OCEAN WAVE MEASUREMENT AND ANALYSIS

PROCEEDINGS OF THE FOURTH INTERNATIONAL SYMPOSIUM WAVES 2001

Volume Two

September 2–6, 2001
San Francisco, California

SPONSORED BY
Coasts, Oceans, Ports, and Rivers Institute
Coastal Zone Management Committee
Waves and Wave Forces Committee
Rubble Mound Structures Committee

EDITED BY
Billy L. Edge
J. Michael Hemsley

**US Army Corp
of Engineers**

**American Society
of Civil Engineers**
1801 Alexander Bell Drive
Reston, Virginia 20191-4400

Abstract: These proceedings, *Ocean Wave Measurement and Analysis,* consist of 180 papers presented at WAVES 2001: the Fourth International Symposium, which was held in San Francisco, California, September 2–6, 2001. The symposium explored the major advances in wave measurement and quantification of ocean and lake waves, including technical knowledge and applications in wave theory, characteristics, design, and techniques. The topics addressed in these proceedings are national and international in scope and include practical examples and case histories on wave transformation, data analysis and reliability, wave modeling, design applications, long waves and tides, wave measurement techniques and instruments, extreme wave statistics, and other topics relating to wave research over the years since the conference in 1997. These proceedings will provide anyone involved with coastal technology a primary reference to the latest information in the field of wave measurement and analysis.

Library of Congress Cataloging-in-Publication Data

International Symposium on Ocean Wave Measurement and Analysis (4th : 2001 : San Francisco, Calif.)
    Ocean wave measurement and analysis : proceedings of the fourth international symposium, WAVES 2001 : September 2-6, 2001, San Francisco, California / sponsored by Coasts, Oceans, Ports, and Rivers Institute ... [et al.] ; edited by Billy L. Edge, J. Michael Hemsley.
        p. cm.
    Includes bibliographical references and index.
    ISBN 0-7844-0604-9
    1. Ocean waves—Measurement—Congresses. 2. Water waves—Measurement—Congresses. 3. Lakes—Congresses. I. Edge, Billy L. II. Hemsley, J. Michael (James Michael), 1944- III. Coasts, Oceans, Ports and Rivers Institute (American Society of Civil Engineers) IV. Title.

GC206 .I58 2001
551.47'02'0287--dc21   2002018276

Foreword

WAVES 2001, the Fourth International Symposium on ocean wave measurement and analysis, was held in San Francisco, California, USA, September 2–6, 2001. The measurement and quantification of ocean and lake waves for verifying wave theory, understanding wave characteristics, and producing economically and environmentally sensitive design are important needs in modern coastal technology. This symposium provided a forum for designers, researchers, and instrument manufacturers to discuss the latest innovations in method and measurement strategies. It is hoped that the symposium will help promote and improve communications, technology transfer, theory, and design. It is anticipated that this publication will become a primary reference in the field of wave measurement and analysis research.

This proceedings, *Ocean Wave Measurement and Analysis,* includes 180 papers amounting to about 2000 pages. This publication includes papers on wave transformation, theory and statistics, measurement, numerical and physical modeling, and application, as well as on long waves, remote sensing, and data analysis. It contains information on measurement projects and programs and provides practical national and international examples and case histories.

To this end, *Ocean Wave Measurement and Analysis* continues the forum for exchange of wave measurement and analysis research and development that was begun 27 years ago. The First International Symposium on Ocean Wave Measurement and Analysis was held in 1974 in New Orleans. It was organized by Orville T. Magoon and Dr. Billy Edge and produced a then state-of-the-art publication on waves. Because of advances in wave measurement and analysis in the 1980s and 1990s and a need to extend the scope to include the formulation of wave statistics for design, a second symposium on ocean wave measurement and analysis was essential. Orville T. Magoon and J. Michael Hemsley organized the Second International Symposium on Ocean Wave Measurement and Analysis, **WAVES 93,** which was held in 1993 in New Orleans. That symposium expanded the focus of the first symposium and became the first of what is intended to be a series of symposia scheduled every four years. The third international symposium, **WAVES 97,** was held in Virginia Beach, Virginia, USA, in November 1997.

The next scheduled international symposium, **WAVES 2005,** is currently being planned for Madrid, Spain. It will continue the exchange of ideas and data in modern coastal wave theory, measurement and analysis, and technology. For more information, please contact Professor Billy L. Edge, Ocean Engineering Program, Department of Civil Engineering, Texas A&M University, College Station, Texas 77843-3136 USA or b-edge@tamu.edu.

The **WAVES 2001** conference co-chairs would like to acknowledge the diligent work of the Organizing and Technical Committees. Their work lead to a very successful program and a smoothly functioning conference. The ASCE and COPRI staff helped to organize the conference to provide as much opportunity for interaction as possible including a pre-conference short course and a post-conference fieldtrip to local sights of technical interest around San Francisco Bay.

Conference Co-Chairs:

Billy L. Edge, Professor
Ocean Engineering Program
Civil Engineering Department
Texas A&M University
College Station, Texas 77843-3136

J. Michael Hemsley, Chief
Engineering Branch
National Data Buoy Center
National Oceanic & Atmospheric Adm.
Stennis Space Center, Mississippi

Acknowledgments

LOCAL ORGANIZING COMMITTEE

Co-Chairs:
Billy L. Edge, *Texas A&M University*
J. Michael Hemsley, *National Oceanic & Atmospheric Administration*

Organizing Committee:
Robert A. Dalrymple, *University of Delaware*
George W. Domurat, *U.S. Army Corps of Engineers, SPD*
Lesley Ewing, *California Coastal Commission*
George Z. Forristall, *Shell International Exploration and Production*
Harald E. Krogstad, *Norwegian University of Science and Technology*
K. Todd Holland, *Naval Research Laboratory*
Robert C. Hamilton, *Evans-Hamilton, Inc.*
James M. Kaihatu, *Naval Research Laboratory*
Orville T. Magoon, *Coastal Zone Foundation*
Jane McKee Smith, *U.S. Army Corps of Engineers, CHL*
Louise Wallendorf, *U.S. Naval Academy*
Philip J. Valent, *Naval Research Laboratory*

TECHNICAL COMMITTEE

Chair: James T. Kirby (USA)
Luc Hamm (France)
Masahiko Isobe (Japan)
Paul A. Hwang (USA)
Lucy R. Wyatt (UK)
Etienne Mansard (Canada)
Noriaki Hashimoto (Japan)
Leo Holthuijsen (Netherlands)
Saleh Abdalla (Turkey)
Alberto Lamberti (Italy)
Phillip L.-F. Liu (USA)
Frederic Raichlen (USA)
Hemming Schaffer (Denmark)
Hajime Mase (Japan)
William F. Baird (Canada)
N.W.H. Allsop (UK)
Kenneth E. Steele (USA)

FINANCIAL SUPPORT

The Conference Organizing Committee expresses its sincere gratitude to the sponsors for their financial contribution to the conference.

Sponsors of Excellence
Hydraudyne B.V.
Shell Global Solutions
U.S. Office of Naval Research
Sponsors of Distinction
AXYS Environmental Systems
InterOcean Systems, Inc.
OCEANOR of Norway ASA
RD Instruments, Inc.
SonTek, Inc.
USACE Coastal and Hydraulics Laboratory
Contributing Sponsors
DHI Water & Environment
Neptune Sciences, Inc.

SUPPORTING ORGANIZATIONS

Coastal Zone Foundation
U.S. Section, PIANC
Marine Technology Society
Office of Naval Research
Naval Research Laboratory
National Data Buoy Center
Texas A&M University
Corps of Engineers
International Association of Hydraulic Engineers
Conference on Offshore Mechanics and Artic Engineering
Japan Society of Civil Engineers
American Geophysical Union

Contents

Wind Wave Generation II

Wave Hindcasting and Climate

Wave Hindcasting and Climate I

Wave Hindcasting and Climate II

Volume Two

A Wave Spectra Study of the Typhoon Across Taiwan

Dong-Jiing Doong[1] Li-Hung Tsai[2] Chia-Chuen Kao[3] Laurence Zsu Hsin Chuang[4]

Abstract: Wave spectra pattern is important information for the input in hydraulics experiments and design of offshore structures. The purpose of this study is to investigate the wave spectral model specifically applicable for typhoon-generated waves. For this, analysis of wave spectra is carried out on measured data obtained by COMC data buoys. This study found that the spectral parameters were related to typhoon states. The typhoon wave spectral model based on JONSWAP formulation but as a function of distance and intensity for the specific track in this study was outlined.

INTRODUCTION

The typhoon-generated waves which have large spatial and temporal variations are significantly different from those observed during normal climate because the input source of energy. The rate of change of wind speed in the moving path of a typhoon is much faster than the ordinary days. The time duration of a given wind speed during typhoon is extremely short in contrast to seas generated by ordinary winds blowing continuously for several hours with constant speed. These cause were responded to the evolution of typhoon wave energy that can be repressed as the shape of wave spectra. It's the motivation of this paper to study the typhoon wave spectra. However, due to the laboratory experiment of simulating severe sea state and the demand of coastal engineering design, the directly purpose of this study is to find the typhoon wave spectral model as a representation by using some simple parameters.

The tracks of the typhoons that approach Taiwan are classified into seven paths as shown in Figure 1. The track of path II display the typhoons move across the Taiwan

---

1 Graduate student, Department of Hydraulics & Ocean Engineering, National Cheng-Kung University, Tainan, Taiwan 701, ROC, doongdj@hotmail.com

2 Graduate student, Department of Hydraulics & Ocean Engineering, National Cheng-Kung University, Tainan, Taiwan 701, ROC, ali@ihmt.gov.tw

3 Professor, Department of Hydraulics & Ocean Engineering and Director, Coastal Ocean Monitoring Center, National Cheng-Kung University, Tainan, Taiwan 701, ROC, kaoshih@mail.ncku.edu.tw

4 Deputy Director, Coastal Ocean Monitoring Center, National Cheng-Kung University, Tainan, Taiwan 701, ROC, laurence@mail.ncku.edu.tw

Island and the Formosa Strait and into Mainland China. Although the occurrence probability of path II is 12% but increases to 71% when including the similar path I and III, extensively. The forward velocity of typhoons by this track is always very fast and stirs much severe sea state to cause the serious disaster. The high occurrence probability and raised huge waves of this type typhoons are the main reasons to choose them in this study. On the other hand, discussion can be simplified without the factor on different typhoon path. There were three typhoons (named AMBER, OTTO and BILIS) moving across Taiwan which had very similar tracks and born days in 1997, 1998 and 2000 respectively. The typhoon paths and some information are shown in Figure 2 and listed in Table 1. The US Joint Typhoon Warning Center (JTWC) stated that typhoon AMBER developed within the monsoon trough during an eight day period (23-30 August 1997). The typhoon intensified to its maximum intensity of 110 kt as it began to approach Taiwan. Reports from CWB indicated that typhoon AMBER began to move across Taiwan with an intensity of 95kt maintaining typhoon intensity as it crossed the island's central mountains, some of which range from 2600 to 4000 m. Land interaction weakened the typhoon. Finally, the system subsequently made landfall in Mainland China with an intensity of 66 kt. In 1998, typhoon OTTO developed on 1 August. OTTO reached a maximum intensity (100 kt on 040000Z August) just prior to making landfall on the southeastern coast of Taiwan. After the mountainous terrain lowered the cyclone's maximum winds to 60 kt, the typhoon continued to move northwestward and made a second landfall in China. The typhoon BILIS, which was classified as a super typhoon was the strongest typhoon of the 2000 Season in West Pacific Ocean. With a diameter close to 600 km and packing gusting winds of 63 m/s, BILIS moved towards Taiwan at a speed of 22 km/hr, and made landfall in the southeastern portion of Taiwan on 23 August 2000.

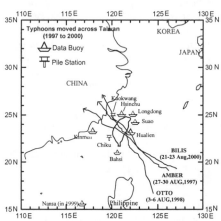

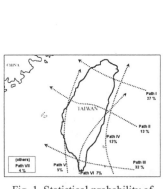

Fig. 1. Statistical probability of typhoon paths

Fig. 2. Typhoon paths and locations of COMC wave measurements network

TYPHOON WAVES MEASUREMENT NETWORK

Typhoon waves measurement is conducted for several purposes, such as provide real time data for disaster prevention and the port operations, calibration and validation of wave models, engineering constructions and determination of local wave climate and for the scientific research of the marine phenomena. In response to the demand for high quality meteorological and oceanographic typhoon data, an enhanced monitoring network comprising buoys, pile stations and other automatic observation systems is planned and is being set-up in nearshore areas as well as in the deep ocean around Taiwan by Coastal Ocean Monitoring Center (COMC) at National Cheng Kung University that was assigned by CWB (Central Weather Bureau) and WRB (Water Resources Bureau) since 1997. The typhoon waves measurement stations operated by COMC are shown in Figure 2.

Both of Data Buoy and Pile Station systems that can provide directional wave data as well as the meteorological information, such as wind, barometric pressure, temperature, etc. are the main instruments in COMC (Kao et al. 1999). The data buoy is the most frequently applied instrument in deep-sea area. To consider the operational requirements of low cost of manufacture and refurbishment, lightweight, land and sea transportability, wave-following characteristics, and reliability, the buoy was designed as a 2.5 meters, as shown in Figure 3. The data buoy can be applied in any water depth, making the site selection more flexible. The system is powered by solar energy, making it suitable for long-term operation. Batteries are used to store the energy for continuous operation during the night and bad weathers. These wave measurements are carried out every 2 hours during normal weather and hourly observation when typhoon comes. In order to get the real-time data for the forecast purposes, data can be transmitted via radio telemetry or GSM communication, in case of far remote station via satellite, immediately after each measurement. In the laboratory, data are quality checked to ensure the veracity before it is sent to the government offices (Doong et al. 1997). The data flow is shown in Figure 4.

The COMC Hualien Buoy station where locate in the nearshore coast was chosen to study the typhoon wave climate of eastern Taiwan, however the Hsinchu Buoy station was used to describe the typhoon wave characteristics of western Taiwan coast (the Formosa Strait). Typhoon BILIS stirred up waves of up to 11 m at Suao Buoy (locate in the Northeast Taiwan) and 7 m at Hualien Buoy (nearly on the track of BILIS). In addition, COMC data buoy had ever recorded the significant wave height up to 14 meter at Hualian station during the prior typhoon AMBER. Detail in-situ measurements for the three typhoons are listed in Table 2.

Table 1. The typhoons that moved across Taiwan (1997~2000)

Typhoon name / No.	AMBER (9717)	OTTO (9802)	BILIS (0010)
Duration	23-30,AUG,1997	2-5,AUG,1998	20-24,AUG,2000
Central pressure (HPa)	948	968	930
Max wind speed during the whole duration	110 kt (55 m/s) at near 280000Z	100 kt (50 m/s) at 040000Z	NA
Max wind speed during the typhoon close to Taiwan	95kt (48m/s)	65 Kt (30 m/s)	105 Kt (53m/s)
Radius of the 30 kt winds, R_7	250 km	150 km	300 km
Forward speed	18 km/hr	20 km/hr	22 km/hr

Table 2. In-situ measurements information

Site	Location	Instruments	Water depth	Max. measurements during typhoons wave height / period / mean wind speed (m / sec / ms^{-1})	
Hualien	121^{0}37'53" E 24^{0}02'05" N	2.5m Data Buoy	30.0	14.75 / 12.3 / 9.4	AMBER
				3.90 / 7.1 / 10.4	OTTO
				7.11 / 9.7 / 17.6	BILIS
Hsinchu	120^{0}56'43" E 24^{0}54'46" N	2.5m Data Buoy	23.0	3.13 / 7.2 / 4.8*	AMBER
				loss	OTTO
				2.02 / 6.9 / 20.7	BILIS
Longdong	121^{0}5'41" E 25^{0}5'41' N	2.5m Data Buoy	32.0	NA	AMBER
				NA	OTTO
				5.38 / 9.0 / 19.0	BILIS

* records weren't complete

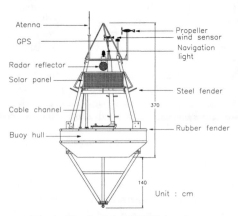

Fig. 3. Sketch of the COMC data buoy

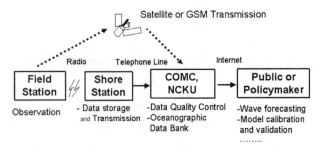

Fig. 4. Data flow of observation system

DISTRIBUTION OF TYPHOON WAVE ENERGY AND ELECTION OF SPECTRAL MODEL

There are two reasons to determine the suitable spectral model for typhoon waves, the distribution of typhoon wave energy and the exponent n in the expression $S(f) \propto f^n$ for the high frequency region. One may expect the shape of wave spectra during typhoons to be extremely random. The evolutions of 1D spectra at eastern and western coast during typhoons moved across Taiwan were analysis. The results on BILIS spectra in Figure 5a and 5b show that there is a consistent uni-modal spectral form, only very few bi-modal spectra before the typhoons arrive in-situ stations. It can be also seen that the peak frequency of wave energy continuously decreases and tends to keep constant for hours right before and after the arrival of typhoons. The wave energy is concentrated primarily in the neighborhood of the peak frequency during the typhoons. The powerful and easier JONSWAP formulation is adopted in this study depend on the most uni-modal typhoon spectra. Ochi and Chiu (1982) fit the shape of typhoon wave spectra by JONSWAP spectral formulation. Again, Young (1997) concluded that either of the JONSWAP or Donelan form (Donelan et al 1985) could be used to represent the typhoon spectra. Based on the results carried out by previous studies, it is concluded that the shape of wave spectra during typhoons can be assumed to the JONSWAP spectral formulation.

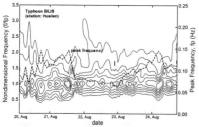

Fig. 5a Evolution of typhoon wave energy   Fig. 5b Evolution of typhoon wave energy
at eastern Taiwan (Hualien Buoy)   at western Taiwan (Hsinchu Buoy)

THE JONSWAP TYPHOON WAVE SPECTRAL MODEL

The JONSWAP spectrum has been extensively employed in coastal and ocean engineering. It may be expressed as

$$S(f) = \alpha \frac{g^2}{(2\pi)^4} \exp\left\{-1.25(f_p / f)^4\right\} \times \gamma^{\exp\left\{-(f-f_p)^2 / 2(\sigma f_p)^2\right\}}$$

(1)

where γ is a peak-enhancement factor, the effect of which is to increase the peak of the PM spectrum, σ is a relative measure of the width of the peak. Recommended values are 0.07 on the low frequency side and 0.09 on the high one and are adopted in this study. f_p is the spectral peak frequency. The parameters involved in the spectral formulation are determined so that the sum of the squared differences between the mathematical formulation and the observed spectra are minimal.

Figure 6 shown the value γ of the JONSWAP form as a function of the inverse wave age, U_{10}/C_p , where C_p is the phase speed of components at the spectra peak frequency and U_{10} is the wind speed at 10m above sea level. As with fetch limited results, there appears to be no systematic dependence on U_{10}/C_p . The mean value obtained from the data set in this study is $\gamma = 1.19$ at Hualien and $\gamma = 1.35$ at Hsinchu. The mean value of $\gamma = 1.9$ was obtained by Young (1997) from Australia typhoon data and $\gamma = 2.2$ obtained by Whalen and Ochi (1978) from their North American data set. The values are much smaller than the mean JONSWAP value of $\gamma = 3.3$ with the normal sea state in the North Sea. Young (1999) explain that although much of the recorded energy has presumably propagated to the measurement stations as swell, it has been transformed locally by a combination of atmospheric input and nonlinear interactions into relatively broad wind-sea spectra. The average JONSWAP spectra of typhoon and monsoon waves are also plot in Figure 7 for comparison.

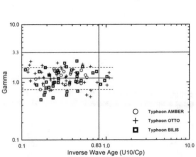

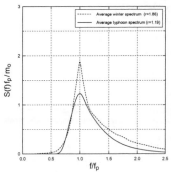

Fig. 6. JONSWAP spectral parameters (γ)   Fig. 7. Comparison of average JONSWAP
for typhoon waves   spectrum between typhoon and monsoon

The demarcation between wind-sea and swell by wave age, $U_{10}/C_p = 0.83$ proposed by Donelan et al. (1985) is also shown in Figure 6. The typhoon waves that recorded at Hualien station are mostly classified to the swell. The ratio of swell energy to total wave energy is analysis by the separation algorithm of wind-sea and swell that presented by Wang and Hwang (2001) to know the proportion of swell components in the directional typhoon waves. More than 60% wave energy is contributed from swell in large typhoon records as shown in Figure 8. The swell components increase with the relative typhoon wave height.

Figure 9 shows the other parameter α of JONSWAP spectrum. Hasselmann et al. (1973) found a relationship for α expressed in terms of U_{10}/C_p (also shows in the figure). Young (1997) identify the typhoon data follow the relationship even there is a significant scatter within his data. The α of typhoon waves present in this study extend the swell data set that presented by Young (1997). Although the good fit with JONSWAP relationship can be used to explain the spectral shape stabilization provided by non-linear interactions even then wind fields are spatially and temporarily variable during typhoon period. However, the authors are going to find the reason of the unconformable data with JONSWAP fetch limited relationship in this study as shown in Figure 9.

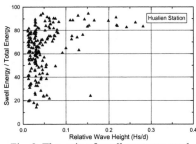

Fig. 8. The ratio of swell energy to total energy of typhoon waves

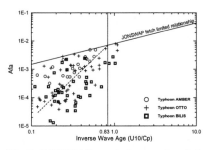

Fig. 9. JONSWAP spectral parameters (α) for typhoon waves

MODEL THE SPECTRAL FORM BY TYPHOON PARAMETERS

In the original work of JONSWAP, the spectral parameters were coupled to the wind speed and fetch length. However, the parameters are functions of dimensionless fetch length which is extremely difficult to evaluate during typhoons since the wind speed is continuously changing. Many studies derived the parameters as the function of wave height and period. However, Shemdin (1980), Young (1998) and Young & Burchell (1996) have demonstrated that the typhoon wave height is a function of the position, forward velocity and the intensity of the typhoon. It seems unlikely that a spectral form could be parameterized in terms of wave height alone or includes the period. In additional, the behavior of typhoon (i.e. typhoon track) is also an important factor to influence the in-situ wave measurements besides the typhoon parameters listed above, although only the data from similar typhoon path is used in this study.

Figure 10 shows the relationship between JONSWAP parameter α and the in-situ relative wave height as Eq. (2). However the relative wave height is a function of the distance from in-situ station to typhoon center as shown in Figure 11 and listed in Eq. (3). Furthermore, the coefficient a_2 in Eq. (3) relates to the typhoon intensity (i.e. max. wind speed of typhoon) as shown in the small diagram in Figure 11. Although there are only three data, the formulation is shown in represent format. In fact, the other coefficient a_1 in Eq. (3) is a meaning of the forward velocity of typhoon that can be found in Figure 11. In addition to parameter α, the parameter γ also has a relationship with position of typhoon as shown in Figure 12 and formulation listed in Eq. (4).

$$\alpha = 0.025\frac{H_s}{d} + 2.5\times10^{-5} \tag{2}$$

$$H_s/d = a_1 e^{a_2 \cdot DD} \tag{3}$$

where $a_1 = f(V_f)$, $a_2 = f(I)$

$$\ln(\gamma) = b_1 \cdot DD + b_2 \tag{4}$$

By using these relationships, we may derive a wave spectral formulation specifically applicable for typhoon-generated seas in the form of JONSWAP formulation and as a function of typhoon parameters as typhoon distance (DD) and intensity (I). The forward velocity (V_f) of typhoon and the location of station should be considered by analysis extensive data.

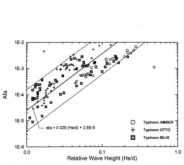

Fig. 10. JONSWAP α is a function of relative wave height at in-situ station

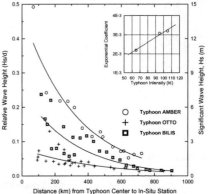

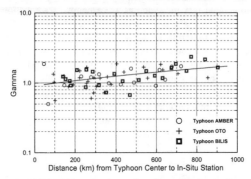

Fig. 11. Relationship between in-situ wave heights and typhoon distances and intensity

Fig. 12. JONSWAP γ is also a function of typhoon distances

CONCLUSIONS

Current engineering practice favors the JONSWAP spectrum to represent the frequency distribution of the incoming waves when computing the action of these waves. In addition, the JONSWAP spectrum is often used to simulate random waves in wave flume in laboratory experiments. This paper presents the wave spectra characteristics and fit them to the JONSWAP formulation to get the parameters according to typhoon waves. A typhoon wave spectral model then was derived as a function of the distance from in-situ station to typhoon and the typhoon intensity. The parametric typhoon spectral model values only for the type of typhoons move toward Taiwan. There maybe have inconsistency results in the other location because of the significant difference on

JONSWAP parameters compared with other data sets in this study. Diverse typhoon data sets should be collected and analysis to strength the results in this study.

ACKNOWLEDGEMENTS

This study was supported by Central Weather Bureau and Water Resource Bureau of Taiwan. The authors would like to express their appreciation to the offices.

REFERENCES

Donelan, M.A., Hamilton, J., and Hui, W.H. 1985. Directional Spectra of Wind Generated Waves. *Philos. Trans. R. Soc. Lond.*, A 315, 509-562.

Doong, D.J., Chuang, L.Z.H., and Kao, C.C. 1997. Development of Quality-Checking System on Oceanographical Observation Data. *Proc. 19th Conf. on Ocean Engineering in Taiwan*, 477-484 (in Chinese)

Hasselmann, K. et al. 1973. Measurements of Wind-Wave Growth and Swell Decay during the Joint North Sea Wave Project (JONSWAP). *Deutsche Hydrographische Zeitschrift*, A, 8, No. 12, 95 p.

Kao, C. C., Chuang, L.Z.H., Lin, Y.P. and Lee, B.C. 1999. An Introduction to the Operational Data Buoy System in Taiwan, *Proceedings of Int. MEDCOAST Conference*, Antalya, Turkey, 33-39.

Ochi, M.K. and Chiu, M.H. 1982. Nearshore Wave Spectra Measured During Hurricane David. *Proc. 18th Int. Conf. Coastal Engineering*, ASCE, Cape Town, 77-86.

Shemdin, O.H. 1980. Prediction of Dominant Wave Properties Ahead of Hurricanes. *Proc. 17th Int. Conf. Coastal Engineering*, ASCE, Sydney, 600-609.

Wang, D.W. and Hwang, P.A. 2001. An Operational Method for Separating Wind Sea and Swell from Ocean Wave Spectra. Submitted to *Journal of Atmospheric and Oceanic Technology*.

Whalen, J.E. and Ochi, M.K. 1978. Variability of Wave Spectral Shapes Associated with Hurricanes. *Proc. 10th Offshore Tech. Conf.*, 1515-1522.

Young, I.R., 1997. Observations of the Spectra of Hurricane Generated Waves. *Ocean Engineering*, 25, 261-276.

Young, I.R., 1998. An Intercomparison of GEOSAT, TOPEX and ERSI Measurement of Wind Speed and Wave Height. *Ocean Engineering*, 26, 67-81.

Young, I.R., 1999, *Wind Generated Ocean Waves*, Elsevier Science, 288 p.

Young, I.R. and Burchell, G.P. 1996. Hurricane Generated Waves as Observed by Satellite. *Ocean Engineering*, 23, 761-776.

SYSTEM IDENTIFICATION TECHNIQUES FOR THE MODELLING
OF IRREGULAR WAVE KINEMATICS

Witold Cieślikiewicz [1] and Ove T. Gudmestad [2]

Abstract: This paper presents the development of a parametric model linking the free-surface elevations with the orbital velocities field under an irregular wave. The *system identification* procedures are applied to estimate the parameters of the model based on data recorded in the wave tank. The free-surface time series are taken as input data and the output data are components of the particle velocity vector. A linear time-invariant model with the static nonlinearities at the input side incorporated is assumed. In the study different system identification procedures are applied for two cases—firstly when orbital velocities are modelled at points that are always submerged in the water and, secondly, in the surface zone where measurement points are not continuously submerged. This paper demonstrates the results of modelling the horizontal component of the orbital velocity in comparison with wave kinematics data taken in a wave flume using Laser Doppler Velocimetry specifically designed to give accurate measurements also in the vicinity of the mean water level. Comparison between the modelled and observed velocity time series presents a sufficiently good agreement to prove the effectiveness of the applied approach. The SI techniques applied to determine from the free-surface records the orbital velocities as well as other hydrodynamic parameters such as particle acceleration and forces may be very important and useful in engineering and oceanography.

INTRODUCTION

System identification (SI) is concerned with constructing *dynamic models* of a *system*. A system in which variables of different kinds interact and produce observable signals that are called *outputs* are of interest to us. The system is stimulated by external signals that are called *inputs*. Usually the inputs can be manipulated by the observer (to the extent of the tolerances of the observer's equipment). Other external signals are called *disturbances* or *noise*.

This concept can be applied when one considers for example a water wave propagating in a wave tank. In such a case the measured input is the wavemaker's horizontal displacement, while the output may be the free-surface elevation measured at a certain distance from the wavemaker or the hydrodynamic pressure and the orbital velocity measured at a selected z-elevation. In this situation it is quite clear that the wavemaker's movement is the only stimuli of the system which fully determines the wave

---

[1] University of Gdańsk, Institute of Oceanography, Al. Pilsudskiego 46, 81-378 Gdynia, Poland, ciesl@univ.gda.pl

[2] Statoil, 4035 Stavanger, Norway, otg@statoil.no

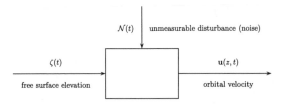

Fig. 1. Basic input-output configuration of wave system being modelled.

field, i.e. the output of the system, if disturbances like those caused by the mechanical limitations of the wave generator or vibrations are not taken into account.

In this study we will apply for water waves the system concept in a context that is somewhat different and, in a sense, wider than this mentioned above. Consider a physical problem described by a number of variables. Those variables are linked together by certain physical relationships that can be defined in terms of mathematical expressions like difference or differential equations. In a sense, we can identify the system with its mathematical description. Let us consider such a selection of variables that it is possible to define a subset of them, which determines via mathematical relations the remaining variables. We will call this subset of variables the input variables while the remaining variables will be referred to as the output variables. This concept of the system, and the input and output variables is more abstract and wider than previous one. We do not need the input variables to be directly controlled by the observer. Moreover, the input variables in this approach are not necessarily physical stimuli in a common sense, i.e. they need not to be the causative "driving force" of the system.

The modelling problem is then to describe how all the variables relate to each other and we will not put the philosophical question how well the constructed model reflects the underlying "true system" given by the Mother Nature. Our approach will be more pragmatic and we will judge our model by usefulness rather than "truth." We will accept the model built if it works, i.e. if it gives accurate enough output for a given input (in terms of comparison with measurement) and if it let us make reliable predictions and simulations of the output variables given the appropriate input.

Selecting the input variables is a part of the modelling problem in our approach. Not always can we be sure that the selected variables are *complete* in such a sense that they uniquely determine the output variables. If we do not know the "true system" or, in other words, if we do not know the "right mathematics" linking all the system variables we can never be sure if our selection of input variables is complete. It is possible this selection is too small and underdetermines the output or may be there are too many input variables and the output is overdetermined. Again, unless we have a full knowledge of the physical system and its mathematical description, the only possible judgement of the completeness of the input variables set is by testing whether the model is "good enough," that is, whether it is valid for its purpose.

In this study the main objective is to answer the question if it is possible to create a mathematical model for the water wave problem which links the free-surface elevations, treated as the input variable, with the orbital velocities at a given z-elevation as the output variables. We will assume the system, which will be referred to as *wave system*, is *dynamic*, which means that the current output values depend not only on the current input but also on their earlier values.

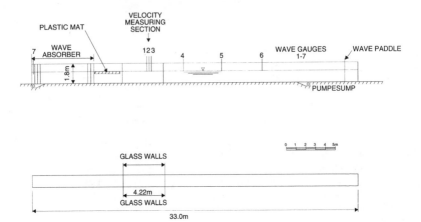

Fig. 2. Wave flume.

The 2-dimensional wave system and basic input-output configuration may be symbolically depicted as in Fig. 1. We assume in this study that this wave system can be described by a linear time-invariant model. The output data across the z-elevations are both components of the orbital velocities $\mathbf{u}(z,t)$. The free-surface elevations $\zeta(t)$ compose the input data vectors. There is also some unmeasurable random disturbance $\mathcal{N}_t$ that influences the output.

In the study we apply the SI techniques for the wave system to two cases. Firstly, when orbital velocities are modelled at points that are always submerged in the water. Secondly, we propose a modification of the SI procedures to model the wave kinematics in the free surface zone where measurement points are not continuously submerged and where so called "emergence effect" (Cieślikiewicz & Gudmestad 1993) should be taken into account.

The SI procedures utilised in this study to model the wave kinematics were created and verified based on experimental data collected in a wave tank. The experimental arrangement is described below.

EXPERIMENT

The wave data utilised in this study and the experimental arrangement are described in detail in paper by Skjelbreia et al. (1991). The experiments were carried out in a tank illustrated in Fig. 2. The irregular wave generator of this tank is hydraulically driven and the control signal was constructed from a JONSWAP target wave spectrum. The flow velocity was measured at several different elevations by a two-component Laser Dopler Velocimeter (LDV). The LDV was specifically designed for studies in which special attention was paid to measuring orbital velocity components within the surface layer.

Each run in the test program had a duration of 819.2 s. Digitizations of the free surface elevation and velocity time series were carried at a rate of 40 Hz and samples of 32 768 measuring points were collected. In this paper we discuss the wave conditions for sea state Case 5 (measurement series I18) from Skjelbreia's measurements in the Norwegian Hydrotechnical Laboratories. These conditions were given by the significant wave height $H_S = 0.21$ m, and the peak period $T_p = 1.8$ s, and water depth $h = 1.3$ m. Measurements of the particle velocity $\mathbf{u} = [u, w]$ near the mean water level have varying degrees of intermittency behaviour depending upon which level they were obtained. This occurs when the probe volume of the LDV emerges from the water. Time series of particle velocities u_i and w_i were therefore modified numerically such that their values during *drop-out* periods were set equal to zero:

$$[\overline{u}_i, \overline{w}_i] = [u_i, w_i]\mathcal{H}(\zeta_i - z) \tag{1}$$

in which z is the level of the LDV and $\mathcal{H}(\cdot)$ is the Heaviside unit step function. The time series ζ_i, $\overline{u}_i$ and $\overline{w}_i$ were subjected to statistical and spectral analysis (see Cieślikiewicz & Gudmestad 1993) and were used as input and output data in the SI modelling.

SYSTEM IDENTIFICATION—BACKGROUND

Let $X(t)$ denotes the output data sequence of the system, i.e., it is one of two components of the orbital velocity vector $\mathbf{u}(z, t)$ at given z-elevation. The input data vectors $\mathbf{Q}_t$ of N_Q dimension are composed of the free-surface elevation sequence $\zeta(t)$. We include the random disturbance $\mathcal{N}_t$ as an additive filtered white noise $\mathcal{E}_t$. We assume in this study that the wave system can be described by a linear time-invariant model which is specified by the sequence of impulse response series $g_q(k)$, $q = 1, \ldots, N_Q$ and the weighting function $h(k)$ of the random additive disturbance, $k = 0, 1, \ldots, \infty$, and, possibly, the probability density function of the white-noise $\mathcal{E}_t$.

It is worth noting, that despite the assumption of a linear model we are still able to incorporate in a system the nonlinearities that have the character of a static nonlinearity at the input side, while the dynamics itself is linear. In case the nonlinearity is known, say as a function F, a certain input variable $Q(t)$ can be transformed as $Q_q(t) = F(Q(t))$ and the system can be treated as linear. We will have such a situation in this study.

A complete model is given by the following relationship (see e.g. Ljung 1987)

$$X(t) = \mathbf{G}(f)\mathbf{Q}(t) + \mathcal{N}(t) \tag{2}$$

in which f is the *forward shift operator* defined by $fQ(t) = Q(t + 1)$, $\mathbf{G}$ is the transfer function of the system and $\mathbf{G}(f)\mathbf{Q}(t)$ is short for $\sum_{q=1}^{N_Q} G_q(f)Q_q(t)$ and for any $q = 1, 2, \ldots, N_Q$

$$G_q(f) = \sum_{k=0}^{\infty} g_q(k)f^{-k}; \qquad f^{-1}Q_q(t) = Q_q(t - 1) \tag{3}$$

Then

$$\mathbf{G}(f)\mathbf{Q}(t) = \sum_{q=1}^{N_Q} G_q(f)Q_q(t) = \sum_{q=1}^{N_Q} \sum_{k=0}^{\infty} g_q(k)Q_q(t - k) \tag{4}$$

As mentioned above, we assume that the disturbance $\mathcal{N}$ can be described as filtered white-noise, so

$$\mathcal{N}(t) = H(f)\mathcal{E}(t) \tag{5}$$

where

$$H(f) = 1 + \sum_{k=1}^{\infty} h(k)f^{-k} \tag{6}$$

Within SI we work with structures that permit the specification of $\mathbf{G}$ and H in terms of a finite number of numerical values. As it is common, we assume that $\mathcal{E}(t)$ is Gaussian, in which case the PDF is specified by the first and second moments. Thus, a particular model (2) is entirely determined in terms of a number of numerical coefficients which are included as parameters to be determined. The purpose of SI is to determine the values of those parameters. If we denote the parameters in question by the vector θ, and if we take into account equation (5), the basic description for the modelled system becomes

$$
\begin{aligned}
X(t) &= \mathbf{G}(f,\theta)\mathbf{Q}(t) + H(f,\theta)\mathcal{E}(t) \\
p_\mathcal{E}(\cdot,\theta), &\quad \text{the PDF of } \mathcal{E}(t); \quad \mathcal{E}(t) \quad \text{white-noise}
\end{aligned}
\tag{7}
$$

which is a set of models, each of them associated with a parameter value θ.

A commonly used way of parameterising G_q and H is to represent them as rational functions of f^{-1} and specify the numerator and denominator coefficient in some way (see e.g. Ljung 1987). Such model structures, which are known as *black-box models*, were utilised in this study. A few different model structures were tested. However, in the frame of the present report we will show the modelling results for one of them. The examples demonstrating the results of SI with the simplest ARX model structures for the horizontal component of orbital velocity will be presented. Those models were estimated using *least-squares methods*.

ARX model structure

If in (7) we assume

$$
G_q(f,\theta) = \frac{B_q(f)}{A(f)} \quad \text{for} \quad q = 1,\ldots,N_Q, \qquad H(f,\theta) = \frac{1}{A(f)}
\tag{8}
$$

where

$$
A(f) = 1 + a_1 f^{-1} + \cdots + a_{N_A} f^{-N_A}
\tag{9}
$$

and for $q = 1, 2, \ldots, N_Q$

$$
B_q(f) = b_0^q + b_1^q f^{-1} + b_2^q f^{-2} + \cdots + b_{N_{B_q}}^q f^{-N_{B_q}}
\tag{10}
$$

we obtain one of the simplest model structures, i.e., the autoregressive with extra input model (ARX). If (7) is rewritten as $A(f)X(t) = \mathbf{B}(f)\mathbf{Q}(t) + \mathcal{E}(t)$, $A(f)X(t)$ is the autoregressive part while $\mathbf{B}(f)\mathbf{Q}(t)$ is the extra input of the ARX model.

The vectorial parameter θ to be determined is in this case

$$
\theta = [a_1, a_2, \ldots, a_{N_A}, b_0^1, b_1^1, \ldots, b_{N_{B_1}}^1, b_0^2, b_1^2, \ldots, b_{N_{B_2}}^2, \ldots, b_0^{N_Q}, b_1^{N_Q}, \ldots, b_{N_{B_{N_Q}}}^{N_Q}]
\tag{11}
$$

where N_A, N_{B_1}, N_{B_2}, $\cdots$, $N_{B_{N_Q}}$ are the *orders of the multi-input ARX model*.

By substituting (8) into (7) the following input-output relationship is obtained

$$
\begin{aligned}
X(t) + a_1 X(t-1) + \cdots + a_{N_A} X(t-N_A) &= b_0^1 Q_1(t) + b_1^1 Q_1(t-1) + \cdots + b_{N_{B_1}}^1 Q_1(t-N_{B_1}) + \cdots \\
&+ b_0^{N_Q} Q_{N_Q}(t) + b_1^{N_Q} Q_{N_Q}(t-1) + \cdots + b_{N_{B_{N_Q}}}^{N_Q} Q_{N_Q}(t-N_{B_{N_Q}}) + \mathcal{E}(t)
\end{aligned}
\tag{12}
$$

which is a *linear difference equation*. The ARX model represented by (12) is sometimes called an *equation error model* because of the way in which the white-noise term $\mathcal{E}(t)$ enters the difference equation (12).

If we denote by N the length of the time series $X(t)$ and write

$$V(\theta) = \sum_{t=t_{min}}^{N} \mathcal{E}^2(t, \theta) \tag{13}$$

where $t_{min} = \max(N_A, N_{B_1}, N_{B_2}, \ldots, N_{B_{N_Q}})$, then the least-squares estimate of the parameter vector θ^* is defined as

$$\theta^* = \arg\min_{\theta} V(\theta) \tag{14}$$

Here arg min means "the minimising argument of the function."

APPLICATION

In this section we present the modelling results for the horizontal component of the orbital velocity, i.e. we take $X(t) = u(t)$. First, we describe the SI procedures for the wave system when orbital velocities are modelled at points which are always submerged in the water, i.e. the emergence effect need not to be taken into account.

Results for selected z-elevation below the wave troughs—no emergence effect

Here we demonstrate modelling results for the Run I18_16 where the LDV recorded the velocity at $z = -0.25\,\mathrm{m}$. In order to select the structure of the model. i.e., to set up the input-vectors and decide about the combination of model orders, a *cross-validation* procedure was utilised. Namely, the time series of 32 768 data points were split into independent *working* and *validation* parts. The first $N_w = 24\,576$ data points were taken as the working data and were used for the estimation, while the last $N_v = 8\,192$ data points (1/4 of the data) were used as the validation data to evaluate an estimated model's properties. That evaluation was done mainly by comparison of the simulated and measured output time series and by computing the variance of the prediction errors (for the validation data):

$$V_v(\theta^*) = \sum_{t=1}^{N_v} \mathcal{E}_\sigma^2(t, \theta^*) \tag{15}$$

where the prediction error $\mathcal{E}$ is

$$\mathcal{E}(t, \theta^*) = X_s(t, \theta^*) - X_m(t) \tag{16}$$

in which $X_s(t, \theta^*)$ is the output value at time t simulated with the parameter vector θ^* that was estimated with use of the working (estimation) data according to (13) and (14). The model structure resulting in the smallest variance (15) was selected. However, in the case when only a small decrease in the variance was achieved by a significant increase of the model order, for practical reasons, an optimum choice may be to take the model structure of a lower order even though it may result in a bit larger variance of prediction errors (15).

First, a very simple model structure with the input-vector $\mathbf{Q}(t) = \zeta(t)$ was tested. This choice of the model structure resulted in a very poor quality of the ARX model, independently of the model orders N_A and N_B. In order to make possible for the model to better control the absolute values of the input signal, a square of ζ was introduced into the input-vector $\mathbf{Q}$. By this we were able to take into account some nonlinear effects, namely, the static nonlinearity. The model structure with the input-vector of dimension $N_Q = 2$

$$\mathbf{Q}(t) = [\zeta(t), \zeta^2(t)] \tag{17}$$

with the orders:

$$N_A = 0; \qquad N_{B_1} = N_{B_2} = N \tag{18}$$

appeared to work very well. In Table 1 the variances of the prediction errors are shown for a number of selected values N. It can be noticed that there is no significant decrease in the modelling quality for smaller model orders. It was found that even the model with the orders $N_A = 0$, $N_{B_1} = N_{B_2} = 1$ produced relatively good results. It appears that quite important findings of this study was the inclusion of the square of the free-surface elevation into the input-vector (17).

Table 1. Variances of prediction errors.

N	$V_v(\theta^*)$ $(\mathrm{m}^2/\mathrm{s}^2) \times 10^{-4}$
1	5.20
5	4.49
40	4.16
80	3.96
160	3.96

Finally, the model structure with the input-vector of the form $\mathbf{Q}(t) = [\zeta(t), \zeta^2(t), \zeta^3(t)]$ with $N_Q = 3$ and with various model orders was verified. It was found that the increase of the modelling quality was insignificant.

In Fig. 3 (a) we show the comparison between the modelled and measured time series of horizontal velocity for the validation part of data with the model structure given by (17) and $N = 160$. There are five time windows containing the six highest wave crests found in the validation time series. In Fig. 3 (a) the free-surface elevation time series is also marked. The Fig. 3 (b) presents the same comparison in a form of a scatter plot. The Fig. 3 (c) shows a scatter plot of the modelled versus measured values for the whole time series of Run I18_16, i.e. for its both estimation and validation parts.

Results for selected z-elevation in the free-surface zone—with emergence effect

In this section we describe the modification of the SI procedures due to the emergence effect. The modelling results for Run I18_25 with the LDV fixed at $z = 0.00$m will be presented here.

The application of the SI procedures as described in the preceding section for the LDV at $z = -0.25$m resulted in a very poor agreement of measured and modelled values of orbital velocity for the LDV positioned at $z = 0.00$ m. A few different approaches have been checked out and it appeared difficult to construct the input vector $\mathbf{Q}$ that allows for modelling of the velocity $\overline{\mathbf{u}}$ modified by the emergence effect as in (1). At this stage of the study, in order to solve the problem the ARX modelling scheme, which is usually applied for the continuous time series of the input and output data, has been modified in such a way that modelling over a number of separate time windows was possible. This can be represented by the following modification of (13):

$$V(\theta) = \sum_{\sigma=1}^{N_c} \sum_{t=t_{min}}^{N_\sigma} \mathcal{E}_\sigma^2(t, \theta) \tag{19}$$

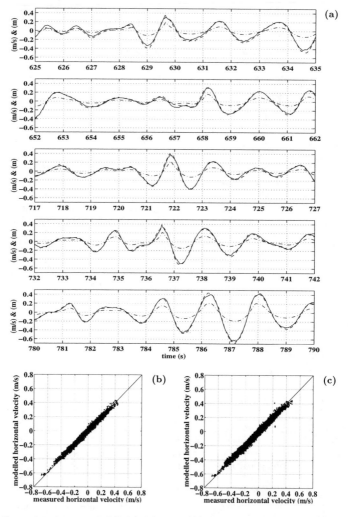

Fig. 3. Comparison between modelled (solid line in (a)) and measured (dashed) time series of horizontal velocity at $z = -0.25$ m; dot-dashed line indicates free-surface elevation. The scatter plot (b) is for the validation part of the data; (c) is for the whole time series.

where N_c is a number of the wave crest windows taken as the model working data, N_σ, $\sigma = 1, 2, \ldots, N_c$ is the length (number of data points) of the wave crest window σ and t_{min} is defined as in (13). The estimate θ^* is determined by the minimisation of $V(\theta)$ using least-squares method.

In order to model the velocity in wave crests a number of time windows covering the highest wave crests was selected using a program prepared for this purpose. Among the input parameters of this program there are the threshold value ζ_{thrld} which should be reached by the free-surface elevation and the minimum value of the free-surface elevation ζ_{min} in the selected time window. The value of ζ_{thrld} decides whether a given crest is selected and included into the model working (estimation) data, while ζ_{min} determines the time window's length.

Below we present the modelling results for the horizontal component of orbital velocity with $\zeta_{thrld} = 0.15$ m which resulted in $N_c = 25$, i.e. 25 of the highest wave crests were selected. $\zeta_{min} = 0.05$ m has been set. In order to present the quality of the proposed parametric modelling technique in an objective way we estimated the model parameters θ^* based on the first 20 of 25 selected time windows, taken as the working data and leaving the remaining 5 as the validation data. The model structure with the input-vector of the form (17) was taken with the orders (18). Since the model is estimated based on the data covering the highest wave crests it may result in a worse modelling of smaller values of the horizontal velocity. It should be emphasised that the small values of the horizontal velocity at $z = 0$ m correspond usually to the time periods in which the observation point (LDV) is not well submerged and one can expect some chaotic fluctuations of the particle velocity in such time periods. These chaotic fluctuations are quite difficult to model.

The modelling results for $N = 6$ are presented on Fig. 4 (a). This figure shows the comparison between the modelled and measured time series of the horizontal velocity for the 5 time windows. These time windows cover the wave crests selected as the validation data and include also waves with the smaller crests. Although some discrepancies in the case of smaller velocity oscillations can be noticed, the modelled data agree quite well with the measured data. It is also demonstrated by Fig. 4 (b) showing the same comparison in the form of a scatter plot for both the 20 evaluation windows and the 5 validation windows.

CONCLUSIONS

It is shown that SI technique may be successfully applied to determine the wave kinematics from the free-surface records. The modelled data agree very well with the measured data for points that are always submerged in the water. For $z = 0$, the modelling is quite difficult and slightly worse agreement between modelled and observed time series has been found. Further research is being directed to improve the modelling in the vicinity of the mean water level. On the other hand, for $z = 0$, the model estimated based on the data covering the highest wave crests is, in a sense, dedicated to predict extreme positive values of horizontal velocity especially well. This model property may be very important and useful in engineering applications.

The authors would like to emphasise that when the parameters of the model are estimated, the measured time series of only the free-surface elevation is used for the construction of wave kinematics. This may be very important for many maritime engineering applications, where measuring of kinematics (or hydrodynamic loading) is more difficult and expansive than the recording of free-surface elevations. For example, it is possible to collect in a relatively easy way the accurate measurements of water surface elevation by using the remote sensing techniques with down-looking laser device installed on an offshore platform. In such situations, it is sufficient to install the underwater measuring equipment for a limited period of time only, allowing for collection of kinematics data needed to estimate the model parameters. When these parameters are established, the laser altimeter, while recording surface elevation, is also able to provide the wave kinematics via parametric model developed.

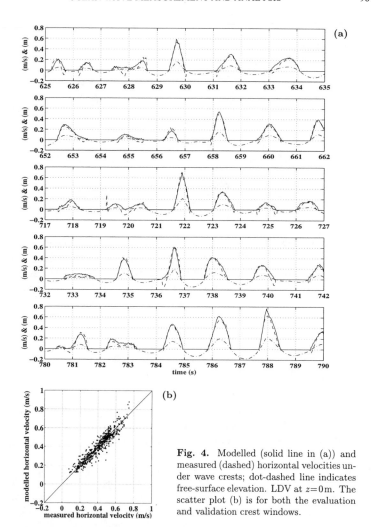

Fig. 4. Modelled (solid line in (a)) and measured (dashed) horizontal velocities under wave crests; dot-dashed line indicates free-surface elevation. LDV at $z=0$m. The scatter plot (b) is for both the evaluation and validation crest windows.

This study was carried out under Statoil contract No. 65163 and was partially founded within University of Gdańsk project BW 1330-5-0059-1.

REFERENCES

Cieślikiewicz, W. & Gudmestad, O. T. 1993 Stochastic characteristics of orbital velocities of random water waves. *Journal of Fluid Mechanics*, **255**, 275–299.

Ljung, L. 1987. *System identification: theory for the user*. Prentice-Hall, Inc., Englewood Cliffs, New Jersey.

Skjelbreia, J. E., Berek, E., Bolen, Z. K., Gudmestad, O. T., Heideman, J. C., Ohmart, R. D., Spidsøe, N. and Tørum, A. 1991. Wave kinematics in irregular waves. Proc. 10th Intern. Conf. on Offshore Mechanics and Arctic Engng, Stavanger, Norway, 223–228.

WAVE EFFECTS ON SHIPS MOORED AT FIGUEIRA DA FOZ HARBOUR

João Alfredo Santos[1], Maria da Graça Neves[2]
and Conceição J. E. M. Fortes[3]

Abstract: The Figueira da Foz harbour is located at the mouth of the Mondego River in the West coast of Portugal. Two shore-connected breakwaters that, due to their layout, have a quite poor performance shelter it. In fact, some of the incident waves cause large movements in the ships moored at the general cargo quay. The breakage of mooring lines is a quite common accident there.

This paper presents the methodology employed and summarizes the results obtained in both the wave penetration study when the harbour basin is dredged up to −7.00 m C.D. and the study of the wave effects on the behaviour of a ship moored at the general cargo quay.

INTRODUCTION

The Figueira da Foz harbour, Fig. 1, is located at the mouth of the Mondego River on the West coast of Portugal. It is protected by two shore-connected breakwaters: the so--called North (776 m long, crest level +6.00 m C.D.) and South (980 m long, crest level +6.00 m C.D.) breakwaters. The harbour entrance is 250 m wide and faces west. The port facilities are distributed on both margins of the Mondego River and include a general cargo quay, a marina and a fishing dock. Nowadays, the seabed depth at the outer-harbour and at the harbour channel is around −5.00 m C.D.. Close to North breakwater head the seabed depth reaches −14.00 m C.D..

The wave regime at the harbour entrance, Fortes et al. (2000), is characterized by wave directions ranging from SW (225°) to W-60°-N (330°), the most frequent wave directions being those of the sector W to NW. The values of the zero crossing period, T_z, range from 6 to 14 s. The significant wave heights vary between 0.16 and 8.4 m, the most common values being in the interval 0.5 to 3.0 m. Records from the Port Authority

[1] Research Officer, jasantos@lnec.pt, [2] Research Officer, gneves@lnec.pt,
[3] Research Assistant, jfortes@lnec.pt
Hydraulics Department, L.N.E.C., 1700-066 Lisboa, Portugal

(*Instituto Portuário do Centro – IPC*) show that for some wave conditions, mainly Southwest storms, the wave heights in the harbour basin can reach high values, which impair the navigation across the outer basin and make the mooring of ships close to the general cargo quay very difficult. When those storms occur broken mooring lines and damaged fenders are common.

Prior to dredging of the main channel down to –7.00 m C.D., which will allow larger ships to call at Figueira da Foz, *IPC* decided to characterize the wave conditions inside the harbour basin, for this new harbour depth, and to analyse, for the same situation, the behaviour of the largest ship that calls the port today moored at the general cargo quay.

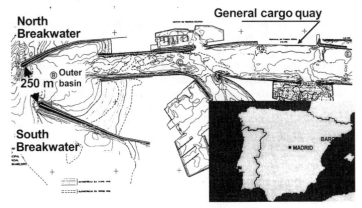

Fig. 1. Figueira da Foz harbour

The wave penetration study (Neves and Fortes 2000) was carried out for several wave conditions in front of the harbour entrance by means of the linear refraction-diffraction-reflection model DREAMS.

The numerical model for the moored ship behaviour, MOORNAV, is based on the linearity of the ship-waves system and assembles and solves, in the time domain, the equations of motion of the moored ship. M.S. "CASSIOPEIA" was the design ship in this study. The fenders considered were those already installed at the quay while several mooring lines, made of polypropylene alone and polypropylene and steel wire, were tested.

After the description of the numerical models employed, the paper proceeds with the presentation of the results obtained in the wave penetration study. The results of the moored ship behaviour study are then presented. At the end of the paper the conclusions of the whole work are presented.

BRIEF DESCRIPTION OF THE NUMERICAL MODELS USED

Numerical model DREAMS

DREAMS numerical model, (Fortes 1993), performs harbour resonance and wave disturbance computations for harbour and sheltered zones. DREAMS is a two-dimensional finite element numerical model based on the elliptic mild-slope equation, Berkhoff (1972). This equation describes the combined refraction-diffraction and back scattering of surface waves propagating over mild-slope seabeds, such as those of ports, harbours and costal regions. The DREAMS model can be applied either to the study of short wave penetration into a harbour or bay, or to the resonance study of long incident waves into a harbour or sheltered zone. It has no limitation related with the incident wave direction at the entrance of the domain and so the same computational domain can be used for a wide range of incident wave directions.

The boundary conditions used in the model are: *radiation boundary condition*, which allows the scattered waves to propagate out of a finite-extent domain; *generation-radiation condition*, which enables the wave generation and the radiation of the outgoing waves and is applied to open boundaries; *total or partial reflection condition,* which simulates the effects of beaches, breakwaters or any other kind of physical boundaries.

The mild-slope equation is solved by the Finite Element Method (FEM), which is based upon a symmetric weak formulation associated to the mild-slope equation. The spatial discretisation can be made of bilinear quadrilaterals or linear triangular elements.

The DREAMS numerical model outputs are the wave amplification factors (H/H_0, the ratio between wave height, H, at any given point of the study domain and the wave height at its entrance, H_0) and the wave propagation directions. A Galerkin weighted residual statement computes the horizontal velocity field. The model's input and output are visualised with the ACE/GREDIT package, Baptista and Turner (1992).

Numerical model MOORNAV

The analysis of the moored ship behaviour is carried out by means of the impulse response approach, Cummins (1962). The implementation of this approach implies the solution, in the frequency domain, of the radiation and diffraction problems of the free--floating ship. While the radiation problem corresponds to the forced oscillation of the ship in otherwise calm waters, the diffraction problem corresponds to the fixed ship withstanding the incoming sea waves. The solution of the linearised radiation and diffraction problems is obtained by a numerical model, which evaluates the potential associated to each of those problems by means of a panel method. The hull's wetted surface is discretised by rectangular and/or triangular panels.

The impulse-response functions (also known as retardation functions) as well as the infinite frequency added mass coefficients associated to each of the six degrees of freedom of the free-floating ship are obtained by Fourier transforming the frequency domain results of the radiation problems. The time series of the wave forces acting on the ship result from the superposition of the forces due to each of the incident wave components.

The time series of the wave forces, together with the impulse response functions, the added mass coefficients and the parameters of the mooring system (mooring lines and fenders) enable the assembly and solution of the equations of motions for the moored ship. The model results consist of time series of both the motions along each of the six degrees of freedom and the forces in the mooring lines and fenders. These procedures are implemented in LNEC's package MOORNAV, Santos (1994).

WAVE PENETRATION STUDY
The main objectives of DREAMS calculations were the characterization of the wave conditions inside Figueira da Foz harbour, namely at the outer basin, the harbour channel and the general cargo quay.

The computational domain and the bottom depth contours considered in those computations are presented in Fig. 2. The information from the hydrographic chart had to be adapted in order to account for the new water depth at the harbour channel, which is to be dredged down to -7.00 m C.D.. Because this study was carried out for short period waves, the boundary condition on the coastline was partial reflection. The reflection coefficients varied with both the coastline characteristics and the period of the incident wave. These coefficients were evaluated by means of the method presented in Seeling and Ahrens (1995), which takes into account the boundary characteristics (porosity, slope, toe depth, among others) and the Iribarren number.

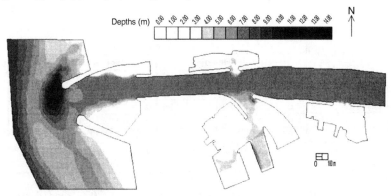

Fig. 2. Computational domain and seabed depth contours of the study region

DREAMS computations were performed for a mean water level equal to +2.00 m C.D.. The computations were carried out for incident waves at the study domain entrance (seabed depth around −10 m C.D.) with directions ranging from SW (225°) to W-30°-N (300°) while the periods varied between 6 and 15 s. For each period and incident wave direction, the output of the computations consisted of one diagram of the wave amplification factors inside the harbour. Fig. 3 presents one of those outputs for an incident wave with a 10 s period and W-30°-S direction.

To get characteristic values for the wave amplification factors at the more important zones of the Figueira da Foz harbour, namely, the outer harbour and the general cargo quay, several points were selected at each of those zones. The maximum values of the wave amplification factor in the set of points representative of the region in front of the general cargo quay, for the several test conditions, are presented in Table 1.

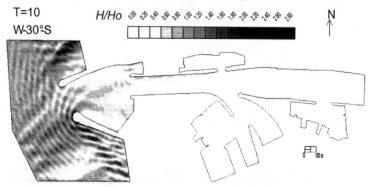

Fig. 3. Wave amplification factors for an incident local wave direction W-30°-S (240°) and wave period of T=10 s

Table 1.　Maximum Values of the Wave Amplification Factors

Incident Wave Direction (°)	Period (s)									
	Outer Basin					General Cargo Quay				
	6	8	10	12	15	6	8	10	12	15
SW .... (225°)	1.22	1.41	1.42	1.20	1.33	0.12	0.14	0.22	0.35	0.24
W-30°-S .. (240°)	1.01	0.78	0.90	0.84	0.95	0.12	0.30	0.34	0.35	0.24
W-15°-S .. (255°)	0.89	1.15	1.13	0.96	1.12	0.41	0.38	0.37	0.32	0.26
W .... (270°)	1.06	1.11	1.16	0.94	1.17	0.38	0.30	0.29	0.25	0.19
W-15°-N .. (285°)	0.95	0.89	1.12	0.93	0.96	0.11	0.13	0.20	0.20	0.13
W-30°-N .. (300°)	0.92	0.89	0.96	1.13	0.94	0.07	0.10	0.15	0.17	0.10

The results from the several computations, Neves and Fortes (2000), showed that, in general, the harbour is not protected from short wave penetration coming from incident directions varying between W (270°) and SW (225°). In front of the general cargo quay, as the incident wave direction goes from W (270°) to SW (225°), there is an increase in the wave period for which the highest wave amplification factor occurs: for an incident wave direction of W (270°), the highest values of the wave amplification factors

(H/H_0=0.38) correspond to a wave period of 6 s while for the direction SW (225°) they correspond to 12 s (H/H_0=0.35).

So, the occurrence of significant wave action inside the harbour is related with the short wave penetration from the third quadrant (incident wave direction between W and SW). In this analysis, the limitations of the DREAMS model, namely those associated with the linear formulation, should be taken into account. In fact, this model does not account for wave breaking, wave overtopping or the interaction between waves and currents, which can be significant at the Figueira da Foz entrance. In addition to that, irregular wave propagation should be studied also, since the directional dispersion can cause important changes in the wave height inside the harbour.

THE SHIP MOORED AT THE GENERAL CARGO QUAY

Due to the small amplitude of the waves radiated by the ship that reach the harbour boundaries distant from the ship, it is quite common to limit the radiation problem of the free-floating ship inside an harbour to the ship and the nearby quay wall. The same reasoning does not apply always to the diffraction problem. However, the results from the wave penetration into the Figueira da Foz harbour show that close to the general cargo quay waves do tend to propagate along the quay and so wave reflection at the marginal protection under the quay is not very important. This means that the diffraction problem can be studied by considering only the ship and the nearby slope, as in the radiation problem.

The ship simulated in this study was the M/S "CASSIOPEIA" which has a length overall of 98.60 m, a beam of 15.90 m and a draught of 4.80 m. Starting with the ship's body plan, supplied by *IPC*, the hull's wetted part was discretised into 764 panels. The ship's reference system has the (x-y) plane coincident with the waterplane, the z-axis points upwards and the x-axis is directed along the ship's longitudinal axis. The origin is amidships.

The marginal protection under the quay has a slope of 2:3 and was modelled as a second body (a four sided ring with three vertical walls and a sloping wall). The ring's reference system has the (x-y) plane coincident with the waterplane, the z-axis points upwards. In this case the x-axis coincides with the longitudinal axis of the ring's intersection with the undisturbed mean water level. This intersection is 150 m long and 20 m wide. The distance between the ring's x-axis and the ship's x-axis is 32.5 m. The ring's slope got the finest discretisation, 500 panels out of a total of 1200 panels. Fig. 4 shows a perspective of the panel dicretisation of the ship's hull and the nearby slope.

The quality of the frequency domain results was verified by means of a symmetry check, for the added mass, a_{ij}, and damping, b_{ij}, coefficients, since a_{ij} must be equal to a_{ji} and b_{ij} must be equal to b_{ji}, for any value of i or j between 1 and 6. For the wave forces acting on the ship, the forces from the diffraction problem were compared with the forces from the Haskind relations. Finally the good behaviour of the whole package was confirmed by comparing the amplitude of the time domain response of the free-

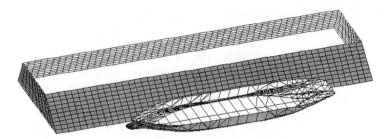

Fig. 4. Panel discretisation employed in the radiation and diffraction problems

-floating ship subject to regular waves with the amplitude of the frequency domain response.

The mooring system was composed of six fenders and eight mooring lines deployed as shown in Fig. 5. Two different mooring configurations were tested, although the mooring lines geometry at rest was kept the same. The second mooring configuration resulted from applying a pre-tension to some of the mooring lines of the initial configuration.

Only the x-coordinates of the contact points of the fenders are listed in Table 2, since for all the fenders the other coordinates are $y = 8.01$ m and $z = +1.00$ m. The reference system used coincides with the ship reference system at rest. All the fenders in the general cargo quay are VREDESTEIN type ALF S1. A limit of 0.5 m was imposed to the fender deflection because values larger than that may imply the collision of the ship's hull to the general cargo quay.

Table 3 lists the coordinates of the mooring line points on the quay and on the ship. The reference system for the quay points is the same as for the fender's contact points. The coordinates of the ship points are on ship fixed reference system.

In spite of the choice of mooring lines being a shipmaster's matter, it was assumed that ship's mooring lines are made of two polypropylene cables with a diameter of 40 mm whose breaking load, per cable, is 201 kN. It was considered that, when a storm

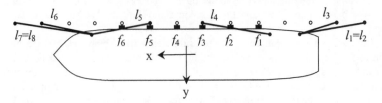

Fig. 5. Mooring system tested

Table 2. x-Coordinates (m) of Fenders Contact Points

f_1	f_2	f_3	f_4	f_5	f_6
-27.00	-17.00	-7.00	+3.00	+13.00	+23.00

Table 3. Coordinates of Mooring Line Points

Mooring Line	Quay Point (m)			Ship Point (m)		
	x	y	z	x	y	z
l_1	-67.00	-9.50	+3.20	-40.79	-7.00	+3.70
l_2	-67.00	-9.50	+3.20	-40.79	-7.00	+3.70
l_3	-57.00	-9.50	+3.20	-40.79	-7.00	+3.70
l_4	-7.00	-9.50	+3.20	-29.29	-8.00	+2.20
l_5	+13.00	-9.50	+3.20	+35.00	-7.00	+4.20
l_6	+53.00	-9.50	+3.20	+35.00	-7.00	+4.20
l_7	+63.00	-9.50	+3.20	+35.00	-7.00	+4.20
l_8	+63.00	-9.50	+3.20	+35.00	-7.00	+4.20

occurs, the port authority can make available stronger mooring lines made of wire rope, with 30 mm diameter and a breaking load of 778 kN, with a tail of polypropylene 10 m long, with 64 mm diameter and a breaking load of 472 kN. The load-elongation curves of the several mooring lines are presented in Fig. 6 and are based on the results presented by Rita (1984).

In the first mooring configuration studied, designated as I, the mooring lines l_1, l_4, l_5 and l_8 are made of two 40 mm polypropylene ropes, 25 m long. The mooring lines l_2 and l_7 are made of one 15 m-long steel wire of 36 mm with a 10 m-long polypropylene tail of 64 mm, and the mooring lines l_3 and l_6, are made of one 5 m-long steel wire of 36 mm with a 10 m-long polypropylene tail of 64 mm. In this configuration the mooring lines have no pre-tension installed when the ship is at rest in the initial condition. In configuration II the same mooring lines are used but it is assumed that the mooring lines supplied by the Port Authority (mooring lines l_2, l_3, l_6 and l_7) have pre-tension of 200 kN.

The numerical tests with the moored ship were carried out with incident regular head waves. Five periods (6 to 15 s) and seven wave heights (0.10 to 1.50 m) were tested. It should be noted that these were wave heights close to the ship. The objective was to gain some insight on the behaviour of the moored ship. All the tests started with the ship at rest and the wave forces intensity grew gradually to their maxima in the first eight wave periods. The tests duration was 240 incident wave periods, unless there were problems with the mooring system due to large movements of the moored ship.

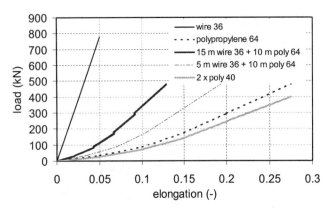

Fig. 6. Load-elongation curve of tested mooring lines

Table 4. Maximum Amplitude of Sway Motion

Wave Height (m)	Period (s)									
	Mooring Configuration I					Mooring Configuration II				
	6	8	10	12	15	6	8	10	12	15
0.10	0.01	0.07	0.67	2.60	2.16	0.03	0.09	0.04	0.64	0.21
0.25	0.03	0.23	2.36			0.04	0.20	0.53	1.19	
0.50	0.07	0.33				0.11	0.49	1.20		
0.75	0.11	2.63				0.12	0.86			
1.00	0.15	0.41				0.13	1.06			
1.25	0.18	2.08				0.35	1.21			
1.50	0.22					0.62				

Table 4 displays the amplitudes of the sway movement for the two mooring configurations tested and for all the test conditions. No results are presented for the test conditions where the simulation had to stop, usually due to fender deformation larger than 0.5 m. Only the sway motion was chosen because it was the one in which large amplitudes occurred.

The results illustrate clearly the non-linear behaviour of the moored ship. In fact, for the same incident wave period, doubling the incident wave height does not imply doubling the sway amplitude.

The results show also that 6 s period waves do not cause any problems to the mooring system. For longer waves, the wave height range without problems decreases as the incident period increases. A pre-tension of 200 kN in the mooring lines l_2, l_3, l_6 and l_7 produces a decrease in the amplitude of the moored ship motions. Actually, by comparing the results of configurations I and II, it may be concluded that for 10 and 12 s period waves the wave height range where no problems occur does increase.

CONCLUSIONS

The results from the studies on both the wave penetration into the Figueira da Foz harbour and their effects on the behaviour of the ships moored at the general cargo quay were presented.

The first study showed that the harbour is not protected from short wave penetration coming from local directions varying between W (270°) and SW (225°). From the second study it may be concluded that introducing a pre-tension of 200 kN in some of the mooring lines produces a notable decrease in the amplitude of the sway motion. In spite of the limitations of the models it is clear that a reduction of the wave height close to the general cargo quay is necessary in order to reduce the excessive motions of the ships moored there.

AKNOWLEDGMENTS

The permission granted by Instituto Marítimo-Portuário to publish the results of these studies and the financial support of the Portuguese Research Council, PDCTM/P/MAR/15239/1999 are gratefully acknowledged.

REFERENCES

Baptista, A.M. and Turner, P. 1992. *ACE/GREDIT User's manual. Software for semi-automatic generation of two-dimensional finite element grids*. Center for Coastal and Land-Margin Research. Oregon Graduate Institute of Science and Technology, 84 p.

Berkhoff, J.C.W. 1972. Computation of Combined Refraction-Diffraction. *Proceedings of 13rd International Conference in Coastal Engineering*, ASCE, 471-490.

Cummins, W.E. 1962. Impulse Response Function and Ship Motions. *Schiffstechnik*, 9, 101-109.

Fortes, C.J.M., Capitão, R.P. and Carvalho, M.M. 2000. *General and Extreme Sea Wave Regime in the Approaches to Figueira da Foz Harbour* (in Portuguese). Relatório 12/00 – NPP, LNEC, 96 p.

Fortes, C.J.M. 1993. *Mathematical Modelling of the Combined Refraction and Diffraction of Sea Waves* (in Portuguese). MSc. Dissertation, IST, 126 p.

Neves, M.G.O. and Fortes, C.J.M. 2000. *Sea Wave Characterization at Figueira da Foz Harbour* (in Portuguese). Relatório 98/00 – NPP, LNEC, 55p.

Rita, M.M. 1984. *On the Behaviour of Moored Ships in Harbours – Theory, Practice and Model Tests*. Relatório 9/85 – NPP, LNEC, 261 p.

Santos, J.A. 1994. *MOORNAV – Numerical Model for the Behaviour of Moored Ships*. Relatório 3/94-B Projecto NATO PO-WAVES, IH-LNEC, 30 p.

Seeling, W.N. and Ahrens, J.P. 1995. Wave Reflection and Energy Dissipation by Coastal Structures. *Wave Forces on Inclined and Vertical Wall Structures*, edited by N. Koboyashi, ASCE, 28-55.

EVOLUTION OF BUBBLY FLOW IN THE SURF ZONE

Yasunori Watanabe[1], Junichi Ohtsuka[2] and Hiroshi Saeki[3]

Abstract: In situ visual observations concerning the entrainment of air bubbles after wave breaking were carried out (Kojohama, Ohkishi and Furubira coasts in Japan). This paper presents a discussion of the behavior of air bubbles under breaking waves as well as the effects of these bubbles on the fluid and sediment based on the results of the observations. Typical bubble cloud structures are identified and classified according to the breaker type. Furthermore, the major differences between the three-dimensional eddy structures containing an obliquely descending eddy (ODE) found in the surf zone relative to those in experiments were investigated by tracking the bubbles involved in the vortices.

INTRODUCTION

The onset of breaking waves forms a very complex turbulent structure in the surf zone. The air bubbles entrained into the water during the wave breaking process add a great deal of complexity to the turbulent field. The local fluid motion, which receives additional stress from the entrained air bubbles (e. g. drag force and buoyancy), tends to develop disturbances in the gas-fluid two-phase turbulent flow. The mechanical contributions to fluid motion associated with the presence of bubbles promote either the production or the dissipation of turbulence, depending on the relative velocity fluctuation of the fluid particles and bubbles. In particular, the collapsing and advecting processes of the bubbles may yield additional turbulence, and, in general, the drag force

1 Assistant Professor, Hydraulic Engineering Laboratory, School of Engineering, Hokkaido University, N-13 W-8, Sapporo 060-8628, Japan, yasunori@eng.hokudai.ac.jp
2 Technical Engineer, Division of Civil Engineering, Penta-Ocean Construction Co., Ltd., Koraku 2-2-8, Bunkyo-ku, Tokyo 112-8576, Japan, Junichi.Ootsuka@mail.penta-ocean.co.jp
3 Professor, Hydraulic Engineering Laboratory, School of Engineering, Hokkaido University, N-13 W-8, Sapporo 060-8628, Japan, h-saeki@eng.hokudai.ac.jp

(on the basis of Stokes's law of resistance) is involved with the energy dissipation of turbulence. The processes involving the bubbly turbulent flow via the bubble-turbulence interactions in shear fields are not yet fully understood.

In terms of sediment transport in a surf zone, it is very important to give attention to air bubbles carried to the vicinity of the seabed because they can easily disturb the sediment due to drag and buoyancy. Therefore, it is essential to determine the mechanisms that produce and transport bubbles in order to develop an accurate model for a global evaluation of the movement of sediment in the surf zone.

In addition, the surf zone is a major supplier of oxygen (O_2) into the seawater. Aeration of the water by the bubbles entrained by breaking waves is essential for marine-resources such as vegetation and fish and for maintaining water quality in near-shore areas. The physical properties of bubbles such as the number and the size distribution of the entrained bubbles have to be estimated in order to assess the effect on coastal environment.

It appears that a major mechanism for transporting the bubbles generated by deep-water wave breaking to greater depths is Langmuir circulation (e.g. Langmuir, 1938; Monahan and Lu, 1990), in terms of bubble transport in very long time scales. Also, Thorpe (1982) characterized the bubble clouds (i. e. "columnar clouds" and "billow clouds") in deep water by thermal conditions. However, in the near-shore breaking case, the local velocity distribution and turbulent structure become predominant factors in specifying the formation of bubble clouds composed of large numbers of bubbles.

The discussions in this paper of the mechanisms that produce the air bubbles, the local bubble behavior and the structure of bubble clouds are based on in situ visual observations. The correlation between the structural organization of the bubble clouds and the vortices caused by wave breaking as well as that between the bubble motion observed in situ and in experiments are also presented.

OBSERVATION SITE AND EQUIPMENT

In August 1999, field (in situ) tests were conducted on three different coasts: Furubira coast, Kojohama beach and Ohkishi beach, all in Hokkaido, the northernmost of the main islands of Japan. The Furubira coast faces the Japan Sea; the Kojohama and Ohkishi beaches face the Pacific Ocean (see Fig. 1). The wave conditions and the type of coast are shown in Table 1. In this paper, we mainly discuss the results from the Furubira coast because of the much clearer visual data resulting

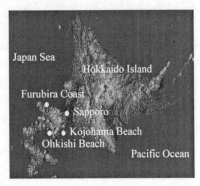

Fig. 1. Observation sites

from the higher water transparency in comparison to the other sites.

A small submerged camera (diameter, 6 cm) was fixed to a steel weight of 12-Kg (see Fig. 2). The Furubira coast features a shelf-like configuration of the seabed (see Fig. 3); the rocky bed spans the entire surf zone, which has a plentitude of marine-resources such as vegetation and fish. The camera was positioned at two recording points (P_a and P_b, see Fig. 3) on the seabed in the on-shore side of the breaking zone behind a shoulder of the shelf. The movements of entrained bubbles were recorded in offshore, on-shore and cross-shore views by the video recording system shown in Fig.

Table. 1. The wave conditions and type of coast

	Significant wave period	Mean breaking wave height	Breaker type	Type of coast (bed material)
Furubira Coast (A)	3.9 sec	34 cm	Spilling	Rock
Ohkishi Beach (B)	8.5 sec	75 cm	Plunging	Sand
Kojohama Beach (C)	10.0 sec	90 cm	Plunging	Sand

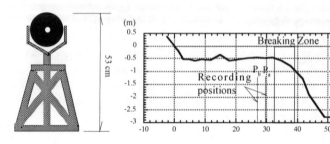

Fig. 2. Submerged camera   Fig. 3. Configlation of the seabed in the Furubira coast

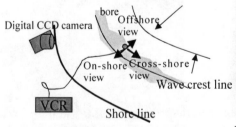

Fig. 4 Recording system of the visual observation

Fig. 5 Incidental breaking waves recorded by the camera on the shore

4. The surface waves were also recorded simultaneously by another digital CCD camera that was set on the shore in order to synchronize the underwater images with the phase of the surface waves (see Fig. 5).

RESULTS

Columnar bubble clouds

The evolution of typical bubble clouds in a bore region is shown in Fig. 6; the interval between successive pictures is 1/15 sec, and the recording point is P_a. An organized structure of bubble clouds has clearly formed under the bore front (note its development inside the circle in phase b), which covers the surface of the entire water domain. The bubble clouds, in which numbers of the entrained bubbles are locally gathered, are regularly arranged in a direction that is parallel to the wave crest line.

Figure 7 shows a sequence of cross-shore images of the bubble clouds at P_a. It can clearly be seen that bubble clouds are produced underneath the front face of the bore; relatively small clouds are organized along the front line of the progressing bore. A typical size distribution consists of small clouds near the front (in the range of several cms), expanding towards the rear of the bore (up to about 50-cm) (see also Fig. 8). The individual bubbles in the clouds tend to disperse in the expanding clouds with the propagation of the bore. In addition, the clouds are frequently observed hitting the seabed, disturbing the sediment during this evolving process, especially under the onset of a large-scale breaking wave.

A sequence of the offshore images of the clouds at P_b is shown in Fig. 9. A typical structure organization can also be seen here, in which numbers of small bubble clouds

Fig. 6. Bubble cloud structure in a bore region (interval: 1/15 sec, camera direction: on-shore, camera position: Pa)

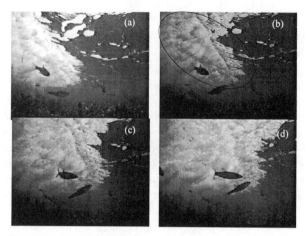

Fig. 7. Bubble cloud structure in a bore region (interval: 1/15 sec, camera direction: cross-shore, camera position: Pa)

packed underneath the front face of bore become larger clouds developing downward over time. Thorpe (1982) was able to detect organized patterns of bubble clouds caused by deep-water wave breaking using sonar. His results are analogous to the structure observed in this field test, suggesting that, in a spilling breaker, the mechanism forming the bubble clouds in a near-shore surf zone is identical to that which exists in deep water. Therefore, we may also call this cloud structure organization "columnar clouds", in the same manner.

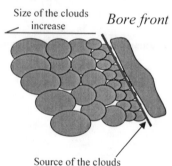

Fig. 8 Size distribution in the bubble cloud structure in a bore region

In addition, a tendency was also found for individual bubbles in each cloud tend to spirally rotate on a vertical axis since they may be involved in a vortex caused by breaking. This local rotational motion of many bubbles forms a typical column-shaped cloud because the bubbles trapped inside the vortex are never ejected out of the cloud (see Fig. 15). This suggests that the three-dimensional eddy structure developed under the bore has an important role in forming the bubble clouds.

Another important feature is that the staying duration of the bubbles in seawater is much longer than in freshwater. The bubbles transported to a deeper region never rise in spite of their buoyancy; they tend to stagnate until the next onset of a breaking wave (see Fig. 10). Since the bubbles entrained in the actual surf zone are very small (according to Su et al. (1988), the peak in the size spectrum appears around 20 μm), buoyancy, which is proportional to the cube of the diameter, is much smaller than the drag force, which is proportional to the square of the diameter. Therefore, the drag force inhibits upward movement while inducing movement following the fluid motion, whereas buoyancy has relatively little effect. The retention of the bubbles may enhance the coastal environment and marine-resources by serving as a source of O_2. On the other hand, when freshwater is used in experiments, the bubbles completely rise to the surface within a short time after the entrainment because their sizes are sufficiently large. The surface tension of water, an important factor in determining bubble size, should be taken into account during investigations of entrained bubbles after wave breaking.

Band structure of the clouds

The structure of bubble clouds inherent in a plunging breaker occurs just after the plunging phase. Figure 11 shows a sequence of cross-shore images at a plunging point (P_a). An air tube is formed in phase a, and then several streaks of bubble clouds gradually occur circumferentially around the air tube in phase b. These clouds, wrapped around and arrayed along the air tube, further develop in thickness to form a well organized band structure (phase d). The mechanism that forms the band structure via the wrapping process of the clouds is believed to develop as follows. Initially, the bubbles are generated in the thin zones on both sides of the overturning jet as it enters the surface (see Fig. 12 c). Then the rotational fluid motion in the region surrounding the air tube causes the generated bubbles to wrap around the air tube. This explanation contradicts a simple hypothesis where the entrained bubbles originate from the col-

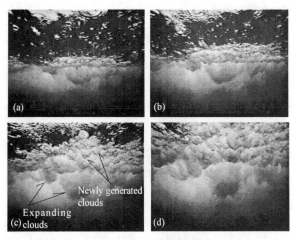

Fig. 9. Bubble cloud structure in a bore region (interval: 1/15 sec, camera direction: offshore, camera position: Pb)

lapse of the entrapped air tube via an expelling process (see Fig. 13). In large eddy simulations of plunging breaking waves (Watanabe and Saeki, 1999), several pairs of helical (stream-wise) and vertical vortices arranged in a direction parallel to the wave crest were found to occur immediately after the plunging phase. They are responsible for inducing vortical fluid motion in a direction along the rotational flow surrounding the air tube (see Fig. 12). Therefore, the bubbles are trapped within each streak of the cloud by the vortical fluid motion in the pairing vortices, forming the band structure.

Obliquely descending eddy (ODE)

The presence of an ODE in three-dimensional coherent eddy structures was confirmed in laboratory experiments by Nadaoka et al. (1989). The downward growth of the ODE is responsible for transport of the bubbles trapped in the ODE to greater depths. The generation of an ODE at the test site is discussed below, based on a development process of bubble clouds that correlates to the presence of vortices.

Figure 14 shows typical paired bubble clouds developing downward behind the bore. The particular shape and the direction of stretch of the clouds (see Fig. 14 d) are analogous with those of an ODE; Figure 15 shows the vertical vortical motion of individual bubbles within the spindle-like columnar cloud involved in an ODE. As a result, it is believed that the spindle-like columnar clouds observed in situ indicate the presence of an ODE. Although several investigations concerning the occurrence of ODE have been conducted in an experimental surf zone in freshwater, the present results clearly demonstrate that ODE occurs in the field and that the production of ODE is an intrinsic feature of wave breaking.

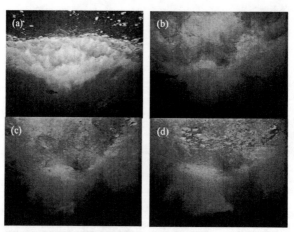

Fig. 10. Bubbles stagnated in a deep region for long time (interval: 1/3 sec, camera direction: offshore, camera position: Pb)

CONCLUSIONS

The structure of the bubble clouds caused by near-shore wave breaking can be characterized in terms of the breaker type. Columnar bubble clouds are predominant in spilling wave breaking and in the bore region for both spilling and plunging breakers (see Fig. 16a). In a plunging breaker, a band structure is formed immediately after the plunging phase via the wrapping process of the clouds, which is associated with the eddy structure after breaking (see Fig. 16b).

In addition, it was found that large numbers of the bubbles, trapped inside the vortex under a breaking wave, spirally rotate within each bubble cloud. The bubble entrapped in the vortex form the bubble clouds. With the dissipation of the vortex, the bubbles tend to disperse and the clouds expand in time because of the reduction of rotational fluid motion.

The field tests have confirmed the presence of ODE. The bubble clouds involved in the ODE have a spindle-like columnar configuration.

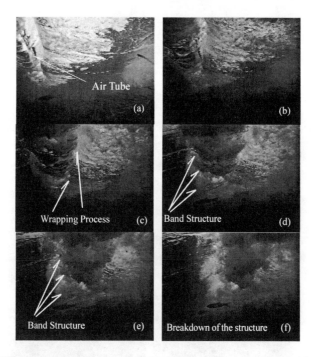

Fig. 11. Typical band structure of bubble clouds in a plunging breaker (interval: 1/15 sec, camera direction: cross-shore, camera position: P_b)

The major difference between experiments and in situ field tests is the staying time of bubbles in the water resulting from differences in bubble sizes. The size of air bubbles in the field is clearly much smaller than in experiments, and the number of bubbles in an actual surf zone is much larger than in an experimental one. These results suggest that experiments in freshwater possibly provide results that are inapplicable to the actual sea.

These numerous fine air bubbles that feature a long-staying time may contribute possibly to the vegetation and the overall ecological system in a surf zone as well as causing a suspension of sediment.

Fig. 12. Mechanism to form wrapped clouds in a plunging breaker

Fig. 13. Convetional hyposis of a generatiing bubbles after breaking

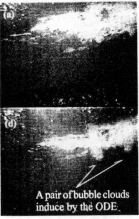
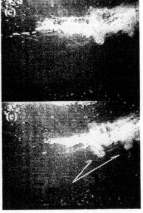

Fig. 14. Evolution of the spindle-like clumnar clouds involved in the ODE (interval: 1/15 sec, camera direction: on-shore, camera position: Pb)

ACKNOWLEDGEMENTS

The authors wish to thank Prof. Philip L- F Liu and Prof. Eiji Suenaga for helpful advises. The work was supported by the Japanese National Scientific Foundation for young researchers.

REFERENCES

Thorpe, A. 1982. On the Clouds of Bubble Formed by Breaking Wind-Waves in Deep Water, and their Role in Air-Sea Gas Transfer, *Phil. Trans. Roy. Soc.*, A304, 155 - 210.

Thorpe, S. A. and Hall, A. J. 1983, The Characteristics of Breaking Waves, Bubble Clouds and Near-Surface Currents Observed Using Side-Scan Sonar, *Continental Shelf Res.*, 1 (4), 353 - 384.

Monahan E. C. and Lu N. Q. 1990, Acoustically Relevant Bubble Assemblages and their Dependence on Meteorological Parameters, *IEEE J. Ocean. Eng.*, 15 (4), 340 - 345

Nadaoka, K., Hino, M. and Koyano, Y. 1989, Structure of the Turbulent Flow Field under Breaking Waves in the Surf zone, *J. Fluid Mech.*, 204: 359 - 387.

Watanabe, Y., Saeki, H. 1999, Three-Dimensional Large Eddy Simulation of Breaking Waves, *Coastal Eng. J.*, 41(3&4): 281 - 301.

Lungmuir I. 1938, Surface Motion of Water Induced by Wind, *Science*, 87 (2250), 119 - 123.

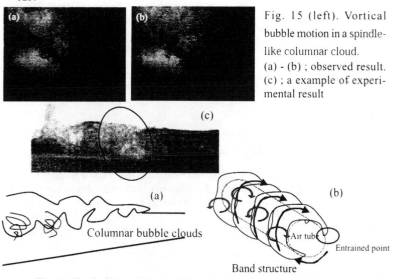

Fig. 15 (left). Vortical bubble motion in a spindle-like columnar cloud. (a) - (b) ; observed result. (c) ; a example of experimental result

Fig. 16. Typical type of the bubble cloud structure (a: in a bore region, in a spilling breaker, b: in a plunging breaker)

LABORATORY STUDY OF BREAKING WAVE INDUCED NOISES

S.A. Sannasiraj[1] and Chan Eng Soon[2]

Abstract: This paper presents the results from an experimental work to characterize the underwater sound field radiated by breaking waves. Measurements of the ambient noise in the frequency range 20Hz – 20kHz, generated by laboratory breaking waves, are presented. It has been observed that the sources of sound in spilling breakers are primarily in the frequencies ranging from 300 Hz to 20 kHz. The plunging waves, however contributes to the noise spectrum from less than a few hundred hertz to 20 kHz. Attempts have been made to correlate breaking wave acoustics with the dynamics of breaking waves. The acoustic radiated energy has been found to be proportional to energy dissipation and the square of the maximum wave slope parameter.

INTRODUCTION

It is well known that wave breaking is a major contributor to free surface ocean ambient noise. Since the noise levels in the ocean surface are linked to the wind speed, early works have been focused on the correlation of wind speed to the sound intensity. The pioneering outcome is the Knudsen spectrum for ambient noise (Knudsen et al. 1948). The Knudsen spectrum, specified in the frequency range 500 Hz to 50kHz, has a typical spectral slope of about -5 to -6 dB/octave. Spectral power in this frequency range is shown to increase with wind speed. In general, the spectral features of the Knudsen spectrum are in agreement with the vast laboratory and field observations.

[1]Research Fellow, tmssas@nus.edu.sg; [2]Director, tmsdir@nus.edu.sg;
Tropical Marine Science Institute, National University of Singapore
Singapore 119 223

In more recent years, ambient noise levels have been correlated with the breaking wave characteristics and the bubble sizes. This is a logical development since underwater sound is predominantly influenced by the bubbles generated by the breaking waves. Most of the recent studies relate the generated sound levels with the structure of bubbles formed by the breaking waves. One of the main hypotheses for noise production is that the oscillation of individual bubbles and bubble clouds could contribute to the noise spectrum in the high and low frequency ranges, respectively (Medwin and Beaky 1989, Medwin and Daniel 1990). The demarcation between the low and high frequencies is difficult although it is reported to range from 300 Hz to 1kHz. This demarcation has been defined based on the maximum possible size of the bubbles formed in the ocean. Farmer and Vagle (1989) and Vagle (1989) reported that breaking waves contribute to the ambient sound field at frequencies as low as 50 Hz. Hollet (1989) and Carey et al. (1993) have verified the existence of lower frequency noise in field experiments. It has been proposed that low frequency noise emissions may be attributed to single or collective mode oscillations of bubble clouds. This observation has been supported by many studies.

In contrast to bubble correlation studies, relatively few studies relate the ambient noise intensity to breaking wave characteristics. Field measurements have shown that the ambient noise did not correlate well with significant wave heights (Farmer and Vagle 1989). Significantly better correlation has been obtained between the mean acoustic intensity, p, and the mean breaking wave speed, C ($p \propto C^{5.83}$) (Ding and Farmer 1994). The mean square acoustic pressure has been found to be proportional to power of energy dissipation (D^n with n in the range of 0.6-0.8) and root mean square wave slope ($s^{3.4}$) (Felizardo and Melville 1994). In laboratory experiments, the radiated acoustic energy has correlated well with the pre-breaking wave slope and the energy dissipation (Melville et al. 1988, Loewen 1991). It was shown that the radiated acoustic power was proportional to energy dissipation (D).

In retrospect, the correlation of ambient noise characteristics to wave steepness and dissipation may be anticipated. However, much of the correlation details have yet to be properly understood. It should be noted, for example, that wave breaking could range from spilling to plunging, each having its own characteristics. This paper aims to provide a better understanding of the sound intensity level of different breaking events and to determine the relationship between acoustic radiated energy, wave energy dissipation and the wave slope parameter. Selected spectral distributions, that is, constant amplitude, constant slope and a Gaussian (Pierson-Moskowitz) spectrum were used in the study to yield a variation of the breaking characteristics.

EXPERIMENTS

The experiments were conducted in a wave flume at the Department of Civil Engineering, National University of Singapore. The ferro-cement wave flume was 36 m long, 2 m wide, 1.3 m height, and filled with fresh water to a depth of 0.8 m. The flume was equipped with a wave signal generator and a piston-type wave board, 1.98 m wide and 1.3 m high. The ability to generate a repeatable single breaking event was important for the highly transient breaking process. The technique adopted here has been used previously by Chan and Melville (1988) and, found to be reliable and accurate. The constructive interference of wave components in a frequency and amplitude modulated wave packet was used for the breaking wave simulation. The breaking wave packet was synthesized from N sinusoidal components. For a linear wave field, the free surface displacement $\eta(x,t)$ of a breaking wave can be specified by

$$\eta(x,t) = \sum_{i=1}^{N} a_i \cos[\ k_i(x - x_b) - 2\pi f_i(t - t_b)] \tag{1}$$

where a_i and k_i were the amplitude and wave number of the frequency component, f_i., (x_b, t_b) were the distance and time of wave breaking.

Three wave packets with differing frequency spectra were generated. The packets had components with i) constant amplitude, a, ii) constant slope, ak, and iii) amplitude a that follows Pierson-Moscowitz spectrum. These are denoted as CA, CS and PM, respectively in this paper. The plunging wave parameters are presented in Table 1. The spilling waves for each wave packet were obtained by reducing the gain factor (τ). Details of the wave characteristics associated with each of the breaking wave packet were reported by Kway et al. (1998). Fig.1 shows the photograph tracings of plunging waves simulated from the CA, CS and PM spectrum. It depicts the difference in the overall length scales of CS, CA and PM plungers.

Table 1. Plunging wave parameters under different wave packets

Wave packets	CS Plunger	PM Plunger	CA Plunger
Centre frequency, f_c (Hz)	0.83	0.83	0.83
Frequency bandwidth, δf (Hz)	0.54	0.54	0.54
Gain factor, τ	0.825	0.20	0.31
Amplitude at f_c, a (m)	0.0111	0.0238	0.0100
Centre frequency wave number, k_c	2.834	2.834	2.834
Slope parameter, γ (= $\tau N a k_c$)	0.73	0.38	0.25
Location of wave breaking, $x_b k_c$	39.7	39.7	39.7

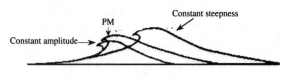

Fig. 1 Photographic tracings of plunging waves

The acoustic levels were measured using a Benthos Modified molded hydrophone with built in pre-amp. The hydrophone was positioned at 0.5m downstream of breaking location and 0.3m below the still water level. The acoustic signals were conditioned using the Precision Model 602 dual filters and recorded using a system comprising a 12 bit IOTech Wavebook (A/D) and an IBM Compatible Pentium computer. The signals were band pass filtered within 20 Hz to 20 kHz. The sampling rate was 50 kHz. The specifications of the hydrophone are presented in Table 2. To synchronise measurements, the acoustic channels were triggered from a reference wave signal voltage.

Table 2. Specifications for the Benthos Modified Molded Hydrophone

Open Circuit Sensitivity	-205 dBV re 1 µPa
Linear Frequency Range	10 Hz to 80 kHz (±3.6 dB)
Omni directional Beam Pattern	± 3 dB
Pre-Amp (AQ-501-1-3) Gain	32 dB

RESULTS AND DISCUSSION

Acoustic Pressure Time Series

The acoustic pressure time histories of incipient, spilling and plunging waves were recorded. Each of these was simulated from CS, PM and CA spectra. Altogether, nine wave packets were generated, i.e. three waves each from three amplitude distributions. A typical acoustic pressure time series for the PM plunger is shown in Fig. 2. The pressure time history was band pass filtered in the low frequency range, 20-500 Hz (Fig. 2a) and in the high frequency range, 500Hz-20kHz (Fig. 2b). A 50 Hz ground noise was observed (Fig. 2a) at the front and tail ends of the time history and hence subsequent spectral analysis was carried out for the frequencies greater than 100 Hz. The envelope maximum at t = 0.55s in the low band pass time history corresponded to a time immediately after the initial impact of the wave crest on the free surface as it plunged over. It can be seen from Fig 2(b) that the contribution to the higher frequencies was earlier than at t = 0.55s. This might be due to the early wave breaking near the flume walls (Kway 2001). The second envelope maximum at 0.62s in the high band pass time series corresponded to the presence of large number of bubbles due to the break up of the air tube. The visualization studies also showed that the time difference between the initial impacts of plunging jet to the break up of the air tube was approximately 0.08s. The higher frequency contribution persisted for longer time compared to lower frequency noise. Based on visualization studies, the duration between the formation of air tube to the stage when the bubbles were widely distributed was about 0.24s. Soon after this stage, the bubbles would become acoustically inactive. The bubbles, thereafter, would rise to the surface by its own buoyancy.

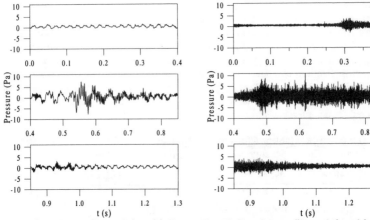

Fig. 2a Band pass filtered time history of noise level from PM plunger in the frequency range of 20 Hz to 500 Hz

Fig. 2b Band pass filtered time history of noise level from PM plunger in the frequency range of 500 Hz to 20kHz

Spectrograph

The spectrograph, a time-frequency plot, was computed with a bandwidth resolution of 48 Hz. The time series was divided into 0.02048s segments and cosine tapered at both ends of the segments. The spectrographs for PM spilling and PM plunging waves are shown in Fig. 3 and Fig. 4, respectively. The frequency is presented in logarithmic scale. Fig. 4 corresponds to the time series shown in Fig. 2. The spectral peak frequency increased from 200 Hz to 925 Hz (logf = 2.30 to 2.97) as the breaker intensity decreased from plunging to spilling. Based on single bubble resonant oscillation, the bubble sizes corresponding to the characteristics frequencies of 200 Hz and 925 Hz were 16.3 mm and 3.5 mm, respectively (Minnaert 1933). The maximum size of bubbles observed in the laboratory, however, was less than 10mm similar to Lamarre and Melville (1994). The lower acoustic frequency contributions, approximately less than 500 Hz could therefore be due to bubble cloud oscillation rather than single bubble oscillations.

In the PM plunging case, there was a significant increase in acoustic energy level around 200 Hz at t = 0.55s, corresponding to the envelope maximum in the low band pass filtered time series (Fig. 2a). The presence of significant low frequency acoustic energy between t = 0.55s and 0.62s, indicated that most of the entrapped air would be released to the atmosphere in the form of larger bubbles as soon as the air tube breaks.

The sound level from the PM spilling breaker was distributed over a narrower frequency band at the time of spilling at t = 0.6s when compared to the plunging waves. The maximum acoustic pressure occurred at about 0.17s after the initiation of spilling. The corresponding spectral peak frequency was 925 Hz. The maximum ambient noise level of the spilling wave was lower than the plunging sound levels by about 13 dB. The low frequency energy attributed to the entrapped air tube in the plunging case was absent in the spilling case, hence supporting the earlier deduction.

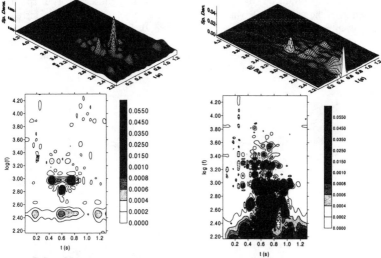

Fig. 3 Spectrograph of PM spiller   **Fig. 4 Spectrograph of PM plunger**

From the time series and contour plots, it was noted that the plunger was acoustically dynamic for about 0.7s and the spiller for about 0.5s. Ding and Farmer (1994) showed that the acoustic duration was about one half of breaking wave period. In the present study, the breaking wave period was 1.2s.

Pressure Spectrum Level

The averaged Pressure Spectrum Level (PSL) has been defined as,

$$PSL = 10\,log_{10}\left(\frac{4S(f)}{\left(1x10^{-6}\right)^2}\right) \quad \text{dB re 1 } \mu Pa^2/Hz \tag{2}$$

where $S(f)$ is the auto-spectral density function of acoustic pressure time series. A 15-point running average smoothing was applied to the spectral estimates. The PSL is presented with logarithmic frequency scale.

The ambient noise spectra for waves simulated with a constant-slope spectral distribution are presented in Fig. 5. There are three spectra plotted in the figure and these correspond to incipient, spilling and plunging waves. The spectrum for the incipient wave was the lowest spectrum and could be considered as the background noise level. PSL of the plunging wave was about 20 to 30 dB higher than the incipient wave. The acoustic energy of the plunging wave covered the entire frequency spectrum considered in the bandwidth of 100 Hz to 20 kHz. The low frequency sound produced by plunging waves was significant. The spectral contribution of the spilling case was dominant above 300 Hz. There was an increase in PSL of up to 20 dB in higher frequencies above 2kHz. The higher energy in the low frequency band associated with plunging waves may be attributed to the dynamics of the entrapped air tube and the resulting larger volume of bubble clouds. In a spiller, the bubble clouds were relatively smaller.

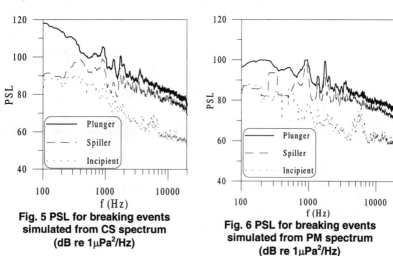

Fig. 5 PSL for breaking events simulated from CS spectrum (dB re 1μPa²/Hz)

Fig. 6 PSL for breaking events simulated from PM spectrum (dB re 1μPa²/Hz)

Fig. 6 depicts the PSL of an incipient, spilling and plunging waves simulated from PM spectrum. PSL of the plunger was 10 to 20 dB more than that of incipient wave. The noise level rose from 10 dB near lower frequency bands less than 600 Hz to 15-20 dB above 2kHz. It was noted that the contribution by the spillers to the frequency spectrum starts from 600 Hz. The pressure energy spectral level of the breaking events simulated from a CA spectrum is presented in Fig. 7. The variation of PSL was similar to the breaking events simulated from CS and PM spectra. The lower frequency contribution of the CA plunger was less than the CS and PM plungers. CA plungers were weak compared to CS and PM plungers and thus, the void fraction in the water column would be less. However, there was a significant contribution by the plungers in the frequency range 300 Hz to 1 kHz compared to the spiller.

Correlation with Wave Dynamics

The variation of sound intensity from different types of breaking waves showed that there was a systematic correlation to the breaking wave dynamics. In the experiments, the CS, PM and CA spectra tested were effectively of three different length scales physically. The corresponding energy dissipation derived from the upstream and downstream surface elevation also varied accordingly. In physical scales, the noise intensity increases with the length scales as expected. Based on the results of Loewen (1991), one could expect a dependence on the relative slope of the plunging wave and also on the wave energy dissipated. From the experimental measurements, it has been found that, the square of the possible maximum slope parameter (γ_{max}) was proportional to the ratio of the acoustic radiated energy (p^2) to the energy dissipation (D). The relation may be given by

$$p^2 = \gamma_{max}^2 (D - D_i) \qquad (3)$$

where, D_i is the energy dissipation due to flume characteristics such as sidewall friction. The radiated energy, p^2 was computed from the first order moment of the pressure spectrum, S(f). The maximum possible slope parameter, γ_{max} corresponded to the steepest wave in a given wave packet, one that yielded a single strong plunger. The normalized acoustic radiated energy (ε) is plotted versus energy dissipation in Fig. 8. The acoustic energy was normalized with the square of the maximum possible slope parameter according to Eq. (3). The acoustic measurements from the nine wave packets discussed in the earlier sections are labeled CS, PM and CA in the plot. Acoustic radiated energy from CS wave packet in a scaled up experiments with a water depth of 1.85m is also plotted. Results obtained in other studies were also examined. These include the experimental measurements by Loewen (1991) for the three CS wave packets P_1, P_2 and P_3 presented in his work (Fig. 2.26, p.121 of Loewen 1991). In overall, the proposed correlation was found to be in good agreement with the experimental data.

CONCLUSIONS

In this paper, breaking wave induced noise intensities have been studied through a series of controlled laboratory experiments. The spilling and plunging waves were simulated with wave spectrum comprising three amplitude distributions: constant slope, PM and constant-amplitude spectra. The characteristics of noise levels in the different frequency bands were discussed. The noise spectra associated with plungers were more broad banded than those associated with spillers.

The ambient noise level has been found to be dependent on the intensity of wave breaking as well as the overall length scales of breaking waves. A strong correlation was obtained between the energy dissipation, radiated acoustic energy and the maximum slope parameter.

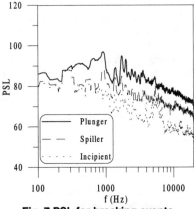

Fig. 7 PSL for breaking events simulated from CA spectrum (dB re 1 μPa²/Hz)

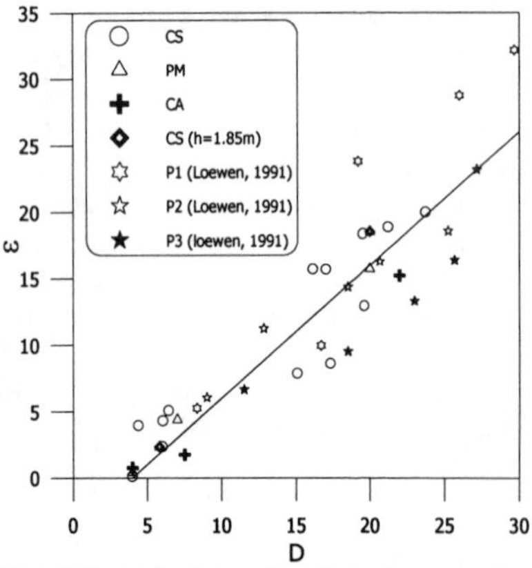

Fig. 8 Correlation of radiated acoustic energy with wave slope and wave energy dissipation. $\varepsilon = p^2 / \gamma_{max}^2$

ACKNOWLEDGEMENTS

This study was the part of the research project "The ambient noise and structure of water column in Singapore waters", sponsored by a research grant, GR6414.

REFERENCES

Carey, W.M., Fitzgerald, J.W. and Browning, D.G. 1993. Low Frequency Noise from Breaking Waves. *Natural Physical Sources of Underwater Sound*, ed. by B.R. Kerman, Kluwer, Dordrecht, 277-304.

Chan, E.S. and Melville, W.K. 1988. Deep-water plunging wave pressures on a vertical plane wall. *Proc. Royal Soc. Of London*, A417, 95-131.

Ding, L. and Farmer, D.M. 1994. Acoustical observation and Numerical simulation of breaking wave statistics. *Sea Surface Sound 1994, Proceedings of the Third International Meeting on natural physical processes related to sea surface sound*, edited by M.J. Buckingham and J.R. Potter, World Scientific, 185-201.

Farmer, D.M. and Vagle, S. 1989. Waveguide propagation of ambient sound in the ocean surface bubble layer. *Journal of the Acoustical Society of America*, 86, 1897-1908.

Felizardo, F.C. and Melville, W.K. 1994. Ambient surface noise and ocean surface waves. *Sea Surface Sound 1994, Proceedings of the Third International Meeting on natural physical processes related to sea surface sound,* edited by M.J. Buckingham and J.R. Potter, World Scientific, 202-213.

Hollet, R. 1989. Underwater sound from whitecaps at sea. *Journal of the Acoustical Society of America,* 85(S1), S145.

Knudsen, V.O, Alford, R.S. and Emling, J.W. 1948. Underwater Ambient Noise. *Journal of Marine Research,* 7, 410-429.

Kway, J.H.L., Loh, Y.S. and Chan, E.S. 1998. Laboratory study of deep-water breaking waves. *Ocean Engineering,* 25 (8), 657-676.

Kway, J.H.L. 2001. *Laboratory study of the kinetics and dynamics of deep-water wave breaking. Ph.D. thesis,* NUS, Singapore.

Lamarre, E.L. and Melville, W.K. 1994. Void-Fraction and Sound-Speed Measurements Near the Ocean Surface, *Sea Surface Sound 1994, Proc. Third International Meeting on natural physical processes related to sea surface sound,* edited by M.J. Buckingham and J.R. Potter, World Scientific, 304-311.

Loewen, M.R. 1991, *Laboratory measurements of the sound generated by breaking waves.* Ph.D. thesis, Mass. Institute of Technology.

Medwin, H. and Beaky, M.M. 1989. Bubble sources of the Knudsen sea noise spectra. *Journal of the Acoustical Society of America,* 86, 1124-1130.

Medwin, H. and Daniel, A.C. 1990. Acoustical Measurements of Bubble Production by Spilling Breakers. *Journal of the Acoustical Society of America,* 88, 408-412.

Melville, W.K. 1995. Oceanographic Applications of Natural Sea Surface Sound. *Sea Surface Sound 1994, Proceedings of the Third International Meeting on natural physical processes related to sea surface sound,* ed. by M.J. Buckingham and J.R. Potter, World Scientific, 3-19.

Melville, W.K., Loewen, M.R., Felizardo, F.C., Jessup, A.T. and Bukingham, M.J. 1988. Acoustic and microwave signatures of breaking waves. *Nature,* 336, 54-59.

Minnaert, M. 1933. On musical air bubbles and the sound of running water. *Phil. Mag.,* 16, 235-248.

Vagle, S. 1989, *An acoustical study of the upper-ocean boundary layer.* Ph.D. thesis, University of Victoria & Institute of Ocean Science, B.C. Canada.

IMPLEMENTATION AND VALIDATION OF A BREAKER MODEL IN A FULLY NONLINEAR WAVE PROPAGATION MODEL

Stéphan N. Guignard [1] and Stéphan T. Grilli[1], M.ASCE

Abstract: A spilling breaker model is implemented in a two-dimensional fully nonlinear coastal wave propagation model. A maximum surface slope breaking criterion is used to identify breaking waves within the incident wave train. Energy dissipation is achieved by specifying an absorbing surface pressure over breaking wave crest areas. The pressure is proportional to the normal particle velocity on the free surface. The instantaneous power dissipated in each breaking wave is specified proportional to the dissipation in a hydraulic jump of identical characteristics. Computations for a periodic wave shoaling and breaking over a plane slope are compared to laboratory experiments. The agreement is quite good, although more work remains to be done in refining the breaker model parameters.

INTRODUCTION

Coastal wave propagation models have been developed, in recent years, following three main approaches : (i) the solution of Fully Nonlinear Potential Flow (FNPF) equations, typically, in an Eulerian-Lagrangian formulation based on a Boundary Integral Equation representation of the solution (e.g., Grilli et al. 1989; Ohyama and Nadaoka 1991); (ii) the derivation and (usually finite difference) solution of approximate long wave equations (such as Boussinesq equations with improved dispersion and/or nonlinear characteristics, e.g., Schäffer et al. 1993; Wei et al. 1995); (iii) the solution of Euler/Navier-Stokes equations, using a Volume of Fluid (VOF) or a Mark And Cell (MAC) method (e.g., Lin and Liu 1998; Guignard et al. 1999).

Models based on approach (i) have proved very accurate for modeling highly nonlinear waves shoaling over arbitrary bottom topography in two-dimensions (2D) (e.g., Grilli et al. 1994; Ohyama et al. 1994; Grilli and Horrillo 1999). In these models, the overturning of one wave can be accurately modeled, up to impact of the breaker jet on the free surface (e.g., Grilli et al. 1997; Li and

---

[1]Dept. of Ocean Engng., University of Rhode Island, Narragansett, RI 02882, USA. E-mail: grilli@oce.uri.edu.

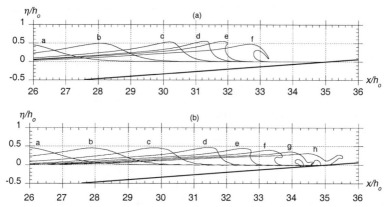

FIG. 1. (a) FNPF and (b) coupled VOF/FNPF computation of the breaking of a solitary wave on a 1:15 slope (based on Guignard et al. 1999).

Raichlen 1998; Grilli et al. 1998; Fig. 1a). [Note that similar FNPF computations of three-dimensional overturning nearshore waves have also recently been performed by Grilli et al. 2001.] Computations, however, break down beyond jet impact, unless waves are "numerically" prevented from breaking. Since potential flow models do not normally include energy dissipation terms (e.g., representing bottom friction and wave breaking effects), wave breaking is typically prevented by dissipating incident wave energy in so-called Absorbing Beaches (AB), using a surface pressure and/or actively absorbing boundaries (e.g., Clément 1996; Grilli and Horrillo 1997). This is further detailed below.

FNPF equations are only approximately solved in models based on approach (ii), in a depth integrated formulation (precluding the modeling of steep bottom obstacles), in which wave nonlinearity and dispersion are only represented to a certain order. These models, however, are less computationally demanding than in approach (i), which makes it possible to perform computations over horizontal 2D domains of meaningful size. Also, energy dissipation terms can be directly introduced in the equations (e.g., Karambas and Koutitas 1992; Schäffer et al. 1993; Skotner and Apelt 1999; Kennedy et al. 2000).

In approach (iii), full dynamic equations are solved, either for the mean fields, on a coarser global grid, together with a turbulence model representing dissipation at sub-grid scales (Lin and Liu 1998), or directly on a finer grid (Guignard et al. 2001). In the latter case, very detailed shape and kinematics of breaking waves can be obtained. However, computational cost rapidly becomes prohibitive, even for small size vertical 2D domain. In that respect, a more efficient solution, proposed by Guignard et al. (1999), has been to combine approaches (i) and (iii), i.e., to couple a FNPF model for representing shoaling waves, to a VOF model for representing breaking waves (Fig. 1b).

In the present work, we use the 2D-FNPF model, initially developed by Grilli et al. (1989) (GSS), to study coastal wave shoaling and breaking over slopes. Grilli and Subramanya (1996) implemented more accurate discretization methods and a node regridding technique in this model, and were able to accurately calculate solitary waves breaking over mild and steep slopes (Grilli et al. 1997). In particular, they were able to predict whether and how (i.e., spilling, plunging, or surging) waves break on a plane slope. Grilli and Horrillo (1997) (GH) implemented exact periodic wave generation (streamfunction wave solution), and numerical energy absorption in the model. Absorption of energy was achieved through a combination of a surface pressure, working against waves, and an open active absorbing boundary, within an AB. They were able to calculate numerically-exact fully nonlinear properties of periodic waves shoaling over mild monotonous slopes, such as wave height and celerity variations (Grilli and Horrillo 1996), and wave transformations over barred-beaches (Grilli and Horrillo 1999). Various comparisons of these numerical results with laboratory experiments showed a very good agreement.

In a FNPF model, periodic waves, shoaling over a sloping bottom, can thus be effectively absorbed in an AB, before they start overturning (Grilli and Horrillo 1997; AB in Fig. 2). For non-periodic incident waves and/or irregular bathymetry, however, it is desirable to have a means of both preventing waves from overturning during shoaling, while gradually dissipating the energy of such individual breaking waves, in relation with the physical rate of energy dissipation in actual waves. This approach is followed in some Boussinesq models, in which empirical eddy viscosity terms are added in the momentum equation, combined to a breaking criterion, such as a maximum angle on the wave front face (Schäffer et al., 1993). The eddy viscosity is calibrated based on laboratory experiments. Here, a spilling breaker model is implemented in the NWT (from the breaking point $x = x_b$ to x_a in Fig. 2), in which the instantaneous rate of energy dissipation for each broken wave is assumed to be that of a hydraulic jump. This analogy was suggested by Svendsen et al. (1978), based on experiments, to estimate the rate of energy dissipation in surf-zone waves.

A maximum/minimum front slope criterion, similar to Schäffer et al.'s, is used to determine whether a wave starts or stops breaking. Following the method used in GH's AB, an absorbing surface pressure distribution is specified over each breaking wave crest area, from the point where normal velocity changes sign behind the crest, to the similar point on the wave front face. The (negative) work produced by this pressure against the wave is calibrated in real time to be proportional to the energy dissipated in an inverted hydraulic jump. This requires knowing values of instantaneous wave characteristics such as wave height H, celerity c, and depth below crest $h_c = h + H$, and trough, h_t (Fig. 3). Hence, a wave tracking algorithm is developed, in which individual waves are identified and followed throughout their shoaling and breaking in the model, while the breaking creterion is being checked. The spilling breaker model is calibrated by comparing results to laboratory experiments for mean wave height, mean-water-

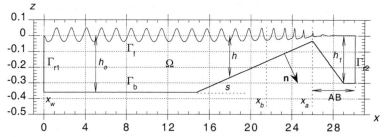

FIG. 2. Sketch of FNPF model for periodic wave shoaling and breaking computations.

level (MWL), and wave celerity variations, during shoaling of periodic waves generated by a piston wavemaker (Hansen and Svendsen 1979).

Note that, in recent years, other criteria have been proposed to detect and/or remove breaking in wave models. Thus, Subramani et al. (1998) developed a maximum surface curvature criterion, to identify deep water breaking waves ($\kappa H \leq 0.7$, with κ the crest curvature), and Gentaz and Alessandrini (2000) used a criterion based on a threshold vertical pressure gradient at the free surface ($\frac{\partial p}{\partial z} \geq \rho g$). Finally, note the method introduced by Wang et al. (1995) to suppress breaking in their computations, in which water is simply peeled away from wave crests reaching a limiting height (this, however, violates mass conservation).

THE NUMERICAL WAVETANK

Governing equations and boundary conditions

Equations for the 2D-FNPF model can be found in the references (GSS,GH). The velocity potential $\phi(x, z, t)$ is used to describe inviscid irrotational flows in the vertical plane and the velocity is defined as $\boldsymbol{u} = \boldsymbol{\nabla}\phi$ (Fig. 2). Continuity equation in the fluid domain $\Omega(t)$ with boundary $\Gamma(t)$ is a Laplace's equation for the potential. On the free surface $\Gamma_f(t)$, ϕ satisfies fully nonlinear kinematic and dynamic boundary conditions. Along the stationary parts of the boundary such as bottom Γ_b and Γ_{r2}, a no-flow condition is prescribed. For comparing results with experiments, periodic waves are generated on boundary $\Gamma_{r1}(t)$ using a solid piston wavemaker ($x = x_w$), moving according to a first-order cnoidal wave solution.

Numerical model

At each time t, the model solves a Boundary Integral Equation (BIE), representing continuity equation, using a higher-order Boundary Element Method (BEM) (see GSS, and Grilli and Subramanya 1996, for details). Free surface boundary conditions are time integrated based on two second-order Taylor series expansions expressed in terms of a time step Δt and the Lagrangian time derivative of free surface position and potential. Detailed expressions for the Taylor series are given in GSS. The optimal time step for each time is selected based on a mesh Courant number $\mathcal{C}_o(t) \simeq 0.45$.

In computations involving finite amplitude waves, mean drift currents occur which continuously move discretization nodes/Lagrangian markers forward in the model. To counteract this effect, Grilli and Subramanya (1996) developed regridding methods in which nodes can be redistributed at constant arclength intervals over specified regions of the free surface. Here, to limit the number of nodes on the free surface and the computational cost, the initial horizontal node spacing Δx_o is gradually reduced over the slope, to match the reduction in wavelength due to shoaling and maintain a node density of at least 15 nodes per waves (from an initial 20 nodes per wavelength over the constant depth part of the NWT). Hence, the method of constant arclength regridding can not be applied. Instead, a new regridding method was used in which the initial ratio of each BEM element length to the total length of the free surface is maintained for all times.

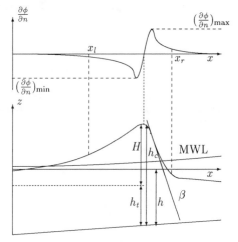

FIG. 3. Definition of geometric parameters for the breaker model. [Note, H is defined as the height between a crest and the previous trough.]

Wave energy absorption

The same method as in GH's AB is used to selectively absorb energy from waves, identified to break for $x \leq x_a$, i.e., an absorbing surface pressure is specified in the dynamic free surface condition (with $z = \eta$), proportional to the normal particle velocity on the free surface. An AB is still used to absorb residual wave energy exiting at the top of the slope, for $x > x_a$ (Fig. 2). [To create additional wave reduction through de-shoaling, the bottom geometry within the AB is specified somewhat similar to a natural bar, with a depth increasing to $h = h_1$.] A wave tracking algorithm (detailed below) identifies breaking waves, within the incident wave train, based on a breaking criterion. The breaker model is assumed to extend from the crest of each such wave to two points on each side of the crest, where $|\frac{\partial \phi}{\partial n}/(\frac{\partial \phi}{\partial n})_{\text{min,max}}| > \varepsilon$ ($x_l \leq x \leq x_r$), with ε is a small threshold value (Fig. 3). $[|(\frac{\partial \phi}{\partial n})_{\text{min,max}}|$ are defined as the maximum absolute normal velocity for each side of the wave.] Over each breaker, the absorbing pressure is defined as,

$$p_{bm}(x,\eta,t) = \nu_{bm}(x)\frac{\partial \phi}{\partial n}(\eta(x,t)) \tag{1}$$

in which $\nu_{bm} = \nu_{bo}\, S(x)$, with $S(x)$ a breaker shape function providing a smooth transition from areas without the absorbing pressure, to the breaker regions over each breaking wave.

The instantaneous power dissipated by each breaking wave is given by,

$$\mathcal{P}_b = \int_{x_l}^{x_r} p_b \frac{\partial \phi}{\partial n} d\Gamma = \nu_{bo} \int_{x_l}^{x_r} S(x)\left(\frac{\partial \phi}{\partial n}\right)^2 d\Gamma \tag{2}$$

and is assumed to be proportional to the power dissipated in a turbulent hydraulic jump (e.g., Lamb 1932, p 280),

$$\mathcal{P}_h = \rho\, g\, c\, \frac{h\, H^3}{4\, h_c\, h_t} \tag{3}$$

where H denotes the wave height, h_t the water depth below trough, $h_c = h_t + H$ the water depth below crest, and (here) c is the absolute wave crest phase speed. Following Svendsen et al. (1978), we define, $\mathcal{P}_{bm} = \mu\, \mathcal{P}_h$, with $\mu \simeq 1.5$. All calculations done, the instantaneous value of the breaker coefficient ν_{bo} for each breaking wave is found based on the values of both wave and breaker parameters $(H, h, h_t, \mu, \varepsilon)$, and the wave shape and kinematics in between x_l and x_r.

Wave tracking algorithm and breaking criteria

For each time t, (x, z) locations of local maxima and minima in surface elevation $\eta(x, t)$ are identified. The crest of each wave is calculated as the highest elevation in between two successive minima. A wave is defined only if its height H is greater than a specified fraction of the incident wave height H_o (typically one-tenth). This avoids including secondary wave crests created during shoaling.

For each wave defined this way, the algorithm finds which wave $i = 1, \ldots$ it corresponded to at the previous time $t - \Delta t$. The search for the right wave is accelerated by extrapolating the crest position of each earlier wave i, to time t,

$$\tilde{x}_c^i(t) \simeq x_c^i(t - \Delta t) + c^i\, \Delta t \tag{4}$$

where, $c^i \simeq \frac{\partial \phi}{\partial t}(x_c^i)/\frac{\partial \phi}{\partial x}(x_c^i)$ is the i's wave crest celerity at time $t - \Delta t$ (assuming a permanent form for the wave over time Δt), and comparing it to the crest position $x_c(t)$ found for the wave currently under consideration. Wave heights $H^i(x, t)$ are stored for each wave after final identification is made as well as other geometric parameters needed to calculate the hydraulic jump power dissipation

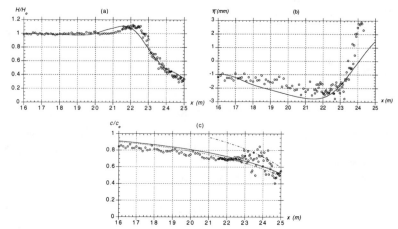

FIG. 4. Calculated (—-) : (a) mean wave height; (b) MWL; and (c) mean celerity. (o) are experimental data. In (c) : NSW (– - –) and LWT (- - -).

using Eq. (3). Crest trajectories $x_c^i(t)$ are calculated for each wave $i = 1, \ldots$ in the incident wave train, and celerities are calculated as the time derivatives of these. Wavelengths are readily found as $L^i = c^i T$, assuming a constant wave period.

The breaking criterion (a simple wave front angle threshold criteria, $\beta > \beta_{max}$) is checked for each wave i identified at time t, to decide whether it breaks or not. For those waves j that break, the procedure described in the previous section is applied to calculate the interval $x_l^j \le x \le x_r^j$, in between which an absorbing pressure p_{bm}^j should be applied (Fig. 3), according to Eq. (1).

APPLICATIONS

One computation is presented here, representing laboratory experiments by (Hansen and Svendsen 1979), for periodic waves with : $H_o = 0.095$ m (at the toe of the slope), $T = 1$ s, and $h_o = 0.36$ m, shoaling over a $s = 1/34.26$ plane slope for $x \ge 14.78$ m (Fig. 2). In the experiments, waves were generated by a piston wavemaker at $x = 0$, using a second-order wave generation method. Variations of wave-averaged, wave height H, mean-water-level (MWL) $\bar{\eta}$, and celerity c, were measured as a function of x. The same tank geometry and wave characteristics are used in the model. Wave generation is performed using a piston wavemaker moved according to a first-order cnoidal wave solution, with wave height 0.095 m. It was observed that generated waves slightly adjusted their shape and height as they propagated over constant depth down the modeled tank, likely due to nonlinear effects. The incident wave finally reached a stabilized height $H_o = 0.085$ m at the toe of the slope i.e., a smaller value than in the corresponding experiments (which used a wave height 0.1 m at the wavemaker). This initial comparison of numerical results with experiments was carried out using those slightly smaller generated waves, after scaling by the wave height.

An AB, with water depth deepening to $h_1 = 0.3$ m, is specified at the top of the slope for $x \geq x_a = 25.96$ m and $x \leq 30$, with $\nu_{ao} = 0.01$. The minimum depth on the slope at $x = x_a$ is thus $h_a = 0.034$ m. Breaking is assumed to occur when the maximum wave front angle reaches $\beta_{\max} = 37^o$, with $\varepsilon = 10^{-5}$. Finally, we select $\mu = 1.5$.

BEM discretization and time step parameters are selected as described before. There is a total of $N = 672$ nodes on the boundary and the initial time step is $\Delta t_o = 0.01$ s. A total of 5000 time steps were run in these computations. The maximum relative numerical error, with respect to the initial volume of the tank ($V_o = 8.322 \, \mathrm{m}^3/\mathrm{m}$) was only 0.011% after 3500 iterations (time $t = 21$ s). At this stage, the model reached an almost steady state in which larger waves kept entering the AB from the top of the slope. This led to somewhat larger numerical errors. After 5000 iterations, at time $t = 30.04$ s the volume error increased to 0.087%, which is still quite small.

Fig. 4 show results of computations averaged over 6 successive waves, after computations reach an almost steady state. Breaking occurs in average for $x_b \simeq 21.5$ m (Fig. 4a). There is a shoaling region for $x < x_b$, where wave height $H(x)$ eventually slightly increases, and a surf-zone region beyond breaking, where H decreases. The agreement with experiments is quite good although breaking occurs slightly too early (i.e. for too small a x) in the model. The agreement of computed MWL with experiments is also good for $x < 24$ m (Fig. 4b). Fig. 4c shows the calculated celerity, together with the celerity predicted by linear wave theory, $c_{\mathrm{LWT}} = c_o \tanh kh$ (with $c_o = gT/2\pi = 1.56$ m/s, wavenumber $k = 2\pi/L$, and wavelength $L = cT$), and that of the Nonlinear Shallow Water equations, $c_{\mathrm{NSW}} = \sqrt{g(H + h)}$ (using the FNPF results for H). The agreement of c and c_{LWT} with experiments is quite good during the shoaling part ($x < 21.5$ m), considering the difficulty in accurately measuring celerities reported by Hansen and Svendsen, and the small difference in H_o value. Beyond breaking, experimental results show a large variance, maybe due to difficulties in identifying crests in the experiments. Experimental results however quite well illustrate the effects of amplitude dispersion, which lead to a larger celerity for larger waves. This is also seen in larger FNPF results as compared to linear wave theory. [The marked increase in celerity seen beyond breaking in experiments, and also predicted in FNPF computations reported by Wei et al. 1995, is not modeled here since the formation of a fast forward moving jet is prevented.] Nonlinear shallow water equations overpredict wave celerity in most places.

CONCLUSIONS

Results show, the spilling breaker model implemented in the FNPF model correctly accounts for the overall physics of periodic waves breaking over a sloping bottom, i.e., the approximate location and height of breaking, and the rate of energy dissipation in breakers, leading to a reduction in wave height and an increase in MWL in the surf-zone. The good agreement of numerical and experimental results confirms the relevance of the hydraulic jump analogy, with $\mu = 1.5$. More work is required to see whether this μ value is general.

The β_{max} value in the breaking criterion is larger than typically used in Boussinesq models, likely because the FNPF model produces steeper waves and thus delays the onset of breaking. Since breaking still occurs too soon, it would be of interest to try and increase β_{max} further. A limit, however, is put on this increase in the sense that, for too large a β_{max} value, it may be difficult to absorb enough (or quickly enough) of the wave energy to prevent wave overturning from occurring in the model.

ACKNOWLEDGMENT

This research was supported by the Office of Naval Research, under grant N00014-99-1-0439 from the US Department of the Navy, Office of the Chief of Naval Research. The information reported in this work does not necessarily reflect the position of the US Government.

REFERENCES

Clément, A. (1996). "Coupling of two absorbing boundary conditions for 2d time-domain simulations of free surface gravity waves." *J. Comp. Phys.*, 26, 139–151.

Gentaz, L. and Alessandrini, B. (2000). "Detection of wave breaking in a 2d viscous numerical wave tank." *Proc. 10th Offshore and Polar Engng. Conf.*, Vol. 3. 263–270.

Grilli, S., Guyenne, P., and Dias, F. (2001). "A fully nonlinear model for three-dimensional overturning waves over arbitrary bottom." *Intl. J. Numer. Methods in Fluids*, 35(7), 829–867.

Grilli, S. and Horrillo, J. (1996). "Fully nonlinear properties of periodic waves shoaling over mild slope." *Proc. 25th Intl. Conf. on Coastal Engng.*, Vol. 1. ASCE Edition, 717–730.

Grilli, S. and Horrillo, J. (1997). "Numerical generation and absorption of fully nonlinear periodic waves." *J. Engng. Mech.*, 123(10), 1060–1069.

Grilli, S. and Horrillo, J. (1999). "Shoaling of periodic waves over barred-beaches in a fully nonlinear numerical wave tank." *Intl. J. Offshore and Polar Engng.*, 9(4), 257–263.

Grilli, S., Skourup, J., and Svendsen, I. (1989). "An efficient boundary element method for nonlinear water waves." *Engng. Analysis with Boundary Elements*, 6(2), 97–107.

Grilli, S. and Subramanya, R. (1996). "Numerical modeling of wave breaking induced by fixed or moving boundaries." *Computational Mech.*, 17, 374–391.

Grilli, S., Subramanya, R., Svendsen, I., and Veeramony, J. (1994). "Shoaling of solitary waves on plane beaches." *J. Waterway Port Coastal and Ocean Engng.*, 120(6), 609–628.

Grilli, S., Svendsen, I., and Subramanya, R. (1997). "Breaking criterion and characteristics for solitary waves on slopes." *J. Waterway Port Coastal and Ocean Engng.*, 123(3), 102–112.

Grilli, S., Svendsen, I., and Subramanya, R. (1998). "Breaking criterion and characteristics for solitary waves on slopes – closure." *J. Waterway Port Coastal and Ocean Engng.*, 124(6), 333–335.

Guignard, S., Grilli, S., Marcer, R., and Rey, V. (1999). "Computation of shoaling and breaking waves in nearshore areas by the coupling of BEM and VOF methods." *Proc. 9th Offshore and Polar Engng. Conf.*, Vol. 3. 304–309.

Guignard, S., Marcer, R., Rey, V., Kharif, C., and Fraunié, P. (2001). "Solitary wave breaking on sloping beaches : 2D two-phase flow numerical simulation by sl-vof method." *Eur. J. Mech. B - FLuids*, 20, 57–74.

Hansen, J. and Svendsen, I. (1979). "Regular waves in shoaling water. Experimental data." *Series paper No. 21*, Technical University of Denmark, Institute of Hydrodynamics and Hydraulic Engng.

Karambas, T. and Koutitas, C. (1992). "A breaking wave propagation model based on the Boussinesq equations." *Coastal Engng.*, 18, 1–19.

Kennedy, A., Chen, Q., Kirby, J., and Dalrymple, R. (2000). "Boussinesq modeling of wave transformation, breaking, and runup. I: 1D." *J. Waterway Port Coastal and Ocean Engng.*, 126(1), 39–47.

Lamb, H. (1932). *Hydrodynamics*. Dover Publications, New York, 6th edition.

Li, Y. and Raichlen, F. (1998). "Breaking criterion and characteristics for solitary waves on slopes — discussion." *J. Waterways Port Coastal Ocean Engng.*, 124(6), 329–333.

Lin, P. and Liu, P. (1998). "A numerical study of breaking waves in the surf zone." *J. Fluid Mech.*, 359, 239–264.

Ohyama, T. and Nadaoka, K. (1991). "Development of a numerical wave tank for analysis of nonlinear and irregular wave fields." *Fluid Dyn. Res.*, 8, 231–251.

Ohyama, T., S., B., Nadaoka, K., and Battjes, J. (1994). "Experimental verification of numerical model for nonlinear wave evolution." *J. Waterway Port Coastal and Ocean Engng.*, 120(6), 637–644.

Schäffer, H., Madsen, P., and Deigaard, R. (1993). "A Boussinesq model for wave breaking in shallow water." *Coastal Engng.*, 20, 185–202.

Skotner, C. and Apelt, C. (1999). "Application of a Boussinesq model for the computation of breaking waves. Part 1 : Development and verification." *Ocean Engng.*, 26, 907–926.

Subramani, A., Beck, R., and Schultz, W. (1998). "Suppression of wave breaking in nonlinear water wave computations." *Proc. 13th Intl. Workshop Water Waves and Floating Bodies.* Delft University of Technology, 139–141.

Svendsen, I., Madsen, P., and Hansen, J. (1978). "Wave characteristics in the surf zone." *Proc. 16th Intl. Coastal Engng. Conf.* ASCE Edition, 521–539.

Wang, P., Yao, Y., and Tulin, M. (1995). "An efficient numerical tank for nonlinear water waves, based on the multi-subdomain approach with BEM." *Intl. J. Num. Methods in Fluids*, 20, 1315–1336.

Wei, J., Kirby, J., and Grilli, S. (1995). "A fully nonlinear Boussinesq model for surface waves. Part 1. Highly nonlinear unsteady waves." *J. Fluid Mech.*, 294, 71–92.

BREAKING IN A SPECTRAL WAVE MODEL

Jane McKee Smith[1], M.ASCE

Abstract: Spectral wave breaking parameterizations are evaluated using 15 days of field measurements from the Duck94 study at the FRF. Measurements include 14 pressure gauges from 8-m depth to the shoreline and 120 wave events (wave heights of 0.5 to 4 m and periods of 4 to 16 s). The Battjes and Janssen model applied with a full Rayleigh distribution to calculate the percentage of wave breaking gave the smallest errors. Bore-type models like Battjes and Janssen degrade significantly as grid spacing increases (resolution greater than 15 m for the FRF comparisons). For applications with coarser resolution, a limit must be placed on dissipation or a simpler breaker model should be applied. Use of spectral transformation models in the surf zone can reduce RMS error by more than a factor of three compared to a monochromatic model defining breaker height as 0.78 times depth. Including the Battjes and Janssen breaking relationship in a monochromatic model reduces error by a factor of two. Spectral breaking models accurately represent wave height in the surf zone, but do not reproduce changes in spectral shape for multi-peaked incident spectra.

INTRODUCTION

Breaking waves are a principle driving force for longshore sediment transport. A typical approach to modeling wave transformation to force longshore transport includes three steps: 1) Generate the offshore wave climate or storm sequence using a spectral generation model. The wave generation model is applied over a large domain (often in a series of grid nests) and driven by wind fields. 2) Transform the offshore waves to near-breaking depth using a spectral wave transformation model. The transformation model is applied at a finer resolution than the generation model (25 to 200 m) to resolve the nearshore shoaling and refraction. 3) Transform the pre-breaking wave parameters (height, period, and direction) to breaking using a one-dimensional monochromatic model. The breaker model used in this case may be a simple 0.78 times the water depth. The breaking wave height and direction are then used to estimate the sediment transport rate. The model approaches

[1] Engineer Research and Development Center, Coastal and Hydraulics Laboratory, 3909 Halls Ferry Road, Vicksburg, MS 39180-6199, USA, Jane.M.Smith@erdc.usace.army.mil

become much simpler from offshore to nearshore, and much of the spectral information is neglected in the final breaking wave calculation. The motivation for this paper is to evaluate how well spectra breaking models perform in the surf zone and to investigate the required model complexity. Spectral models have the advantages of including wave randomness, but they neglect nonlinearities.

The purpose of this paper is to evaluate breaking parameterizations applied in spectral transformation models. First the selected breaking parameterizations are described. Then a field data set used to evaluate the models is presented. The measurements were taken during the Duck94 field experiment at the US Army Engineer Research and Development Center, Field Research Facility (FRF) in Duck, North Carolina. Wave measurements were made along a cross-shore transect from depths of 8 m to the shoreline. A fifteen-day period of Duck94 is selected for simulation because it covers a range of incident wave conditions. Breaking parameterizations are evaluated based on errors in predicted wave height, model efficiency, and model robustness. A short discussion of spectral shape in the surf zone is also included.

BREAKER PARAMETERIZATIONS

Five breaker parameterizations were selected for evaluation in this study. These parameterizations were selected because they have been applied in spectral transformation models. The five parameterizations are:

- Modified Miche criterion (Miche 1951, Smith et al. 1997):

$$H = 0.1L \tanh(kd) \tag{1}$$

where H = significant wave height (m); L = wave length (m); k = wave number (m^{-1}); and d = water depth (m). This criterion is applied by limiting the significant wave height by the value given in Eqn. 1.

- Battjes and Janssen (1978):

$$D = 0.25 Q_b f_m (H_{max})^2 \tag{2}$$

$$H_{max} = 0.14L \tanh(kd) \tag{3}$$

where D = energy dissipation (m^2); Q_b = percentage of waves breaking based on a truncated Rayleigh distribution of wave heights; and f_m = mean frequency (s^{-1}).
Equation 2 is applied as a source term in the wave energy balance equation with the maximum wave height given by Eqn. 3. The Battjes and Janssen model is denoted as BJ. The following three parameterizations also apply the bore-type dissipation given by Eqn. 2, but redefine the maximum wave height.

- Booij et al. (1999):

$$H_{max} = 0.73d \tag{4}$$

The Booij et al. model is denoted as SWAN, after the model in which it is implemented.

- Nelson (1994, 1997):

$$H_{max} = [0.55 + 0.88 \exp(-0.012 \cot \beta)]d \tag{5}$$

where β = bed slope.

- Baldock et al. (1998):

$$H_{max} = [0.39 + 0.56 \tanh(33S_0)]d \tag{6}$$

where S_0 = deepwater wave steepness. Baldock et al. define Q_b based on a full Rayleigh distribution instead of the Battjes and Janssen truncated distribution, with Q_b

calculated explicitly. For the comparisons to data, Q_b will be calculated using both the truncated and full Rayleigh distributions for the Battjes and Janssen, Booij et al., Nelson, and Baldock et al. parameterizations.

DUCK94 DATA SET

The Duck94 field study was conducted at the FRF in Duck, North Carolina on a natural, Atlantic coast beach. One element of this multi-investigator experiment was collection of wave data along a cross-shore transect using 14 pressure gauges from the depth of 8 m to the shoreline. During Duck94, daily surveys were taken of the nearshore bathymetry. The beach profiles are gently sloping in the outer profile (~1/100) with a single nearshore bar at a depth of approximately -2 m NGVD. At the 8-m depth contour there is also a pressure gauge array that provides high-resolution directional spectra used as input to the transformation calculations. The 15-day period 10-24 Oct 1994 was selected for evaluation of the breaking criteria (Fig. 1). Four different types of conditions exist during this period: prior to the storm, waves are moderate in height and relatively steep ($H \sim 2$ m, $T_p \sim 7$ sec); at the storm peak, waves are large, steep, and are breaking along most of the profile ($H \sim 4$ m, $T_p \sim 11$ sec); as the storm wanes, height decreases and period increases as swell dominates ($H \sim 1.5$ m, $T_p \sim 14$ sec); and post-storm, the wave height is small and two wave trains are present ($H \sim 0.5$ m, $T_p \sim 9$ and 14 sec). The Duck94 wave and bathymetry data are available on the FRF web site (http://www.frf.usace.army.mil/).

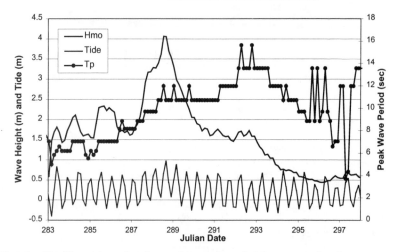

Fig. 1. Significant wave height, peak period, and tide measured at 8-m array (10-24 Oct 1994)

MODEL-DATA COMPARISONS

Comparisons of the five breaking parameterizations are illustrated in Figs. 2-4. Fig. 2 is representative of the steep pre-storm waves. The Battjes and Janssen parameterization gives the greatest dissipation offshore of the bar and the best representation of the measurements. The Baldock and modified Miche parameterizations give the least dissipation and the poorest fit to the measurements. Fig. 3 shows results at the peak of the storm. Again, the modified Miche and Baldock et al. parameterizations underestimate dissipation. The other three parameterizations give similar results, with Battjes and Janssen slightly overestimating dissipation and Booij et al. and Nelson agreeing closely with the measurements. Fig. 4 is representative of post-storm swell conditions. Baldock et al. gives the best fit for these low steepness waves. The results from Battjes and Janssen, Booij et al., and Nelson are similar and underestimate dissipation. The modified Miche provides even a greater under estimate. The full Rayleigh wave height distribution was used for the examples shown in Figs. 2-4.

For the 120 cases simulated (15 days, measurements at 3 hr intervals), the Battjes and Janssen breaker parameterization using the full Rayleigh distribution gave the smallest root-mean-square (RMS) and mean errors, 3.4% and 0.8%, respectively. The Booji et al. and Nelson parameterizations gave only slightly larger errors than Battjes and Janssen. The modified Miche and Baldock parameterizations gave RMS errors of 5.8% and 6.2%, respectively. The Baldock maximum wave height overestimated steep wave conditions, but provided better agreement with the data for swell. Results with the truncated Rayleigh distribution are similar for each of the parameterizations (errors within a couple percent). Errors were slightly smaller for the full Rayleigh distribution and the computation time for Q_b was approximately 50% shorter for the full Rayleigh distribution. A summary of the error statistics is given in Table 1.

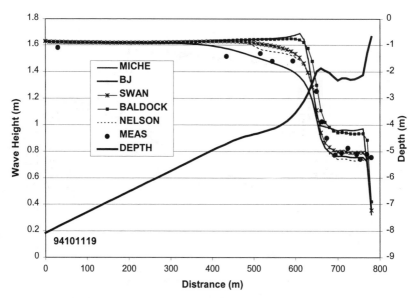

Fig. 2. Model comparison for 11 Oct 1994 at 1900 hours

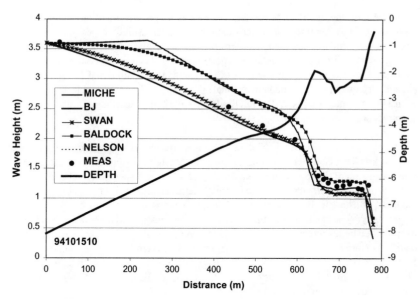

Fig. 3. Model comparison for 15 Oct 1994 at 1000 hours

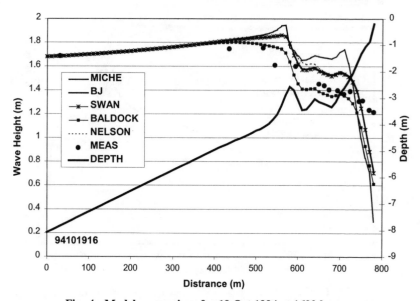

Fig. 4. Model comparison for 19 Oct 1994 at 1600 hours

Table 1. RMS and Mean Error for Breaking Parameterizations

Breaker Model	Truncated Rayleigh Distribution RMS Error, %	Mean Error, %	Full Rayleigh Distribution RMS Error, %	Mean Error, %
Mod. Miche	5.81	0.91	5.81	0.91
Battjes & Janssen	5.09	1.51	3.36	0.80
Booij et al.	4.05	1.11	3.67	0.99
Nelson	3.90	1.06	3.63	0.97
Baldock et al.	6.67	1.55	6.23	1.42

Monochromatic Implementation

Monochromatic wave transformation is more efficient for repeated calculations, but how does the accuracy compare to spectral models? To address this question, the same measurements were used to compare three models: single wave component with a limiting wave height of 0.78 times depth; a single wave component with the Battjes and Janssen breaker parameterization (Eqs. 2 and 3 with the full Rayleigh distribution); and the full spectral transformation with the Battjes and Janssen breaker parameterization. Figures 5-7 illustrate the results for the same cases shown in Figs. 2-4. Summary statistics are given in Table 2. The wave heights in the surf zone are over estimated by up to 50% using the 0.78 single component model compared to the Battjes and Janssen spectral approach. Simplifying the Battjes and Janssen from a spectrum to single component doubles the RMS error. Using the 0.78 depth breaking parameterization for a single component model doubles the error again. The spectral transformation model gives significantly smaller error in wave height estimates compared to the simple 0.78 depth model.

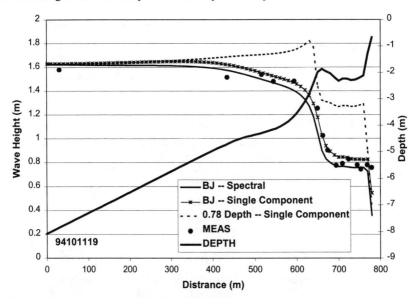

Fig. 5. Model comparison for 11 Oct 1994 at 1900 hours (monochromatic v. spectral)

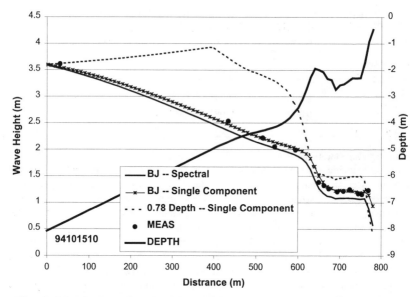

Fig. 6. Model comparison for 15 Oct 1994 at 1000 hours (monochromatic v. spectral)

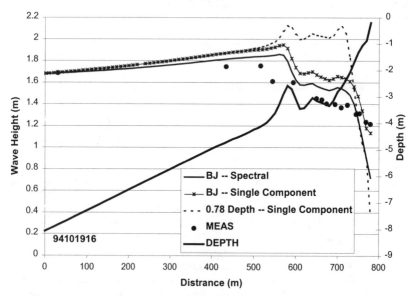

Fig. 7. Model comparison for 19 Oct 1994 at 1600 hours (monochromatic v. spectral)

Robustness

Nearshore spectral wave modeling requires that the breaking solution be robust. To investigate robustness, the Duck94 spectral simulations were repeated, varying the grid spacing from 1 to 50 m. Increasing spacing degrades the solution in two ways: 1)

Table 2. Error for Monochromatic v. Spectral Breaking Parameterizations

Breaker Model	RMS Error %	Mean Error %
0.78 Depth (mono)	11.55	2.77
Battjes & Janssen (mono)	5.75	1.05
Battjes & Janssen (spec)	3.36	0.80

bathymetry is more coarsely resolved, and 2) the dissipation calculation is poorly resolved. For the modified Miche parameterization, dissipation is not directly calculated, so the solution changes only in terms of bathymetric resolution. For the parameterizations based on Eq. 2, total dissipation is significantly overestimated for grid spacing greater than approximately 15 m for this application. Figure 8 shows RMS error as a function of grid resolution. For models with coarse resolution, the more robust Miche model is preferable. For bore-type models, error increases with grid spacing and a limit is required on the total dissipation within one grid cell (as done in the SWAN numerical model). Little is presently known about the physics or length scales of such a limiter.

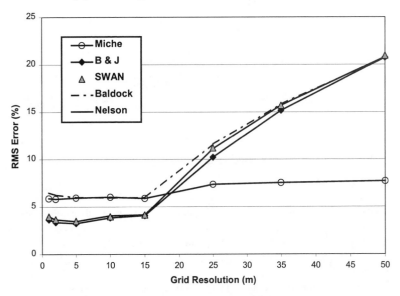

Fig. 8. Increase in RMS model error as a function of grid resolution

Spectra

The spectral transformation provides significantly better results for wave height in the surf zone than a single component approach. The next logical question is then how good is the representation of spectral shape in the surf zone? The models discussed in this paper are linear, so breaking does not change the spectral shape in the surf zone. Parameterizations of nonlinear three-wave interactions have generally given poor results in the surf zone (Vink 2001, Wood et al. 2001). For cases with single peaked spectra, linear models give reasonable agreement with the measured spectra in the surf zone, although the model tends to over predict the energy at the peak and under predict energy at high and low frequencies. For multiple incident wave trains, spectral shape changes significantly in the surf zone. Energy is maintained in the low frequency peak and dissipated from the high frequency peak (Smith and Vincent 1992). Example measured and modeled two-peaked spectra from Duck94 are given in Fig. 9. Although the linear model estimate of wave height is good, the mean and peak frequency and the mean wave direction may have significant errors.

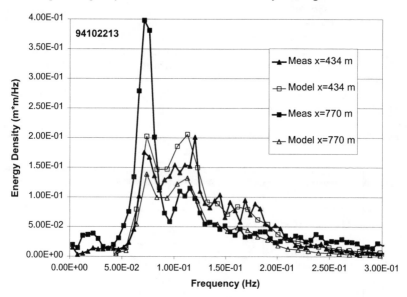

Fig. 9. Measured and modeled spectral shape for 22 Oct 1994 at 1300

CONCLUSIONS

Of the five breaker parameterizations examined in this study, the Battjes and Janssen model gave the smallest errors when applied using the full Rayleigh distribution to estimate percentage of wave breaking. Errors for the Booij et al. and Nelson models were only slightly larger. Using the full Rayleigh distribution as suggested by Baldock et al. (1998) reduced model error and reduced computational time for Q_b by 50%. Bore-type models of wave breaking give best results for high-resolution simulations (grid resolution less than 15 m), but error increased significantly for larger grid spacing. These models require a limiter if applied at larger resolutions. Little is know about the correct form for such a limiter. The simpler modified Miche model provides a more robust solution, but at the cost of increased RMS error (6% for modified Miche versus 3% for Battjes and Janssen at high resolution).

The single component (monochromatic) breaker model (0.78 depth) gave an RMS error for wave height nearly four times larger than the error for the Battjes and Janssen spectral model (12% versus 3%). The RMS error was reduced to 6% using the Battjes and Janssen breaker parameterization with the single component model. Thus, if computational demands limit modelers to a monochromatic approach, application of a breaker parameter that includes wave randomness through percentage of wave breaking can significantly improve wave height estimates. Potential sediment transport rates are reduced up to 30% by replacing the breaker height from the monochromatic 0.78 depth model with the maximum wave height from the spectral model. Additional work is needed to determine the most appropriate wave height from the spectra transformation to use for sediment transport calculations.

Linear spectral transformation models provide good estimates of wave height in the surf zone. But for complicated input spectra shapes (e.g., two wave trains), the spectral shape in the surf zone may not be accurately represented. More computationally rigorous solution methods, e.g., Boussinesq models, or improved parameterizations are required to represent surf zone spectral shapes.

ACKNOWLEDGEMENTS
This project was supported by the Transformation-Scale Waves Work Unit of the Engineer Research and Development Center, Coastal and Hydraulics Laboratory (CHL). Permission to publish this paper was granted by the Chief of Engineers, U.S. Army Corps of Engineers. Tom Herbers (Naval Postgraduate School), Steve Elgar (Woods Hole Oceanographic Institution), Bob Guza (Scripps Institution of Oceanography (SIO)), and Bill O'Reilly (SIO) collected and analyzed the Duck94 nearshore wave data. The FRF supplied wave measurements from the 8-m array (Chuck Long) and bathymetry data. Mark Gravens (CHL) and Ernie Smith (CHL) provided helpful review comments.

REFERENCES
Baldock, T.E., P. Holmes, S. Bunker, and P. Van Weert. 1998. Cross-shore hydrodynamics within an unsaturated surf zone. *Coastal Engineering*, 34, 173-196.

Battjes, J.A. and J.P.F.M. Janssen. 1978. Energy loss and set-up due to breaking of random waves. *Proceedings 16th International Conference on Coastal Engineering*, ASCE, 569-587.

Booij, N., R.C. Ris, and L.H. Holthuijsen. 1999. A third-generation wave model for coastal regions 1. Model description and validation. *Journal of Geophysical Research*, 104(C4), 7649-7666.

Miche, M. 1951. "Le pouvoir reflechissant des ouvrages maritimes exposes a l' action de la houle." *Annals des Ponts et Chaussess*, 121e Annee, 285-319 (translated by Lincoln and Chevron, University of California, Berkeley, Wave Research Laboratory, Series 3, Issue 363, June 1954).

Nelson, R.C. 1994. Depth limited wave heights in very flat regions. *Coastal Engineering*, 23, 43-59.

Nelson, R.C. 1997. Height limits in top down and bottom up wave environments. *Coastal Engineering*, 32, 247-254.

Smith. J.M., D.T. Resio, and C.L. Vincent. 1997. "Current-Induced Breaking at an Idealized Inlet," *Proceedings Coastal Dynamics '97*, ASCE, 993-1002.

Smith, J.M., and C.L. Vincent. 1992. "Shoaling and Decay of Two Wave Trains on a Beach," *Journal of Waterway, Port, Coastal and Ocean Engineering*, 118(5), 517-533.

Vink, A.S. 2001. Transformation of wave spectra across the surf zone. M. Sc. Thesis, Delft University of Technology, The Netherlands.

Wood, D.J., M. Muttray, and H. Oumeraci. 2001. "The SWAN model used to study wave evolution in a flume." *Ocean Engineering*, 28, 805-823.

KINEMATICS AND TRANSFORMATION OF NEW TYPE
WAVE FRONT BREAKER OVER SUBMERGED BREAKWATER

Takehisa Saitoh[1] and Hajime Ishida[2]

Abstract: This study presents the discovery of a new type of wave breaking over a submerged breakwater. Our findings were made through laboratory experiments. In this new type, wave breaking first occurs at a wave's front part, not at the top of the wave crest. Next, the wave crest passes over the breakwater's crown without breaking. In this study, we call this wave breaker a *wave front breaker*, and the process of wave transformation and the conditions for the *wave front breaker* to occur are clarified. The wave kinematics of the *wave front breaker* is also investigated by PIV and LDV measurements. Consequently, it is shown that the occurrence of the *wave front breaker* strongly depends on the incident wave length, and the wave transmission coefficients of the *wave front breaker* become larger. It is also found that an inverse layer of the mass transport velocity exists near the seabed against the main region of the mass transport velocity. Furthermore, the direction of the mass transport velocity near the toe in the inverse layer is onshore, while that of the maximum acceleration is offshore and the value of the acceleration increases as the measuring points get closer to the seabed and away from the toe.

1. INTRODUCTION

Submerged breakwaters or artificial reefs are expected to play an important part in controlling wave transformations, preventing beach erosion, and maintaining coastal zone environments. In Japan, the 1999 revision of the seacoast law recommended a new coastal management system. This is based on not only protection by seawalls

1 Assistant Professor, Department of Civil Engineering, Kanazawa University, Kanazawa, 920-8667, Japan,   saitoh@t.kanazawa-u.ac.jp
2 Professor, Department of Civil Engineering, Kanazawa University, Kanazawa, 920-8667, Japan, hishida@t.kanazawa-u.ac.jp

but also environmental protection and creation of coastal amenity by artificial reefs, headlands and sand fills. Therefore, design requirements make it necessary to investigate and estimate accurately the wave transformations over the submerged breakwaters or artificial reefs as well as the characteristics of internal velocities around them. In the past, however, many research works on wave barriers have focused on vertical wall breakwaters, rubble mounted breakwaters, and composite breakwaters. Only a few research works (Smith and Kraus 1991) have studied submerged breakwaters or artificial reefs. Therefore, there is still a lack of enough knowledge about the interaction between waves and submerged breakwaters or artificial reefs.

From this point of view, we have conducted fundamental laboratory experiments in order to clarify wave transformation over an impermeable submerged breakwater due to regular waves and the wave kinematics around it. A new type of wave breaking over the submerged breakwater was found from the experimental results. In this new type, wave breaking first occurs at the wave's front part not at the top of the wave crest. Next, the wave crest passes over the breakwater's crown without breaking. In this paper, we call this wave breaker a *wave front breaker*. The process of wave transformation and the condition for the *wave front breaker* to occur are then examined in detail. Furthermore, the wave kinematics of the *wave front breaker* is investigated by using the time variation of velocity vector plots, the mass transport velocity, and the acceleration, which are obtained from PIV (Particle Image Velocimetry) and LDV (Laser Doppler Velocimeter) measurements.

2. EXPEIMENTAL SET-UP
The experimental apparatus is shown in Fig.1. The flume is 14 m long and 0.8 m wide. Monochromatic waves are generated by a piston-type absorbing wave-maker located at the right end of the flume. Another piston-type absorbing wave-maker was located at the left end of the flume to minimize wave reflection, and another wave absorber was also installed on this side to absorb the high-frequency wave components generated after wave breaking. An impermeable submerged breakwater made of Plexiglas plates with steel frames was settled in this flume. The height, length and slope of the submerged breakwater were fixed at 20 cm, 1 m, and 1/3, respectively. Three water depths h were employed: h = 22, 26, 30 cm (corresponding crown depth R = 2, 6, 10 cm). Incident waves were regular waves, and wave heights H and periods T were 2-10 cm and 1-2 s (h/L: relative water depths were 0.078-0.219; L: wave length), respectively. A total of 150 combinations of water depth, wave height, and wave period were tested over the range of conditions possible for the equipment. For each condition tested in this experiment, we started to take the measurements after the wave generator had been in continuous operation for 200 s. The process of wave breaking over the submerged breakwater was recorded by a high-speed video camera at a rate of 1125 frames/s. The wave transmission coefficients were calculated as the ratio of the incident wave heights to the transmitted wave heights. The incident wave heights were separated from the data, which were sampled from two wave gauges in front of the submerged breakwater at a rate of 100 Hz for a duration of 100 s. The transmitted wave heights

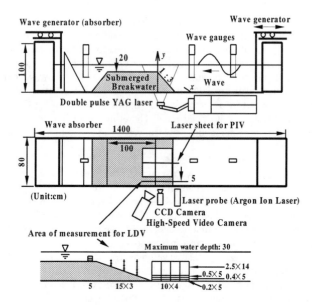

Fig. 1.   Schematic of experimental apparatus

were obtained from one wave gauge behind the submerged breakwater at the same rate. The measurements of time variations for internal velocities were conducted by PIV and LDV. Moreover, the mass transport velocities and accelerations around the submerged breakwater were calculated from the above time variations of velocities. In the PIV system (Saitoh and Ishida 2000), a laser sheet about 1 mm in thickness was created by frequency-doubled dual Nd:YAG lasers with a –6.35 mm cylindrical lens combined with a 500 mm spherical lens. The lasers had a 12 mJ/pulse maximum output energy and 5 ns pulse duration at the 532 nm wavelength. A high-resolution digital cross-correlation CCD camera (1008×1018 pixels) connected to a Pentium PC with frame straddling (Dabiri and Gharib 1997) in it was utilized with the built–in synchronizer. The camera was equipped with a 25 mm focus lens set at f/1.4. The time interval between two laser pulses was 2 ms (Chang and Liu, 1996), and a seeding particle made of nylon had a mean diameter of 86 μ m and a specific gravity of 1.02. After the images were captured, they were analyzed by performing cross-correlation (Adrian 1991). The interrogation area for the velocity calculations was 64×64 pixels, and the time interval between two vector plots in the sequence was 15 Hz. On the other hand, in the LDV system, a 4W argon ion laser was employed, and the measurements were conducted at 184 points on the vertical plane located 5 cm from the flume side (Fig. 1) for each condition. The vertical points for the measurement were located at a 2 mm space from the seabed up to $z = 1$ cm, a 4 mm space from $z = 1$ cm to $z = 3$ cm, a 5 mm space from $z = 3$ cm to $z = 5$ cm and a 2.5 cm space from $z = 5$ cm to $z = 20$ cm. The data were sampled at a rate of about

400 Hz.

3. RESULTS AND DISCUSSIONS

3.1 Wave transformation of the *wave front breaker*

First, we present the process of the *wave front breaker*. Figure 2 shows a typical example of the process of the *wave front breaker* recorded by the high-speed video camera, where $H/h = 0.27$, $h/L = 0.092$ and $R/h = 0.33$. After a backrush occurred (Fig. 2 (a)), a wave crest approached the slope of the submerged breakwater and wave breaking occurred at a wave's front part of the wave crest (Fig. 2 (b) and Fig. 2

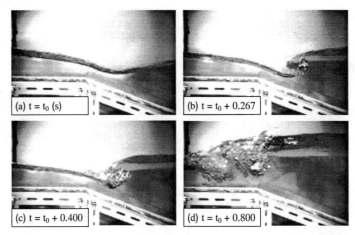

(a) t = t₀ (s) (b) t = t₀ + 0.267
(c) t = t₀ + 0.400 (d) t = t₀ + 0.800

Fig. 2. **Process of *wave front breaker* ($H/h = 0.27$, $h/L = 0.092$, $R/h = 0.33$)**

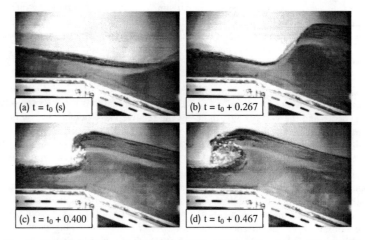

(a) t = t₀ (s) (b) t = t₀ + 0.267
(c) t = t₀ + 0.400 (d) t = t₀ + 0.467

Fig. 3. **Process of plunging breaker ($H/h = 0.32$, $h/L = 0.219$, $R/h = 0.33$)**

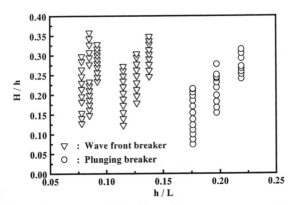

Fig. 4.   Conditions of occurrence for *Wave front breaker* ($H/h - h/L$)

(c)). Next, the wave crest passed through the crown with the bore formed at the first breaking of the wave's front part (Fig. 2 (d)). The wave crest kept its own height on the crown before gradually decreasing. In this process, the breaking point was not at the top of the wave crest, unlike such typical breaking types as a plunging breaker and a spilling breaker. Furthermore, the process of the *wave front breaker* did not necessary correspond to that of a collapsing breaker (Galvin 1968), which occurred when the wave crest remained unbroken and relatively flat while the lower part of the front face became steep and then fell with a minimal air pocket, forming an irregular turbulent water surface that slid up a beach (constant bottom slope) without developing a bore-like front.

In order to confirm the difference between the process of the *wave front breaker* and that of other breaking types, a comparison was made with a typical example of the process of a plunging breaker where $H/h = 0.32$, $h/L = 0.219$ and $R/h = 0.33$ (Fig. 3). It was clearly found that after a backrush occurred, the wave crest approached the slope and wave breaking happened at the top of the wave crest. Then, the wave crest height decreased rapidly on the crown, and a splash was generated.

Figure 4 shows the conditions for the occurrence of the *wave front breaker* compared with these of a plunging breaker. The horizontal axis indicates the relative water depth and the vertical axis indicates the relative wave height. As can be seen from this figure, regardless of the relative wave height, the *wave front breaker* occurs when h/L is smaller than 0.15. This means that the occurrence of the *wave front breaker* strongly depends on the incident wave length. In general, a breaker type is governed by a surf similarity parameter (Battjes 1974) in the case of a constant beach slope. This is a single similarity parameter that embodies the effects of both the beach slope angle and the incident wave steepness (Galvin 1968). In order to clarify the dependence on the wave steepness for the occurrence of the *wave front breaker*, the condition for this occurrence of the *wave front breaker* is rearranged by using the wave steepness in Fig. 5. As can be seen from the figure, the breaker type cannot be

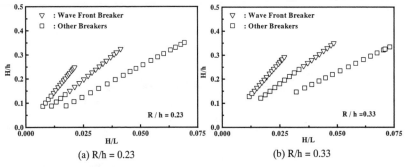

(a) R/h = 0.23   (b) R/h = 0.33

Fig. 5.   Conditions of occurrence for *Wave front breaker* (*H/h − H/L*)

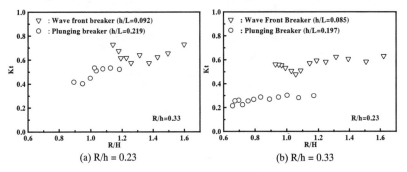

(a) R/h = 0.23   (b) R/h = 0.33

Fig. 6.   Transmission coefficients

classified simply on the basis of the wave steepness, even when we model the submerged breakwater with slope of 1:3. Therefore, it is even more difficult to classify the *wave front breaker* as well as other breakers by only using the surf similarity parameter in the case of a submerged breakwater.

Figure 6 compares the wave transmission coefficients of the *wave front breaker* with the plunging breaker. In this figure, the horizontal axis indicates the ratio of the crown depth to the wave height, and the vertical axis indicates the transmission coefficient. In the case of $R/h = 0.33$, the transmission coefficients of the *wave front breaker* are almost over 0.6, and the transmission coefficients of the plunging breaker are not so small. On the other hand, in the case of $R/h = 0.23$, it was found that the transmission coefficients of the plunging breaker decrease to 0.3 as the relative crown depth decreases. However, the transmission coefficients of the *wave front breaker* remained large at around 0.6 regardless of the change in crown depth. This means that energy dissipation through the submerged breakwater is not expected in the occurrence of a *wave front breaker*.

3.2 Wave kinematics of the *wave front breaker*
In this section, we focus on the wave kinematics of the *wave front breaker*. Figure 7

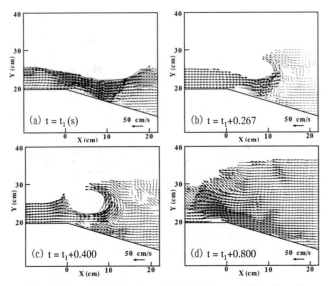

Fig. 7. Time variation of vector plots for *wave front breaker*
($H/h = 0.27$, $h/L = 0.092$, $R/h = 0.33$)

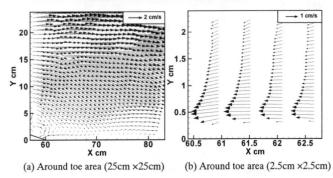

(a) Around toe area (25cm ×25cm)   (b) Around toe area (2.5cm ×2.5cm)

Fig. 8.   Mass transport velocity of *wave front breaker* by PIV
($H/h = 0.28$, $h/L = 0.092$, $R/h = 0.33$)

shows a typical example of the time variation of the internal velocity (vector plots) for the *wave front breaker* by PIV where $H/h = 0.27$, $h/L = 0.092$, and $R/h = 0.33$. Maximum offshore velocity appeared on the top of the slope as backrush velocity (Fig. 7 (a)), and wave breaking occurs at the wave's front part of the waver crest (Fig. 7 (b) and Fig. 7 (c)), while the wave crest keeps its velocity distribution without mixing (Fig. 7 (d)). On the other hand, we also conducted measurements of the internal velocity (vector plots) for the *wave front breaker* by PIV around the toe of the submerged breakwater. The velocity distribution on the space at each time was

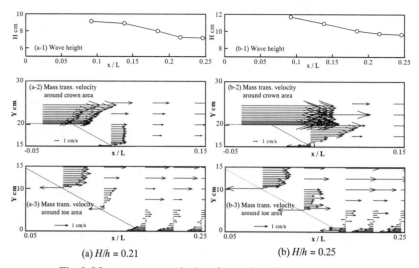

Fig. 9. Mass transport velocity of *wave front breaker* by LDV
($h/L = 0.092$, $R/h = 0.33$)

quite similar to, for example, the case of a partially standing wave without breaking. Consequently, we could not find any effects on the velocity distribution from the occurrence of the *wave front breaker*. Furthermore, the time variations of the vector plots around the toe seemed similar for the conditions tested in this study.

In addition to the time variation of the internal velocity, the characteristics of the mass transport velocity and the acceleration is strongly related to the stability of the armor blocks or scouring phenomena around the toe of the submerged breakwater. Figure 8 shows a typical example of the mass transport velocities for the *wave front breaker* by PIV where $H/h = 0.28$, $h/L = 0.092$, and $R/h = 0.33$. The measurement area in Fig. 8 (a) was 25 cm × 25 cm and the resolution was 0.28 mm/pixel; in Fig. 8 (b), the measurement area was 2.5 cm × 2.5 cm, which was expanded, and the resolution was 0.028 mm/pixel. As can be seen from Fig. 8 (a), it was found that an inverse layer of the mass transport velocity existed near the seabed against the main region of the mass transport velocity with offshore direction. The direction of the mass transport velocity in the inverse layer shown in Fig. 8 (a) was clearly onshore, as can be seen from Fig. 8 (b).

In order to more accurately estimate the mass transport velocity and acceleration which is calculated as the rate of change between two velocities in sequence, we also measured time variations for the internal velocities by LDV, since the time interval between two vector plots was 15 Hz in the PIV system. Figure 9 shows a typical example of the mass transport velocity for the *wave front breaker* by LDV. As can be seen from Fig. 9 (a) where $H/h = 0.21$, $h/L = 0.092$, and $R/h = 0.33$, the mass transport velocities on the crown became larger in the offshore direction, while in the

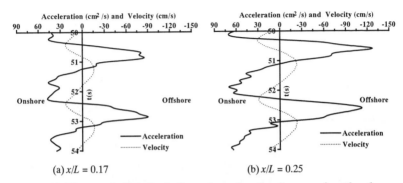

Fig. 10. **Time variation of velocity and acceleration for *wave front breaker***
by LDV (*H/h* = 0.25, *h/L* = 0.092, *R/h* = 0.33, *y*=0.2cm)

Table 1.   **Maximum accelerations for *wave front breaker* by LDV**
(*H/h* = 0.25, *h/L* = 0.092, *R/h* = 0.33)

	$x/L = 0.17$		$x/L = 0.20$		$x/L = 0.23$		$x/L = 0.25$	
	Onshore	Offshore	Onshore	Offshore	Onshore	Offshore	Onshore	Offshore
$y = 1.0$(cm)	53.7	-89.2	64.7	-103.5	68.5	-112.3	75.2	-116.1
$y = 0.8$	57.0	-90.9	62.8	-103.9	65.7	-110.2	73.6	-118.8
$y = 0.6$	52.5	-92.1	66.1	-105.7	68.3	-112.7	73.5	-116.6
$y = 0.4$	57.9	-99.1	62.7	-104.2	72.9	-113.8	76.0	-120.6
$y = 0.2$	73.1	-106.4	69.0	-109.1	78.6	-116.1	71.9	-122.5

(Unit: cm/s²)

inverse layer, the mass transport velocity of the offshore direction occurred around
the seabed. Moreover, the mass transport velocities on the crown and in the inverse
layer increased as the relative wave height increased as can be seen from Fig. 9 (b) in
where H/h = 0.25, h/L = 0.092, and R/h = 0.33. Although the existence of an inverse
layer around the seabed was presented by Sumer and Fredsøe (2000) under standing
waves, we found that the inverse layer occurred in the case of the submerged
breakwater under partially standing waves.

Furthermore, the time variations of velocity and acceleration in this inverse layer
were measured at y = 0.2 cm, x/L = 0.17, 0.23 and 0.25 for the *wave front breaker*
where H/h = 0.25, h/L = 0.092, and R/h = 0.33 (Fig. 10). Here, if the standing wave
occurs due to a vertical wall located at x/L = 0, the location of x/L = 0.25 indicates
the point of a node. From this figure, it was found that the direction of the maximum
acceleration is offshore. In addition, the maximum absolute values of acceleration (-:
offshore direction; +:onshore direction) are shown in Table 1. The direction of the
mass transport velocity was onshore, while that of the maximum acceleration was
offshore and the value of acceleration increased as the measuring points got closer to
the seabed and away from the toe, which suggests that light and small sediments
around the toe would move onshore due to the mass transport velocity, while heavy

and large sediments would move offshore due to the maximum acceleration.

4. CONCLUSIONS

A new type of wave breaking, *wave front breaker* over a submerged breakwater was found from laboratory experiments. The wave transformation and kinematics of the *wave front breaker* clarified for the experimental conditions used in this study.

The occurrence of the *wave front breaker* strongly depends on the incident wave length, and the *wave front breaker* occurs when h/L is smaller than 0.15. The transmission coefficients of the *wave front breaker* become larger than those of other breakers, so the energy dissipation through the submerged breakwater is not expected for the *wave front breaker*.

In the case of the *wave front breaker*, the mass transport velocities on the crown become larger in the offshore direction. Furthermore, the inverse layer of the mass transport velocity exists near the seabed against the main region of the mass transport velocity with offshore direction. The direction of the mass transport velocity in the inverse layer is onshore, while that of the maximum acceleration is offshore. The value of acceleration increases as the measuring points get closer to the seabed and away from the toe.

ACKNOWLEDGEMENTS

The authors thank Dr. H. Mase (Disaster Prevention Research Institute of Kyoto University) for helpful suggestions and discussions.

REFERENCES

Adrian, R.J. 1991. Particle-imaging techniques for experimental fluid mechanics, *Annu.Rev.Fluid Mech.*, 23, 261-304.

Battjes, J.A. 1974. Surf similarity, *Proc. 14th Int. Conf. Coastal Engineering, ASCE*, 466-480.

Galvin, C.J. Jr. 1968. Breaker type classification on three laboratory beaches, *Journal of Geophysical Research*, 73, 3651-3659.

Chang, K.-A. and Liu, P. L.-F. 1996. Measurement of Breaking Wave Using Particle Image Velocimetry, *Proc. 25th Int. Conf. Coastal Engineering, ASCE*, 527-536.

Dabiri, D. and Gharib, M. 1997. Exprimental invesigation of the vorticity generation within a spilling water wave, *Journal of Fluid Mech.*, .330, 113-139.

Saitoh, T. and Ishida, H. 2000. Vortex formation behind a vertical slender plate and wave force, *Proc. 27th Int. Conf. Coastal Engineering, ASCE*, 1773-1786.

Smith, E.R. and Kraus, N.C. 1991. Laboratory study of wave-breaking over bars and artificial reef, *Journal of Waterway Port Coastal and Ocean Engineering, ASCE*, 117-4, 307-325.

Sumer, B.M. and Fredsøe, J. 2000. Experimental Study of 2D Scour and its Protection at a Rubble-Mound Breakwater, *Coastal Engineering*, 40, 59-87.

EFFECT OF REFLECTIVE STRUCTURES ON UNDERTOW DISTRIBUTION

Md. Azharul Hoque[1], Toshiyuki Asano[2], M.ASCE,
and Mir Ahmed Lasteh Neshaei[3]

Abstract: The study investigates the effects of reflective seawalls on wave breaking processes and mean flow distribution throughout the surf zone. The presence of reflective structures results in a significant change of wave breaking characteristics and turbulence structure in the surf zone. The organized large vortexes, which cause onshoreward mass transport, will also be changed thus changing the undertow considerably. The physical processes and mechanisms of wave breaking and turbulence structures in presence of a seawall have been investigated using visualization techniques. A model for evaluating the distribution of undertow in front of a reflective structure is also proposed on the basis of the change in wave fields due to the presence of reflected waves in the surf zone. The model is in reasonable agreement with the experimental results. It has been found that the presence of reflected waves due to reflective condition in the surf zone changes the wave breaking characteristics and reduces the magnitude of undertow. The results of the present work can be implemented for predicting cross-shore sediment transport models and beach evaluation models where reflective conditions exist.

INTRODUCTION

Under normal sea conditions the seawall might provide little interactive effects on wave fields. But under stormy condition, the sea level surges up and waves may attack the wall directly. As a result, not only the wave fields but also the wave breaking characteristics and turbulence structure in the surf zone will be much altered. The First

[1]Ph.D. Student, Department of Ocean Civil Engineering, Kagoshima University, Korimoto 1-21-40, Kagoshima 890-0065, JAPAN. hoque@oce.kagoshima-u.ac.jp
[2]Professor, Department of Ocean Civil Engineering, Kagoshima University, Korimoto 1-21-40, Kagoshima 890-0065, JAPAN. asano@oce.kagoshima-u.ac.jp
[3]Assistant Professor, Department of Civil Engineering, Gilan University, Gilan Blvd., Golshar, Rasht, 41668, IRAN. MALN@eng.gu.ac.ir

objective of this study is to investigate the physical processes and mechanisms of wave breaking and turbulence structures in the surf zone for monochromatic waves in the presence of a seawall. As the wave fields are highly complex due to the presence of broken incident waves and interaction of reflected waves, the visualization techniques were employed to investigate the physical transformation processes through observation and phenomena interpretation. The initiation of wave breaking, breaker type, formation and movement of eddy area, splash up region and generated turbulence were carefully studied through a series of laboratory experiments for different incident wave parameter. Significant differences in the wave breaking characteristics in terms of above parameters were found with and without wall.

The mass flux due to wave breaking causes the undertow, which is the offshore-directed steady current below the trough level. The mass transport by breaking waves is induced by wave motion and organized large vortexes. Due to the change in breaking processes and generated turbulence in presence of a reflective structure, the mass flux in the surf zone will also be changed with the consequence of changing the undertow velocity. Moreover reflected wave causes an offshore-directed mass flux, which accordingly reduces the mean undertow below the trough level. Although some models were presented to estimate the undertow distribution in normal beach condition, but there is no complete work on the undertow distribution in reflective beaches. There have been only few experimental works on the undertow in case of reflective beaches. Holmes and Neshaei (1996) measured the vertical and horizontal distributions of undertow in front of a partially reflective seawall in a series of random wave experiments. Their investigation revealed that the magnitude of the undertow is reduced in the presence of partially standing waves, which is in agreement with the work of Rakha and Kamphuis (1997) indicating a reduction in undertow by reflected waves.

The second objective is to investigate the effects of reflective structures on the distribution of undertow under monochromatic waves through performing physical hydraulic model test and formulating a numerical model for estimating the undertow distribution throughout the surf zone. In the theoretical analysis the effects of reflected waves have been considered in terms of the process of wave transformation, mass flux transported by broken waves and resultant wave set-up.

EXPERIMENTAL PROCEDURES

Laboratory experiments on wave breaking were carried out in a two-dimensional wave flume having dimensions 17.5 m long, 1.5 m deep and 1.0 m wide. Waves were generated at one end of the tank. A plane beach profile with a constant slope of 1:10 was built at the other side of the tank. Experimental arrangements are shown in Figure 1(a). The measurement points were in the middle of the flume. Water surface elevations have been measured at different locations with 20 cm spacing up to a distance of 6 m from the shoreline. Two wave gauges were mounted further offshore to measure the incident waves. The physical transformation mechanisms of breaking waves were observed using a video camera. The cross-shore and vertical positions were identified by 5x5 cm fluorescent grid placed on the glass wall of the flume. Several trials were made to enhance the visibility of wave breaking phenomenon in the surf zone. A black lamp lighting system was placed both above the tank and in front of the glass wall in order to enhance the visibility of the wave breaking phenomena through the 5x5 cm grid frames. Wave breaking structures for seven cases of monochromatic waves have been investigated for the condition of without wall and with wall at shoreline. Characteristics of the incidental waves are shown in Table 1

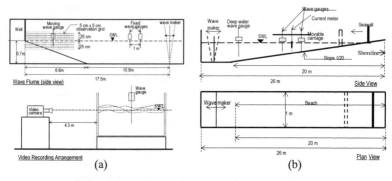

(a)   (b)

Figure 1: Experimental set-up of (a) wave breaking
experiments and (b) undertow experiments

Table 1: Characteristics of incident waves used in wave breaking experiments

Case	H_0 (cm)	T (sec)	L_0 (m)	$\dfrac{H_0}{L_0}$	$\dfrac{\tan \beta}{\sqrt{H_0/L_0}}$	Breaker Type
B1	20.5	1.25	2.44	0.084	0.345	Spilling
B2	21.8	1.20	2.25	0.097	0.322	Spilling
B3	14.0	1.46	3.32	0.042	0.487	Spilling
B4	15.0	2.00	6.24	0.024	0.645	Plunging
B5	18.6	2.00	6.24	0.030	0.612	Plunging
B6	20.0	1.50	3.50	0.057	0.444	Plunging
B7	17.6	1.50	3.50	0.050	0.474	Spilling

Experimental investigations on undertow distribution were conducted in another wave flume having dimension 26 m long, 1.5 m deep and 1.0 m wide as shown in Figure 1(b). A plane beach profile with a constant slope of 1:20 has been build by concrete and reinforcing wire mesh. The velocity measurements were taken in a cross-shore directed vertical plane up to a distance of 500 cm from the shoreline at different horizontal locations in 25 cm intervals. Again at each location, vertical points have been selected with a spacing of about 1 cm in the surf zone starting from 0.8 cm above the bottom up to 85% of the water depth. Water surface elevation data have also been recorded at each location synoptic with the velocity measurement. Another wave gauge was mounted further offshore to measure the deep-water incident wave. Data were acquired and analyzed using a personal computer with a sampling rate of 20 Hz for each channel. The recording length was 3.42 minutes allowing approximately 200 waves in account.

Measurements were taken for the condition of without wall and with wall at 1 m and 2 m distance from the shoreline. The incident wave parameters and breaking characteristics of the experimental wave cases used in undertow experiments are shown in Table 2.

Table 2: Characteristics of incident waves used in undertow experiments

Case	H_o (cm)	T (sec)	L_o (m)	H_o/L_o	$\tan \beta \big/ \sqrt{H_o/L_o}$	Breaker Type	h_b (cm)	x_b (m)
U1	10.0	2.00	6.24	0.016	0.395	Spilling	12.3	2.5
U2	12.5	1.50	3.51	0.036	0.264	Spilling	16.2	3.2
U3	11.8	1.48	3.42	0.034	0.271	Spilling	15.0	3.0

EXPERIMENTAL FINDINGS

Observational investigations of video records of wave breaking characteristics for one spilling (case -B1) and one plunging (case B4) type of breaking are shown in Figure-2 and 3 respectively. Figure-4 illustrates the comparison of wave height distribution for the conditions of without wall and with wall at the shoreline for case-B1. It can be seen that wave-shoaling pattern has been changed and the breaking point has been shifted seaward. The earlier formation and growth of eddies can be observed in presence of reflected waves. From different stages of splash up formation zone as shown in the figure, it can be observed that upon mixing with reflected water mass, strong turbulence was generated with high splash up of water. In case of B3, having breaker index near to the transition for spilling and plunging type of breaking, the breaker type changed from spilling to plunging due to the presence of the seawall. It has also been found that for the same position of seawall, the effects are more significant for milder waves.

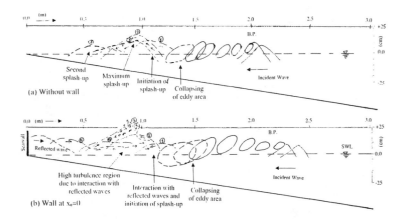

Figure 2: Comparison of wave breaking characteristics (case-B1)

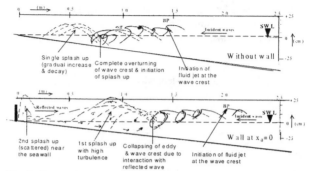

Fig. 3: Comparison of wave breaking characteristics (Case-B4)

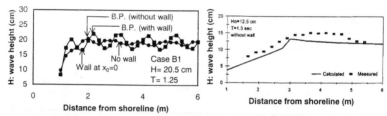

Fig. 4: Experimental wave height distribution with and without wall (case- B1)

Fig. 5: Calculated and measured wave attenuation (case-U2)

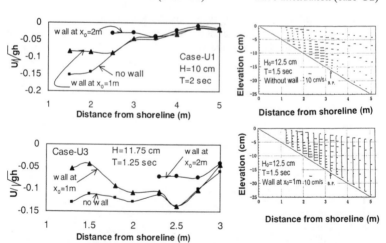

Fig. 6: On-offshore distribution of measured undertow velocity

Fig. 7: Distribution of time averaged current(Case-U2)

Figure 5 illustrates the comparison of calculated and measured wave attenuation for case-U2. On-offshore distributions of undertow at the height of z=0.8 cm from the bottom for case-U1 and U3 are shown in Figure-6. As the reflectivity increases, the magnitude of undertow velocity decreases. The reduction of undertow velocity is found to be more pronounced in the inner surf zone areas. The time average current distributions without seawall and using the model seawall at x_0=1m for case-U2 of undertow experiments are shown in Figure 7. It can be observed that the shape of undertow profiles in the surf zone is significantly different in presence of the seawall.

MODELING OF UNDERTOW IN PRESENCE OF REFLECTED WAVES

A theoretical model is proposed to predict undertow velocities under partially reflective waves. This model is basically an extension of Okayasu's (1990) energetic approach model to include reflective waves. The average undertow velocity U_m under wave trough level in the surf zone can be expressed as a compensating flow of the total mass flux comprised of mass flux by wave motion, M_w, and mass flux by organized large vortex M_v. If d_t is the water depth below the wave trough level, U_m is given by

$$U_m = -\frac{1}{d_t}\left(M_w + M_t\right) \tag{1}$$

The incident and reflected water surface variation, η_i, η_r is given as follows:

$$\eta_i = A_i(x)e^{i\left[k(x-x_0)-\sigma t\right]} \tag{2a}$$

$$\eta_r = A_r(x)e^{i\left[k(x-x_0)+\sigma t\right]} \tag{2b}$$

Where A is wave amplitude, k is wave number, σ is angular frequency, x_0 is the distance of sea wall from shoreline and subscripts i and r represent incident and reflected waves.

For the condition in which reflected waves exist, the mass flux by the reflected wave is subtracted from the total mass flux by the incident waves to modify the mass balance in the surf zone. In consequences of reduction of total mass flux, the mean undertow velocity U_m will be reduced according to Eq. (1). Thus the present model presumes that the reflected waves reduce the total mass flux in the surf zone as follows.

$$M_t = \left(M_w + M_v\right)_i - \left(M_w\right)_r \tag{3}$$

Following Okayasu's model, the mass flux due to incident wave can be found as:

$$\left(M_w\right)_i = \frac{1.6c}{\rho gh}\frac{\rho gA_i^2}{4}, \tag{4a}$$

$$\left(M_v\right)_i = \frac{3}{8}\rho\frac{H^2}{L}c \tag{4b}$$

where, c is wave celerity, H is wave height, ρ is the water density and g is the gravitational acceleration. For reflected waves, mass flux due to wave motion, $(M_w)_r$ is calculated by using A_r instead of A_i in Equation (4a).

Wave set-up $\bar{\eta}$ in the surf zone can be evaluated from the momentum balance equation in the onshore direction,

$$\frac{dS_{xx}}{dx} + \rho gh\frac{d\bar{\eta}}{dx} = 0 \tag{5}$$

where, h is the total water depth comprised of still water dept d and wave set-up $\bar{\eta}$, S_{xx} is the radiation stress tensor normal to the shore which can be calculated based on the linear wave theory as:

$$S_{xx} = \frac{\rho g \overline{\left(\eta\right)^2}}{2} + \rho \overline{\int_{-h}^{\eta} \left(u_w^2 - w_w^2\right) dz} , \qquad (6)$$

in which, u_w and w_w are horizontal and vertical component of water particle velocity.

In presence of reflected waves the radiation stress component should be modified adding extra term due to reflected wave, which will result in a modified wave set-up in the surf zone. The water surface elevation η and water particle velocity u_w are given by

$$\eta = \eta_i + \eta_r \qquad (7a)$$

$$u_w = (u_w)_i + (u_w)_r \qquad (7b)$$

Substituting Eqs. (2) and (7) into Eq. (6) and after integration, radiation stress component S_{xx} can be written as:

$$S_{xx} = E_i \left\{2n - (1/2)\right\} + E_r \left\{2n - (1/2)\right\} - \rho g C_{ir} A_i A_r \cos 2k(x - x_0)/2 , \qquad (8)$$

where, $2n = 1 + 2kh/\sin(2kh)$; E_i and E_r is total wave energy as $E_i = \rho g A_i^2/2$, $E_r = \rho g A_r^2/2$, C_{ir} is the phase correlation between incident and reflected waves.

The vertical distribution of undertow velocity can be evaluated if the mean shear stress acting on the horizontal plane at the trough level, τ, and the distribution of mean eddy viscosity v_e are given. Following Okayasu's model, these quantities are assumed to vary linearly with the vertical elevation from the bottom, z' as: $\tau = \alpha_\tau z' + \beta_\tau$, $v_e = \alpha_v z'$ and undertow U is expressed in a first order linear differential equation as:

$$(\alpha_\tau z' + \beta_\tau) = (\alpha_v z' + v) \frac{\partial U}{\partial z'} , \qquad (9)$$

in which, v is kinematic viscosity, α_τ, β_τ, α_v are factors independent of z', but related to the rate of energy dissipation by wave breaking D_B. The energy dissipation rate is calculated based on the linear wave theory on a constant slope following Battjes and Janssen (1978) as:

$$D_B = \frac{5}{6} \rho g^{\frac{3}{2}} \tan \beta \gamma_H^2 h^{\frac{3}{2}} , \qquad (10)$$

where, $\tan\beta$ is bottom slope and γ_H is the ratio of wave height to water depth.

The vertical distribution of undertow is found from the solution of equation (9) as:

$$U(z) = \alpha_1\left(z' - \frac{d_t}{2}\right) + \alpha_2\left(\alpha_3 + \log f_1(z'/d_t)\right) + U_m \qquad (11)$$

where, α_1, α_2, α_3 are constants.

The first term represents a linear distribution to have zero value at the mid depth $z' = d_t/2$, and the last term represents uniform velocity of which magnitude is given by Equation (1). Existence of reflected waves alter the turbulence intensity, but in this treatment it is only accounted for the change of breaking point, which is indirectly related to turbulence intensity. Thus the effects of reflected waves are only considered for calculating the mean undertow velocity U_m.

For the cases using wall, wave heights at different location of the sloping beach is found by adding the contribution of reflected waves with that of incident waves. The composite wave fields, comprised of after broken waves and their reflected waves, are difficult to treat theoretically. Newly advanced numerical simulation by direct tracing fluid element might be useful tools to describe the flow field, however such methods are only able to provide universal properties after accumulating a number of reliable numerical results. In this study, the wave field is treated by simple superposition of the onshore directed incident waves including shoaling, breaking, after breaking, and off-shore directed reflected waves of which wave height is reduced by a certain reflection coefficient and phase shift is assumed to be zero. No non-linear characteristics between incident and reflected waves are considered. The reflection coefficient was analyzed by comparing the experimental wave height with theoretical wave height distribution for different reflection coefficients. The reflection coefficient, R, varies over a wide range (0.2 to 0.8) depending on the location of the seawall and wave properties. The cases that the seawall was located at $X_0=200$ cm is found to have larger reflection coefficients. For judging the breaking in this composite wave field, the criterion of Iwata and Kiyono (1985) was used. However, this criterion does not always predict the breaking point in the experimental results, thus the ordinary breaking criterion on a wave height and depth ratio $H_b/h_b=0.7$ is evoked. Since this study deals with the regular waves, the breaking idealistically should occur at a certain point. In that situation, the undertow velocity changes discontinuously at the breaking point. But in reality the breaking points fluctuate to and fro for each wave.

COMPUATIONAL RESULTS AND COMPARISON WITH EXPERIMENTS

Comparisons between calculated and measured undertow profiles for case-U1 and case-U2 are shown in Figure 8. It is clearly observed that undertow velocity is reduced due to the presence of reflective seawall and the reduction is more significant when the seawall was located at $x_0=2$m compared to $x_0=1$m. It can also be seen that the calculated result well reproduces the amount of reduction obtained in the measurements. Around the breaking point, the profiles are found to be nearly vertical and closely spaced. This is because the profiles greatly depend on the energy dissipation rate D_b, which is very small at the breaking point. Though the undertow around the breaking position is of small magnitude, the profiles are still in clear agreement with that by the present model. Figure 9 illustrates the comparisons of the on-offshore undertow distribution at height $z=0.8$ cm from the bottom. The results also indicate that the calculations, as a whole, reproduce the measured data showing the reduction of undertow velocity due to the presence of the seawall. Moreover, it can be observed that the reduction of undertow for a given reflective condition is more significant in the inner surf zone areas.

The wave set-up of case U2 is shown in Figure 10. The calculated wave set-up and set-down well reproduce the experimental results, though set-down is found to be slightly over-estimated near the breaking point and set-up is under-estimated at outer surf zone. This is because, in the original model of Okayasu the wave set-down is slightly over-estimated at the breaking points due to over estimation of radiation stress by the small amplitude wave theory. Again the set-up is under estimated in outer region.

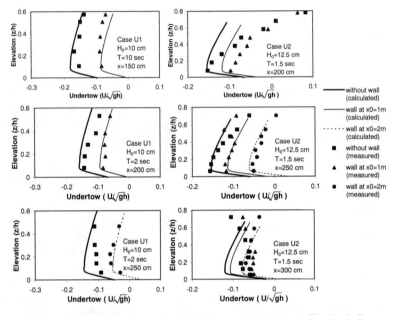

Figure 8: Comparisons between calculated and measured undertow profiles (vertical)

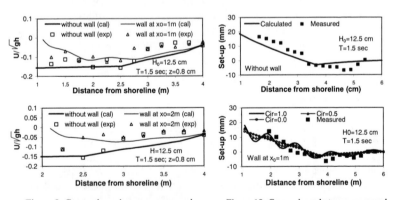

Figure 9: Comparisons between measured and calculated undertow along the beach

Figure 10: Comparisons between measured and calculated wave set-up and set-down

Finally, some parametric calculations of the model are performed varying the reflection coefficients and incident wave parameters as shown in Figures 11 and 12, which are found to be in good accordance with the experimental findings.

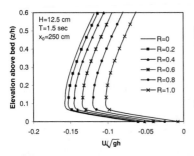

Figure 11: Vertical distribution of undertow
for different reflection coefficients

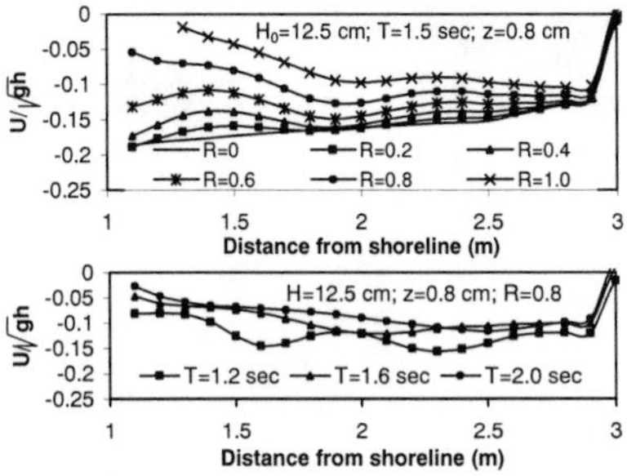

Figure 12: Cross-shore undertow profiles
for different reflections and beach slopes

CONCLUSIONS

Following conclusions are revealed from the experimental investigations as well as computational results of the present study:

(1) The presence of reflective structure changes breaking pattern, breaker type and turbulence structure. If the wave breaker type remains the same for both cases of with and without seawall, the breaking point will be shifted seaward.

(2) Existence of reflective structure in the surf zone results in the reduction of undertow velocity and modifies its distribution across the surf zone. The reduction of undertow is more pronounced in the inner region of the surf zone.

(3) The level of beach reflectivity is an important parameter to control the magnitude and distribution of undertow. Higher reflectivity causes higher reduction in undertow.

(4) The present undertow model including reflection effects is capable to reproduce the experimental undertow profiles both in vertical and cross-shore direction.

REFERENCES

Batjes, J. A. and Janssen, J. P. 1978. Energy loss and set-up due to breaking of random waves, Proceedings of the 16th ICCE, ASCE, pp. 569-588.

Holmes, P. and Neshaei, M. A.L. 1996. The effect of seawalls on coastal morphology, Proceedings of the Second IAHR Symposium, On Habitats Hydraulics, Ecohydraulics 2000, Vol. A, pp.525-530.

Iwata, K. Kyono, H. 1983. Breaking of standing two-component composite and irregular waves, Coastal Engineering in Japan, Vol. 28, pp.71-87.

Longuet-Higgins, M. S. 1953. Mass transport in water waves, Phil. Trans. Royal Soc., London, Vol.245, pp. 535-581.

Okayasu, A.1989. Characteristic of turbulence structure and undertow in the surf zone, Ph.D thesis, Univ. of Tokyo, 119p.

Rakha, K. A. and Kamphuis, J. W. 1997. Wave- induced currents in the vicinity of a seawall, Coastal Eng., Vol. 30., pp.23-52.

WAVE AND TURBULENCE CHARACTERISTICS IN NARROW-BAND IRREGULAR BREAKING WAVES

Francis C. K. Ting[1], Associate Member, ASCE

Abstract: Water surface elevations and fluid velocities were measured in a laboratory surf zone created by the breaking of narrow-band waves on a 1/35 plane slope. The incident waves formed wave groups that were strongly modulated. The wave characteristics inside the surf zone were influenced by low-frequency waves. The waves with the smaller initial heights break closer to shore and have higher wave-height-to-water-depth ratios at breaking than the waves with the larger initial heights. When normalized by the period-averaged water depth, the wave height has a narrow variation in the inner surf zone. The standard deviation of the wave period increases as the water depth decreases due to the effects of long-wave motion. Turbulent kinetic energy and Reynolds stress were estimated at two different levels under the wave trough level by ensemble averaging. The distributions of the period-averaged turbulent kinetic energy and Reynolds stress for narrow-band and broad-band waves are different in the outer surf zone, but similar in the inner surf zone. It is suggested that the wave and turbulence characteristics in an irregular wave surf zone are determined by the breaker types, which in turn, are determined by the wave height and wave period distributions of the incident waves, and the beach slope.

INTRODUCTION

This study was concerned with a narrow-band irregular wave train breaking on a plane slope. The main objective of this experimental investigation was to study the wave and turbulence characteristics inside the surf zone. The percentage of breaking waves in the flow region investigated ranged from 63% to 96%. Longitudinal and vertical components of water particle velocity were measured simultaneously with wave elevation at three cross-shore locations in this zone.

Detailed experimental data of the breaking and decay of irregular waves in the surf zone, and of the generation and evolution of the related turbulent flow fields are

---

[1] Associate Professor, Department of Civil and Environmental Engineering, South Dakota State University, Brookings, South Dakota 57007, USA. Francis_Ting@sdstate.edu

important for improving detailed models of surf zone breaking processes. To date, only a few studies of fluid turbulence under irregular breaking waves are available in the literature. George et al. (1994) analyzed turbulence generated by wave breaking on a natural beach using hot-film anemometer data. Sultan (1995) measured the flow velocities in a laboratory surf zone for irregular waves with predominantly spilling breakers and predominantly plunging breakers. Recently, Ting (2001) presented the probability distributions of wave-averaged turbulent kinetic energy and Reynolds stress obtained in a laboratory surf zone created by the breaking of broad-band waves on a plane slope.

This article examines the wave and turbulence characteristics in the surf zone of a narrow-band incident wave spectrum. A basic difference between a narrow-band wave spectrum and a broad-band wave spectrum is that a narrow-band spectrum may produce wave and turbulence characteristics in the surf zone that are similar to those found in a regular wave train, whereas the surf zone flow field of a broad-band spectrum may show a high level of variation from wave to wave, depending upon the distributions of wave heights and periods in the incident waves.

EXPERIMENTAL PROCEDURE

The experiments were conducted in a 37 m long, 0.91 m wide and 1.22 m deep glass-walled wave flume equipped with a hinged-flap programmable wave maker. A 1 on 35 sloped false bottom built of marine plywood was installed in this flume to create a plane beach. The still water depth at the toe of the beach was 45.72 cm (Fig. 1). One 10-min irregular wave time series was used in all the test runs. The time series was developed from the TMA spectrum, with a spectral significant wave height H_{m0} of 0.1229 m, a spectral peak period T_p of 2.0 s and a peak enhancement factor γ of 100.

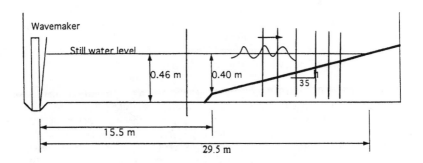

Fig. 1. Experimental arrangement

Water surface elevations were measured in the horizontal portion of the wave flume and at six locations on the plane slope sequentially by using a resistance-type gage. The gage positions are shown as vertical lines in Fig. 1. A video camera recorded the breaker types at each wave gage location. A laser-Doppler anemometer measured the longitudinal and vertical velocity components below the wave trough level at three wave gage positions near the shoreline. Titanium dioxide particles were used to "seed" the water to provide particles for scattering the laser light. Typical signal dropout rate was 4% where the laser beams were never above the water surface. The uncertainty in velocity measurements was ±0.5 cm/s. The sampling frequency was 100 Hz per channel.

At five measuring points, the experiment was repeated ten times with the exact same incident wave time series. The average of all ten runs represents an ensemble average. Instrumental data were collected for 8.192 min from the start of wave generation, but only the last 5.12 min of data was analyzed to minimize transient effects.

To reduce the effect whereby small waves riding on long waves may be carried entirely above or below the mean water level, the long waves were filtered out before computing the short wave statistics. The filtering was done by a Fourier filter at a cutoff frequency of 0.2 Hz. Individual waves in the short-wave time series were defined using the zero-downcrossing method. The measured velocities were partitioned into a low-frequency and a high-frequency time series in the same manner as the wave elevations. The ensemble average of the low-frequency time series is consisted of the undertow and the long-wave-induced velocity. The ensemble average of the high-frequency time series represents the short-wave-induced velocity. The difference between the short-wave-induced velocity and the high-frequency time series is defined as turbulence.

RESULTS

Wave Characteristics
Figure 2 shows the probability distributions of the measured wave heights and periods in the horizontal portion of the wave tank. As shown in Fig. 2, the frequency content of the incident waves was concentrated over a very narrow range. The interaction of waves of slightly different frequencies produced wave groups that were strongly modulated. Thus, the wave height has a wide variation as shown in Fig. 2.

Figure 3 shows the distributions of wave-height-to-water-depth ratios at the six wave gage positions on the plane slope. Because the local water depth varied from wave to wave due to long waves generated by wave breaking, the wave height H_i has been normalized by the period-averaged water depth h_i ($i=1,2...N$), where N is the total number of waves in the wave record. Wave breaking occurred at all six wave gage positions on the slope. The percentage of breaking and broken waves at the still water depth $d = 27.04$, 22.77, 18.23, 13.72, 9.39 and 6.25 cm was 2%, 10%, 31%, 63%, 82% and 96%, respectively. It may be observed from Fig. 3 that there is a consistent increase in the kurtosis of the H_i/h_i ratio from offshore to onshore. In addition, the maximum value of H_i/h_i increases as the water depth decreases until close to the shoreline. The large H_i/h_i values in the distributions are from waves that were near their break points. The small H_i/h_i values are from secondary waves in the wave profiles that were interpreted as individual waves by the zero-downcrossing method. For the broken waves, the H_i/h_i ratio has a narrow variation inside the surf zone.

A wave-by-wave analysis was carried out to examine the transformation of the wave groups through the surf zone. The analysis showed that the H_i/h_i ratio for the individual waves increases to a maximum value of 0.7 to 1.1 at breaking, and then decreases steadily to a constant value of about 0.5 inside the surf zone. The order in which the individual waves break follows closely their initial wave heights in the groups. In general, the waves with the smaller initial heights break closer to shore and have higher H_i/h_i ratios at breaking than the waves with the larger initial heights. Similar results have been observed in regular waves. It is well known that for regular waves, the wave-height-to-water-depth ratio at breaking, $(H/h)_b$, increases as the deep-water wave steepness H_0/L_0 decreases, i.e., when the breaker type changes from spilling to plunging (Hansen 1990). Furthermore, spilling breakers usually produce a wider surf zone than plunging breakers.

Figure 4 presents the distributions of wave periods on the plane slope. The wave periods have been normalized by $T_{1/3}$, the average period of the highest one-third wave. Figure 4 shows that the standard deviation of the wave period increases as the water depth decreases. The low-frequency waves generated by wave breaking create a time-varying water depth and current that becomes significant at small water depths. The long-wave motion changes the wave propagation speed of the short waves, causing the period of the short waves to increase or decrease. Thus, the short-wave period has a wide variation near the shoreline.

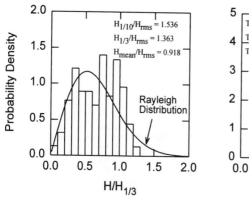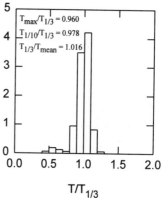

Fig. 2. Distributions of wave heights and periods in the constant-depth region

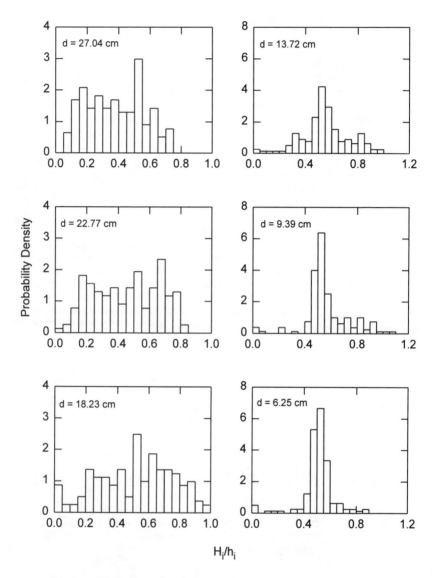

Fig. 3. Distributions of wave-height-to-water-depth ratios on the plane beach

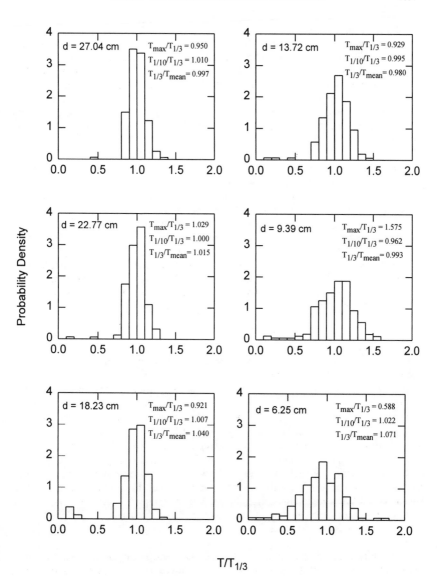

Fig. 4. Distributions of wave periods on the plane beach

Turbulence Characteristics

Figure 5 presents the probability distributions of the period-averaged turbulent kinetic energy $[k/(gh)]^{1/2}$ at the last three wave gage positions. The measurements shown were taken at two different levels below the still water level $z = 0$. The instantaneous turbulent kinetic energy k' is defined as $1/2$ $(u'^2 + w'^2)$, where u' and w' are the longitudinal and vertical components of the turbulence velocity, respectively. The ensemble average of k' was determined from ten realizations of k'. The ensemble average was averaged over the wave period for the individual waves and normalized by gh, where h is the time-averaged water depth.

Figure 5 shows that not all the waves were turbulent in the outer and middle surf zone. At $d = 13.72$ cm, non-breaking waves were found at the head and rear of the wave groups. These were the waves with the smaller initial heights. Since adjacent groups of broken waves were separated in between by non-breaking waves, the distribution of $[k/(gh)]^{1/2}$ has a large standard deviation. Most of the waves were broken at $d = 6.25$ cm, and $[k/(gh)]^{1/2}$ has a narrow variation. Compared to the narrow-band spectrum, a broad-band spectrum (Ting 2001) has a smaller range of $[k/(gh)]^{1/2}$ values at $d = 13.72$ cm. This is because wave breaking was more random in the broad-band waves. The turbulence generated by one breaker did not die out completely before the arrival of the next breaker. Thus, the turbulence energy generated by wave breaking was distributed more uniformly among the breaking and non-breaking waves in the broad-band waves.

In reference to Ting (2001), the distributions of $[k/(gh)]^{1/2}$ for the narrow-band and broad-band waves are similar at $d = 6.25$ cm. These results indicate that changing the shape of the incident wave spectrum may not affect the distribution of turbulent kinetic energy in the inner surf zone; other factors are also important. Experimental studies on regular spilling and plunging breakers showed that the turbulent flow field in the inner surf zone depends on a particular breaker type and it is not similar for different breaker types (Ting and Kirby 1994). Breaker type is characterized by the surf similarity parameter $\xi_0 = \tan\beta/(H_0/L_0)^{1/2}$, where β is the beach slope, and H_0 and L_0 are the deep-water wave height and wave length, respectively (Battjes 1974). When calculated based on H_{m0} and T_p, ξ_0 has a value of 0.20 for the narrow-band spectrum and 0.18 for the broad-band spectrum. Visual observations showed that the surf zone of the narrow-band spectrum and broad-band spectrum were both dominated by spilling breakers. It is suggested that the distributions of turbulent kinetic energy for the narrow-band and broad-band waves were similar in the inner surf zone because the turbulence was produced by similar breaker types.

Figure 6 presents the distributions of the period-averaged Reynolds stress correlation coefficient for the narrow-band spectrum. The Reynolds stress correlation coefficient is defined as $<-u'w'>/(<u'^2> <w'^2>)^{1/2}$, where the angular brackets denote an operator to take an ensemble average. The period average, τ_i (i=1,2...N), is in the range of -0.2 to 0.4. Comparing to a broad-band spectrum (Ting 2001), the mean and the highest one-third value of τ for the narrow-band spectrum are significantly lower in the upper layer. In the lower layer, the distributions of τ for the narrow-band spectrum and broad-band spectrum are similar.

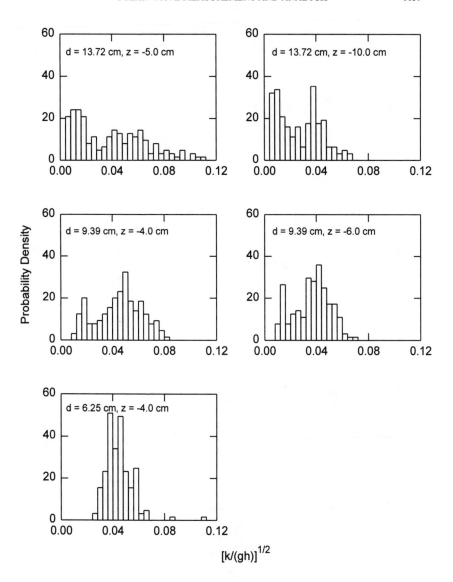

Fig. 5. Distributions of wave-averaged turbulent kinetic energy

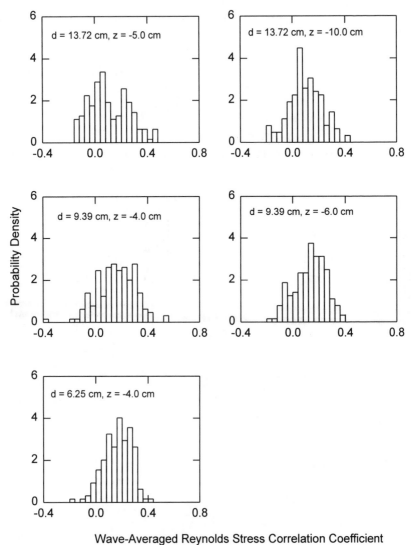

Wave-Averaged Reynolds Stress Correlation Coefficient

Fig. 6. Distributions of wave-averaged Reynolds stress correlation coefficient

CONCLUSIONS
The following main conclusions can be drawn from this study:

1. The wave-height-to-water-depth ratio for narrow-band waves has a narrow variation inside the surf zone when the wave height of the individual waves is normalized by the period-averaged water depth. The waves with the smaller initial heights break closer to shore and have higher wave-height-to-water-depth ratios at breaking than the waves with the larger initial heights.

2. The standard deviation of the wave period increases as the water depth decreases. The long-wave motion generated by wave breaking creates a time-varying water depth and current that changes the propagation speed of the short waves riding on the long waves. This causes the wave period of the short waves to increase or decrease inside the surf zone.

3. In the outer and middle surf zone, the period-averaged turbulent kinetic energy has a wider range of values for narrow-band waves than for broad-band waves. After most of the waves have broken, the distributions of the period-averaged turbulent kinetic energy for the narrow-band and broad-band waves are similar. It is suggested that the wave and turbulence characteristics in the inner surf zone are determined by the breaker types, which in turn, are determined by the distributions of wave heights and wave periods of the incident waves, and the beach slope.

ACKNOWLEDGEMENTS
This study was sponsored by the Office of Naval Research through Grant N00014-00-1-0461. Their support is gratefully acknowledged.

REFERENCES
Battjes, J. A. 1974. Surf Similarity. *Proceedings of the 14th International Coastal Engineering Conference*, ASCE, 466-480.

George, R., Flick, R. E., and Guza, R. T. 1994. Observations of Turbulence in the Surf Zone. *Journal of Geophysical Research*, 99 (C1), 801-810.

Hansen, J. B. 1990. Periodic Waves in the Surf Zone: Analysis of Experimental Data. *Coastal Engineering*, 14, 19-41.

Sultan, N. J., 1995, *Irregular Wave Kinematics in the Surf Zone*. Ph.D. dissertation, Department of Civil Engineering, Texas A&M University, College Station, Texas, 135 p.

Ting, F. C. K. 2001. Laboratory Study of Wave and Turbulence Velocities in a Broad-Banded Irregular Wave Surf Zone. *Coastal Engineering*, 43, 183-208.

Ting, F. C. K. and Kirby, J. T. 1994. Observation of Undertow and Turbulence in a Laboratory Surf Zone. *Coastal Engineering*, 24, 51-80.

WAVE DAMPING AND SPECTRAL EVOLUTION AT ARTIFICIAL REEFS

Matthias Bleck[1] and Hocine Oumeraci[2]

Abstract: Artificial reefs are increasingly used as an active shore protection measure. Nevertheless, the physical processes occuring at these structures are still not fully understood. Existing design formulae usally take into account only the change in wave height over the reef, i.e. the energy transfer within the wave spectrum is not considered. Physical model tests on artificial reefs have been performed at Leichtweiß-Institute for Hydraulic Engineering, focussing on the energy transfer and on local effects occuring at the reef. First results are discussed in this paper.

INTRODUCTION

Due to the increase of storminess and sea level rise observed in the last decades as a result of global climate changes and owing to the increasing pressure for the use of coastal zones it is expected that the demand for coastal protection will increase. Particularly due to the increasing activities and infrastructure for recreating, the traditional shore protection structures are getting more and more controverse. Hard structures such as revetments, sea walls and breakwaters represent a passive protection measure which substantially affect the marine landscape. Therefore, active protection measures such as artificial reefs constitute a better alternative. In contrast to the passive protection measures, active protection measures do influence the waves further seaward before they reach the shore. As a result, the protection works at the shore - if still necessary - do not need to resist large wave loads and overtopping, so that they can be built much less massive. In addition, the life time of the structures at the shore will substantially be increased as less severe wave conditions prevail behind the active structure.

[1] teaching and research assistant, Leichtweiß-Institute for Hydraulic Engineering; Beethovenstr. 51a; 38106 Braunschweig, Germany; M.Bleck@tu-bs.de

[2] Professor and Head of Leichtweiß-Institute for Hydraulic Engineering; H.Oumeraci@tu-bs.de

The best known and most investigated active shore protection structure is the artificial reef also known as submerged breakwater. An advantage of this type of structure is, that it is invisible from the coastline not disturbing the natural landscape. Moreover, artificial reefs are also used to create new habitats for marine flora and fauna or to create an appropriate surf beach for wave riders. Both concepts may also have a strong economical impact associated with the touristic use of the reef either as a surfing site or as a diving and fishing spot.

MOTIVATION

For artificial reefs as a shore protection measure most existing design concepts are based on a global approach. This means, the local effects occuring at the reef (Fig.1) such as wave breaking, vortex sheeding and non-linear interactions are not explicitly addressed. Only the waves in front and behind the reef are compared. Thereby, two changes can be observed afterwards addressed as global effects (Fig.1). In the time domain the waves behind the reef seem to be shorter and smaller than in front of the reef. In the frequency domain this would be expressed by a decrease in the integral of the energy density spectrum (wave height reduction) and a deformation of the spectrum (change in wave form / period).

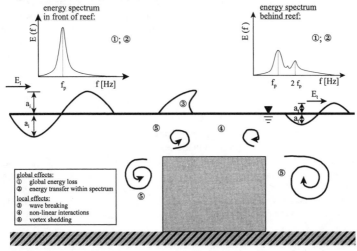

Fig.1: Effects Occuring at Artifical Reefs

Existing design rules commonly rely on wave energy coefficients; i.e. the energy of the wave spectrum in front of the artifical reef is compared to the energy of the spectrum behind the reef by means of the transmission coefficient k_t:

$$k_t = \frac{H_t}{H_i} = \sqrt{\frac{E_t}{E_i}} \qquad (1)$$

with: H_i = incoming wave height, H_t = transmitted wave height, E_i = incoming wave energy and E_t = transmitted wave energy.

The link between wave height and wave energy is achieved by the zero order moment of the spectrum m_0. The H_{m0} wave height is then expressed by:

$$H_{m0} = 4\sqrt{m_0} \tag{2},$$

All wave components in the record of the surface elevation are taken into account, including the higher harmonics. One should remind that this is not necessarily the case when applying a zero-crossing approach in the time domain analysis which is very sensitiv to the selected mean level.

Using the energetic approach the deformation of the energy density spectrum is not explicitly considered. Nevertheless, this change of spectral shape can not be neglected for the design of coastal structures, because the frequency of a wave (component) and thus its period is a measure for the wave celerity, which is determinant for the calculation of the energy flux.

Beside the fact that the change in spectral shape is implicitly taken into account by using the zero order moment (and thus the whole energy contained in the wave) most authors also use to mention a change in wave form or wave period. Nevertheless, no quantitative consideration of this fact has yet been reported. The existing design concept based on transmission coefficients can therefore be regarded as insufficient to describe the hydraulic performance of reefs. A comprehensive overview of the existing design formulas is provided by BLECK (1997).

EXPERIMENTAL SET-UP AND PROCEDURE

In the wave flume (length = 100m; width = 2m) of the Leichtweiß-Institute for Hydraulic Engineering (LWI) physical model tests at an idealised artificial reef (Fig.2) have been conducted. A rectangular box with different heights (h = 0.4m; 0.5m; 0.6m) and widths (B = 0.5m; 1.0m) placed on the flume bottom has been employed to represent the reef. In total six alternatives were investigated using both regular and irregular (JONSWAP spectra) waves (Table 1). Water level elevations were recorded in front and behind the reef. The water depth was kept constant at $d_f = 0.70$m.

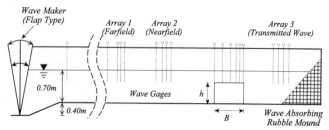

Fig.2: Model Set-Up at LWI

Table 1. Test Conditions

T bzw. T_p [s] / H bzw. H_s [m]	1,1	1.5	2.0	2.5	3.0	3.5	4.0	4.5	6.0
0.08	×	×	×	×	×	×	×	×	×
0.12	×	×	×	×	×	×	×	×	
0.16	×	×	×	×	×	×			
0.20	×	×	×	×	×				

FIRST EXPERIMENTAL RESULTS
General
The generation of secondary wave crests can be clearly observed while the waves pass the reef. Because of the irregularities of the transmitted waves, all wave recordings, including regular waves, are analysed using Fourier spectral analysis. Besides a decrease of the total wave energy, a broadening of the spectrum can be observed behind the reef (Figs.3 & 4). Unlike the time series which is altered by a different superposition of bound and free waves depening on the position there is no further evolution of the spectrum behind the reef (e.g. ELDEBERKY; 1996).

incident amplitude spectrum

transmitted amplitude spectrum

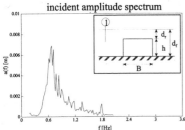

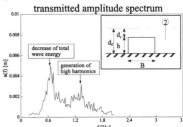

Fig.3: Spectrum in Front of Reef

Fig.4: Spectrum Behind Reef

Wave Heights
Transmission coefficients k_t are calculated using the wave height H_{m0} directly in front of the reef (array 2) and behind the reef (array 3).

In general the transmission coefficient depends on the following parameters (Fig.5):

$$k_t = f(H_i; T_i; d_f; h; B; g; \rho_w; v) \qquad (3)$$

with: H_i = incident wave height [m], T_i = incident wave period [s], d_f = water depth at toe of structure [m], h = reef height [m], B = reef length [m], g = acceleration of gravity [m/s^2], ρ_w = water density [kg/m^3] and v = viscosity of water [m^2/s].

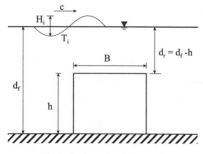

Fig.5: Influencing Parameters

These dimensional parameters are combined to non-dimensional parameters (Table 2). The relation between the non-dimensional parameters and the transmission coefficients are investigated. The transmitted wave energy shows a clear dependency of the non linearity parameter for shallow water d_r/H_i, the Ursell number and two other universal non linearity parameters proposed by BEJI (1995) and GODA (1983).

Table 2. Investigated Dimensionless Parameters

notation	description of parameter	definition
d_r/H_i	relative water depth; non-linearity parameter for shallow water (breaker criterion)	d_r/H_i
S	wave steepness; non-linearity parameter for deep water (breaker criterion)	$\dfrac{H}{L}$
Ur	Ursell parameter; universal non-linearity parameter	$\dfrac{H \cdot L^2}{d^3}$
Π	universal non-linearity parameter following GODA (1983)	$\left(\dfrac{H}{L}\right) \cdot \coth^3\left(\dfrac{2\pi}{L} \cdot d\right)$
ε	universal non-linearity parameter following BEJI (1995)	$\dfrac{g \cdot a}{c^2}$
B/L	relative reef length	B/L
B/H	relative reef length	B/H

The relative water depth d_r/H_i proved to be the best parameter for describing the wave transformation. The following relationship has been established for the transmission coefficient k_t (Fig.6):

$$k_t = 1.0 - 0.83 \cdot \exp\left(-0.72 \cdot \frac{d_r}{H_i}\right) \tag{4}$$

having a coefficient of determination of $r^2=0.90$ and a coefficient of variation $\sigma`=6.7\%$.

Comparing the results of the own tests with existing design formulas (for details see OUMERACI and BLECK; 2001) the method of NAKAMURA et al. (1966), which is comparable to Eq.4, shows the best fit (Fig.7).

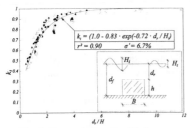

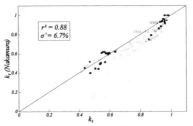

Fig.6: Transmission Coefficient as Function of Relative Water Depth

Fig.7: Own Results Compared to NAKAMURA (1966)

Wave Period

Considering the evolution of the Fourier based energy spectrum before and behind the reef, no change of the peak period can be observed. Actually, this only indicates that the highest energy remains at the same frequency. An information about the distribution of the energy is not provided. Nevertheless, the shape of the spectrum definitly changes while the waves pass the reef (compare Figs.3 and 4). To quantify this change a method to describe the spectral shape has to be applied. As will be shown the peak period is of no use for this purpose. Useless are also the spectral shape parameters ε and μ proposed by LONGUET-HIGGINS (1975):

$$\varepsilon = \sqrt{1 - \frac{m_2^2}{m_0 \cdot m_4}} \tag{5}$$

$$\mu = \frac{m_0 \cdot m_2}{m_1^2} - 1 \tag{6}$$

In fact both include higher order spectral moments defined as:

$$m_n = \int f^n \cdot S(f) \cdot df \tag{7}$$

which are mathematically quite unstable, because of a strong weighting of the frequency. A small amplitude which may be caused by measurement noise (not representing the signal) at high frequencies will highly influence the calculation of the higher order moments m_n.

The best results in describing the evolution of the spectral shape are achieved by using the wave periods T_{01} and T_{0-1}:

$$T_{01} = \frac{m_0}{m_1} = \frac{\int S(f) \, df}{\int f \, S(f) \, df} \tag{8}$$

and
$$T_{-10} = \frac{m_{-1}}{m_0} = \frac{\int f^{-1} S(f)\ df}{\int S(f)\ df} \qquad (9).$$

Both periods are calculated as a weighted mean for the frequency of the energy density spectrum comparable to the centre of gravity for a plane. While the period T_{01} is weighting more the high frequency components of the spectrum the lower components are more weigthed for the period T_{-10}. In addition the period T_{-10} has been used by VAN GENT (1999) to parameterize real sea spectra for the prediction of wave overtopping at sea dykes.

The wave periods T_{02} and T_{-20} are less appropriate due to the second order moment which is more sensitive to measurement noise and thus mathematically less stable.

For the incident and transmitted energy spectra shown in Figs.3 and 4 these parameters are given in Table 3. As already mentioned the energy of the spectrum expressed by the wave height H_{m0} decreases while the peak period T_p remains constant. For the other parameters a change can be observed.

Table 3. Spectral Parameters for Example Case (Figs. 3 and 4)

parameter	H_{m0}	T_p	T_{01}	T_{-10}	T_{02}	T_{-20}	ε	μ
incident spectrum	0.102m	1.515s	1.349s	1.421s	1.303s	1.453s	0.570	0.72
transmitted spectrum	0.085m	1.515s	1.098s	1.247s	1.029s	1.320s	0.437	0.14

The change of the wave period T_{-10} as the waves pass the reef can be predicted as a function of the relative water depth d_r/H_i by the following equation:

$$\Delta T_{-10} = 1 - \frac{T_{-10,t}}{T_{-10,i}} = -0.24 \cdot \exp\left(-0.63\ \frac{d_r}{H_i}\right) \qquad (4)$$

with $r^2 = 0,77$ and $\sigma' = 41$ %, showing that the correlation is relativly low and the uncertainty too high.

As the transmission coefficient k_t is a squared ratio of the zero order moment before and behind the reef (comp. Eq.1) and the wave period a ratio of two spectral moments (Eq.9), the change in wave period can also be expressed by:

$$\Delta T_{-10} = 1 - \frac{m_{-1,t} \cdot m_{0,i}}{m_{0,t} \cdot m_{-1,i}} = 1 - k_t^{-2} \cdot \frac{m_{-1,t}}{m_{-1,i}} = 1 - \frac{k_{m-1}}{k_t^2} \qquad (11)$$

with: $k_t = \sqrt{m_{0,t}/m_{0,i}}$ = transmission coefficient = zero order moment coefficient,
$k_{m-1} = m_{-1,t}/m_{-1,i}$ = minus first moment coefficient.

The wave transformation at artificial reefs can therefore be described by using the transmission coefficient (zero order moment coefficient) and the higher order moment coefficients. The moment coefficient k_{m-1} expressed by the relative water depth d_r/H_i will lead to:

$$k_{m-1} = 1.0 - 1.17 \cdot \exp\left(-0.55 \cdot \frac{d_r}{H_i}\right) \qquad (12)$$

Combining Eqs.4, 11 and 12 the period T_{-10} of the transmitted waves can be predicted by the following relationship:

$$T_{-10,t} = \frac{k_{m-1}}{k_t^2} \cdot T_{-10,i} = \frac{1.0 - 1.17 \cdot \exp\left[-0.55 \cdot \frac{d_r}{H_i}\right]}{\left(1.0 - 0.83 \cdot \exp\left[-0.72 \cdot \frac{d_r}{H_i}\right]\right)^2} \cdot T_{-10,i} \qquad (13)$$

This approach results in a high correlation for the measured data (Fig.8) which is expressed by a regression coefficient of $r^2 = 0.99$ and a relative standard deviation of $\sigma' = 3.0\%$.

SUMMARY AND CONCLUDING REMARKS

Based on the own experimental data a predictive formula for the energy coefficients describing the wave transformation at artificial reefs is developed. In addition, the change in the spectral shape for waves passing the reef is quantified, i.e. a predictive formula is derived for the period of the transmitted waves.

As a first result a new design concept for artificial reefs is proposed which is sketched in Fig.9. The concept is based on spectral moment coefficients to describe the spectral evolution at artificial reefs completely. The first part of the new concept is identical with the standard methods yet available:

•   The wave height behind the reef will be calculated applying energy coefficients such as the transmission coefficient.

The second (new) part of the proposed design concept aims at describing the deformation of the wave spectra:

•   The change in spectral shape expressed by the wave periods T_{01} and T_{-10} is quantified by using first and minus first order moment coefficients in addition to the energy coefficients (equivalent to zero order moment coefficients).

As the proposed design concept for artificial reefs represents a first result to describe the wave transformation at the reef entirely, the proposed relationships can still be improved. In addition, the data base will be extended for other reef geometries and porosities. Moreover the applicabilty of the concept for natural multi peak spectra, which have a different form than theoretical, single peak spectra, will be investigated.

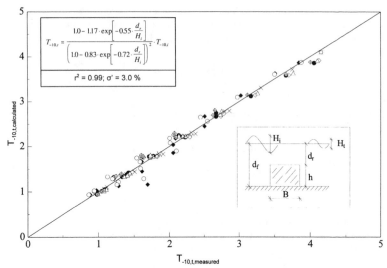

Fig.8: Accuracy of Proposed Formula for Transmitted Wave Period

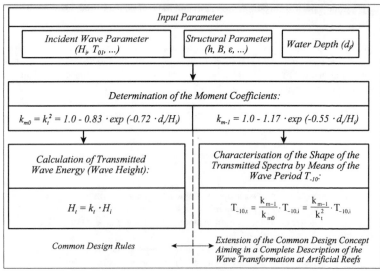

Fig.9: Proposed Design Concept

Finally, the local effects at the structure which are the source of the global effects described here have to be addressed both qualitatively and quantitatively. Especially the term "non-linear interactions" which is very often mentioned in the literature to explain the effect of the reef on the waves is an indication that the physical processes are not completely understood.

ACKNOWLEDGEMENTS
This project is supported by the German Research Council (DFG) within the basic research project DFG OU 1/6-1 („Hydrodynamic Efficiency of Artificial Reefs with Particular Consideration of the Energy Transfer within the Wave Spectrum"). This support is gratefully acknowledged.

REFERENCES
BEJI, S: (1995) Note on a Nonlinearity Parameter of Surface Waves. Coastal Engineering 25

BLECK, M. (1997) Hydraulische Wirksamkeit von "künstlichen Riffen" im Küstenschutz. (Hydraulic Performance of Artificial Reefs for Coastal Protection). Master Thesis at the Leichtweiss-Institute. (in German)

ELDEBERKY, Y. (1996) Nonlinear Transformation of Wave Spectra in the Nearshore Zone. Communications on Hydraulic and Geotechnical Engineering, Delft University of Technology, Report No. 96-4.

VAN GENT, MRA (1999) Physical Model Investigations on Coastal Structures with Shallow Foreshores - 2D Model Test with Single and Double-Peaked Wave Energy Spectra. Delft Hydraulics Report H3608

GODA, Y. (1983) A Unified Nonlinearity Parameter of Water Waves. Report of the Port and Harbour Research Institute, Vol.22 No. 3

LONGUET-HIGGINS (1975) On the Statistical Distribution of the Periods and Amplitudes of Sea Waves. Journal of Geophysical Research 80

MASON, A. and KEULEGAN, G.H. (1944) A Wave Method for Determining Depths over Bottom Discontinuities

NAKAMURA, M.; SHIRAISHI, H. and SASAKI, Y. (1966) Wave Dampening Effect of Submerged Dike. Coastal Engineering 24

OUMERACI, H. and BLECK, M. (2001) Hydraulische Wirksamkeit von künstlichen Riffen unter besonderer Berücksichtigung des Energietransfers im Wellenspektrum („Hydrodynamic Efficiency of Artificial Reefs with Particular Consideration of the Energy Transfer within the Wave Spectrum"). LWI report No. 863 (in German).

THREE-DIMENSIONAL NUMERICAL MODEL FOR FULLY NONLINEAR WAVES OVER ARBITRARY BOTTOM

Philippe Guyenne [1], Stéphan T. Grilli [2], M.ASCE, and Frédéric Dias [3]

Abstract: We present an accurate three-dimensional (3D) Numerical Wave Tank (NWT) solving the full equations in the potential flow formulation. The NWT is able to simulate wave propagation up to overturning over an arbitrary bottom topography. The model is based on a high-order 3D Boundary Element Method (BEM) with the Mixed Eulerian-Lagrangian (MEL) approach. The spatial discretization is third-order and ensures continuity of the inter-element slopes. Waves can be generated in the tank by wavemakers or they can be directly specified on the free surface. A node regridding can be applied at any time step over selected areas of the free surface. Results are presented for the computation of overturning waves over a ridge and their kinematics.

INTRODUCTION

Many numerical wave models solving Fully Nonlinear Potential Flow (FNPF) equations have been developed, mostly in two dimensions (2D), which have been shown to accurately simulate wave overturning in deep and intermediate water (Dommermuth et al. 1988) as well as wave shoaling and breaking over slopes (Grilli et al. 1997). In most recent 2D models, incident waves can be generated at one extremity and reflected, absorbed or radiated at the other extremity (Grilli and Horrillo 1997). In three dimensions (3D), only a few attempts have been reported of solving FNPF problems, for arbitrary transient nonlinear waves in a general propagation model, with the possibility of modeling overturning waves. Xu and Yue (1992) and Xue et al. (2001) calculated 3D overturning waves in a doubly periodic computational domain with infinite depth (i.e. only the free surface was discretized). In their case, progressive Stokes waves were led to breaking

---

[1] Dept. of Math. and Stat., McMaster Univ., Hamilton, ON L8S 4K1, Canada. E-mail: guyenne@math.mcmaster.ca.

[2] Ocean Engrg. Dept., Univ. of Rhode Island, Narragansett, RI 02882, USA. E-mail: grilli@oce.uri.edu.

[3] CMLA, ENS de Cachan, 94235 Cachan cedex, France. E-mail: dias@cmla.ens-cachan.fr.

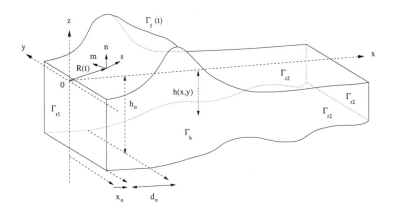

FIG. 1.   Sketch of the computational domain for the 3D BEM.

by specifying an asymmetric surface pressure. Sawtooth instabilities eventually developed near the wave crests and were eliminated by smoothing. Broeze (1993) developed a numerical model similarly to Xu and Yue's but for non-periodic domains and finite depth. He was able to produce the initial stages of wave over-turning over a bottom shoal. Numerical instabilities were also experienced which limited the computations.

In the present study, we propose a new 3D nonlinear surface wave model (Fig. 1), solving FNPF equations based on a high-order 3D Boundary Element Method (BEM) and a mixed Eulerian-Lagrangian time updating of the free surface Γ_f. The methods used for both spatial and temporal discretizations are direct 3D extensions of those in Grilli and Subramanya (1996). The model is applicable to nonlinear wave transformations up to overturning and breaking from deep to shallow water of arbitrary bottom topography Γ_b. This, in fact, constitutes a Numerical Wave Tank (NWT), where arbitrary waves can be generated by wavemakers on Γ_{r1} or they can be directly specified on the free surface. If needed, absorbing boundary conditions can be simulated on lateral boundaries Γ_{r2} (Grilli and Horrillo 1997). In addition, techniques are developed for regridding nodes at any time step, over selected areas of the free surface.

MATHEMATICAL FORMULATION

Equations for the FNPF formulation with a free surface are summarized below. The fluid velocity is expressed as $\boldsymbol{u} = \boldsymbol{\nabla}\phi = (u, v, w)$, with $\phi(\boldsymbol{x}, t)$ the velocity potential.

Continuity equation in the fluid domain $\Omega(t)$, with boundary $\Gamma(t)$, is Laplace's equation for the velocity potential,

$$\nabla^2\phi = 0 . \qquad (1)$$

The 3D free space Green's function for Eq. (1) is defined as

$$G(\boldsymbol{x},\boldsymbol{x}_l) = \frac{1}{4\pi r} \quad \text{and} \quad \frac{\partial G}{\partial n}(\boldsymbol{x},\boldsymbol{x}_l) = -\frac{1}{4\pi}\frac{\boldsymbol{r}\cdot\boldsymbol{n}}{r^3}\,, \tag{2}$$

with $r = |\boldsymbol{r}| = |\boldsymbol{x} - \boldsymbol{x}_l|$ the distance from the source point $\boldsymbol{x}$ to the field point $\boldsymbol{x}_l$ (both on boundary Γ) and $\boldsymbol{n}$ the outward unit normal vector at $\boldsymbol{x}$ on Γ.

Green's second identity transforms Eq. (1) into the Boundary Integral Equation (BIE)

$$\alpha(\boldsymbol{x}_l)\,\phi(\boldsymbol{x}_l) = \int_\Gamma \left\{ \frac{\partial\phi}{\partial n}(\boldsymbol{x})\,G - \phi(\boldsymbol{x})\,\frac{\partial G}{\partial n} \right\} d\Gamma\,, \tag{3}$$

where $\alpha(\boldsymbol{x}_l) = \theta_l/(4\pi)$ with θ_l the exterior solid angle at point $\boldsymbol{x}_l$.

The boundary is divided into various sections, with different boundary conditions (Fig. 1). On the free surface $\Gamma_f(t)$, ϕ satisfies the nonlinear kinematic and dynamic boundary conditions

$$\frac{D\boldsymbol{R}}{Dt} = \boldsymbol{u} = \boldsymbol{\nabla}\phi\,, \tag{4}$$

$$\frac{D\phi}{Dt} = -gz + \frac{1}{2}\boldsymbol{\nabla}\phi\cdot\boldsymbol{\nabla}\phi - \frac{p_a}{\rho}\,, \tag{5}$$

respectively, with $\boldsymbol{R}$ the position vector of a fluid particle on the free surface, g the acceleration due to gravity, p_a the atmospheric pressure, ρ the fluid density and D/Dt the Lagrangian time derivative.

Various methods can be used in the NWT for wave generation. When waves are generated by a wavemaker at the "open sea" boundary $\Gamma_{r1}(t)$, motion and velocity $[\boldsymbol{x}_p(t), \boldsymbol{u}_p(t)]$ are specified over the wavemaker as

$$\overline{\boldsymbol{x}} = \boldsymbol{x}_p \quad \text{and} \quad \overline{\frac{\partial\phi}{\partial n}} = \boldsymbol{u}_p\cdot\boldsymbol{n}\,, \tag{6}$$

where overlines denote specified values.

Along the bottom Γ_b and other fixed parts of the boundary referred to as Γ_{r2}, a no-flow condition is prescribed as

$$\overline{\frac{\partial\phi}{\partial n}} = 0\,. \tag{7}$$

The solution within the domain can be easily evaluated from the boundary values. For instance, the internal velocity and local acceleration are given by

$$\boldsymbol{\nabla}\phi(\boldsymbol{x}_l) = \int_\Gamma \left\{ \frac{\partial\phi}{\partial n}(\boldsymbol{x})\,Q - \phi(\boldsymbol{x})\,\frac{\partial Q}{\partial n} \right\} d\Gamma\,, \tag{8}$$

$$\boldsymbol{\nabla}\frac{\partial\phi}{\partial t}(\boldsymbol{x}_l) = \int_\Gamma \left\{ \frac{\partial^2\phi}{\partial t\partial n}(\boldsymbol{x})\,Q - \frac{\partial\phi}{\partial t}(\boldsymbol{x})\,\frac{\partial Q}{\partial n} \right\} d\Gamma\,,$$

respectively, where

$$Q(\boldsymbol{x}, \boldsymbol{x}_l) = \frac{1}{4\pi\, r^3}\, \boldsymbol{r} \quad \text{and} \quad \frac{\partial Q}{\partial n}(\boldsymbol{x}, \boldsymbol{x}_l) = \frac{1}{4\pi\, r^3}\, \{\boldsymbol{n} - 3\,(\boldsymbol{e}_r \cdot \boldsymbol{n})\,\boldsymbol{e}_r\},$$

with $\boldsymbol{e}_r = \boldsymbol{r}/r$.

Note, results presented here only have no-flow conditions on lateral boundaries Γ_{r1} and Γ_{r2}. For the use of a "snake" flap wavemaker and an absorbing piston at extremities of the tank, the reader can refer to Brandini and Grilli (2001).

TIME INTEGRATION

Following the method implemented in Grilli and Subramanya's 2D model (1996), second-order explicit Taylor series expansions are used to express both the new position $\overline{\boldsymbol{R}}(t + \Delta t)$ and the potential $\overline{\phi}(\boldsymbol{R}(t + \Delta t))$ on the free surface, in the MEL formulation, as

$$\overline{\boldsymbol{R}}(t + \Delta t) = \boldsymbol{R} + \Delta t \frac{D\,\boldsymbol{R}}{D\,t} + \frac{(\Delta t)^2}{2} \frac{D^2\boldsymbol{R}}{Dt^2} + \mathcal{O}[(\Delta t)^3], \tag{9}$$

$$\overline{\phi}(\boldsymbol{R}(t + \Delta t)) = \phi + \Delta t \frac{D\,\phi}{D\,t} + \frac{(\Delta t)^2}{2} \frac{D^2\phi}{Dt^2} + \mathcal{O}[(\Delta t)^3], \tag{10}$$

where all terms in the right-hand sides are calculated at time t.

Coefficients in these Taylor series are expressed as functions of the potential, its partial time derivative, as well as the normal and tangential derivatives of both of these along the free surface. Thus, the first-order coefficients are given by Eqs. (4) and (5), which requires calculating $(\phi, \frac{\partial\phi}{\partial n})$ on the free surface. The second-order coefficients are obtained from the Lagrangian time derivative of Eqs. (4) and (5), which requires also calculating $(\frac{\partial\phi}{\partial t}, \frac{\partial^2\phi}{\partial t\partial n})$ at time t.

As in Grilli and Svendsen (1990), the time step Δt in Eqs. (9) and (10) is adapted at each time as a function of the minimum distance between two nodes on the free surface and a constant mesh Courant number $\mathcal{C}_o \simeq 0.45$.

The advantages of this time stepping scheme are of being explicit and using spatial derivatives of the field variables along the free surface in the calculation of values at $(t + \Delta t)$. This provides for a better stability of the computed solution and makes it possible to use larger time steps, for a similar accuracy, than in Runge-Kutta or predictor-corrector methods, which only use point to point updating based on time derivatives and thus are more subject to sawtooth instabilities. Hence, this also makes the overall solution more efficient for a specified numerical accuracy of the results.

BOUNDARY DISCRETIZATION AND REGRIDDING

The BIEs for ϕ and $\frac{\partial\phi}{\partial t}$ are solved by a BEM. The boundary is discretized into collocation nodes and M_Γ high-order elements are used to interpolate in between m of these nodes. Within each element, the boundary geometry and the field variables (denoted by $u = \phi$ or $\frac{\partial \phi}{\partial t}$, and $q = \frac{\partial \phi}{\partial n}$ or $\frac{\partial^2 \phi}{\partial t \partial n}$, for simplicity) are discretized using polynomial shape functions $N_j(\xi, \eta)$ as

$$\boldsymbol{x}(\xi, \eta) = N_j(\xi, \eta)\, \boldsymbol{x}_j^k$$

$$u(\xi, \eta) = N_j(\xi, \eta)\, u_j^k \quad , \quad q(\xi, \eta) = N_j(\xi, \eta)\, q_j^k$$

where $j = 1, \ldots, m$ denotes the nodes within each element $k = 1, \ldots, M_\Gamma$. The summation convention is applied to repeated subscripts.

Isoparametric elements can provide a high-order approximation within their area of definition but only offer C_0 continuity of the geometry and field variables at nodes in between elements. Based on the experience in modeling overturning waves in 2D NWTs, for producing stable accurate results one needs to define elements which are both higher-order within their area of definition and at least locally C_2 continuous in between elements. Here, the elements are defined using an extension of the so-called Middle-Interval-Interpolation (MII) method introduced by Grilli and Subramanya (1996). The boundary elements are 4-node quadrilaterals with cubic shape functions defined using both these and additional neighboring nodes in each direction for a total of $m = 16$ nodes.

The discretized boundary integrals are calculated for each collocation node by numerical integration. When the collocation node does not belong to the integrated element, a standard Gauss-Legendre quadrature method is used. When it belongs to the element, the distance r in the Green's function and in its normal gradient becomes zero at one of the nodes of the element (Eq. (2)). It can be shown that the integrals including G are weakly singular whereas the integrals including $\frac{\partial G}{\partial n}$ are non-singular. For the former integrals, a method of "singularity extraction", well-suited to MII elements, is applied based on polar coordinate and other transformations.

The linear algebraic system resulting from the discretization of Eq. (3) for ϕ (and $\frac{\partial \phi}{\partial t}$) is in general dense and non-symmetric. Since the number of nodes N_Γ can be very large in 3D, the solution by a direct method of order $\mathcal{O}(N_\Gamma^3)$ is prohibitive. As in Xu and Yue (1992) and Xue et al. (2001), a preconditioned GMRES (Generalized Minimal Residual) algorithm is used to iteratively solve the linear system.

Two types of regridding methods for the free surface are included in the model. When the free surface is still single-valued, a 2D horizontal regridding to a finer resolution can be performed in selected areas of the free surface. It consists in a reinterpolation of nodes for equally spaced MII elements in the x and y directions. In addition, we developed a local regridding technique similar to that in Grilli and Subramanya (1996) which redistributes the nodes in regions of flow convergence like in the breaker jet. When the distance between 2 nodes on grid lines along the direction of wave propagation becomes too small in comparison with the distance between neighboring nodes, the nodes are locally regridded to make these distances equal. The purpose is to limit the occurrence of quasi-singular integrals in the BIEs, resulting from the node convergence. Note, for regridding, the same

(a)

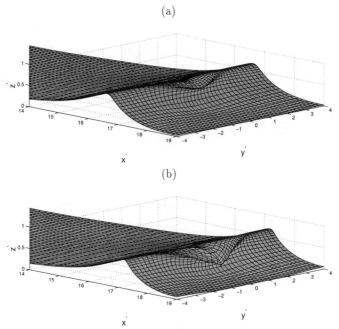

(b)

FIG. 2. Wave profiles over a sloping ridge, at $t' =$ (a) 8.577 and (b) 8.997.

interpolation functions as in the BEM are used to recalculate the solution at the new nodes, without modifying the solution obtained at the old nodes.

The reader can refer to Grilli et al. (2001) for more details on the model (description, validation, etc.).

RESULTS: SOLITARY WAVE SHOALING AND BREAKING OVER A SLOPING RIDGE

A domain of depth h_o and width $8h_o$ in the y direction is considered, with a sloping ridge at its x extremity. The ridge starts at $x' = 5.225$ and has a 1:15 slope in the middle ($y' = 0$), tapered in the y direction by specifying a depth variation in the form of a sech^2 modulation. The ridge is truncated at $x' = 19$ where the minimum depth is $h' = 0.082$ in the middle part ($y' = 0$) and the maximum depth $h' = 0.614$ on the sides ($y' = \pm 4$). [Dashes indicate non-dimensional variables based on the long wave theory, i.e. lengths are divided by h_o and times by $\sqrt{h_o/g}$.] The initial condition is an exact FNPF solitary wave of height $H'_o = 0.6$, with its crest located at $x' = 5.7$ for $t' = 0$. Such a wave is obtained using the numerical method proposed by Tanaka (1986).

The initial BEM discretizations on the bottom and the free surface have 50 by 20 quadrilateral elements in the x and y directions, respectively ($\Delta x'_o = 0.38$

(a)

(b)

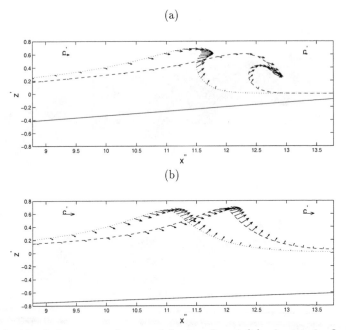

FIG. 3. Vertical cross-sections at (a) $y' = 0$ and (b) $y' = \pm 4$. Surface velocity field at $t' = 8.259$ (....) and $t' = 8.997$ (- -).

and $\Delta y'_o = 0.40$). The total number of nodes and of quadrilateral MII elements are $N_\Gamma = 2862$ and $M_\Gamma = 2560$, respectively. The initial time step is set to $\Delta t'_o = 0.171$ ($C_o = 0.45$). Maximum numerical errors of 1 % on wave mass and energy conservation are considered acceptable in this application. Computations are first performed in the initial discretization up to reaching these maximum errors. The 2D regridding of part of the NWT to a finer discretization is then specified at an earlier time $t' = 5.769$ for which errors are very small (0.012 % and 0.032 % for wave mass and energy respectively). At this stage, the wave crest is located at $x' = 13.2$ with $H' = 0.64$. The regridded discretization is increased to 60 by 40 quadrilateral elements on the free surface and bottom boundaries for $x' = 8.075$ to 19 ($\Delta x'_o = 0.182$, $\Delta y'_o = 0.20$, $N_\Gamma = 6022$, $M_\Gamma = 5600$) and computations are pursued, up to $t' = 8.577$.

Fig. 2(a) shows the wave computed at this time. Errors on wave mass and energy are still small (0.026 % and 0.054 % respectively) but the time step has considerably reduced, to $\Delta t' = 0.0016$. Wave overturning has already started in the middle of the NWT and has not yet reached the sidewalls. However, computations cannot be pursued much beyond this stage due to the node convergence at the wave crest. This problem is overcome by using the local adaptive regrid-

(a)

(b)

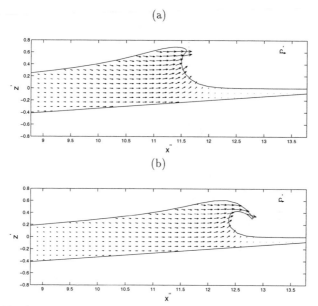

FIG. 4. Vertical cross-sections at $y' = 0$. Internal velocity field at $t' =$ (a) 8.259 and 8.997.

ding technique for nodes in the breaker jet. As a result, the solution in Fig. 2(b) exhibits a well developed plunging jet at $t' = 8.997$. One clearly sees that the overturning process tends to propagate laterally to the sidewalls. Grilli et al. (2001) evaluated the lateral mean speed of propagation of wave overturning.

Fig. 3 depicts the velocity field (u', w') in vertical sections in the middle of the NWT $(y' = 0)$ and at the sidewalls $(y' = \pm 4)$ for fluid particles on the free surface. The results are qualitatively in good agreement with those obtained by New et al. (1985) for overturning waves in 2D. In Fig. 4, we show the internal velocity field (u', w') for $(y' = 0)$ at $t' = 8.259$ and $t' = 8.997$. This was computed by using Eq. (8). For comparison, the celerity of a linear wave in shallow water $c' = \sqrt{g'h'_o}$ is given on the figures. Finally, the horizontal internal velocity field (u', v') is shown in Fig. 5, at depth $h' = -0.2$ for $t' = 7.911$ and $t' = 8.997$. Curves represent the bottom cross-sections. Focusing of the flow by the ridge can be seen on the figures, illustrating 3D breaking effects.

CONCLUSIONS

A 3D computation of wave shoaling and overturning over an arbitrary bottom topography was presented. Our results show a better stability and numerical accuracy than in previous attempts reported in the literature for calculating such strongly nonlinear 3D surface waves. Regridding techniques are developed to

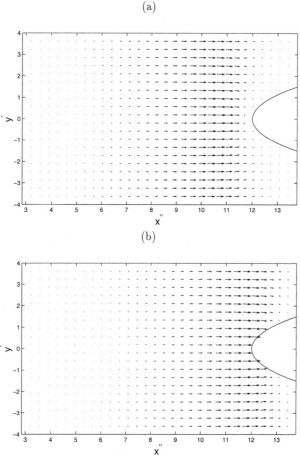

FIG. 5. Horizontal cross-sections at $h' = -0.2$. Internal velocity field at (a) the breaking point $t' = 7.911$ and (b) $t' = 8.997$.

describe the solution far beyond the breaking point. To our knowledge, this was never attempted before in a general 3D-NWT. The model can be applied to a wide range of problems such as the modeling of freak waves (Brandini and Grilli 2001) and the modeling of wave impact against structures (Guyenne et al. 2000).

ACKNOWLEDGMENTS

The first and third authors acknowledge support from the Délégation Générale pour l'Armement (France). The second author acknowledges support from the US Office of Naval Research, under grant N-00014-99-10439 of the Coastal Dynamics

Division (code 321 CD). Computations carried out in this research project were performed in part on the NEC SX5 supercomputer located at IDRIS, under CNRS funding.

REFERENCES

Brandini, C. and Grilli, S. T. (2001). "Modeling of freak wave generation in a 3D NWT." *Proc. 11th Offshore and Polar Engrg. Conf.*, Stavanger, Norway. 124–131.

Broeze, J. (1993). "Numerical modelling of nonlinear free surface waves with a 3D panel method," Phd dissertation, Enschede, The Netherlands.

Dommermuth, D. G., Yue, D. K. P., Lin, W. M., Rapp, R. J., Chan, E. S., and Melville, W. K. (1988). "Deep-water plunging breakers: a comparison between potential theory and experiments." *J. Fluid Mech.*, 189, 423–442.

Grilli, S. T., Guyenne, P., and Dias, F. (2001). "A fully nonlinear model for three-dimensional overturning waves over arbitrary bottom." *Int. J. Numer. Meth. Fluids*, 35, 829–867.

Grilli, S. T. and Horrillo, J. (1997). "Numerical generation and absorption of fully nonlinear periodic waves." *J. Engrg. Mech.*, 123(10), 1060–1069.

Grilli, S. T. and Subramanya, R. (1996). "Numerical modeling of wave breaking induced by fixed or moving boundaries." *Computational Mech.*, 17, 374–391.

Grilli, S. T. and Svendsen, I. A. (1990). "Corner problems and global accuracy in the boundary element solution of nonlinear wave flows." *Engrg. Analysis with Boundary Elements*, 7(4), 178–195.

Grilli, S. T., Svendsen, I. A., and R., S. (1997). "Breaking criterion and characteristics for solitary waves on slopes." *J. Waterways Port Coastal and Ocean Engrg.*, 123(3), 102–112.

Guyenne, P., Grilli, S. T., and Dias, F. (2000). "Numerical modeling of fully nonlinear 3D overturning waves over arbitrary bottom." *Proc. 27th Int. Conf. on Coastal Engrg.*, Sydney, Australia. 1–12.

New, A. L., McIver, P., and Peregrine, D. H. (1985). "Computation of overturning waves." *J. Fluid Mech.*, 150, 233–251.

Tanaka, M. (1986). "The stability of solitary waves." *Phys. Fluids*, 29(3), 650–655.

Xu, H. and Yue, D. K. P. (1992). "Computations of fully nonlinear three-dimensional water waves." *Proc. 19th Symp. on Naval Hydrodynamics*, Seoul, Korea. 1–24.

Xue, M., Xu, H., Liu, Y., and Yue, D. K. P. (2001). "Computations of fully nonlinear three-dimensional wave-wave and wave-body interactions. Part 1. Dynamics of steep three-dimensional waves." *J. Fluid Mech.*, 438, 11–39.

USING A LAGRANGIAN PARTICLE METHOD FOR DECK OVERTOPPING

Robert A. Dalrymple[1], F. ASCE, Omar Knio[2], Daniel T. Cox[3],
Moncho Gesteira[4], and Shan Zou[5]

Abstract: Smoothed Particle Hydrodynamics (SPH) methods have
recently been used to examine water wave motions. Here we apply
SPH to the problem of water waves impinging on a deck of an
offshore platform. Laboratory experiments by Cox and Ortega (2001)
are used to motivate the numerical modeling efforts. Realistic
comparisons are found, although no direct one-one comparisons are
made as the resolution of the numerical model was insufficient at that
time.

INTRODUCTION

Smoothed Particle Hydrodynamics (SPH) was developed in astrophysics to study
the behavior of large numbers of celestial bodies, such as solar systems and galaxies
(see Betz, 1990, and Monaghan, 1992, for review articles). Recently Monaghan
(1994) and Monaghan and Kos (1999, 2000) have applied the technology to free
surface flows, including long waves. Fontaine *et al.* (2000) examine a jet impacting
with a fluid, while Fontaine (2000) has examined the possible use of SPH for wave
impact studies on offshore structures. Comparisons between SPH models and BEM
model of waves are shown to be very good. Dalrymple and Knio (2001) show

---

1   E.C. Davis Professor, Center for Applied Coastal Research, University of Delaware, Newark, DE
    19716, rad@udel.edu
2   Assoc. Professor, Mechanical Engineering, Johns Hopkins University, Baltimore, MD 21218,
    knio@flame.me.jhu.edu
3   Assoc. Professor, Ocean Engineering Program, Civil Engineering Department, Texas A&M
    University , College Station, Texas 77843-3136, dtc@eddycat.tamu.edu
4   Professor, Faculty of Science, University of Vigo, Campus de Ourense, 32004 Ourense, Spain,
    mggesteira@uvigo.es
5   Graduate Student, Center for Applied Coastal Research, University of Delaware, Newark, DE
    19716, szou@coastal.udel.edu

shoaling solitary waves in a numerical wave tanks and present a new boundary condition for solid boundaries.

EXPERIMENTS

Here we use experiments by Cox and Ortega (2001) to motivate an SPH study of waves overtopping a flat deck, representing a fixed deck of an offshore structure. As indicated by Buckner (1995), the overtopping velocities of a wave on a deck are not the same as either the orbital motions that would be present in the absence of the deck nor equal to the phase velocity of the wave. Cox and Ortega noticed that both the velocities on the top of the deck and those under the deck can at times be much greater than those occurring in the absence of a deck.

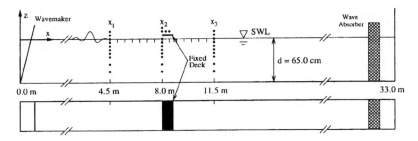

Figure 1 Laboratory Setup (from Cox and Ortega, 2001)

Their experiment setup consisted of fixing flat plate horizontally across a 36 m long and 0.95 m wide wave tank, with the plate bottom 5.25 cm above the water surface. The plate is 8 m away from a flap wavemaker and it is 0.61 m long and 1.5 cm thick. The wavemaker was driven by a drive signal that provides for two

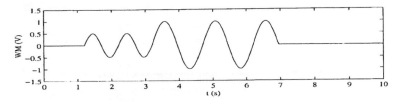

Figure 2 Wavemaker Drive Signal (Ortega and Cox, 2001)

cycles of one second period followed by 2.5 cycles of 1.5 second period, with a larger amplitude. In the wave tank, this leads to a large single wave striking the deck (located at $x=8\ m$, as a freak wave. The water surface at a variety of locations down the tank is shown in Figure 3. Velocity profiles taken with a laser doppler anemometer are shown in Figure 4, comparing the Case 1 situation (no deck) to the case when a deck is present, Case 2. In the figure, the velocity profiles over the water column are shown at the plate location for different phases of the freak wave. In the figure, the short heavy line refers to the instantaneous free surface in the Case 1, while the short, light line is the same for Case 2. The horizontal dashed lines indicate the crest and trough positions of the wave (14.6 and -8 cm respectively) and the dotted line corresponds to 5.25 cm, the deck bottom.

The conclusions of the experiments are that the presence of deck causes an 20% increase of wave height at front of deck. The maximum velocity just under the deck can exceed 2.5 times the velocity in the absence of the deck. The maximum velocities over the deck can also exceed 2.5 times the velocity in the absence of the deck. The experimental data are available from Dan Cox.

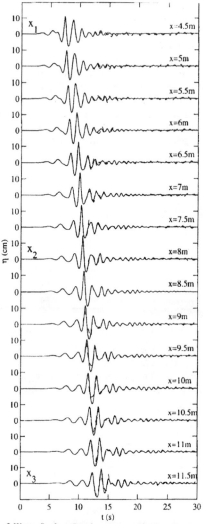

Figure 3 Water Surface Displacement at Various Location in Tank (Cox and Ortega, 2001)

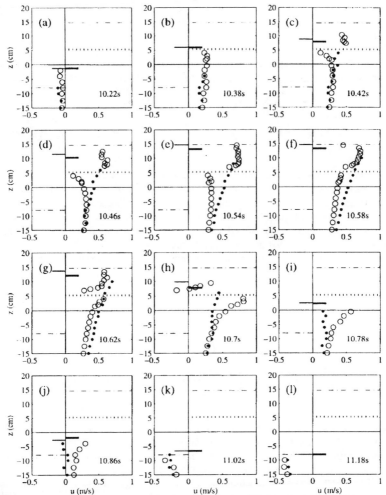

Figure 4 Measured Horizontal Velocities With (open circles) and Without Deck (Cox and Ortega, 2001)

SPH METHODOLOGY

The numerical development of SPH schemes is based on a particle representation of field quantities. Consider the approximate identity:

$$\zeta (r) \approx \int_{r'} \zeta (r) W (r - r', h) \, dr \quad (1)$$

Here, $\zeta(r)$ is a generic field quantity, r is the position vector, and W is a rapidly-decaying, radial smoothing function that approximates a Dirac delta function as h goes to zero.

The particle representation of $\zeta(r)$ is then obtained by approximating the integral in Eq. 1 using numerical quadrature with a cell size h, resulting in:

$$\zeta(r) = \sum_{i=1}^{N} \zeta_i h^n W(r - r_i, h) \quad (2)$$

where N is the total number of particles, ζ_i is the value of $\zeta(r_i)$ for the i-th particle, and r_i is the position of the i-th particle.

In SPH as applied to fluid flows is based, this particle representation is applied to density and velocity fields, by using a distribution of Lagrangian particles that are described in terms of their positions r_i, densities ρ_i, masses m_i, and velocities v_i. The evolution of the solution away from the initial condition is then described in terms of a coupled system of ordinary differential equations that express the mass and momentum conservation laws. As discussed in Gingold and Monaghan (1977), Monaghan (1982), and Monaghan (1992), this system can be expressed as:

$$\frac{d m_i}{d t} = 0$$

$$\frac{d r_i}{d t} = v_i$$

$$\frac{d \rho_i}{d t} = \sum_{j=1}^{n} m_j (v_i - v_j) \cdot \nabla_i W(r_i - r_j) \quad (3)$$

$$\frac{d v_i}{d t} = -\sum_{j=1}^{N} m_j \left(\frac{p_i}{\rho_i^2} + \frac{p_j}{\rho_j^2} + P_{ij} \right) \nabla_i W(r_i - r_j) + F_i$$

where t is time, P_{ij} is the viscosity term, and F_i is a body force term (including gravity). The quantity p_i is the pressure at the i th particle. The first equation states thats that the mass of every particle is conserved and the second defines the particle velocity. The last two equations represent the conservation of mass and momentum in SPH format. These equations show the advantage of the kernel smoothing functions W as derivatives of field quantities are found by using derivatives of the kernel function.

The fluid pressure is obtained from an equation of state (Monaghan, 1994):

$$p = B \left[\left(\frac{\rho}{\rho_0} \right)^\gamma - 1 \right] \quad (4)$$

where B and γ are constants, and ρ_o is the density at the free surface; here we use $\gamma = 7$, and $\rho_o = 1000$ kg/m^3. Monaghan uses $B = 200 \, \rho \, g \, H \, / \, \gamma$, where H is a representative depth. This modified equation of state actually gives a sound speed $(c^2 = d P / d \rho)$, which is much slower than the actual speed of sound in the water. This is important as the numerical time stepping is determined by the apparent speed of sound in the fluid.

One of the drawbacks of this equation of state is the extreme sensitivity of the pressure to fluctuations, due to the power of 7 in Eq. 4. Repeatability of calculations depends on the precision of the computations.

Monaghan and Lattanzio (1985) suggest a cubic spline kernel function:

$$
W \, (r,h) = \frac{10}{7 \, \pi \, h^2}
\begin{cases}
1 - \dfrac{3}{2}(\dfrac{r}{h})^2 + \dfrac{3}{4}(\dfrac{r}{h})^3 \; ; \, 0 \le (\dfrac{r}{h}) \le 1 \\[2mm]
\dfrac{1}{4}\Big(2 - (\dfrac{r}{h})\Big)^3 \; ; \quad 1 \le (\dfrac{r}{h}) \le 2 \\[2mm]
0 \, ; \qquad\qquad 2 < (\dfrac{r}{h})
\end{cases}
\qquad (5)
$$

This kernel has continuous first and second derivatives, and vanishes at distances greater than 2 h. This means the interactions between particles, given by the summations in Eqns. 3 are only important in the neighborhood of a given particle. By restricting the summations to these nearest neighbors leads to immense computational time savings.

The time stepping of the governing equations (Eqns. 3) is done with a predictor-corrector scheme utilized by Monaghan (1994).

NUMERICAL SIMULATIONS
The numerical modeling is done through the use of a numerical wave tank. The model consists of a wavemaker at one end of the tank and a sloping beach is located at the other. The bottom, the wavemaker, and the beach are composed of a double row of particles that have either no motion (and are not subjected to the equations of motion above or, in the case of the wavemaker, the motion is prescribed (Dalrymple and Knio, 2001).

Due to computational time considerations, the number of particles used in these simulations is small (~5000), with the particle sizes in the order of 5 cm, the distance from the water surface to the bottom of the deck. This size is clearly larger than desired.

Two different cases are examined: (1) initiating the wave in the basin as a solitary wave of the same height as the single freak wave striking the deck or (2) using the Cox and Ortega drive signal (unfortunately, while the drive signal is easily replicated, the correspondence between the drive signal and the paddle displacement is not known). Further, we use a piston motion rather than a flap. Nevertheless, in front of the paddle a wave similar to that in the laboratory is generated.

Case 1: solitary wave initialization. This technique utilizes theoretical expressions for the velocities and displacement of the particles to initialize the wave motion in a fixed box (no wavemaker). While the use of a solitary wave in a finite length box is fraught with theoretical difficulties, the model runs fine, producing a wave moving towards the platform deck and passing over and under it. The deck is represented by a single row of closely spaced fixed particles.

While the wave action is accentuated on the front of the deck as expected, the behavior of the wave at the trailing edge of the deck was interesting. For large waves, the pressure on the underside of the deck causes an upwards flow that impinges with that flow going over the top of the deck. The collision of these two separated flows at the rear of the deck can be dramatic. This is illustrated by Figure 4.

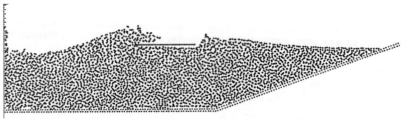

Figure 5Solitary Wave Overtopping Deck (note free surface at rear of deck)

Case 2: the wavemaker in the end wall was used to generate the wave using the Cox and Orgega drive signal. The wavemaker was moved closer to the deck than in the laboratory experiment to save computational time. As is shown in Figure 5, the large wave is being generated at the wavemaker and some overtopping has occurred on the previous wave. The deck here is represented by two rows of fixed particles-- which provides too great a thickness to the deck. In Figure 6, the horizontal velocities obtained by SPH for the same times as in the laboratory experiment, Figure 3, are shown.

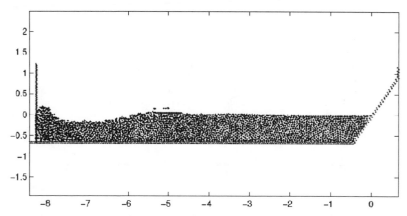

Figure 6 Wavemaker Case

CONCLUSIONS

The numerical technique, Smoothed Particle Hydrodynamics (SPH), is a viable method to model wave motions, as has been illustrated earlier by Monaghan and Kos (1999, 2000), Fontaine (2000), and Dalrymple and Knio(2001).

The separated flow over a platform deck is modeled well by SPH; however, in this study, an insufficient number of particles was used and the resolution is not adequate to compare with data.

SPH velocities under and over the deck do not show the faster velocities observed in the laboratory--this however is likely due to the resolution problem above and perhaps the computational thickness of the deck.

At the rear of the deck SPH predicts that the rejoining of the flows over and under the deck can lead to large impact forces.

ACKNOWLEDGMENTS

The efforts of Dalrymple and Szou were supported by the U.S. Department of Commerce Sea Grant Program at the University of Delaware. Gesteira was supported by the Ministerio de Educacion, Cultura Y Deporte, under grant PR2001-0175.

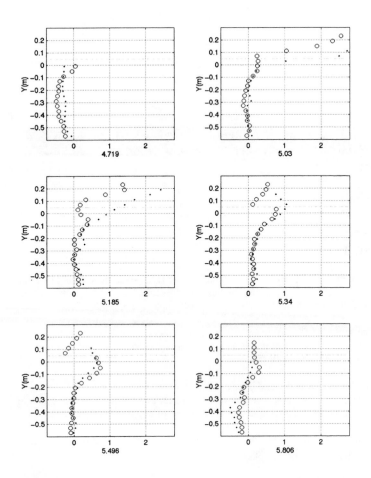

Figure 7 Horizontal Velocities From SPH With (open circles) and Without the Deck

REFERENCES

Bea, R.G., T. Xu, J. Stear, and R. Ramos, ``Wave forces on decks of offshore platforms,'' *J. Waterways, Port, Coastal, and Ocean Engineering, 125 (3), 136-144,* 1999.

Betz, W., ``Smoothed particle hydrodynamics: a review,'' in **The Numerical**

Modelling of Nonlinear Stellar Pulsations, J.R. Buchler, ed., 29-288, Kluwer Acad. Publ., 1990.

Buckner, B., ``The impact of green water on FPSO design,'' *Proc. Offshore Technology Conf.*, Paper 7698, 45-57, 1995.

Cox, D.T. and J.A. Ortega, ``Laboratory measurements: green water overtopping a fixed deck,'' ms., 2001.

Cox, D.T. and C.P. Scott, ``Exceedance probability for wave overtopping on a fixed deck,'' *Ocean Engineering, 28, 707-721,* 2001.

Dalrymple, R.A. and Omar Knio, ``SPH modeling of water waves,'' *Proc. Coastal Dynamics,* Lund, 2001.

Fontaine, E., ``On the use of smoothed particle hydrodynamics to model extreme waves and their interaction with structures,'' *Proc. Rogue Waves 2000,* Brest, France, 2000. www.ifremer.fr/metocean/conferences/wk.htm

Fontaine, E., M. Landrini, and M. Tulin, ``Breaking, splashing, and ploughing phases,'' *Intl. Workshop on Waterwaves and Floating Bodies, 34-38,* 2000.

Gingold, R.A. and J.J. Monaghan, ``Smoothed particle hydrodynamics: theory and application to non-spherical stars,'' *Mon. Not R. Astro. Soc,* 181, 375-389, 1977.

Gudmestad, O.T. and K. Hansen, "On some research issues related to requalification of fixed steel jacked structures," *Proc. 10th Intl. Offshore and Polar Engineering Conference,* 1, 191-195, 2000.

Monaghan, J.J., ``Why particle methods work,'' *SIAM Journal of Scientific and Statistical Computing,* 3, 4, 422-433, 1982.

Monaghan, J.J., ``Smoothed particle hydrodynamics,'' *Ann. Rev. Astron. Astrophys.* 30, 543-574, 1992.

Monaghan, J.J., ``Simulating free surface flows with SPH,'' *J. Comput. Physics,* 110, 399-406, 1994.

Monaghan, J.J. and A. Kos, ``Solitary waves on a Cretan Beach,'' *J. Waterway, Port, Coastal, and Ocean Engineering,* ASCE, 125, 3, 1999.

Monaghan, J.J. and A. Kos, ``Scott Russells' wave generator,'' *Phys. Fluids,* 12, 3, 622-630, 2000.

Monaghan, J.J. and J.C. Lattanzio, ``A refined particle method for astrophysical problems,'' *Astronomy and Astrophysics,* 149, 135-143, 1985.

FULLY NONLINEAR WAVES AND THEIR KINEMATICS: NWT SIMULATION VS. EXPERIMENT

Weoncheol Koo[1] and Moo-Hyun Kim[2], F.ASCE

Abstract: A 2D fully nonlinear NWT is developed based on the potential theory, mixed Eulerian-Lagrangian (MEL) time marching scheme, and boundary element method (BEM). Wave profiles and wave kinematics of highly nonlinear waves are calculated using the NWT and the results are compared with linear, Stokes second-order theory and experimental values. It is confirmed that the spatial variation of intermediate-depth waves along the direction of wave propagation is caused by the unintended generation of second-order free waves, which was originally investigated both experimentally and theoretically (3rd order perturbation theory) by Goda. The various phenomena observed by Goda are clearly reproduced by the present fully nonlinear NWT. It is shown that the wave kinematics above mean water level can be significantly different from the perturbation-based prediction. It is also found that small mean positive or negative flows can be generated below mean water level depending on the water depth and wave condition.

INTRODUCTION

Nonlinear waves have higher/sharper crest and lower/flatter trough compared to linear waves and their kinematics can be significantly different from those of linear waves particularly above the mean water level (MWL). The mean velocities of steady flows generated by nonlinear waves can play an important role for various mass transport problems. The interactions of such nonlinear free-surface waves with compliant and fixed structures are also of vital importance in various ocean engineering applications. Although linear wave-body interaction theory is still very useful, it is precisely the conditions of large motions and extreme loads for which high performance, safety, and ultimate survivability are of concern that nonlinear effects become critical. The essential

[1] Research Assistant, Ocean Engineering Program, Civil Engineering Department, Texas A&M University, College Station, Texas 77843-3136, kwc@tamu.edu
[2] Associate Professor, Ocean Engineering Program, Civil Engineering Department, Texas A&M University, College Station, Texas 77843-3136, m-kim3@tamu.edu

characteristics of fully nonlinear free-surface waves and their kinematics can accurately be simulated by using potential-theory-based numerical wave tanks (NWTs). When a body is present, the validity of the potential theory can vary case by case.

In this paper, a 2D fully nonlinear NWT is developed based on the potential theory, mixed Eulerian-Lagrangian (MEL) time marching scheme, and boundary element method (BEM). The use of fully nonlinear free-surface time-stepping method for 2D waves by MEL technique was first introduced by Longuet Higgins and Cokelet (1976). At each time step, the procedure requires (i) solving the Laplace equation in the Eulerian frame, and (ii) updating the moving boundary points and values in Lagrangian manner. Subsequently, the MEL scheme has been used by many researchers for various fully nonlinear wave-wave or wave-body interaction problems. The 2D-NWT examples include Dommermuth et al(1988). Cointe et al(1990), Cao et al.(1991), Clement(1996), Grilli et al(1989), Tanizawa(1996). There are also several fully nonlinear 3D-NWTs. A complete review on this topic is given, for example, in KIM et al(1999).

The fully nonlinear wave simulation is still computationally very intensive and requires meticulous treatment of free-surface time marching, inflow/outflow boundaries, and removal of possible saw-tooth instability caused either by variable mesh size/high-order aliasing or inherent singular behavior near the moving-body and free-surface intersection. In addition, the relative effectiveness and accuracy of various absorbing/open boundary conditions is still in debate.

Despite relative abundance in simulations for fully nonlinear waves, the numerical results for their kinematics are surprisingly scarce. The accurate prediction of nonlinear wave kinematics above MWL is very important for various ocean engineering applications, and so far, most of the published results have been from laboratory measurement. In some cases, however, the lab-generated waves vary spatially in the direction of wave propagation due to the presence of undesirable second-order spurious free waves. When this kind of wave is present, the spatial variability of the experimental results has to be carefully checked. The physics of the spatial variation was studied both experimentally and theoretically (3rd –order perturbation theory) by Goda, and the essential physics explained by Goda are accurately reproduced in this paper by using fully nonlinear 2D NWT simulations.

BOUNDARY-VALUE PROBLEM

An ideal, irrotational fluid is assumed so that the fluid velocity can be described by the gradient of velocity potential ϕ. A Cartesian coordinate system is chosen such that the z=0 corresponds to the calm water level and z is positive upwards.

Then the governing equation of the velocity potential is given by

$$\nabla^2 \phi = 0 \tag{1}$$

and the boundary conditions consist of
1) Fully nonlinear dynamic free surface condition

$$\frac{\partial \phi}{\partial t} = -g\eta - \frac{1}{2}|\nabla \phi|^2 - \frac{P_a}{\rho} \quad \text{satisfied on the exact free surface} \tag{2}$$

where P_a is the pressure on the free surface, and we assume that it is zero from now on.

2) Fully nonlinear kinematic free surface condition

$$\frac{\partial \eta}{\partial t} = -\nabla \phi \cdot \nabla \eta + \frac{\partial \phi}{\partial z} \qquad \text{satisfied on the exact free surface} \qquad (3)$$

3) Rigid boundary condition

$$\frac{\partial \phi}{\partial n} = 0 \qquad (4)$$

on the rigid bottom, the end of numerical beach, and other rigid boundaries.

4) Input boundary condition: At the inflow boundary, either feeding a theoretical particle velocity profile along the fixed input boundary or actual wavemaker condition along the moving boundary is used. For example, when a linear regular wave is prescribed, the following equation is used.

$$\frac{\partial \phi}{\partial n} = -\frac{\partial \phi}{\partial x} = -\frac{gAk}{\omega} \frac{\cosh k(z+h)}{\cosh kh} \cos(kx - \omega t) \qquad (5)$$

where A, w, k, and h are wave amplitude, frequency, wave number, and water depth respectively.

At each time step, the velocity potential is obtained by solving the discretized form of the following integral equation.

$$\alpha \phi_i = \iint_\Omega (G_{ij} \frac{\partial \phi_j}{\partial n} - \phi_j \frac{\partial G_{ij}}{\partial n}) ds \qquad (6)$$

where G is a Green function for Laplace equation and α is solid angle ($\alpha = 0.5$ on the boundary). For two-dimensional problems, the simple source G is given by

$$G(x, z, x_i, z_i) = \ln R_1 \qquad (7)$$

where, R_1 is the distance between source and field points (Brebbia and Dominguez, 1992).

Time Marching for Fully Nonlinear Free Surface Conditions

To update the fully nonlinear kinematic and dynamic free-surface conditions at each time, Runge-Kutta 4th order scheme was used and the MEL (Mixed Eulerian-Lagrangian) approach was adopted. In the present calculation, the free-surface node is allowed to move only in the vertical direction to maintain the uniform nodal distance throughout entire simulation i.e. nodal velocity $\vec{v} = (0, \frac{\delta \eta}{\delta t})$. Then considering $\frac{\delta}{\delta t} = \frac{\partial}{\partial t} + \vec{v} \cdot \nabla$, the fully nonlinear free-surface conditions can be modified as follows in the Lagrangian frame

$$\frac{\delta \phi}{\delta t} = -g\eta - \frac{1}{2} |\nabla \phi|^2 + \nabla \phi \cdot \vec{v} \qquad (8)$$

$$\frac{\delta \eta}{\delta t} = \frac{\partial \phi}{\partial z} - (\nabla \phi - \vec{v}) \cdot \nabla \eta \qquad (9)$$

Compared to material-node approach where $v=\nabla\phi$, the free-surface equations become more complicated but the horizontal uniformity of initial grid system can be preserved.

Ramp Function

When the simulation is started, the following ramp-function was used. The ramp function was used for preventing impulse-like behavior of wave maker so as to reduce the corresponding transient wave amplitude.

$$r(t) = \begin{cases} 1 & \text{, for } t > 2T \\ \{1 - \cos(\pi \dfrac{t}{2T})\}/2 & \text{, for } t \le 2T \end{cases} \tag{10}$$

Numerical Beach (Artificial Damping Zone)

Toward the end of the computational domain, an artificial damping zone was applied for absorbing wave energy gradually in the direction of wave propagation. The length of the damping zone (l_d) was determined to be at least 2 wavelengths in this study after comprehensive tests. In general, the longer l_d is needed for more nonlinear waves. In this paper, both ϕ_n & η -type damping terms were added to the fully nonlinear free surface condition. The damping is designed to grow in a gradual manner to the target constant value to minimize wave reflection from the entrance of the damping zone. Through linear stability analysis, the appropriate damping coefficients were adopted ($\mu_{01} = 1.5$ and $\mu_{02} = k\mu_{01}$), which minimize the dispersion error. The results were numerically confirmed.

$$\frac{\delta\phi}{\delta t} = -g\eta - \frac{1}{2}|\nabla\phi|^2 + \nabla\phi \cdot \vec{v} + \mu_1 \frac{\partial\phi}{\partial n} \tag{11}$$

$$\frac{\delta\eta}{\delta t} = \frac{\partial\phi}{\partial z} - (\nabla\phi - \vec{v}) \cdot \nabla\eta + \mu_2\eta \tag{12}$$

where, $\mu_i = \begin{cases} \mu_{0i}[1 - \cos\{\dfrac{\pi}{2}\left(\dfrac{x-l}{l_d}\right)\}] & \text{for } x > l \\ 0 & \text{for } x \le l \end{cases}$

l is the length of computational domain and l_d is the length of damping zone.

Smoothing Scheme

It is well known that the so-called saw-tooth instability may occur on the free surface during the simulation of highly nonlinear waves. It is caused either by variable mesh size/high-order aliasing or inherent singular behavior at the wavemaker and free-surface intersection. To avoid the numerical problem associated with it, a Chebyshev 5-pts. smoothing scheme was used along the free surface during time marching. The smoothing scheme was applied at every 5 time step. It is confirmed that the smoothing scheme does not affect the higher-order components up to third order. An evenly spaced-node for Chebyshev 5-pts. smoothing was first developed by Longuet-Higgins & Cokelet (1976) and it was modified for variable-node-space case (Sung, 1999).

NUMERICAL RESULTS AND DISCUSSIONS

Nonlinear Wave evolution (Spatial Variation) for Intermediate Depth

Waves generated by wave makers would undergo nonlinear interactions between each harmonics. In particular, the second-order free waves generated due to the mismatch between the particle velocity profile and actual wave maker motion propagate with different phase velocity compared to the primary wave and thus cause some spatial variation of wave profile. Goda (1998) observed this phenomenon in a 2D physical wave tank and succeeded to reproduce/explain it using the third-order perturbation theory. He claimed that the interaction between the primary wave and the second-order free wave causes the spatial variation of the first- and third-harmonic components. The nonlinear spatial wave evolution investigated by Goda (1998) is reproduced in the present study using fully nonlinear NWT simulations.

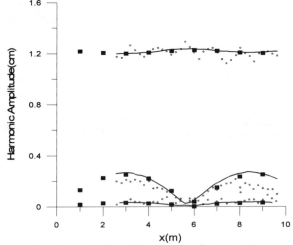

Fig. 1. Comparison of spatial variation of Fourier Amplitudes with numerical results of inputting velocity potential profiles (rectangle), Goda's theoretical (solid line) and experimental measurement (small circle).

Figure 1 shows the spatial variation of each harmonic amplitude when linear wave of period T=1.697s and height H=2.5cm is fed along the input boundary. Computational parameters from figure 1 to 4 are water depth=0.25m, wavelength=2.5m, free surface (computational domain)=9m, damping zone (artificial beach zone)=6m, delta x on the free surface=0.1m(first 3m) and 0.05m(rest 6m) and delta t is T/64. Figure 2 shows the same case when the piston type wave maker and moving boundary is used instead of feeding velocity profile at the fixed input boundary. The piston-wavemaker results show slightly more spatial variation than the feeding case. The difference between the smoothing and non-smoothing cases is shown in Figure 3. The maximum of 1st harmonic amplitude is located around 6m, where the minima of 2nd and 3rd harmonic amplitudes occur.

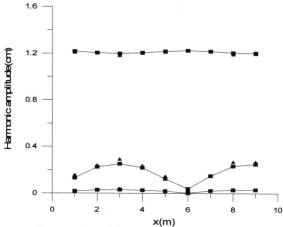

Fig. 2. Comparison between feeding velocity profiles (rectangle) and piston type wave maker (triangle). (same case as Fig. 1).

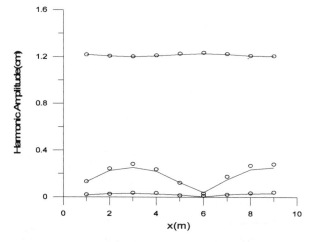

Fig. 3. Comparison of spatial variation with smoothing (solid line) and without smoothing (white circle). (same case as Fig. 1).

Figure 4 shows similar plots when wave heights are doubled (H=5.0cm). As wave height increases from 2.5cm to 5.0cm, the 2nd and 3rd harmonic amplitudes grow rapidly and the fluctuation of the 1st harmonic amplitude increases as well. For both wave heights, the present nonlinear NWT results are in good agreement with Goda's theoretical and experimental results. Figure 5 shows the change of wave profiles and corresponding heights at various locations and the appearance/deformation of secondary crest at the trough. The overall pattern is similar to that of experimental measurement by Goda

(1998). The secondary crest appears behind the main crest, propagates with slower speed than the main crest, and is overtaken by the next crest as the wave propagates along the wave tank.

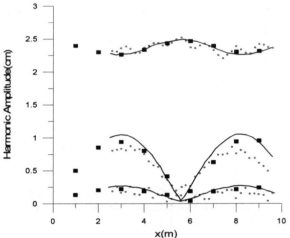

Fig. 4. Case of H=5cm. (legend and parameters same as Fig. 1)

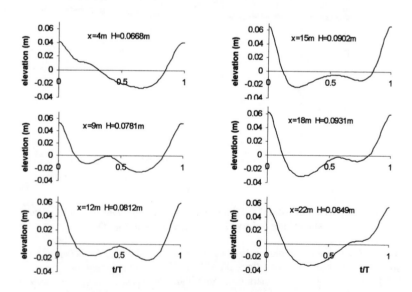

Fig. 5. Examples of wave profiles with secondary crests at various locations.
(water depth=0.32m, T=3.07s, dt=T/64, dx=0.1m)

The spatial variation of wave profiles along the propagation distance is mainly caused by the generation of second-order free waves due to the mismatch of wave particle velocity profile and actual wave maker motions. If it is true, the phenomenon should disappear by matching the wave maker motion as close as possible to the actual velocity profile of the generated waves. To confirm this fact more clearly, the second-order Stokes wave velocity profile resembling the kinematics of the actual nonlinear wave more closely is fed along the input boundary. The results for H=5cm are shown in Figure 6 and the spatial variation is greatly reduced as expected. Care needs to be taken in interpreting experimental/simulation data when undesirable second-order free waves are present.

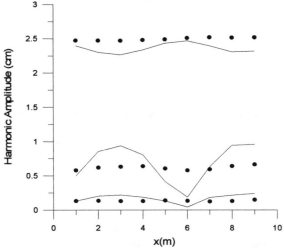

Fig. 6. Comparison of spatial variation with linear wave input (solid line) and Stokes 2nd –order wave input (black circle) (parameters same as Fig. 1.).

Using the fully nonlinear NWT, the wave kinematics of Figure 1 under wave crest are calculated and they are presented in compared in Figure 7 with linear and second-order Stokes wave results. For the NWT computation, the 2nd-order Stokes waves are fed along the input boundary. The actual maximum horizontal velocity turned out to be greater than that of Stokes second-order waves. The difference is more pronounced above mean water level. Below mean water level, the maximum horizontal velocity is greater than minimum value and it causes some mean transport flow in the direction of wave propagation (Figure 7).

Finally, the wave kinematics of deepwater waves of H=9cm and T=0.887 are calculated using the fully nonlinear NWT and the results are compared in Figure 8 with linear theory. The experiment was conducted at Texas A&M University (Choi et al. 2001). It is seen that the NWT simulations agree more reasonably with the lab measurement than the linear theory. It is noticeable that the nonlinear NWT simulations produce small mean negative flow below mean water level (i.e. negative minimum value is larger than positive maximum value) and the phenomenon is also confirmed in the experiments.

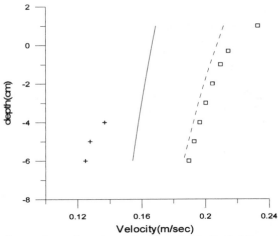

Fig. 7. Comparison of maximum horizontal velocity (white rectangle, cross=minimum horizontal velocity) with Stokes 2nd order wave velocity (dashed line) and linear wave velocity (solid line).

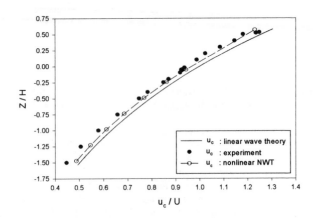

Fig. 8. Comparison with wave crest horizontal velocities (U is measured at z=0)

CONCLUDING REMARKS

A fully nonlinear 2D NWT is developed based on the potential theory, MEL approach, and boundary element method. A numerical beach using artificial damping both in kinematic and dynamic free-surface conditions is devised and its performance is found to be satisfactory. Both wave profiles and wave kinematics of highly nonlinear waves are calculated using the NWT. It is confirmed that the spatial variation of intermediate-depth waves along the direction of wave propagation is caused by the unintended generation of second-order free waves, which was originally investigated both experimentally and theoretically (3rd order perturbation theory) by Goda. The various phenomena observed by Goda are clearly reproduced by the present fully nonlinear NWT. His explanation of pertinent physics is also demonstrated by comparing the simulation results for different wave maker motions. The wave kinematics of nonlinear NWT simulations are compared with the linear, Stokes second-order theory and experimental values. It is shown that the wave kinematics above mean water level can be significantly different from the perturbation-based prediction. It is also found that small mean positive or negative flows can be generated below mean water level depending on the water depth and wave condition.

REFERENCES

Brebbia, C.A. and Dominguez, J. 1992. *Boundary elements: an introductory course,* Computatinal mechanics publications, Southampton, U.K. McGraw-Hill.

Cao, Y., Schultz, W.W., and Beck, R.F. 1991. Three-Dimensional Desingularized Boun-Dary Integral Methods for Potential Problems. International Journal of Numerical Methamatics Fluids, 12, 785-803.

Choi, H.J., Cox, D., Kim, M.H, and Ryu, S.S. 2001. Laboratory Investigation of Nonlinear Irregular Wave Kinematics. 4th International Symposium on Ocean Wave Measurement and Analysis, WAVES 2001.

Clement, A.H. 1996. Coupling of Two Absorbing Boundary Conditions for 2-D Time-Domain Simulations of Free Surface Gravity Waves. *Journal of Computational Physics,* 126, 139-151.

Cointe, R., Geyer, P., King, B., Molin, B., and Tramoni, M. 1990. Nonlinear and Linear Motions of a Rectangular Barge in a perfect fluid, *Proceedings 18th Symposium on Naval Hydrodynamics.* 85-99.

Dommermuth, D.G., and Yue, D.K.P. 1987. Numerical Simulation of Nonlinear Axisymmetric Flows with a Free Surface. *Journal of Fluid Mechanics,* 178, 195-219.

Goda, Y. 1998. Perturbation analysis of nonlinear wave interactions in relatively shallow water, *Procdings of the 3rd International Conference on Hydrodynamics,* 33-51.

Grilli, S.T., Skourup, J., and Svendsen, I.A. 1989. An Efficient Boundary Element Method for Nonlinear Water Waves. *Engineering Analysis with Boundary Elements,* 6(2), 97-107.

Kim, C. H., Clement, A. H., and Tanizawa, K. 1999. Recent Research and Development of Numerical Wave Tanks-A Review. *International Journal of Offshore and Polar Engineering,* 9(4), 241-256.

Longuet-Higgins, M. S., and Cokelet, E.D. 1976 The Deformation of steep surface waves on Water: I. A Numerical Method of Computation, *Proceedings Royal Society London.* A 350, 1-26.

Sung, H.G. 1999. A Numerical Analysis of Nonlinear Diffraction Problem in Three Dimensions by Using Higher-Order Boundary Element Method. *Ph.D. Dissertation,* Seoul National University.

Tanizawa, K. 1996. Nonlinear Simulation of Floating Body Motions. *Proceedings 6th International Offshore and Polar Engineering Conference,* Los Angeles, ISOPE, 3, 414-420.

THREE-DIMENSIONAL WAVE FOCUSING IN FULLY NONLINEAR WAVE MODELS

Carlo Brandini [1] and Stéphan T. Grilli [2], M.ASCE

Abstract: Wave frequency focusing has been used in two-dimensional (2D) laboratory wave tanks to simulate very large waves at sea, by producing large energy concentration at one point of space and time. Here, three-dimensional (3D) frequency/directonal energy focusing is simulated in a fully nonlinear wave model (Numerical Wave Tank; NWT), and shown to produce very large waves. This method alone, however, cannot explain why and how large waves occur in nature. Self-focusing, i.e., the slow growth of 3D disturbances in an initially regular wave train, is shown to also play a major role in the formation of "freak waves". Self-focusing is studied in a more efficient space-periodic nonlinear model, in which long term wave propagation can be simulated. The combination of directional/frequency focusing and self-focusing, and resulting characteristics of large waves produced, could be studied within the same NWT.

INTRODUCTION

The existence of abnormally large waves at sea and the understanding of physical phenomena creating them have received increasing attention in recent years. Early reports by Mallory (1974) described a long series of naval accidents caused by unexpectedly large waves. Since then, many authors have given a considerable attention to the study of large transient waves, with the aim of understanding possible physical mechanisms determining when and how these are generated. The goal is to calculate kinematics and dynamics of such wave events and, eventually, to provide models for better designing vessels and off-shore structures. So far, a number of mechanisms have been proposed for the generation of such steep wave events, but it seems that these are still poorly understood. The consensus, however, is that all of these mechanisms require to model nonlinear behavior of ocean waves.

---
[1] Dept. of Civil Engng., University of Firenze, Via di S.Marta, 50135 Firenze, Italy. E-mail: brandini@dicea.unifi.it
[2] Dept. of Ocean Engng., University of Rhode Island, Narragansett, RI 02882, USA. E-mail: grilli@oce.uri.edu.

In some situations, the occurrence of giant waves can be explained by the focusing of wave energy due to the presence of ocean currents or the bottom topography. This is typical of some areas around the world (such as the famous 'Agulhas Current', which is responsible for the formation of freak waves off the South-East Coast of Africa). Why giant waves are generated in the open ocean, far away from non-uniform currents or bathymetry, however, is still very much an open problem. In past research efforts, the concept of 'phasing' was often used as an explanation, whereas a short-lived large wave occurred when waves in an irregular sea combined their phase at one spatial point and at a particular time. Although this concept has been used in fairly ingenious ways (Boccotti 1981), it is still usually defined within the limits of linear wave theory. Many experimental and observational results, however, have shown that, when considering rarely occurring waves, the processes are far from being stationary Gaussian ones. In other words, the Rayleigh distribution of wave heights (based on a linear representation of the sea surface) does not predict that waves as large as 2.2 to 2.4 the significative wave height—as observed at sea—will normally occur (Wolfram 2001). Such steep wave events appear to belong to a non-Gaussian distribution of rare events (Skourup et al. 1997; Haver and Andersen 2000). Therefore, when dealing with extreme waves, full nonlinearity should, in principle, be kept in the equations, since nonlinear (i.e., non-Gaussian) wave interaction processes will likely play a dominant role.

More specifically, nonlinerarity seems to act in two main ways : (i) a trivial nonlinear superposition mechanism (Dean 1990), and (ii) a more complex nonlinear instability process. The latter way will be detailed in a following section. The former way has been widely used, mainly by naval architects and off-shore engineers, who developed (mostly experimental) techniques referred to as "wave focusing", to produce extremely large waves. In practice, in a laboratory or in a numerical wave tank, wave phases are calculated to produce a large 'design' wave at a given location and, sometimes, to study the interaction of a huge wave with vessels, piles, or other mobile or fixed structures. Nonlinearity usually makes it hard to produce the highest wave at a pre-determined point, due to amplitude dispersion effects. In traditional wave focusing techniques, waves having different frequencies are focused to produce a single large wave at one prescribed time and location : the basic idea is to first generate shorter waves, followed by longer ones which, due to frequency dispersion, are faster and catch up with the shorter waves over a some small area of space, thus producing a particularly high and steep wave through superposition (Chaplin 1996). This focusing method, however, is mostly limited to unidirectional situations and used in laboratory wave tanks to simulate all kinds of conditions, from slightly spilling breakers (Schlurmann et al. 2000) to violent plunging breakers (Dommermuth et al. 1988). Three-dimensional effects, which can be very important, have often been neglected, mainly because 3D wave tanks are very costly to operate, and 3D wave generation is also a difficult task. Observations show, however, directional focusing effects associated with 3D features of the wave field, such as a continuous curvature of the wave front,

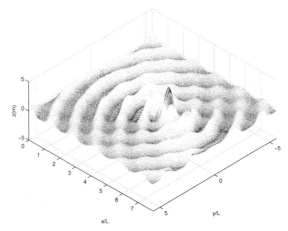

FIG. 1. Example of second-order directional wave focusing

cannot be neglected in the analysis of extreme waves at sea (She et al. 1997, Nepf et al. 1998). Experiments have shown that curved wave fronts lead to 3D breaking waves, and that the shape and kinematics of 3D breaking waves may greatly differ from those of 2D breakers. The degree of angular spreading is found to have great effects on wave breaking characteristics and kinematics, and, hence, non-directional wave theories are demonstrated to be insufficient to describe the kinematics of 3D waves.

LOW-ORDER 3D WAVE FOCUSING

A ready-to-use solution for wave focusing may be easily obtained using low-order wave theories. Curved wave fronts are generated by using a number of wave fronts of the same height and frequency, equally spaced within an (horizontal) angular range. The phases φ of the fronts are calculated so that energy becomes focussed at a predetermined point

$$\varphi = k(x - d_f) \cos\theta + ky \sin\theta - \omega t \tag{1}$$

where d_f is the focal distance (in the x direction) and $-\alpha \leq \theta \leq \alpha$. To accelerate focusing, one can impose an additional frequency-focusing, by adjusting the wave frequency as a function of the angle of incidence θ, thus increasing the curvature of wave fronts. For mild incident waves, this can approximately be done based on the linear dispersion relationship. [Thus, if k denotes the wavenumber for $\theta = 0$ and frequency ω, and $k_\theta = k \cos\theta$ is the wavenumber for angle θ, then frequency may be slightly changed to satisfy the linear dispersion relationship, $\omega_\theta^2/g = k_\theta \tanh k_\theta h$.]

Up to second-order, it is possible to obtain an analytical solution for directional focusing, considering all 2nd-order interaction terms, as given, e.g., by Hu

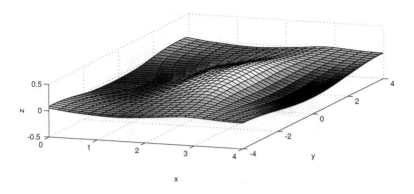

FIG. 2. Directional focusing in a NWT : a non-breaking case

(1996) (Fig. 1). Nonlinear wave-wave interactions become only important near the focusing point while far from this point, linear theory and second-order theory give nearly the same results (Brandini and Grilli 2000,Brandini and Grilli 2001a).

FULLY NONLINEAR 3D WAVE FOCUSING

To generate a focussed signal in a wave tank, such as shown in Fig. 1, one needs a wave generation system that can specify wave propagation from many directions. This is achieved using directional or "snake" wavemakers, which are long articulated wavemakers consisting of numerous wave paddles, that can be moved independently from one another. The linear solution for directional wave focusing in a wavetank equipped this way, and having impermeable (reflective) lateral walls, was derived by Dalrymple (1989). A 3D fully nonlinear potential flow model (i.e., NWT), based on a Boundary Element Method (BEM) with an Eulerian-Lagrangian flow representation, was recently developed by Grilli et al. (2001a) (also see, Grilli et al. 2001b). Extension of this NWT to model 3D directional wave focusing, including the additional possibility of frequency-focusing, was done by Brandini and Grilli (2001b). A snake wavemaker, similar to those used in laboratory facilities, was modeled at one extremity of the 3D-NWT, and a new open boundary condition, based on a snake absorbing wavemaker, was modeled at the other extremity. The snake wavemaker motion was prescribed according to linear wave theory (Dalrymple 1989), such as to generate curved wave fronts and focus wave energy at a specified distance d_f away from the wavemaker. The image method and symmetrical properties of the solution were implemented, to reduce the size of the computational domain in the BEM, and hence the computational cost. Figs. 2 and 3 illustrate such computations. In both cases, we see the generation of a curved wave front having a very high elevation around $y = 0$. In Fig. 3, the wave starts spilling breaking at the crest, which interrupts computations.

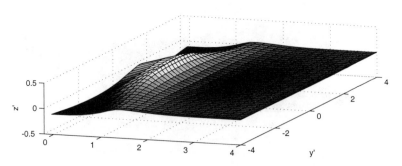

FIG. 3.   Directional focusing in a NWT : a breaking case

3D SELF-FOCUSSED WAVES

Wave instability to small modulations

Although many solutions are theoretically possible, based on a 'phasing' concept, the probability of finding a large enough number of waves, in a random wave field, whose phases would match at the same wave crest, is extremely low. Thus, research was done in recent years to discover which other physical mechanism might be responsible for the generation of large wave energy concentration at one point in the ocean. Many researchers concentrated their efforts on wave instability phenomena. The pioneering work of Benjamin and Feir (1967) on the instability of periodic waves of finite amplitude, caused a small revolution at the time. In water of sufficiently depth, Benjamin-Feir's (BF) theory predicts that a slightly modulated 2D periodic wave train will evolve into strongly modulated wave groups, where the wave of maximum amplitude may be much larger than that of the original wave train. Since even a relatively regular ocean swell contains many frequencies, according to BF's theory, a perfectly regular time-harmonic wave train can therefore never exist. The phenomenon of 'natural' evolution of periodic waves into a series of wave groups has been referred to as 'self-focusing'. Many authors suggested that BF instability is the mechanism explaining the formation of waves much larger than expected. However, other instability phenomena have been identified. McLean (1982) theoretically predicted a type of wave instability (called type II), which is predominantly 3D, while BF (called type I) is only 2D. Su et al. (1982) experimentally confirmed this prediction by showing how a steep 2D wave train can evolve into 3D spilling breakers.

Type I and II instabilities involve nonlinear effects. In fact, they can only be identified by developing evolution equations at least to the third-order (such as the nonlinear Shrödinger equation or its modifications, e.g., Henderson et al. 1999, Trulsen and Dysthe 1999). Henderson et al. (1999) also performed fully nonlinear calculations to study the behaviour of 2D uniform wave trains of moderate steepness, perturbed by a small periodic perturbation. After a large time of propagation (typically over 100 wave periods), it is observed that a large steep wave (i.e., a "freak wave"), may emerge from the initial wave train, and break or recede, and periodically reappear.

Self-focusing in a 3D model

In previous numerical studies, 3D effects were not usually addressed because, either it was not possible to generalize the method of solution to 3D, or the computational effort in a 3D model was too high. In the present work, we adopt the computationally efficient Higher Order Spectral (HOS) method, independently developed by West et al. (1987) and Dommermuth and Yue (1987). This rapidly convergent method represents the sea surface as a modal (Fourier) superposition, by way of a perturbation expansion. Doubly periodic boundary conditions are specified in the horizontal plane. HOS allows computations at any desired order in nonlinearity. Here, we use a fourth-order method. We start with an initial quasi-2D wave train (modeled as a streamfunction wave), with small initial periodic perturbations in both the longitudinal x direction (as it has been done to show the BF instability) and the lateral one y. Cases are characterized by the initial steepness of the wave train ak, and two characteristic modulation wavelengths (l_x and l_y). Computations are carried out for many wave periods, because a strong growth of instabilities only appears after about 100 wave periods.

The evolution of a modulated wavetrain with $ak = 0.14$, $l_x = 5$ and $l_y = 10$ (hence with a lateral modulational wavelength twice the longitudinal one) is shown in Fig. 4. While at earlier stages of evolution waves are essentially 2D, at later stages, the growth of transverse perturbations causes a 3D structure to develop. At final stages, both a longitudinal and a transverse growth of such modulations is observed. Fig. 4a shows the evolution at time $t/T = 90$ (with T the wave period). We see the combination of two effects :

- In the longitudinal direction, a BF-like mechanism causes the wave group to shorten ahead and to lengthen behind, with a wave energy concentration in the middle of the wave envelope.
- In the lateral direction the growth of transverse perturbations affects the highest wave and its first predecessor. Lateral features in the form of standing waves across the (periodic) wavetank appear.

The combination of these two effects gives rise to a fully 3D structure of the wave group. Fig. 4b shows the evolution after just one more wave period, at time $t/T = 91$. The observed wave evolution is clearly a truly directional self-focusing process. Finally, the appearance of curved wave fronts is an important feature of such 3D waves (Fig. 5). These wave groups are characterized by skewed wave patterns that qualitatively agree with Su's experiments.

According to the existing theory, instabilities of type II only affect the steepest waves. The combination of lateral effects with the BF instability, however, has not yet been properly studied in laboratory experiments for the highest waves. Based

(a)

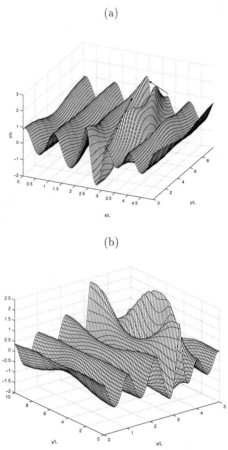

(b)

FIG. 4. Evolution of a doubly modulated wave train after: (a) 90 and (b) 91 wave periods, with $ly = 2lx$

on our limited computational results, our analysis is that the BF-like mechanism produces a short wave group of increasing height and steepness, and it is within such a group that the lateral instability significantly manifests itself, provided the modulational wavelength in the lateral direction is long enough. In fact, not all longitudinal perturbations produce a BF instability, as well as not all lateral perturbations are able to produce instabilities of type II. For instance, the evolution of a modulated wavetrain having the same initial steepness $ak = 0.14$ and $l_x = 5$, but a shorter $l_y = 4$ (so that the lateral modulational wavelength is 0.8 times the longitudinal one) is shown in Fig. 6 at time $t/T = 90$. In this case,

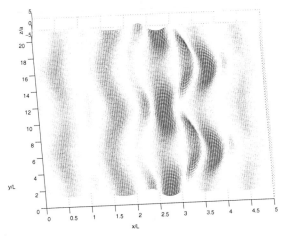

FIG. 5. Spatial structures of the doubly modulated wave train in Fig. 4

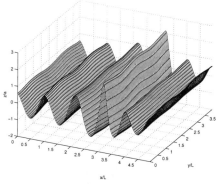

FIG. 6. Evolution of a doubly modulated wave train after 90 wave periods, $ly = 0.8lx$

only the longitudinal modulation grows significantly, according to a classical BF modulational mechanism. The modulation growth observed in 3D modulations should be limited by wave breaking, which cannot be described by a single-value free surface representation such as used in the HOS method. Breaking will not happen uniformly along a wave crest, and a 3D self-focused breaking wave is expected to appear at some stage of the modulation.

CONCLUSIONS

We simulated fully nonlinear 3D focused and self-focused waves, in nonlinear wave models, with the goal of understanding the kinematics and dynamics of 3D large transient waves, as they occur in nature.

In future work, the computationally efficient HOS method could be used to calculate the initial stages of the self-focusing modulation (i.e., the longer duration ones, on the order of 100 wave periods). Then, free surface elevations and potential $[\eta(x, y, t), \phi(x, y, t)]$, found in the HOS solution, could be used to initialize a 3D-NWT having doubly periodic boundary conditions specified on lateral boundaries.

Therefore, 3D self-focusing cases producing extreme, possibly breaking (the worst scenarios for engineering applications), waves could be studied in the NWT. This would be quite difficult to do in a laboratory, due to the long distances of propagation required for the instabilities to grow.

REFERENCES

Benjamin, T. and Feir, J. (1967). "The disintegration of wave trains on deep water. Part 1. Theory." *J. Fluid Mech.*, 27, 417–430.

Boccotti, P. (1981). "On the highest waves in a stationary Gaussian process." *Atti Acc. Lig.*, 27, 45–73.

Brandini, C. and Grilli, S. (2000). "On the numerical modeling of extreme highly nonlinear deep water waves." *Proc. IABEM 2000 Symp.*, Intl. Assoc. for Boundary Element Methods, Brescia, Italy. 39–42.

Brandini, C. and Grilli, S. (2001a). "Evolution of 3D unsteady water wave modulations." *Proc. Rogue Waves 2000 Workshop (in press)*, Brest, France.

Brandini, C. and Grilli, S. (2001b). "Modeling of freak wave generation in a 3D NWT." *Proc. 11th Offshore and Polar Engrg. Conf.*, Stavanger, Norway. 124–131.

Chaplin, J. (1996). "On frequency-focusing unidirectional waves." *Intl. J. Offshore and Polar Engng.*, 6, 131–137.

Dalrymple, R. (1989). "Directional wavemaker theory with sidewall reflection." *J. Hydraulic Res.*, 27, 23–34.

Dean, R. (1990). *Water Wave Kinematics*, chapter Freak waves: a possible explanation, 609–612. Kluwer.

Dommermuth, D. and Yue, D. (1987). "A higher-order spectral method for the study of non linear gravity waves." *J. Fluid Mech.*, 184, 267–288.

Dommermuth, D. G., Yue, D. K. P., Lin, W. M., Rapp, R. J., Chan, E. S., and Melville, W. K. (1988). "Deep-water plunging breakers: a comparison between potential theory and experiments." *J. Fluid Mech.*, 189, 423–442.

Grilli, S. T., Guyenne, P., and Dias, F. (2001a). "A fully nonlinear model for three-dimensional overturning waves over arbitrary bottom." *Int. J. Numer. Meth. Fluids*, 35(7), 829–867.

Grilli, S. T., Guyenne, P., and Dias, F. (2001b). "Three-dimensional numerical model for fully nonlinear waves over arbitrary bottom." *Proc. WAVES 2001 Conf. (in press)*, San Francisco, USA. ASCE.

Haver, S. and Andersen, O. J. (2000). "Freak waves: rare realizations of typical population or typical realization of a rare population ?." *Proc. 10th Offshore and Polar Engrg. Conf.*, Vol. 3, Seattle, USA. 123–130.

Henderson, K., Peregrine, D., and Dold, J. (1999). "Unsteady water wave modulations: fully non linear solutions and comparison with the non linear Schrödinger equation." *Wave motion*, 9, 341–361.

Hu, S.-L. (1996). *Computational Stochastic Mechanics*, chapter Nonlinear random waves, 519–543. Comp. Mechanics Pub., Southampton.

Mallory, J. (1974). "Abnormal waves on the south east coast of South Africa." *Intl. Hydrog. Rev.*, 51, 99–129.

McLean, J. (1982). "Instabilities and breaking of finite amplitude waves." *J. Fluid Mech.*, 114, 331–341.

Nepf, H., Wu, C., and Chan, E. (1998). "A comparison of two- and three-dimensional wave breaking." *J. Phys. Oceanography*, 28, 1496–1510.

Schlurmann, T., Lengricht, J., and Graw, K.-U. (2000). "Spatial evolution of laboratory generated freak waves in deep water depth." *Proc. 10th Offshore and Polar Engrg. Conf.*, Vol. 3, Seattle, USA. 54–59.

She, K., Greated, C., and Easson, W. (1997). "Experimental study of three-dimensional breaking wave kinematics." *Appl. Ocean Res.*, 19, 329–343.

Skourup, J., Hansen, N., and Andreasen, K. (1997). "Non-Gaussian extreme waves in the central north sea." *J. Offshore Mechanics and Artic Engng.*, 119, 146–150.

Su, M., Bergin, M., Marler, P., and Myrick, R. (1982). "Experiments on nonlinear instabilities and evolution of steep gravity-wave trains." *J. Fluid Mech.*, 124, 45–72.

Trulsen, K. and Dysthe, K. (1999). "Note on breather type solutions of the nls as model for freak waves." *Phys. Scripta*, 82, 45–73.

West, B., Brueckner, K., Janda, R., Milder, D., and Milton, R. (1987). "A new numerical method for surface hydrodynamics." *J. Geophys. Res.*, 92, 11803–11824.

Wolfram, J. (2001). "Some experiences in estimating long and short term statistics for extreme waves in the north sea." *Proc. Rogue Waves 2000 Workshop (in press)*, Brest, France.

ON THE VALIDITY OF THE SHALLOW WATER EQUATIONS FOR VIOLENT WAVE OVERTOPPING

Stephen R Richardson[1], David M Ingram[1],
Clive G Mingham[1] and Derek M Causon[1],

ABSTRACT

This paper reports results of numerical wave flume predictions against experimental data for violent overtopping. The numerical scheme solves the non-linear shallow water equations which, for violent overtopping should not be valid as they neglect vertical accelerations. The results presented show relatively good agreement with experimental results under certain conditions. The paper reviews and identifies the conditions under which engineers can use the numerical model as a viable design tool.

INTRODUCTION

Wave overtopping of coastal structure has been extensively investigated over the last few decades (Goda et al. 1975; Owen 1982; Franco et al. 1994; Besley et al. 1998; Pearson et al. 2001). Originally many sea walls were designed for nominal run-up, based on simple tests with regular waves. Under most of these tests the waves were shown to reflect back out to sea without breaking (pulsating conditions). These conditions are found where the water depth is far in excess of the wave length. Where this isn't the case, violent overtopping can often occur as structures interact with breaking waves. These are known as impacting conditions. Currently, in the UK, most predictions for mean overtopping are based on empirical formulae from laboratory measurements (Besley 1999). These measurement cover pulsating wave conditions, yet sudden violent wave conditions are experienced on different types of sea walls i.e. composite vertical walls. It is therefore imperative that these design methods are investigated to determine whether they can safely be used under very impulsive conditions (Pearson et al. 2001). In addition, reliable design methods are not currently available to calculate peak overtopping volumes, for a significant range of structures. The Violent

---

[1]Centre for Mathematical Modelling and Flow Analysis, Manchester Metropolitan University, Manchester, M1 5GD, United Kingdom, E-mail: s.r.richardson@mmu.ac.uk

Overtopping of Waves at Sea walls (VOWS) project funded by the Engineering and Physical Science Research Council (EPSRC) is looking into these phenomena. The project incorporated both physical and numerical modelling to develop new or improved prediction formulae for mean and peak overtopping events. Physical modelling at small scale was undertaken at Edinburgh University (Bruce et al. 2001).

The recent advances in solution techniques has led to numerous computer packages which solve the Non-Linear Shallow Water Equations (NLSWE) for wave runup and overtopping (Gent 1994; Watson et al. 1996; Titov and Synolakis 1998; Dodd 1998; Hu et al. 2000). Numerical modelling, undertaken by the authors, uses advanced numerical techniques to solve the NLSWE. Bed slope terms are used to model beach geometry and near-vertical walls. Care is needed in formulating bed slope terms to avoid non-physical water surface levels. This has been accomplished here by using the Surface Gradient Method (SGM) (Zhou et al. 2001).

The NLSWE are derived by assuming that the vertical velocity is negligible in comparison to the horizontal velocity and this assumption is obviously broken during overtopping events. An alternative, though more computationally expensive approach is to solve the full Navier-stokes equations using a solver capable of resolving the free surface, e.g. the Volume of Fluid (VoF) method (Hirt and Nichols 1981). Nevertheless, the successful application of the NLSWE to overtopping has been reported (Hu et al. 2000) and these equations may yet prove a viable tool.

A high resolution finite volume solver (Causon et al. 2000), which represents breaking waves as bore waves and uses the SGM is implemented to represent the numerical flume. Initially the Cartesian cut cell method, which admits moving boundaries, is used as a wave generator in the numerical flume and is driven by the demand signal from the experimental wave generator.

Using these techniques the authors have undertaken detailed comparisons between predictions of the numerical flume and experimental observations.

The paper will hope to provide guidance on the regimes under which the NLSWE model performs successfully as a design tool for engineers.

NUMERICS

The Shallow Water Equations

The one-dimensional integral form of the shallow water equations is:

$$\frac{\partial}{\partial t} \int_A \mathbf{U} dA + \oint_s \mathbf{F}.d\mathbf{s} = \int_A \mathbf{\Omega} dA \tag{1}$$

where

$$\mathbf{U} = \begin{pmatrix} \phi \\ \phi u \end{pmatrix}, \quad \mathbf{F} = \begin{pmatrix} \phi u \\ \phi u^2 + \frac{1}{2}\phi^2 \end{pmatrix}, \quad \mathbf{\Omega} = \begin{pmatrix} 0 \\ g\phi\frac{\partial H}{\partial x} \end{pmatrix}$$

and where A is the area enclosed by the the control surface S and s is the outwardly pointing normal vector. $\phi = gh$, is the geopotential; h is the water depth; g is the acceleration due to gravity; u is the depth-averaged velocity in the x-direction; H is the partial depth between a fixed reference level and the bed surface; $\mathbf{F}$ is the convection flux; and Ω is the vector of source terms. Here the source term is bed slope.

The equations are of hyperbolic type and admit discontinuous solutions as bore waves, which are difficult to resolve numerically. Therefore, any numerical scheme applied to (1) must be able to satisfactorily deal with this behaviour.

To solve the integral form of the shallow water equations (1), a MUSCL-Hancock finite volume scheme (van Leer 1984) is used by the authors. This two-step, high resolution, upwind scheme of Godunov type, uses an HLL approximate Riemann solver (Harten et al. 1983) to provide solutions at each cell interface. Full details of the scheme have previously been published by the authors (Mingham and Causon 1998)

Surface Gradient Method (SGM)

To permit the modelling of real problems it is often necessary to include source terms, such as bed topography and bed shear stress. Conservative schemes require special treatment of source terms, with Bermudez and Vázquez (1994) proposing an upwind scheme for the treatment of the bed slope term for unsteady flow. A wider range of flow problems including steady state flow, were looked into by Vázquez-Cendón (1999), but the method has drawbacks in its complexity. Both schemes improved the accuracy of earlier methods such as the fractional step method (Toro 1997), which provided poor solutions to both steady and quasi-steady problems. LeVeque (1998) developed a method for quasi-steady problems, though the scheme is reported to be less accurate for steady transcritical flows with a shock.

Zhou et al. (2001) developed an accurate, simple and robust scheme called the Surface Gradient Method (SGM). The method is a general scheme for treating source terms in the shallow water equations and, unlike conventional data reconstruction methods, the water surface level is chosen as the basis for data reconstruction. Only the bed slope term has been incorporated into this paper though bed friction has been treated using this method (Zhou et al. 2001). Figure 1 shows a 1D sketch of the flow domain for the Surface Gradient Method.

To solve the continuity equations, fluxes based on the conservative variables are required at cell interfaces.

The values of ϕ to the left and right of the cell interface $\left(i - \frac{1}{2}\right)$ are

$$\phi_{i-\frac{1}{2}}^{L} = \phi_{i-1} + \frac{1}{2}\Delta\phi_{i-1}\Delta x_{i-1}, \qquad \phi_{i-\frac{1}{2}}^{R} = \phi_{i} - \frac{1}{2}\Delta\phi_{i}\Delta x_{i} \qquad (2)$$

where Δx_{i} is the grid spacing and $\Delta\phi_{i}$ is the gradient of ϕ within the cell i, based on neighbouring cell centre data for ϕ.

For other methods such as the depth gradient method (DGM), the gradient in ϕ, is effectively the gradient in depth. The water depth at cell interfaces will

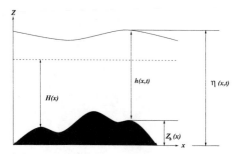

FIG. 1.   Definition sketch for bed topography.

therefore be influenced by bed topography, in addition to the variation of the free surface with time. This will lead to errors introduced by the DGM giving rise to inaccurate fluxes and consequently inaccurate solutions. This is where the new SGM method differs from the DGM. The water surface level $\eta(x,t)$ at the cell interface for the SGM is defined as:

$$\eta(x,t) = h(x,t) + z_b(x) \tag{3}$$

as shown in Figure 1. Following a similar approach as above the water level at the left and right of the cell interface $\left(i - \frac{1}{2}\right)$ are given as

$$\eta^L_{i-\frac{1}{2}} = \eta_{i-1} + \frac{1}{2}\Delta\eta_{i-1}\Delta x_{i-1}, \qquad \eta^R_{i-\frac{1}{2}} = \eta_i - \frac{1}{2}\Delta\eta_i\Delta x_i \tag{4}$$

and $\Delta\eta_i$ is the gradient of η within cell i.

The values of ϕ at the left and right of the cell interface $\left(i - \frac{1}{2}\right)$ are then calculated as

$$\phi^L_{i-\frac{1}{2}} = \left(\eta^L_{i-\frac{1}{2}} - z_{bi-\frac{1}{2}}\right)g, \qquad \phi^R_{i-\frac{1}{2}} = \left(\eta^R_{i-\frac{1}{2}} - z_{bi-\frac{1}{2}}\right)g. \tag{5}$$

The depth related errors in the computation of the fluxes are eliminated by the surface gradient method, giving accurate values of the conservative variable ϕ at the cell interfaces.

The surface gradient method is now incorporated into the MUSCL-Hancock finite volume method for the solution of Equation (1), i.e.

Predictor step

$$\mathbf{U}^{n+\frac{1}{2}}_i = \mathbf{U}^n_i - \frac{\frac{\Delta t}{2}}{\Delta x}\left(\mathbf{F}_R - \mathbf{F}_L - \mathbf{\Omega}^n_i\right) \tag{6}$$

Corrector step

$$\mathbf{U}^{n+1}_i = \mathbf{U}^n_i - \frac{\Delta t}{\Delta x}\left(\mathbf{F}^*_R - \mathbf{F}^*_L - \mathbf{\Omega}^{n+\frac{1}{2}}_i\right) \tag{7}$$

Discretisation of the source terms can be found in (Zhou et al. 2001)

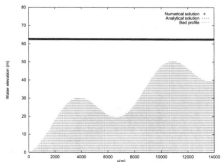

FIG. 2. Comparison of numerical and analytical solution of water surface $\eta(x, t)$ **for tidal wave flow.**

VALIDATION OF THE NUMERICAL METHOD

Tidal wave flow

Coastal engineers often report tidal waves. We consider the test problem that Bermudez and Vázquez (Bermudez and Vázquez 1994) used for the verification of their upwind discretisation.

The bed topography, for this one-dimensional problem, is defined as:

$$H(x) = 50.5 - \frac{40x}{L} - 10\sin\left[\pi\left(\frac{4t}{86400} + \frac{1}{2}\right)\right]$$

where $L = 14000$m is the channel length. The initial conditions are:

$$h(x, 0) = h(x), \qquad u(x, 0) = 0 \tag{8}$$

and the boundary conditions are;

$$h(0, t) = H(0) + 4 - 4\sin\left[\pi\left(\frac{4t}{86400} + \frac{1}{2}\right)\right], \qquad u(L, t) = 0 \tag{9}$$

Under the conditions specified, the tidal wave is relatively short and an analytical solution, for water elevation, is derived in (Bermudez and Vázquez 1994) as

$$h(x, t) = H(x) + 4 - 4\sin\left[\pi\left(\frac{4t}{86400} + \frac{1}{2}\right)\right]$$

The computation has a grid of 200 cells, ie $\Delta x = 70m$ and equations 8-9 were used as the initial and boundary conditions.

A comparison of the numerical water elevation with the analytical solution at $t = 7600$ seconds is shown in Figure 2. The agreement between the results is excellent and suggests that the SGM is accurate for unsteady shallow water flow problems.

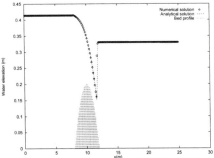

FIG. 3. Comparison of numerical and analytical solution of water surface $\eta(x, t)$ for steady transcritical flow over a bump with a shock.

Transcritical flow with shock

This was one of the test problems used by Vázquez-Cendón (1999). The test employed a 25m long channel with a bump defined by:

$$z_b(x) = \begin{cases} 0.2 - 0.05\,(x - 10)^2 & \text{if } 8 < x < 12 \\ 0.0 & \text{otherwise} \end{cases}$$

and is a classic bench mark test.

A grid of 200 mesh intervals, $\Delta x = 0.125$m was used in this computation. Initial conditions were set to: $h(x, 0) = 0.33$, $u(x, 0) = 0$. A discharge of $q = 0.18$m^2/s was specified at the upstream boundary and $h = 0.33$ imposed as the downstream boundary condition. The steady state solution was obtained at $t = 162.6$ seconds. The numerical results are shown in Figure 3 and again show very good agreement with the analytical solution.

RESULTS

The small scale physical modelling tests for the VOWS project were performed in the 20m wave flume at Edinburgh University. The flume is 0.4m wide and operates at a depth of 0.7m with an absorbing flap-type wave generator, see Figure 4.

The model bed-profile and wall were constructed from perspex, with overtopping discharged directly recorded by a measuring container suspended from a load cell. Individual overtopping events were detected by water closing the connection between two parallel metal strips along the structure crest. Water elevation readings along the flume were taken by 8 wave gauges: 1.00, 2.00, 3.00, 4.25, 5.50, 6.75, 8.00, 11.21 metres from the wall. The 5 tests reviewed here take place on a 10:1 battered wall with a 1:10 uniform bathymetry. Each of the tests were run for approximately 1000 waves of a JONSWAP spectrum with $\gamma = 3.3$. Repeat tests of approximately 100 wave also undertaken and it is these tests that have been numerically simulated. To generate these waves the movement of the wave

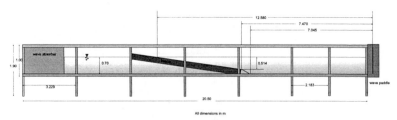

FIG. 4.   Sketch of Edinburgh flume.

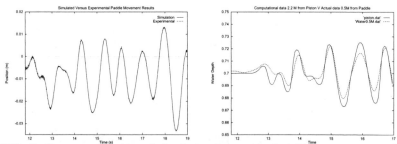

FIG. 5. Comparative view of physical and numerical paddle movement and surface elevation in front of wave generators.

generator was recorded so that the numerical moving boundary could imitate it, creating the same wave train.

As can be seen from Figure 5, the numerical piston performs very well in recreating the paddle movement. When the water surface elevation is reviewed, 0.5m in front of the wave generator, the numerical scheme provides good agreement with the experimental results (Figure 5).

Energy loss occurs in the physical model as the waves propagate along the flume. The shallow water equations do not include a dispersion term, so this energy loss is not observed in the numerical flume, as can be seen in Figure 6.

To compensate for this energy loss, the boundary condition in generating the wave train was moved nearer to the battered wall. The numerical model was able to generate waves from the surface elevation readings, obtained from the physical experiments wave gauges. Generating waves from data obtained 2m from the wall, allowed the water to be shallow enough for energy loss terms to be neglected. Water surface elevation was entered into the boundary condition to generate the waves, with velocities for these waves obtained from the first node inside the flow domain.

Extensive tests were undertaken for this condition, so that the authors could confirm that the reflected waves pass out of the flow domain at the seaward boundary. A very long flow domain was constructed with no bed topography, a

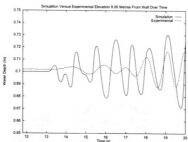

FIG. 6. Comparative view of physical and numerical surface elevation further along flume.

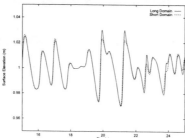

FIG. 7. Comparison of water elevation just inside flow domain for new boundary condition

random wave generator at one boundary and a reflective wall at the other end. Water depth and velocity were recorded at two adjacent nodes in the middle of the flow domain and the simulation run for sufficient time for the reflected waves to return to wave generator. Simulations in a smaller flow domain, half the original length, where then undertaken. The recorded nodes in the middle of the domain were positioned so that they spanned the inflow boundary, i.e. one was used to generate the wave conditions from the previous test while the other was inside the flow domain. Comparisons in the water surface elevation at the internal node can be seen in Figure 7. This test showed a high degree of correlation between the two approaches giving R^2 values of over 99% for the water level and 85% for the velocities.

This boundary condition was then used to generate the wave train in the numerical model. The numerical wave tank is shown in figure 8.

Overtopping processes were shown to be strongly influenced by the form of the incident wave. When water depth is large compared to wave length, waves are generally reflected back to sea off vertical or composite walls. When this is not the case, waves can break onto structures causing significantly more overtopping. The wave breaking parameter, h^*, was formulated by these observations (Allsop

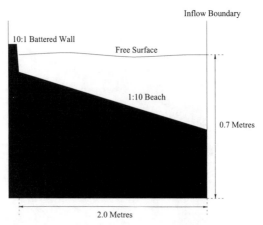

FIG. 8.   Geometry of numerical wave tank

et al. 1995) and is given by,

$$h^* = \frac{h}{H_s}\left(\frac{2\pi h}{gT^2}\right). \tag{10}$$

Reflecting waves predominate when h*> 0.3 and impacting waves are more likely when h*≤ 0.3. New dimensionless discharge (Q_h) and freeboard (R_h) parameters, were established to incorporate h* (Besley 1999),

$$Q_h = \frac{\left(\frac{Q}{(gh^3)^{0.5}}\right)}{h*^2} \tag{11}$$

$$R_h = \left(\frac{R_c}{H_s}\right)h*, \tag{12}$$

where Q is the mean overtopping discharge per metre run.

Tables 1 and 2 show the wave parameters for the physical and numerical tests undertaken. In all tests the wall height = 0.24m, h = 0.09, Rc = 0.15. Besley et al. (1998) undertook an extensive range of measurements for mean overtopping discharge on simple vertical walls. The curve shown in Figure 9 shows the line of best fit described in (Besley 1999) for these measurements. The numerical results, represented by squares and the experimental results (triangles) are plotted in Figure 9.

The proportion of overtopping waves for each of the test are plotted in Figure 10 (numerical - squares, experimental - triangles) as well as the predicted line from (Besley 1999).

TABLE 1. Physical Parameters/Results

Tests	1	2	3	4	5
Hs	0.069	0.071	0.078	0.062	0.066
Tz	1.25	1.50	1.48	0.97	0.98
h^*	0.05	0.03	0.03	0.09	0.08
Q	6.38E-05	6.71E-05	1.55E-04	3.77E-05	4.88E-05
Q_h	3.26E-01	7.52E-01	1.99E+00	5.63E-02	8.61E-02
R_h	0.10	0.07	0.06	0.22	0.19
N_{ow}	15	19	27	18	19
N_w	48	43	43	62	62
N_{ow}/N_w (%)	31	44	63	29	31

TABLE 2. Numerical Parameters/Results

Tests	15060017	15060018	15060019	15060020	15060021
Hs	0.069	0.071	0.078	0.062	0.066
Tz	1.25	1.50	1.48	0.97	0.98
h^*	0.05	0.03	0.03	0.09	0.08
Q	5.67E-06	1.75E-05	3.04E-05	1.68E-05	1.66E-05
Q_h	2.89E-02	1.96E-01	3.90E-01	2.52E-02	2.93E-02
R_h	0.10	0.07	0.06	0.22	0.19
N_{ow}	18	34	30	19	19
N_w	48	43	42	62	61
N_{ow}/N_w (%)	38	79	71	31	31

Both Figures 9 and 10 show the numerical results for high values of R_h, and consequently h^*, are close to the experimental results. The numerical results under estimate the discharge (Figure 9) in comparison to the experiments, though the tests around $R_h = 0.2$ show good agreement with Besley (1999). In predicting peak overtopping and statistics of wave-by-wave overtopping volumes, the number of overtopping waves (N_{ow}) is an important parameter. Besley (1999) also obtained a line of best fit for this parameter, divided by the total number of waves in the sequence against R_h^2. Plotting the experimental and numerical results in Figure 10 good agreement is observed between them, especially for higher values of R_h, though these results deviate from Besley's line. Figure 10 also shows that the numerical model over-predicts the number of overtopping waves in these tests.

More numerical simulations of other experimental tests are required. The five tests presented all have h^* values of < 0.1, which, under the definition given previously, makes them highly impulsive conditions. For these conditions,

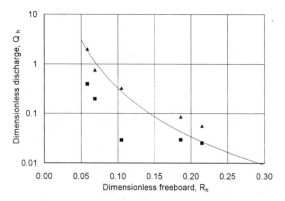

FIG. 9.   Overtopping discharge on a 10:1 battered wall

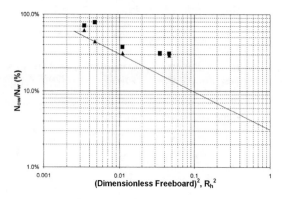

FIG. 10.   Percentage of waves overtopping a 10:1 battered wall

which incorporate broken or breaking waves onto structures, the current one-dimensional scheme should not be valid. It has been shown that for the higher values of R_h and h^* which have been tested, the scheme works reasonably well and could yet be a more viable tool for predicting wave overtopping, than VoF methods which are computationally more expensive.

CONCLUSIONS

A computational method for calculating the non-linear shallow water equations, with bed topography represented by the Surface Gradient Method has been presented. The method is a high order MUSCL-Hancock finite volume scheme, with an HLL approximate Riemann solver and produces high resolution solutions to the shallow water equations.

The scheme has been used to model wave overtopping of steep sea walls and impulsive conditions with good agreement between numerical and physical results. These tests show that reasonable results are obtained for values of R_h of around 0.2. Further tests are require for other values of h^*, particularly $0.1 \leq h* \leq 0.3$. These will help determine the range of values for h^* and R_h, where the numerical model is a viable tool for sea defence design. From the present results it is estimated that the model provides a viable tool for $h^* \geq 0.1$ and $R_h \geq 0.15$.

ACKNOWLEDGEMENTS

As previously stated this work is funded by the UK EPSRC (GR/M42428 and GR/M42312), and supported by the VOWS management Committee. Members included University of Edinburgh (Tom Bruce, Jon Pearson), University of Sheffield (William Allsop), Posford Duvivier (Dick Thomas, Keming Hu), Bullen & Co (Mark Breen, Dominic Hames), HR Wallingford (Philip Besley, Tim Pullen), of whose input and support is greatly appreciated.

The authors are especially grateful to Tom Bruce, Jon Pearson and William Allsop for providing the experimental data used in this paper.

REFERENCES

Allsop, N., Besley, P., and Madurini, L. (1995). "Overtopping performance of vertical walls and composite breakwaters, seawalls and low reflection alternatives." *Paper 4.7 in MCS Final Report, University of Hannover.*

Bermudez, A. and Vázquez (1994). "Upwind methods for hyperbolic conservation laws with source terms." *Computers and Fluids,* 23, 1049.

Besley, P. (1999). "Overtopping of Seawalls - design and assessment manual. R & D Technical Report W178, ISBN 1 85705 069 X, Environment Agency, Bristol.

Besley, P. B., Stewart, T., and Allsop, N. (1998). "Overtopping of vertical structures: new methods to account for shallow water conditions." *Proceedings of Int. Conf. on Coastlines, Structures & Breakwaters '98, Institution of Civil Engineers.* Thomas Telford, London, 46–57.

Bruce, T., Allsop, N., and Pearson, J. (2001). "Violent overtopping of seawalls - extended prediction methods." *Proc. Coastlines, Structures and Breakwaters 2001, ICE.* Thomas Telford, London.

Causon, D., Ingram, D., and Mingham, C. (2000). "A cartesian cut cell method for shallow water flows with moving boundaries.." *Water Resources Research,* Submitted.

Dodd, N. (1998). "A numerical model of wave run-up, overtopping and regeneration." *Proc ASCE, J. Waterway, Port, Coast & Ocean Eng.,* Vol. 124(2). ASCE, New York, 73–81.

Franco, L., de Gerloni, M., and van der Meer, J. (1994). "Wave overtopping on vertical and composite breakwaters." *Proc. 24th Int. Conf. Coastal Eng., Kobe,* publn. ASCE, New York.

Gent, v. M. (1994). "The modelling of wave action on and in coastal structures." *Coastal Engineering*, 22, 311–339.

Goda, Y., Kishira, Y., and Kamiyama, Y. (1975). "Laboratory investigation on the overtopping rates of seawalls by irregular waves." *Ports and Harbour Research Institute*, 14(4), 3–44. PHRI, Yokosuka.

Harten, A., Lax, P., and vanLeer, B. (1983). "On upstream differencing and godunov-type schemes for hyperbolic conservation laws." *SIAM Review*, 25(1), 35–61.

Hirt, C. and Nichols, B. (1981). "Volume of fluid (vof) methods for the dynamics of free boundaries." *Journal of Computational Physics*, 39, 201–225.

Hu, K., Mingham, C., and Causon, D. (2000). "Numerical simulation of wave overtopping of coastal structures using the nlsw equations." *Coastal Engineering*, In press.

LeVeque, R. (1998). "Balancing source terms and flux gradients in high-resolution godunov methods: The quasi-steady wave-propagation algorithm." *Journal of Computational Physics*, 146, 346.

Mingham, C. and Causon, D. (1998). "High-resolution finite-volume method for shallow water flows." *Journal of Hydraulic Engineering*, 605–614.

Owen, M. (September 1982). "Overtopping of sea defences." *Proc. Conf. Hydraulic Modelling of Civil Engineering Structures, BHRA, Coventry*, 469–480.

Pearson, J., Bruce, T., and Allsop, N. (2001). "Prediction of wave overtopping at steep seawalls - variabilities and uncertainties." *Proc. Waves '01*, ASCE, San Francisco.

Titov, V. and Synolakis, C. (1998). "Numerical modelling of tidal wave runup." *ASCE Journal of Waterways, Port, Coastal and Ocean Engineering*, 124(4), 157–171.

Toro, E. (1997). "Riemann solvers and numerical methods for fluid dynamics." *Springer-Verlag, Berlin, Heidelberg.*

van Leer, B. (1984). "On the relationship between the upwind-differencing schemes of godunov, engquist-osher and roe." *SIAM Journal on Scientific and Statistical Computing*, 5.

Vázquez-Cendón, M. (1999). "Improved treatment of source terms in upwind schemes for shallow water equations in channels with irregular geometry." *Journal of Computational Physics*, 148, 497.

Watson, G., Barnes, T., and Peregrine, D. (1996). "Numerical modelling of solatary wave propagation and breaking on a beach and runup on a vertical wall." *In: Long Wave Runup, Yeh H, Liu P and Synalakis J (eds), Word Scientific*, 291–297.

Zhou, J., Causon, D., Mingham, C., and Ingram, D. (2001). "The surface gradient method for the treatment of source termsin the shallow water equations." *Journal of Computational Physics*, 168, 1–25.

APPENDIX I. NOTATION

The following symbols are used in this paper:

U	$=$	vector of conserved quantities (geopotential and momentum)
F	$=$	convection flux
Ω	$=$	vector of source terms
S	$=$	control surface
s	$=$	outwardly pointing normal vector
ϕ	$=$	geopotential
h	$=$	water depth
g	$=$	acceleration due to gravity
u	$=$	depth-averaged velocity in x-direction
H	$=$	partial depth between a fixed reference level and bed surface.
η	$=$	water surface elevation including bed topography
z_b	$=$	height of bed topography
h	$=$	water depth at toe of structure (m);
Rc	$=$	freeboard (m);
Hs	$=$	significant wave height at structure (m)
Tz	$=$	mean wave period at structure (s)
h*	$=$	wave breaking parameter
Q	$=$	mean overtopping discharge $(\text{m}^3/\text{s/m})$
Qh	$=$	dimensionless discharge
Rh	$=$	dimensionless freeboard

LARGE-EDDY SIMULATION OF LOCAL FLOWS
AROUND THE HEAD OF A BREAKWATER

Masaya Kato[1], Yuhki Okumura[2], Yasunori Watanabe[3] and Hiroshi Saeki[4]

Abstract: In this study, three-dimensional fluid motion in waves around the head of a semi-infinite vertical breakwater was investigated using large-eddy simulation (LES). Wave deformation, separation, and evolution of vortices were investigated to obtain basic data in order to develop a reliable model for predicting local scouring and for estimating local fluid forces acting on the structure. Moreover, the processes by which separated eddies are generated and evolved were discussed in detail in order to determine the mechanisms by which the separated vortices cause local failures of armor blocks and the mound of a breakwater. Two significant vertical separations appeared around both edges of the breakwater head, especially when the wave front passed each edge. The velocity field around the head of the breakwater became significantly three-dimensional because secondary horizontal and helical eddies arose due to the vertical gradient of shear associated with the vertical separation. Furthermore, during interactions among these eddies, a complex three-dimensional eddy structure was formed around the head of the breakwater. It is possible that the vertical velocity caused by the three-dimensional eddy structure is associated with the suspension of sediment around the structure.

INTRODUCTION

A composite breakwater consists of a concrete caisson as the vertical wall and a rubble mound foundation. There are several reports concerning damages to the breakwater. Takahashi et al. (2000) classified typical failures of composite breakwaters

---

1 Senior Research Engineer, Nippon Data Service Co., Ltd., Kita-16 Higashi-19, Sapporo 065-0016, JAPAN, kato@kowanws2.hyd.eng.hokudai.ac.jp
2 Graduate Student, Offshore & Coastal Engineering Lab., Graduate School of Eng., Hokkaido University, Kita-13 Nishi-8, Sapporo 060-8628, JAPAN, okumura@kowanws2.hyd.eng.hokudai.ac.jp
3 Assistant Professor, Offshore & Coastal Engineering Lab., Graduate School of Eng., Hokkaido University, Kita-13 Nishi-8, Sapporo 060-8628, JAPAN, yasunori@eng.hokudai.ac.jp
4 Professor, Offshore & Coastal Engineering Lab., Graduate School of Eng., Hokkaido University, Kita-13 Nishi-8, Sapporo 060-8628, JAPAN, h-saeki@eng.hokudai.ac.jp

in Japan, and they reported that local scouring of the rubble mound foundation around the breakwater head is one major causes of failure of composite breakwaters. Kimura et al. (1994) and Kimura et al. (1998) presented the stability of armor units for composite breakwaters. Sumer et al. (1997) investigated local scouring of the seabed around the head of a vertical-wall breakwater. These studies showed that a formation of separated vertical eddies induce the damage to the armor layer and local erosion around the breakwater head. However, previous studies have focused on relationships of measured or computed flow velocity with block weight and scouring depth, and there have been few studies in which the characteristics of spatial and temporal variations in the formation, advection and dissipation of eddies around the head of a breakwater were examined. Therefore, it is necessary to clarify the mechanisms in which such a complex flow structure is formed around the breakwater head in order to be able to design a more reliable and cost-effective breakwater. The aim of this study is to determine the basic properties of local velocity field around the head of a breakwater in terms of the three-dimensional eddy structure.

METHOD OF COMPUTATION

In this study, the evolution of the three-dimensional flow structure around the head of a breakwater was computed by the numerical model via a three-dimensional large-eddy simulation (hereinafter called LES), which has been proposed Watanabe et al. (1999). For grid-scale quantities, the filtered Navier-Stokes equation was solved numerically by a fractional two-step method. The numerical method based on the MAC method was used for Eulerian terms in the filtered momentum equation. The CIP method was used for an advection phase. As a sub-grid scale stress model, we used a nonlinear sub-grid viscosity model that was corrected for a low Reynolds number, which was based on a renormalization group theory. The density function method was used to determine a free surface. Figure 1 shows the computational domain and the coordinate system, and Table 1 shows the conditions of calculation, where H is wave height of the incident wave, T is the period of the incident wave, h is water depth, L is the wavelength on the basis of small amplitude wave theory, B is width of the breakwater, Ws is width of the opening, Δx, Δy, Δz are grid spacing in each axial direction, and Δt is an interval between each time step. On the bottom and the surface of structure, a non-slip condition was imposed. For the side boundary, a slip condition was used as a symmetry condition. As an open boundary condition, the energy absorbing zone and the free outflow condition, in which the velocity gradient is zero, were used in combination. The second-order cnoidal wave for case 1 and case 2 and the third-order Stokes wave for case 3 were numerically generated at the incident boundary.

RESULTS AND DISCUSSION
Comparison of computational and experimental results

The method for computation described here is basically similar to that used by Watanabe et al. (1999) as mentioned above. Its validity has already been proven by results of computations for wave breaking. In the present study, water levels around the head of a model breakwater under the conditions of case 3 were measured to determine the validity of this numerical method with a structure and an open boundary. Considering the symmetry in the center of the opening of the breakwater, calculations were carried for only half of the domain in order to reduce calculation load. Figure 2 shows a comparison of the computational and experimental results of water surface elevation at the vicinity of the breakwater head. As can be seen in the figure, the results of calculation coincide satisfactorily with the experimental results, demonstrating that each boundary treatment used in this calculation was valid in general.

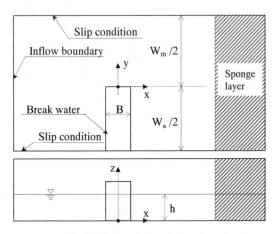

Fig. 1. Computational domain.

Table 1. Computational conditions.

	H/h	h/L	B/L	Ws/L Wm/L	Δx/h, Δy/h Δz/h	Δt/T
Case 1	0.4	0.083	0.07	0.4		
Case 2	0.4	0.083	0.07	1.0	0.04 – 0.2 0.04	1/512
Case 3	0.33	0.192	0.15	1.0		

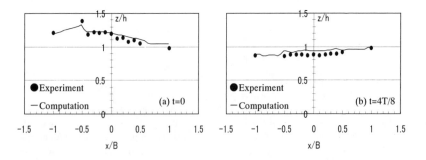

Fig. 2. Comparison of surface elevations near the breakwater head ($y/B=0.025$).

Distribution of flow velocity near the bottom surface

The flow structure around the head of a breakwater changes slightly even in each period due to the influence of the residual eddy from the previous period and other factors, and it is therefore impossible to achieve a perfectly steady state. However, since the purpose of the present study was to determine the basic properties of the structure of a flow around the head of a breakwater, we looked over the computational results for several periods and used the results for one period that were representative of the results for several periods for analysis in this study. Moreover, although the fluid velocity in the vicinity of the head of the breakwater was slightly smaller in case 2 than in case 1 due to the difference in the widths of the opening of the breakwater, the flow structures in case 1 and case 2 were qualitatively similar, and therefore case 1 and case 3 are mainly discussed below.

Figure 3 shows the velocity distribution in the vicinity of the bottom surface (z/h = 0.02) in case 1. The vector in the figure refers to the distribution of fluid velocity within the x–y plane, and the black-and-white refers to the vertical velocity. The velocity is indicated dimensionless by the wave velocity of long wave. The phase when the wave crest passes over the opening of a breakwater is referred as t = 0. The upper figure shows the phase when the wave crest is just passing over the opening of the breakwater. Two significant vertical separations appear around both edges of the breakwater head. The mechanisms of these separations are completely different. The offshore-ward separation is caused by a cross-shore flow (in the y-direction) induced by the cross-shore gradient of the free surface. On the other hand, the onshore-ward separation develops in the dead-water region behind the breakwater due to the shoreward flow (in the x-direction) of a progressive wave. The lower figure shows the velocity distribution when the wave trough is just passing over the head of the breakwater. Two reverse separations appear around both edges of the breakwater head.

These eddies observed in the vicinity of the bottom are formed all the way from the bottom to the water surface. However, their structures change greatly over time, and these eddies differ in characteristics of development. For example, the central location and the size of the eddy differ depending on the water depth. Furthermore, a locally faster fluid velocity occurs in the eddy, and the vertical fluid velocity forms a complex three-dimensional structure. A descending flow is formed around the eddy, and an ascending flow is formed near the center of the eddy.

The maximum horizontal velocity in the vicinity of the bottom (z/h = 0.02) calculated in this study was 2.3 to 3.4-times greater than the horizontal velocity of the progressive wave without a structure. The maximum vertical velocity in the vicinity of the bottom surface (z/h = 0.02) was 10.7 to 30.7-times greater than the vertical velocity of the progressive wave without a structure. However, the vertical velocity in the case of z/h = 0.02 calculated in this study was roughly 1/10 of the horizontal velocity.

Figure 4 shows the velocity distribution in the vicinity of the bottom under the conditions of case 3. There is no counterclockwise separation from the offshore side when the wave trough passes over the opening of the breakwater. However, for the onshore side, a very small eddy is formed 90 degrees behind this phase, but it disappears immediately. On the other hand, there are relatively large clockwise eddies. As is also clear from the direction of rotation, these eddies are residuals of the separation eddies formed during the passage of the wave crest.

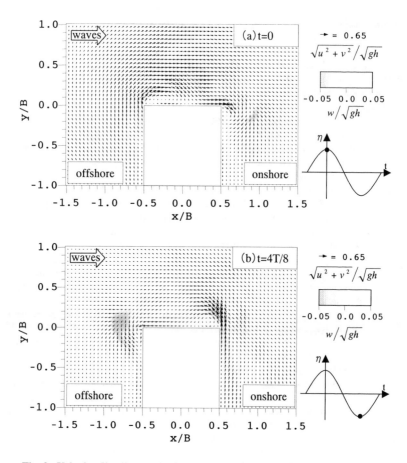

Fig. 3. Velocity distribution in the vicinity of the bottom (case 1, z/h=0.02).

Figure 5 shows the mean flow velocity per period in the vicinity of the bottom ($z/h = 0.02$) for case 1 and case 3. The separated eddy causes formations of a flow along the structure, a flow away from the structure and large circulation currents in the vicinity of the structure, the velocities of which are much greater than those of flows in other regions. The circulation currents in case 3 are asymmetry because of the asymmetry of vertical separation. From the nonuniformity in the distribution of vertical fluid velocity, it is also clear that the mean flow is given a complex three-dimensional structure in the vicinity of the head of a breakwater.

Distribution of flow velocity in a y-z cross section
Figure 6 shows the distribution of fuid velocity in a y-z cross section near the center

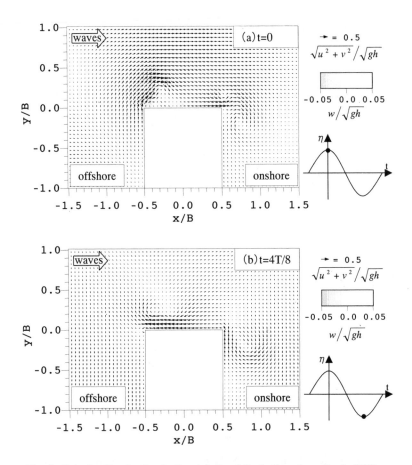

Fig. 4. Velocity distribution in the vicinity of the bottom (case 3, z/h=0.02).

$(x/B = 0.725)$ of the vertical eddy where $t = T/8$ in case 1. From Figure 6, it is clear that a locally strong ascending flow and an eddy (helical eddy) having its axis in the progressive direction of the incident wave are formed near the center of the vertical eddy behind the breakwater generated by the flow moving toward the shore and that they have complex three-dimensional structures. It is thought that spatial variations in the local flow near the bottom mentioned above cause scattering of the materials covering the mound and local scouring.

Figure 7 shows the distribution of fluid velocity in a y-z cross-section near the center of the opening $(x/B = 0.0)$ where $t = 5T/8$ in case 1. With the formation of a vertical eddy caused by the flow moving in the offshore direction, a larger helical eddy with its center near the middle of the depth of water is formed outside of the vertical eddy.

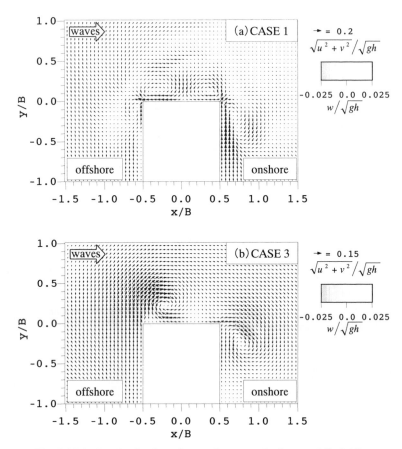

Fig. 5. Velocity distribution of mean flow near the bottom (z/h=0.02).

Distribution of flow velocity in an x-z cross section

Figure 8 shows the distribution of fluid velocity in an x-z cross-section near the opening of a breakwater ($y/B = 0.225$) where t = $6T/8$ in case 1. An eddy (horizontal eddy) with its axis in the direction of the normal to the breakwater is formed by the strong flow in the offshore direction accompanying a vertical eddy that is generated by the edge of the shoreward side of the breakwater in the phase of $t = 4T/8 - 5T/8$.

Figure 9 shows the distribution of fluid velocity in an x-z cross-section near the head of a breakwater ($y/B = -0.225$) where $t = 3T/8$ in case 1. The phase shown in Figure 9 refers to the reversion from the current moving in the onshore direction to that moving in the offshore direction, and the vertical eddy separated in the onshore direction is disappearing. In this phase, various eddies with complex three-dimensional flow structures, such as a horizontal eddy accompanying a fast flow along the breakwater and a horizontal eddy in the vicinity of the corner where the bottom surface and the wall surface of the breakwater meet, are formed on both sides of the breakwater due to the interaction between fluid motion and the surface of the breakwater or the bottom.

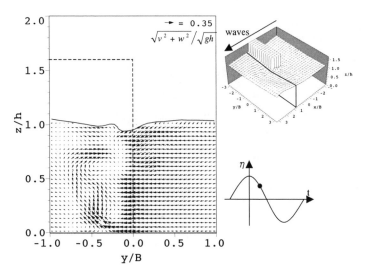

Fig. 6. Velocity distribution in a y-z cross section (case 1, x/B=0.725, $t=T/8$).

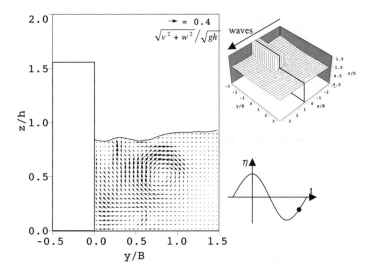

Fig. 7. Velocity distribution in a y-z cross section (case 1, x/B=0.0, $t=5T/8$).

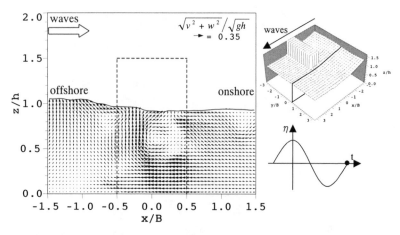

Fig. 8. Velocity distribution in an *x-z* **cross section (case 1,** *y/B*=**0.225,** *t*=6*T*/8**).**

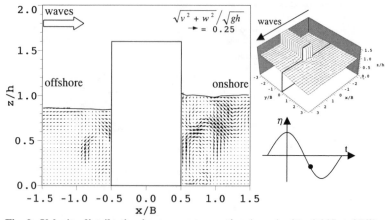

Fig. 9. Velocity distribution in an *x-z* **cross section (case 1,** *y/B*=**-0.225,** *t*=3*T*/8**).**

CONCLUSIONS

The basic characteristics of flow structures around the head of a breakwater were elucidated with LES. The local flows involved with separated eddies (vertical eddy) generated at the edges of the breakwater are predominant in wave field containing the head of a breakwater. The secondary horizontal and helical eddies are formed due to the vertical gradient of shear associated with the vertical separation, which in turn develop a spatially complex three-dimensional structure. This structure itself also underwent a major change over time. Moreover, differences of the size of the eddy and the generation/dissipation process of the eddy greatly depend on the incident wave and the configurations of the breakwater. The horizontal eddy and the helical eddy are generated noticeably when a wave of longer period acts on the head of a breakwater. The vertical velocity caused by the three-dimensional eddy structure possibly promotes a rolling-up process of the sediment around the structure, which causes local scouring. The velocity of the local flow in the vicinity of the bottom, which is associated with a formation of an eddy structure, was estimated in this study to be 2.3 to 3.4-times greater for horizontal velocity and 10.7 to 30.7-times greater for vertical velocity than the fluid velocity in the wave field without a structure. These results suggest that it is necessary to use the intensified fluid velocity due to separations when designing armor layer of mound around the head of a breakwater. Significant circulating mean flows appear around the head of the breakwater. The mean flows tend to go away from the edges of the head of the breakwater, possibly causing sediment to be transported away from the structure.

REFERENCES
Kimura, K., Takahashi, S., and Tanimoto, K.   1994.   Stability of Rubble Mound Foundation of Composite Breakwaters under Oblique Wave Attack. *Proceedings of Coastal Engineering '94*, ASCE, 1227-1240.

Kimura, K., Mizuno, Y., and Hayashi, M.   1998.   Wave Force and Stability of Armor Units for Composite Breakwaters. *Proceedings of Coastal Engineering '98*, ASCE, 2193-2206.

Sumer, B. and Fredsøe, J.   1997.   Scour at the Head of a Vertical-wall Breakwater. *Coastal Engineering*, Vol.29, 201-230.

Takahashi, S., Shimosako, K., Kimura, K., and Suzuki, K.   2000.   Typical Failures of Composite Breakwater in Japan. *Proceedings of Coastal Engineering '00*, ASCE, 1899-1910.

Watanabe, Y. and Saeki, H.   1999.   Three-dimensional Large Eddy Simulation of Breaking Waves. *Coastal Engineering Journal*, JSCE, Vol.41, 281-301.

SPATIAL REGULAR WAVE VELOCITY FIELD MEASUREMENTS NEAR SUBMERGED BREAKWATERS

Francisco Taveira-Pinto, M. Fernanda Proença & F. Veloso-Gomes[1]

Abstract: Submerged breakwaters are a possible solution for the prevention of coastal erosion problems. They guarantee increased protection between themselves and the coastline, diminishing the risk of coastline erosion and assisting sand accretion or retention. Their efficiency depends and it's normally evaluated through the knowledge of the incident, transmitted and reflected wave heights. A different approach of the problem can be done through the measurement of incident, reflected and transmitted wave energies using a full energy concept, i.e., through the measurement of the velocity flow field and sea surface elevation for a regular wavelength. The results presented show the variation of the measured kinetic, potential, total and turbulent kinetic energies, as a function of the distance to the submerged breakwater, calculated through the phase-dependant measured velocity profiles and water surface elevation. Laser Doppler anemometry technique was used for measuring flow velocities and experiments were carried out at the wave tank of the Hydraulics Laboratory of the University of Porto's Faculty of Engineering.

INTRODUCTION

The interference of the wave flow with a submerged breakwater causes dissipation of energy. The incident wave energy is partially reflected, partially transmitted and partially absorbed by the structure. The breakwater's interference with the wave flow leads to the generation of a reflected wave, and gives rise to a standing or partially standing wave, different from the one corresponding to the non-existence of the structure. The resulting water surface elevation is schematically shown in figure 1, as well as the oscillatory flow field pattern. Node and anti-node sections arise and the velocity flow field variation is also sketched on the same figure, where T represents the wave period (s), L the wavelength (m) and η the water elevation (m).

[1] Institute of Hydraulics and Water Resources, Faculty of Engineering of the University of Porto, Rua do Dr. Roberto Frias, 4200 – 465 Porto, Portugal, Tel/Fax: 00 351 22 508 1952; E-mail: fpinto@fe.up.pt

The total wave energy, E_t, is the sum of the kinetic, E_K, and the potential, E_p, wave energies. The potential energy results from the displacement of the free water surface in relation to its still level and the kinetic energy is due to the movement of the water particles throughout the fluid.

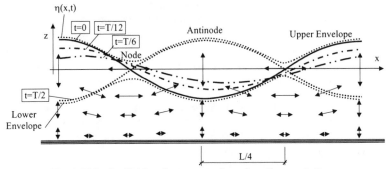

Fig. 1. Velocity field and water surface elevation variations in partially standing waves

The assessment of kinetic (E_k), potential (E_p) and turbulent kinetic (E_k') energies (N.m/m) requires the measurement of the wave velocity components, u and v (m/s) and the corresponding water surface elevations, η (m). The measurements after been analyzed gives the mean velocity components ($\bar{u}$ and $\bar{v}$) and the root mean square (rms) its fluctuation components (u' and v'). So, it's possible do determine the related energies, through the following equations,

$$E_k = \frac{\gamma}{2g} \int_0^L \int_{-d}^{\eta} (\bar{u}^2 + \bar{v}^2)\, dz\, dx \tag{1}$$

$$E_p = \frac{\gamma}{2} \int_0^L \eta^2\, dx \tag{2}$$

and

$$E_k' = \frac{\gamma}{2g} \int_0^L \int_{-d}^{\eta} ((u')^2 + (v')^2)\, dz\, dx \tag{3}$$

According to previous work of Taveira Pinto *et al.*, 1998, 1999, 2000, flow energy values can be obtained through the phase dependent measured velocity profiles at each section of the flow.

The design of submerged breakwater with characteristics that can provide the required energy dissipation and meet the environmental requirements oblige to the definition of the loads acting on the structure. The forces imposed by waves on impermeable slopes, as smooth submerged breakwaters, where percolation effects do not participate, depend on the energy transformation process. The wave loads act over the slope in the form of pressure forces (impacts) and as shear forces acting tangentially to the structure slope, due to the wave motion.

Additional phenomena should also be taken into account in the energy dissipation process. In fact the computation of wave loads on structures are based on the incident wave characteristics which act immediately before the structure and that should be reliably computed from the deep-water wave parameters, modified by diffraction, refraction (due to underwater morphology), shoaling effects and by breaking process which affects it during its propagation from offshore to nearshore areas.

Progressive wave characteristics are transformed by the variation in the bottom configuration and by the existence of obstacles such as submerged breakwaters, until they become unstable and break at a certain water depth. When a wave enters shallow water, its characteristics, namely its height, H (m), length, L (m), and celerity, c (m/s), change. Considering that the slope of the bottom has a negligible effect on the wave characteristics, the waves at any location on a sloping bottom can be described by small amplitude wave theory (not valid for steep waves), being the wavelength and wave celerity as follows,

$$L = \frac{gT^2}{2\pi} tanh \frac{2\pi d}{L} \quad , \quad c = \frac{gT}{2\pi} tanh \frac{2\pi d}{L} \tag{4}$$

which in deep water are, respectively equal to,

$$L_0 = \frac{gT^2}{2\pi} \quad , \quad c_0 = \frac{gT}{2\pi} \tag{5}$$

The relation between them is then equal to,

$$\frac{L}{L_0} = \frac{c}{c_0} = tanh \frac{2\pi d}{L} \Rightarrow \frac{d}{L_0} = \frac{d}{L} tanh \frac{2\pi d}{L} \tag{6}$$

That is to say that the wavelength L at the water depth d can be obtained from the water depth and deep-water wavelength values, using equation (5).

In order to achieve a better understanding of the effect of a submerged breakwater on the wave propagation let us analyse the same effect with a single uniform slope. On a uniform slope, if breaking occurs, at the breaking point the wave height increases to the maximum breaker height, with a certain crest elevation and breaking depth. Breaking process will occur when the velocity of the water particles of the crest becomes higher than the wave celerity itself. During breaking, a large amount of wave energy is dissipated by a very complicated turbulent mixing processes connected with an intensive air entrainment, figure 2.

For the spilling breaker (predominant on very flat slopes of natural beaches) the wave energy in a surf zone is dissipated gradually over a distance of several wavelengths. The air entrainment is only near the surface and very little wave energy is reflected. On the plunging breaker the energy dissipation takes place over a very short distance, of the order of one wavelength, with strong air entrainment through the water depth. A part of the wave energy is reflected. For the surging breaker (predominant on steep slopes) at the breaking point the water mass forms a nearly vertical front, which flattens producing a strong, up-rush swell with a high reflection of the wave energy.

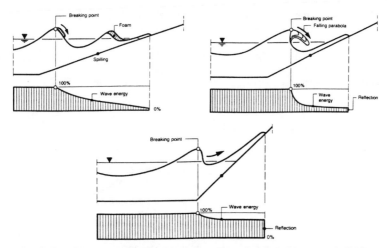

Fig. 2. Breaking modes and related energy variation (Abbott *et al.* 1990)

The high up-rush and down-rush velocities on the slope, cause high tangential forces (shear stresses). The energy reflection affects the energy dissipation. Very low reflection is found for the spilling breaker (less than 10%), while in the range of the plunging breaker the reflection increases up to 60% - 80% at the limit for surging breakers.

Assuming conservation of the mean wave energy, $\overline{W}$, through a vertical section with unit crest width and per unit time, i.e., considering that no breaking phenomena occurs and that there are no energy losses, and taking as reference section the one located at deep-water, the definition of the wave height at a given water depth, can be done the following way,

$$\overline{W} = \overline{W}_0 \Rightarrow \overline{E} c n = \overline{E}_0 c_0 n_0 \Rightarrow \frac{1}{2} \rho g H^2 c n = \frac{1}{2} \rho g H_0^2 c_0 n_0 \tag{7}$$

and finally,

$$K_S = \frac{H}{H_0} = \sqrt{\frac{1}{2n} \frac{c_0}{c}} = \sqrt{\frac{\cosh^2 \frac{2\pi d}{L}}{senh \frac{2\pi d}{L} \cosh \frac{2\pi d}{L} + \frac{2\pi d}{L}}} \tag{8}$$

being,

$$n = \frac{1}{2}\left[1 + \frac{\frac{4\pi d}{L}}{senh \frac{4\pi d}{L}}\right] \tag{9}$$

and K_S the shoaling coefficient.

K_S and n are functions of d/L and consequently of d/L_0, figure 3. In the range of $d/L_0 < 0.01$, the following approximations can be used,

$$\frac{H}{H_0} \approx \left(8\pi \frac{d}{L_0}\right)^{-1/4} \quad , \quad n \approx 1 \quad , \quad \frac{c}{c_0} \approx \left(2\pi \frac{d}{L_0}\right)^{1/2} \quad , \quad \frac{d}{L} \approx \left(\frac{1}{2\pi}\frac{d}{L_0}\right)^{1/2} \tag{10}$$

where H represents the local wave height, H_0, the wave height in deep water and d the water depth.

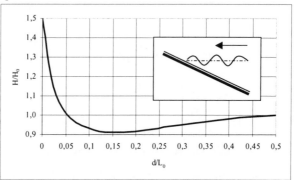

Fig. 3. Wave height theoretical variation as a function of the relative depth

In which regards energy dissipation, wave energy should be dissipated gradually due to the effects of fluid viscosity. This dissipation is negligible in relatively deep water, but cannot be neglected in the shallow region, where a part of the wave energy is transformed due to bottom friction, and the waves are damped in the course of wave propagation. The shear stress on the sea bottom, τ_0, is given by,

$$\tau_0 = f \rho u_0^2 \tag{11}$$

where f is a friction coefficient (0.01–0.02), ρ the seawater density and u_0 the horizontal velocity component of a fluid particle in the vicinity of the bottom.

Taking two sections I and II, Δx (m) apart, and assuming a uniform water depth, the following energy equation can be established based on the shear stress definition,

$$\frac{d(\overline{E}cn)}{dx} = -D_f \quad , \quad D_f = \frac{4\pi^2}{3} \frac{\rho f H^3}{T^3 senh^3\left(\dfrac{2\pi d}{L}\right)} \tag{12}$$

where D_f is the dissipated energy at the bottom, per unit area and per unit time.

Solving the above equation, the damping coefficient, K_f, will be equal to,

$$K_f = \frac{H_2}{H_1} \approx \left[1 + \frac{64}{3}\frac{\pi^3}{g^2}\frac{f H_1 \Delta x}{d^2}\left(\frac{d}{T^2}\right)^2 \frac{K_S^2}{senh^3\left(\dfrac{2\pi d}{L}\right)}\right]^{-1} \tag{13}$$

where H_1 and H_2 are the wave heights at sections I and II respectively and K_S the shoaling coefficient.

The process of wave transformation could lead with the possibility of a certain potential energy increasing, during the process of shoaling and run-up, whilst the kinetic energy could decrease or not. The relation between these two energies depends on the degree of energy dissipation that occurs on the slope.

In the present work detailed measurements of the velocity field due to regular wave action, near different types of submerged breakwaters has been made. Those measurements allowed wave energy values calculation as a function of the distance to the submerged breakwater, whose analysis has been made in order to have a better understanding of the wave-structure interference phenomena.

EXPERIMENTAL SET-UP

The measurements were carried out in the unidirectional wave tank of the Faculty of Engineering of the Porto University, figure 4, with 4.8 m wide, 24.5 m long, a water depth of 0.40 m and a piston type wave generator, figure 5. Wave probes were used in order to register instantaneous water surface elevations. An Argon-Ion Laser Spectra-Physics Stabilité 2017S operating in single-mode with 2 Watts of power was used for the velocity measurements. The optical system consisted of 55X modular LDA optics based on a Dantec fibre optic system, with a 60 mm probe, working in a backscatter configuration. A frequency shift of 0.1 MHz was imposed on the Bragg Cell system and a 600 mm front lens was used, with an expander 55x12 (Dantec) placed before the front lens. The scattered light was collected by a photomultiplier and a TSI 1990C counter processed the signal. A 1400A Dostek card and an analogue digital converter module were respectively interfaced with the counter and the wave probe monitor of the probe located at the measuring section.

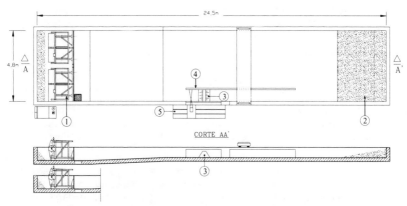

Fig. 4. Plant and section of the wave tank (LH-FEUP)
(1 – Wave generator; 2 – Dissipation beach; 3 – Submerged breakwater
model; 4 – Thin dividing wall; 5 – Table of LDA coordinates)

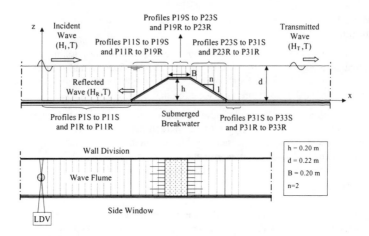

Fig. 5. Wave tank test section and location of measured profile

Both signals (velocity and water level) were acquired simultaneously, whenever the counter validated the velocity. The main characteristics of the LDA system were: wavelength – 514.5 nm; measured half angle of beams in air – 3.487°; dimension of control volume in air – major axis equal to 1.507 mm and minor axis equal to 91.9 µm; fringe spacing – 4.229 µm and frequency Shift – 0.1 MHz.

Using the Laser Doppler Anemometry technique, measurements of horizontal and vertical velocities were made, together with the simultaneous recordings of the corresponding water levels. Phase dependent velocity flow fields have bean measured, which enabled the calculation of the different energy values.

Two types of submerged breakwater models have been tested. Those models had similar dimensions, which are shown in figure 5, but they had different characteristics in which concerns surface roughness. One model was built with a smooth surface and the other with scaled rough blocks.

RESULTS AND DISCUSSION

The usual measured velocity profiles generated by the submerged breakwater, in a particular fixed section do not show the total flow evolution, neither how the transmission is processed (Taveira Pinto et al., 1999).

For this reason measurements in several sections were made, according to schema of figure 5, to allow the analysis of the wave energy propagation, generated by the flow per unit of wavelength, considered as the lake of time equal to a wave period on a fixed section and not a forward wavelength. The location of those profiles is indicated on table 1. Figure 6 shows some examples of phase dependent velocity measurements on different sections, which allow energy calculations.

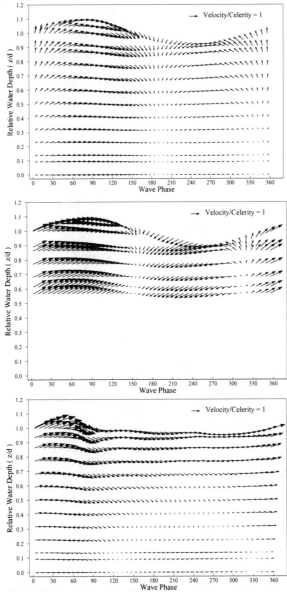

**Fig. 6. Phase-dependant regular wave velocity field
(Smooth model – profiles P10S, P16S and P33S)**

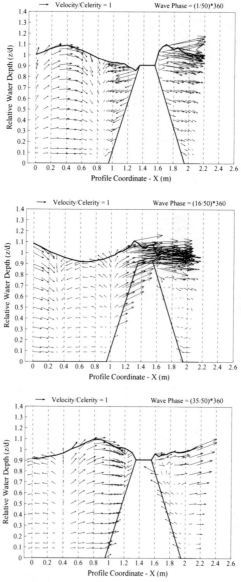

Fig. 7. Spatial phase-dependant regular wave velocity field (smooth model)

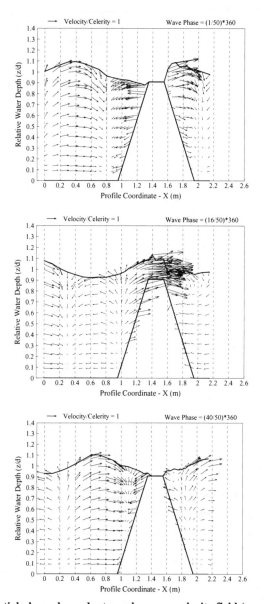

Fig. 8. Spatial phase-dependant regular wave velocity field (rough model)

Table 1. Location of measured profiles

Profile Number	Measured Profile (Smooth Model)	Measured Profile (Rough Model)	Profile Distance ΔX (m)
1 to 10	P1S to P10S	P1R to P10R	0.10
10 to 32	P10S to P32S	P10R to P32R	0.05
32 to 33	P32S to P33S	P32R to P33R	0.10

The results of the phase dependent velocity measurements of all analyzed sections allowed the definition of the flow field around the submerged breakwater in 50 different situations, corresponding to 50 different phase values. In figure 7 and 8 spatial and phase dependent velocity fields for smooth and rough breakwater models are shown.

The measurements were made for a water depth of 0.22 m and as mentioned with a smooth and a rough model of 0.20 m height, 0.20 m of crest width and a 1:2 slope on both sides. The regular waves had a 0.035 m height and a period of 1.20 s. The figure 7 and 8 show the measured mean phase dependent velocity flow field normalized by the wave celerity equal to 1.3 m/s.

The analysis of these wave flow fields shows the existence, in certain moments, of high velocities over the submerged breakwater crest, which lead to a high wave energy concentration in small water depths. In other hand there are also some moments, when over the breakwater slope occurs a reverse flow and high values of vertical velocities, as well as the formation and growing of a turbulent eddy zone. On the smooth model in particular, this region moves upwards the breakwater slope, increasing the turbulence at the water surface.

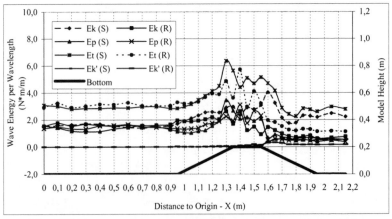

Fig. 9. Kinetic, potential, total and kinetic turbulent energies variation

This information could have a wide importance in the stability analysis of the submerged breakwater blocks. For each model and for each measuring section the kinetic, potential, total and kinetic turbulent energies were determined, figure 9.

The kinetic energy at the seawards side, on the beginning of the crest is slightly constant and similar for the two models. In reality there is an oscillatory variation due to the partial standing wave field, more evident in the smooth model. When the flow reaches the crest and as for the used tests conditions (d=0.22 m, h=0.20 m, H=0.035 m; T=1.20 s) breaking does not occur the reduction of the flow field section lead to a significant increasing of the wave velocities. So, the kinetic energy reaches a relative maximum at the beginning of the crest structure, in different sections of the smooth and rough models. The kinetic energy in this region is greater for the smooth model than for the rough one, due to the fact of a greater energy dissipation caused by the surface roughness. The kinetic energy variations verified are a result of the interaction between the incident oscillatory flow and the related reflux over the crest, verified on the model. This phenomena is stronger on the smooth model, producing a maximum at the end of the crest, due to the greater acceleration achieved by the flow when it reaches the seaward slope.

Comparing the variation of the kinetic energy values for both model tests it can be observed that, for the last landward section analyzed, the kinetic energy value for the smooth model is greater than the seaward kinetic energy value, being reduced to approximately 54% of this value for the model with a rough surface. The analysis of the potential energy shows the existence of an oscillation corresponding to the nodes and antinodes, seawards the structure, defined respectively by the maximum and minimum values. Through the location of these sections and considering a linear analysis, it's possible to calculate the relative phase of the reflected wave, θ. So, considering that $X = 0$ at the beginning of the structure slope, as at the node the total phase $2kX+\theta$ is equal to π, as $X_{ROUGH} \cong 0.35$ m and $X_{SMOOTH} \cong 0.25$ m, then for $L=1.554$ m, $\theta_{ROUGH} \cong 18°$ and $\theta_{SMOOTH} \cong 65°$, what shows the greater reflection ability of the smooth submerged breakwater model.

When the flow reaches the top of the seaward slope, there is a significant increasing of the potential energy, due to the shoaling phenomena, as explained before, related with the corresponding wave height growing and with the no existence of breaking. That potential energy increase is smaller for the rough model due to the greater loose of energy on its surface. For this reason the maximum potential energy occurring over the crest model is greater for the smooth model and do not correspond to the same crest section.

The variations observed on the potential and kinetic energies over the crest, could be explained not only by the measuring process, but also by the interaction over the crest between the incident wave flow and the reflux of the flow from landward, added by the different roughness of the models. For the potential energy there's a mean reduction, regarding the potential energy observed seawards, of 63% for the smooth model and of 65% for the rough one, that corresponds in terms of wave height, to a mean reduction of 38% and 41%, respectively, having only as reference the analyzed landward sections.

The variation of the total energy, being the sum of the two components previously analyzed, presents for that reason the characteristics already mentioned. In mean terms it's verified a reduction of 24% for the smooth model and 61% for the rough one, that corresponds in terms of wave height to a mean reduction of 2% and 38%, respectively.

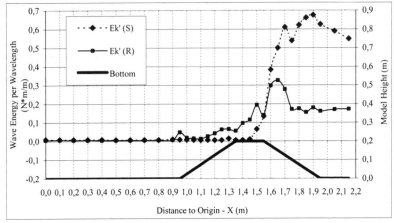

Fig. 10. Kinetic turbulent energy variation

Finally the analysis of the kinetic turbulent energy, figure 10, shows that it is slightly constant in the seaward area before the model, and similar for both of them. It stands increasing over the slope for the rough model and over the crest for the smooth one. At landward, the kinetic turbulent energy increases and is higher on the smooth model than the rough one, due to the smaller energy dissipation and consequent greater volume of water that reaches this area, achieving a greater turbulence.

To note also that, in both models, there's a maximum of the kinetic turbulent energy, value that follows another smaller relative maximum, with approximately the same lag. Taking in consideration that the kinetic turbulent energy in the rough model reaches its maximum before the maximum of the smooth one, sooner it remains constant at landward, what is not possible to conclude in the smooth one.

CONCLUSIONS

Spatial wave flow velocity measurements were taken in order to determinate the variation of the potential, kinetic, total and kinetic turbulent energies.

The analysis of the velocity flow fields showed the existence of high values over the slope and on the submerged breakwater crest, which could be related with the block stability analysis. The differences between smooth and rough models are important in what concerns the wave energy dissipation.

This aspect can also be verified through the wave energy components showing other important phenomena, like shoaling effect and energy transfer between potential and kinetic energies. It's possible to quantify the mean transmission coefficient and the related difference according to the roughness of the models. The roughness of models leads also to the generation of turbulence, mainly in the landward side of the model. The behavior of the two models is similar, but the turbulence generated by the rough model gives a quicker dissipation of the incident wave energy.

The presented work allows not only the analysis of breakwater energy dissipation but can provide data for calibration of numerical models, in order to study other submerged breakwater configurations and characteristics.

ACKNOWLEDGEMENTS

The authors wish to thank to FCT (Foundation for Science and Technology) for its support under the project PRAXIS/C/ECM/ 11303/1998.

REFERENCES

Abbott, M. B. and Price, W. A. Editors, 1993, *Coastal, Estuarial and Harbour Engineers' Reference Book*, E&FN SPON, London, UK.

Taveira-Pinto, F., Proença, Maria Fernanda & Veloso-Gomes, F. 1998. Energy dissipation study of submerged breakwaters using velocity measurements. *Proc. of Ninth International Symposium on Applications of Laser Techniques to Fluid Mechanics*, Lisbon, Portugal.

Taveira Pinto, F., Proença, Maria Fernanda e Veloso Gomes, F, 1999. Dissipation Analysis on Submerged Breakwaters Using Laser Doppler Velocimetry, for Regular and Random Waves, *Proc. of the 8th Int. Conf. On Laser Anemometry Advanced and Applications*, Roma, Italy, 253-265.

Taveira Pinto, F.; Proença, M.F. e Veloso Gomes, F., 2000. Experimental analysis of the energy reflected from submerged breakwaters, Iñigo J. Losada, Balkema, *Proc. of the Int. Conf. Coastal Structures '99*, Santander, Espanha, Vol. 2, 683-688. ISBN 90 5809 092 2.

LABORATORY MODEL STUDIES OF WAVE ENERGY DISSIPATION IN HARBORS

Steven J. Wright[1], M. ASCE, Donald D. Carpenter[2], A.M. ASCE
and Amy L. Cunningham[3], S.M. ASCE

Abstract: Wind waves may be amplified in many Great Lakes harbors due to reflection and resonance phenomena. Although the short period waves do not excite the harbors at their fundamental periods, wave amplification in many situations has been sufficient to pose navigational concerns. It is useful to dissipate a portion of the wave energy in order to reduce wave heights to acceptable levels. We have performed numerous laboratory studies at the University of Michigan Hydraulics Laboratory utilizing stone to provide energy dissipation in order to reduce wave activity to an acceptable level. These studies have considered a wide array of stone size and use from large armor stone that would be statically stable under design waves down to much smaller stone confined within gabions. It is likely that a significant portion of this energy dissipation occurs due to fluid motion in the pore spaces. We have systematically varied the stone size in laboratory models for both enclosed harbors and jettied harbor channel entrances. These experiments have indicated that the reduction in wave amplitudes is independent of the stone size. It is therefore concluded that the results of the small-scale laboratory model tests may be applied to estimate prototype harbor behavior for typical Great Lake harbor settings. General results of these studies are discussed with respect to the general effectiveness of different configurations for reducing wave amplitudes.

INTRODUCTION

A number of existing small craft harbors on the Great Lakes have experienced difficulties with excess wave activity posing navigational hazards and other problems associated with harbor function. The forcing is primarily due to wind

Professor of Civil and Environmental Engineering, University of Michigan, Ann Arbor, Michigan, 48109-2125, sjwright@engin.umich.edu
Assistant Professor of Civil Engineering, Lawrence Technological University, Southfield, Michigan 48075-1058, carpenter@ltu.edu
Undergraduate Student, Civil and Environmental Engineering, University of Michigan, alcunnin@engin.umich.edu

waves in the range of 4-8 seconds. Two primary types of harbor configurations have been investigated in various studies to mitigate wave activity. One is an enclosed harbor typically lined with sheet piling. The reflective surfaces lead to resonant interactions at certain periods. Although the short period forcing resonates at high harmonic modes, the wave amplification may be sufficient to hinder use of the harbor under moderate incident waves. A second situation involves a typical configuration at many of the Great Lakes navigational harbors. These harbors are frequently located in river mouths and provide sufficient internal damping of waves. However, these harbors provide access to the lake through a jettied channel, often several hundred meters in length. If these jetties are constructed of sheet pile or other reflective surface, waves the jetty channel may provide difficult navigational conditions within the jetty channel.

A primary solution to the wave problems in both of these situations is to provide for energy dissipation. A cost-effective solution can often be accomplished utilizing strategically placed stone in order to dissipate wave energy. One problem utilizing loose stone is that it must be placed on a maximum slope on the order of 1V:2H for stability purposes. For even moderate depths, the stone has a large footprint on the harbor, leading to concerns for small craft potentially impacting on the stone. This has led to the use of two different configurations to address this issue. We refer to one configuration as pocket wave absorbers; stone is placed in the recess of a jetty or harbor wall thereby limiting exposure during navigational use of the harbor. The second utilizes stone-filled gabions to confine the stone to vertical faces in unobtrusive locations within the harbor. This paper reports on the experiences gained in a variety of studies performed with one or the other of these two wave mitigation solutions. One particular issue was the effect of stone size on wave energy dissipation. Results are included to indicate this influence as well as other factors that play a role in wave height reduction.

JETTIED CHANNEL ENTRANCES

The use of pocket wave absorbers in the Great Lakes region was initiated by the Detroit District of the U.S. Army Corps of Engineers and has been implemented at five Great Lakes harbors. The typical approach is to use recessed pockets on the order of 60 m long and 12 m deep. Most installations have pockets on both of the jetty walls directly opposite each other. Stone size has been on the order of 2-4 tons. A laboratory study was conducted of a typical jettied with a length of 500 m, a width of 60 m, and a depth of 5 m. The hydraulic model was based on a scale ratio of 1:50 with Froude Number scaling to define the dynamic similarity requirement. The laboratory model was constructed of parallel rows of concrete blocks that could be easily adjusted to examine various geometries. Figure 1 provides a schematic of a typical test layout. Wave absorbing material was placed at the back of the jettied walls to simulate the wave absorbing conditions in a typical river mouth harbor found at these sites.

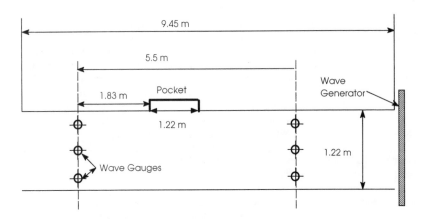

Figure 1. Basic Pocket Wave Absorber Model Layout.

Monochromatic waves were generated with a plunger-type wave generator and most experiments were conducted with the wave rays parallel to the channel axis as this produced the least amount of wave energy dissipation. In order to avoid difficulties in interpreting results due to wave breaking, wave heights were set to just below breaking and were held constant through any particular set of experiments. Wave heights were measured with capacitance wave gauges. Initially, only one probe was placed on either side of the pocket. However, the results of initial investigations showed a considerable variation in wave height across the channel on the downstream side of the pocket and subsequent experiments were performed with three wave gauges spaced across the width of the pocket. The wave heights from these three gauges were averaged to provide a more accurate representation of the wave energy flux. The wave height computed for each gauge is actually the standard deviation of the water surface displacement in order to potentially account for any out-of-phase reflected wave components. Wave records indicated progressive waveforms with nonlinear effects of steep crests and flattened troughs but no reflected or standing wave components were apparent. Results are generally presented as a percent dissipation that is the difference between the incident and downstream waves normalized by the incident amplitude. A reduction in wave amplitude was observed even in the absence of a pocket, presumably due to wall and bottom friction. For the standard configuration, this reduction was approximately 4-5 percent.

ENCLOSED HARBOR STUDIES

This paper only presents results for a generic rectangular basin 1.8 m deep by 0.5 m wide by 0.11 m deep. The harbor opening was the full width of the basin. The same wave generation and measurement techniques described above were employed in this study with the waves entering the harbor parallel to the longitudinal axis. The frequency on the wave generator was initially varied continuously to define those wave periods that resulted in high wave amplification within the harbor.

Most of the experiments were performed for a period of 0.512 s, corresponding to a fifth longitudinal resonant mode for the harbor. Testing was performed to determine the effect on wave amplitude at this and other resonant conditions. Results are presented in a *before* (wave heights measured at a given location with the unmodified harbor) and *after* (some manipulation such as the placement of a stone pocket or gabion) comparison.

EFFECT OF STONE SIZE

A major effort was made to investigate the effect of stone size on the reduction in wave amplitudes for both harbor configurations. Stone of a particular size was prepared by mechanical sieving of crushed limestone. Stone size is reported as the mesh size on which the stone was retained. In particular studies, various mixtures of the different stone sizes were prepared in order to determine the role of size gradation. A set of consistent experiments was performed with the jetty entrance model in which all wave and geometric variables were held constant except for the stone size in the pocket. Figure 2 presents the results for four different uniform stone sizes as well as a mixture of equal portions of the three smallest stone sizes. Within the uncertainty of the measurement, the percentage of incident energy dissipated was the same for all the uniform stone sizes and slightly higher for the mixture. Experiments were also performed over a range of incident wave periods for two different stone sizes; results are presented in Figure 3 with the dissipation percentage presented as a function of the pocket length to wave length ratio. The results in Figure 3 are for a single downstream probe at the center of the channel and so do not reflect the true total energy dissipation. The lack of dependence on stone siae is consistent with the earlier study by Bishop (1987) who investigated wave propagation through rubble lined jettied channels.

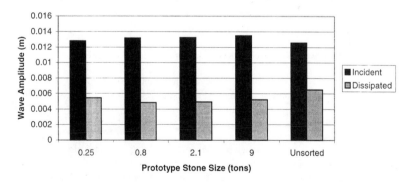

Figure 2. Incident and Dissipated Wave Amplitudes for Various Stone Sizes.

A similar result was found in the confined harbor study where four different uniform stone sizes were investigated. A stone pocket was placed on the sidewall at the back of the harbor and filled with stone placed on a 1V:2H slope to above the still water level. The pocket length was systematically varied at equal intervals of approximately 0.2 m up to a maximum length of 0.61 m. Figure 4 indicates the

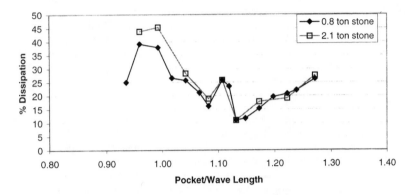

Figure 3. Dissipation for a 0.81 m long pocket over a range of frequencies.

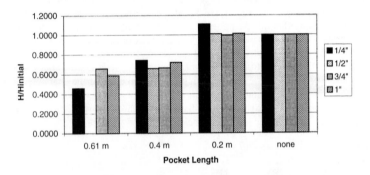

Figure 4. Wave Height Reduction as Function of Stone Size.

experimental results for these cases. The resonant condition introduced is sensitive to small changes in harbor geometry and small differences in experimental conditions can have a significant impact on the measured wave amplitude. There is no trend to the variations in wave response with stone size, again indicating a lack of dependence on stone size.

There may be a concern in extrapolating small-scale laboratory measurements on dissipative processes to field applications due to the enhancement of viscous effects in a Froude scaled model. There are various potential mechanisms for wave height reduction in these experiments including dissipation due to surface roughness, internal dissipation within the pore spaces in the stone, flow separation at the front edge of the pocket and scattering of the wave energy due to diffraction into the pocket. It is difficult to isolate the relative importance of these individual effects. Since the results in Figure 2 indicated a potential difference in dissipation for the mixture of stone, it was considered that the porosity of the stone revetment might be an important factor in energy dissipation. A series of experiments were performed to systematically vary porosity. The uniform stone sizes all had a porosity on the order of 0.49. By various packing arrangements, the porosity could be varied over the range of 0.39 to 0.59. These involved mixtures of stones as well as special arrangements to create greater pore space by placing wire mesh screens between layers of stone. Figure 5 presents the results obtained at three different wave frequencies. The zero porosity result is for a sheet of Plexiglass placed on a 1:2 slope at the location of the stone interface. There is a clear trend of increased dissipation with increasing porosity indicating that motion within the pore spaces plays a significant role on energy dissipation. Wave frequency over the range tested did not have a major effect on energy dissipation, although for jettied channel experiments with a single pocket, energy dissipation increased somewhat with increasing wave frequency while two pocket configurations showed a slight decrease in dissipation with increasing frequency.

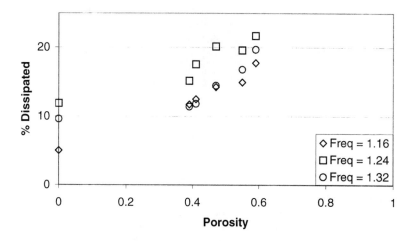

Figure 5. Effect of Porosity on Wave Dissipation at Various Frequencies.

Flow in porous media should be dependent on the characteristics of the pore spaces. The permeability for steady flow is generally assumed to be proportional to the square of a characteristic pore dimension, indicating a size effect (Bear, 1979). However, turbulent flow within the pore spaces is expected to prevail when Reynolds numbers based on the Darcy velocity and pore dimension are greater than about 100-300. Estimating a characteristic velocity as the maximum horizontal orbital velocity and using the stone diameter as the characteristic dimension yields Reynolds numbers in excess of 1200 for the smallest stone size studied. Since pore velocities will decrease with distance into the stone, Reynolds numbers should decrease until viscosity influences the pore velocities. An additional set of experiments was performed examining the effect of wave penetration into the stone. For a single uniform stone size, sheets of flexible plastic were placed into the stone section at different distances from the surface to limit water motion at that location. According to the results presented in Figure 6, after a penetration between one to three stone layers, the presence of the plastic sheet did not affect energy dissipation. Therefore the wave motion is restricted to a thin layer close to the stone surface. Additional tests for different stone sizes are planned to confirm this result. It is concluded that viscosity does not play a significant role in the wave energy dissipation and that the laboratory results are consistent with what would be observed at the field scale. Field observations at jettied harbor entrances with pockets similar to those studied in this investigations yielded wave height dissipation percentages consistent with those observed in the laboratory experiments (Carpenter, 2001).

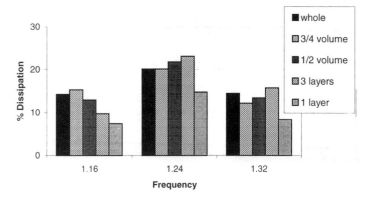

Figure 6. Effect of Reducing Stone Volume on Wave Energy Dissipation.

GABION STRUCTURES

Stone-filled gabions have been installed in several small craft harbors to mitigate wave amplification due to resonance at relatively shortwave periods. A common theme in most of these installations was the retrofitting of an existing harbor with tight space constraints. Typical solutions have involved gabions of 1 m by 1 m cross-section and two or three horizontal gabions stacked vertically have been used to dissipate wave energy. Common arrangements have been to hang the gabions on existing sheet pile walls or to place them beneath dock structures projecting into the harbor. The latter arrangement has been found to be more efficient in dissipating wave energy. Confirmation can be obtained from the data presented in Figures 7 and 8. Figure 7 presents results for the jettied channel configuration for four cases. The results for a standard pocket configuration are included as well as observations when no energy dissipation system is present. Two additional cases are presented with gabion structures affixed to a channel wall. In one configuration, the gabion was placed in a very small recess so that the outside of the gabion is just flush with the channel wall; the energy dissipation is only slightly greater than if the stone is not present. Placing the gabions along the back wall at the standard pocket depth increases the dissipation indicating that the presence of the pocket has a positive impact on wave height reduction. Figure 8 presents results when the gabions are projecting from the back wall of the pocket. Two gabions with a length equal to the pocket depth were placed within the pocket at the one-third points along the length. Results for three different wave periods are presented in Figure 8 and indicate comparable dissipation rates to the standard pocket configuration. The variation in results at different periods is primarily due to the fact that there is substantially more wave reflection off the vertical projections.

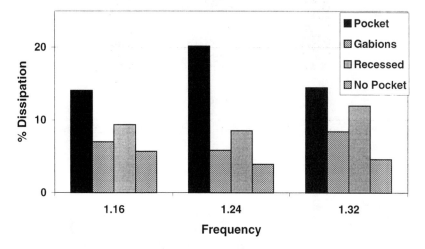

Figure 7. Dissipation for Gabions Installed Along Solid Boundaries.

The effect of projecting gabions can also be observed in Figure 9; this figure presents the findings for a single gabion one-sixth of the harbor width projecting from one wall of the confined harbor model. The single variable is the longitudinal location of the gabion along the harbor. Dissipation rates vary considerably with location compared to the pocket wave absorber results, and can be explained on the basis of the gabion location relative to longitudinal nodes and anti-nodes within the harbor. Figure 10 indicates the location of the gabions for the various experiments relative to the estimated locations of nodes and anti-nodes. A clear pattern of greater dissipation at the anti-nodes where horizontal water particle motions are concentrated as compared to the nodal locations where motions are constrained to the vertical is indicated. This finding also supports the previous conclusions that interactions within the stone pore spaces are important for wave height reductions.

The significance of the gabion investigation is that reductions in wave amplitudes comparable to those observed with the pocket configurations can be achieved with much smaller stone volumes provided that the gabions project into the harbor as opposed to placement along harbor walls. The smaller stone volumes may result in economic benefits as well as more efficient space utilization if the gabions are located beneath existing docks or other similar locations that do not interfere with harbor functions.

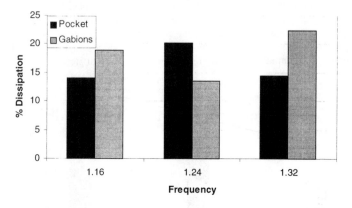

Figure 8. Dissipation for Pocket Wave Absorber and Projecting Gabions.

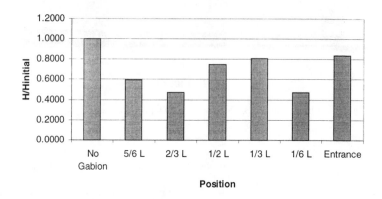

Figure 9. Wave Height Reduction for Projecting Gabions at Various Locations.

CONCLUSIONS

Additional laboratory results as well as the findings from a field investigation at several harbors that have had pocket wave absorbers installed may be found in Carpenter (2001). Several conclusions may be drawn from the findings presented in this paper, including:

- Stone size does not play a role in the reduction of wave amplitudes for the experimental configurations investigated. This conclusion is also supported by the findings of a limited number of prior studies by other researchers.
- The stone porosity has a clear effect on wave dissipation with increasing dissipation as the stone porosity is increased. Practical limitations on

achievable porosities in field installations controls design considerations, but one implication is that a single size of stone would be preferable to a graded stone due to the larger achievable porosity.

• Gabion structures are capable of achieving wave amplitude reductions comparable to the pocket wave absorbers with an order of magnitude less stone required. However, the gabions should project from harbor walls as opposed to placement along the walls.

• In confined harbors subject to resonant conditions, placement of gabions at the locations of anti-nodes is significantly more effective in reducing wave amplitudes than if placed at nodal locations. This indicates the necessity of a detailed investigation of resonant interactions within the harbor at typical wave periods.

These conclusions were made on the basis of studies on relatively short period wind waves and no basis for extrapolating the findings to longer period wave phenomena is available.

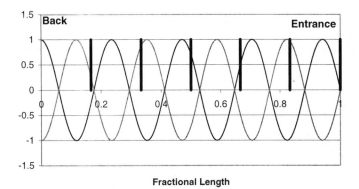

Figure 10. Location of Standing Wave Nodes and Antinodes.

ACKNOWLEDGEMENTS

The Detroit District of the U.S. Army Corps of Engineers supported a portion of the research conducted on pocket wave absorbers; James Selegean was the contact person on this contact. Vivian Lee performed the experiments on the confined harbors and the following undergraduate students assisted with the research: Katherine Sultani, Russell Sieg and Justin Voss.

REFERENCES

Bear, J. 1979. *Hydraulics of Groundwater*, McGraw Hill, 567 pp.

Bishop, C.T. 1987. Wave Attenuation by Rubble-lined Channels, *Canadian Journal of Civil Engineering*, 14:828-836.

Carpenter, D.D. 2001, *Wave Energy Dissipation in Great Lake Harbor Entrances*, Ph.D. dissertation presented to the University of Michigan, Ann Arbor Michigan.

EXPERIMENTAL INVESTIGATIONS OF WAVE PROPAGATION OVER THE SWASH ZONE

Renata Archetti[1] and Francisco Sancho[2]

Abstract: The paper will present the experimental set up and description of the spatial evolution of waves propagating in the swash zone. Wave and velocity measurements performed in the LIM laboratory of the UPC University are described. Velocity profiles are acquired with a Doppler Velocimeter Profiler, which enables simultaneous high frequency velocity profiles measurements over the whole swash zone. Advantages of this new technique with respect to more traditional ones are presented. Velocity profiles are used to estimate bed shear stresses in the swash and to estimate the friction factors.

INTRODUCTION

The swash zone (hereinafter, SZ) is that part of the beach over which the instantaneous shoreline moves back and forth as waves meet the shore. For many incident wave conditions it is also the region of maximum sediment transport and bed changes. Experimental and field measurements aimed to a detailed study and description of hydrodynamics and waves in the SZ are less frequent than experimental studies on the inner surf zone (Baldock et al., 1997). In most cases, the *set-up* is the only measured quantity. Difficulties of acquiring velocity profiles in this zone are due to the intrusion of available instruments, such as Electromagnetic Current Meters (ECM), Acoustic Doppler Velocimeters (ADV) and Laser Doppler Velocimeters (LDV) on the SZ thin film of water (fluid field). In general, they allow to measure at a defined single position, therefore, some assumptions must be made about the vertical distribution of the velocity in the swash. Furthermore, Laser Doppler Velocimetry need clear water while acoustic instrumentation have the advantages to work properly in presence of seeding and or small bubbles.

---

1 Research Assistant, D.I.S.T.A.R.T. University of Bologna. Viale Risorgimento, 2, 40136 Bologna ITALY. Renata Archetti@mail.ing.unibo.it.
2 Assistant Professor, Dept. Civil Eng., University of Coimbra, Polo II, Pinhal de Marrocos, 3030-290 Coimbra, PORTUGAL. fsancho@dec.uc.pt

Recent SZ measurements are due to Petti & Longo,(1998) and Cox & Hobensack (2000),they performed some measurements of the surface elevation, mean and turbulent velocities in the swash with a Laser Doppler Velocimeter (LDV), and assumed a logarithmic vertical velocity profile in their analysis.

The subject matter of this study is concerned with the spatial evolution of waves (or bores) propagating over the swash. Laboratory experiments were performed in the CIEM Large Wave Flume at the Polytechnic University of Catalonia (Sancho et al., 2001): an extensive program of measurements of instantaneous velocity profiles, surface elevation and turbulence levels in the swash zone was performed using an Acoustic Doppler Velocimeter Profiler, (ADVP hereinafter). The innovative characteristics of this set up relies upon the simultaneous high frequency velocity profiles measurements over the whole swash zone, which were obtained with probes encapsulated in cubes of PVC installed in the bottom of the flume (see Figure 1) in order to have non-intrusive measurements. An ADV was used for the calibration of the ADVP signal. The position of the instruments during the tests is shown in Figure 2.

This paper presents the experimental set up and description of the spatial evolution of waves propagating in the SZ. In section 1, the experimental set-up is described. In section 2, velocity measurements acquired simultaneously with an ADVP and an ADV are compared in order to validate the data. In section 3, vertical velocity profiles are shown and bed shear stresses and friction coefficients are estimated.

EXPERIMENTAL SET-UP DESCRIPTION

The test were performed at the CIEM wave flume, which is 100 m long, 3 m wide and 5m deep. A hard beach was constructed by filling the flume in (from the bottom up to the desired height) with compacted sand and then by pouring mortar (soft concrete) on the top. The final surface resulted in a non-smoothed mortar layer. The surface roughness is estimated to be equal to that of coarse sand grains.

The rigid bottom profile was designed to match an equilibrium bar (Sancho, 1999). The cross-shore bottom profile is shown in Figure 2, and waves propagate from right to left (the wave-maker is positioned at x=86 m).

The complete experimental programme covered four wave conditions (for details, see Sancho et al. 2001). For the present analysis, we selected the regular wave condition C, which originated the wider swash zone, with an incoming wave height, H=0.38 m, and wave period, T=3.5 s.

Fig. 1 – View of the wave flume at LIM

Several instruments were placed during each test, being moved from one location to another, after each run. The following is a list of all the instruments used during the experiment for the present analysis, and described in detail below:

- Wave gauges (WG);

- Acoustic Doppler Velocity Meters (ADV);

- Acoustic Doppler Velocity Profilers (ADVP);

Data were acquired with the following acquisition frequency WG: 8 Hz; ADV: 25-50 Hz; and ADVP: 20-60 Hz

The DOP1000 ADVP, by Signal processing S.A. (www.signal-processing.com), was used to measure the fluid velocity at several sampling volumes along the beam of a piezoelectric transducer which works as both emitter and receiver. This technique offered the possibility of collecting spatio-temporal information of the flow field over a single line. In the present setup, this measurement line made an angle of $\alpha = 60°$ with the local bottom surface. Further details of the instruments specifications are given in Sancho et al., 2001.

A run-up wave gauge, seen at the foreground in Figure 1, was positioned at the bottom of the flume in the swash zone. Wave gauges were placed along the flume and

ADVP probes were positioned at the bottom of the wave flume, along a longitudinally-centred trench, 0.14 m wide. Position of instrumentation during the tests is sketched in Figure 2

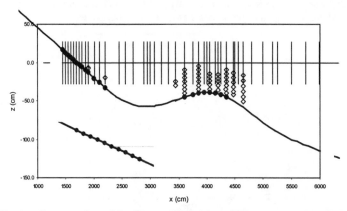

Fig. 2 – Cross section of the flume and position of Wave gauges (+) ADV (◊) and ADVP probes (•). Little figure identify the swash zone.

The ADVP probes were encapsulated in a PVC support. The support for the 4 MHz probes, used in the swash zone, were open at the upper face so the probes were directly exposed to the water, avoiding any refraction effect of the acoustic beam. In figure 3, it can be seen the set up of the ADVP probes in the trench of the flume.

Fig. 3– Trench at the bottom of the flume, the ADVP probes are encapsulated in the red blocks

VALIDATION OF ADVP MEASUREMENTS

Simultaneous measurements of velocity were gathered by the ADV and the ADVP instruments, at the slope of the beach, as closest as possible to the swash zone (x=19 m). The velocity magnitudes projected in the direction of the ADVP measurement line were compared and the correlation between data is r=0.89 during back-wash.

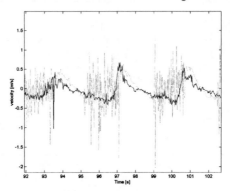

Fig. 4– Simultaneous measurements of velocity gathered by the ADV (dotted line) and ADVP (solid line)

For all others phases, the ADV data could not be considered for the comparison, signal shows a high frequency noise, probably arising from air entrainment and /or from the vibration of the instrument (Figure 4). Others comparisons between ADV and ADVP data were performed at another section (position x=22 m, Figure 1) and the correlation was good with a high correlation coefficient for all phases.

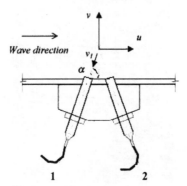

Fig. 5 – ADVP probes reference system

ADVP probes measure the velocity projected in the direction of the probe axis. In order to calculate the velocity components of the flow, two probes were coupled as shown in figure 5.

Velocity along the beam axes are:

$$v_1 = -u \cos\alpha - v \sin\alpha \tag{1}$$
$$v_2 = u \cos\alpha - v \sin\alpha \tag{2}$$

In the swash zone the water level was so low that the intersection between the beam axes was at a position higher than the water level. In this case, to evaluate the flow velocity, some assumptions had to be taken regarding the direction of the flow.

Direction of the flow was measured at the lower limit of the swash, where the water depth was larger, with the two ADVP crossing probes setup, as described above. In Figure 6 we see the flow direction θ (defined as the angle between the bottom surface and the 2DV flow direction) at the lower limit of the swash: except when the flow swaps from up-rush to back-wash, flow direction is for all phases $-5° < \theta < 5°$.

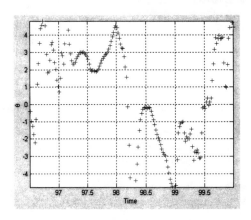

Fig. 6 – Mean direction of the flow in the swash during a wave cycle

As conclusion of the preliminary analysis we assume to have valid data acquired by the ADVP and that the flow is parallel to the bottom except during transition between up-rush and back-wash; flow velocity in the swash was calculated as (see figure 7):

$$u = v_1 / \cos\alpha \tag{3}$$

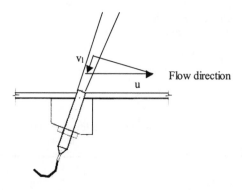

Fig. 7 – Reference system in the swash zone

VELOCITY PROFILES AND BED STRESSES

In order to estimate the bed stresses and to estimate the velocity profiles in the swash zone, measurements in section x=22 m to x=15 m were used. Velocity was measured by an ADVP by a spatial resolution of 0.9 mm and a sampling rate of 24 Hz.

Our measurements indicate that the logarithmic profile, valid in the bottom boundary layer of steady flows, can also represent the velocity profiles of unsteady flows like those occurring in the SZ. These occurred for most all flow-phases with the only exception occurring in the transition between up-rush and backwash. For these phases of flow reversal, the velocity profile cannot be adequately represented by a logarithmic law. Similar findings have also been obtained by Cox and Hobensak (2001), while for a detailed analysis of the vertical structure of both mean and turbulent flow properties in the SZ we refer the reader to Petti and Longo (2001).

Two vertical profiles of the horizontal velocity at the lower limit of the swash during backwash and up rush are respectively shown in Figure 8.

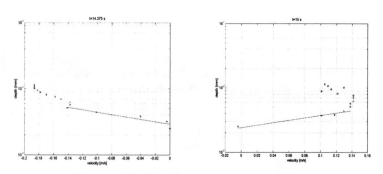

Fig. 8 – Vertical velocity profiles at the lower limit of the swash

The above result suggests that the shear velocity u^* can be computed by means of the procedure summarized below. The logarithmic velocity for a rough bottom is expressed as:

$$u = \frac{u^*}{k} \ln\left(\frac{z}{z_0}\right) \tag{4}$$

where u^* is the shear velocity, z is the vertical coordinate z_0^* the bottom roughness and k is the von Karman constant (~ 0.4).

The shear velocity and the bottom roughness were computed at each phase using a least square procedure. These analysis follows that of Cox *et al.* (1996) while modeling frictional force in the surf zone. The temporal variation of the bed stress was estimated by

$$\tau_b(t) = \rho \cdot u^*(t) \cdot |u^*(t)| \tag{5}$$

with ρ being the fluid density. This can also be expressed in the quadratic form suggested by Grant and Madsen (1979) as:

$$\tau_b(t) = \frac{1}{2} f \cdot \rho \cdot u(t) \cdot |u(t)| \tag{6}$$

where u is the free stream velocity, which in the thin fluid depth of the swash zone does not differ too much from the depth-averaged velocity and *f* is an empirical factor, which if we assume constant for all phases, may be estimated by a least squares fit of (5) and (6). The value of *f* is then obtained as:

$$f = 2 \frac{\sum_{i=1}^{N} \left(u_i^* \cdot |u_i^*|\right) \cdot \left(u_i \cdot |u_i|\right)}{\sum_{i=1}^{N} (u_i)^4} \tag{7}$$

where N is the number of measured phases (we excluded in the calculation phases during transition from up-rush to back-wash). Our results indicate that *f* is equal to 0.02. This range of variation agrees well with that found by Puleo and Holland (2001), who estimated the friction coefficient in the swash zone under a large number of flow conditions.

The quadratic law (6) seems to be well supported by figure 8 in which the temporal variation within the swash zone of both measured and modeled (by eq. 6) seabed stresses are illustrated. Please, notice that moving from the lower boundary of the swash zone (Figure 9) to the final end of the swash zone, dry conditions cover a larger and larger portion of the period, hence the lack of data between the 2s to 3s time period. The alternating wet and dry conditions also seem to introduce a larger data scatter for the stress estimated by means of the shear velocity.

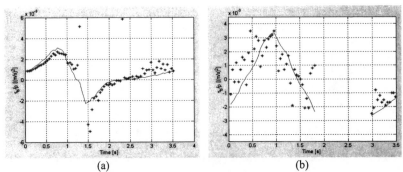

(a)   (b)

Fig. 9 – Temporal variation of sea bed stresses in the swash zone: at the lower limit of the swash zone (a) into the swash zone (b).

CONCLUSIONS

Free-surface and ADVP measurements in the swash zone are presented. The new technique allows to obtain measurements of velocity clean from any instrumental presence.

Simultaneous measurements with an ADVP and ADV allowed to validate the ADVP data and determine the flow direction. The flow direction in the swash is considered parallel to the bottom of the flume. The logarithmic velocity profile, typical of steady flows, represents adequately the instantaneous velocity profiles near the bottom, in the swash zone. The quadratic law gives reasonably good estimates of the measured bed shear stress and the estimated friction factors are in agreement with those found in the literature.

ACKNOWLEDGEMENTS

Experimental tests were funded by the *European Commission, Training and Mobility of Researchers Programme – Access to Large-scale Facilities*, under contract no. ERBFMGECT9500073, and by the *"Laboratorio de Ingenieria Maritima"* (LIM) of the *"Universitat Politècnica de Catalunya"* (UPC). Thanks are extended to Prof. Alberto Lamberti for providing the Acoustic Velocimeter Profiler DOP1000, to Paulo Mendes (from the Univ. of Coimbra) for helping with the ADV installation and to Dr. Maurizio Brocchini for the useful discussions.

REFERENCES

Baldock, T. E., Holmes, P., Horn, D.P. 1997. Low frequency swash motion induced by wave grouping. *Coastal Eng.* 32, 197 – 222.

Cox, D. T. , Hobensack, W. A. 2001. Temporal and spatial bottom shear variations in the swash zone. *J, geophys. Res.* (under peer review).

Cox, D. T., Kobayashi, N., Okayasu, A. 1996. Bottom shear stress in the surf zone. *J. Geophys. Res.* 101, 14337 – 14348.

Petti, M., Longo, S. 2001. Turbulence experiments in the swash zone. *Coastal Eng.* 43, 1-24.

Puleo, J. A., Holland, K. T. 2001. Estimating swash zone friction coefficients on a sandy beach. *Coastal Eng.* 43, 25-40.

Sancho, F.E. 1999. Perfil de equilíbrio de uma praia com barra-fossa. *Proc. of 1st Portuguese Conference on Harbour and Coastal Eng., Porto (in Portuguese).*

Sancho F.E., Mendes, P.A., Carmo, J.A., Neves, M.G., Tomasicchio G.R., Archetti R., Damiani, L., Mossa, M., Rinaldi, A., Gironella, X., Sanchez Arcilla, A. 2001. Wave hydrodynamics over a barred beach. *Presented at WAVES'01 – ASCE.*

WAVE HYDRODYNAMICS OVER A BARRED BEACH

F. Sancho[1], P.A. Mendes[1], J.A. Carmo[1], M.G. Neves[2], G.R. Tomasicchio[3],
R. Archetti[4], L. Damiani[5], M. Mossa[5], A. Rinaldi[5], X. Gironella[6], A. S.-Arcilla[6]

Abstract: This paper reports the experimental Project "SPANWAVE-SPPORITA" carried out at the "Canal de Investigación y Experimentación Marítima" of the Polytechnic University of Catalonia (Barcelona, Spain). The Project goals were to obtain detailed and accurate measurements of turbulent and mean velocities over the bar and trough regions, for regular and random waves breaking over the bar. Four wave conditions were simulated, and both surface elevation and velocity measurements were carried out at a large number of locations. The experiments are considered successful and provide a unique data set on surfzone hydrodynamics over a barred beach. Preliminary data results reveal quite interesting aspects, deserving further investigation.

1. INTRODUCTION

There have been several extensive experiments addressing the mean flow hydrodynamics over barred beaches (e.g., Kraus *et al.*, 1992; Wu *et al.*, 1994), which did not address the turbulence generated by the breaking waves. On the other hand, other experiments covered wave-induced turbulence over laboratory planar beaches, (e.g., Stive, 1980; Nadaoka and Kondoh, 1982; Hattori and Aono, 1985; Okayasu, 1989; Cox *et al.*, 1995) and field monotonic profiles (e.g., George *et al.*, 1994). These experiments comprised a wide range of bottom slopes and wave conditions, including both spilling and plunging breakers.

1 Dpt. Civil Eng., University of Coimbra, Pólo II, Pinhal de Marrocos, 3030-290 Coimbra, Portugal. (fsancho@dec.uc.pt, pamendes@dec.uc.pt, jsacarmo@dec.uc.pt)

2 LNEC, Dpt. Hydraulics, Av. do Brasil, 101, 1700-066 Lisboa, Portugal. (gneves@lnec.pt)

3 Dpt. Civil and Environm. Eng., University of Perugia, via G. Duranti, 93 - 06125 Perugia, Italy. (tomas@unipg.it)

4 D.I.S.T.A.R.T. University of Bologna, Viale Risorgimento, 2, 40136 Bologna, ITALY. (renata.archetti@mail.ing.unibo.it)

5 Dpt. Civil and Environm. Eng., Bari Polythechnic, Via E. Orabona, 4; 70125 Bari, Italy. (l.damiani@poliba.it, mossa@poliba.it)

6 Laboratori d' Enginyeria Marítima, Polytechnical University of Catalonia, Jordi Girona, 1-3, Edif. D1, 08034 Barcelona, Spain. (xavi.gironella@upc.es, agustin.arcilla@upc.es)

None of the above studies, however, analysed turbulence from breaking waves over a bar and wave reforming over the trough, as often happens in nature (e.g., Birkemeyer *et al.*, 1997). Rodriguez *et al.* (1995, 1999) analysed wave-induced macro-turbulence over a barred beach at a field location, but their study suffers from non-simultaneity of the measurements at different cross-shore positions. Moreover, it appears that the sea-state generated a single surfzone, where waves did not reform after breaking over the bar.

In this paper we present the experimental Project "SPANWAVE-SPPORITA" carried out at the "Canal de Investigación y Experimentación Maritima" (hereafter referred as CIEM wave flume) of the Polytechnic University of Catalonia (Barcelona, Spain). The Project goals were to obtain detailed and accurate measurements of turbulent and mean velocities over the bar and trough regions, for regular and random waves breaking over the bar. These measurements provided a unique set of data, allowing one to estimate important hydrodynamic parameters, such as energy dissipation and shear stresses, and to better understand the surfzone dynamics.

Section 2 contains a description of the experimental setup, instrumentation and test conditions. A review of the data analysis parameters is given in Section 3, and preliminary results are presented in Section 4. Finally, in Section 5 we provide a summary and conclusions of the present study.

2. EXPERIMENTAL SETUP
2.1. Facility
The tests were performed at the CIEM wave flume, which is 100 m long, 3 m wide and 5 m deep. A barred beach was built in the flume, topped by a non-smoothed soft-concrete layer, with roughness nearly equal to that of coarse sand grains.

The rigid bottom profile was designed to match an equilibrium bar. This was accomplished by scaling-down prototype profiles at Duck (North Carolina, USA), taking into account the SUPERTANK (Kraus and Smith, 1994) and DELTA-flume (Sanchez-Arcilla *et al.*, 1995) movable-bed experiments, and also by tuning the final "equilibrium-bar" shape with the assistance of a numerical Boussinesq-type wave model (Kennedy *et al.*, 2000) adapted to provide tendencies for onshore/offshore sediment transport (Sancho, 1999). From the above, and comparing with the conditions commonly found at Duck (North Carolina, USA), we consider the present experiment a 1:5 scale of field conditions.

The cross-shore bottom profile is shown in Fig. 1, and the wave-maker is positioned at x=86 m. The following parameters characterize the bottom profile and water depth:
- still water depth at wave maker, $h_0 = 2.05$ m;
- depth at the bar-crest, $h_c = 0.39$ m;
- depth at the bar-trough, $h_t = 0.575$ m;
- still water shoreline position, $X_{shoreline} = 17.0$ m;
- bar-crest to shoreline distance, $X_c = 23.0$ m;
- bar-trough to shoreline distance, $X_t = 12.0$ m;
- beach-face slope=1:15;
- beach-toe slope=1:8;
- mean slope=1:25.

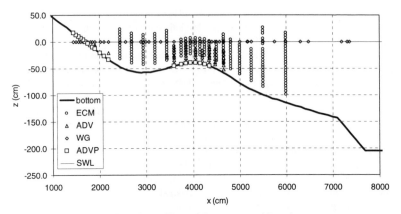

Fig. 1. Beach profile and instruments' locations.

The wave motion is set forward by a wedge-type hydraulic wave generator. A paddle slides up and down in a thirty-degree inclined plane and is controlled by a PC-based wave generating-absorption system, able to eliminate the spurious re-reflections of the wave paddle, for the most energetic wave periods. During the present experiments, the wave absorption system worked quite well for that period range, but was not able to eliminate the *seiching* motions. As an example, Fig. 2 shows the time series of the runup motion for one test, where the low-frequency oscillation is quite visible. *Seiching* periods were identified as $T_1 \approx 55.0$ s, $T_2 \approx 25.6$ s, and $T_3 \approx 19.0$ s.

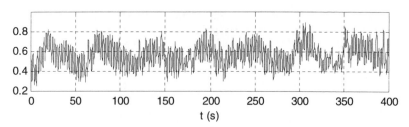

Fig. 2. Time series of the surface elevation at the swash region.

2.2. Instrumentation
A combination of six different equipments was adopted. All instruments were placed primarily in the breaking and post-breaking regions, although other positions were covered as well. Fig. 1 shows the locations for each instrument type. The following is a list of all the instruments, described in detail below:
   - Wave Gauges (WG);
   - Pressure Transducers (PT);
   - Electromagnetic Current Meters (ECM);

- Acoustic Doppler Velocity Meters (ADV);
- Acoustic Doppler Velocity Profilers (ADVP);
- Video and photographic equipment.

The free-surface elevation was registered at 49 different locations by a combination of up to eight resistance-type wave gauges (WG). There were six "standard" one-meter long wave gauges, and three others, specially built to measure the wave conditions nearer the shoreline. Three surface elevation sensors remained the whole time of the experiment at fixed positions, in front of the wave paddle, for repeatability and quality control of the tests. All sensors were mounted from vertical masts standing at the bottom of the flume. At the top of the base-plates of 5 different masts we installed pressure transducers.

Seven spherical S-type Electromagnetic Current Meters (ECM) were used to collect flow velocities at the same sampling frequency as the WG measurements (8 Hz). The sensors were mounted on circular masts, three at each vertical, 20 cm apart. Due to intrinsic limitations, the ECMs could not be placed nearer than 15 cm from the bottom. Thus, the ECM measurements cover the vertical range between the mean surface elevation and 15 cm above the bottom, every 5 cm apart.

Due to physical constraints, both the WG and ECMs were positioned off-centered the flume. This caused, in some situations, the flow to be 3-dimensional, which was visible by the wave crest not being fully perpendicular to the flume axis. Care was taken in recognizing these effects and identifying the correspondent data files.

Two "Nortek ADV Lab" ADVs were used to measure the 3-component flow velocities, mostly within the surf region, at both 25 and 50 Hz sampling frequencies. As the ADV uses acoustic sensing techniques, the sampling volume (located 5 cm away from the probe tip) is not disturbed by the presence of the probe. In the present experiment it resulted clear that the probe orientation along a longitudinal vertical plane was quite difficult to obtain, which induced cross-flume velocity readings larger than expected. Therefore, the velocity data needs to be corrected through rotation of the coordinate system.

An Ultrasound Doppler Velocity Profiler DOP1000 (by *Signal Processing S.A.*), hereinafter ADVP, was used to gather velocity measurements along the wave flume. The probes were fixed in PVC supports, located in a longitudinal trench at the bottom (14 cm wide), and running along the beach profile. This setup allowed to obtain near-simultaneous high-frequency velocity profile measurements, over the water column, and undisturbed from any intrusive equipment. The ADVP signal was sampled at frequencies ranging from 3.8 to 141.4 Hz, depending on the number of simultaneous probes, the spatial resolution along each beam, and the local water depth.

The water surface elevation was always measured simultaneously and at the same transect of the ADVP sensors. In the breaking region, two probes were set-up in pairs. In the swash zone, where the water depth was very shallow, the probes were only installed individually. Furthermore, the transducers were installed with its beam oriented 60°-70° with respect to the bottom, so that the measured velocities correspond to the flow velocities projected along that oblique axis.

Video imaging was used to record one full test for each wave condition. The video camera was both located near the swash zone and at the surfzone, and helped to identify the regions corresponding to initiation of wave breaking and wave reforming. The video cameras were also setup aiming vertically, downwards, towards the water surface. It is expected that the analysis of the digital images will enable to estimate several surfzone parameters.

2.3. Test Conditions

Wave conditions were chosen such that prototype measurements over a fixed bed beach simulated those that happen when a near-equilibrium profile condition is attained. Several preliminary runs were performed in order to select a few, most adequate, wave conditions. Therefore, four types of wave conditions were chosen, such that waves broke on the seaward slope of the bar and reformed into the trough region, breaking secondly nearer the shoreline. The 4 wave conditions (3 monochromatic and 1 irregular sea state) were repeated consecutively, giving rise to nearly 230 independent tests. During each test, the measuring instruments were fixed at a single position being then moved for the next repetition.

Table 1 – Summary of Test Parameters

Wave condition	H, H_{rms} (m)	T_p (s)	H_{rms}/L	No. Ursell, $H/L/(h/L)^3$	x_b (m)	H_b (m)	h_b (m)	Breaking type
A (regular)	0.21	2.50	0.024	1.89	42.0	0.30	0.41	Spilling
B (regular)	0.21	3.50	0.015	4.71	43.5	0.35	0.45	Plunging
C (regular)	0.38	3.50	0.027	8.52	46.5	0.58	0.56	Plunging
D (irregular)	0.21	2.50	0.024	1.89	–	–	–	–

Table 1 summarizes the four types of wave conditions analysed here, where H and H_{rms} are the target regular and root mean square wave height in front of the wave maker, T_p is the peak wave period, L is the computed wavelength at the wave-maker (using linear wave theory), x_b, H_b and h_b are the approximate breaking location, breaking height (defined as the maximum wave height from wave height measurements) and depth, respectively. The values for the random wave condition D correspond to those associated with the peak period, satisfying a Jonswap spectrum with a peak enhancement factor of $\gamma=3.3$.

For each wave condition, at least 56 independent tests were performed, with repetitions being performed for some tests, if any abnormal event occurred. Generally, the wave conditions A, B, C, and D were run sequentially for each test number, with about 6 minutes of rest between each test. In order to achieve stationarity of each sea-state, the data acquisition was started 360 seconds after the start of the wave-maker for conditions A, B and C, and 240 seconds for condition D. Each data acquisition lasted 400 seconds for wave conditions A, B and C and 1250 seconds for wave condition D.

3. DATA ANALYSIS TECHNIQUE

Frequency and time domain analysis were performed on the data, both during and after the data acquisition. The analysis performed through the experiments helped to detect faults and to improve the setting of the apparatus' parameters. This was particularly helpful for the newer instruments used in this environment, such as the ADV and ADVP sensors.

Due to the fact that the ADV and ADVP contain a lot more noise than the surface elevation data, we pre-processed all the velocity data, whereas the surface records were not. In the case of ADV measurements, it has been assessed the level of the auto-correlation and of the signal-to-noise ratio (SNR) levels, which are an integrant part of the ADV readings. For most of the sampled time series these resulted to be larger than 90% and 20 dB, respectively, yielding quite acceptable velocity readings. For the present pre-processing data analysis, it has been admitted that two consecutive readings are affected by "noise" whenever the correspondent acceleration is larger than two times the gravity ($dv/dt > 2g$). For uniformity between the ADV and the ECM data statistics, we applied the same pre-processing procedure to all point-velocity records.

For both the surface elevation and velocity data we followed the same time series data analysis as carried out in the SUPERTANK laboratory Project (Kraus and Smith, 1994). This was performed with a zero-upcrossing definition of a wave. We note that a few data acquisition signals corresponded to the paddle horizontal position. Hence, these signals were processed as if they were surface elevation records, i.e., the surface elevation should be interpreted as the paddle position, and the wave height as to the paddle stroke. In the following we list the parameters calculated from time series analysis, for both the surface elevation and velocity records:
- mean, standard deviation, skewness and kurtosis of surface elevation;
- mean, root-mean-square, significant, one-tenth, maximum and minimum wave heights;
- mean, significant, one-tenth, maximum and minimum wave periods;
- mean, standard deviation, skewness and kurtosis of the point-velocity components;
- maximum and minimum velocity magnitudes.

4. RESULTS

Along this Section we present a few preliminary results of the experimental data collected within the Project. We first analyze the data with respect to quality parameters, and then show a few significant results.

Regarding quality control, and since the experiments reflect over-54 repetitions of the same wave condition, for four distinct situations, a major concern along the whole experiment was the repeatability of the tests. Therefore, we have used the input signals corresponding to the paddle position to assess test repeatability. Fig. 3 shows the stroke root-mean-square, based on the paddle position measurement (the "feedback" signal), for all tests of wave condition A. The solid-thick lines represent the average of the measured values of all tests and the dashed lines represent the average plus or minus 5%. Therefore, the band within the ±5% of the average values is portrayed, and this allows us to reject any test (we assume it is not a repetition of the same stochastic process) whose stroke$_{rms}$ do not fall within the ±5% band.

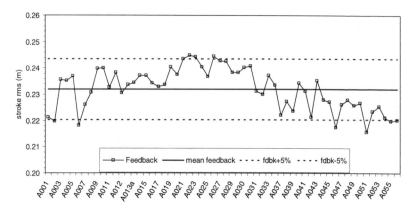

Fig. 3. Stroke root-mean-square for all tests of wave condition A.

Using the repeatability procedure outlined above for all wave tests, we conclude that 96% of them are accepted. Similar conclusions are drawn from the surface elevation records measured at the first three wave gauges that remained fixed during the experiment.

A second concern regarding data quality was maintaining stationarity of the processes during each run. This was verified by means of acquiring a wave record, for each wave condition, much longer than the other standard data records. A detailed analysis of these records (Sancho *et al.* 2001) allowed to conclude that the process is considered to achieve stationarity approximately 360 s past the start of the wavemaker. This condition was satisfied for all acquired wave records.

Most data analysis is underway, but next, we show a few results for the regular wave condition C (see table 1, for details). For this wave condition, the root-mean-square wave height, H_{rms}, and wave setup variation along the flume are presented in Fig. 4. From right to left, we observe wave shoaling up to the breaker height, and then a fast decay correspondent to an intensive plunging breaker. The setup is initiated only past (about 2 m) the start of the wave breaking, as observed in several other previous studies. Afterwards, as waves break, the wave height remains nearly constant all through the first surfzone ($32 < x < 40$ m), over the bar crest, and then waves shoal again over the bar trough and break secondly nearer the shoreline. Interestingly, the wave setup remains nearly constant slightly past the bar crest ($x < 40$ m). Nearer the still water shoreline (at $x = 17$ m), since the sensors were initially at dry conditions, the wave setup measurement is poorly defined, although they should tend to zero as shown.

The mean hydrodynamic flow field generated by the wave condition C is portrayed in Fig. 5. The depicted currents correspond to those measured only by the ECMs; therefore, the 15 cm layer immediately above the bottom has no measurements. Also, the ECMs were not deployed for $x < 24.5$ m because the experiment focused on the wave breaking and reforming regions over the bar. Hence, all the following analysis is preliminary and reports solely to the plotted data.

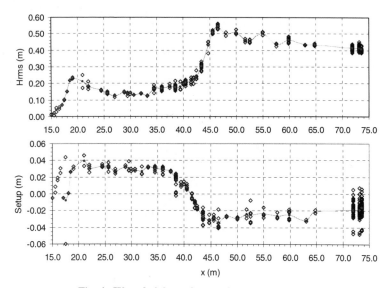

Fig. 4. **Wave height and setup for wave condition C.**

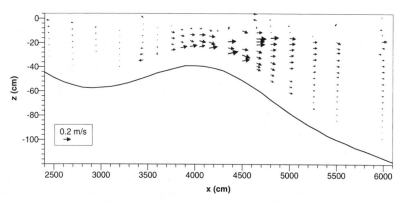

Fig. 5. **Mean currents for wave condition C.**

Firstly, we note that the maximum velocity magnitude is 0.28 m/s and occurs at the breaking region. Despite the fact that the bottom layer is lacking data, the velocity profiles are in agreement with those presented by other authors for different bottom configurations, both inside and outside the surfzone (*e.g.*, Okayasu, 1989; Putrevu and Svendsen, 1993). It is further interesting to point out that, in the region above the 15 cm layer shown here, we note a flow-direction reversal around $x=36$ m, which falls still within the surfzone. For $x< 36$ m, all velocity measurements point towards the shoreline, meaning that, in order to satisfy mass conservation, we either have a 3-dimensional flow (thus, the flow at the flume can not be considered 2-dimensional), or the 15 cm region lacking data will show offshore-directed velocities. Further data analysis will provide light on this issue.

Finally, a few results of the turbulent velocities are promising (Archetti *et al.*, 2000). Fig. 6 shows the spectrum of the horizontal velocity from an ADV record. The dashed line has the $-5/3$ slope of the power law used to describe fully turbulent flows. This indicates that the high frequency energy is mostly turbulent.

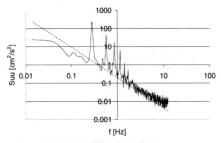

Fig. 6. Horizontal velocity spectrum from ADV measurements, wave condition C, $x=36$ m, $z=25$ cm.

5. CONCLUSIONS

In the present paper we have presented a comprehensive experimental Project, targeted towards understanding the hydrodynamics over a fixed-bed, barred beach. Four wave conditions were generated and both surface elevation and velocity measurements were carried out at a large number of locations. Several different runs of the same wave condition were performed, and were considered to represent the same stochastic process. Flume *seiching* was evident and could not be eliminated by the present wave-absorption system, which worked well in main wave-frequency range. Preliminary velocity results are shown and reveal quite interesting aspects, needing further investigation.

ACKNOWLEDGEMENTS

This work was partly funded by the *European Commission, Training and Mobility of Researchers Programme – Access to Large-scale Facilities*, under contract no. ERBFMGECT9500073, and by the *"Laboratorio de Ingenieria Maritima"* (LIM) of the *"Universitat Politècnica de Catalunya"* (UPC). The success of this Project is the result of the work of a large, innumerable, team. Special acknowledgement is given to Messrs. Andreu Fernández, Oscar Galego and Joaquim Sospreda (from UPC) for assisting with the instrumentation and helping overcoming all experimental problems. We are also particularly grateful to Prof. J.M. Redondo for providing one the ADV units and to Prof. A. Lamberti for making available the ADVP. Finally, the authors greatly appreciated the discussions and valuable comments provided by Prof. N. Kobayashi.

REFERENCES

Archetti, R., Damiani, L., Lamberti, A., Mossa, M., Rinaldi, A., and Tomasicchio, G.R. (2000). Indagine sperimentale su di una spiaggia con barra. *IDRA 2000 – XXVII Convegno di Idraulica e Costruzioni Idrauliche*, Genova (in Italian).

Birkemeyer, W.A., Donoghue, C., Long, C.E., Hathaway, K.K., and Baron, C. (1997). 1990 DELILAH nearshore experiment: summary report. *Tech. Rep. CHL-97-24*, U.S. Army Corps of Engineers, Waterways Experiment Station.

Cox, D.T., Kobayashi, N., and Okayasu, A. (1995). Experimental and numerical modeling of surf zone hydrodynamics. *Res. Report CACR-95-07 (Ph.D. Dissertation* of 1st author), Center for Applied Coastal Research, Univ. of Delaware, pp. 293.

George, R.A., Flick, R.E., and Guza, R.T. (1994). Observations of turbulence in the surf zone. *J. Geophys. Res.*, 99 (C1), 801-810.

Hattori, M. and Aono, T. (1985). Experimental study on turbulence under spilling breakers. In Y. Toba and H. Mitsuyasu (Eds.), *The Ocean Surface*, 419-424, D. Reidel Publishing Company.

Kennedy, A.B., Chen, Q., Kirby, J.T., and Dalrymple, R.A. (2000). Boussinesq modeling of wave transformation, breaking and runup. I: One dimension. *J. Waterways, Ports, Coastal and Ocean Engng.*, 126 (1), 39-47.

Kraus, N.C., Smith, J.M., and Sollitt, C.K. (1992). SUPERTANK laboratory data collection project. *Proc. 23rd Int. Conf. Coastal Engng.*, Vol. 3, Venice, ASCE, 2191-2204.

Kraus, N.C. and Smith, J.M. (1994). SUPERTANK Laboratory Data Collection Project. *Tech. Rep. CERC-94-3*, US Army Corps of Engineers, Waterways Experiment Station.

Nadaoka, K. and Kondoh, T. (1982). Laboratory measurements of velocity field structure in the surf zone by LDV. *Coastal Engng. in Japan*, 25, 125-145.

Okayasu, A. (1989). Characteristics of turbulence structure and undertow in the surf zone. *Ph.D. Dissertation*, Univ. of Tokyo.

Putrevu, U., and Svendsen, I.A. (1993). Vertical structure of the undertow outside the surf-zone. *J. Geoph. Res.*, 98, C12, 22707-22716.

Rodriguez, A., Sánchez-Arcilla, A., Gomez, J., and Bahia, E. (1995). Study of surf-zone macroturbulence and mixing using DELTA'93 field data. *Proc. Coastal Dynamics '95*, Gdansk, ASCE, 305-316.

Rodriguez, A., Sánchez-Arcilla, A., and Redondo, J.M. (1999). Macroturbulence measurements with electromagnetic and ultrasonic sensors. *Experiments in Fluids*, 27, 31-42.

Sanchez-Arcilla, A., Roelvink, J.A., O'Connor, B.A., Reniers, A., and Jimenez, J.A. (1995). The Delta Flume'93 experiment. *Proc. Coastal Dynamics '95*, Gdansk, 488-502.

Sancho, F.E. (1999). Equilibrium barred-beach profile. *1st Portuguese Conference on Harbour and Coastal Eng.*, Porto (in Portuguese).

Sancho, F.E., Mendes, P.A., Carmo, J.A., Neves, M.G., Lamberti, A., Tomasicchio, G.R., Archetti, R., Damiani, L., Mossa, M., Rinaldi, A., Gironella, X., and S.-Arcilla, A. (2001). Wave induced turbulence and undertow over barred beaches. *Technical report* (in progress).

Stive, M.J.F. (1980). Velocity and pressure field of spilling breakers. Proc. 17th Int. Conf. *Coastal Engng.*, Vol. 1, Sydney, ASCE, 547-566.

Wu, Y., H.-H. Dette and H. Wang (1994). Cross-shore profile modelling under random waves. *Proc. 24th Int. Conf. Coastal Engng.*, Vol. 3, Kobe, ASCE, 2843-2855.

A LOW-COST WAVE-SEDIMENT-TOWING TANK

Patrick J. Hudson[1], M.ASCE, Michael E. McCormick[2], F.ASCE,
and Shannon T. Browne[3], Visitor

Abstract: The design and construction of a small wave-sediment tank, built and instrumented for under $10,000, is described. The tank is equipped with a piston-type wavemaker, capable of producing both small-amplitude and solitary waves. The middle section of the tank contains a sediment bed appropriate for wave-soil-structure interaction and sediment transport experiments. Wave energy is absorbed by a gravel beach located at the opposite end of the tank from the wavemaker. Instrumentation includes three capacitive wave height gauges. The wavemaker is controlled by a desktop computer, which also provides data-logging capability. Applications of this wave-sediment tank to research into grounded ship migration and barge added-mass are presented.

INTRODUCTION

A wave tank with a sediment bed was constructed primarily to conduct experiments into the wave-induced migration of grounded ships. The center section of the tank started life as a sand seepage tank in the soil mechanics laboratory at Johns Hopkins University. When a replacement was acquired by the Civil Engineering Department, the old seepage tank was slated for disposal. It was, however, salvaged to become the core of the wave-sediment-towing tank. This paper describes the design, construction, and testing of the tank and associated equipment, and discusses examples of its application to experimental research.

[1] Naval Architect, David Taylor Model Basin, 9500 MacArthur Boulevard, West Bethesda, Maryland 20817-5700, HudsonPJ@nswccd.navy.mil
[2] Research Professor, Civil Engineering Department, The Johns Hopkins University, 3400 North Charles Street, Baltimore, Maryland, 21218, memccormick@jhu.edu
[3] Research Assistant, Civil Engineering Department, The Johns Hopkins University, 3400 North Charles Street, Baltimore, Maryland, 21218, STBrowne@jhu.edu

TANK DESIGN AND CONSTRUCTION

The original sand tank was built in 1983 by the Johns Hopkins Civil Engineering Department for conducting seepage testing in the soil mechanics laboratory. This tank, sketched in Figure 1, was 244-cm long, 61-cm deep, and 30-cm wide. The bottom and ends of the tank were constructed of aluminum plate with 13-cm thickness. The sides of the tank were formed from 13-cm-thick clear acrylic sheeting, which were supported by an aluminum structural frame. The tank rested on a wooden base mounted on 15-cm-diameter wheels. The tank bottom had a screened drain at both ends, which were plumbed through the wooden base to ball valves. These valves could be opened to drain the tank.

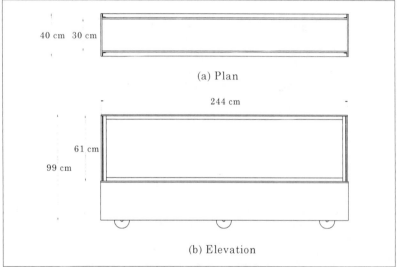

Fig. 1. Original Johns Hopkins sand seepage tank before modification into a wave-sediment-towing tank (Hudson, 2001).

Conversion from Seepage Tank to Wave Tank

The sand seepage tank was well-suited to modification into a small wave tank with a sediment bed, and its only disadvantage was a short length relative to its depth. As discussed in Dean and Dalrymple (1991), a wavemaker produces a system of evanescent standing waves in addition to the desired radiated waves, and these evanescent waves are negligible at a distance greater than three water depths from the wavemaker. A wave tank with a wavemaker, then, should be long enough so that the generated waves are fully developed by the time they reach the test section. Tank length also determines the maximum time that waves may be generated before wave reflection becomes significant. The length of the original seepage tank was doubled at the waterline without extending the sediment bed or foundation structure.

The wave tank design includes a 122-cm-long by 30-cm-deep extension at either end, increasing the total length to 488 cm. The two end plates were cut in half, and the top half of each was place at the end of the extension. Two 10-mm-thick aluminum braces were added at either end to support the weight of the extensions. On one side, the original acrylic sheet was replaced with a new scratch-resistant acrylic to improve visibility. A 1.3 cm-thick by 244-cm-long aluminum plate was provided to cover the sediment bed and form a continuous impermeable bottom. The final wave tank design is sketched in Figure 2.

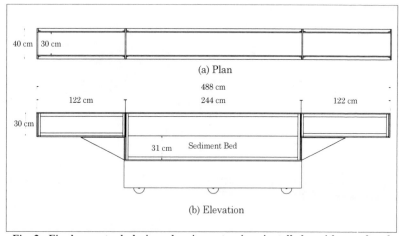

Fig. 2. Final wave tank design, showing extensions installed at either end and the added sediment bed (Hudson, 2001).

A unique feature of the wave tank is the sediment bed in its middle section. This bed is 31-cm deep, 30-cm wide, and 244-cm long, and can be filled with sand, clay, silt, or other marine sediments. For the grounded ship experiments described later in this paper, the sediment bed was filled with 0.4 mm to 0.5 mm silica sand. This sand, although washed, contained enough fines to cause significant water turbidity when disturbed. For this reason, a circulation and filtration system was added to the tank.

A 1/15-hp, 3,400-rpm aquarium pump draws water from one end of the tank, pumps it through a 5-micron residential-type water filter, and then discharges it at the opposite end of the tank. All piping is 25-mm polyvinyl chloride (PVC) with cemented fittings. The intake and discharge pipes can be easily removed for testing to eliminate reflection and diffraction affects. For sediment transport studies, the pump can generate an depth-averaged current over the bed of approximately 4 cm/s. This system has proven to be very effective in maintaining photographically-clear water which is free of algae growth. The tank is filled with chlorinated tap water.

Wavemaker System

A piston-type wavemaker at one end of the tank can generate linear and nonlinear waves, including solitary waves. The piston is an acrylic plate which is 29.8-cm wide, 26.0-cm high, and 1.3-cm thick. When installed in the tank, it maintains a clearance of 1 mm along each side and the bottom. The piston-plate is supported at its top edge by an aluminum plate bolted to two shuttles, which ride on a 15-mm wide anodized aluminum rail. This linear guide is manufactured by Igus, Inc., of Providence, Rhode Island, under their Drylin-T® product line. The two shuttles are bolted to a 15-cm long aluminum plate, and that plate is bolted to an aluminum flange which supports the wavemaker piston. The piston is centered between the two shuttles, which minimizes any moment-induced rotation of the piston.

A gravel beach with a 1:4 slope is located at the end of the tank opposite the wavemaker. This beach absorbs wave energy both by causing the incident waves to break, and by allowing percolation of water through the gravel after breaking to minimize reflection. The beach is 60-cm long and 30-cm wide, and rises from the tank bottom to a maximum depth of 15 cm.

To generate small amplitude, sinusoidal waves with a desired period and wave height, the required stroke of the wavemaker is given by

$$S = H \frac{\sinh 2kh + 2kh}{2(\cosh 2kh - 1)} \tag{1}$$

where H is the wave height, k is the wave number, and h is the water depth. Equation (1) is derived from linear wavemaker theory as presented, for example, in Dean and Dalrymple (1991). The stroke is equal to twice the harmonic amplitude of the piston motion.

Wavemakers are typically driven by hydraulic, pneumatic, or electromagnetic systems. For this application, the electromagnetic option was chosen for its accuracy, simplicity, and ease of operation. The first actuator installed was an electromagnetic voice coil, driven by a DC brushless motor driver. This voice coil developed a force proportional to the applied current with a maximum stroke of 2.5 cm. However, this small stroke greatly limited the height of waves which could be generated. Since equation (1) is based on wavemaker displacement and not force, use of a force actuator required a closed-loop feedback control system to accurately match the necessary sinusoidal motion. To increase the wave height generation capability, and to eliminate the need for a feedback control system, the voice coil was replaced by a stepper motor and lead-screw combination.

The driving rod of a stepper motor linear actuator moves a fixed distance when an electrical pulse is applied to it. As long as the motor is operated within its torque limits, the system may be run without the need of a feedback loop. The Johns Hopkins wavemaker uses a 1.8° per-step, 1.54 amp-per-coil, 6-lead stepper motor linear actuator manufactured by Eastern Air Devices, Inc., of Dover, New Hampshire. The lead screw on this actuator moves 1.0 cm for every 630 pulses applied to the stepper motor. A 7006opto™ High-Speed Chopper Drive, manufactured by AMSI Corporation of Palos Hills, Illinois, controls the linear actuator. This drive has a maximum effective stepping rate of about 5 kHz, which translates to a linear velocity of the piston of 7.9 cm/s.

To generate a sinusoidal wave, the controlling software calculates the required linear velocity for the wavemaker, and converts that speed to a pulse frequency. This frequency is output to the chopper drive, which then steps the motor at the appropriate rate. This pulse frequency is updated every 0.01 s, which results in a *quasi*-smooth operation of the wavemaker piston.

The linear actuator can generate a maximum torque of 0.28 J, but this torque drops off rapidly to about 0.10 J at linear speeds above 1.5 cm/s. If the combination of required period and stroke results in a higher torque than the stepper motor can produce, then the motor will slip and the generated wave will be unpredictable. Limits on wave period and height for given depths are, thus, established to prevent such motor slippage. Generation of solitary waves requires torques which greatly exceed the capability of the stepper motor; so, a gravity-pull system is used in place of the linear actuator.

According to Wiegel (1964), solitary waves may be generated in the laboratory using several methods, including an impulsively-loaded piston, rapid addition of a specified volume of additional water, and dropping a body into the water. All of these methods normally produce a dispersive *tail* which follows the desired solitary wave. Goring and Raichlen (1980) describe a method for minimizing this tail by matching the wavemaker plate velocity to the solitary wave water particle velocity. The presence of the dispersive tail was not of significant concern in the grounded-ship experiments described in this paper. The solitary wave traveled faster than the tail, and its affect on the hull model was recorded prior to the arrival of the tail. The torque limitation of the stepper motor prevented use of the Goring and Raichlen method for generation of waves higher than 1 cm. Instead, the same wavemaker piston used for linear wave generation is used for solitary waves, but it is driven by a weight-and-pulley system instead of the stepper motor.

The driving system for solitary wave generation consists of a nylon line connecting the wavemaker piston through two small pulleys to a variable weight. These pulleys are used to guide the line only and do not apply any mechanical advantage. To generate a solitary wave, the piston is moved to its zero position and secured. The combination of driving weight and stroke distance used depends on the desired height of the solitary wave. Once the water in the tank in calm, the wavemaker is carefully released and travels its pre-determined stroke before impacting at a rail stop. The waves generated using this method are of consistent height and form from test to test, and closely match that of a theoretical solitary wave.

Wave Gauges and Data Acquisition

Three capacitive wave height gauges are used to measure water surface elevation in the wave tank. They are model WG-50 gauges with 30-cm probes, manufactured by Richard Brancker Research, Ltd., of Ottawa, Ontario, Canada. Capacitive wave gauges measure the change in electrical discharge time of an immersed, insulated wire. This measured capacitance is linearly proportional to the immersion depth of the wire, and is output by the gauge as a ± 5.0 v signal. The gauges may be easily positioned in the tank as needed. Figure 3 shows the wave gauge configuration for the grounded ship studies discussed later in this paper. The gauges are calibrated by adjusting two trimmer potentiometers. The first potentiometer sets the zero point for the gauge, and the second potentiometer controls the gain in units of centimeters per volt. Once calibrated, the gauges can operate for several weeks without significant drift. The voltages output from the wave gauges are converted to integer values by an analog-to-digital converter card located in a desktop computer.

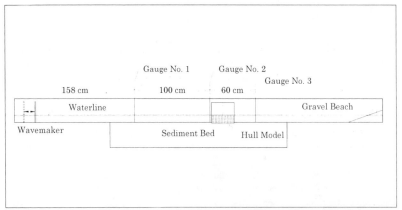

Fig. 3. Locations of the three wave height gauges. The configuration shown was used in an experimental study of the wave-induced migration of grounded ships.

Wavemaker Performance

In order to quantify the performance of the linear wavemaker system, waves were generated while there was no model in the tank. Five wave trains were generated for each of 31 combinations of wave height, period, and water depth. For each period, the number of individual waves per train was limited so that reflected waves did not interfere with the incident waves. The resulting wave height-to-stroke ratios from each of these 155 wave tests are plotted against the relative depth in Figure 4. The relative depth is the wave number, k, multiplied by the water depth, h.

The measured wave height-to-stroke ratios in these 155 wave tests were approximately 20% lower than those predicted by wavemaker theory. There was better agreement in the shallower water than in deeper water. The linear waves generated in the grounded ship experiments described later in this paper had relative

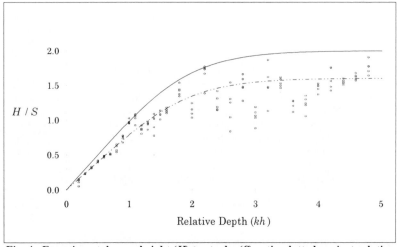

Fig. 4. Experimental wave height (*H*) to stroke (*S*) ratio plotted against relative depth (*kh*). The solid line represents the theoretical ratio, while the dashed line represents a 20% reduction from theory (Hudson, 2001).

depths between 0.286 and 1.298. The disagreement between theoretical and measured wave heights may be partly due to damping caused by the sediment bed and viscous losses along the tank walls. Eddy losses around the edges and bottom of the wavemaker plate may have also contributed to the lower measured wave heights.

The solitary waves generated using the impulsive weight-and-pulley system include a dispersive tail. Since the solitary wave travels faster than its tail, the influence of this tail decreases with distance from the wavemaker. An experimental solitary wave, measured at the closest gauge to the wavemaker, is compared with its theoretical prediction in Figure 5. This wave was generated using a pull force of 26.6 N and a wavemaker stroke of 15 cm. As is evident in that figure, the generated wave profile is very close to that of the theoretical solitary wave. The dispersive tail is evident starting at time t = 1.8 s.

Gravity-Towing System
The wave tank includes a removable gravity-towing system to pull models through the water with a constant force. A 2-m long linear rail is installed along the top frame of the tank, centered transversely. A shuttle plate, which is bolted to the model, rides along the rail on self-lubricating bearings. An electromagnetic latch-release mechanism is used to hold the shuttle at its desired start point. A nylon line, attached at one end to the shuttle plate, runs along the length of the rail, through a 1-cm-diameter hole in the end plate of the tank, around a pulley, and finally attaches to a weight-carrier plate.

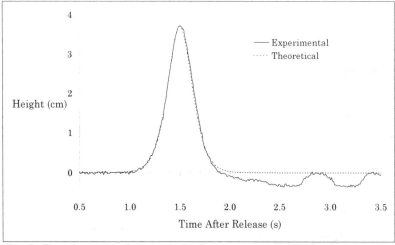

Fig. 5. Experimental and theoretical solitary wave profiles in the time domain. The wavemaker piston was released at time t = 0s.

Prior to a towing experiment, the model is latched at the starting point and the required towing weight is added to the weight-carrier plate. Once the water in the tank is calm, the electromagnetic release is activated and the model begins accelerating down the tank. Data collection begins simultaneously with model release. At the end of the rail, the model impacts a cushioned rail-stop which prevents further motion. An accelerometer is installed on the shuttle plate to record model acceleration. Model velocity over the run is then calculated by integrating this accelerometer data.

EXPERIMENTAL APPLICATIONS

The wave-sediment-towing tank described in this paper has so far been successfully used in two experimental studies. The first investigated the wave-induced migration of grounded ships, and is described in detail in Hudson (2001). A sectional model of a ship hull was placed on the sediment bed, embedded into the sand at varying depths, as shown in Figure 3. A series of small-amplitude and solitary waves were generated, and the oscillatory motion and migration of the model was recorded. The small-amplitude waves ranged in period from 0.6 s to 2.0 s, with wave heights from 0.8 cm to 1.0 cm. The solitary waves had wave heights from 3.0 cm to 4.0 cm, and wave height-to-depth ratios between 0.3 and 0.5. A digital camera was used to record model and sand displacements. The results of this study showed very good correlation to an analytical method to predict grounded ship migration.

Kraemer *et al* (2001) discusses the second study using this wave-sediment-towing tank. That study experimentally determined the added-mass of a barge approaching a quay wall. The wave gauges were removed from the tank and the gravity-towing system was installed. Several model runs were completed with different towing forces and model drafts, both with the quay wall in place and with it removed. The acceleration and velocity data from these runs was used to calculate the affect of the quay wall on the added mass.

CONSTRUCTION AND EQUIPMENT COSTS

The wave-sediment-towing tank was designed and constructed on a limited budget. Table 1 lists all direct costs actually expended to build and equipment the wave-sediment-towing tank. The original sand seepage tank was designated for disposal, and as such was obtained at no cost. However, its estimated replacement cost is included in Table 1. All labor was performed by graduate students and by the staff machinist of the Johns Hopkins Civil Engineering Department, and was not a direct cost. An estimated labor cost, based on 120 hours of labor at an average rate of $50/hour, is also included in Table 1.

Table 1. Actual and Estimated Costs for Wave-Sediment-Towing Tank

Cost Item	US Dollars
Extension to Seepage Tank	900
Wavemaker	
Stepper Motor	95
Motor Controller	175
Acrylic Plate and Miscellaneous Hardware	50
Linear Guide System	250
Capacitive Wave Gauges (3)	2,300
Pentium III Desktop Computer	1,800
Analog-to-Digital Converter	295
Gravity towing system	90
Sand	75
Filtration System	250
Total Acutally Expended:	**6,280**
Replacement Value of Seepage Tank	2,000
Estimate of Labor Cost (not a Direct Cost)	6,000
Total Cost Estimate Including Indirect Costs:	**14,280**

CONCLUSIONS

The wave-sediment-towing tank described in this paper, although primarily designed for a specific study and built with a limited budget, resulted in a robust and versatile experimental tool. Although small compared to most wave and towing tanks, it is well-suited to educational and basic research studies. Larger tanks require a staff and funding to operate and maintain them, and these overhead costs often prohibit their use in small-scale experiments. Instead of reserving a block of time well in advance to conduct experiments, the researchers using this small tank only need to turn around in their chairs to investigate a physical behavior or try out an idea. These initial studies can then be refined and eventually expanded to more complex experiments in large wave tanks or flumes.

REFERENCES

Dean, R.G., and Dalrymple, R.A., 1991, *Water Wave Mechanics for Engineers and Scientists*, World Scientific, 353 pp.

Goring, D.G. and Raichlen, F., 1980, "Generation Of Long Waves In The Laboratory," *Proceedings of the 17th International Coastal Engineering Conference*, American Society of Civil Engineers,

Hudson, P.J., 2001, *Wave-Induced Migration of Grounded Ships*, Doctoral Dissertation, Johns Hopkins University, 212 pp.

Kraemer, D., McCormick, M.E., Hudson, P.J., and Noble, W., 2001, *Analysis of the Added-Mass of a Barge in Restricted Waters, Final Report – Phase 2*, U.S. Army Corps of Engineers.

Wiegel, R., 1964, *Oceanographical Engineering*, Prentice-Hall, 532 pp.

HIGH-QUALITY LABORATORY WAVE GENERATION
FOR FLUMES AND BASINS

Ap van Dongeren[1], Gert Klopman[2], Ad Reniers[3] and Henri Petit[1]

Abstract: A second-order wave generation algorithm called "Delft-Auke/generate" which includes online Active Reflection Compensation is presented. Results are shown from various experimental and numerical verification tests which confirm the software is capable of generating very accurate regular and irregular waves. The short-wave ARC performance tests show that the reflection coefficient (in terms of wave height) is in the order of 5% which is much less than can be achieved for "conventional" long wave ARC.

INTRODUCTION

A wave generation routine which suppresses spurious and absorbs reflected waves is essential in order to conduct high-quality laboratory experiments. In this paper an algorithm called "Delft-Auke/generate" is presented which can be applied to wave flumes and basins, and to "piston" and "flap"-type wave generators with a hinge position at an arbitrary depth above or below the bottom. With this software regular, bichromatic and irregular (short- and longcrested) waves according to well-known or user-defined energy density functions can be generated. Second-order (Stokes) wave generation for longcrested waves is included. The software has additional features such as the reproduction of a wave record at a prescribed location, and the selection of certain wave groups from a longer record which reduces tank time.

In addition, a realtime (online) Active Reflection Compensation (ARC) algorithm has been developed which compensates the wave board motion for reflected short waves

1 WL|DELFT HYDRAULICS (mailing address), P.O Box 177, 2600 MH Delft, The Netherlands. Ap.vanDongeren@wldelft.nl
2 Albatros Flow Research, P.O. Box 85, 8325 ZH Vollenhove, The Netherlands.
3 Naval Postgraduate School, Monterey, CA, USA.

from an arbitrary angle with the use of a wave height meter which is mounted flush on the wave boards (Klopman *et al.*, 1996). ARC dramatically improves the wave field because reflected waves are suppressed effectively. In addition, ARC quickly damps out residual waves after an experiment has been concluded which means that the downtime between experiments is reduced.

To this date, the software has been installed at the new Seakeeping and Offshore Basins at the Maritime Research Institute Marin, The Netherlands, at Delft University of Technology's new research flume and at WL|Delft Hydraulics' own flumes and basin. The construction of a new ocean basin has been commisioned by UFRJ/Coppe in Rio de Janeiro, Brazil, and the implementation of wave generation software is expected in May 2002.

SECOND-ORDER WAVE GENERATION

In order to generate waves accurately, the wave board motion should match the water motion under the (progressive) waves as well as possible. Any discrepancy between the motions will result in the generation of disturbance (or spurious) waves which contaminate the wave field. These spurious waves can be greatly suppressed when one realizes that a large contribution to the mismatch between the wave board motion and the water wave motion is caused by higher and lower harmonic "bound" waves which travel at the phase speed of the base harmonic. These bound waves are automatically generated by the primary wave because of the physics of wave motion. If the bound wave motion is not accounted for in the wave board motion, the boundary conditions at the wave board dictate that free, or spurious, waves are also generated at the higher or lower harmonic frequencies. These spurious waves travel at phase speeds given by the dispersion relation for free waves, wheras the higher harmonic bound waves travel at the phase speed of the base harmonic, and the lower harmonic bound waves at the group velocity. Except in the shallow water limit, their phase speeds, and hence their wave lengths, are different. The spurious waves result in a wave field which is contaminated as is shown in the top panel of Figure 1. The bottom panel shows a wave record when using second-order wave theory, the derivation of which is outlined below. The figure shows that the wave field is much more uniform.

DERIVATION OF SECOND-ORDER WAVE MAKER THEORY

The present algorithm is essentially an extension of the work by Van Leeuwen and Klopman (1996) who derived the second-order wavemaker theory for a piston-type wave flume generator. The new software is extended for generic wave makers where the hinge position may vary from the free surface to a position infinitely far below the bed. The generic theory thus includes conventional piston and flap theories. The software can be used in flumes as well as basins with one, two or more wave generator boundaries.

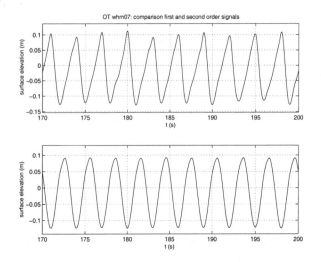

Figure 1: Time series of measured regular wave train with first order wave generation (top panel) and second order wave generation (bottom panel)

The wave maker theory is a solution to the classical boundary value problem with a wave maker condition on one lateral boundary. The derivation is based on the perturbaton method of multiple scales where the scaling parameter is the wave steepness $\varepsilon = ka$. This is done under the assumption of a narrow-banded spectrum which means that the water motion can be described as an oscillation with a slowly-varying amplitude and frequency. We refer to Van Leeuwen and Klopman (1996) for details, but in essence the expansion is as follows:

$$(\varsigma^n, X^n) = \sum_{m=-n}^{n} (\varsigma^{n,m}, X^{n,m}) e^{-im\omega t} \tag{1}$$

where ς is the surface elevation, X is the horizontal paddle movement at the still water line, ω is the radial frequency, n is the order and m is the harmonic. The derivation shows that the following corrections to first order $(n,m) = (1,1)$ or linear theory are important:

- the $(n,m) = (2,2)$ contribution of the second-order waves, second-harmonic waves;
- the $(n,m) = (2,0)$ contribution of the second-order waves, zeroth or lower-harmonic waves. These are the long bound waves;
- the $(n,m) = (2,1)$ contribution of the second-order waves at the base frequency, which correct for the paddle excursion and for the slowly-varying amplitude of the wave envelope.

NUMERICAL VERIFICATION

The software has been verified numerically using the HYPAN model, a highly accurate numerical wave propagation model which is based on a panel method (De Haas, 1997; De Haas & Dingemans, 1998). Figure 2 shows an instantaneous picture of the improved wave field characteristics in the case of second-order wave generation as compared to linear (first-order) wave generation. In the top panel spuriously generated waves are visible between the waveboard at x=0 m. and x=30 m. These waves are generated because the motion of the wave board does not match the motion in the water exactly. Second-order wave generation ensures a closer match between the board and the water motion and hence the spurious waves are suppressed, which can be seen in the bottom panel. The wave field is much more homogeneous. The wave generation algorithm was tested thoroughly using the HYPAN model which proved to be a very cost-efficient way to test the newly developed code.

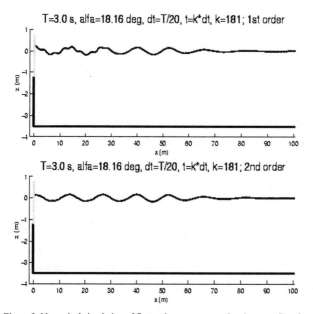

Figure 2: Numerical simulation of first-order wave generation (top panel) and second-order wave generation (bottom panel).

LABORATORY VERIFICATION

The wave generation and absorption algorithms have been tested thoroughly in both the Scheldtflume and the large Deltaflume at WL|Delft Hydraulics. Figure 1 already showed the comparison between measured first and second-order wave theory. Figure 3 shows the comparison of a measured wave train in the Deltaflume and the higher-order theory by Rienecker and Fenton (1981). The agreement is very good which is an indication that the second-order effects are properly accounted for in the wave generation software.

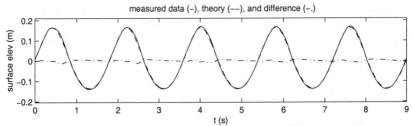

Figure 3: Comparison between measured data (solid line), theory (dashed line) and the difference wave (dash-dotted line) at a distance of 20 meters from the wave board.

Figure 4 shows the comparison a measured Jonswap spectra at three locations in the Scheldt wave flume as compared with the target spectral shape with parameters water depth $h = 0.80$ m., peak period $T_p = 1.5$ s, significant wave height $H_{m0} = 0.10$ m and a record length of 30 minutes. The agreement is generally quite good, expect for some loss of energy in the peak, which is partially due to some breaking in the wave tank.

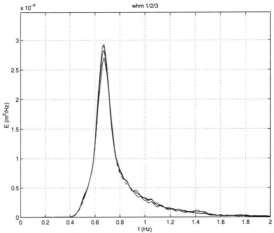

Figure 4: Jonswap spectra (measured at three locations and target)

An additional feature of the wave generation software is that a measured wave record in the field can be reproduced at a certain location in the laboratory wave basin. Figure 5 shows the scaled field-measured wave record (solid line) and the laboratory reproduction (dashed line) at a distance of 40 meters (or about four peak-period wavelengths) from the wave boards in the Delta Flume. We see that the agreement is good and that especially the phases match well.

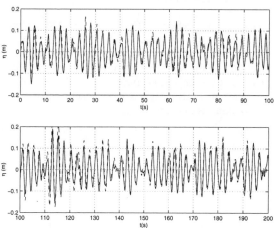

Figure 5: Field-measured (solid) and laboratory-reproduced (dashed) wave records at a distance of x=40 m. from the wave boards.

Another feature which has been included in the algorithm is the option to select certain wave groups from a long-time record. The motivation behind this is that for normative design under irregular wave conditions only wave groups with a certain wave group length and or steepness are important. It is then time consuming to run a full length record of an irregular wave train while only a fraction of the wave groups are important. In order to save time a selection criterion has been developed with which wave groups can be selected. The criterion is based on the wave group length parameter κ as developed by Battjes and Van Vledder (1984) and Van Vledder (1992) and the wave steepness parameter S_t as derived by Haller and Dalrymple (1995).

This feature was experimentally verified in the Delta Flume. The first case was a full-length simulation of an irregular wave train based on a Pierson-Moskowitz spectrum with parameters water depth $h = 3.5$ m., peak period $T_p = 3$ s, significant wave height $H_{m0} = 0.20$ m. The simulation was rerun for a selection of wave groups and a lenght of about 10% of the original length. The comparison of the group length and steepness parameters for both realisations is give in Table 1, which shows that the shorter duration test accurately reproduces the mean characteristics of the full-length simulation.

Table 1: Grouplength and steepness of the full-length simulation and the selected simulation

test	grouplength κ	steepness S_t
full-length	0.4156	2.14 e-5
selection	0.4052	2.00 e-5

ACTIVE REFLECTION COMPENSATION

A realtime (online) Active Reflection Compensation (ARC) algorithm has been developed which compensates the wave board motion for reflected short waves from an arbitrary angle with the use of a wave height meter which is mounted flush on the wave boards (Klopman et al., 1996). ARC dramatically improves the wave field because spuriously reflected waves are suppressed effectively. In addition, ARC quickly damps out residual waves after an experiment has been concluded which means that the downtime between experiments can be reduced.

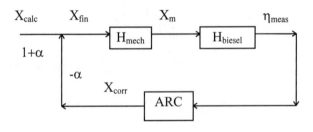

Figure 6: Schematic diagram of ARC loop

In principle, the online ARC loop works as sketched in Figure 6. Here X_{calc} is the a priori calculated wave generation signal, X_{fin} is the generation signal when the presence of ARC is taken into account. H_{mech} and H_{biesel} are the transfer functions of the mechanics of the wave board and the transfer from the paddle motion to the water surface elevation, respectively. X_m is the actual motion of the paddle, and η_{meas} is the measured water surface elevation at the wave board. This measured signal is used as input in the ARC routine to calculate online the correction signal X_{corr} to the wave generation signal. This corrected signal is multiplied by a factor α and added to $(1+\alpha)*X_{calc}$ which yields X_{fin}. The online calculation needs to be performed in a matter of milliseconds, because any time delay will cause a phase shift in the correction signal. Large phase shifts decrease the performance of the ARC dramatically. The ARC routine is based on full-spectrum wave theory, as opposed to long wave theory only. In essence this means that the long wave filter is multiplied by a digital filter which in the frequency domain reads:

$$ARC = \sqrt{\frac{g}{h}} \frac{1}{i\omega} \frac{\sum\limits_{j=0} b_j e^{-i\omega j \Delta t}}{1 + \sum\limits_{j=i} a_j e^{-i\omega j \Delta t}} \qquad (2)$$

where the first part is the long wave filter and the second part is the short wave correction filter. In this equation, ω is the frequency, and a_j and b_j are the filter coefficients. The choice of the parameters is limited due to stability constraints, see Schäffer and Klopman (2000) for details.

EXPERIMENTAL RESULTS

The performance of the ARC was tested in the new research flume at Fluid Mechanics Laboratory at Delft University of Technology. The flume is 38 meters long and a still water depth of 0.70 m. was used. Relatively high waves with three periods were propagated in a short group from the wave board. The wave train reflects off a vertical wall at the end of the flume and propagates back towards the wavemaker. In Figure 7 the ARC on the wavemaker is turned off to create a reference case. The top panel shows the recorded surface elevation at about 15 meters from the waveboard. The second panel shows the recorded velocity at a depth of 18 cm below still water level. The third and fourth panels show the surface elevation of the wave propagating away and towards the wavemaker. These signals are computed from the measured signals using a simple separation technique based on linear theory. The third panel shows a wave group propagating in the positive direction which is fully reflected off the wall and appears in the bottom panel. This wave train is again fully reflected at the wavemaker (since the ARC is turned off) and reappears in the third panel, etc.

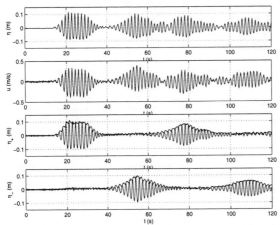

Figure 7: Timeseries of wave motion in the case of no ARC. First panel: total surface elevation. Second panel: total velocity. Third panel: Surface elevation of waves propagation away from the waveboard. Fourth panel: Surface elevation of waves propagating toward the wave board.

Figure 8 shows the same case when the ARC is turned on. The wave train can still be seen propagating away from the wavemaker in the third panel, and after reflection in the bottom panel, but there appears very little re-reflection from the waveboard which would have been visible around t=70-90 s.

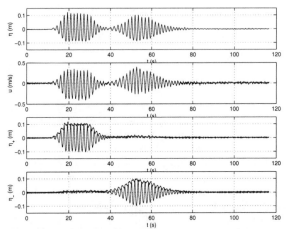

Figure 8: Timeseries of wave motion in the case of no ARC. First panel: total surface elevation. Second panel: total velocity. Third panel: Surface elevation of waves propagation away from the waveboard. Fourth panel: Surface elevation of waves propagating toward the wave board.

The reflection coefficients of the ARC (in terms of wave height) can be estimated from the energy flux in the wave groups. The results (including the case of longwave ARC, which means that no shortwave filter in (2) has been used) show that the reflection coefficients are of the order 5-10% for the case of shortwave ARC and about a factor 40% higher in the case of longwave ARC. The presence of shortwave ARC causes a large reduction in the reflection coefficients.

Table 2: Computed ARC reflection coefficients for three periods using longwave and shortwave ARC

T (s)	H(m)	R_H (longwave ARC)	R_H (shortwave ARC)
1.2	0.1	0.13	0.09
1.6	0.2	0.07	0.04
2.0	0.2	0.06	0.05

CONCLUSIONS

A second-order wave generation algorithm called "Delft-Auke/generate" which includes online Active Reflection Compensation, is presented. This software can be applied to wave flumes and basins, and to "piston" and "flap"-type wave generators with a hinge position at an arbitrary depth above or below the bottom. Regular, bichromatic and irregular (short- and longcrested) waves according to well-known or user-defined energy density functions can be generated. The software also includes additional features such as the reproduction of a wave record at a prescribed location, and the selection of certain wave groups from a longer record which reduces tank time.

Results are shown from various experimental verification test which confirm the software is capable of generating very accurate regular and irregular waves. The short-wave ARC performance tests show that the reflection coefficient (in terms of wave height) is in the order of 5% which is much less than can be achieved for conventional "long wave" ARC.

ACKNOWLEDGMENTS

The work has been funded in part by the Netherlands Ministry of Economic Affairs. Marin Research Institute Netherlands is gratefully acknowledged for the use of the laboratory test results.

REFERENCES

Battjes, J.A. and G.Ph. Van Vledder (1984). "Verification of Kimura's theory for wave group statistics", *Proc. 19th ICCE*, Houston TX, USA, pp. 642-648.

De Haas, P.C.A. (1997). "Numerical simulation of nonlinear waterwaves using a panel method; domain decomposition and applications." *Ph.D Dissertation*, Twente University, Netherlands, 191 p.

De Haas, P.C.A. and M.W. Dingemans (1998). "Simulation of nonlinear wave deformation by a shoal in 3D." *Proc. 26th ICCE*, Copenhagen, Denmark, pp. 670-681.

Haller, M.C. and R.A. Dalrymple (1995). "Looking for wave groups in the surf zone." *Proc. Coastal Dynamics*, Gdansk, Poland, pp. 81-92.

Klopman, G., Reniers, A.J.H.M., Wouters, J. and De Haan, Th. (1996). "Active Multi-Directional Wave Absorption". *Abstract no. 415, 25th ICCE*, Orlando, Fla, USA.

Rienecker, M.M. and J.D. Fenton (1981). "A Fourier approximation method for steady water waves." *J. Fluid Mech*, vol. 104, pp. 119-137.

Schäffer, H.A. and G. Klopman (2000). "Review of multidirectional active wave absorption methods." *J. of Wat., Port, Coastal and Oc. Eng.*, vol. 126, No. 2, pp. 88-97.

Van Leeuwen, P.J. and G. Klopman (1996). "A new method for the generation of second-order random waves." *Ocean Engng*. Vol. 23, No. 2., pp. 167-192.

Van Vledder, G.Ph. (1992). "Statistics of wave group parameters", *Proc. 23rd ICCE*, Venice, Italy, pp. 946-959.

ACTIVE WAVE ABSORPTION IN FLUMES AND 3D BASINS

Hemming A. Schäffer[1]

Abstract: The theory of 2D and 3D active wave absorption is outlined in Fourier space and the principles of approximate time-space realisation are given. An analysis method by which the system checks its own performance is derived and the usage is demonstrated for a wave flume case. For active absorption of oblique waves a qualitative test shows promising results.

INTRODUCTION

Wavemakers equipped with a control system for simultaneous wave generation and active wave absorption are now widely used in hydraulic laboratories. These are often called absorbing wavemakers. Several features make active absorption attractive. First of all spurious re-reflection of outgoing waves is largely avoided. This improves control of the incident wave field, especially in situations with large reflections. Another important feature is that active absorption helps suppressing wave flume resonance. This is true despite the fact that any active absorption system has a quite limited performance at low frequencies to avoid slow drift of the wave paddle. Even so, their effect is important since their absorption ability is typically much better at low frequencies than that of passive absorbers.

This paper builds on two decades of accumulated experience at DHI Water & Environment in the field of active absorption. No attempt is made to give credit to the many other contributors to the field of active wave absorption. We only mention the pioneering work of Milgram (1970) and refer to a review of the various approaches to 2D and 3D active absorption given by Schäffer and Klopman (2000).

Active absorption needs a hydrodynamic feedback from the flume or basin in order to detect the waves to be absorbed. For this purpose we only consider the well-proven and commonly used method, where the hydrodynamic feedback is provided by surface elevation gauges integrated in the paddle front.

---

[1] DHI Water & Environment, Agern Allé 11, DK-2970 Hørsholm, Denmark, www.dhi.dk, has@dhi.dk

THEORY IN FOURIER SPACE

Based on linear wavemaker theory, see e.g. Dean and Dalrymple (1991), we can write up the governing equations for the active absorption problem in Fourier space. Time series of surface elevation and paddle position along the wave paddle are denoted $\eta(t, y)$ and $X(t, y)$, respectively, while $A(\omega, k_y)$ and $X_a(\omega, k_y)$ denote the equivalent complex Fourier amplitudes:

$$X(t, y) \quad \overset{\text{2D Fourier Transorm}}{\Leftrightarrow} \quad X_a(\omega, k_y) \tag{1}$$

$$\eta(t, y) \quad \overset{\text{2D Fourier Transform}}{\Leftrightarrow} \quad A(\omega, k_y) \tag{2}$$

Here t is time, ω is angular frequency and k_y is wave number component along the wavemaker. For the wave flume case k_y vanishes and only a one-dimensional Fourier transform is needed. The quantities A and η carry the following subscripts: "I" for the target, progressive, incident waves, "0" for waves measured right at the paddle front, "R" for progressive reflected waves and "RR" for progressive re-reflected waves, see Figures 1 and 2 for definitions. All surface elevations and equivalent amplitudes are taken at $x = 0$, the mean position of the paddle. Using linear wavemaker theory and further assuming full re-reflection on the wave paddle when the wavemaker is at rest, the following three equations apply

$$A_I = i e_0 X_a + A_{RR} \tag{3}$$

$$A_0 = i X_a \left(e_0 + \sum_{j=1}^{\infty} e_j \right) + A_R + A_{RR} \tag{4}$$

$$A_R = A_{RR} \tag{5}$$

see below for further explanations. Here

$$e_j = \frac{k_j}{k_{xj}} c_j \tag{6}$$

is the oblique-wave transfer function for paddle position to surface elevation. Furthermore, k_j satisfies the linear dispersion relation generalised to complex wave numbers

$$\omega^2 = g k_j \tanh k_j h \tag{7}$$

and k_0 is the ordinary real wave number for the progressive wave, while k_j is purely imaginary for $j \geq 1$ corresponding to evanescent modes. The x-component k_{xj} of k_j is given by

$$k_{xj}^2 = k_j^2 - k_y^2 \tag{8}$$

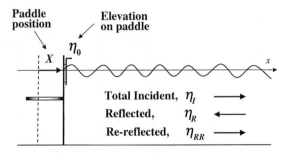

Figure 1. Vertical cross-section perpendicular to the wavemaker. Definition sketch

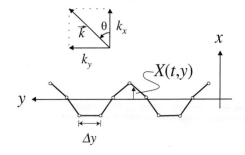

Figure 2. Horizontal cross-section. Definition sketch

For $\theta = 0$, e_j reduces to c_j. Here c_0 is a real transfer function often denoted the Biésel transfer function (Biésel, 1951), and c_j ($j \geq 1$) is the purely imaginary transfer function for the j'th evanescent mode. For a piston-type wavemaker, we have

$$c_j = \frac{4\sinh^2 k_j h}{2k_j h + \sinh 2k_j h} \tag{9}$$

In traditional linear wave generation only the progressive mode is relevant and we note that the influence of wave direction θ comes in as

$$e_0 = \frac{1}{\cos\theta} c_0 \tag{10}$$

taking $\theta = 0$ for normally emitted waves. Details of the complex formulation can be found in Steenberg and Schäffer (2000).

If no reflections occur then $A_R = A_{RR} = 0$ and (3)-(4) are consistent with standard linear wavemaker theory; (3) gives the amplitude of the incident waves in terms of the paddle amplitude (with the imaginary unit giving a 90 degree phase shift) while (4) gives the amplitude including evanescent modes. Conversely, if $X_a = 0$ then (3) expresses that incident waves are due to re-reflections alone, while (4) indicates that both reflected and re-reflected waves contribute to the amplitude measured at the wave paddle. Finally (5) is an assumption of full reflection from a wave paddle at rest.

Solving (3)-(5) with respect to X_a yields

$$X_a = (2A_l - A_0)F \tag{11}$$

where

$$F = -\frac{i}{e_0 - \sum_{j=1}^{\infty} e_j} \tag{12}$$

This is the Fourier-domain recipe for active absorption. The directional dependence comes in through e_j, primarily for $j = 0$ and to a smaller extent for $j > 0$.

In the present formulation we have neglected the inevitable delay in the position servo loop of the wavemaker control. While it is quite important to compensate for this to get good results at high frequencies, we have chosen to leave it out here for simplicity.

As the system has to work on the fly it is necessary to transform this Fourier-domain recipe to the time domain. This is done using recursive, digital filters. For the wave flume case only a one-dimensional time-domain filter is needed. Multidirectional wavemakers also call for a spatial filter variation along the wavemaker to make the system sensitive to wave directionality. Quasi-3D active absorption can be made neglecting this feature.

REALISATION OF ACTIVE ABSORPTION IN TIME AND SPACE

For practical realisation of active absorption the desired paddle position must be available for each paddle segment at every instant of time. For this purpose a 2D digital filter of the form

$$v_{n,m} = \sum_{l=-M_2}^{M_2} \sum_{k=0}^{M_1} a_{k,l} u_{n-k,m-l} + \sum_{l=-N_2}^{N_2} \sum_{k=1}^{N_1} b_{k,l} v_{n-k,m-l} \tag{13}$$

is used. Here $(a_{k,l}, b_{k,l})$ are filter coefficients, and (M_1, N_1) and (M_2, N_2) are the respective filter orders in time and space. In line with (11), the filter input is $2\eta_l - \eta_0$, and the output is X at discrete points in time and space

$$\left. \begin{array}{l} v_{n,m} = X \\ u_{n,m} = 2\eta_l - \eta_0 \end{array} \right\} \quad @ \quad (t, y) = (n\Delta t, m\Delta y) \tag{14}$$

where Δt is the time increment and Δy is the paddle width of the segmented wavemaker. The second term in (13) is the recursive part of the filter. Present output is a function of *previous output* as well as present and previous input. We have chosen to limit the recursive element to the time variable by specifying $N_2 = 0$. For the wave flume case we further have $M_2 = 0$. The general transfer function for (13) is

$$\tilde{F}(\omega, k_y) = \frac{\displaystyle\sum_{l=-M_2}^{M_2} \sum_{k=0}^{M_1} a_{k,l} z_1^{-k} z_2^{-l}}{1 - \displaystyle\sum_{l=-N_2}^{N_2} \sum_{k=1}^{N_1} b_{k,l} z_1^{-k} z_2^{-l}}; \qquad \begin{array}{l} z_1 = e^{i\omega \, \Delta t} \\[2mm] z_2 = e^{ik_y \, \Delta y} \end{array} \tag{15}$$

and the coefficients $\left(a_{k,l}, b_{k,l}\right)$ are determined by matching $\tilde{F}(\omega, k_y)$ with the active absorption target transfer function (12). This must be done under the stability constraints of a recursive filter. With $N_2 = 0$, the stability condition states that the poles of (15) must be within the unit circle in the z_1-plane, see e.g. Antoniou (1979). Another constraint is that the response must be limited in the zero-frequency limit. This is necessary to avoid slow drift of the wave paddle.

SYSTEM PERFORMANCE AUTO-CHECK

The constraints of the filter design result in a less than perfect system. In order to evaluate the system performance we now repeat the equations (11) and (3)-(5), but now for the actual system, meaning that $F = F(\omega, k_y)$ is now replaced by $\tilde{F}(\omega, k_y)$. We still assume that the water behaves according to linear wavemaker theory and that full reflection occurs from a paddle at rest. Signifying actual quantities by a tilde, we have

$$\tilde{X}_a = \left(2A_I - \tilde{A}_0\right)\tilde{F} \tag{16}$$

$$\tilde{A}_I = ie_0\tilde{X}_a + \tilde{A}_{RR} \tag{17}$$

$$\tilde{A}_0 = i\tilde{A}_a \left(e_0 + \sum_{j=1}^{\infty} e_j\right) + \tilde{A}_R + \tilde{A}_{RR} \tag{18}$$

$$\tilde{A}_R = \tilde{A}_{RR} \tag{19}$$

Note that we distinguish between the desired incident wave amplitude, A_I, and the one we actually expect to obtain, $\tilde{A}_I$. It is instructive to look at the ratio λ defined as

$$\lambda = \frac{\tilde{A}_I - \tilde{A}_R}{A_I - \tilde{A}_R} \tag{20}$$

Further defining RR as

$$RR = 1 - \lambda = \frac{\tilde{A}_I - A_I}{\tilde{A}_R - A_I} \tag{21}$$

we see that in case of pure absorption, where $A_I = 0$, RR is the re-reflection coefficient of the absorbing wavemaker. From (16)-(19) we may obtain

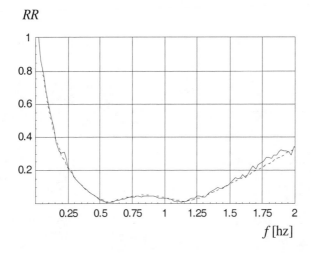

Figure 3. "Re-reflection" coefficient. Dotted curve: expected value, full curve: experimental value

$$\lambda = \frac{2e_0}{-\dfrac{i}{\tilde{F}} + e_0 + \sum_{j=1}^{\infty} e_j} \tag{22}$$

which shows how expected values of λ and RR can be determined theoretically given the characteristics of the system through $\tilde{F}$. It also shows that the expected value of λ and thereby of RR is independent of both A_I and $\tilde{A}_R$. Thus we may regard RR as a "re-reflection" coefficient of the active absorption system in any case.

Multiplying both numerator and denominator of (22) by $\tilde{X}_a$ and eliminating $\tilde{F}$ by (16), we get

$$\lambda = \frac{2e_0 \tilde{X}_a}{-i\left(2A_I - \tilde{A}_0\right) + \tilde{X}_a\left(e_0 + \sum_{j=1}^{\infty} e_j\right)} \tag{23}$$

This form provides a means of obtaining λ and RR from a practical experiment. Any systematic deviation between the theoretical values (22) and experimental values (23) of λ is due to unexpected behaviour of the control system. Thus comparison between (22) and (23) provides an auto-check of the system performance. Figure 3 shows an example. The water depth in the experiment was 0.7m and a fully reflective plate was placed in the far end of the wave flume. The incident waves were synthesised from a JONSWAP spectrum with a peak period of 2.0s and a significant wave height of 0.05m.

Figure 4. New wavemaker at DHI Water & Environment

QUALITATIVE BASIN TESTS FOR OBLIQUE WAVES

The Active Wave Absorption Control System DHI AWACS[3], was developed as an important part of a new multidirectional wavemaker at DHI Water & Environment, see Schäffer et al. (2000). Figure 4 shows a photograph of the new wavemaker. The wavemaker is of piston-type with vertical hinges between the segments providing a step-wise linear segmentation of the paddle front. *Two* wave gauges are integrated in each segment for providing a hydrodynamic feedback with *reduced* aliasing effects. The paddle width is $\Delta y = 0.5\text{m}$, the maximum stroke is 0.6m and the maximum nominal wave height is 0.55m at period 2.3s. With a paddle height of 1.2m usual water depths would range from 0.2 to 0.9m. Precision control of each actuator is achieved using a brushless AC servomotor with a ball screw transmission and encoder feedback. This ensures a highly accurate control of paddle position. Novel Digital Servo Controller (DSC) design allows for cross-communication between neighbouring controllers. The DSC's receive real-time information on the desired incident waves from a host PC and operate the active absorption procedures. The cross-communication is required for fully 3D active absorption corresponding to $M_2 > 0$ in (15). We have chosen a hardware limitation that allows for $M_2 \le 2$, which means that up to the five nearest DSC's are involved in each paddle position update. The rate of this update is 1000 Hz.

Figure 5. Oblique waves absorbed by the 3D wavemaker to the left. The waves are generated by the long-crested wavemaker to the right

Figure 6. Oblique waves reflected on the 3D wavemaker at rest

Schäffer and Skourup (1996) tested the active absorption of oblique waves in a numerical wave tank based on the Boundary Integral Equation Method. Their quantitative results agreed quite well with the theoretically expected performance. The active absorption system is now ready for use with the new multidirectional wavemaker. Qualitative tests of the absorption performance were made by letting a long-crested wavemaker send oblique waves onto the 3D wavemaker. Figure 5 shows the resulting progressive wave field being absorbed by the 3D wavemaker. The wave direction is $\theta = 45°$. For reference Figure 6 shows the same situation but with the active absorption turned off. Due to reflection at the wavemaker a cross pattern results.

CONCLUSIONS
The general theory and realisation procedures have been outlined for simultaneous generation and active absorption of multidirectional waves. A method by which the system checks itself has been derived and the use of it has been shown for a wave flume example. A qualitative test of the active absorption of oblique waves by a multidirectional wavemaker has shown promising results. Although there is no doubt that the performance of the system is good, quantitative tests are still required for checking the system in detail.

ACKNOWLEDGEMENTS
J.U. Fuchs, P. Hyllested, K.P. Jakobsen, N. Mathiesen, and B. Wollesen are acknowledged for their important role in the development of wavemaker technology at DHI Water & Environment.

REFERENCES

Antoniou, A. (1979). *Digital Filters: Analysis and Design.* McGraw-Hill, New York, New York, USA.

Biésel F. (1951). Etude théorétique d'un type d'appareil à houle. *La Houille Blanche* (6), pp 152-165.

Dean, R. and R.A. Dalrymple (1991). *Water Wave Mechanics for Engineers and Scientists.* World Scientific, Singapore.

Milgram, J.H. (1970). Active water-wave absorbers. *J. Fluid Mech.* 42(4), pp 845-859.

Schäffer H.A., J.U. Fuchs, P. Hyllested, N. Mathiesen and B. Wollesen (2000). An absorbing multidirectional wavemaker for coastal applications. *Proceedings of 27th Int. Conf. Coastal Engng.* Sydney, Australia, pp 981-993.

Schäffer, H.A. and G. Klopman (2000). Review of multidirectional active wave absorption methods. *Journal of Waterway, Port, Coastal and Ocean Engineering*, 126(2), ASCE, pp 88-97.

Schäffer, H.A. and J. Skourup (1996). Active absorption of multidirectional waves. *Proceedings of 25th Int. Conf. Coastal Engng.* Orlando, Florida, USA, pp 55-66.

Steenberg C.M. and H.A. Schäffer (2000). Second-order wave generation in laboratory basins. *Proceedings of 27th Int. Conf. Coastal Engng.* Sydney, Australia, pp 994-1007.

THE DEVELOPMENT OF A
NEW SEGMENTED DEEPWATER WAVE GENERATOR

Lambert N.G. Romijnders[1], B.Sc

Abstract: For MARIN's new Offshore Basin, an L-shaped segmented wave generator has been developed and installed that allows high and multidirectional deepwater waves while simultaneously allowing layered surface current in one direction. Other features are a low profile above water, very accurate drive and control system, second-order wave steering, selectable wave groupiness, and active absorption of reflected waves. The system can produce a new wave field accurately from scratch with only one single correction.

INTRODUCTION

In June 1997, the Maritime Research Centre Netherlands (MARIN) commissioned a consortium of Hydraudyne Systems & Engineering (now Rexroth Hydraudyne) and WL|Delft Hydraulics with the design and supply of two multidirectional wave generator systems, for their new facilities Seakeeping and Manoeuvring Basin and Offshore Basin. The contract was won in a world-wide competition against major wave generator manufacturers. This paper focuses on the Offshore Basin wave generator.

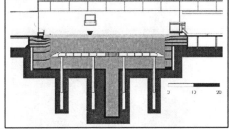 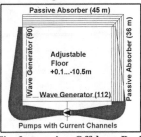

Fig.1: cross-section of MARIN's Offshore Basin Fig. 2: top view Offshore Basin

1 Senior System Engineer, Systems & Control Department, Rexroth Hydraudyne B.V., P.O.Box 32, NL 5280 AA Boxtel, The Netherlands, l.romijnders@rexroth-hydraudyne.com

The Offshore Basin layout is shown in fig.1 and fig.2: an L-shaped 202-segment wave generator with opposing passive wave absorbers, in a 10.5 m deep 45x36m basin with extensive current capabilities and a stiff adjustable floor. The segments are wet-back hinged flaps.
The layout is described in detail by MARIN's Buchner, Wichers and de Wilde (1999).

REQUIREMENTS

The type of wave generation specified was: longcrested random and regular waves under any angle, and shortcrested waves. The longcrested waves were specified as 1st or 2nd order waves, with or without absorption of reflected waves from any angle.
Design wave was: JONSWAP, Tp=3.0 s, Hsig=0.3 m. Maximum wave frequency was 2 Hz. Wave quality was mainly specified as an allowed deviation of a measured regular wave time trace from the corresponding theoretical wave, and as an allowed deviation of the measured random wave spectrum from the theoretical spectrum.
The above was specified with the current system switched off.

The segment width was specified as 0.4 m, and the hinge depth as only 1.2m. The latter was minimised to create maximum space below the hinge, for current to reach the water surface (hinge as high as possible). Motion quality of the segments was mainly specified with maximum amplitude errors and with maximum super- and subharmonics. The result (in combination with the design wave) was the requirement of a huge segment rotation amplitude (30 degree). To enable good quality waves with this, a special steering algorithm was applied.

Fig. 3: wave generator in final construction phase

DESIGN DESCRIPTION
Mechanical:

The hinged stainless steel segments each have an arc-segment attached to the backside, stiffened with spokes and provided with a running surface for a timing belt. This black toothed belt is looped around two reversing wheels and a pulley, that is mounted on the output shaft of a planetary helical gearbox with an AC-servomotor. The belt is fastened at both arc-segment ends, and pretensioned. So if the motor rotates the pulley, the entire segment is rotated. Reversing wheels and pulley are combined into a stainless steel drive housing. This novel concept allows:

- a small hinge depth, suitable for high frequencies but –with the big amplitude- also suitable for high and accurate waves at low frequencies.
- the current outlet opening to be close to the surface.
- to keep the total transmission ratio constant over the full displacement of the segment.
- to drive the segment in the beneficial area of high speed and low force (construction is lightweight and stiff).
- to bring in sufficient lateral flexibility to avoid alignment problems (with their associated lifetime reduction).
- to keep the test area as large as possible, because a short distance to the rear wall behind the wave generator is sufficient.
- to maintain a low profile above water, facilitating towing carriage freedom.

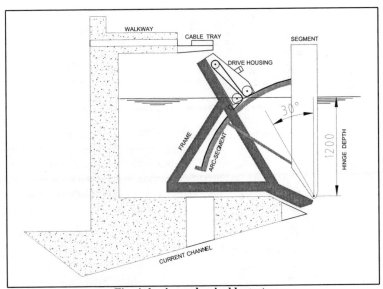

Fig. 4: basic mechanical layout

Electric:

Main power requirement is approx. 250 kVA; the power is fed in via a 10 kV transformer. The electric equipment is mounted in standard cabinets, located on the walkway behind the segments, see fig.12 (in the background).

A control strategy is designed, that implements on each segment a 'water level' control loop called '3D Active Reflection Compensation' (3D-ARC), around an accurate position control loop. It absorbs reflected waves from any angle.

To keep things workable and flexible towards any future demands, the control system is designed as a 3-level distributed digital control system, using modular hardware. This hardware is selected mainly as Commercial Off The Shelf items. In total some 350 microprocessors work in concert, and synchronous within 1 ms.

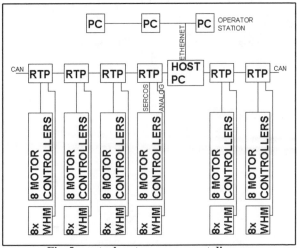

Fig. 5: control system component diagram

On the bottom control level are located the AC-servomotors and controllers, that control each segment in the complex multipath position loop (setpoint and feedback directly in degrees segment angle). The interface with the medium control level was selected to be the fibre-optic based SERCOS interface, designed for extremely accurate and extremely synchronous control of multiple axes. Controllers are grouped as 16 in a cabinet.

Wave Height Meters (WHM) along the front of each segment measure the local dynamic water level. They use the 2-wire capacitive principle. The wires are looped back under water, so 4 wire parts are piercing the surface. The wires are isolated, so the WHM's are galvanically separated from the water and from each other.

On the medium control level are located the Real Time Processors (RTP), that each have 8 segments connected. The RTP is a multiprocessor single-board computer, developed by WL|Delft Hydraulics. The main processor is a single SHARC DSP, the board design however has space for four of these to avoid bottlenecks in computational power for future projects.
The RTP's are accompanied by a real-time Host PC for setpoint distribution.

Fig. 6: Real Time Processor

The RTP software has several clever features:

- State machine for bumpless start-up, motion and shutdown. All states and any errors are reported back to the operator.
- ARC executed as long wave ARC or as 3D-ARC, using wave information from adjacent segments.
- Automatic check of offset and gain for all WHM's simultaneously. Any WHM exceeding a pre-set threshold deviation is identified.
- Filters for suppression of basin resonances.
- Filters for compensation of drive system dynamics.
- Smooth clipping of long strokes in ARC, that are required for long wave absorption.
- Smooth clipping of extreme (out of design range) setpoints in random spectra.
- Full protection from system overload.

At the top control level are located three MSWindows PC's in a network, that perform different tasks. The Operator Station has the graphical user interface for wave computation and for wave generation. The other PC's perform auxiliary functions.

Important software features are the possibilities for:

- second-order waves and for shortcrested waves.
- the pre-selection of waves with certain wave group characteristics: without actually using the basin, a project manager can select a suitable wave with desired groupiness to arrive at his test goal, using the least possible basin time.
- compensation of the dynamic behaviour of both wave generator and basin, to aid in obtaining a good fitting spectrum at first time.
- quality control: every parameter set by the operator is stored, so the same wave and/or wave event can be repeated, even after years.

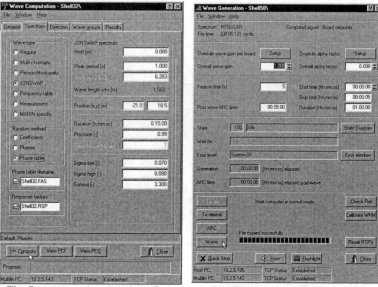

Fig. 7: computation user interface   **Fig. 8: generation user interface**

The computation user interface is a friendly shell around WL|Delft Hydraulics' 'Auke/generate' wave computation software package that runs on an auxiliary PC.

DEVELOPMENT

Rexroth Hydraudyne chose not to adopt a so-called concurrent engineering approach. Beside the technical performance and risks, also the economy was very important in view of the large number of segments. The step by step approach seemed safer and sounder, so the development sequence was:

- a Concept Test, to get quick and firsthand experiences with the drive concept.
- an Engineering Model Test (4 segments in a water-filled sea-container) to get an impression of the wave performance and of the effects of the nearby rear wall.
- detail design, using full finite element analysis.
- a Prototype Test: a 12-segment pre-production prototype (pilot series) was constructed and thoroughly tested in a wave flume at WL|Delft Hydraulics, to confirm the wave performance and as a final check on the mechanical and control system with its key software.
- series production: after approval of the prototype, series production was undertaken, guided by Factory Acceptance Tests
- erection at the building site, commissioning and Site Acceptance Test. This whole process was guided by a dedicated site manager.

TEST RESULTS
Segment drive system:
Due to the careful design and production, the accuracy of the segment motion is very good. Putting a hand on a segment top during motion, reveals no irregularity, not at motion turnaround nor at midstroke. Gearbox backlash is not noticeable.
Analysis of measured segment motion shows near-perfect results. The attenuation of the higher harmonics is at least 50 dB, for the lower even more.

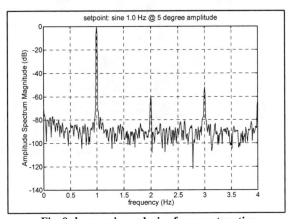

Fig. 9: harmonic analysis of segment motion

Drive system Bode plot, measured by sinusoidal motion corresponding to a wave steepness of 7%, shows the -3dB point at 9.2 Hz and the 90 degree phase lag point at 5.6 Hz. Such a generous bandwidth allows for the proper generation of 2nd order waves, since 2nd order waves largely require double bandwidth (for a 2 Hz wave, 4 Hz bandwidth is required).

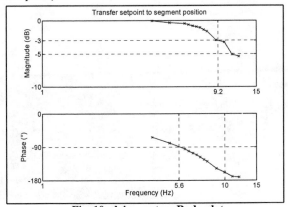

Fig. 10: drive system Bode plot

Wave performance:

The wave performance and 3D-ARC performance is elaborated by WL|Delft Hydraulics' Van Dongeren, Klopman, Petit and Reniers (2001).

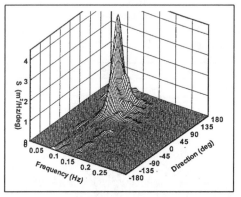

Fig. 11: sample shortcrested spectrum, generated and measured by MARIN

Electromagnetic compatibility:

Electromagnetic compatibility was found to be excellent. It was measured to conform to the strict European standard EN55011 class B (for residential areas and small industries). The wave generator does not emit significant noise to the laboratory sensors, and vice-versa is not influenced by noise from outside.

Fig. 12:wave generator in operation

RELIABILITY ISSUES

After manufacture of the equipment in the 1st Q. 1999, there was a one-year delay in the civil works. Commissioning was started on 10 April 2000, and there were some failing items in that period, most of them already at the first power-on. This was quickly resolved, the failing items can be considered normal for this kind of project size. Transfer to MARIN was on 07 July 2000. At this moment, power-on hours are ~12000 (control system); operation hours with waves are ~650.

Items that failed:
- 8 motor controllers (out of 202)
- 1 hard disk
- 1 rotating light
- 2 WHM's

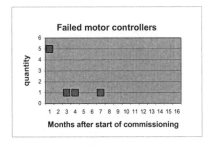

Fig. 13: reliability vs. time: start of a bathtub curve

ASPECTS FOR THE FUTURE

There is always something that could be developed further, such as:
- Reduction of phase lag. The -90 degree point can be moved to above 10 Hz. It has already been tested, but not yet implemented.
- Remote maintenance and support, to look inside a remote system's computers and controls. Under development.
- Reduction of bypass flow between the segments, by sliding or gap seal.
- Introduction of surface current as a parameter for waves. Under development.

CONCLUDING REMARKS

Rexroth Hydraudyne has successfully developed and built a new wave generator system to MARIN's full satisfaction. The state-of-the-art technology and the economy of this system make it suitable for many laboratories, for newbuilding or for upgrade.

ACKNOWLEDGEMENTS

Rexroth Hydraudyne wishes to thank MARIN for the confidence that MARIN has put in them. The excellent co-operation with WL|Delft Hydraulics during the realisation of this large project is greatly appreciated.

REFERENCES

Buchner, B., Wichers, J.E.W. and Wilde, J.J. de, 1999. Features of the State-of-the-Art Deepwater Offshore Basin, OTC 10841,

Van Dongeren, A.R., G. Klopman, H. Petit, A. Reniers, 2001. High-Quality Laboratory Wave Generation for Flumes and Basins, WAVES2001.

LONGSHORE SEDIMENT TRANSPORT AS A FUNCTION
OF ENERGY DISSIPATION

Ernest R. Smith[1], M.ASCE, and Ping Wang[2]

Abstract: Experiments to measure waves, currents, and sediment transport rate for two breaker types, plunging and spilling, were conducted in a large-scale three-dimensional physical model. A large difference in cross-shore distribution and total sediment transport rate was found between the two breaker types. Existing predictive equations for total transport rate did not predict the data well. With the exception of the Kamphuis (1991) equation, which included a dependence on wave period, the predictive equations did not differentiate between breaker types. Sediment transport increased where dissipation was high for plunging waves, but high dissipation did not necessarily produce increased transport rates with spilling waves.

INTRODUCTION

Total longshore sediment transport rate and its cross-shore distribution in the surf zone are essential to many coastal engineering studies. The U.S. Army Engineer Research and Development Center recently completed a large-scale Longshore Sediment Transport Facility (LSTF) to study longshore sediment transport. The facility has the capability of simulating wave heights that are comparable to annual averages along many low-wave energy coasts, such as many of the beaches along the Gulf of Mexico and the Great Lakes in the U.S. This paper describes the capabilities of the LSTF, presents comparisons to existing predictive equations for longshore sediment transport, and examines sediment transport as a function of energy dissipation.

LONGSHORE SEDIMENT TRANSPORT FACILITY

The LSTF was designed to generate waves and currents and conduct sediment transport experiments at a large scale. The facility consists of a 30-m wide, 50-m long, 1.4-m deep basin, and includes four wave generators, a sand beach, a recirculation system, and an

1 Research Hydraulic Engineer, Engineering Research and Development Center, Coastal and Hydraulics Lab, 3909 Halls Ferry Road Vicksburg, MS 39180-6199, smithe@wes.army.mil
2 Department of Geology SCA 528, University of South Florida, Tampa, FL 33620, pwang@chmua1.cas.usf.edul

instrumentation bridge (Figure 1). The following paragraphs describe the equipment used in the facility.

Fig. 1. An experiment in progress at the LSTF

Four wave generators were used to produce waves in the LSTF. The generators are synchronized to produce unidirectional long-crested waves at a 10-deg angle to shore normal. A digitally controlled servo electric drive system controls the position of the piston-type wave board, and produces waves with the periodic motion of the board. The system allows a variety of regular and irregular wave types to be produced. The generators were oriented at a 10-deg wave angle for the present study, but they can be positioned to produce waves from 0 to 20 deg from shore normal.

The beach was constructed using approximately 150 m^3 of fine quartz sand having a median grain diameter, d_{50}, of 0.15 mm. It was desired to obtain an accurate rate of longshore sediment transport and its cross-shore distribution with minimal longshore variation and boundary influences. Therefore, straight and parallel contours were maintained throughout the model to maximize the length of beach over which longshore uniformity of waves and currents exist in the basin. Beaches having "three-dimensionality" affect incident waves and, subsequently, the longshore currents and sediment transport associated with the waves.

The model beach was of finite length and bounded at the upstream and downstream ends. To minimize adverse laboratory effects created by the boundaries and to produce uniform longshore currents across the beach, it was necessary to supplement wave-driven currents. A recirculation system was installed which consisted of 20 independent vertical turbine pumps placed in the cross-shore direction at the downdrift boundary. Flow channels were placed upstream of each pump to direct flow to the pump, which externally re-circulated water to the upstream end of the facility, where it was discharged through flow channels onto the beach. The objective of the system was to maximize the length of beach over which waves and longshore currents were uniform by continually re-circulating wave-driven longshore current through the lateral boundaries of the facility. Each pump included a variable speed motor to control discharge rates. The variable speed motors were adjusted to control the cross-shore distribution of longshore current based on the measured distribution of waves.

A 21-m instrumentation bridge spanned the entire cross-shore dimension of the beach and served as a rigid platform to mount instruments and observe experiments. Each end of the bridge is independently driven on support rails by drive motors, which allow it to travel the entire length of the wave basin. The bridge could be driven either at the bridge or remotely by entering the desired longshore (Y) location on a PC in the LSTF control room.

INSTRUMENTATION

Time series of water surface elevations were measured using single-wire capacitance-type wave gauges. Ten gauges were mounted on the instrumentation bridge to provide wave heights as they transformed across the beach. The cross-shore location of the gauges could be repositioned on the bridge depending on the wave conditions. Additionally, a gauge was placed in front of each wave generator to measure offshore wave characteristics.

Ten acoustic doppler velocimeters (ADVs) were used to measure orbtal wave velocities and unidirectional longshore currents. The ADVs were positioned at the same cross-shore position on the bridge with the wave gauges, but separated by approximately 40 cm in the longshore direction to prevent interference between the two instrument types. As with the wave gauges, the ADV cross-shore location could be repositioned on the bridge.

The eight most offshore ADVs were down-looking three-dimensional (3D) sensors, which sampled velocities in the x, y, and z directions 5 cm below the sensor. The two most inshore ADVs were two-dimensional (2D) side-looking sensors, which sampled velocities in the x and y directions, also 5 cm from the sensor. The 2D sensors were used to allow measurements in water too shallow for the 3D current meters.

All of the ADVs were mounted on vertical supports that allowed the vertical position of the sampling volume to be adjusted. Typically, the ADVs were positioned vertically to sample at a location that gave the average velocity in the water column (an elevation equal to one third of the water depth from the bottom). However, some experiments were conducted in which the vertical position of the ADVs was varied to obtain the velocity distribution through the water column.

An automated beach profiler mounted to the instrumentation bridge was used to survey the beach. A mechanical spring-wheel system attached to a vertically mounted rod followed the sand elevation as the system moved cross-shore along the bridge. The vertical movement of the rod produced a voltage, which was recorded and converted to elevation. The vertical resolution of the system was ±1 mm. Horizontal positioning of the profiler was controlled by the bridge position and a cross-shore motor mounted on the bridge. The profiling system was amphibious to allow the entire beach to be surveyed without draining the basin. Survey data were obtained every 5mm in the cross-shore direction and every 0.5 or 1.0 m in the longshore direction. The middle portion of the beach remained uniform and the higher resolution was not required. Higher irregularity in the bathymetry occurred near the upstream and downstream boundaries, and denser profile lines were required.

Twenty traps were installed in the downdrift flow channels to collect sand transported through the downdrift boundary. Seventeen traps were placed in the flow channels of the first 17 pumps, and one trap was placed in the flow channel of pump 19. Traps were omitted from channels 18 and 20 because sediment transport was expected to be low in the offshore region. The remaining two traps were placed in the swash zone.

To determine the cross-shore distribution of longshore sediment transport, each sand trap was equipped with three load cells to weigh the amount of trapped sand. The total capacity of the 20 traps is 2500 kg. As experiments progressed, the updrift end of the beach became depleted of sand. It was necessary to dredge the traps to replenish the updrift portion of the beach, and rebuild the beach to uniform and parallel contours.

LONGSHORE TRANSPORT EXPERIMENTS

Longshore transport experiments were conducted for two breaker types with wave conditions given in Table 1. Zero-moment wave height, H_{mo}, water depth, h, and incident wave angle at the generator, θ, were the same in both cases, and only peak wave period, Tp, varied.

Table 1. Longshore Sediment Transport Experiment Wave Conditions

Breaker Type	H_{mo} m	T_p sec	h m	θ deg
Spilling	0.247	1.5	0.9	10
Plunging	0.247	3.0	0.9	10

Uniformity of Longshore Currents

The first step of the experiments was to determine the distribution of wave-induced longshore current. Visser (1991) determined from laboratory experiments that if the re-circulated currents either exceeded or were less than the wave-driven currents, an internal current would develop and re-circulate within the offshore portion of the basin. Visser also found that as the pumped currents approached the proper current, the internally re-circulated current was minimized. Therefore, it was desired to match wave-driven currents with pumped currents. Initial pump settings were based on results of the numerical model NMLONG (Kraus and Larson 1991), and the iterative approach described by Hamilton and Ebersole (2001) and Hamilton, et al. (2001) was used to determine the optimum pump settings. After the beach had reached an equilibrium or quasi-equilibrium profile, and the pumped currents matched measured velocities, experiments on longshore transport rate were initiated. Spilling breakers required 1,330 minutes to reach equilibrium in the model and plunging breakers reached equilibrium after 280 minutes of wave action.

Results

Figure 2 shows the transformation of waves from offshore to nearshore. Spilling waves show a fairly gradual decrease in wave height across the beach. Plunging breakers shoal farther shoreward and the break point is evident where wave height sharply decreases. The corresponding equilibrium beach profiles formed by the two wave types are shown in

Figure 3. The profile associated with spilling breakers fit the theoretical shape of Dean (1977) reasonably well. The offshore portion of the profile is controlled by the constructed profile. A breakpoint bar and trough formed under plunging waves, and slight erosion occurred compared to the spilling wave case.

Fig 2. Wave transformation of spilling and plunging breakers

Depth-averaged longshore current measured in the center portion of the beach for both wave cases is shown in Figure 4. Both cases show a subtle peak immediately inshore of the

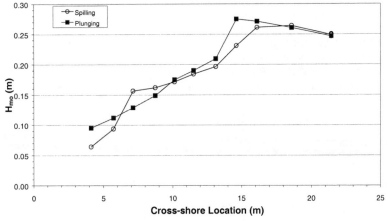

breaker line, and velocities increase near the shoreline. Although, measured velocities are slightly higher with plunging waves, the two cases have similar cross-shore distributions of longshore velocity.

Sediment flux was calculated from the rate of sand collected in each trap. Although the wave heights were identical for the two wave cases, and the resulting wave-driven longshore currents were similar, the sediment flux of the two breaker types is significantly different (Figure 5). The plunging wave case has peaks in transport at the break point and in the swash. A single peak occurs in transport with the spilling wave case in the swash zone. Total transport for spilling waves is approximately a third less than that of the plunging waves. Throughout the rest of the surf zone, sediment flux is similar between the cases.

It is not surprising that the plunging waves show greater transport over spilling waves near the breaker line. Turbulence associated with spilling breakers remains close to the surface in the bore. The jet associated with plungers penetrates deep into the water column and causes sand to be suspended and transported by the longshore current. Through the mid-surf zone, turbulence associated with broken plunging waves remains near the surface,

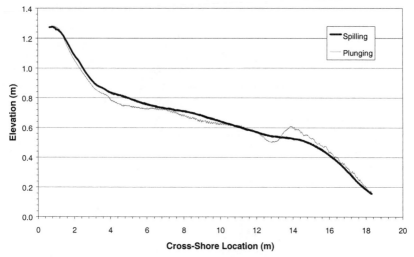

Fig 3. Equilibrium profiles formed by spilling and plunging breakers

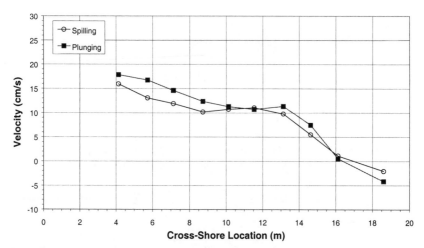

Fig. 4 Distribution of longshore currents produced by spilling and plunging waves

and transport is similar between breaker types. However transport in the swash zone with plunging waves is influenced by more energetic conditions associated with the longer wave period, and is substantially higher.

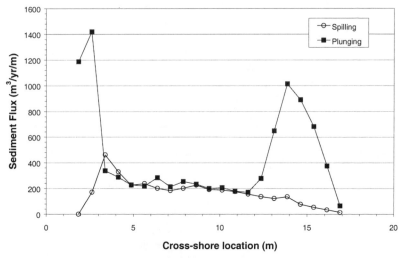

Fig. 5. Sediment flux distribution of spilling and plunging waves

Comparison to Predictive Equations

Measured transport rates for the spilling and plunging case were compared to the CERC formula (Shore Protection Manual 1984), and predictive equations of Kamphuis, et al. (1986), Kamphuis (1991), and Kraus (1988). The CERC formula is given by:

$$I = \frac{K}{16\sqrt{\dfrac{H_b}{h_b}}}\rho g^{\frac{3}{2}} H_b^{\frac{5}{2}} \sin(2\theta_b) \tag{1}$$

in which I is the submerged weight transport rate, K is an empirical coefficient recommended by the Shore Protection Manual to be 0.39, H_b, h_b, and θ_b are significant wave height, water depth, and wave angle at breaking, respectively, ρ is fluid density, and g acceleration due to gravity.

Kamphuis, et al. (1986) developed an empirical equation based on field data as:

$$Q = 1.28\frac{H_b^{3.5}m}{d}\sin(2\theta_b) \tag{2}$$

in which Q is the total volume transport rate in kg/s, d is sediment grain size, and m is beach slope. . Kamphuis (1991) modified Equation 2 based on laboratory data and re-analysis of existing field data to include the influence of wave period and give Q in m^3/yr:

$$Q = 6.4x10^4 H_b^2 T_p^{1.5} m^{0.75} d^{-0.25} \sin^{0.6}(2\theta_b) \tag{3}$$

Kraus, et al. (1988) used a different approach and assumed that the total rate of longshore sediment transport in the surf zone is proportional to the longshore discharge of water. They found:

$$Q \propto K_d (R - R_c) \tag{4}$$

where K_d is an empirical coefficient that may relate to sediment suspension, R_c is a threshold value for significant longshore sand transport, and R is the discharge parameter, which can be accurately measured in the LSTF and calculated in the field as:

$$R = nV_{ls} x_b H_b \tag{5}$$

in which n is a constant, V_{ls} is the average longshore current velocity, and x_b is the surf-zone width. Based on field data, Kraus, et al. suggest $K_d = 2.7$ and $R_c = 3.9$ m^3/s.

Table 2 lists measured transport rates and the values of the predictive equations. The CERC equation over-predicts the total rate for spilling waves by over 700 percent and plunging waves by nearly 250 percent. It is interesting to note that the other equations may predict transport rate for one breaker type well, but none predict values that estimate both types well. In fact, with the exception of Kamphuis (1991), which includes wave period, the predictions do not reflect differences between breaker types.

Table 2. Comparison of Measured and Predicted Transport Rates

Transport Rates	Spilling Case m^3/yr	Plunging Case m^3/yr
Measured	2,660	7,040
CERC formula	22,030	23,850
Kamphuis (1986)	10,760	9,100
Kamphuis (1991)	2,200	5,360
Kraus (1988)	2,670	3,150

Measured sediment flux and an energy dissipation parameter were plotted as a function of cross-shore location for plunging and spilling breakers in Figures 6 and 7, respectively. The dissipation parameter was calculated by multiplying longshore velocity by the difference in energy, i.e., the difference in H^2, divided by distance between cross-shore measurement locations. Figure 6 shows a sharp increase in sediment flux corresponding to a peak in energy dissipation for plunging waves. Although dissipation is less for the spilling waves, Figure 7 shows that sediment flux does not increase with a significant increase in energy dissipation.

The figures indicate that turbulence associated with plunging breakers penetrates to the sand bed and suspends sediment, which is transported by the longshore current. Regions characterized by higher energy dissipation with spilling waves did not produce an increase in sediment flux.

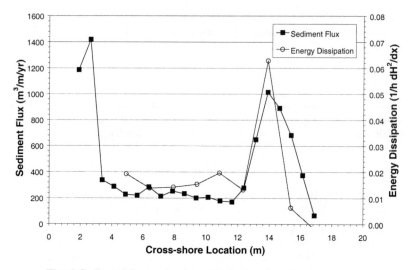

Fig. 6. Sediment flux and energy dissipation for plunging waves

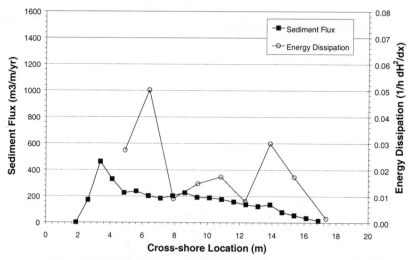

Fig. 7. Sediment flux and energy dissipation for spilling waves

SUMMARY

Waves, currents, and sediment transport rate were measured in a large-scale physical model. Large differences were found in the cross-shore distribution and total sediment transport rates between spilling and plunging breakers. Existing predictive equations did not predict the total transport well and, with the exception of the Kamphuis (1991) equation, generally didn't differentiate between breaker types.

The data show that transport increases where wave energy dissipation is high with plunging breakers. High dissipation didn't necessarily produce increased transport rates with spilling breakers. Further study will be given to sediment transport as a function of breaker type, including additional experiments and comparisons to field data.

ACKNOWLEDGEMENTS
David Hamilton, William Halford, David Daily, and Tim Nisley provided technical support for this study. Ping Wang was jointly funded by the U.S. Army Engineer Research and Development Center and the Louisiana Sea Grant College Program. Permission to publish this paper was granted by the Headquarters, U.S. Army Corps of Engineers.

REFERENCES
Dean, R. G., 1977. Equilibrium Beach Profiles: U.S. Atlantic and Gulf Coasts. *Ocean Engineering Report No. 12*, Department of Civil Engineering, University of Delaware, Newark, DE.

Hamilton, D.G., Ebersole, B.A., Smith, E.R., and Wang, P., 2001. Development of a Large-Scale Laboratory Facility for Sediment Transport Research. *Technical Report*, U.S. Army Engineer Research and Development Center, Vicksburg, MS.

Hamilton, D.G., and Ebersole, B.A., 2001. Establishing Uniform Longshore Currents in a Large-Scale Laboratory Facility. *Coastal Engineering*, 42, 199-218.

Kamphuis, J.W., 1991. Alongshore Sediment Transport Rate. *Journal of Waterway, Port, Coastal and Ocean Engineering*, 117(6), ASCE, 624-641.

Kamphuis, J.W., Davies, M.H., Nairn, R.B., and Sayao, O.J., 1986. Calculation of Littoral Sand Transport Rate. *Coastal Engineering*, 10, 1-21.

Kraus, N. C., Gingerich, K.J., and Rosati, J.D., 1988. Toward an Improved Empirical Formula for Longshore Sand Transport. *Proceedings of 21st Coastal Engineering Conference*, ASCE, 1183-1196.

Kraus, N.C., and Larson, M., 1991. NMLONG: Numerical Model for Simulating the Longshore Current – Report 1: Model Development and Tests. *Techincal Report DRP-91-1*, U.S. Army Engineer Research and Development Center, Vicksburg, MS.

Shore Protection Manual, 1984. U.S. Army Engineer Waterways Experiment Station, U.S. Government Printing Office, Washington, D.C.

Visser, P. J., 1991. Laboratory Measurements of Uniform Longshore Currents. *Coastal Engineering*, 15, 563-593.

WAVE SCALING IN TIDAL INLET PHYSICAL MODELS

William C. Seabergh[1], M.ASCE, and Jane McKee Smith[2], M.ASCE

Abstract: Two cases of scaling waves at tidal inlets with physical models are examined. One case was for a study of wave breaking over an ebb shoal in the presence of ebb currents (1:50 scale). Wave heights were compared for two different model scales. The other case compared physical model and prototype currents adjacent to a jetty resulting from a combination of tidal and wave-generated currents using a 1:75 scale model. Results indicated tidal inlet models at these scales reproduce the nearshore wave heights and currents well.

INTRODUCTION

Many physical model studies of tidal inlets must be conducted at scales of 1:50, 1:60 and 1:75 (ratio of model to prototype, or field, dimensions) because of the large horizontal extent of the study region, perhaps exceeding several square kilometers. This paper examines two cases of wave processes modeled in the typical 1:75 to 1:50 range of scales. The first compares laboratory results of wave breaking on a current at an inlet to larger scale experiments. The second case compares model results of wave and tide-generated currents in the vicinity of Grays Harbor north jetty to field measurements. The comparisons allow evaluation of the performance of scaled physical models of tidal inlets.

WAVE BREAKING ON A CURRENT

As part of the U.S. Army Corps of Engineers' Coastal Inlets Research Program, laboratory experiments were conducted to measure wave breaking on an ebb shoal for a typical dual-jettied tidal inlet (Smith and Seabergh 2001). The objective was to provide information to parameterize breaking for numerical wave transformation models. The modeling test basin contained an inlet configuration that included ocean, inlet and bay

1 Research Hydraulic Engineer, U.S. Army Engineer Waterways Experiment Station, Coastal and Hydraulics Laboratory, 3909 Halls Ferry Road, Vicksburg, MS 39180-6199 USA. William.C.Seabergh@erdc.usace.army.mil
2 Research Hydraulic Engineer, U.S. Army Engineer Waterways Experiment Station, Coastal and Hydraulics Laboratory, 3909 Halls Ferry Road, Vicksburg, MS 39180-6199 USA. Jane.M.Smith@erdc.usace.army.mil

regions (Figure 1). The inlet was idealized in shape, with simple contours for the adjacent beach, ebb shoal, and inlet channel. The inlet model could be considered approximately a 1:50 scale model, and this scale was assumed for determining model parameters defining wave height and period based on typical field values. See Seabergh (1999) for more detail. The model wave heights ranged from 2 to 8 cm (which represented 1 to 4 m), and periods were 0.7 to 1.7 s (which scaled 5- to 12-s waves). Also, model ebb current speeds varied from 0 to 32 cm/s. To confirm the use of Froude scaling, the experiments included a series of runs with a 1.45 scaling factor (or 1:34 scale) of the original 1:50 scale. The full model bathymetry was not altered for the scale tests, but the model water level was adjusted to give the correct scaled depth on the shallowest portion of the ebb shoal, where the most intense breaking occurs.

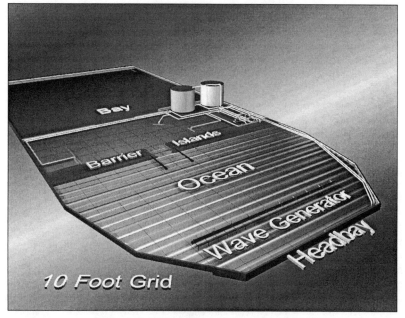

Figure 1. CIRP Idealized Inlet model facility

The laboratory setup included a wave generator located in 30-cm depth, parallel to the shoreline and perpendicular to the ebb current flowing from the inlet. Measurements were made with a co-linear gauge array of six capacitance wave rods and five acoustic Doppler velocimeters, spaced 30 cm apart and alternating along their placement line. These instruments were mounted on a portable rack.

Small-scale physical models are designed to replicate prototype processes in controlled laboratory settings. The premise is that the physical model behaves similar to the prototype, and the model results can be "scaled up" to estimate prototype results. Surface gravity wave processes are scaled using the Froude Number, which is the ratio of inertial forces to gravity forces. The Froude Number, F, is given by

$$F = \frac{U}{\sqrt{g\ell}}$$
(1)

where U is a characteristic velocity, g is gravitational acceleration, and ℓ is a characteristic length. To achieve similitude, the Froude number must be the same for model and prototype. For constant gravitational acceleration, the scaling for velocity is given by

$$\frac{U_m}{U_p} = \sqrt{\frac{\ell_m}{\ell_p}}$$
(2)

where the subscript m denotes a model parameter and p denotes a prototype parameter. The scaling for wave period, T, is given by

$$\frac{T_m}{T_p} = \sqrt{\frac{\ell_m}{\ell_p}}$$
(3)

Froude scaling is applicable for processes in which inertial forces are balanced primarily by gravitational forces, as is the case in most gravity wave problems. Additional information on physical model similitude is given by Hughes (1993).

Stive (1985) conducted a scale comparison of wave breaking on a 1:40-slope beach and found no significant deviation from Froude scaling for a wave height range of 0.1 to 1.5 m with periods of 1.6 to 5.4 s. The wave heights used in the present study range from 2 to 8 cm with periods of 0.7 to 1.7 s. The wave heights in the study are smaller than those presented by Stive, the bathymetry is more complex, and the waves are breaking on a strong ebb current. To confirm the applicability of Froude scaling, a series of model runs was scaled up by a factor of 1.45 and repeated. The wave height, water depth, wave period, and current speed were all scaled (wave height and water depth by a factor of 1.45, and wave period and current speed by a factor of $\sqrt{1.45}$). The full model bathymetry was not altered for the tests, but the model water level was adjusted (raised up) to give the correct scaled depth on shallowest portion of the ebb shoal, where the most intense breaking occurs.

Figure 2 shows a cross-section view of the model depth and gauge positions. Figs. 3 to 5 show results of the wave height variation over the ebb shoal and into the inlet channel for incident wave height of 5.5 cm, peak period of 0.7 s, and current speed of 0, 16, and 27 cm/s. The scaled wave heights are also plotted. The results show good agreement in the wave height across the ebb shoal (cross-shore distance 300-800 cm). The heights in the flat channel (cross-shore distance > 800 cm) show poorer agreement because the depth is not scaled and the current distribution in the channel varies somewhat between cases.

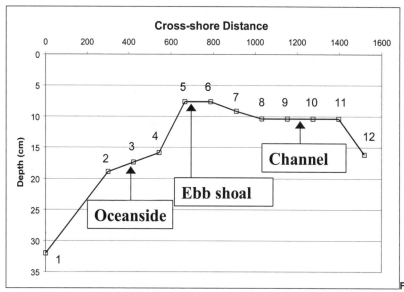

Figure 2. Laboratory cross-shore still-water depth profile and
Gauge Locations (1-12)

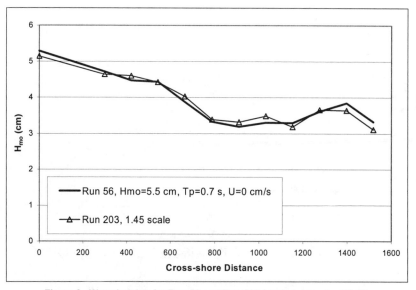

Figure 3. Wave heights for Run 56 and Run 203 (scaled down by 1/1.45)

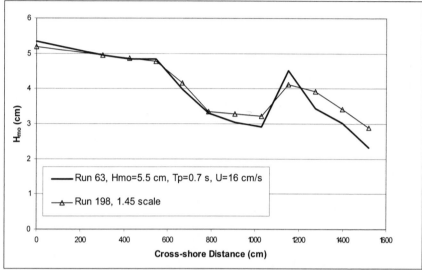

Figure 4. Wave heights for Runs 63 and Run 198 (scaled down by 1/1.45)

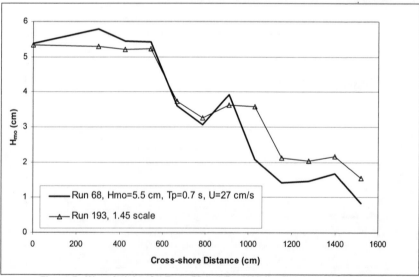

Figure 5. Wave heights for Runs 68 and Run 193 (scaled down by 1/1.45)

Scaling Wave-Generated Currents

Many physical model studies of tidal inlets are concerned with modeling the wave-generated currents created along the shoreline adjacent to the inlet. These currents originate as waves break and are responsible for sediment transport towards the inlet channel, and any modification to a jetty structure must include an examination of effects on wave-generated currents. Also in the vicinity of the inlet the tidal currents are accelerating as they approach the inlet entrance during a rising tide (flood flow) and will contribute to sediment transport. As part of a study effort for the Corps of Engineers' Seattle District, a comparison of field measurements and physical model measurements was performed. The study site was located at Grays Harbor, Washington, and involved the adjacent shoreline of the Ocean Shores community and the Grays harbor Inlet, and the north jetty separating the two (Figure 6). Figure 6 also shows the 1:75 scale physical model limits, representing approximately 1.6 km of Ocean Shores' beach and a portion of the Grays Harbor entrance channel. A flow manifold system was set-up to create tidal current flood flow across the study region as shown in numerical model work of Cialone and Kraus (2001). A snapshot of maximum flood flow for a spring tide is presented in Figure 7. With baffling and aligning flow guides from the manifold into the modeled region, flow distribution was adjusted.

Waves were created with a 24-m long plunger-type wave generator. The generator could be moved to accommodate different wave angles. The generators were programmed with actual prototype wave spectrum information to recreate the scaled waves. Waves selected were based on an approach angle about in agreement with the setup of the generator at 296 deg azimuth. Table 1 shows the conditions selected for simulation in the physical model. Also shown are results from physical model measurements. Figure 8 illustrates a segment of the field data collected by Pacific International Engineering, under contract to the Corps of Engineers. The location of gauge OS5 is shown in Figure 6.

Figure 8 shows the cyclical nature of the longshore current adjacent to the north jetty. The tidal current component can be noted in most cases for the five day period shown. Peak currents typically occur at mid-tide, when peak flood current occurs in the inlet. During large wave events the wave-generated currents dominate and peak currents do not necessarily occur at rising mid-tide. An accoustic-doppler velocimeter was located in the physical model at the location of gauge OS-5 and measurements were collected for the five wave heights and tidal elevations shown in Table 1. The tidal flow system was adjusted for the stage of tide. Figure 9 shows a plot of prototype versus physical model current magnitudes with a line of perfect agreement. Four of the five current conditions were reproduced very closely. One condition, not closely matched, was most likely related to some affect to the flow field, such as wind or a larger scale disturbance/circulation. In light of the complex interactions of the flow field with waves and strong bathymetric variation in the north jetty and shoreline region, the comparison of model and field currents was very good.

Table 1. Verification Conditions at Gauge OS-5*

Case number	Time period	Offshore wave height (m), period, direction	OS-5 prototype current, cm/s	OS-5 model current, cm/s	Tide Elevation, m, mean lower low water
1	5/9/01 - 0430 hr	2.8, 12.5, 288	74.7	83.0	1.02
2	5/9/01 - 1400 hr	2.2, 11.8, 288	60.9	56.0	0.23
3	5/10/01 - 0630 hr	1.8, 9.1, 293	39.2	44.7	1.71
4	5/20/01 - 0430 hr	1.8, 9.1, 293	40.8	47.6	2.15
5	5/25/01 - 1600 hr	2.0, 9.2, 299	54.7	32.0	-0.05

* Location of Gauge OS-5 shown in Figure 6

Figure 6. 1999 photograph showing Grays Harbor Inlet and the boundary for the 1:75 scale physical model of Ocean Shores/Grays Harbor.

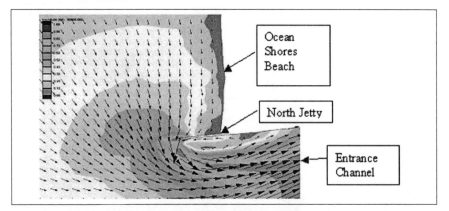

Figure 7. Maximum flood tidal currents flowing into Grays Harbor entrance channel from the north. (from Cialone and Kraus 2001)

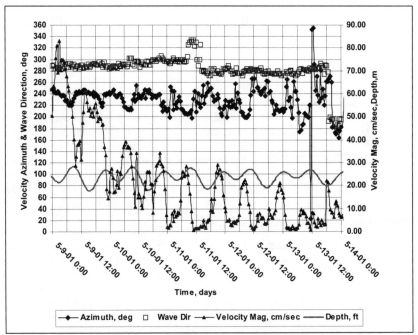

Figure 8. Field measurements of combined wave-generated and tidal currents, 100 m up-coast of submerged portion of north jetty.

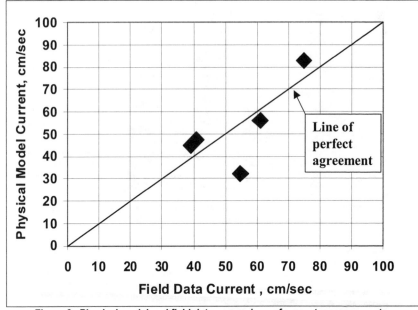

Figure 9. Physical model and field data comparison of current measurements at Gauge OS-5.

CONCLUSIONS

The experiments discussed in this paper were designed to examine whether Froude scaling for three-dimensional physical models in the 1:50 to 1:75 scale range produces reliable simulations of complex flow phenomena at tidal inlets. In the first case, wave breaking on a current over a tidal inlet ebb shoal was reproduced at a larger scale than the original 1:50 scale experiments. A series of scaling runs was made with a geometric scaling factor of 1:45. The appropriate scaling was applied to the wave height, wave period, and water depth across the ebb shoal. The scaling runs showed good agreement in wave height and energy dissipation across the ebb shoal.

The second case considered a comparison of a 1:75 scale model current measurements to field data at Grays Harbor Inlet, Washington. The measurement location was in a region of complex bathymetry, adjacent to the inlet's north jetty, where flood tidal currents and breaking-wave generated currents were comparable in magnitude. Model-prototype comparisons were found to be in reasonable agreement in four out of five cases examined. The discrepancy in the one case is attributed to larger scale circulations in the prototype, possibly due to wind.

This paper has demonstrated that scale physical models perform well in simulating complex coastal flow fields in the vicinity of tidal inlets. The capability to model coastal phenomena at the scales discussed makes this type of model a cost-effective tool for coastal design.

ACKNOWLEDGEMENTS

This work was conducted under the Inlet Laboratory Investigations work unit, Coastal Inlets Research Program and the Grays Harbor North Jetty Study, Corps of Engineers, Seattle District. Thanks to Dr. Nicholas Kraus and Ms. Julie Rosati for review comments, and Dr. Phil Osborne, Pacific International Engineering, Inc., for providing of field data. Permission to publish was granted by the Office, Chief of Engineers.

REFERENCES

Cialone, M. A. and Kraus, N. C. 2001. "Engineering Study of Inlet Entrance Hydrodynamics: Grays Harbor, Washington, USA," *Proc 4th Coastal Dynamics Conf.*, ASCE, 413-422.

Hughes, S. A. 1993. Physical Models and Laboratory Techniques in Coastal Engineering, World Scientific, Singapore, 568 pp.

Seabergh,W. C. 1999. "Physical Model for Coastal Inlet Entrance Studies," Coastal Engineering Technical Note, CETN-IV-19, Mar 99, U.S. Army Engineer Waterways Experiment Station, Coastal and Hydraulics Laboratory, Vicksburg, MS.

Smith, J.M. and Seabergh, W.C. 2001. "Wave breaking on a current at an idealized inlet with an ebb shoal," Technical Report ERDC/CHL TR-01-7, U.S. Army Engineer Waterways Experiment Station.

Stive, M. J. F. 1985. "A scale comparison of wave breaking on a beach," *Coastal Engineering*, 9, 151-158.

LABORATORY STUDY OF SHORT-CRESTED BREAKING WAVES

Koh Yih Ming[1], S.A. Sannasiraj[2] and Chan Eng Soon[3]

Abstract: A series of laboratory experiments was conducted to understand the kinematics and dynamics of short crested breaking waves. The simulation of wave breaking was achieved through wave-wave interactions of a frequency and amplitude modulated wave packet. Directionality and short crestedness were added to the breaking wave packet by varying the amplitude distribution of the wave paddles. The experiments have highlighted the basic features of a short-crested breaker through a series of still and movie pictures. Surface elevations at prescribed locations within the breaking wave field were monitored using wave gauges. Flow visualizations, incorporating digital imaging illuminated by a laser light sheet, were used to elucidate the mixing process.

INTRODUCTION

Most studies on breaking waves have concentrated on the two-dimensional features of the process, focusing on issues such as the causes of breaking, the wave energy dissipation and the generation of turbulence and mixing (Rapp and Melville 1990, Banner and Peregrine 1993, Melville 1996, Kway et al. 1998). While progress has been achieved in the two-dimensional studies, investigations of unsteady, three-dimensional and short-crested breaking waves are scarce. Examples of these three-dimensional investigations include the studies by She et al. (1991, 1994 & 1997), Kolaini and Tulin (1998), Nepf et al. (1998) and Wu (1999).

In many of the three-dimensional studies, the short crested waves were primarily generated using the method of spatial focusing or wave convergence as it generates a stronger three-dimensional breaker. It has been mentioned that such breakers have

1 Research Scholar, Department of Civil Engineering
2 Research Fellow, Tropical Marine Science Institute, tmssas@nus.edu.sg
3 Director, Tropical Marine Science Institute, tmsdir@nus.edu.sg
National University of Singapore, Singapore 119 223

higher energy losses and thus larger turbulence intensities (Wu 1999). Such waves may occur in the field when obliquely traveling wave packets meet, for example after passing around an island. However, in an open ocean, the constructive interferences of dispersive waves and their directional distributions could also cause the diverging waves to break. The kinematics and dynamics of these short crested breakers are expected to vary due to the different directional effect. However, little is known about the physics of the short crested breaker.

The intention of this paper is to elucidate the kinematics and dynamics of a short crested breaking wave. A controllable unsteady breaking wave was simulated using the methodology of wave-wave interaction. By systematic variation of the gain factor in the paddles of the wave generator, we were able to simulate a three-dimensional directional breaking wave. Combinations of flow visualization techniques, incorporating the use of high-speed camera, digital video camera and laser system, were integrated to provide the qualitative delineation of the breaking processes. Laboratory instrumentation including capacitance-type wave gauges and micro-Acoustic Doppler Velocimeters were used to quantify the kinematics of the breaking field and provide a greater understanding of the kinematics and mixing. The sequences of simulation process, the wave breaking process, energy re-distribution during the wave propagation and the mixing regions are discussed in this paper.

The experiments were carried out in a wave basin at the Hydraulics Laboratory of the Department of Civil Engineering, National University of Singapore. The wave basin was 6.6m wide and 24m long and filled to a depth of 0.5m with fresh water. A beach with a vertical height of 0.7m and a slope of 1:8 extended from the far end of the wave basin towards the wave generator to reduce the influence of reflections.

GENERATION OF THREE-DIMENSIONAL BREAKING WAVE PACKET

The breaking waves were generated through the methodology of wave focusing (Chan and Melville 1988, Rapp and Melville 1990, Kway et al. 1998). In this study, the methodology was translated to the simulation of modulated wave packets based on the summation of 28 sinusoidal wave components of discrete frequencies. Their relative phases were chosen such that within linear context all wave crests arrived simultaneously at a point in space and time.

Without the directional effect, the free surface displacement, $\eta(x,t)$, can be expressed by

$$\eta(x,t) = \sum_{n=1}^{N} a_n \cos[k_n(x-x_b) - 2\pi f_n(t-t_b)] \tag{1}$$

where N is the number of wave components, a_n is the amplitude of the nth frequency component, f_n. x_b and t_b are the theoretical wave breaking location and the time of occurrence of wave breaking respectively. The wave number, k_n and frequency, f_n are related by the linear dispersion relation

$$(2\pi f_n)^2 = k_n g \tanh(k_n d), \tag{2}$$

where g and d are the gravitational constant and water depth respectively. Nonlinear amplitude dispersion and directionality were not included.

The discrete frequency, f_n, for each wave component was equally spaced across a bandwidth of $\Delta f = 0.72$Hz centered at frequency $f_c = (f_1 + f_{28})/2 = 1$Hz. The amplitude of each wave component, a_n was chosen to produce a constant wave steepness; that is,

$$a_n = G / k_n \tag{3}$$

where G can be considered as the constant gain factor used to vary the overall intensity of the wave packet and it would be determined during the fine-tuning stage. The theoretical surface displacements generated based on the formulation above was converted to input signals by incorporating the wave maker transfer function. To eliminate any abrupt paddle movement, both ends of the input time series were tapered with a half cosine bell function for two wave periods. Details of the wave packet parameters and inputs are presented in Table 1 and the wave packet is shown in Fig. 1.

Table 1. Wave parameters and input

Wave packet input			Wave parameter		
Δf	frequency bandwidth (Hz)	0.72	T	Characteristic wave period (s)	1
f_{max}	highest frequency (Hz)	1.36	L	Characteristic wave length (m)	1.51
N	number of components	28	C	Characteristic phase speed (m/s)	1.51
d	water depth (m)	0.5	k_c	Characteristic wave number (m⁻¹)	4.15
x_b	breaking distance (m)	8.7	w_c	Characteristic angular frequency (s⁻¹)	6.28
t_b	Breaking time (s)	8.5			

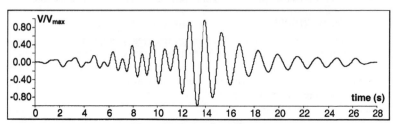

Fig. 1 Time series of input signal

To generate a three-dimensional wave, a lateral variation in the wave amplitude was introduced. This was done by varying the gain in each paddle, with the maximum gain in the middle paddle and tapered down to about 11% at both ends using a cosine square window. By defining the mean paddle position to be x = 0, the desired surface displacement at the mth paddle was given by

$$\eta(x = 0, y = mb, t) = \sum_{n=1}^{N} \frac{G}{k_n} \cos[-k_n x_b - 2\pi f_n(t - t_b)] \left\{ \frac{2}{3} - \frac{1}{3}\cos\left[\frac{\pi}{5}(m-1)\right]\right\}^2 \tag{4}$$

where b is the width of each paddle. With this function, each paddle produced a similar modulated wave packet but with different amplitude. As a result, there was a passive directional spreading during the wave evolution. It was noted that the procedure could also be refined to yield a straight crest. However, this was not done since we were focusing on the possibility of generating a short-crested breaking wave.

THREE-DIMENSIONAL WAVE PLUNGING PROCESS

Figs. 2(a)-(d) show the sequence of wave plunging process. Fig. 2(a) captured the moment when the wave crest at the center was almost vertical, but towards the side, the crest had already started to overturn. It can be seen in Fig. 2(c) and Fig. 2(d) that plunging at the side occurred prior to the plunging at the center.

As depicted by Fig. 2(c), the surface of overturning crest could be quite rough. With careful examination, it was noted that roughness occurred on the inner side of the overturning crest rather than on the outer surface. From Figs. 2(b)-(d), it can be seen that the surface of the crest was still smooth, confirming the above observation. The roughness seemed to be quite regular, resembling vertical capillary waves. It is clear from Fig. 2(c) that the wave had yet to plunge at this stage. Thus, the vertical capillary waves like features could not have been caused by the impingement of the plunging jet.

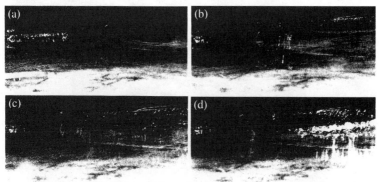

Fig. 2 The evolution of three-dimensional plunging breaker

An interesting feature observed in the study was the vertical spray along the boundary between the plunging crest and the splash-up (Fig. 3). Similar features had been documented for two-dimensional plungers (Kway 2001) and were evident in the results by Rapp and Melville (1990) although it was not discussed. It was clear from the video pictures that the back spray first occurred near the side (Fig. 3a) and then at the middle (Fig. 3b). The intensity of the back spray at the middle was significantly higher than that at the side (Figs. 3c-d). These observations were consistent with the fact that breaking occurred at the side first even though there was higher energy in the middle.

Fig. 4 shows the plunging action viewed from a position under the water surface. A stretch of curved 'wall' could be seen penetrating the free surface. On careful observation, this 'wall' was actually the roll down of the entrained air tube. From the front view (Fig. 4), it looked like a 'wall'. At this instant, the tip of the 'wall' was full of bubbles suggesting that air could be entrained. The air tube eventually lost its energy, broke down into bubbles, and emerged to the surface. Through a series of synchronized videos taken simultaneously from different angles, it was noted that, when the entrained air tube started to break down, it corresponded to the time when back spray started.

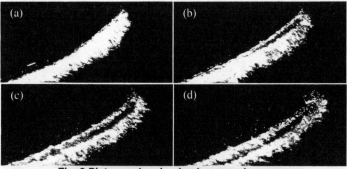

Fig. 3 Pictures showing back spray phenomenon

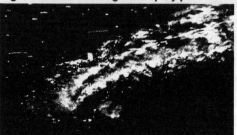

Fig. 4 Picture showing the 'wall' phenomenon

LATERAL VARIATION OF SURFACE ELEVATION

The coordinate system in the wave basin was defined as follows: x was positive in the direction of wave propagation and the y-axis was parallel to the mean position of the wave paddles. The origin (0,0) was located at intersection of the centerline of the basin and the breaking point. Wave breaking was symmetrical about the centerline (y=0). The surface elevation time histories for transects at $x = 0$cm are presented in Fig. 5. The distinctive crest shape of the waves at each transect could be observed by comparing the surface displacement records. From these plots, it could be observed that the breaking wave crest was led by the crest in the centerline and there was a successive phase lag across the entire crest length. This depicted a directional-divergence of the breaking wave generated. It was also clear that along the centerline and the transect y =100cm, the wave crests were locally very steep, resulting in wave plunging. The presence of high frequency contents in the center region of the basin was an important contributing factor to the steepness formation. Away from the plunging region, there was a reduction in the wave steepness. This was correlated to the lack of high frequency wave components at the side. Hence, only spilling and incipient waves were observed in those regions.

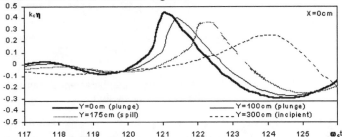

Fig. 5 Surface displacement time histories at x=0cm

Lateral Transfer of Momentum Flux

Fig. 6 shows the variation of the normalized wave momentum flux across the width of the basin for different sections. From this figure, it is clear that as the waves propagated downstream, the momentum flux in the center region decreased whereas the flux at the sides increased. This showed that despite of the flux losses due to viscosity and wave breaking, there was an energy transfer from the center to the side as the waves propagated. Based on the cosine-shaped input signals in the wave paddles, one would expect the momentum flux level to diminish from the center to the side. However, it was interesting to note that the lowest flux level was at the transects ranging from y = 150cm to 220cm, instead of sidewall. It was not entirely clear how this came about, but one possibility was that it might be due to the wave reflections from the sidewall, which provided an additional source of momentum flux to the area near the side of the basin. From Fig. 6, it could also be noted that the minimum flux level shifted closer to the centerline as the measurement point moved further downstream. This might imply that the influence of the reflected waves increased as waves propagated downstream. That being the case, the results at the sides would not be indicative of the momentum flux (i.e. the estimate would have been erroneous if there were reflected waves).

Fig. 7 presents the surface elevation time histories upstream and downstream of breaking at transects y = 0cm and 300cm. It can be observed that along the centerline (y = 0cm), the dispersive characteristic of the wave component was as expected, that is, short waves were ahead upstream and long waves were ahead downstream. At the side (y = 300cm), such characteristic was not obvious. Instead, the wave packet was dominated by the lower frequency waves. It appeared that, as the wave propagated downstream, the lateral transfer of the wave energy was associated with the lower frequency wave components.

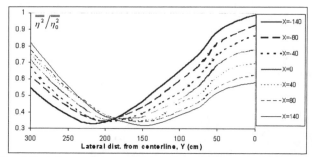

Fig. 6 Variations of normalized surface displacement variances across different lateral sections

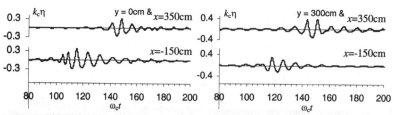

Fig. 7 Surface elevation time histories at y=0cm and y=300cm

EVOLUTION OF MIXING DUE TO WAVE BREAKING

Using flow visualizations obtained with the use of laser and fluorescence, the surface mixing due to wave breaking at selected transects was studied. Details of the measurements can be found in Koh (2001). It was observed that, as the wave broke, the plunging jet and the turbulent eddies quickly initiated the mixing. The subsequent passing waves then carried the dye down the water column. This resulted in a high penetration of dye when the wave group propagated past. After approximately three wave periods, the waves had passed and further deepening was due to turbulent diffusion and advection. The log-log plot of the maximum depth of dye excursion, D, mixing area, A, and mixing volume, V, against time are presented in Fig. 8. The length and time scales were normalized by k_c and f_c respectively. Note that one wave height corresponded approximately to $0.83\ k_cD$.

Three periods after breaking, the evolution of the penetration depth could be estimated by a power law

$$k_cD = \gamma[w_c(t-t_b)]^s \qquad (5)$$

where s was the growth rate due to turbulent diffusion and advection and was the given by the slope of the log-log plot. γ could be interpreted as the initial penetration due to the plunging jet and turbulent eddies, and $\log_{10}\gamma$ was the y intercept. From the plot, it was observed that a slope of $s = \frac{1}{4}$ fits all the data reasonably well, for both the plunging and spilling regions. In the plunging region, the parameter γ ranged from 0.32 to 0.50. Though there was no distinct trend in the variation of γ among the different section, the common growth rate had shown a consistency in the behaviour of mixing due to wave plunging. This slope of ¼ also agreed with the results from a two-dimensional plunger (Rapp and Melville 1990), but the γ range in this study was smaller. This suggested that the initial sources of turbulent mixing were weaker in this study.

It should be noted that the mixing area was obtained from the total area of the dye patches in each transect. From Fig. 8(b), it was apparent that the evolution of the mixing area in the plunging region follows a similar power law with $s = \frac{1}{2}$ and γ ranging from 0.25 to 0.45. It was interesting to note that the growth rate of $s = \frac{1}{2}$ was similar to the results from the two-dimensional plunger (Rapp and Melville 1990) and spatially focused three-dimensional breaker (Wu 1999).

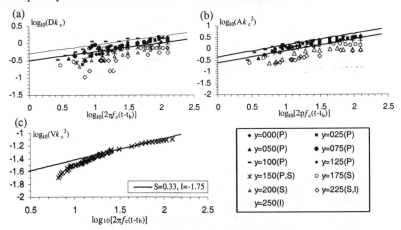

Fig. 8 Log-log plot of (a) depth of penetration, (b) mixing area and (c) mixing volume against time

Assuming that the mixing was symmetrical along the centerline, the temporal variation of the total three-dimensional mixing volume, V(t), was calculated by integrating the mixing area in each transect laterally. The variation of the mixing volume after 3 wave periods could be estimated by

$$k_c^3 V(t) = 0.0178[w_c(t-t_b)]^{1/3} \qquad (6)$$

Table 2 shows a comparison of the slope, s, in the two- and three- dimensional breaking. It was clear that mixing of the directionally-diverging breaker was the slowest. It should be noted that this was partly due to the slow growth rate in the spilling region.

Using the ensemble-average method (Rapp and Melville 1990), the turbulence generated by wave breaking was derived from the fluctuating velocity. The three turbulence components were accounted for in the turbulent kinetic energy, TKE, defined as

$$TKE = \frac{1}{2}\left[< u_t'^2 > + < v_t'^2 > + < w_t'^2 >\right] \qquad (7)$$

where $u_t'^2$, $v_t'^2$ and $w_t'^2$ were the fluctuating velocity component due to turbulence in the longitudinal, lateral and vertical direction respectively.

In our study, the intensity level within the plunging region ranged from $7 \times 10^{-5}C^2$ to $1 \times 10^{-4}C^2$. In terms of the r.m.s. velocities, it was about half that of two-dimensional breaker (Rapp and Melville 1990) and a quarter that of spatially focused three-dimensional breaker (Wu 1999). The lower intensity level compared to two-dimensional plunger was consistent with the fact that the plunging wave studied was a directionally-diverging wave.

Table 2. Comparison of the slope in $w_c(t\text{-}t_b)^s$ power law in 2D & 3D wave breaking

Breaker type	Author	Slope (s)
Single 2D plunger	Rapp and Melville, 1990	0.5
	Wu, 1999	0.52
3D plunging breaker, spatially focused	Wu, 1999	0.64
3D plunging breaker, directional-diverging	Present experiment	0.33

CONCLUSION

By varying the input amplitudes in the paddles of the three-dimensional wave making system, we have successfully extended the methodology of wave focusing to the generation of an unsteady short crested plunging wave. When the wave steepened, the crest tip started to overturn, followed by the impingement of the plunging tip. It has been noted that plunging started at the side and then continued to the center. Before impingement, undulations resembling vertical capillary waves were observed on the inner surface of the overturning crest. After plunging, the entrained air tube rolled into the water column, yielding a penetrating 'wall' when viewed from the front. Similar to the observation in two-dimensional breaking waves (Kway 2001), vertical sprays were obtained. In addition, it was noted that the occurrence of the vertical spray was synchronous with the collapse of the entrained air tube.

As anticipated, there was a lateral transfer of wave energy from the center to the side. This energy transfer was associated with the lateral spreading of the lower frequency wave components. The high frequency components were observed to remain within the middle of the wave basin, which was an important factor in forming a plunger in the middle. Based on flow visualizations on mixing, the maximum depth of the mixing area was about two to three times the wave amplitude. The growth of the penetration depth, mixing area and volume were found to follow power laws of $t^{1/4}$, $t^{1/2}$ and $t^{1/3}$ respectively.

On the whole, these results have elucidated the mechanism of short crested wave breaking and the resulting mixing characteristics. The overall wave breaking process, vertical sprays and mixing of the directionally-diverging short crested breaking waves were similar to those observed in two dimensional frequency-focused breaking waves. Additional features such as the formation of three-dimensional air tube, lateral spreading of wave energy and the initial breaking at the sides were also observed. It should be noted, however, that wave breaking could range from spilling to plunging, and from short crested breaking to long crested breaking.

REFERENCES

Banner, M. L. and Peregrine, D. H. 1993. Wave breaking in deep water. *Annual Review of Fluid Mechanics*, 25, 373-397.

Chan, E. S. and Melville, W. K. 1988. Deep water plunging wave pressures on a vertical plane wall. *Proceedings of the Royal Society,* A417, 95-131.

Koh, Y. M. 2001, *Laboratory study of three-dimensional breaking waves.* Masters thesis in communication, NUS, Singapore.

Kolaini, A. and Tulin, M. 1998. Laboratory measurements of breaking inception and post-breaking dynamics of steep short-crested waves. *International Journal of Offshore and Polar Engineering*, 5, 212-218.

Kway, J. H. L. 2001, *Laboratory study of the kinetics and dynamics of deep-water wave breaking.* Ph.D. thesis in communication, NUS, Singapore.

Kway, J. H. L., Loh, Y. S. and Chan, E. S. 1998. Laboratory study of deep-water breaking waves. *Ocean Engineering*, 25(8), 657-676.

Melville, W. K. 1996. The role of surface-wave breaking in air-sea interaction. *Annual Review of Fluid Mechanics*, 28, 279-321.

Nepf, H. M., Wu, C. H. and Chan, E. S. 1998. A comparison of two and three dimensional wave breaking. *Journal of Physical Oceanography*, 28, 1496-1510.

Rapp. R. and Melville, W. K. 1990. Laboratory measurements or deep water breaking waves. *Philosophical Transactions of the Royal Society of London*, Series A, 331, 735-800.

She, K., Greated, C. A. and Easson, W. J. 1991. Effects of three-dimensionality on wave kinematics and loading. *Proceeding of the first international offshore and polar engineering conference*, Edinburgh, UK, 3, 1-5.

She, K., Greated, C. A. and Easson, W. J. 1994. Experimental study of three-dimensional wave breaking. *Journal of Waterway, Port, Coastal and Ocean Engineering*, 120, 20-36.

She, K., Greated, C. A. and Easson, W. J. 1997. Experimental study of three-dimensional breaking wave kinematics. *Applied Ocean Res.* 19, 329-343.

Wu, C. H. 1999, Laboratory measurements of three-dimensional breaking waves. Ph.D. thesis, MIT, US.

LABORATORY "FREAK WAVE" GENERATION FOR THE STUDY OF EXTREME WAVE LOADS ON PILES

Uwe Sparboom[1], Jan Wienke[2] and Hocine Oumeraci[3]

Abstract: Large-scale experiments of breaking wave attack on slender cylinders were performed in the LARGE WAVE CHANNEL at the Coastal Research Centre in Hannover, Germany. Using large-scale physical models the influence of scale effects on the real physical mechanisms can be widely minimized. Very steep waves (similar to "freak waves") were generated to investigate extreme breaking wave impacts.

INTRODUCTION

Slender cylindrical structures (piles) are needed for various coastal and ocean structures. Until now, for the attack of very steep breaking waves ("freak waves") design procedures for slender cylinders are not very reliable (Sarpkaya and Isaacson, 1981; Kjeldsen, 1981). Valuable contributions of breaker induced forces on piles were given by Wiegel (1982). Kjeldsen et al. (1986) performed laboratory tests with vertical piles attacked by plunging breakers. Field and large-scale laboratory tests with a 0.7 m diameter vertical pile were carried out by Sparboom (1987) in order to measure real wave-loads in shallow water under storm surge conditions (high Reynolds-numbers).

Some results of small-scale tests on breaking wave impact at slender cylinders were reported by Chan et al. (1995). The importance of scale effects in physical wave modeling was pointed out by Oumeraci (1984).

The occurrence of "freak waves" with high damage potential was described by Kjeldsen (1997). Most important for this paper was the laboratory generation of so-called "freak-waves" which start to break at a predetermined location at the test

[1] Dr.-Ing., Senior Researcher, FORSCHUNGSZENTRUM KÜSTE (FZK)/"Coastal Research Centre", Merkurstrasse 11, D-30419 Hannover, Germany; sparboom@fzk.uni-hannover.de
[2] Dr.-Ing. Dipl.-Phys., Researcher, FORSCHUNGSZENTRUM KÜSTE
[3] Full Professor, Dr.-Ing., LEICHTWEISS-INSTITUTE FOR HYDRAULICS; Technical University of Braunschweig

cylinder in the wave flume (Clauss and Bergmann, 1986; Clauss and Kühnlein, 1997).

Concerning extreme wave loadings of a structure or of a structural component it is of particular interest to analyse single wave effects in a wave flume. Single wave tests have the advantage that very high sampling rates for the data acquisition can be used. In the present study, pressure impact time histories were resolved in the range of milliseconds and below. Local loads or pressures were determined exactly and total forces as well as wave kinematics were measured simultaneously (Wienke et al., 2000).

TEST SET-UP

The experiments were carried out in the LARGE WAVE CHANNEL of the Forschungszentrum Küste (Coastal Research Centre) in Hannover, Germany. This channel has an effective length of 309 m, a width of 5 m and a depth of 7 m. For the tests, the still water level was around 4 m (Fig. 1).

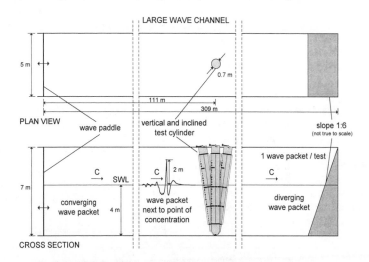

Fig. 1. Test set-up in the LARGE WAVE CHANNEL

A steel cylinder with a diameter of 0.7 m was installed on the horizontal channel bottom in a distance of 111 m from the wave paddle. The top of the test cylinder was fixed at a traverse structure at the top edge of the channel. At the bearings strain gauges were installed to measure the total force as the sum of the forces at both bearings. Furthermore, 55 pressure transducers were installed in the cylinder. Some were installed in the front line and some others around the circumference of the cylinder (Fig. 2).

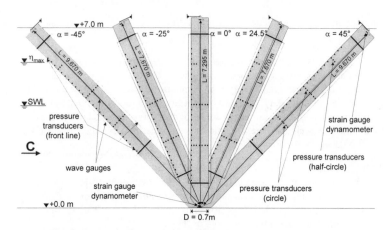

Fig. 2. Test cylinder with installed measuring instruments

WAVE GENERATION AND WAVE LOAD CLASSIFICATION

There are two different opportunities for the generation of single waves in the LARGE WAVE CHANNEL. On the one hand solitary waves can be generated by the piston-type wave maker. Using a typical still water level of 4.75 m the maximum elevation of a solitary wave is around 0.8 m (Fig. 3). On the other hand wave packets can be simulated. Focussing the point of concentration of a wave packet at the structure the elevation of the water surface corresponds to a single wave. In this way maximum wave elevations of about 2 m above SWL can be generated at the structure (Fig. 4).

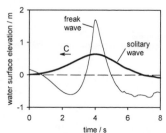

Fig. 3. Single waves in the LARGE WAVE CHANNEL

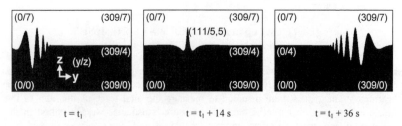

Fig. 4. Gaussian wave packet in the LARGE WAVE CHANNEL

The wave packet is significantly steeper than the solitary wave (Fig. 3). If the maximum elevation of the wave packet is enlarged moreover the wave packet starts to break (Fig. 5). Focussing the breaking location in front of the structure very extreme loads act on the structure and can be examined.

All generated wave packets were quite similar at the location of breaking and only plunging breaker occurred. The time history of the wave elevation was detected with a wave gauge next to the location of breaking.

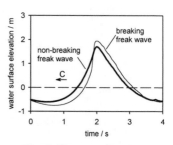

Fig. 5. Wave packets in the LARGE WAVE CHANNEL

Neglecting the small variation during one wave period the shape of the water surface is obtained by multiplying the time history with the wave celerity. In this way values for the parameters which characterise the breaking waves are determined (Fig. 6).

Parameter	Definition	Values given by Kjeldsen, (1990)*	Measured values
wave height	H		2.2 m - 2.8 m
maximum water surface elevation	η'		1.7 m - 2.0 m
wave period	T		4.11 s - 4.28 s
wave celerity	C		5.8 m/s - 6.2 m/s
steepness	$s = \dfrac{2 \cdot \pi \cdot H}{g \cdot T^2}$		0.08 - 0.10
crest front steepness (in space)	$\varepsilon_{X,B} = \dfrac{\eta'}{L'}$	0.32 – 0.78	0.55 - 0.80
crest front steepness (temporal)	$\varepsilon_t = \dfrac{2 \cdot \pi \cdot \eta'}{g \cdot T' \cdot T}$		0.50 - 0.75
vertical asymmetry factor	$\lambda = \dfrac{L''}{L'}$	0.90 - 2.18	1.9 – 2.7
horizontal asymmetry factor	$\mu = \dfrac{\eta'}{H}$	0.84 – 0.95	0.71 – 0.77

* field measurements

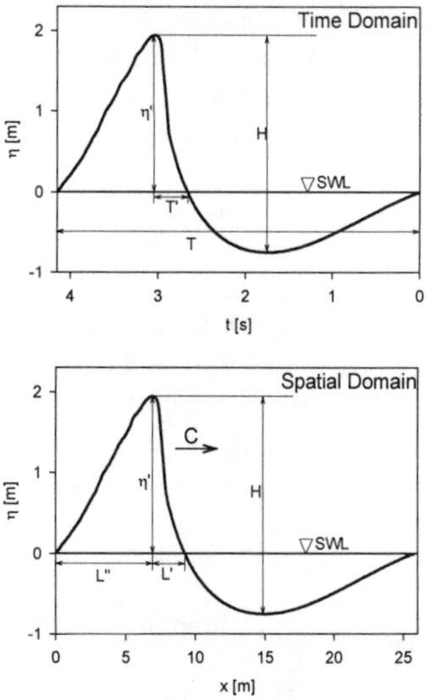

Fig. 6. Breaking wave parameters measured in the LARGE WAVE CHANNEL: Definition and comparison with field data

Acting on the cylinder these breaking waves of similar shape showed obvious differences concerning the loading records. Therefore, the distance between breaking location and cylinder was varied systematically. Five loading cases were defined (Fig. 7). With increasing number of the loading case the breaking location was shifted closer to the cylinder.

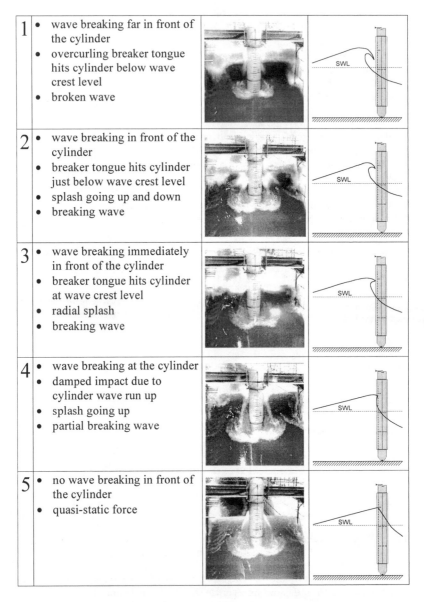

1	• wave breaking far in front of the cylinder • overcurling breaker tongue hits cylinder below wave crest level • broken wave		
2	• wave breaking in front of the cylinder • breaker tongue hits cylinder just below wave crest level • splash going up and down • breaking wave		
3	• wave breaking immediately in front of the cylinder • breaker tongue hits cylinder at wave crest level • radial splash • breaking wave		
4	• wave breaking at the cylinder • damped impact due to cylinder wave run up • splash going up • partial breaking wave		
5	• no wave breaking in front of the cylinder • quasi-static force		

Fig. 7. Classification of different loading cases

Waves of loading case 1 break far in front of the cylinder and the breaker tongue hits the cylinder in an angle of around 45° related to the horizontal plane. Loading case 2 means that the waves break evidently in front of the cylinder and the hitting breaker tongue is inclined regarding the horizontal plane by an angle of around 25°. Waves of loading case 3 break immediately in front of the cylinder and the breaker tongue is still moving along the horizontal plane when the cylinder is attacked. Loading case 4 describes waves breaking at the cylinder. Waves of loading case 5 do not break in front of the cylinder but at the rear. This definition of the five loading cases is applied for each of the five investigated yaw angles (Fig.8).

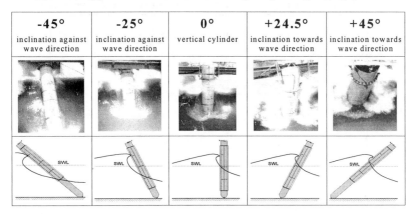

-45°	-25°	0°	+24.5°	+45°
inclination against wave direction	inclination against wave direction	vertical cylinder	inclination towards wave direction	inclination towards wave direction

Fig. 8. Loading case 3 for the different investigated yaw angles

IMPACT LOADING

The total force acting on the cylinder has been determined by adding the loads measured at the two bearings at the top and at the bottom. The maximum force caused by a typical solitary wave is below 1 kN. Attacked by a wave packet the total force acting on the cylinder is around 7 kN. Since the wave packet is steeper than the solitary wave the increase of force takes place faster. Very much larger values of the total force are measured for wave packets breaking in front of the cylinder (Fig. 9).

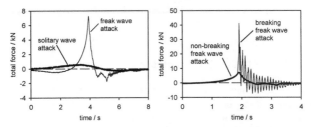

Fig. 9. Measured forces acting on the cylinder for different wave attack

The large forces due to breaking waves are restricted to a small extension at the cylinder and these forces act only during an extremely short time. Therefore this force due to breaking waves is called slamming force. It can be detected by pressure measurement at the cylinder. Comparing the pressure time history with the time history of the water surface elevation the different values of the wave period and the slamming duration are illustrated (Fig. 10). The small extension of the slamming force is confirmed by the pressure distribution along the height of the cylinder.

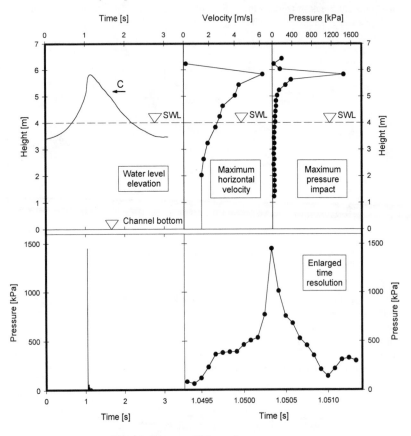

Fig. 10. Measured slamming pressures

The total force measured at the bearings of the test cylinder was separated into a quasistatic part varying in time with the water surface elevation and a dynamic part due to the slamming force. The experimentally obtained values for the dynamic force are plotted in Fig. 11. For each angle of inclination the mean, the maximum and the minimum values are plotted. The highest maximum value of the dynamic force is obtained for the yaw angle of -25° (inclination against wave direction).

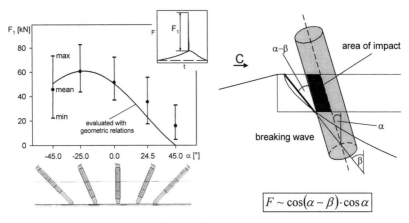

$$F \sim \cos(\alpha - \beta) \cdot \cos \alpha$$

Fig. 11. Dynamic force versus yaw angle

The dynamic force is related to an area of impact at the cylinder. It is assumed, that the slamming force is only acting in this area and that the force is equal at the different levels of the area of impact. In this way a height of the impact area at the cylinder can be deduced. Dividing this length by the maximum elevation of the breaking wave, the curling factor λ is obtained. In Fig. 12 this curling factor is plotted, namely the mean, the maximum and the minimum values for each angle of inclination are shown. For the vertical cylinder a maximum value of 0.5 is obtained. This is in agreement to values given in literature (e.g. Wiegel, 1982). If the cylinder is inclined against wave direction, the height of the impact area increases. However, the height of the area of impact decreases, if the cylinder is inclined towards the wave direction (Wienke et al., 2001).

CONCLUDING REMARKS

It is shown that wave packets are a suitable tool for the simulation of steep breaking waves in a wave flume. The location of wave breaking can accurately be varied by the predetermination of the concentration point of the wave packet. So waves breaking in front of a test structure can be generated. The shift of the test structure can be equalized by the variation of the point of concentration of the wave packet. In this way slamming forces on a test cylinder with different yaw angles were investigated. For each yaw angle maximum values of the slamming force were obtained by varying the point of concentration over a range of nearly 10 m. The results show that the maximum force is acting on a cylinder which is slightly inclined against wave direction.

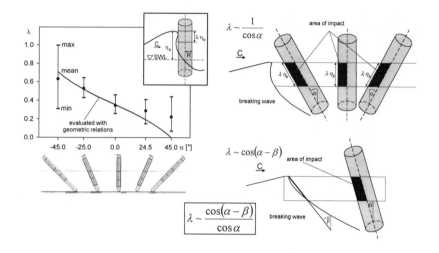

Fig. 12. Curling factor versus yaw angle

ACKNOWLEDGEMENTS

The DEUTSCHE FORSCHUNGSGEMEINSCHAFT (DFG) is gratefully acknowledged supporting the research project on breaking wave impact of slender cylinders (Contract-No. Ou1/4-1).

The authors thank the research team at the LARGE WAVE CHANNEL of the FORSCHUNGSZENTRUM KÜSTE (FZK) for valuable help during the research program (Dipl.-Ing. Grüne, Dipl.-Ing Schmidt-Koppenhagen, Dipl.-Ing. Irschik, Dipl.-Ing. Bergmann, Mr. Junge and Mr. Malewski).

REFERENCES

Chan, E.-S.; Cheong, H.-F. and Tan, B.-C. (1995). "Laboratory study of plunging wave impacts on vertical cylinders." *Coastal Engineering*, 25, 87-107.

Clauss, G.F. and Bergmann, J. (1986). "Gaussian wave packet – A new approach to seakeeping tests of ocean structures." *Applied Ocean Research*, 8, No. 4.

Clauss, G.F. and Kühnlein, W.L. (1997). "Simulation of design storm wave conditions with tailored wave groups." *Proceedings of the Seventh International Offshore and Polar Engineering Conference*, Honolulu, 3, 228-237.

Kjeldsen, S.P. (1981). "Shock pressures from deep water breaking waves." *Proceedings of the International Symposium on Hydrodynamics*, Trondheim, 567-584.

Kjeldsen, S.P.; Tørum, A. and Dean, R.G. (1986). "Wave forces on vertical piles caused by 2- and 3-dimensional breaking waves." *Proceedings of the 20th International Conference on Coastal Engineering*, Taipei, 1929-1943.

Kjeldsen, S.P. (1997). "Examples of heavy weather damages caused by giant waves." *Techno Marine*, Bulletin of the Society of Naval Architects of Japan, 820, No. 10.

Oumeraci, H. (1984). "Scale effects in coastal hydraulic models." Kobus (ed.), *Symposium on Scale Effects in Modelling Hydraulic Structures*, Technische Akademie Eßlingen.

Sarpkaya, T.and Isaacson, M. (1981). "Mechanics of wave forces on offshore structures." Van Nostrand Reinhold, New York.

Sparboom, U. (1987). "Real wave-loads on small cylindrical bodies." *Proceedings of the 2nd International Conference on Coastal and Port Engineering in Developing Countries*, Beijing.

Wiegel, R.L. (1982). "Forces induced by breakers on piles." *Proceedings of the 18th International Conference on Coastal Engineering*, Cape Town, 1699-1715.

Wienke, J., Sparboom, U. and Oumeraci, H. (2000). "Breaking wave impact on a slender cylinder." *Proceedings of the 27th International Conference on Coastal Engineering*, Sydney, 2, 1787-1798.

Wienke, J., Sparboom, U. and Oumeraci, H. (2001). "Large-scale experiments with slender cylinders in breaking waves." *Proceedings of the 11th International Offshore and Polar Engineering Conference*, Stavanger, 4, 358-362.

WAVE TRANSFORMATION AND QUASI-3D NEARSHORE CURRENT MODEL OVER BARRED BEACH

Masamitsu Kuroiwa[1], Chang Bae Son[2] and Hideaki Noda[3]

Abstract: The purpose of this study is to develop a quasi-3D numerical model that is applicable to nearshore currents over barred beaches. The presented model is composed of two sub-modules of wave field and nearshore current field. The wave filed was determined using wave by wave approach to consider the irregularity of waves. The current field was determined from the Q-3D model, taking account of the effects of wave decay and recovery processes in the surf zone. The present Q-3D model was calibrated and verified by comparing with data measured over barred beaches in the field. The applicability of the Q-3D model was discussed.

INTRODUCTION

Nearshore currents have previously been predicted by using two-dimensional models in the horizontal plane (2DH model). However, in the surf zone, the direction of current vectors near the water surface is different from that at the sea-bottom because of the effect of undertow velocities. In order to accurately predict the changes of beach profile, it is very important that the three-dimensional distribution of nearshore currents is determined.

Some models for determining the vertical distribution of nearshore currents have previously been proposed. De Vriend et al. (1987) presented a semi-analytical model, and then Svendsen et al. (1989) proposed an analytical model composed of cross-shore and longshore current velocities. Q-3D numerical models by combining 2DH model and one-dimensional model defined in the vertical direction (1DV model), have also been proposed (e.g. Okayasu et al., 1994). Kuroiwa et al. (1998) proposed a Q-3D numerical model based on the solution method developed by Koutitas et al. (1980) and tried to calculate nearshore currents around a detached breakwater in the experiments

1 Research Associate, Dept. of Civil Engineering, Tottori University, 4-101,Koyama-Minami,Tottori,680-8552,Japan, kuroiwa@cv.tottori-u.ac.jp
2 Lecturer , Korea Maritime University, 1 Dongsam-dong, Yeongdo-ku Pusan 606-791, KOREA
3 Professor, Tottori University of Environmental Studies,1-1,Wakabadai-kita,Tottori-shi Tottori,689-1111, JAPAN, hidenoda@kankyo-u.ac.jp

Furthermore, the presented model was applied to the nearshore current field around a harbor in the field (Kuroiwa et al.,2000). It was clarified that the accuracy of the prediction for nearshore currents in the field was not sufficient because the influence of beach profile, that is, the wave decay, recovery and secondary breaking process in the surf zone, was not considered. And the irregularity of waves was not also considered. Therefore, the purpose of this study is to develop a quasi-3D numerical model that is applicable to nearshore currents over barred beaches and that can take account of the influence of random waves. In this study ,the Q-3D model proposed by Kuroiwa et al. (1998) is modified. The applicability of the modified Q-3D model is investigated by comparing the compute results with data measured in the field.

NUMARICAL MODEL

The presented model consists of two models, that is, a wave module and a nearshore current module. The co-ordinate system used in the present model is shown in Figure 1.

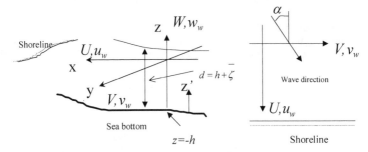

Fig. 1. Co-ordinate system

Wave module

The wave transformation under random wave is computed by using the energy flux balance equation with an additional term to represent the energy dissipation of breaking wave proposed by Dally et al. (1984). Assuming that the bottom topography is uniform in the longshore direction, the energy flux balance equation can be expressed as

$$\frac{\partial E C_g \cos\alpha}{\partial x} = -D_b \tag{1}$$

where E is the wave energy density, C_g is the group velocity, and α is the wave angle. The angle α is determined by Snell's law. D_b is the energy dissipation rate. D_b is proportional to the difference between the local energy flux and the stable energy flux of breaking wave:

$$D_b = \frac{K}{h} C_g (E - Es) \tag{2}$$

where h is a still water depth, K is a dimensionless constant, which is set to 0.15 in the surf zone and zero seaward of the wave breaking. Es is the energy associated with the stable wave height Hs, given by

$$Es = \frac{1}{8}\rho g(\Gamma h)^2 \tag{3}$$

where Γ is a dimensionless constant, which is set to 0.4. The breaking points are determined by the breaking criteria of Goda(1975). The criteria is given by

$$H_b = AL_0 \left\{ 1 - \exp\left[-1.5 \frac{\pi h}{L_0} \left(1 + 15 \tan^{4/3} \beta\right) \right] \right\} \tag{4}$$

where $\tan \beta$ is bottom slope, and A is a dimensionless constant set at 0.1 in this model.

In order to determine the distribution of wave heights under random waves, the wave-by-wave approach used by Kuriyama et al. (1999) was employed. The method is as follows. A JONSWAP spectrum is determined from a significant wave height and period at deep water. From the spectrum, a time series of water surface elevation is numerically computed. The time series is divided into approximately two hundred individual waves using the zero-down crossing method. Then the distribution of individual wave height in the cross-shore direction is numerically computed from Eq.(1), and then the root-mean-square and significant wave heights at each grid point of the computation are determined. The root-mean square wave heights and peak period of the JONSWAP spectrum are used to compute nearshore current velocities.

Nearshore current module
Governing Equations

The equations of motion for Q-3D nearshore currents derived by Svendsen et al. (1989) can be expressed as

$$\frac{\partial U}{\partial t} + U \frac{\partial U}{\partial x} + V \frac{\partial U}{\partial y} + W \frac{\partial U}{\partial z} = -g \frac{\partial \overline{\zeta}}{\partial x} - \frac{\partial (\overline{u_w^2} - \overline{w_w^2})}{\partial x} - \frac{\partial \overline{u_w v_w}}{\partial y}$$
$$+ \frac{\partial}{\partial x}\left(v_h \frac{\partial U}{\partial x} \right) + \frac{\partial}{\partial y}\left(v_h \frac{\partial U}{\partial y} \right) + \frac{\partial}{\partial z}\left(v_v \frac{\partial U}{\partial z} \right) \tag{5}$$

$$\frac{\partial V}{\partial t} + U \frac{\partial V}{\partial x} + V \frac{\partial V}{\partial y} + W \frac{\partial V}{\partial z} = -g \frac{\partial \overline{\zeta}}{\partial y} - \frac{\partial (\overline{v_w^2} - \overline{w_w^2})}{\partial y} - \frac{\partial \overline{v_w u_w}}{\partial x}$$
$$+ \frac{\partial}{\partial x}\left(v_h \frac{\partial V}{\partial x} \right) + \frac{\partial}{\partial y}\left(v_h \frac{\partial V}{\partial y} \right) + \frac{\partial}{\partial z}\left(v_v \frac{\partial V}{\partial z} \right) \tag{6}$$

where U, V and W are steady current velocities in the x, y and z directions, respectively, as shown in Figure 1. $\overline{\zeta}$ is the mean water level. $\overline{u_w^2} - \overline{w_w^2}$, $\overline{v_w^2} - \overline{w_w^2}$, $\overline{u_w v_w}$ and $\overline{v_w u_w}$ represent the excess momentum fluxes due to waves and are estimated by linear wave theory.

v_v and v_h represent the turbulent eddy viscosity coefficients in the vertical and horizontal direction, respectively. v_h is estimated by using the method presented by Longuet-Higgins (1970), that is,

$$v_h = N\left(h + \overline{\zeta}\right) / \tan \beta \sqrt{g\left(h + \overline{\zeta}\right)} \tag{7}$$

where N is a dimensionless constant of 0.01. v_v is estimated by Tsuchiya model(1986), that is given by

$$v_v = A_v C_p H_{rms} \tag{8}$$

where C_p is the wave celerity with peak period of spectrum. H_{rms} is the root-mean-square wave height and A_v is a dimensionless constant, set at 0.01.

The continuity equation is expressed as

$$\frac{\partial U}{\partial x} + \frac{\partial V}{\partial y} + \frac{\partial W}{\partial z} = 0 \qquad (9)$$

and the depth-integrated continuity equation is

$$\frac{\partial \overline{\zeta}}{\partial t} + \frac{\partial \widetilde{U}(h + \overline{\zeta})}{\partial x} + \frac{\partial \widetilde{V}(h + \overline{\zeta})}{\partial y} = 0 \qquad (10)$$

where $\widetilde{U}$ and $\widetilde{V}$ are the depth-averaged steady currents. The steady current velocity W in the vertical direction and the mean water level $\overline{\zeta}$ are calculated from Eq.(9) and Eq.(10), respectively.

Boundary conditions and solution method

Cross-shore current velocities at the shoreline and offshore boundaries are set to zero and longshore current velocities at lateral boundaries are uniform flows. These conditions can be written as

$$U\big|_{shore} = U\big|_{offshore} = 0 \;, \quad \partial V / \partial y = 0 \qquad (11)$$

Boundary conditions at the mean water surface and sea-bottom are needed in order to determine the vertical distribution of the neashore currents. In general, the boundary condition at the free surface is the no-flux condition. However, in the case where mass transport due to wave breaking is dominant in the surf zone, shear stress caused by water surface rollers must be considered. In previous study(Kuroiwa et al,1998), the shear stress was given by the model based on Svendsen et al.(1989). In this study, in order to consider the effect the wave decay, recovery and secondary breaking processes in the surf zone, the stresses in x and y directions are rewritten as

$$\tau_{xx} = A_{Sx} \rho^{1/3} D_b^{2/3} \left(\frac{h}{L_p} \right) \cos \alpha \qquad (12)$$

$$\tau_{sy} = A_{Sy} \rho^{1/3} D_b^{2/3} \left(\frac{h}{L_p} \right) \sin \alpha \qquad (13)$$

where ρ is the density of seawater, h is the water depth , L_p is the wave length with peak frequency, and α is the wave direction. D_b is given by Eq.(2). A_{sz} and A_{sy} are constant values determined empirically by comparing computed nearshore currents with measured data. The boundary conditions at the mean water level are given by

$$v_v \frac{\partial U}{\partial z}\bigg|_{z=\overline{\zeta}} = \tau_{xx} / \rho \;, \quad v_v \frac{\partial V}{\partial z}\bigg|_{z=\overline{\zeta}} = \tau_{sy} / \rho \qquad (14)$$

where α is the wave direction. The boundary conditions at the sea-bottom level are given as

$$v_v \frac{\partial U}{\partial z}\bigg|_{z=-h} = \tau_{bx} / \rho \;, \quad v_v \frac{\partial V}{\partial z}\bigg|_{z=-h} = \tau_{by} / \rho \qquad (15)$$

in which τ_{bx} and τ_{by} are the shear stresses caused by bottom friction and include the effects of interaction between the steady current and wave oscillatory motion. The bottom shear stresses are determined from the model proposed by Nishimura(1988).

The governing equations mentioned above are solved numerically. In the actual computation, a grid size of 5m in the horizontal direction, 6 nodal points in the vertical direction and a time increment of 0.01s were employed. The detailed method has been reported by Kuroiwa et al.(1998).

VERIFICATION OF MODEL
Comparisons with HORS data(Kuriyama et al,1999)

First, the results of field observation conducted by Kuriyama et al. (1999) were used in this study to calibrate and verify the modified Q-3D model. The field observations were curried out from January 29 to February 3,1997 at the Hazaki Oceanographical Research Station (HORS) of Port and Airport Research Institute (PARI), Independent Administrative. Table 1 shows the wave conditions. Four data sets of the fourteen wave conditions reported by Kuriyama et al.(1999) were used to calibrate and verify the modified model.

Table 1. Input wave conditions for HORS data(Kuriyama et al.,1999)

Case	Date	$H_{1/3}$	$T_{1/3}$	α	T_p
1	Jan 31	2.00	9.69	24.5	10.4
2	Jan 32	2.11	9.63	16.5	10.3
4	Jan 33	2.91	11.81	19.0	12.6
7	Feb.1	2.37	12.16	6.0	13.0

$H_{1/3}$:significant wave height(m)
$T_{1/3}$:significant wave period(sec)
T_p :peak period (sec)

In this model, A_{sx} and A_{sy} in Eq. (12) and (13) are key coefficients for determining the magnitude of cross-shore and longshore current velocities. From the comparisons of the computed data with the measured data, A_{sx} and A_{sy} were set to 2.4 and 0.4, respectively.

Figure 2 shows comparisons between the computed results and measured data of the significant wave heights, the cross-shore and longshore current velocities near bottom. In this figure, z'/d represents dimensionless height from the sea-bottom, and d is the mean water depth, as shown in Figure 1. These comparisons show that the computed results of wave heights and undertow velocities approximate the measured data well, particularly in Cases 2 and 4. The computed longshore current velocities for Case 2 are also good agreement with the measured data. The computed results overestimated the measured values in Case 1 and 4. On the other hand, the compufed results for Case 7 underestimated the measured data. It is thought that the computed results in Case 7 are lower because the deep-water wave angle is less than in the other cases. Hence, the longshore current velocities computed by the present model are dependent on the wave angle. In this study, the computed longshore currents for Case 2 are the most accurate predictions of the measured data. The computed vertical distributions of the cross-shore and longshore current velocities for Case 2 are shown in Figure 3. It is found that the computed results for Case 2 are good agreement with the measured data.

Comparisons with DELILAH data(Smith et al.,1992,1993)

Secondly, the present model was applied to the field data for DELILAH (Duck Experiment on Low-frequency and Incident band Longshore Across-shore Hydrodynamics) project. Table 2 shows the input wave conditions for DELIAH data. Figure 4 shows the comparisons between the computed results and measured data of the root mean square wave heights and the longshore current velocities in the lower layer.

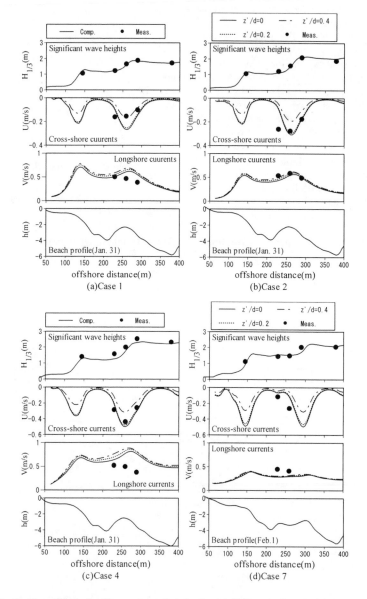

Fig. 2. Computed significant wave heights, cross-shore currents and longshore currents at HORS

It is found that the computed root-mean square wave heights coincide with meausred data. On the other hand, the peak velocities of the computed longshore currents are different from measured data. The longshore currents are not predicted well. Figure 5 shows the comparisons of computed cross-shore currents with data measured at 0.35m height from sea bottom. From this figure, it is found that the present model can well compute the peak velocities of cross-shore currents over barred beach, and that the present model can apply to the prediction of undertow over barred beach. Figure 6 shows the computed vertical distribution of cross-shore

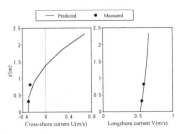

Fig.3. Vertical distribution of cross-shore and longshore current (Case 2)

currents on the longshore bar for Case Oct19. In this case, the present model can well predict the vertical distribution of cross-shore current velocities.

Table 2. Input wave conditions for DELIAH data in October of 1990

Case	Date	Time	H_{rms}	T_p	α	Reference
Oct14-0100	Oct.14	0100	0.94	9.7	32	Smith et al.
Oct14-1900	Oct.15	1900	0.83	12	18	(1993)
Oct17	Oct.17	1000	0.54	9.7	15	Smith et al.
Oct19	Oct.19	1200	0.65	7	24	(1992)

*H_{rms} : root-mean square wave height (m),T_p:peak period(s)

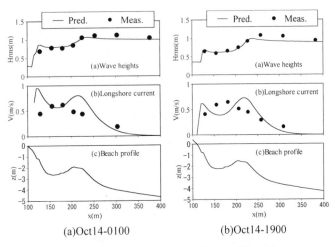

(a)Oct14-0100   (b)Oct14-1900

Fig.4 . Computed root-mean square wave heights and longshore currents

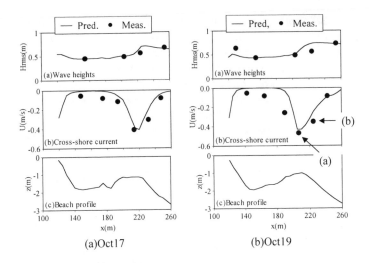

(a)Oct17   (b)Oct19

Fig.5. Computed root-mean square wave heights and cross-shore currents

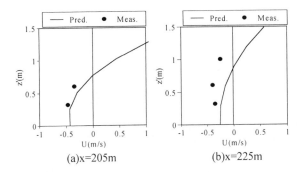

(a)x=205m   (b)x=225m

Fig.6. Vertical distribution of cross-shore currents on longshore bar (Oct 19)
((a) x=205m, (b) x=225m, see Figure.5(b))

Prediction errors

Figures 7(a),(b) and (c) show the computed vs. measured values of the significant and root mean square wave heights, undertow and longshore current velocities, respectively. In order to evaluate the accuracy of the computation, the root-mean-square relative errors used by Smith et al. (1993) were calculated. The error is defined as

$$Er = 100 \times \sqrt{\sum (Meas - Pred)^2 \Big/ \sum Meas^2} \tag{16}$$

where *Meas* and *Pred* represent the measured and computed data, respectively. The relative errors Er of the wave height, undertow and longshore current velocity are 8.7%, 37.7% and 44.0%, respectively. The prediction of wave heights under random waves is sufficiently to be used in practice. The computed results of undertow and longshore current velocities are qualitatively agreement with measured data. The present model has two tendencies to underestimate the cross-shore current velocities and overestimate the lognshore current velocities. In this study, it is assumed that the key coefficients Asx and Asy are constant values, and that these coefficients are independent on wave conditions such as wave height and period and wave direction.

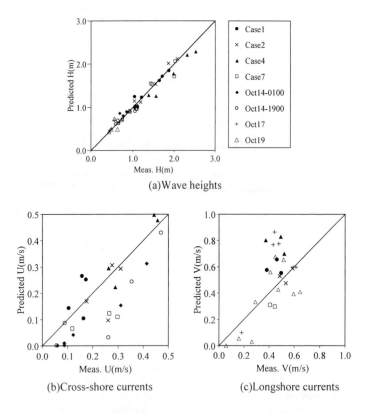

(a)Wave heights

(b)Cross-shore currents   (c)Longshore currents

Fig. 7. Computed vs. measured values of the wave heights(a), cross-shore(b) and longshore current velocities(c)

CONCLUSIONS

This paper presented a quasi-three dimensional numerical model for estimating the distribution of nearshore currents on barred beaches. The computed results were compared with data measured in the field sites.. The applicability of the Q-3D model to nearshore currents was investigated. It was found that computed wave heights were quantitatively agreement with the measured data, with a relative computational error 8.7%. The computed cross-shore and ongshore current velocities were qualitatively agreement with the measured data. The errors in these computations are 37.7% and 44.07%, respectively. Although these errors are large, it was confirmed that the cross-shore and longshore current velocities on a barred beach could be computed by using a modified Q-3D numerical model.

ACKNOWLEDGEMENTS
We would like to thank Dr. Yoshiaki Kuriyama, Head of Littoral Drift Laboratory, Independent Administrative Institution, Port and Airport Research Institute(PARI), for providing the field data from HORS. We would also like to thank Dr. Takao Ohta, Department of Social System Engineering, Faculty of Engineering, Tottori University, for advising in the computation of random waves.

REFERENCES
Dally, W. R., R. G.Dean and R. A. Dalrymple. 1984. A Model for Breaker Decay on Beaches, Proc. 19th ICCE, pp.82-97

De Vriend, H.J. and Stive, M.J.F. 1987. Quasi-3D modelling of neashore currents, Coastal Engineering, 11, no.5/6, pp.565-601.

Kuriyama,Y and T. Nakatsukasa 1999. Undertow and Longshore current on a Bar-Trough Beach,Field Measurements at HORS and Modeling-, Rep. Port and Harbor Res. Inst., Vol.38,No.1

Kuroiwa M, H.Noda and Y. Matsubara. 1998. Applicability of a quasi-three dimensional numerical model to nearshore currents, Proceedings of 26th ICCE, pp.815-828.

Kuroiwa,M., H.Noda,C.B.Son, K.Kato and S.Taniguchi. 2000. Numerical prediction of bottom topographical change around coastal structures using quasi-3D nearshore current model, Proceedings of 27th ICCE,pp.2914-2927.

Longuet-Higgins, M.S. 1970. Longshore currents generated by obliquely Incident sea waves(1 and 2). J.Geophys. Res., 75,pp6778-6801.

Nishimura, H. 1988. Computation of nearshore current, Neashore Dynamics and Coastal Process -Theory, Measurements and Predictive Models-(ed. Horikwa, K), University of Tokyo Press, pp.271-291.

Okayasu, A., K.Hara and T. Shibayama. 1994. Laboratory experiments on 3-D nearshore currents and a model with momentum flux by breaking wave, Proc. 24th Int. Conf. Coastal Eng., pp.2461-2475.

Smith,J.M.,I.A.Svendsen and U.Putrevu, 1992. Vertical structure of the nearshore current at DELILAH : Measured and Modeled, Proc. 22th Int. Conf. Coastal Eng., pp.2825-2838

Smith,J.M., M. Larson and N.C. Kraus, 1993. Longshore current on a barred beach : field measurements and calculation, J. of Geophs. Res.,Vo.98, C12,pp22,717-22,731.

Svendsen, I.A. and R.S. Lorenz. 1989. Velocities in combined undertow and longshore currents, Coastal Engineering, Vol.13, pp.57-79.

Tsuchiya,Y., T.Yamashita and M.Uemoto. 1986. A model of undertow in the surf zone, Proc. of 33rd Conf. of Coastal Eng., JSCE, pp.31-35.(in Japanese)

VISUALIZATION OF SEEPAGE DURING VORTEX RIPPLE FLOW OVER A SANDBED

Kojiro Suzuki[1] and Mostafa Foda[2]

Abstract: The action of large waves causes the disappearance of vortex ripple flow and produces sheet flow, although numerous experimental studies have not clarified why this occurs. The shape of a vortex ripple indicates that the flow velocity over the crest of the ripple is higher than the velocity over the trough, with the resultant flow field inducing low pressure above the crest of the ripple such that upward seepage inside ripples will occur. An oscillating water tunnel and numerical simulations were accordingly employed to elucidate the phenomenon; ultimately showing that the oscillatory flow-produced excess pore pressure due to low pressure around the ripple crest is larger than the overburden soil pressure such that ripple fluidization occurs. The effect of wave height, permeability of sand, and ripple shape are also investigated using numerical simulations.

INTRODUCTION

Three types of sea floor sand transport are known to occur (Fig. 1), i.e., initial sand movement, called bedload, suspension of sand particles over vortex ripple flow produced by small height amplitude waves, and larger-scale flow of sand across a flat bed produced by the action of large waves, called sheet flow. The shape, size, and sediment transport rates of vortex

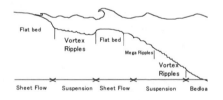

Figure-1 Bedform and sediment transport

ripple and sheet flows have both been investigated (e.g., Dibajnia and Watanabe (1992)), yet only the bottom shear stress due to oscillatory flow, or Shields number, has been considered. In fact, the reason as to why ripple flow disappears remains unclear although numerous experimental studies have been carried out (e.g., Horikawa (1988)).

1 Research Engineer, Port and Airport Research Institute, 3-1-1 Nagase, Yokosuka, Japan , 239-0826, suzuki_k@pari.go.jp
2 Professor, 412 O'Brien Hall, Civil and Environmental Eng., University of California at Berkeley, CA USA.

The seabed configuration, which includes no-motion, bed-load, or suspension of sand in ripple/sheet flow, is thought to be related to the fluidization of a sandbed produced by wave-induced dynamic water pressure. Zen and Yamazaki (1990) reported that liquefaction of a sandbed is due to wave-associated dynamic pressures (Fig. 2(a)), while Sakai et al. (1996) used an oscillating water tunnel incorporating a cylinder system for controlling water pressure to experimentally show that seabed configuration is affected by the coexistence of oscillatory flow and dynamic water pressure, ultimately finding that changes in water pressure effect the ripple geometry, or flatness of ripples.

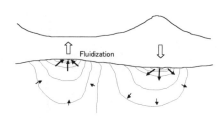

Figure-2(a) Fluidization of sandbed due to wave pressure

When the shape of a vortex ripple is observed, it appears that the dynamic pressure is driven by wave pressure in conjunction with wave-induced oscillatory flow. That is, flow velocity over the crest of the ripple is faster than the velocity over the trough such that resultant flow field induces low pressure above the crest of the ripple (Fig. 2(b)). Accordingly, we surmised that it

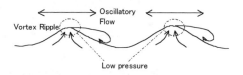

Figure-2(b) Fluidization of sandbed due to oscillatory flow

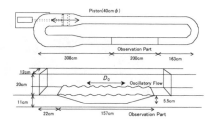

Figure-3 Oscillating water tunnel

is this low pressure which induces upward seepage flow around the crest such that fluidization or boiling of sandbed occurs around the crest. Under this scenerio, vortex ripple flow disappears and sand movement transitions into sheet flow.

As a means to elucidating the phenomenon of fluidization of ripples, seepage flow is visualized here using an oscillating water tunnel. In addition, fluidization depth is estimated by numerical simulation. A lot of studies has been done about vortex ripple and sheet flow. The shape, size and the sediment transport rates over vortex ripples and sheet flow seem like almost been clarified. But these studies are classified based only on the bottom shear stress due to the oscillatory flow, or Shields number.

FLOW VISUALIZATION
(1) Oscillating water tunnel

Figure 3 shows the oscillating water tunnel used for seepage flow visualization. Oscillatory flow of water molecules across the sandbed is produced by piston-driven waves that allow test section observation of sediment transport from the bed load to sheet flow. The tunnel can respectively produce symmetric oscillatory and asymmetric flows. The test section is 2.0 m long, 0.12 m wide, and 0.31 m deep including a 0.055 m deep sandbed.

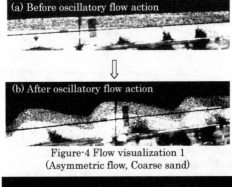

Figure-4 Flow visualization 1
(Asymmetric flow, Coarse sand)

Figure-5 Flow visualization 2
(Symmetric flow, Coarse sand)

Figure-6 Flow visualization 3
(Symmetric flow, Coarse sand)

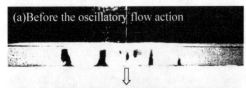

Figure-7 Flow visualization 4
(Symmetric flow, Mixed sand)

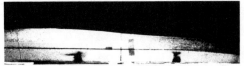

Figure·8 Flow visualization 5
(Asymmetric flow, Mixed sand, Sheet flow)

(2) Results

Figure 4 shows a photo of upward seepage flow due to asymmetric oscillatory flow using a sand diameter of 0.87 mm, oscillatory flow period T of 3 s, and water particle orbital diameter D_0 of 0.53 m. Dye is initially injected at the bottom of the sandbed every 10 cm, and after 2 min oscillatory flow action produces ripples and dye moves toward the crest of the ripples.

Asymmetric oscillatory flow is oscillatory flow in which the onshore and offshore velocities are different. In Fig. 4, the left side represents onshore where the onshore velocity is 2.3 times larger than offshore velocity.

Figures 5 and 6 are photos that more clearly show upward seepage due to symmetric oscillatory flow (T = 3 s and D_0 = 0.53 m). Note that the dye is transported to the crests of the ripples.

Figure 7 is a photo of seepage when mixed sand is used (50% 0.87 mm and 50% 0.18 mm). Although the seepage velocity is small because of the low permeability of sand, dye still reaches the crest after 10 min of oscillatory flow (T = 3 s, D_0 = 0.35 m).

As shown in Fig. 8, increasing oscillatory flow (T = 3 s, D_0 = 0.4 m) causes ripples to disappear and the sand bed to flatten, i.e., sand movement transitions into sheet flow in which upward seepage flow does not occur. In other words, upward seepage flow is only present in vortex ripple flow conditions, i.e., inside the ripples.

NUMERICAL METHOD

Seepage flow inside ripples is quantified using a numerical method to simulate turbulent flow over them.

(1) Flow over the ripples

The oscillatory flow over ripples is turbulent flow which can be modelled using Eqs. (1)–(6) below. Equations (4) and (5) are transport equations of the k-ω turbulence model developed by Wilcox (1988) and later successfully used by Fredsoe et al.(1999) to model oscillatory turbulent flow over a ripple. After transformation to curvilinear coordinates, Eqs. (2)–(5) are discretized using a hybrid scheme, and the governing equations are solved using the SIMPLE algorithm (Patankar (1980)).

$$\frac{\partial(\rho u)}{\partial x}+\frac{\partial(\rho v)}{\partial y}=0 \qquad (1)$$

$$\frac{\partial(\rho u)}{\partial t}+\frac{\partial(\rho uu)}{\partial x}+\frac{\partial(\rho uv)}{\partial y}=-\frac{\partial p}{\partial x}+\frac{\partial}{\partial x}\left\{(\mu+\mu_T)\frac{\partial u}{\partial x}\right\}+\frac{\partial}{\partial y}\left\{(\mu+\mu_T)\frac{\partial u}{\partial y}\right\} \qquad (2)$$

$$\frac{\partial(\rho v)}{\partial t}+\frac{\partial(\rho uv)}{\partial x}+\frac{\partial(\rho vv)}{\partial y}=-\frac{\partial p}{\partial y}+\frac{\partial}{\partial x}\left\{(\mu+\mu_T)\frac{\partial v}{\partial x}\right\}+\frac{\partial}{\partial y}\left\{(\mu+\mu_T)\frac{\partial v}{\partial y}\right\} \qquad (3)$$

$$\frac{\partial(\rho k)}{\partial t} + \frac{\partial(\rho u k)}{\partial x} + \frac{\partial(\rho v k)}{\partial y} = \frac{\partial}{\partial x}\left\{(\mu + \sigma^* \mu_T)\frac{\partial k}{\partial x}\right\} + \frac{\partial}{\partial y}\left\{(\mu + \sigma^* \mu_T)\frac{\partial k}{\partial y}\right\} + G - \rho\beta^* k\omega \qquad (4)$$

$$\frac{\partial(\rho \omega)}{\partial t} + \frac{\partial(\rho u \omega)}{\partial x} + \frac{\partial(\rho v \omega)}{\partial y} = \frac{\partial}{\partial x}\left\{(\mu + \sigma \mu_T)\frac{\partial \omega}{\partial x}\right\} + \frac{\partial}{\partial y}\left\{(\mu + \sigma \mu_T)\frac{\partial \omega}{\partial y}\right\} + \gamma\frac{\omega}{k}G - \rho\beta k\omega^2 \qquad (5)$$

$$G = \mu_T\{2[(\frac{\partial u}{\partial x})^2 + (\frac{\partial v}{\partial y})^2] + (\frac{\partial u}{\partial y} + \frac{\partial v}{\partial x})^2\} \qquad (6)$$

where u and v are the horizontal and vertical mean velocities, p the pressure, k the kinetic energy, ω the dissipation of kinetic energy, ρ the density of water, μ the viscosity, and μ_T the eddy viscosity. From k and ω, μ_T is calculated using

$$\mu_T = \rho\gamma^* \frac{k}{\omega} \qquad (7)$$

where the closure coefficients are taken as

$$\gamma^* = 1, \gamma = \frac{5}{9}, \beta^* = \frac{9}{100}, \beta = \frac{3}{40}, \sigma^* = \frac{1}{2}, \sigma = \frac{1}{2} \quad . \qquad (8)$$

Momentum Eqs. (2) and (3) are solved using the best guess pressure determined from Eqs. (9) and (10) representing analytical pressure from small amplitude theory. Briefly, after substituting assumed velocities into the pressure-correction equation, correct pressure and velocities are obtained. Here, Eq. (9) is used as the estimated pressure for progressive waves and Eq. (10) as the estimated pressure for standing waves.

$$p = \frac{1}{2}\rho g\frac{H}{\cosh(kh)}\cos(kx + \frac{\pi}{2})\cos(\sigma t) \approx -\frac{1}{2}\rho g\frac{kH}{\cosh(kh)}\sin(\frac{\pi}{2})\cos(\sigma t) \qquad x \approx 0 \quad (9)$$

$$p = \frac{1}{2}\rho g\frac{H}{\cosh(kh)}\cos(kx + \frac{\pi}{2} - \sigma t) \approx -\frac{1}{2}\rho g\frac{kH}{\cosh(kh)}(\sin(\frac{\pi}{2})\cos(\sigma t) + \cos(\frac{\pi}{2})\cos(\sigma t)) \quad x \approx 0 \quad (10)$$

Figure 9 shows the numerical grid and boundary conditions. Periodic conditions are used at the side boundaries, a symmetric condition at the top, and fixed condition at the surface of the ripples as described next (Eqs. (11)–(23)). Three or five ripples are located at the middle of the grid. As the calculation becomes unstable when a large separation passes the side boundaries, a flat part is located at both sides of the ripples to avoid flow separation passing the side boundaries.

Periodic conditions (Side boundaries)

$$u_{1,j} = u_{n-1,j}, u_{n,j} = u_{2,j} \qquad (11)$$

$$v_{1,j} = v_{n-1,j}, v_{n,j} = v_{2,j} \qquad (12)$$

$$p_{1,j} = p_{n-1,j}, \ p_{n,j} = p_{2,j} \qquad (13)$$

$$k_{1,j} = k_{n-1,j}, k_{n,j} = k_{2,j} \qquad (14)$$

$$\omega_{1,j} = \omega_{n-1,j}, \omega_{n,j} = \omega_{2,j} \qquad (15)$$

Fixed condition (Bottom boundary)

$$u = 0, \ v = 0 \qquad (16)$$

$$k = 0, \ \omega = U_f^2 S_R / \nu \qquad (17)$$

$$S_R = (50/k_N^+)^2 \quad k_N^+ < 25 \qquad (18\text{-}1)$$

$$S_R = 100/k_N^+ \quad k_N^+ \geq 25 \qquad (18\text{-}2)$$

$$k_N^+ = k_N U_f / \nu \qquad (19)$$

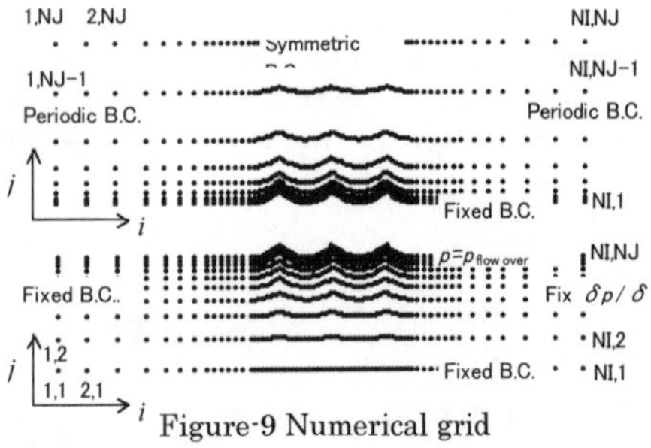

Figure-9 Numerical grid

in which k_N is Nikuradse's equivalent roughness height (grain roughness) and U_f the local friction velocity.

Symmetric condition (Top boundary)

$$u_{i,NJ} = u_{i,NJ-1} \qquad (20)$$

$$v_{i,NJ} = u_{i,NJ-1} \qquad (21)$$

$$p_{i,NJ} = p_{i,NJ-1} \qquad (22)$$

$$k_{i,NJ} = k_{i,NJ-1}, \; \omega_{i,NJ} = \omega_{i,NJ-1} . \qquad (23)$$

(2) Flow inside ripples

The governing equation for seepage flow inside ripples is

$$k\left(\frac{\partial^2 p}{\partial x^2} + \frac{\partial^2 p}{\partial y^2}\right) = (m_v + n_p \beta)\frac{\partial p}{\partial t} \qquad (24)$$

After overlying turbulent flow is solved for, turbulent flow pressure is given as the input pressure of seepage flow at the ripple surface, i.e.,

$$p_N(\text{flow in ripple}) = p_1(\text{flow over ripple}) \qquad (25)$$

Other boundaries are fixed boundaries, i.e.,

$$\frac{\partial p}{\partial n} = 0 \qquad (26)$$

in which p is the pressure, k the permeability, n_p the porosity, m_v the coefficient of the volume compressibility of sand, and β the coefficient of the volume compressibility of water.

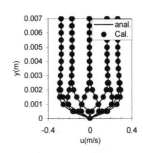

Figure-10 Comparison of numerical and analytical velocity profiles in laminar oscillatory boundary layer

(3) Verification of numerical simulation

To validate the accuracy of the numerical simulation, the analytical velocity in laminar oscillatory flow in the boundary layer is compared with the experimental flow velocity profiles over the ripples (Fredsoe et al. (1999)). Figure 10 compares the numerical and analytical velocity profiles in the laminar oscillatory boundary layer above a flat plate using a wave height H of 0.13 m, wave period T of 2.5 s, and depth of 0.43 m, where good agreement is apparent.

Figure 11 compares the experimental and numerical velocity profiles over ripples for a wave height H of 0.13m, wave period T of 2.5s, depth of 0.43m, and ripple height η

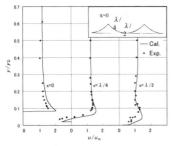

Figure-11 Experimental and numerical velocity profiles over ripples

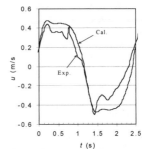

Figure-12 Time series of the horizontal velocity 1cm over the crest of a ripple

and length λ of 0.22 and 0.035m, respectively. Again note the good agreement present.

Figure 12 shows a time series of the horizontal velocity 1cm above the crest of the ripple, where except for negative velocity, agreement is apparent. It is surmised that disagreement exists because the experiment was conducted in an open channel whereas the numerical simulation considers a closed channel.

NUMERICAL RESULTS

Table-1 Calculation cases

Case		T (s)	D_0 (m)	H (m)	k (cm/s)	λ (cm)	η (cm)
1	Oscillatory flow	2.5	11.2	0.13	0.01	22.0	3.5
2	//	//	5.2	0.06	//	//	//
3	//	//	17.2	0.20	//	//	//
4	//	//	11.2	0.13	0.1	//	//
5	//	//	11.2	0.13	0.001	//	//
6	Progressive wave (Oscillatory flow + Pressure)	//	11.2	0.13	0.01	//	//
7	//	//	11.2	0.13	//	//	1.5

Table 1 shows calculation cases and conditions.

(1) Flow field over and inside a ripple (only oscillatory flow): Case-1

Figure 13 shows the calculated velocity, turbulent kinetic energy, and pressure fields over and inside ripples during a half wave period. Separation occurs at $t = 0$ above the crest of ripples, then the vortex expands up the lee side of ripples. Note that the pressure around the crest is always low and that seepage flow moves toward this low pressure area. The upward seepage flow is especially large when $t = 0.05$ to 0.1 T.

Figures 14 and 15 show pore pressure p and excess pore pressure $p\text{-}p_b$ profiles under the crest of the ripple, where p_b is the pore pressure at the ripple surface and p and $p\text{-}p_b$ are divided by water head ρg. Around the surface of ripples, p can be seen to immediately decrease at $t = 0.05$ T.

The broken line in Fig. 15 is the overburden soil pressure such that when the excess pore pressure $p\text{-}p_b$ is larger then boiling (fluidization) of sand occurs (Zen and Yamzakaki(1990)). In this case, fluidization occurs 0.015 m below the ripples' surface.

Figure 16 shows the mean velocity and pressure field during one wave period, where at the both sides of the crest of the ripple, vortices arise and pressure is lowest at the crest of the ripple. Seepage flow merges up to the crest and this seepage flow field looks like that in Figs. 5–7.

Based on these results, fluidization of a sandbed is believed to be caused by the low pressure around the crest of the ripple produced by oscillatory flow. Ripples disappear and the sandbed becomes flat due to this fluidization of ripples.

(2) Effect of wave height: Cases 1–3

Figure 17 shows the excess pore pressure profiles when the wave height is varied from 0.06 to 0.2 m, where effect of increasing wave height increases the boiling depth. Note that the ripple is completely fluidized when the wave height is 0.2 m.

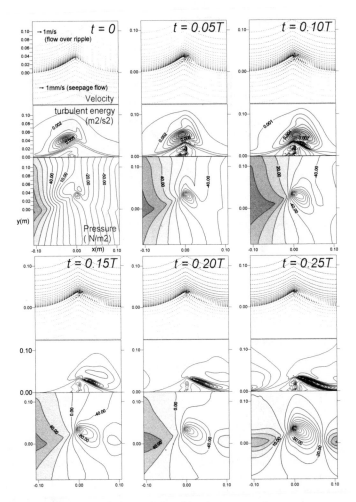

Figure-13 velocity, turbulent kinetic energy and pressure

(3) Effect of permeability: Cases 1,4 and 5

Figure 18 shows the pressure profiles when the permeability of sand is changed from 0.001 to 0.00001 m/s, with only a slight difference resulting. Since the excess pore pressure profiles also show no difference in fluidization depth, permeability does not affect fluidization.

(4) Effect of wave-associated dynamic pressure: Case 6

When no ripple is present, as indicated by Eq. (9) there is no pressure difference along the vertical direction of oscillatory flow. The action produced by a progressive wave (Eq. (10)), however, results in a pressure difference. Figure 19 shows the pore pressure

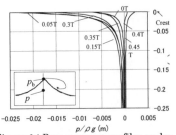

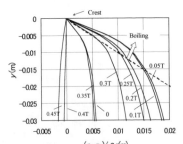

Figure-14 Pore pressure profiles under
the crest of ripple

Figure-15 Excess pore pressure profiles

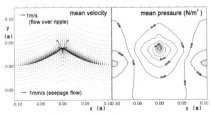

Figure-16 Mean velocity and pressure
field during 1 wave period

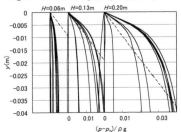

Figure-17 Effect of wave height

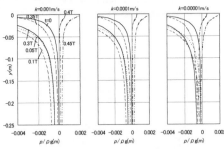

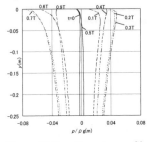

Figure-18 Effect of permeability

Figure-19 Pore Pressure profiles
under progressive wave

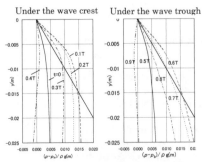

Figure-20 Excess pore pressure profiles
under progressive wave

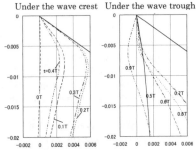

Figure-21 Excess pore pressure profiles
when ripple height is half that of Fig-20

profiles using Eq. (10), where the dynamic pressure due to a progressive wave is transmitted deeply into the sandbed. Around the ripple on the surface of the sandbed, the pressure profiles tend to be in a negative direction. Figure 20 shows excess pore pressure under the wave crest and trough. Fluidization is thought to occur under the wave crest (left side) rather than under the wave trough (right side). This phenomenon is different from the phenomenon predicted under a flat bed (no ripple) condition by Zen and Yamazaki (1990) who predicted that fluidization likely occurs under the wave trough (see Fig. 2(a)). Accordingly, this phenomenon is unique because the fluidization of sandbed is not only affected by the wave but also by the shape of sandbed.

(5) Effect of ripple height: Case 7

Figure 21 shows the excess pore pressure profiles when the ripple height is the half that of cases 1–6. Note that excess pore pressure is much less than that present in Fig. 20 and fluidization is no longer predicted. This indicates that fluidization is strongly related to ripple shape. Also, the effect of sandbed flattening due to large waves (Fig. 8) stops upward seepage flow.

CONCLUSIONS

The results of our numerical simulations and oscillating water tunnel observations are summarized as follows:

1) Low velocity over the crest of the ripple is confirmed to be faster than the velocity over the trough producing a flow field that induces low pressure above the crest of the ripple, which in turn produces upward seepage flow inside a vortex ripple. This seepage flow is only present in vortex ripple flow conditions, disappearing in sheet flow conditions.

2) Numerical simulations indicate that oscillatory flow-produced pore pressure under the crest of the ripple is larger than the overburden pressure such that fluidization of ripples occurs. These ripples disappear and the sandbed becomes flat due to this fluidization of ripples. In addition, increases in wave height were found to increase the fluidization depth.

3) Under progressive wave action, fluidization of the ripples likely occurs under the wave crest rather than under the wave trough; a phenomenon that is different from the phenomenon predicted under a flat bed condition, being unique because the fluidization of a sandbed is not only affected by waves but by the shape of a sandbed as well.

REFERENCES

Dibajnia, M. and Watanabe, A. 1992. Sheet flow under nonlinear waves and currents, Proc. of 23 Int. Conf. On Coastal Engineering : 2015-2028.

Fredsoe, J. Andersen, K.H., Sumer, B.M. 1999. Wave plus current over a ripple-covered bed, Coastal Engineering 38 : 177-221.

Horikawa, K. 1988. Nearshore Dynamics and Coastal Processes, University of Tokyo Press, 522p.

Patankar, S.V. 1980. Numerical heat transfer and fluid flow, Taylor & Francis, 197p.

Sakai, T. and Hitoshi Gotoh, H. 1996. Effect of Wave-Induced-Pressure on Seabed Configuration, Proceedings of 25th Conference on Coastal Engineering : 3155-3168.

Zen, K. and Yamazaki, H. 1990. Mechanism of wave-induced liquefaction and densification in seabed. Soils and Foundations, Vol.30, No.4 : 90-104.

ROLE OF NONLINEAR WAVE-WAVE INTERACTIONS IN BAR FORMATION UNDER NON-BREAKING AND BREAKING CONDITIONS

Cyril Dulou[1], Max Belzons[2], and Vincent Rey[3]

Abstract: This small scale laboratory study deals with the action of monochromatic or bichromatic waves on a sand bed of initial constant slope. The aim was to address the effects of nonlinear wave-wave interactions onto bed evolution and bar formation. Properties of a spurious wave, produced either by the wavemaker or by rapidly varying bathymetry, were used to obtain or not nonlinear interactions. A strong relation between the wave envelope and the bathymetric profile was found in the breaking case as in non-breaking wave conditions.

INTRODUCTION

Many works have been conducted to enhance the knowledge of nonlinear wave propagation in the nearshore zone for more than ten years. Since linear approach was deficient in the surf zone for wave modelling, nonlinear effects, as wave-wave interactions, were taken into account in the recent models (Ohyama *et al.*, 1992; Wei *et al.*, 1995; Madsen *et al.*, 1997).

In a parallel, the prediction of shoreline evolution for a sandy coast was more and more accurate thanks to wave models, but today it was always a challenge. Bar formation was an important process in shoreline evolution and it was not well understood. Several mechanisms were yet proposed and validated in the field or in laboratory, and they are dependent of initial wave and bottom conditions (Dulou *et al.*, 2000). These mechanisms were reviewed by Dean *et al.* (1992) and also by Van Rijn (1998).

I. 1 PhD, Département de Géologie et d'Océanographie, Université de Bordeaux, avenue de l'Université, 33405 Talence, France. c.dulou@epoc.u-bordeaux.fr

II. 2 Lecturer, IUSTI, Technopôle de Château-Gombert, 13453 Marseille, France. belzons@iusti.univ-mrs.fr

III. 3 Lecturer, LSEET, Université de Toulon et du Var, 83957 La Garde, France. rey@lseet.univ-tln.fr

This paper presents a small-scale experimental study of the initiation of bar formation under breaking and non-breaking waves. An experimental artifact, the free wave generation (Hulsbergen, 1974; Schäffer, 1996; Dulou, 2000), was used to simulate nonlinear wave-wave interactions and their effects on the sediment transport. Thanks to the control of a free wave generated by the wavemaker, the role of nonlinear wave-wave interactions on bar formation has been displayed under non-breaking conditions (Dulou *et al.*, 2000). These results were used to analyze the breaking case and to propose some way to explain the mechanism of bar formation by nonlinear wave effects.

EXPERIMENTAL FACILITIES

The glass-walled flume was 4.7 m long, 0.39 m wide and the maximum depth was 0.15 m. A piston-type wavemaker produced a regular wave with a first harmonic frequency $f = 1.5$ Hz and of maximum height $H_0 = 0.02$ m. Generally, a spurious free wave, of small amplitude, having the frequency $2f$, was also generated by the wavemaker (Madsen, 1971). The resulting wave was then bichromatic and the spurious component was revealed through a beating in the envelope of the second harmonic (Dulou *et al.*, 2000). This spurious wave can be suppressed by placing a sill in front of the wavemaker (Hulsbergen, 1974) and a monochromatic wave is obtained.

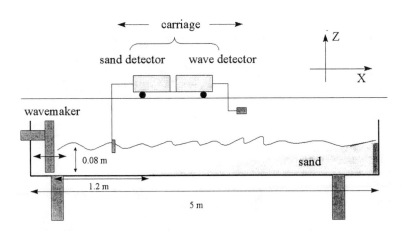

Fig.1. Sketch of experimental facilities

The generated wave was propagated first in a zone of constant depth $h = 0.08$ m, of extend 1.2 m until the toe of a sloping sandy bottom (Fig. 1). For both breaking and non-breaking conditions, the slope of the bed was gentle ($\beta < 0.03$). For the non-breaking case, a solid sloping bed replaced the sand in the upper part ($h < 0.04$ m), so that breaking occurred far downstream of the sandy zone. Eventually, this solid sloping bed could significantly reflect the incident wave leading to a partially-standing wave-field.

An artificial non-cohesive sand of diameter 0.08 mm was used for both cases. The wave envelope and the bottom profile were precisely measured along the flume using ultra-sonic sensors (Dulou *et al.*, 2000).

EXPERIMENTS WITH NON-BREAKING WAVES

A number of laboratory experiments have been devoted to a better understanding of the multiple-bar formation observed in nature, presumably under non-breaking conditions. Additional studies of bar formation were recently carried out by Dulou *et al.* (2000) by considering a gently sloping sandy bed and weakly nonlinear waves. Both monochromatic (regular wave) and bichromatic (regular wave + spurious free wave) waves were used, in order to address the role of nonlinear wave-wave interactions in the simplest situation in a suspension-dominated case.

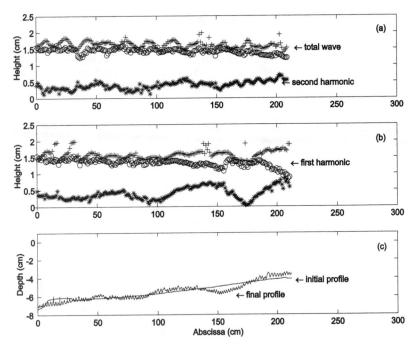

Fig.2. Envelope of the initial wave (a), final (b) and bathymetric profile (c) after 2660 min of wave action under non-breaking conditions (H_0=1.7 cm and β<0.05)

The main finding of this study was that the final bed profile was a replica of the first harmonic envelope. In the case of a monochromatic wave of first harmonic frequency f and local wavenumber k, a spatial modulation of local wavenumber 2k was only observed for partially standing wave. This was explained from linear analysis (Mei, 1983). In the bichromatic case, where the frequency of the parasitic wave was 2f and the wavenumber was K, nonlinear wave-wave interactions are expected. They have been addressed using the Boussinesq approach of Madsen and

Sorensen (1993) (Dulou *et al.*, 2000). Indeed, in addition to spatial modulation of local wavenumber 2k, modulations of local wavenumbers K and K-2k were also obtained. Moreover an important result of this modelling was that in the case of a progressive wave (no reflection), the only spatial modulation was of local wavelength $\Lambda = 2\pi/(K-2k)$. This theoretical result was confirmed by the Figure 2 which represents the final equilibrium modulated profile with a wavelength close to Λ (which is 0.7164 m, with K = 37.89 m^{-1}, and k = 14.56 m^{-1}, for the mean water depth h = 0.05 m). This last result displayed the role of the nonlinear wave-wave interactions in bar formation.

These nonlinear interactions are particularly relevant under breaking wave conditions.

EXPERIMENTS WITH BREAKING WAVES

The suspension-dominated case (D_1 = 0.08 mm) was considered and compared with the non-breaking experiments (Fig.2) . Figure 3 presents both initial and final envelopes and the final bathymetric profile. The final envelope (fig.3b) looks like the non-breaking one (Fig.2b): upstream the breakpoint, there is a beating in the envelope of the second harmonic and a bar is also observed in that zone, as in the non-breaking case. In both cases the configurations of the envelope-bar system over the extend of the beating are very similar. Although, no free wave was initially present in the breaking case, the final envelope of the second harmonic reveals the appearance of an additional free harmonic through the beating observed from x = 50 cm to x = 150 cm. The bar was then of the same length as the beating, as in the non-breaking case were the free wave was however initially present. So the questions are: was it the same mechanism of bar formation as in the non-breaking case ? What was the origin of the beating in the final envelope?

From the above described experimental results, the strongly nonlinear part of the wave just before the breakpoint (decrease of the first harmonic amplitude) was suspected to play an important role in bar formation.

We then modified the initial bottom to preserve highly nonlinear wave conditions but in a zone located far away from breaking and thus poorly affected by the undertow. This was obtained by placing an horizontal sandy zone (of 0.8 m long) prolonged downstream with an inclined solid plate where the breaking occurred.

The breaking occurs only on the solid plate of which slope is about 0.10. After 60 min small bars are formed with a length of half the local wavelength of the wave, according to Bragg conditions (Dulou *et al.*, 2000).

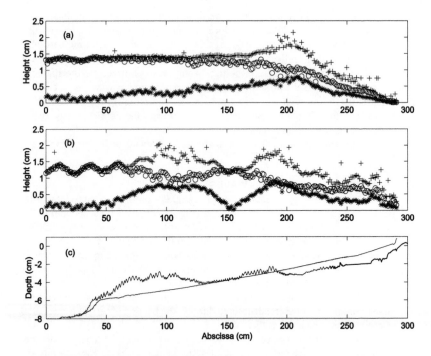

Fig.3. Envelope of the initial wave (a), final (b) and bathymetric profile (c) after 4660 min of wave action under breaking conditions (H₀=1.9 cm)

A beating is observed in the envelope of the second harmonic from x = 100 cm to x = 200 cm suggesting a local production of free harmonics, naturally induced by bathymetric changes, and which interact with the bound ones. The final envelope of the second harmonic (Figure 4a) shows the same length scale beating from x = 120 cm to x = 220 cm. In the final bathymetric profile (Figure 4b), small-scale bars have disappeared and only a larger bar is present from x = 80 cm to x = 170 cm. This bar is then of the same order than the beating length, like in the non-breaking case. The sediment was transported in the offshore direction.

Finally, the bar was formed as in the standard breaking case, under the same nonlinear conditions, but with a strongly diminished action of the undertow. Another mechanism than the undertow is then responsible of the seaward sediment transport.

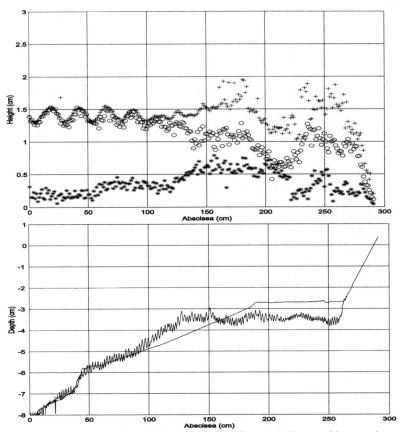

Fig.4. Envelope of (+) the total wave and of (o) first and (*) second harmonics at 1320 min, above the initial bathymetric profile (- -) and at 1320 min (--) (H_0=1.6 cm)

CONCLUSION

The density and the accuracy of the measurements in a small-scale wave flume allowed observations of the spatial and temporal evolution of bar formation resulting from the wave-sand bed interaction under breaking wave conditions. The important result of this study concerns the suspension-dominated case and the role of both the nonlinear wave-wave interactions and the undertow in bar formation.

Indeed, by increasing complexity of the experimental conditions (non-breaking waves ? breaking waves), we observed that the formation of a long bar is associated with a beating in the wave envelope of the second harmonic, under both non-breaking and breaking wave conditions. The last experiment has demonstrated that a bar is formed even under lowered action of the undertow, thanks to the nonlinearities that control the mean flows. This result could explain bar formation far from the breakpoint or the offshore migration of the bars. In any case, it demonstrated that spatial modulations in the wave envelope may be at the origin of bottom evolution, which cannot be predicted by linear wave analysis (contrary to modulations due to partially-standing waves).

The breaking point mechanism of bar formation could be improved by considering the near bed flow, which can be different in direction, and an additional flow in the same direction as the undertow. Nevertheless, these mean flows are related to the position of the breakpoint which means that this position is indeed an important parameter in bar formation. The next stage of this study would be then the issue of the influence of the breakpoint location on bar formation under action of bichromatic waves or irregular wave group.

ACKNOWLEDGMENTS

The authors are grateful to the french scientific program PNEC (Programme National d'Environnement Côtier).

REFERENCES

Dean, R.G., Srinivas, R., and Parchure, T.M., 1992. Longshore bar generation mechanisms, *Proceedings of the 23rd Coastal Engineering Conference,* (ASCE), 1, 2001-2014.

Dulou, C., Belzons, M., and Rey, V., 2000. Laboratory study of wave-bottom interaction in the bar formation on an erodible sloping bed, *Journal of Geophysical Research*, 105, 19745-19762.

Dulou, C., 2000. Interactions houle-sédiments : application à la formation des barres littorales, *thèse, Université de Provence*, 185p.

Hulsbergen, C.H., 1974. Origin, effect and suppression of secondary waves, *Proceedings of the 14th Coastal Engineering Conference*, 392-411.

Madsen, O.S., 1971. On the generation of long waves, *Journal of Geophysical Research*, 76, 8672-8683.

Madsen, P.A. and Sørensen, O.R., 1993. Bound waves and triads interactions in shallow water, *Ocean Engineering*, 20, 359-388.

Madsen, P.A., Sørensen, O.R., and Schäffer, H.A., 1997. Surf zone dynamics simulated by a Boussinesq type model. part 1. Model description and cross-shore motion of regular waves, *Coastal Engineering*, 32, 255-287.

Mei, C.C., 1983. *The applied dynamics of ocean surface waves*, Wiley-Interscience.

Ohyama, T. and Nadaoka , K. 1992. Modeling the Transformation of Nonlinear Waves Passing over a Submerged Dike. *Proceeding of 23rd Coast. Eng. Conf.*, ASCE, 526-539.

Schäffer, H.A., 1996. Second-order wavemaker theory for irregular waves, *Ocean Engineering,* 23, 47-88.

Van Rijn, L.C., 1998. *Principles of coastal morphology*, Aqua Publications, Amsterdam.

Wei, G., Kirby, J.T., Grilli, S.T., and Subramanya, R., 1995. A fully nonlinear Boussinesq model for surface waves. Part 1. Highly nonlinear unsteady waves, *Journal of fluid Mechanic*, 294, 71-92.

EFFECTS OF FREQUENCY-DIRECTIONAL SPREADING ON THE LONGSHORE CURRENT

Matilda Kitou[1] and Nikos Kitou[2]

Abstract: This paper presents a detailed study on the sensitivity of a number of important wave and current parameters (wave height, driving forces, wave-induced current) to the directional and frequency spreading of the random sea. The analysis is carried out using a purpose-built wave and current model. Three forms of input are used, a directional spectrum, a frequency spectrum and a monochromatic wave, corresponding to different sea conditions, short-crested, unidirectional and monochromatic waves. It is concluded that the directional spreading affects radically the wave and current fields. Its effect is much more pronounced on the driving forces and longshore current velocity (differences up to 50%) rather than on the wave heights. Moreover, the obtained results show that the above quantities are not very sensitive to the frequency spreading.

INTRODUCTION

The radical advances in computer capacity as well as recent scientific research have resulted in the development of advanced fully spectral nearshore models (Ris et al, 1994). However, engineers very often introduce simplifying assumptions for the description of the random sea climate (e.g. one–dimensional spectral description or a representative wave) for reasons of model availability and time restrictions. Nevertheless, they should consider whether the simplifications are justified by the physics of the sea state and to what extent they affect the results obtained in wave transformation and nearshore circulation modeling.

As a contribution towards this direction, this paper presents a detailed study on the sensitivity of the energy and momentum related quantities to the frequency and directional spreading and aims to reach solid conclusions based on qualitative and quantitative comparison between various kinds of spectral description in a transparent manner.

1 Assistant Engineer, High-Point Rendel Ltd., 61 Southwark Street, London SE1 1SA, UK, m.kitou@highpointrendel.com
2 Research Associate, Democritus University of Thrace, Dept. of Civil Engineering, , 67100 Xanthi, Greece, kitou@civil.duth.gr

DESCRIPTION OF MODEL AND OUTLINE OF MODEL TESTS

For the purposes of the analysis use is made of a fully discrete spectral wave model and a depth and period averaged nearshore circulation model. Calculations are carried out for a uniform plane beach, steady state conditions and ignoring current refraction and wave-wave interaction effects.

The wave transformation model evaluates the wave quantities by advancing directional energy components, thus working on a spectral grid which is linear in frequencies and refraction adaptive in directions (refraction-wise resolution). Wave and current modules are coupled through the varying depth due to set-up.

The model runs are designed so as to facilitate investigation of the various ways in which frequency and directional spreading influence the wave field. For this purpose, test runs are performed for three different forms of input and two incident wave directions in deep water. The tests are grouped according to the incident wave direction.

The forms of input for each group of tests are as follows:
- 2D form: a two-dimensional spectrum $S(f, \theta)$
- 1D form: an one-dimensional frequency spectrum $S(f)$ and a direction θ_o
- 0D form: a representative wave height H_{rms} with period T_{m01} and a direction θ_o

The 2D cases are compared to the 1D cases to assess the effect of directional spreading and the 1D cases are compared to the 0D cases to assess the effect of frequency spreading.

The test case considered is a plane beach of 1:50 slope. The coordinate system and sign convention for various wave parameters is shown in figure 1. The bed configuration (together with the locations of output reference points) is shown in figure 2. The cross-shore dimension of the modeled area, as figure 2 shows, extends to 2500 m from the offshore boundary (at 50 m depth) to the shoreline (zero depth).

The quantities used to monitor and assess the results are the root-mean-square wave height H_{rms}, the mean longshore current velocity U_y and the S_{xy} component of the radiation stress tensor. We choose to monitor S_{xy} because for the particular domain and imposed coordinate system, its gradient is the only driving force for the longshore mean current. For a different domain and/or coordinate system, the other components of the radiation stress tensor would also have a contribution to the driving force terms. Nevertheless, the physics of the wave-induced current generation process and therefore the relevance of the present discussion will remain the same.

For the set of runs presented here, representative formulations are selected for the wave dissipation and current friction mechanisms. Wave energy dissipation due to breaking is calculated by the Battjes and Janssen (1978) model (with the steepness effect off). Frictional dissipation in the wave model is calculated using the model by Madsen et al (1988). Friction effects in the current model are incorporated using the bed shear stress model by Bijker (1966).

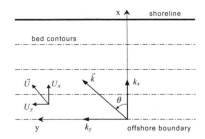

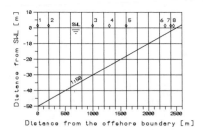

Fig. 1. Coordinate system and sign conventions

Fig. 2. Bed configuration and location of reference points

In all cases where incident wave conditions are given in spectral form, the frequency spectrum $S(f)$ is assumed to be of the Pierson-Moskowitz type, implying a sea of low steepness and of wide frequency spreading. The spreading function D_s used for the 2D runs, is of the $\cos^{2s}$ type, where the spreading parameter s is taken equal to 1 (wide directional spreading) and the directional range is 180^0. The 2-dimensional spectrum in deep water, is thus written:

$$S(f,\theta) = S(f)D_s(\theta) \quad , \quad \int_0^{2\pi} D_s(\theta)\,d\theta = 1 \tag{1}$$

where θ = direction of the spectral component

Two groups of tests are presented with the main wave directions in deep water chosen to be 60^0 and 30^0 to the shore normal.

DESCRIPTION OF RESULTS

Results from the 60^0 and 30^0 runs are presented in figures 3-5 and 6-8 respectively, which show the variation of H_{rms}, U_y and S_{xy} as a function of the cross-shore distance. It can be seen that the results produced by the 0D and 1D runs are close or almost identical whereas the results produced by the 1D and 2D runs deviate significantly.

The ratios between selected values of the 2D to 1D results are as follows:
- 60^0 runs: 88% for the wave height H_{rms} at the offshore boundary, 74% for the peak value of the current velocity U_y and 61% for the radiation stress S_{xy}
- 30^0 runs: 98% for the wave height H_{rms} at the offshore boundary, 57% for the peak value of the current velocity U_y and 51% for the radiation stress S_{xy}

By comparing the results of the two directions, one can notice that the larger deviation of the H_{rms} values appears in the 60^0 runs whereas the larger deviations of the U_y and S_{xy} values appear in the 30^0 runs.

In trying to assess the factors affecting the obtained results, we must consider the wave spectrum in deep water and its evolution within the model area. Figures 9, 10, 11 and 12 show the 2-dimensional spectrum in deep water (at an orientation of 60^0 to the shore normal), at the offshore boundary (Location 1, at 50m depth) and at two inshore locations (Locations 4 and 7 at 24m and 2m depth respectively).

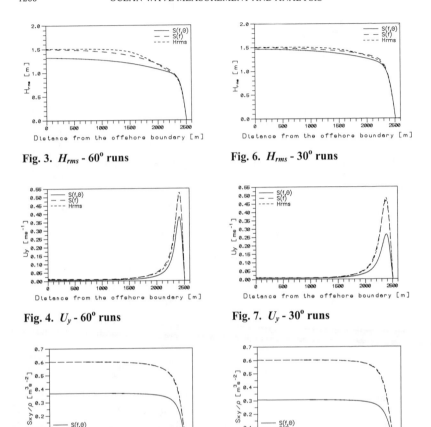

Fig. 3. H_{rms} - 60° runs   **Fig. 6.** H_{rms} - 30° runs

Fig. 4. U_y - 60° runs   **Fig. 7.** U_y - 30° runs

Fig. 5. S_{xy} - 60° runs   **Fig. 8.** S_{xy} - 30° runs

With regard to figures 9, 10, 11 and 12 we note that the directional range becomes more and more narrow as waves approach shallow water. At each depth, as figures 10, 11, and 12 show, the spectrum is confined between boundary curves in the f-θ domain (also shown), which indicate the permissible directional range for the refracted wave components.

It can be shown (Kitou, 1999) that for straight and parallel bed contours the directional bounds for the refracting spectrum are described by equation (2). Figures 13 and 14 show these directional bounds in Cartesian and in polar coordinates for various depths and they clearly demonstrate the contraction of the directional range in shallow water.

$$\theta = \arcsin\left(\pm \tanh\left(kd\right)\right) \tag{2}$$

where k = wave number, and d = mean water depth

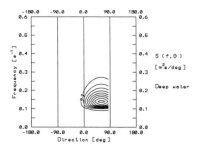

**Fig. 9. 2D Spectrum in deep water
        - 60° runs**

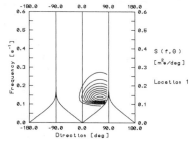

**Fig. 10. 2D Spectrum at 50m depth
        - 60° runs**

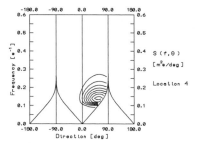

**Fig. 11. 2D Spectrum at 24m depth
        - 60° runs**

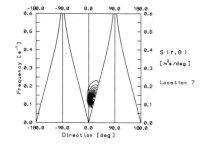

**Fig. 12. 2D Spectrum at 2m depth
        - 60° runs**

Moreover, figure 9 shows that there is a part of the incident spectrum not entering the computational domain, which corresponds to a set of offshore travelling wave components. Due to the presence of these components, the spectral wave energy entering the nearshore zone is less compared to the one associated with the full spectrum. This justifies the differences in H_{rms} values at the offshore boundary (more pronounced in the 60^0 runs), although the wave input parameters were based on the same spectrum at the open sea.

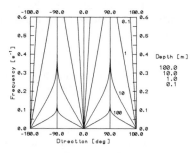

**Fig. 13. Directional bounds at
        various depths (Cartesian)**

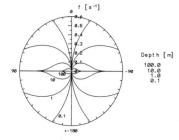

**Fig. 14. Directional bounds at
        various depths (Polar)**

The reduction in incoming energy is an effect of the directional spreading. It is possible to eliminate the differences associated with this particular effect of directionality by adjusting the incoming energy. To achieve this, we take into account only the onshore part of the wave directional spectrum and on this basis calculate the equivalent mean direction θ_m and input wave - $S(f)$ or H_{rms} - in deep water. We can thus produce 1D and 0D runs energetically equivalent to the corresponding 2D runs. The additional runs are named 60^0-adjusted and 30^0-adjusted and their input parameters are shown in Table 1. These runs may also be seen as an attempt to obtain unidirectional parametric descriptions for the directional sea.

Table 1. Model Test Runs and Corresponding Input

CODE NAME OF RUN	INPUT TO 1D & 0D-RUNS			(%) percentage of incoming energy
	θ_o (deg)* Used in 1D & 0D	T_{m01} (s) Used in 0D	H_{rms} (m) Used in 0D	
60^0	$\theta_w = 60.0$	5.7	1.51	100%
60^0 - adjusted	$\theta_m = 48.0$	5.7	1.33	78.3%
30^0	$\theta_w = 30.0$	5.7	1.51	100%
30^0 - adjusted	$\theta_m = 27.7$	5.7	1.48	96.5%

* θ_o denotes the main wave direction in deep water

Results from the 60^0-adjusted and 30^0-adjusted runs are presented in figures 15-17 and 18-20 respectively, which show the variation of H_{rms}, U_y and S_{xy} as a function of the cross-shore distance. It can be seen that the results produced by the 0D and 1D runs are almost identical whereas the results produced by the 1D and 2D runs deviate significantly, with the exception of the H_{rms} values which are in close agreement for all three forms of input.

The ratios between selected values of the 2D to 1D-adjusted results are as follows:
- 60^0-adjusted runs: 100% for the wave height H_{rms} at the offshore boundary, 77% for the peak value of the current velocity U_y and 68% for the radiation stress S_{xy}
- 30^0-adjusted runs: 100% for the wave height H_{rms} at the offshore boundary, 60% for the peak value of the current velocity U_y and 55% for the radiation stress S_{xy}

By comparing the results of the two directions, one can notice that the larger deviations between U_y and S_{xy} appear in the 30^0 runs.

The results obtained from the 60^0-adjusted and 30^0-adjusted runs suggest that, although the directional cut-off has an influence on the calculated quantities, significant differences still exist. Clearly this is not the only factor causing deviations between 2D and 1D/0D runs.

DISCUSSION

With regard to the results obtained we notice that whilst frequency spreading has a minor effect on the wave height, driving forces and longshore current, the directional spreading has a dominant effect on all three quantities.

We also notice that the wave height is less sensitive to directionality than the driving forces and longshore current. This is logical since energy fluxes from individual wave components contribute to the energy level (associated to the wave height) with positive increments regardless of their directions, whilst momentum fluxes from individual wave components depend strongly on their direction and may even result in opposing contributions to the radiation stress tensor (associated to the driving forces).

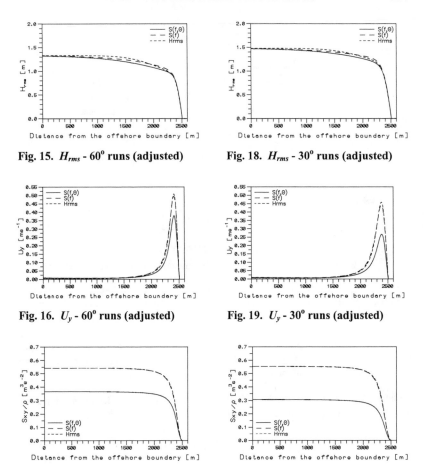

Fig. 15. H_{rms} - 60° **runs (adjusted)**

Fig. 18. H_{rms} - 30° **runs (adjusted)**

Fig. 16. U_y - 60° **runs (adjusted)**

Fig. 19. U_y - 30° **runs (adjusted)**

Fig. 17. S_{xy} - 60° **runs (adjusted)**

Fig. 20. S_{xy} - 30° **runs (adjusted)**

To a certain extend, directionality can affect the amount of the total energy entering the nearshore zone and causes a "cut-off" of the incident wave spectral energy. To examine this particular effect we need to consider the energy flux in deep water. Actually, the directional cut-off of the incident wave energy is a direct consequence of the fact that in most cases the shore is not perpendicular to the main wave direction. A schematic example is shown in figure 21 where the open sea spectrum has a full directional range of 180° (indicated by semicircles) but closer to the area of interest the range is restricted (black areas) by large formations of the shoreline.

A further and more important effect of directionality concerns the fact that momentum fluxes associated with individual wave components depend strongly on their direction. The key point here is the radiation stresses in deep water. This is because the gradients of this quantity (in this particular case of S_{xy}) are proportional to the longshore driving forces and

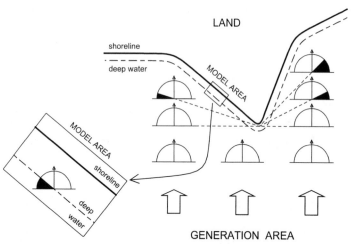

Fig. 21. The effect of large land formations on the directional range

therefore the difference in S_{xy} at two locations represents the total force exerted on the fluid mass between these two locations. The radiation stress component eventually becomes zero at zero depth (due to breaking) and because of that S_{xy} in deep water represents the total force applied on the whole body of water. The dissipation mechanisms, through their intensity, in fact regulate the distribution of this force along x. However due to the fact that dissipation mainly takes place in the surf zone, the S_{xy} component in deep water more or less represents the total longshore driving force acting on the breaker zone.

On the basis of Eq.(1), the S_{xy} radiation stress component in deep water for a multi-frequency and multi-directional wave field can be written as follows:

$$\frac{S_{xy}}{\rho g} = \int_f S(f) \left\{ \int_\theta n_o \cos\theta \sin\theta\, D_s(\theta)\, d\theta \right\} df \tag{3}$$

where n = wave parameter; g = acceleration of gravity, and ρ = water density

We should now focus on the integral along the directions, i.e. the quantity in curly brackets in Eq.(3). Figures 22, 23, 24 and 25 show the functions involved as well as the integrand quantities. They clearly show how the integral along the directions builds up with negative and positive contributions from the partially opposing wave components.

For unidirectional waves travelling at a main direction θ_0, an expression analogous to Eq.(3) can be written for the S_{xy} component in deep water:

$$\frac{S_{xy}}{\rho g} = \int_f S(f) \left\{ n_o \cos\theta_o \sin\theta_o \int_{2\pi} D_s(\theta)\, d\theta \right\} df \tag{4}$$

In the limiting case of zero directional spreading centered on the wind direction θ_w (1D-runs), the distribution function D_s is a delta-function centered on $\theta_o = \theta_w$, whose integral

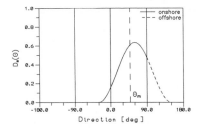

Fig. 22. Spreading function $D_s(\theta)$ and adjusted θ_m - 60° runs

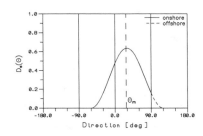

Fig. 24. Spreading function $D_s(\theta)$ and adjusted θ_m - 30° runs

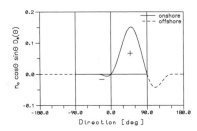

Fig. 23. $n_o \cos\theta \sin\theta D_s(\theta)$ - 60° runs

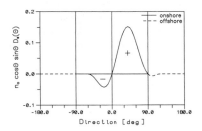

Fig. 25. $n_o \cos\theta \sin\theta D_s(\theta)$ - 30° runs

over the directional range is equal to 1 (because of Eq. 1). Similarly, in the limiting case of zero directional spreading centered on the mean direction θ_m (1D-adjusted runs), the distribution function D_s is a spike centered on $\theta_o = \theta_m$, but in this case its integral over θ is less than 1, because it includes only the onshore travelling wave components, i.e. its integration range extends from $-\pi/2$ to $\pi/2$.

Evaluation of the directional integrals for the 2D, 1D and 1D-adjusted runs (within curly brackets in equations 3 and 4) are shown in Table 2. The reasons for the differences between the 2D and 1D runs and between the 2D and 1D-adjusted runs become clear, as the ratios between the above numbers are exactly the same as the ones found between the values of S_{xy} at the offshore boundary (see figures 5, 8, 17 and 20).

Table 2. Directional Integrals in Computing S_{xy} in Deep Water

RUN	Directional integral	Direction 60°	Direction 30°
2D	$\int_{-\pi/2}^{\pi/2} n_o \cos\theta \sin\theta \, D_s(\theta)\, d\theta$	-0.002+0.134=0.132	-0.024+0.134=0.110
1D	$n_o \cos\theta_w \sin\theta_w \int_{2\pi} D_s(\theta)\, d\theta$	0.217	0.217
1D-adjusted	$n_o \cos\theta_m \sin\theta_m \int_{-\pi/2}^{\pi/2} D_s(\theta)\, d\theta$	0.195	0.199

CONCLUSIONS

The work presented constitutes a detailed study on the sensitivity of a number of wave and current parameters to the directional and frequency spreading of the random sea. The major conclusions drawn can be summarized as follows:

Frequency spreading has a minor effect on all three quantities considered, i.e. wave heights, driving forces and longshore current.

Wave height is rather insensitive to directional spreading. The main influence comes through the possible reduction of the energy entering the model area in cases of wide directional spreading and oblique wave incidence.

The driving forces and longshore current are largely affected by the directional spreading. This is because the longshore current is controlled by the radiation stress in deep water which is proportional to the total longshore force. Energy dissipation mechanisms will only regulate the distribution of the total force acting mainly within the breaker zone.

The effect of directionality on the energy flux (wave height) could be eliminated by appropriately adjusting the incoming energy to the onshore part of the 2-dimensional spectrum. However, it is not possible to eliminate the effect of directionality on the momentum related quantities (driving forces, longshore current) by similar considerations.

In connection to this latter effect, results have shown that ignoring/oversimplifying directionality always leads to an overestimation of the driving forces and the resultant longshore current velocity. This remark is of particular importance when calibrating predictive numerical models against results obtained in real field situations.

The use, in our analysis, of a domain uniform in the longshore direction and a coordinate system that results in S_{xy} being the only component of the radiation stress tensor involved in the wave-induced current generation, does not affect the physics and therefore the generality of the above arguments.

ACKNOWLEDGEMENTS

Part of the work of the first author was funded by the European Marine Science and Technology Program, MAST II (Contract No B/MAST - 900065). The second author was partly funded by Mast-G8 Coastal Morphodynamics (Contract No Mas2-CT92-0027).

REFERENCES

Battjes, J.A. and Janssen, J.P.F.M. 1978. Energy loss and set-up due to breaking of random waves. *Proc. 16th Int. Conference on Coastal Engineering,* ASCE, 569-587

Bijker, E.W., 1966. The increase of bed shear in a current due to wave motion. *Proc. 10th Int. Conference on Coastal Engineering,* ASCE, 746-765

Kitou, M. 1999. Cross-shore modeling of short-crested wind wave transformation and wave-induced circulation in the nearshore zone. *Ph.D. Thesis,* University of Liverpool, UK, 432 pp.

Madsen, O.S., Poon, Y.K. and Graber, H.C 1988. Spectral wave attenuation by bottom friction theory. *Proc. 21th Int. Conf. on Coastal Engineering,* Malaga, ASCE, 492 - 504

Ris, R.C., Holthuijsen L.H. and Booij, N. 1994. A spectral model for waves in the nearshore zone, *Proc. 24th Int. Conference on Coastal Engineering,* ASCE, 224 – 238

RELATIONSHIP BETWEEN VERTICAL WAVE ASYMMETRY AND THE FOURTH VELOCITY MOMENT IN THE SURF ZONE: IMPLICATIONS FOR SEDIMENT TRANSPORT

Ismael Mariño Tapia[1], Paul E. Russell[2], Tim J. O'Hare[3], Mark A. Davidson[4], Andrew N. Saulter[5], Jonathon R. Miles[6], David A. Huntley[7]

Abstract: The commonly used Bailard sediment transport model relates sediment suspension to the fourth velocity moment. However field studies have shown that wave orbital acceleration, which is related to vertical wave asymmetry, is also important for sediment suspension. The general view is that asymmetry is unrelated to velocity moments. Here, incident wave cross-shore velocity data from dissipative beaches shows very similar cross-shore distributions for both vertical wave asymmetry and kurtosis. The implications for sediment transport of this possible relationship are investigated by obtaining the cross-spectra between sediment suspension and (i) a Bailard-type pick up function $(u^4)^{3/4}$, and (ii) vertical asymmetry, as the Hilbert transform of the incident cross-shore velocity. For dissipative conditions, both the Bailard term and asymmetry represent similarly sediment suspension. For reflective conditions the Bailard term fails to describe suspension whilst vertical asymmetry is very strongly correlated to sediment suspension. Accurate representation of sediment suspension in models should include the effects of vertical asymmetry.

INTRODUCTION

The evolution of wave asymmetries due to non-linear wave-wave interactions caused by shoaling and breaking is now reasonably well understood (Elgar and Guza 1985, Elgar et al. 1990, Elgar et al. 1997). In the shoaling zone primary frequencies interact in phase with their first harmonics resulting in skewed (Stokes-like) wave records. As water depth decreases and into the surf zone the biphase of the triad interactions approach $-90°$, the wave profile becomes more symmetrical in the horizontal plane and begins to pitch forward towards a vertically asymmetric sawtooth shape. Vertical

---

1 Graduate Student – imarino-tapia@plymouth.ac.uk;
All authors at: Institute of Marine Studies, University of Plymouth, Drake Circus, Plymouth, UK. PL4 8AA. Fax +44 1752 232406

asymmetry in the wave orbital velocity produces strong fluid accelerations, which are shown to be important for sediment suspension (Hanes and Huntley 1986, Osborne and Greenwood 1993).

On the other hand, some of the most robust physics-based models regard net sediment transport as proportional to velocity moments (e.g. Bailard 1981). The moments dependence of the energetics sediment transport equation reflects the importance of non-linear wave properties for suspending and transporting sediment. The general consensus is that the energetics formula does not account for wave vertical asymmetry, as this process has not been related with velocity moments. The exclusion of vertical asymmetry is suggested as one of the underlying reasons for poor performance in cross-shore transport models based on the energetics formulation, especially when trying to reproduce the slow onshore bar migration observed under low energy wave conditions (Gallagher et al. 1998, Elgar, et al. 2001).

Field data presented here suggests that under certain conditions the normalised fourth order velocity moment, kurtosis, and vertical wave asymmetry are related to each other. The results compare observations of the cross-shore structure of kurtosis with the cross-shore structure of vertical wave asymmetry in order to elucidate the nature of the relationship between the two. The implications of such a relationship for sediment suspension is addressed by investigating if a Bailard-type pick up function, which contains the fourth moment, better represents suspension events than a pickup function based on wave vertical asymmetry.

CHARACTERISTICS OF THE FIELD SITES

The data used in this study was gathered on three beaches across the UK. Figure 1 presents the geographical location of the sites and the characteristics of their profiles. Data was gathered in Llangennith, Wales, under the British Beach And Nearshore Dynamics program, B-BAND (Russell et al. 1991). This is a 5 km long beach with a shallow and concave beach profile (gradient 0.014), consisting on fine to medium grained quartz sands (D_{50}=0.21 mm). It is exposed to high-energy Atlantic swell and locally generated wind waves. Tidal range is up to 9 m. Perranporth, Cornwall, is a 2 km long beach with very similar characteristics to Llangennith. It has a shallow and concave beach profile (gradient 0.013), fine to medium grained quartz sands (D_{50}=0.24 mm) and is also exposed to high-energy Atlantic swell and locally generated wind waves. Tidal range is up to 7 m. Teignmouth, South Devon is a narrow beach 2 km long facing ESE into the English Channel. The beach is backed by a seawall and meets the mouth of an estuary and a prominent headland to the south. It is protected from Atlantic swell and infrequent periods of locally generated wind waves from the east dominate the wave climate. Data was gathered in Teignmouth under the European project COAST 3D. The profile shown on Figure 1 for Teignmouth beach reveals a steep and reflective region at the high water mark (gradient up to 0.110) with a wide variety of grain sizes (0.4 – 5 mm), and a less steep gradient at low water (gradient = 0.057) where dissipative conditions occur. Fine sand (0.2mm) predominates on this low tide terrace.

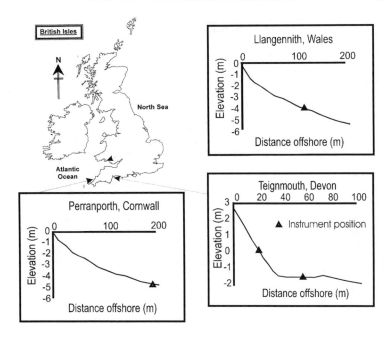

Figure 1. Location of field sites and beach profile characteristics

OBSERVATIONS

Each rig of instruments contained an optical backscatter sensor (OBS) to monitor sediment suspension, a pressure transducer (PT) for recording surface elevation changes, and a 2-axis electromagnetic current meter (EMCM) to measure horizontal currents. The instruments were collocated at 10 to 15 cm above the bed. The large tidal range at the three sites allows the acquisition of data at different cross-shore positions relative to the shoreline with a single rig of instruments. Table 1 summarises the characteristics of the data sets used in this study. The data sets with code **D** indicate dissipative conditions. The dissipative data from Teignmouth was gathered on the low tide terrace under high-energy conditions. Figures 2a and 2b are an example of the hydrodynamic conditions prevailing on the three dissipative data sets. Wave height was linearly dependent on water depth, which means that the surf zone was saturated, hence the breaker index (ratio between wave height and water depth) remained fairly constant across the surf zone. The data sets with code **R** on Table 1 indicate that the beach was reflective. All this data was collected on the steep part of the beach at Teignmouth. Here the hydrodynamic conditions were consistent as well. The narrow surf zones did not allow the wave height to be linearly dependent on water depth, hence the surf zone was not saturated and consequently the breaker index was not constant but increased shorewards. Figures 2c and 2d show an example of such conditions.

Table 1. Data Set Characteristics

Beach	Code	Hb* (m)	Tp** (sec)	Surf width (m)	Max. tidal range (m)	Sampling mode
Llangennith	D1	3	14	350	9	continuous
Perranporth	D2	1.5	11	80	7	continuous
Teignmouth	D3	1	7.6	51	6	burst
Teignmouth	R1	0.6	6	5	6	continuous
Teignmouth	R2	0.8	7.6	7	6	continuous
Teignmouth	R3	0.8	6.25	7	6	continuous

*Breaking wave height from observations on the field, data statistics and photographs
**Peak spectral wave period on the outermost measured position from the shoreline.

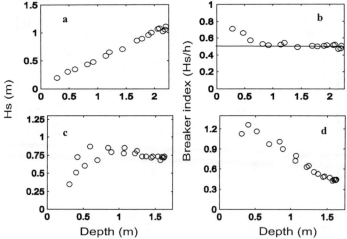

Figure 2. Typical hydrodynamic conditions on the dissipative (panels *a* and *b*) and reflective (panels *c* and *d*) data sets. Figures *a* and *b* are taken from Perranporth (D2) and Figures *c* and *d* from Teignmouth (R3).

DATA REDUCTION AND ANALYSIS
Velocity Moments

The probability density function (PDF) of the orbital velocity time series reveals characteristics of the wave profile shape. This PDF can be characterised by its moments. For example, the time history of swell waves in deep water can be well represented by a linear superposition of Airy waves with independent random phase. This stochastic process has a normal or Gaussian PDF, and can be characterised by the first two moments: mean and variance. By definition, the normalised third moment (skewness) will be zero, and the normalised fourth moment (kurtosis) will be 3.

As waves shoal and become non-linear, the PDF of the orbital velocity will have values of skewness different from zero, and kurtosis will depart from three. As sediment transport has been related to powers of velocity, our variable of interest is the cross-shore velocity time series. Skewness (S) and kurtosis (K) are calculated from 17 min long time series as:

$$S = \frac{\langle (\tilde{u})^3 \rangle}{\langle (\tilde{u})^2 \rangle^{3/2}} \quad (1) \quad ; \quad K = \frac{\langle (\tilde{u})^4 \rangle}{\langle (\tilde{u})^2 \rangle^2} \quad (2)$$

where $\tilde{u}$ is the incident cross-shore orbital velocity obtained after detrending and demeaning the total velocity record (u_t), removing all associated infragravity (IG) energy (using the spectral valley as frequency cut off criteria), and linearly removing the effects of wave reflection (Guza et al. 1984). Standing wave patterns modify the shape of the waves and potentially affect the velocity moments.

Vertical Wave Asymmetry

The observed asymmetries in $\tilde{u}$ are manifestations of non-linearities that grow as the waves shoal and break. The horizontal and vertical asymmetries are usually characterised by the skewness (eq. 1) and asymmetry (A) respectively. Both are third moments, but asymmetry has been defined as the skewness of the Hilbert transform of the cross-shore velocity time series $H(\tilde{u})$ (Elgar and Guza 1985).

$$A = -\frac{\langle H(\tilde{u})^3 \rangle}{\langle H(\tilde{u})^2 \rangle^{3/2}} \quad (3)$$

The difference between $H(\tilde{u})$ and $\tilde{u}$ is the phase relationship between the primary frequency and the phase-locked harmonics (Elgar et al. 1990). $H(\tilde{u})$ and $\tilde{u}$ time series are 90° out of phase. It can also be shown that $H(\tilde{u})$ is related to the slopes of the original time series and hence to acceleration. As asymmetry is commonly defined as a negative quantity, the values of kurtosis presented here will be set negative to aid comparison.

VELOCITY MOMENTS AND KURTOSIS IN THE SURF ZONE

Skewness (Figures 3a and 3b) shows the expected behaviour for all data sets. Outside the surf zone (normalised depth >1) skewness tends to increase shorewards towards the point of breaking. Inside the surf zone skewness decreases shorewards. This behaviour is fairly consistent, but there are important differences in the magnitude of skewness between data sets, which produces a lot of scatter, especially under reflective conditions.

Kurtosis (Figure 3c) shows a more consistent trend when plotted against normalised depth. The observations depart from the Gaussian value of three during shoaling with tendency to decrease in magnitude. Around the mid surf zone kurtosis goes back to normality and increases dramatically in magnitude in the inner surf zone. Reflective beaches (triangles) do not show this increase in negative kurtosis. On steep beaches waves are forced to plunge because they are slope limited, the surf zone is narrow and waves can not develop an equilibrium stage in the form of saturation.

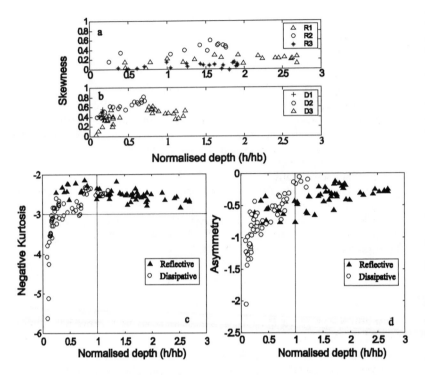

Figure 3. Cross-shore structure of skewness (a and b), kurtosis (c) and vertical wave asymmetry (d), plotted against normalised depth where 1 is the approximate breaking point and 0 the shoreline.

Large values of kurtosis seem to be associated with a zone of saturated spilling bores characteristic of dissipative beaches. Asymmetry (Figure 3d) increases in magnitude towards the shore in accordance with previous observations made on unbarred beaches. This pattern is consistent on both reflective and dissipative conditions.

From Figures 3c and 3d it can be seen that kurtosis has a similar cross-shore distribution to asymmetry suggesting a possible relationship between the two. In order to elucidate the nature of such a relationship, kurtosis was plotted against asymmetry. Figure 4 shows the results. Data for the dissipative case (circles) shows that kurtosis is directly proportional to asymmetry. The correlation coefficient (R^2) is 0.68 for this case. Conversely, data from reflective beaches appears to show a weak inverse relationship between kurtosis and asymmetry. Another intriguing element is that the whole trend is shifted from the neutral point in which linear waves in deeper water should fall.

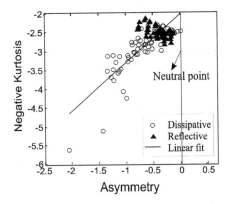

Figure 4. Relationship between kurtosis and vertical asymmetry

IMPLICATIONS FOR SEDIMENT TRANSPORT

On one hand there is some evidence showing that acceleration, and consequently vertical asymmetry, are important variables for sediment suspension (Hanes and Huntley 1986, Elgar et al. 2001), but the conditions under which this occurs or the underlying mechanisms involved remain speculative. On the other hand, we have the energetics Bailard formula which relates sediment transport to moments of velocity:

$$i_t = \rho Cf \, \varepsilon_b \left(\left\langle u_t^3 \right\rangle + \frac{\tan \beta}{\tan \phi} \left\langle |u_t|^3 \right\rangle \right)$$
$$+ \, \rho Cf \, \frac{\varepsilon_s}{W} \left(\left\langle |u_t|^3 u_t \right\rangle + \frac{\varepsilon_s}{W} \tan \beta \left\langle |u_t|^5 \right\rangle \right) \tag{4}$$

where ρ = water density, Cf = Drag coefficient, ε_b and ε_s are efficiency factors, u_t is the instantaneous velocity, β = bed slope, ϕ = sediment angle of repose, and W = sediment fall velocity. This expression does not explicitly include the effects of vertical asymmetry. The third velocity moment describes the bed load transport (first term), and the fourth velocity moment describes the suspended load transport (second term). The third moment includes the effects of Stokes-like waves for moving sediment near the bed, but the fourth moment does not have a straightforward physical meaning.

Considering suspended load only, if we assume that the total velocity field u_t can be decomposed into a mean and an oscillatory component (with short and long waves), the term $|u_s|^3 u_s$ will be contained in the total velocity term, where u_S denotes the short wave component of velocity. Here $|u_s|^3$ represents a pick up function that stirs sediment into suspension, to be subsequently transported by u_S. This pick up function can also be defined in terms of the fourth velocity moment: $|u_s|^3 = (u_s^4)^{3/4}$.

If there is any relationship between a fourth moment and vertical asymmetry, they should relate in a similar way to sediment suspension events. In order to test their individual effects on sediment suspension, a cross-spectral analysis was carried out between the Bailard pick-up term $\left|u_s\right|^3$ and suspended sediment concentration (ssc) time series, and between the $H(\widetilde{u})$ time series, representing asymmetry, and ssc.

The results of this analysis show that both vertical asymmetry and the Bailard pick up function describe similarly sediment suspension for dissipative conditions. Figure 5, left panel, shows the co-spectrum for dissipative beaches integrated over the incident wave band. Both variables show comparable cross-shore distributions, except for some outlying points. The cross-shore distribution of coherence, for this case, is also very similar (not shown). However, the values of coherence were usually below 0.25, suggesting that vertical wave asymmetry is not playing a mayor role on sediment suspension under these conditions. This result is rather unexpected, as it is under dissipative conditions where the largest vertical asymmetries and accelerations are found (see Figure 3d).

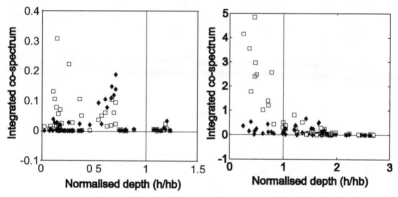

Figure 5. Integrated version of the relationships between $H(u)$ and ssc (squares), and $|u_s|^3$ vs ssc (black diamonds) for the dissipative (left) and reflective (right) cases.

Under reflective conditions (Figure 5, right panel) the differences between a Bailard term and vertical asymmetry are pronounced. Vertical asymmetry is the governing factor for sediment suspension. Values of coherence inside the surf zone are consistently above 0.5 and as high as 0.75 (not shown). The Bailard term, on the other hand, is poorly correlated to ssc with coherence values rarely reaching 0.5.

Figure 6 provides a summary of our findings. The integrated co-spectrum of vertical asymmetry $(H(u_S))$ against ssc is presented for both the dissipative and the reflective beaches. All data is included.

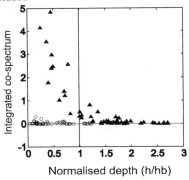

Figure 6. Integrated version of the relationships between $H(u)$ and ssc, including all data sets. Dissipative cases on circles, reflective on triangles.

The trend for the three reflective cases is quite consistent. The in-phase oscillations are increasingly important as the shore is approached. The dissipative cases do not show similar consistency, with values staying near zero all across the surf zone. Figure 6 reveals clearly that the processes that suspend sediments at incident frequencies on reflective beaches are different from those acting on dissipative beaches.

The driving mechanism for sediment suspension is not the magnitude of the asymmetry $(A,$ eq. 3), as waves on dissipative beaches are considerably more asymmetric than waves on reflective beaches (Figure 3d).

Inspection of the $H(\tilde{u})$ time series reveals that inside the surf zone of reflective beaches, where asymmetry driven suspension is more important, the $H(\tilde{u})$ variances are large (e.g. large accelerations and decelerations), whereas on dissipative beaches $H(\tilde{u})$ variances are smaller. This may imply that $H(\tilde{u})$ variance is an appropriate equivalent for sediment suspension.

Conversely, on reflective beaches where waves are usually plunging, the large amount of turbulence transmitted to the bottom will suspend more sediment than the surface limited turbulence produced by spilling breakers commonly found at dissipative beaches.

Further work is needed to test either of the above hypotheses. Nonetheless, a feasible scenario for non-barred beaches could be the following. On a reflective beach, waves dissipate their energy very near to the shore and the surf zone is very narrow, IG energy is weak and high frequency processes (e.g. incident vertical wave asymmetry) dominate sediment suspension. Within this narrow and turbulent surf zone vertical asymmetries are not fully developed and the acceleration field is still oscillating considerably. These oscillations (variance) on the acceleration field might provide an effective pick up mechanism for sediment suspension. In contrast, dissipative beaches have a wide surf zone, low frequency processes tend to dominate the hydrodynamics and turbulence is limited to the surface. When the very asymmetric waves of the inner surf zone reach the beach, IG energy is already dominating the velocity and driving sediment suspension. At the same time, the acceleration field under the more asymmetric waves of the inner surf zone is governed by large onshore pulses (skewness) and has less variability. Under these conditions sediment suspension by asymmetric waves seems to be less effective.

CONCLUSION

Large values of kurtosis (K>3) are associated with wide saturated surf zones with spilling breakers. In contrast, small values of kurtosis (K<3) are associated with non-saturated conditions found in the outer surf zone or in narrow surf zones where plunging breakers are more likely to occur.

For dissipative surf zones, kurtosis and asymmetry are directly proportional. It is also under this conditions where the Bailard type pick up function and asymmetry relate in a similar way to ssc, therefore both are good descriptors of sediment suspension events. In spite of the very large magnitudes of asymmetry found, processes at IG frequencies are likely to dominate the sediment suspension and vertical asymmetry is relatively unimportant in these conditions.

In reflective conditions, where kurtosis is not clearly related to asymmetry, $|u_s|^3$ does not hold a strong nor consistent relationship with ssc, whereas asymmetry does have a very robust relationship with ssc. Coherence shows a consistent cross-shore structure with values above 0.5 and up to 0.75 inside the surf zone. Hence a term including vertical asymmetry, maybe in the form of $H(u)$, will be describing more accurately sediment suspension events in sediment transport models.

ACKNOWLEDGEMENTS

We appreciate the financial support of the National Council for Science and Technology (CONACYT), Mexico, which makes possible this investigation.

REFERENCES

Bailard,J. 1981. An Energetics Total Load Sediment Transport Model For A Plane Sloping Beach. *Journal of Geophysical Research,* 86(C11): 10,938-10,954.

Elgar,S. and Guza,R.T. 1985. Observations of Bispectra of Shoaling Surface Gravity Waves. *Journal of Fluid Mechanics,* 161, 425-448.

Elgar,S., Freilich, M.H., and Guza,R.T.1990. Model-Data Comparisons of Moments of Nonbreaking Shoaling Surface Gravity Waves. *Journal of Geophysical Research,*

95 (C9), 16,055-16,063.

Elgar,S., Guza,R.T., Raubenheimer,B., Herbers,T.H.C. and Gallagher,E. 1997. Spectral Evolution of Shoaling and Breaking Waves on a Barred Beach. *Journal of Geophysical Research*, 102 (C7), 15,797-15805.

Elgar,S., Gallagher,E. and Guza,R.T. 2001. Nearshore Sandbar Migration. *Journal of Geophysical Research*, 106 (C6), 11,623-11,627.

Gallagher,E., Elgar,S., and Guza,R.T. 1998. Observations of Sandbar Evolution on a Natural Beach. *Journal of Geophysical Research*, 103 (C2), 3203-3215.

Guza,R.T., Thornton, B., and Holman,R.A. 1984. Swash on Steep and Shallow Beaches. *Proceedings of International Conference on Coastal Engineering*, ASCE, 708-723.

Guza,R.T. and Thornton, B. 1985. Velocity Moments in Nearshore. *Journal of Waterways, Ports, Coastal and Ocean Engineering*, 111(2), 235-256.

Hanes,D., and Huntley,D. 1986. Continuous Measurement of Suspended Sand Concentration in a Wave Dominated Nearshore Environment. *Continental Shelf Research*, 6, 585-596.

Osborne,P., and Greenwood,B. 1993. Sediment Suspension Under Waves and Currents: Time Scales and Vertical Structure. *Sedimentology*, 40, 599-622.

Pierson,W.J. and Jean-Pierre,A. 1999. Monte Carlo Simulations of Nonlinear Ocean Wave Records with Implications for Models of Breaking Waves. *Journal of Ship Research* , 43(2),121-134.

Russell, P.E., Davidson, M.A., Huntley, D.A., Cramp, A., and Hardisty, J. 1991. The British Beach and Nearshore Dynamics (B-BAND) Programme. *Proceedings of Coastal Sediments,* ASCE, 371-384.

EROSIONAL HOT SPOT PREDICTION THROUGH WAVE ANALYSIS

Christopher J. Bender [1] and Robert G. Dean [2], M.ASCE

Abstract: Construction of beach nourishment projects requires large volumes of sediment, which are usually obtained by the removal from borrow pits located in reasonably shallow water. These modifications of the nearshore bathymetry have the potential to affect wave transformation processes and thus to alter the equilibrium shoreline planforms landward of the borrow pits. Recent cases of erosional hot spots (EHS's) associated with beach nourishment projects have increased interest in the prediction of the mechanisms of borrow pit alteration of the local wave field. A better understanding of the interaction of borrow pit characteristics such as size and depth with the incident wave field is needed to determine the effect of the pits on the nourished beach and to anticipate the effects of various designs.

The present study employs numerical and theoretical approaches. The numerical analysis uses a boundary element approach for a pit of finite dimensions in a uniform shallow water depth as in Williams (1990). An analytical solution to the shallow water pit problem is obtained using the general method of MacCamy and Fuchs (1954) with a circular pit instead of a solid cylinder and allowing a solution inside the pit with appropriate boundary conditions. For both the numerical and analytical solutions, the velocity potential can be determined anywhere in a fluid domain containing a pit, or a shoal, of uniform depth. This allows for the calculation of the free surface elevation as well as other wave related quantities such as wave direction, wave energy flux and wave-induced shoreline change.

INTRODUCTION

Beach nourishment projects require large amounts of sediment, which are usually obtained from offshore borrow areas. The dredging of offshore shoals or the creation of dredge pits in areas of otherwise uniform bathymetry is common practice when nearshore sites are used to obtain the required sediment for a beach

---

[1]Graduate Student and [2]Professor, both at the Civil and Coastal Engineering Dept, P.O. Box 116590, 345 Weil Hall, University of Florida, Gainesville, FL 32611; ph: 352-392-1436; fax: 352-392-3466; email: bender@ufl.edu; dean@coastal.ufl.edu

nourishment project.   Four wave transformation processes can occur when the offshore borrow area significantly alters the local bathymetry: wave refraction, wave diffraction, wave reflection and wave dissipation.   The presence of the altered bathymetry can result in a shoreline with a non-uniform planform, which may contain erosional hot spots (EHS's).   Recent cases of EHS's associated with beach nourishment projects have increased interest in the prediction of the mechanisms by which borrow areas alter the local wave field.

The beach nourishment project at Grand Isle, LA is an often-cited example of a project where the offshore borrow area has resulted in a non-uniform planform containing several EHS's (Figure 1).   Details of the beach nourishment project, borrow area location and dimensions, and the post-nourishment project evolution are presented in Combe and Soileau (1987).   This example demonstrates that in order to anticipate the effects of various project designs a better understanding of the interaction between the incident wave field and the borrow pit characteristics is required.

Fig. 1. Aerial Photograph showing salients and EHS's shoreward of borrow area at Grand Isle, LA (Combe and Soileau, 1987); printed with permission of ASCE.

The overall aim of this study and current research is to develop the capability to predict the shoreline response in the presence of offshore borrow pits and shoals through the development of both simple and more complex and complete models. These models will facilitate the design of successful beach nourishment projects when the borrow area location may impact the shoreline through wave transformation processes.   In this study, numerical and analytic solutions are used to model the wave conditions in and around a pit or a shoal using shallow water wave theory.   These models determine the complex velocity potential found anywhere in the fluid domain, allowing quantities such as the velocity and pressure to be solved. This allows modeling of the shoreline change induced by a pit or a shoal.

PREVIOUS STUDIES

Wave field modification caused by changes in the bathymetry has been studied analytically, numerically and through laboratory and field studies. Studies of the wave field modifications by pits of infinite length have determined the reflection and transmissions coefficients of various pit dimensions and incident wave conditions (Lassiter 1972, Lee and Ayer 1981, Miles 1982, Kirby and Dalrymple 1983, and Ting and Raichlen 1986). A fluid domain defined by one horizontal dimensional and one vertical dimension is used in these models. More recently the wave field modifications by pits of arbitrary shape have been investigated (Williams, 1990, McDougal et al, 1996). The models developed are valid in the shallow water region with the fluid domain defined in two horizontal dimensions.

Motyka and Willis (1974) and Horikawa et al. (1977) describe two studies that have examined the shoreline change induced by offshore pits using mathematical models that only consider the wave refraction caused by the pit. These models reached opposing conclusions with the model of Motyka and Willis indicating erosion shoreward of a pit and the model of Horikawa et al. showing accretion. A fixed bed laboratory model containing a rectangular pit and a moveable sand beach was employed by Horikawa et al. (1977) to investigate the shoreline response due to the presence of a pit. The model showed accretion occurring behind the pit, flanked by two areas of erosion for the case of normally incident, monochromatic waves. Through field studies, Price et al. (1978) and Kojima et al. (1986) have investigated the effect of offshore dredging on the coasts of England and Japan, respectively.

THEORY

The present study employs numerical and theoretical approaches. In the models a uniform wave field encounters a region of otherwise uniform depth containing a pit, or a shoal, of uniform depth. The fluid domain is comprised of two regions; one representing the pit (or shoal) and its projection and the other, the rest of the fluid domain. The fluid is considered incompressible and the flow irrotational allowing the application of potential flow and linearized shallow water wave theory. The boundary conditions consist of a no flow condition on the bottom boundary and two conditions on the pit (or shoal) boundary; one that equates pressure across the boundary and one that equates discharge. A radiation condition is also implemented to ensure the incident solution is retained in the far field.

The numerical analysis uses a boundary element approach for a pit (or shoal) of finite dimensions in uniform depth as in Williams (1990). Using Green's second identity and appropriate Green's functions, the velocity potential and derivative of the potential normal to the pit are found at the interface of the two regions using standard matrix techniques. This is done after the pit is discretized into a finite number of equally spaced points. The velocity potential at any location in the fluid domain can then be determined after a reapplication of Green's Law with the values determined on the pit boundary. The reader is referred to the paper by Williams and to Bender (2001) for further details on the formulation of the model.

An analytical solution to the problem is obtained using the general method of MacCamy and Fuchs (1954) with a circular pit (or shoal) instead of a solid cylinder by allowing a solution inside the pit with appropriate boundary conditions. Three equations for the velocity potential result:

Incident potential: $\phi_I = M_I \left[\sum_{m=0}^{\infty} \beta_m \cos(m\theta) J_m(kr) \right] e^{-i\omega t}$ (1)

Scattered potential: $\phi_S = \left[\sum_{m=0}^{\infty} A_m \cos(m\theta)[J_m(kr) + i Y_m(kr)] \right] e^{-i\omega t}$ (2)

Potential inside pit: $\phi_{ins} = \left[\sum_{m=0}^{\infty} B_m \cos(m\theta) J_m(kr) \right] e^{-i\omega t}$ (3)

where t is the time, (r, θ) is the polar location for a point in the fluid domain with the pit centered at the origin, ω is the angular frequency, k is the wave number at the point of interest, M_I is a coefficient for the incident wave, A_m and B_m are unknown coefficients J_m and Y_m are Bessel functions and $\beta_m = 1$ for m = 0 and $2i^m$ otherwise. The boundary conditions are used to solve for the two unknown coefficients. The analytical solution allows for the determination of the potential at any location in the fluid domain containing a circular pit or shoal.

For both the numerical and analytical solutions, the wave refraction, wave diffraction, and wave reflection caused by the change in bathymetry are accounted for in the solution although there is no consideration of the wave dissipation caused by the pit or shoal. The free surface elevation is obtained based on the known velocity potential at any point. This provides the basis for the development of a diffraction diagram of relative amplitude over the entire field of interest. The velocity potential can also be used with sediment transport and sediment conservation equations to calculate the shoreline change landward of the pit.

RESULTS
Wave Modeling
Many trials were run to verify and analyze the models that were developed. For verification of the numerical model, the results were compared to those of Williams (1990). Good agreement was expected as the same formulation was used for in both models. Figure 2 shows a contour plot of the relative amplitude determined using the numerical model. In this example the incident period is 12.77 s, the water depth and pit depth are 2 and 4 m, respectively, 120 points comprise the pit boundary, 1600 points define the fluid domain and the waves approach from left to right, normal to the pit. This figure can be compared to the Fig. 2a in Williams (1990), which uses the same model conditions. The results of the two models are found be very comparable, as was expected. Fig. 2 clearly shows a shadow zone of decreased relative amplitude shoreward of the pit and a partial standing wave pattern of increased and decreased relative amplitude seaward of the pit. McDougal et al. (1996) contains an analysis of the relationship between the minimum diffraction coefficient and certain pit and incident wave characteristics.

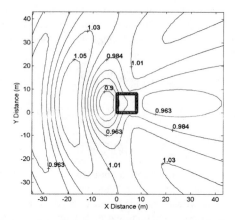

Fig. 2. Contour plot of relative amplitude from numerical model for 8 by 8 m pit; ambient water depth is 2 m and pit depth is 4 m.

With confidence that the numerical model accurately determined the velocity potential and therefore wave amplitude in the presence of a pit, it was next compared with the analytic solution model for further confirmation. Defining a circular pit in the numerical model allowed for the direct comparison of results for the numerical model and the analytical solution model (Fig. 3).

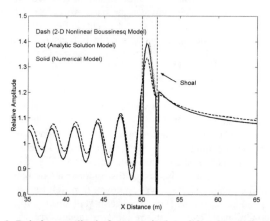

Fig. 3. Relative amplitude for numerical model and analytic solution model for a shoal of radius = 1 m, compared to 2-D fully non-linear Boussinesq model.

For this comparison a transect is taken in the direction of wave propagation, running through the center of a shoal with a radius of 1 m. In this example the incident period is 3.5 s, the water depth and shoal depth are 0.3 and 0.15 m, respectively and the waves approach from the left to right. A standing wave pattern develops seaward of the shoal, as occurred for case of a pit. An area of increased relative amplitude is found in the area where a pit produces a shadow zone as the shoal induces convergence of the wave rays, and a pit causes a divergence of the wave rays. The results of the analytic solution model are seen to match those of the numerical model. A 2-D fully nonlinear Boussinesq model is also compared to the results on the plot, and is found to compare reasonably well with the numerical and analytic solution model for this case.

Shoreline Modeling

The capability to model the wave heights and wave directions at a transect representing the shoreline is necessary to predict the shoreline response due to an offshore borrow area. The shoreline change induced by a pit or shoal was predicted using the modeled wave heights and directions along the transect to calculate the longshore transport using equilibrium beach profile concepts. The continuity equation was used to relate the change in the shoreline position with the change in volume for the profile. The assumptions of straight and parallel bottom contours and no nearshore refraction were made.

Up to this point, only monochromatic incident wave conditions have been considered. This situation is not very representative of a natural shoreline, where waves of many different periods, directions, and wave heights occur. An averaging procedure was developed to produce more realistic wave conditions at the transect representing the shoreline. This was done by averaging the wave conditions at a transect for different wave directions and different periods. The averaged relative amplitude values for the case of a transect that lies 80 m shoreward of a pit of radius equal to 6 m are shown in Fig. 4.

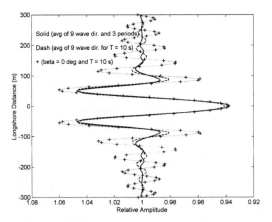

Fig. 4. Relative amplitude versus longshore distance for analytic solution model along a transect for two averaging conditions and monochromatic waves.

A shadow zone of decreased relative amplitude is seen directly shoreward of the pit bordered by two lobes of increased relative amplitude. For a shore-normal, monochromatic wave the relative amplitude oscillates with longshore distance, with large oscillations found even at the end of the transect. Averaging over 9 wave directions centered around 0 degrees is seen to greatly dampen the oscillations occurring along the shore, while maintaining the shadow zone and areas of wave focusing directly shoreward of the pit. By averaging over 3 different periods, and also the wave direction, further dampening occurs at large longshore distances, with little change occurring directly shoreward of the pit. Averaging over only the wave direction was found to provide an adequate reduction in the longshore oscillation and was used in the shoreline evolution model.

The shoreline evolution model determines the wave height and wave direction at each point along a transect that represents the shoreline. Using the wave height and direction at each point, the longshore transport is calculated using the equation:

$$Q = \frac{K_1 H^{2.5} \sqrt{g/\kappa} \sin(\theta - \alpha_b)\cos(\theta - \alpha_b)}{8(s-1)(1-p)} + \frac{-K_2 H^{2.5} \sqrt{g/\kappa} \cos(\theta - \alpha_b)}{8(s-1)(1-p)\tan(\gamma)} \frac{dH}{dy} \quad (4)$$

where H is the wave height, g is gravity, κ is the breaking index, θ is the shoreline orientation, α is the wave angle at the shoreline, γ is the beach slope, s and p are the specific gravity and porosity of the sediment, respectively, and K_1 and K_2 are sediment transport coefficients. The standard value for K_1 is 0.77 while K_2 is usually specified in a range from ½ K_1 to K_1. The first term in the transport equation is driven by the wave angle at the shoreline and has been extensively studied in the field and laboratory. The second transport term is driven by the longshore gradient in the wave height. This term is negligible in most situations such as a long straight beach with parallel offshore contours, but is important in areas with irregular offshore bathymetry, or near structures where significant longshore variation in the wave height can occur.

The shoreline evolution model calculates the change in shoreline position using the gradient in the longshore transport along the transect. After the wave height and direction are calculated at each transect location, the longshore transport is calculated and the shoreline change determined. The updated shoreline position is then used to recalculate the shoreline change for a specified number of iterations before the whole model is run again to calculate the wave height and direction at the current shoreline positions. The averaging procedure introduced earlier is used with the longshore transport values averaged over a specified number wave directions and/or periods. The shoreline evolution using the longshore transport from Eq. (4) for a transect located 80 shoreward of a pit with a radius of 6 m is shown in Fig. 5. For this example the water depth and pit depth are 2 m and 4 m, respectively, the period is 10 s, the incident wave height is 1 m, K_1 and K_2 equal 0.77 and averaging over 5 wave directions is used. The time step is 120 s and 10 iterations of shoreline change are calculated between wave height and direction updates for a total modeling time of 48 hours.   The figure shows an area of accretion directly shoreward of the pit, flanked by two large areas of erosion with oscillations of advancement and erosion occurring with increasing longshore distance.   The shoreline is found to approach an equilibrium condition with the shoreline change approaching zero at the largest time steps.

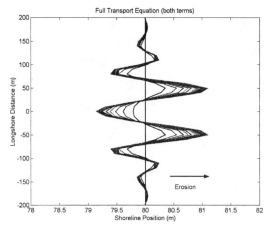

Fig. 5. Shoreline evolution using full transport equation and analytic solution model for transect located 80 shoreward of a pit with a radius = 6 m.

Examining the two transport terms in Eq (4) individually shows the influence of each term on the full transport equation (Fig. 6).   The first transport term leads to a shoreline with a large EHS occurring shoreward of the pit created by the divergence of the wave rays behind the pit.   Two areas of shoreline advancement flank this area of erosion.   The shoreline is seen to approach an equilibrium state with little shoreline change at the largest time steps.   Directly shoreward of the pit the shoreline is found to advance for the first few time steps and then retreat as sand is drawn back into the large area of erosion and moved into the areas where the shoreline is advancing.

This condition is contrasted by the shoreline that results from the second transport term, which is driven by the longshore gradient in the wave height.   The shoreline resulting from this term has a large area of accretion directly shoreward of pit, bordered by two areas of erosion.   The shadow zone that developed behind the pit created a gradient in the wave height that draws sediment into this area, resulting in shoreline advancement.   As can be seen in the figure, the shoreline change due to the gradient in the wave height is much greater than that caused by the wave direction.   This occurs because the shoreline change due to the gradient in the longshore wave height does not approach an equilibrium value.   As time increases the shoreline change directly landward of the pit remains almost constant, with a negative value, resulting in a large advancement.   When the two terms are combined in the full transport equation the diffusive nature of the first transport term modifies the second transport term to produce a shoreline that approaches an equilibrium state, as seen in Fig. 5.

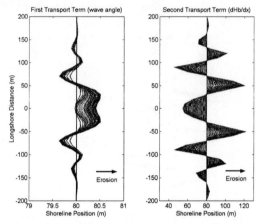

Fig. 6. Shoreline evolution using each transport term individually for transect located 80 shoreward of a pit with a radius = 6 m.

The shoreline evolution occurring in the presence of a shoal was also modeled. For this case an equilibrium shoreline developed with an EHS directly shoreward of the shoal, bordered by two areas of accretion when the full transport equation was used. An equilibrium shoreline developed for the transport driven by only the wave angle with accretion occurring directly shoreward of the shoal due to the convergence of the wave rays. The transport due to the gradient in the longshore wave height resulted in a non-equilibrium shoreline as was found for the case of a pit.

SUMMARY AND CONCLUSIONS
Summary
Two models have been developed and tested to represent the wave transformation resulting from shallow water waves propagating over pits or shoals with steep sides. One of the models is numerical and follows the development by Williams (1990) and the other model is analytical with the solution similar to the MacCamy and Fuchs (1954) problem. The impact of the altered wave field on the shoreline was modeled using equilibrium beach profile theory and the continuity equation. The resulting shoreline evolution has been illustrated for the case of a circular pit.

Conclusions
Application of the models confirms that offshore borrow areas, through the creation of pits and/or the modification of shoals, can cause significant wave transformation and associated shoreline effects. The role and magnitudes of the two

sediment transport coefficients, K_1 and K_2, are significant to the equilibrium shoreline response. More realistic modeling will require removal of the shallow water requirement and a better understanding of the transport due to the longshore variation in the wave setup.

ACKNOWLEDGEMENTS

The first author is grateful for his support through the Alumni Fellowship Program at the University of Florida. Dr. Andrew Kennedy is kindly thanked for the results he provided from his 2-D fully nonlinear Boussinesq model shown in Fig. 3.

REFERENCES

Bender, C. J. 2001. Wave Field Modifications and Shoreline Response Due to Offshore Borrow Areas, Masters Thesis. University of Florida.

Combe, A. J., and Soileau, C. W. 1987. Behavior of Man-Made Beach and Dune: Grand Isle, Louisiana. *Coastal Sediments '87*, 1232-1242.

Horikawa, K., Sasaki, T., and Sakuramoto, H. 1977. Mathematical and Laboratory Models of Shoreline Changes due to Dredged Holes. *Journal of the Faculty of Engineering*, The University of Tokyo(B), (34), No. 1, 49-57.

Kirby, J. T. and Dalrymple, R. A. 1983. Propagation of obliquely incident waves over a trench. *Journal of Fluid Mechanics*, (133), 47-63.

Kojima, H., Ijima, T. and Nakamuta, T. 1986. Impact of offshore dredging on beaches along the Genkai Sea, Japan. *Proceedings, 20th International Conference on Coastal Engineering*, Taipai, (2), 1281-1295.

Lassiter, J. B. 1972. The propagation of water waves over sediment pockets. PhD. Thesis, Massachusetts Institute of Technology.

Lee, J. J. and Ayer, R. M. 1981. Wave propagation over a rectangular trench. *Journal of Fluid Mechanics*, (110), 335-347.

MacCamy, R. C., and Fuchs, R. A. 1954. Wave Forces on Piles: A Diffraction Theory. U.S. Army Corps of Engineers, Beach Erosion Board, Tech. Memo #69.

McDougal, W. G., Williams, A. N., and Furukawa, K. 1996. Multiple Pit Breakwaters. *Journal of Waterway, Port, Coastal, and Ocean Engineering*, ASCE, 122(1), 27-33.

Miles, J. W. 1967. Surface-wave scattering matrix for a shelf. *Journal of Fluid Mechanics*, 28, 755-767.

Motyka, J. M., and Willis, D. H. 1974. The Effect of Wave Refraction Over Dredged Holes. *Proceedings, 14th International Conference on Coastal Engineering*, ASCE Copenhagen, Vol. 1, 615-625.

Price, W. A., Motyka, J. M. and Jaffrey, L. J. 1978. The effect of offshore dredging on coastlines. *16th International Conference on Coastal Engineering*, ASCE, Hamburg, Vol. 2, 1347-1358.

Ting, C. K. F. and Raichlen, F. 1986. Wave interaction with a rectangular trench. *Journal of Waterway, Port, Coastal, and Ocean Engineering*, ASCE, 112(3), 454-460.

Williams, A. N. 1990. Diffraction of Long Waves by a Rectangular Pit. *Journal of Waterway, Port, Coastal and Ocean Engineering*, ASCE, 116, 459-469.

REGIONAL WAVE TRANSFORMATION AND ASSOCIATED SHORELINE EVOLUTION IN THE RED RIVER DELTA, VIETNAM

Nguyen Manh Hung[1], Magnus Larson[2], Pham Van Ninh[1], and Hans Hanson[1]

Abstract: Several stretches of shoreline in the Red River Delta, Vietnam, are suffering from severe erosion, especially an area in the southern part of the delta known as Hai Hau Beach. Commonly found explanations for this erosion is the construction of a large dam upstream the Red River Delta and the closing of a river branch that used to supply sediment to the beach. However, in this study it is shown that the prevailing wave climate during the winter monsoon in combination with the complex topography of the delta create longshore transport gradients that promotes erosion at Hai Hau Beach. A newly developed two-dimensional random wave transformation model based on the mild slope equations was employed for a 20-year long time series of hindcast waves to compute representative nearshore wave properties for the Red River Delta shoreline. These properties were utilized to compute the longshore sediment transport rate along the delta from which shoreline evolution was inferred.

INTRODUCTION

The Red River Delta (RRD) located in northern Vietnam (see Fig. 1) is under constant threat from waves and high water levels in the sea causing erosion, flooding, and intrusion of saline water. Large waves and high water levels are typically associated with storms during the monsoon periods or the occurrence of typhoons, although long-term sea level rise due to climate change is a concern as well (VCZVA 1996). Deposition of river sediment along the coastline combined with sediment transport by nearshore waves and currents induce variability in the shoreline position at many scales in time and space that must be estimated in order manage the delta and associated coastal areas in a rational manner (Zeidler and Nhuan 1997). In spite of the importance of the delta and the other coastal areas in Vietnam relatively few comprehensive studies have been carried out to clarify the dynamics of the coastal regions. A difficulty in carrying out such studies has

---

1 Center for Marine Environment Survey, Research, and Consultation, Institute of Mechanics, 264 Doican, Hanoi, VIETNAM; Fax: +84 4832 7903; nmhung@im01.ac.vn

2 Department of Water Resources Engineering, Lund University, Box 118, S-22100 Lund, SWEDEN.; magnus.larson@tvrl.lth.se.

been the limited amount of data that are available on nearshore waves, currents, sediment transport, and coastal evolution. Thus, in order to estimate these characteristics along the coast numerical models are valuable and necessary tools to apply (Hung 2001, Ninh *et al.* 2001).

The overall aim of the present study was to investigate the coastal evolution along the RRD in response to incident, short-period waves, with special focus on stretches of shoreline suffering from erosion. Particularly, the conditions at Hai Hau Beach located in the southern part of the delta (see Fig. 1) were studied using numerical models for wave transformation and sediment transport. Because of the complex topography of the delta and the associated spatial scale of variation, wave modeling over a fairly large spatial region was required. A newly developed two-dimensional (2D) random wave model based on the mild-slope equations was employed using a 20-year time series of hindcast waves in the offshore as input. Calculations based on this time series yielded nearshore wave properties necessary for estimating local longshore sediment transport rates. A recently proposed longshore sediment transport formula was utilized to determine these transport rates. The results of the present study supported the hypothesis that the erosion at Hai Hau Beach is mainly due to the prevailing waves generated by the winter monsoon in combination with the complex delta topography. Interaction between the topography and the waves produces longshore sediment transport gradients that induce erosion at Hai Hau Beach.

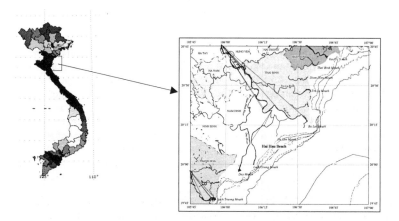

Fig. 1. Red River Delta and its location in northern Vietnam

STUDY AREA

The RRD is located in the subtropical climate region implying a marked influence from monsoon winds and severe storms (*i.e.*, typhoons) on the coastal evolution. Along the delta coastline, the hydrodynamic characteristics are mainly defined by the monsoons, which through persistent wind action generate waves, large-scale coastal circulation, and nearshore currents. Analysis of a 20-year wind time series recorded at the meteorological station Bach Long Vi in the central part of the Tonkin Gulf indicates that there are two main wind directions during the year (NE-N and S-SE) belonging to NE monsoon in winter and SW monsoon in summer, respectively. Long-term statistics show that the average number of typhoons striking the coastline of Vietnam annually is four, and the RRD is in an area that is most frequently subject to typhoon attack. About 30% of the recorded typhoons caused storm surge over 1 m and 4 % over 2 m (Ninh *et al.* 2001).

The Red River Delta plain has an approximately triangular shape extending about 150 km from its apex to the coast, and with a base length at the coast of 130 km. Holocene delta sediments, relatively fine-grained and up to 30 m thick deposited at the present sea level, overlay coarse-grained Pleistocene sequence of braided river and alluvial fan deposits formed during glacial low sea level stands (Mathers and Zalasiewicz 1999). The overall evolution of the delta indicates about equal influence from fluvial, wave, and tidal (meteorological and astronomical) processes. The Tonkin Gulf, to which the Red River discharges its sediment, represents a quite complex environment, both in terms of the forcing and the prevailing geomorphology. Zeidler and Nhuan (1997) as well as Pruszak (1998) have presented schematic sediment budgets for the delta region describing the contribution from the various river branches, the longshore transport pattern, and transport of material to the offshore. About 70% of the sediment from the river is transported to the offshore. Pruszak (1998) investigated the coastal evolution at a few sites in Nam Ha and Thai Binh Provinces including stretches of coast that displayed both severe long-term erosion and significant accretion. The sediment typically found in the Red River Delta consists of sand, silt, and clay, mostly of riverine origin, with grain sizes in the range 0.001 mm to 0.25 mm. Because of the dominance of fine fractions the sediment is very mobile, requiring fairly low water velocities to initiate motion and net transport.

Most of the river sediment is discharged into the sea through six different branches. In general, accumulation occurs in the vicinity of the branches, at a rate depending on the local sediment discharge of each branch. The most intensive accumulation is recorded at Balat River mouth situated to the north of Hai Hau Beach, followed by Day River mouth to the south of this beach. However, simultaneously with this accumulation, high rates of erosion has occurred at Hai Hau Beach (VCZVA 1995, Vinh *et al.* 1996). Several different hypothesis have been put forward to explain this erosion. The construction of Hoa Binh Dam in the Red River has reduced the sediment supply to the delta (Zeidler and Nhuan 1997), possibly affecting Hai Hau Beach. Also, in 1955 one river branch in the delta that discharged sediment at the upstream end of the beach was cut, implying a decrease in the sediment input (Vinh *et al.* 1996). Most likely, these two factors have had some impact on the erosion, but since the shoreline retreat has been observed for at least 100 year, the main cause must be different. Another hypothesis, to be more closely investigated here, is that the prevailing wave climate in connection with the complex topography of the delta induce a gradient in the longshore sediment transport that promotes erosion along Hai Hau Beach.

REGIONAL WAVE TRANSFORMATION

Offshore Waves

The data set encompassing 20 years of measured wind speed and direction at Bach Long Vi Station (6-hr interval) was employed to calculate offshore wave parameters for the RRD coastal area using the wave forecasting method for shallow water recommended in SPM (1984). However, the wave prediction formulas were first validated through comparison with 20 years of visually observed wave heights and periods carried out near Bach Long Vi (20^008 N, 107^043 E). Figs. 2 and 3 display the measured and calculated frequency of occurrence for wave heights and wave periods with respect to direction (time period: 1976-1995). Overall, the computed wave heights, periods, and directions agree well with the observations, giving confidence to the simple method for wave prediction utilized in this study. The results show that the wave regime in Tonkin Gulf is primarily defined by the monsoon wind climate. Thus, there are two predominant wave directions NE and S associated with the winter and summer monsoon, respectively.

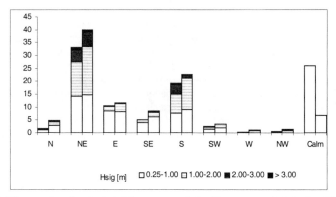

Fig. 2. Measured and calculated frequency of significant wave height for different wave directions (in the central part of Tonkin Gulf)

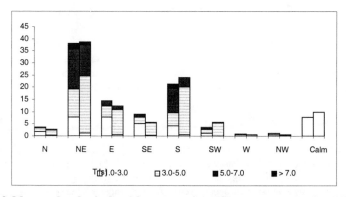

Fig. 3. Measured and calculated frequency of significant wave period for different wave directions (in the central part of Tonkin Gulf)

Table 1 summarizes significant wave height and period, depending on direction, as an annual average in 17-m water depth outside Hai Hau Beach. The total number of wave events during the 20-year period in respective directional band is also given (Hung 2001).

Table 1. Annual average of significant wave height (H_{mo}) and period (T), and total number of events (N), depending on direction, offshore of Hai Hau Beach

	N	NNE	NE	ENE	E	ESE	SE	SSE	S	Calm
H_{mo} [m]	1.22	1.60	2.05	1.47	1.35	1.15	1.54	1.99	1.73	-
T [s]	5.0	6.0	7.0	6.0	6.0	6.0	6.0	8.0	7.0	-
N	1604	3211	7962	1466	3054	972	1850	1415	4922	2028

Nearshore Waves

A recently developed numerical 2D random wave transformation model based on the mild-slope equation (MSE) was employed in the present study to determine the nearshore wave conditions. The one-dimensional version of this model has been validated with a substantial amount of laboratory and field data (Larson 1995) and it is used for computing the cross-shore variation in statistical wave parameters in several numerical models of coastal processes. No similar extensive validation has been performed for the 2D version, although indirect validation has been obtained by using this model in simulations of beach topography evolution (Larson and Hanson 1996). Here, only a brief summary of the main equations are given (see Larson and Hanson 2002 for a more extensive discussion).

Employing the real part of the mild-slope equation with a sink term for depth-limited wave breaking in accordance with Dally (1992), and utilizing a wave-by-wave approach as described by Larson (1995) to simulate the randomness in wave height, the following wave energy conservation equation is obtained,

$$\frac{\partial}{\partial x}\left(F_{rms}\frac{1}{k}\frac{\partial S}{\partial x} \right) + \frac{\partial}{\partial y}\left(F_{rms}\frac{1}{k}\frac{\partial S}{\partial y} \right) = \frac{\kappa}{d}(F_{rms} - F_{stab}) \tag{1}$$

where F_{rms} is the wave energy flux based on the root-mean-square (rms) wave height (H_{rms}), k the wave number, S a phase function, x and y coordinates in a rectilinear system (cross-shore and longshore direction, respectively), κ an empirical dissipation coefficient (about 0.15), and the equivalent stable wave energy flux for random waves is defined as,

$$F_{stab} = \beta F_m + \mu F_r + \alpha(\Gamma/\gamma)^2 F_b \tag{2}$$

where α, β, and μ is the ratio of breaking, unbroken, and reformed waves, respectively, and F_m and F_r is the wave energy flux based on the rms wave height for the unbroken and reformed waves, respectively. The energy flux F_b is based on the wave height corresponding to incipient breaking at the specific depth given by $H_b=\gamma_b d$, where γ_b is the breaker depth index. The phase function S is determined by solving the imaginary part of the MSE, neglecting diffraction. In Eq. 1, F_{rms} is the primary unknown that should be solved for, implying that α, β, μ, F_m, F_r, and F_b must be supplied in the calculation. Once F_{rms} has been determined everywhere, other wave quantities can be derived, for example, the average energy dissipation $P=\kappa/d(F_{rms}-F_{stab})$.

For a beach where the depth is decreasing monotonically in the direction of the approaching waves, α is determined directly from the truncated Rayleigh pdf through,

$$\alpha = e^{-F_B/F_x} \tag{3}$$

where F_x is the local rms wave height neglecting wave breaking. In order to calculate α, F_x has to be obtained everywhere, which is done by solving the wave energy flux equation (Eq. 1) without including energy dissipation due to breaking. Thus, in the solution procedure for the wave field two wave transformation equations have to be solved simultaneously, one with and one without energy dissipation. Wave reforming may occur if the beach topography is complex with non-monotonic sections. In this case α can not be predicted directly from Eq. 3, but α is determined based on how the wave reforming is modeled (not discussed here; see Larson and Hanson 2001). The ratio of unbroken waves is given by $\beta=1-\alpha$, if the beach is monotonic in the direction of the local wave. The energy flux based on the rms height of the unbroken waves may be calculated from the truncated local Rayleigh pdf yielding:

$$F_m = \frac{1}{\beta}\left[F_x - (1-\beta)(F_x + F_B)\right] \tag{4}$$

Calculations were performed with the 2D model for every wave event in the 20-year time series and average wave properties were computed at each grid point. The coordinate system was oriented along the main trend of the shoreline, and grid cells with $\Delta x=50$ m and $\Delta y=600$ m were employed (900 x 165 grid points). Figure 4 and 5 show the calculated spatial distribution of the annual mean rms wave height and mean energy dissipation due to depth-limited wave breaking, respectively. The results indicate that the wave height decay and energy dissipation occur in a narrow strip in the middle and southern part of the RRD, particularly in the area where significant erosion has been observed.

SEDIMENT TRANSPORT AND COASTAL EVOLUTION

Gross Sediment Transport Pattern

As a first step to estimate the gross longshore sediment transport pattern along the RRD the CERC formula (SPM 1984) was employed for the generated time series of offshore waves. The properties at breaking were determined through a one-dimensional wave transformation calculation including refraction and shoaling up to the break point. In these calculations, different shoreline orientations were assumed and the annual net transport rate was determined. Fig. 6 displays the mean annual net longshore transport rate as a function of the shoreline orientation (degree TN). The results indicate that the maximum transport rate occurs for an orientation of about 45 deg, which is approximately the orientation of the shoreline at Hai Hau Beach. The shoreline orientations both to the north and south of Hai Hau Beach are larger indicating smaller annual net transports and gradients that would promote erosion along the beach and accretion downdrift of it.

Regional Sediment Transport Pattern

The simple estimate of the nearshore wave properties and the calculation of the annual net longshore sediment transport, discussed in the previous section, provided some support for the hypothesis that the erosion at Hai Hau Beach is mainly due to the wave climate and its interaction with the nearshore topography. However, more detailed information on the

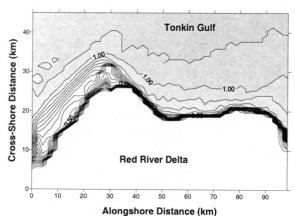

Fig. 4. Spatial distribution of annual mean rms wave height along the RRD coastal zone (alongshore distance goes from N to S)

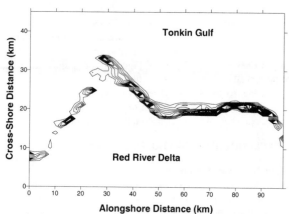

Fig. 5. Spatial distribution of annual mean wave energy dissipation along the RRD coastal zone (alongshore distance goes from N to S)

wave characteristics and this interaction are needed in order to establish that gradients in the longshore sediment transport is the primary cause of the erosion. Thus, the time series of nearshore wave properties obtained from the 2D wave transformation calculations were used to compute local longshore sediment transport rates (q_l). The following formula proposed by Larson and Hanson (1996) were utilized,

$$q_l = \frac{\varepsilon_c}{1-a} \frac{1}{\rho_s - \rho} \frac{1}{gw} VP \tag{5}$$

where ε_c is a coefficient (transport efficiency factor), a the porosity, ρ_s and ρ the density of sediment and water, respectively, w the sediment fall speed, V the longshore current velocity, and P the energy dissipation. The local current velocity was estimated from a balance between the radiation stress gradient and bottom friction (*i.e.*, the alongshore momentum equation with linearized friction). Furthermore, since reliable values for ε_c is lacking at present, ε_c was determined through comparison with the CERC formula. Thus, for a particular beach, if the local transport rate obtained from Eq. 5 is integrated across shore and the total rate obtained is compared with the CERC formula, $\varepsilon_c=0.77c_f K$ results, where c_f is the bottom friction and K the transport coefficient in the CERC formula. In the case of only wave-generated longshore currents (no effects from wind or tide), c_f will cancel out since it appears both in ε_c and V. The coefficient K was assigned the standard value in accordance with SPM (1984).

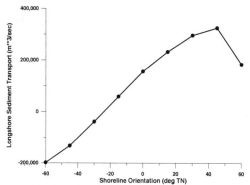

Fig. 6. Mean annual net longshore sediment transport along the RRD shoreline determined based on the CERC formula

Fig. 7 shows the mean annual net longshore transport rate along the RRD shoreline obtained by integration in the cross-shore direction. The results are based on the wave properties computed with the 2D wave transformation model employed for the 20-year time series of offshore waves. Overall, the calculated transport rates are in agreement with previous investigations for the area and the gradients indicate accumulation and erosion in accordance with field observations. The transport rate along Hai Hau Beach increases at a low rate from the northern end, but towards the southern end the increase (and associated gradient) is substantial. Again, the calculated transport rates support the hypothesis that the wave climate in combination with the bottom topography is the cause of erosion at Hai Hau Beach.

Observed and Estimated Future Coastal Evolution

Hai Hau Beach has been eroding for more than one hundred years (see Fig. 8). Although the dam construction in the Red River upstream Hanoi, as well as the cutting of the Ha Lan branch north of Hai Hau Beach, might have contributed to the erosion, calculations here indicated that the main factor behind the erosion is gradients in the longshore sediment transport. These gradients are caused by interaction between the topography and prevailing wave climate. Hai Hau Beach is located downdrift an area of marked accumulation, occurring especially during the NE monsoon. This accumulation creates conditions similar to a barrier for the longshore transport (i.e., a "natural groin") with marked effects on the wave transformation pattern and associated transport gradients. In order for the erosion to cease, a shoreline shape must evolve that is in equilibrium with the wave climate, which in turn is a function of the complex bottom topography. Quantitatively evaluating the equilibrium shape is a non-trivial task considering the complex wave transformation and the feedback between topography and waves. However, some approximate estimates of the equilibrium shape could be made based on the work by Hsu *et al.* (1989).

CONCLUSIONS

Several stretches of coastline in the Red River Delta, Vietnam, are suffering from severe erosion, especially Hai Hau Beach in the southern part of the delta. Different hypotheses have been put forward to explain the erosion at this beach, including the construction of a large dam in the upstream part of the Red River (trapping of sediment), the cutting of a river branch that formerly supplied sediment to the beach, and the imbalance between the prevailing wave climate and the local shoreline orientation. Wave transformation calculations in this study indicated that the most likely explanation for the erosion is the third hypothesis. Wave transformation calculations, taking into account the complex topography of the RRD, together with computed local net transport rates produced gradients in the longshore transport rate consistent with the observed coastal evolution along the RRD delta in general, and at Hai Hau Beach in particular.

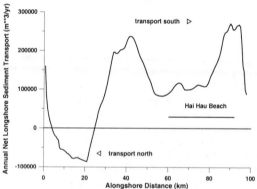

Fig. 7. Mean annual net longshore sediment transport along the RRD shoreline determined based on wave properties from a 2D wave transformation model

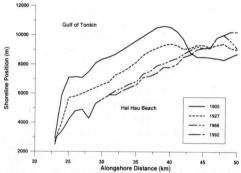

Fig. 8. Measured shoreline location at Hai Hau Beach, Vietnam

ACKNOWLEDGEMENTS

Financial support from the Swedish International Development Cooperation Agency is gratefully acknowledged (SWE-2000-208). The development of the wave transformation model utilized in this study was partly supported by the Coastal and Hydraulics Laboratory, U.S. Army Engineer Development and Research Center, Vicksburg, MS. Also, part of ML's work was carried out during a research visit to the Institute of Environmental Sciences at the University of Tokyo, sponsored by Japan Society for the Promotion of Science. Professor Masahiko Isobe's assistance during this visit is much appreciated.

REFERENCES

Dally, W.R. 1992. Random Breaking Waves: Field Verification of a Wave-By-Wave Algorithm for Engineering Application. *Coastal Engineering*, 16, 369-397.

Hsu, J. Silvester, R., and Xia, Y.M. 1989. Generalities on Static Equilibrium Bays. *Coastal Engineering*, Vol 12, 353-369.

Hung, N.M. 2001. Wave and Sediment Transport Computation. Final report, National Marine Project KHCN-0610, Institute of Mechanics, Hanoi, Vietnam (in Vietnamese).

Larson, M. 1995. "Model for Decay of Random Waves in the Surf Zone," *Journal of Waterway, Port, Coastal and Ocean Engineering*, Vol 121(1)1-12.

Larson, M. and Hanson, H. 1996. Schematized Numerical Model of Three-Dimensional Beach Change. *10th Congress of the IAHR Asia and Pacific Division*, Langkawi Island, Malaysia, Vol 2, 325-332.

Larson, M. and Hanson, H. 2002. Numerical Modeling of Beach Evolution. *Advances in Coastal Modeling*, Elsevier (in preparation).

Mathers, S. and Zalasiewicz, J. 1999. Holocene sedimentary architecture of the Red River Delta. *Journal of Coastal Research*, Vol 15(2), 314-325.

Ninh P.V., Quynh D.N., Lien N.V. 2001. Tidal, Storm Surge, and Nearshore Circulation Computations. Institute of Mechanics, Final Report, Marine Project KHCN-0610, Institute of Mechanics, Hanoi, Vietnam (in Vietnamese).

Pruszak, Z. 1998. Coastal Processes in the Red River Delta Area with Emphasis on Erosion Problems. Internal Report, Institute of Hydroengineering, Polish Academy of Sciences, Gdansk, Poland.

SPM 1984. Shore Protection Manual. 4th Ed. Vol. 1, US Army Engineer Waterways Experiment Station, Coastal Engineering Research Center, US Government Printing Office, Washington, DC.

VCZVA. 1996. Vietnam Coastal Zone Vulnerability Assessment. Final report, Government of the Netherlands (Ministry of Foreign Affairs) and the Socialist Republic of Vietnam (Hydrometeorological Service).

Vinh, T.T., Kant, G., Huan, N.N., and Pruszak, Z. 1996. Sea Dike Erosion and Coastal Retreat at Nam Ha Province. *25th International Coastal Engineering Conference*, American Society of Civil Engineers, 2820-2828.

Zeidler, R.B. and Nhuan, H.X. 1997. Littoral Processes, Sediment Budget and Coast Evolution Vietnam," *Proceedings of Coastal Dynamics '97*, American Society of Civil Engineers, 566-575.

SEDIMENT TRANSPORT RELATED TO
POTENTIAL SAND MINING OFFSHORE NEW JERSEY

Steven M. Jachec[1] and Kirk F. Bosma[2]

Abstract: Interest has developed to use borrow material from resource areas located on the inner continental shelf offshore New Jersey for beach replenishment purposes. A study was conducted to quantify the potential impacts of the offshore dredging and consequent beach nourishment on the local sediment transport patterns. In order to determine the offshore sediment transport patterns at the borrow locations, sediment transport rates were determined by coupling the results of spectral wave modeling with ambient current measurements. Having determined the sediment transport rates and directions at the borrow locations, and by knowing the volume of sediment to be mined, estimates of the borrow location recovery time were determined. These recovery time estimates were utilized to determine the time scale of influence on wave transformation, nearshore sediment transport, and re-mining frequencies.

INTRODUCTION

Beaches are key elements of coastal tourism, and the practice of replenishing beaches with sand from upland and nearshore sources have increased in direct relation to population growth. As coastal and nearshore borrow areas become depleted, alternate sources of beach nourishment must be evaluated. In many cases, sand resource excavation from the outer continental shelf (OCS) may prove to be environmentally preferable to nearshore borrow areas due to potential changes in waves and currents as large quantities of sand are mined. In recent years, the U.S. Department of Interior, Minerals Management Service (MMS) has initiated environmental studies along Atlantic and Gulf coastlines that evaluate OCS borrow locations.

1 Coastal Engineer. The Woods Hole Group 81 Technology Park Dr. E. Falmouth, MA 02536. Phone: (508) 495-6224; Fax: (508) 540-1001; Email: sjachec@whgrp.com
2 Team Leader and Coastal Engineer. The Woods Hole Group 81 Technology Park Dr. E. Falmouth, MA 02536. Phone: (508) 495-6228; Fax: (508) 540-1001; Email: kbosma@whgrp.com

A total of seven sites along the New Jersey coastline were identified as potential sand sources (Figure 1). Although environmental impacts were examined at each of these locations, this paper focuses on only one of the borrow sites and only the offshore sediment transport aspect of the comprehensive environmental study. Borrow site F2 is located offshore along an open-coast region in a depth of approximately 17 m (Figure 2) seaward of the State-Federal OCS boundary. The shoal in this area was a good candidate for sand mining for two reasons: 1) it is composed of beach-compatible material, and 2) it is expected to have minimal potential impact on wave transformation due to the leveling of the shoal. The F2 borrow site will be excavated three meters to produce 2,100,000 m³ of sand.

Although of the overall purpose for the environmental study was to assess the physical, geological, and biological impacts of dredging numerous locations along the New Jersey coastline, this paper focuses on the offshore sediment transport and borrow site recovery time at the F2 site by presenting methodology, results, and conclusions.

INITIATION OF SEDIMENT MOTION UNDER COMBINED WAVE-CURRENT ACTION

The spectral wave model REF/DIF S (Kirby and Özkan 1994) was applied to assess changes in wave propagation across the New Jersey continental shelf relative to the potential sand mining. REF/DIF S simulates the behavior of a random sea by describing the wave energy density as of function of a directional spectrum and frequency spectrum. Through a combination of various wave directions and frequencies input, derived from WIS station 2070 (1976-1995), waves were simulated on a numerical grid used to describe the area and bathymetry (obtained from NOAA). It was key to include as much of the energy as possible. The table below lists the peak directions simulated and the associated energy percentage. For this scenario, over 90% of the wave energy was accounted for within the model runs.

Table 1. Directional Cases Obtained from WIS Station 2070 (1976-1995).

Approach Direction (from)	Percent Contribution (%)
NE	2.3
ENE	3.4
E	26.1
ESE	25.8
SE	19.9
SSE	13.9
CALM	8.6

Wave model simulations were used to assess the changes in wave height and direction, and hence wave-induced bottom velocity. Wave simulations were performed for both existing conditions and potential dredge scenarios. The F2 borrow site was numerically dredged three meters to represent the proposed dredging. Figure 3 shows a difference plot in wave height across the shelf between the post- and pre-dredge cases with waves propagating from a peak easterly direction. As expected, sand mining creates a zone of decreased wave energy behind the borrow site and areas of increased energy adjacent to the

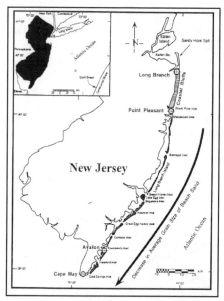

Figure 1. Overview of the project study area located along the New Jersey coastline.

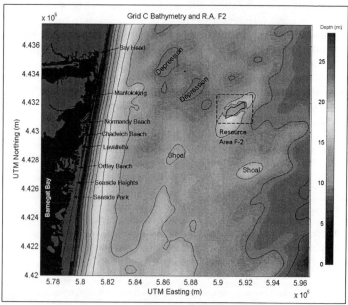

Figure 2. Close-up of the F2 borrow site and surrounding area. Thick black-lined polygon within dashed box represents the F2 borrow site.

borrow site. Waves propagating over the sand borrow site are deflected outward, while a zone of reduced wave height is created directly behind the site. A maximum increase of approximately 0.1 meters (m) and a maximum decrease of approximately 0.2 m result from the sediment excavation scenario for resource area F2. Other directional and dredge locations are similar.

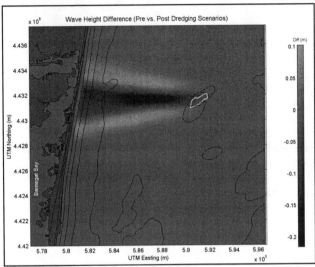

Figure 3. Pre- versus post-dredged wave heights at F2 and surrounding area.

Before sediment can be transported, it must be moved from the seabed by combined wave and current motion. When sufficient stress is applied to the bed, sediment may begin to move. Through dimensional analysis, Shields (1936) derived an expression that identifies the point where bed stress equals bed resistance. The Shields parameter (ψ) results from equating the driving and stabilizing forces. For a flat bed:

$$\psi = \frac{\tau_b}{(s-1)\,\rho g\, d_{50}} \tag{1}$$

where τ_b = maximum bottom shear stress (N/m), ρ = density of the sea water (kg/m^3), s = relative density (equals 2.65 for natural sediment), g = acceleration due to gravity (m/s^2), d_{50} = grain diameter which corresponds to 50% by weight finer (m)

The shear stress at the bed, τ_b, is given by Madsen and Grant (1976) and Raudkivi (1990) as:

$$\vec{\tau}_b = \frac{1}{2}\rho f_{cw}\left|\vec{u}_{cw}\right|\vec{u}_{cw} \tag{2}$$

where f_{cw} = combined wave/current friction factor and u_{cw} = combined wave/current reference velocity (m/s).

In this study, u_{cw} includes the effects of waves and a steady current. A combination of the two creates a more realistic representation of maximum bottom velocity and bed shear stress. Proper combination of wave-induced and ambient currents requires an accurate representation of flow dynamics located directly at the seabed. In most cases, it is difficult to measure ambient current magnitude and direction directly at the seafloor. In the present study, historical current observations were measured a certain distance from the bottom.

The combined wave/current reference velocity, u_{cw}, is a function of the wave-induced bottom orbital velocity and the apparent current velocity at the bottom, U_a, as given by:

$$\bar{u}_{cw} = \left(U_b \cos \omega t + U_a \cos \phi_a, U_a \sin \phi_a \right) \tag{3}$$

where, U_b = wave-induced bottom velocity (m/s), U_a = apparent ambient current bottom velocity, and ϕ_a = the angle between the apparent current and wave-induced current (radians).

Because current observations were not measured at the bottom, they must be translated to the seafloor based on the application of a current profile through the bottom boundary layer. In order to determine the appropriate vertical current profile, the thickness of the bottom wave/current boundary layer (δ_w) must be determined and compared to the observed current location within the water column. Jonsson (1980) presents an equation for the thickness of the wave boundary layer in oscillatory rough turbulent flow, which is most common in nature, as:

$$\delta_w = \frac{2\kappa U_{*m}}{\omega} \tag{4}$$

where δ_w = boundary layer thickness (m), κ = Von Karman's constant (0.4), U_{*m} = maximum current velocity at the seabed (m/s), ω = angular frequency (radians).

If observed currents were measured outside of the bottom boundary layer ($z > \delta_w$), which is usually the case in field measurements, a logarithmic current profile is assumed, as:

$$U_c = \frac{U_{*c}}{\kappa} \ln\left(\frac{30z}{k_{bc}} \right) \tag{5}$$

where U_{*c} = critical bottom velocity (m/s), z = height above the bed (m), U_c = magnitude of the measured current (m/s), and k_{bc} = the apparent bed roughness.

The apparent bed roughness presented in the above equation is defined as:

$$k_{bc} = k_b \left(60\kappa \frac{U_{*m}}{k_b \omega} \right)^\beta \tag{6}$$

where k_b = roughness coefficient, which is assumed to be equivalent to d_{50} of the local sediment, and $\kappa = 1-(U_{*c}/U_{*m})$.

In the present study, the observed current was measured outside of the wave boundary layer at all of the measurement stations; therefore, Equation 5 was applied to translate the observed current data to the seabed for each of the borrow site regions.

Having defined the ambient current velocity at the bottom, the bottom shear stress resulting from combined wave/current interaction can be determined. Maximum bottom shear stress, $\tau_{b,max}$, due to the combined current and wave action can be determined from:

$$\tau_{b,max} = \rho U^2_{*m} = \frac{1}{2}\rho f_{cw}U^2_b\left(1 + 2\varepsilon\cos\phi_a\right) \tag{7}$$

where $\varepsilon = (U_a/U_b)$.

The combined wave/current friction factor, f_{cw}, is provided by Madsen and Grant (1976) as:

$$f_{cw} = \frac{U_c f_c + U_b f_w}{U_c + U_b} \tag{8}$$

where f_c and f_w = friction factors corresponding to ambient current flow and wave-induced flow, respectively. The wave friction factor was presented by Jonsson (1966) and is a function of the wave Reynolds number and (U_b/k_bT).

$$f_w = f_w\left(\frac{U_b^2}{\nu\omega}, \frac{U_b}{k_b\omega}\right) \tag{9}$$

The wave friction factor can be determined using Jonsson's wave friction factor diagram (Jonsson 1966). In a similar manner, the current friction factor can be determined from the standard Darcy-Weisbach approach:

$$f_c = \frac{1}{4}f\left(\frac{U_m 4h}{\nu}, \frac{d_{50}}{4h}\right) \tag{10}$$

where h = water depth (m), ν = viscosity (m²/s), and f = friction factor.

The maximum bottom shear stress under the combined wave/current interaction is then used to calculate the Shields parameter (Ψ_{max}) from Equation 1; recast as:

$$\Psi_{max} = \frac{U^2_{*m}}{g(s-1)d_{50}} \tag{11}$$

Once the Shields parameter has been calculated at points of interest, the resulting values can be compared to a critical Shields parameter (Ψ_{crit}) to determine if sediment initiation occurs at each point of interest. The critical Shields parameter may be determined using a modified Shields diagram developed for sediment transport in the coastal environment (Madsen and Grant 1976).

SEDIMENT TRANSPORT RATES AND BORROW SITE RECOVERY TIMES
Sediment initiation provides valuable insight regarding sediment movement, but it does not provide information relative to the magnitude or direction of sediment transport.

Therefore, sediment transport rates and directions need to be calculated in and around the offshore borrow site. The benefits to the transport calculations are three-fold:

1. Approximate rates and directions of sediment transport
2. Aid in borrow site recovery time
3. Provide directional fluctuations in sediment transport patterns

Offshore sediment transport rates are based on analytical expressions developed by Madsen and Grant (1976). These instantaneous sediment transport rates along the bed in the easting and northing directions respectively are:

$$q(t)_{east} = 40\,\omega\,d_{50} \left[\frac{\frac{1}{2} f_{cw}\left(u(t)^2 + v(t)^2\right)}{(s-1)g\,d_{50}} \right]^3 \frac{u(t)}{\sqrt{u(t)^2 + v(t)^2}} \tag{12}$$

$$q(t)_{north} = 40\,\omega\,d_{50} \left[\frac{\frac{1}{2} f_{cw}\left(u(t)^2 + v(t)^2\right)}{(s-1)g\,d_{50}} \right]^3 \frac{v(t)}{\sqrt{u(t)^2 + v(t)^2}} \tag{13}$$

where $q(t)_{east}$ = instantaneous easting component of sediment transport (m³/s-m), $q(t)_{north}$ = instantaneous northing component of sediment transport (m³/s-m), ω = sediment fall velocity (m/s), f_{cw} = wave-current friction factor, $u(t)$ = instantaneous bottom velocity in the easting direction (m/s), and $v(t)$ = instantaneous bottom velocity in the northing direction (m/s)

Since we are interested in long-term answers, not instantaneous, the above equations are period-averaged and multiplied the respective grid spacing, which yield:

$$\bar{q}(t)_{east} = \left(\frac{1}{T} \int_0^T q(t)_{east}\ dt \right) \Delta x \tag{14}$$

$$\bar{q}(t)_{north} = \left(\frac{1}{T} \int_0^T q(t)_{north}\ dt \right) \Delta y \tag{15}$$

where $\bar{q}(t)_{east}$ = period-averaged sediment transport in the east direction (m³/s), $\bar{q}(t)_{east}$ = period-averaged sediment transport in the east direction (m³/s), Δx = grid spacing in the easting direction (m), Δy = grid spacing in the northing direction (m), and T = wave period (s).

Once the results are period-averaged, they are then annually averaged to determine the yearly sediment transport magnitude and direction. The averaging is based upon the energy contributed from each directional model bin (see Table 1). By knowing the annual average sediment transport rate and dredge volume, the recovery time can be estimated. The recovery time is the time it would take the dredge site to fill to the original volume.

The above methodology does have some limitations and simplifications. The limitations are: a wave-dominated environment, cohesionless sediment, and non-breaking waves. The simplifications are the steady ambient current, and the non-dynamic morphology at the borrow site and surrounding area.

RESULTS

Figure 4 illustrates wave-induced bottom velocities (panel 1), steady near-bottom shelf ambient currents (panel 2), sediment initiation potential (panel 3), and sediment transport results (panel 4). These figure panels are based upon results from the spectral wave modeling in which the input wave direction was from the east: this only represents one of the seven scenarios evaluated at borrow site F2. The thin black lines represent bathymetric contours, while the thick black-lined polygon represents the borrow site. The colorbar corresponding to each panel shows the respective magnitude (velocity, potential, or rates).

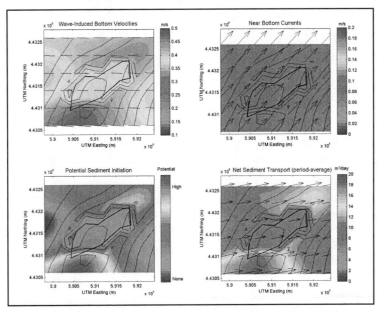

Figure 4. Hydrodynamic and sediment transport results for the eastern directional case at borrow site F2.

Panel 1, which represents the variable wave-induced bottom velocities, show that the currents are oscillatory reaching a peak velocity south of the borrow site while they are reduced within the mined borrow site. Near bottom ambient currents (panel 2) are treated as steady for this directional case in this study, and are flowing in a northeast direction.

Sediment initiation potential and sediment transport rates are presented in panels 3 and 4, respectively. Areas of greater initiation are similar to the areas of larger wave-induced bottom velocities, primarily around the north and south of the borrow site. However, given the increased depth within the mined borrow site and decreased wave height, sediment initiation and transport are reduced within it. Sediment transport rates entering this borrow location, for this particular easterly case, are estimated to 8 to 15 m³/day. A similar type of analysis is carried out from the remaining directional cases at F2.

Based on these calculations, the mean annual sediment transport rate is 28 m³/day towards the east. Given this information and the fact that the F2 borrow site dredge volume is 2,100,000 m³, the recovery time can be calculated as 205 years. Table 2 summarizes the results of all seven offshore borrow sites examined along the New Jersey coastline.

Table 2. Annually-Averaged Sediment Transport Results and Estimated Recovery Times.

Borrow Site	Magnitude of Sediment Transport (m³/day)	Direction of Net Sediment Transport (to)	Mined Sand Volume (x 10⁶ m³)	Recovery Time (yrs)
A1	450	SW	8.8	54
A2	327	SW	7.8	65
C1	55	W	6.1	303
G2 Top	100	NE	3.4	93
G2 Bottom	40	E	0.95	65
G3	110	NE	3.3	82
F2	28	E	2.1	205

In order to compare the sediment transport rates, an independent method was used to verify the model. The independent method consisted of analyzing seafloor change rates offshore New Jersey. The original data was obtained from two USC&GS data sets from 1834-1891 and 1934-1977. The model predicted rates of 10,000 – 160,000 m³/yr, while the analytical seafloor analysis yielded 60,000 – 230,000 m³/yr. This suggests that the two methods produce results of the same order of magnitude. In general, the analytical seafloor change rates are greater than the model. One possible reason could be the fact that storms were not included in the above sediment transport analysis. The storms would tend to increase these rates since wave heights and currents would be greater than "normal" conditions.

CONCLUSIONS

Results from a spectral wave model coupled with ambient near-bottom currents were coupled to simulate offshore sediment transport and estimate borrow site recovery time. This method was applied to seven offshore borrow sites along the New Jersey shelf. The analytical sediment transport model provided reasonable results when compared with a separate, independent seafloor change analysis. However, improvements to model are needed to more accurately portray the situation. This includes: the effects of time-varying bathymetry, nonlinear wave-induced bottom currents, suspended sediment transport, and the effects of storms.

ACKNOWLEDGEMENTS
Funded by the U.S. Department of the Interior, Minerals Management Service (MMS), International Activities and Marine Minerals Division (INTERMAR).

REFERENCES

Applied Coastal Research and Engineering, Inc., Aubrey Consulting, Inc., Barry A. Vittor and Associates, Inc., and Continental Shelf Associates, Inc. 2000. Environmental Survey of Potential Sand Resource Sites: Offshore New Jersey, Final Report. *Prepared for the Mineral Management Service (INTERMAR) under contract number 14-35-01-97-CT-30864.*

Grant W.D. and Madsen, O.S. 1979. Combined Wave and Current Interaction with a Rough Bottom. *Journal of Geophysical Research*, 87, 1797-1808.

Grant W.D. and Madsen, O.S. 1982. Movable Bed Roughness in Unsteady Oscillatory Flow. *Journal of Geophysical Research*, 87, 469-481.

Grant W.D. and Madsen, O.S. 1986. The Continental Shelf Bottom Boundary Layer. *Annual Review of Fluid Mechanics*, 18, 265-305.

Jonsson, I.G. 1966. Wave Boundary Layers and Friction Factors. *Proceedings of the Tenth Conference on Coastal Engineering*, ASCE, Vol. 1. 127-148.

Jonsson, I.G. 1980. A New Approach to Oscillatory Rough Turbulent Boundary Layers. *Ocean Engineering*, 7, ASCE, 109-152.

Kirby, J.T., and Özkan, H.T. 1994. Combined Refraction/Diffraction Model for Spectral Wave Conditions, REF/DIF S v.1.1. *CACR-94-04*, University of Delaware, Newark, NJ.

Madsen, O.S. and Grant, W.D. 1976. Quantitative Description of Sediment Transport by Waves. *Proceedings, Fifteenth International Coastal Engineering Conference*, ASCE, Vol 2. 1093-1112.

Raudkivi, A.J., 1990, *Loose Boundary Hydraulics, Third Edition, Pergamon Press, 538 p.*

Shields, I.A. 1936. Application of Similarity Properties and Turbulence Research to Bedload Movement. *Mitteilunden der Preoss.* Versuchsanst. Fur Wasserbau and Schiffau, Berlin, 26 (in German).

Tanaka, H., and Shuto, N. 1981. Friction Coefficient for a Wave-Current Coexisting System. *Coastal Engineering of Japan*, 24, 105-128.

A COUPLED WAVE, CURRENT, AND SEDIMENT TRANSPORT MODELING SYSTEM

David J.S. Welsh[1], Keith W. Bedford[2], M.ASCE, Yong Guo[3], and
Ponnuswamy Sadayappan[4]

Abstract: The impact of inter-model coupling on regional scale numerical predictions of wind-waves, marine circulation, and sediment transport is investigated. This is done using the COupled MArine Prediction System (COMAPS), which consists of parallel-processing versions of the CH3D-SED circulation and sediment transport model, the WAM wind-wave model, and the WCBL bottom boundary layer model. COMAPS includes physics couplings between wave and current motions at the water surface and wave, current, and sediment motions at the marine bed. Adriatic Sea hindcasts are discussed. These indicate that the coupling mechanisms have significant impacts on coastal water levels, currents, and wave heights, as well as basin-wide sediment transport.

INTRODUCTION

Regional or basin scale numerical models of marine circulation and wind-waves have traditionally been treated as independent entities. Furthermore, regional predictions of sediment transport have typically used current forcing only, neglecting wave-induced transport and related bedforms. These strategies were necessary due to theoretical and computational limitations, but recent research indicates significant interactions exist between wave, current, and sediment motions. Tolman (1991) and Mastenbroek et al. (1993), for example, modeled wave-current interactions at the air-sea boundary layer in North Sea hindcasts. Signell et al. (1990) investigated the effect of wave-induced bottom roughness on the circulation in an idealized bay. More recently, Xie et al. (2001) included both water surface and marine bed wave-current couplings to predict the impact of waves on coastal circulation. This paper concerns the formulation and application of a marine modeling system that accounts for the aforementioned wave-current couplings as well as

---

1 Senior Research Engineer, Dept. of Civil Engineering, Ohio State University, Columbus, Ohio 43210. welsh.3@osu.edu

2 Professor, Dept. of Civil Engineering, OSU. bedford.1@osu.edu

3 Research Associate, Dept. of Civil Engineering, OSU. yguo@superior.ceegs.ohio-state.edu

4 Professor, Dept. of Computer and Information Science, OSU. sadayappan.1@osu.edu

sediment transport interactions. The COupled MArine Prediction System (COMAPS; Welsh et al., 2000) consists of coupled, parallel-processing versions of the CH3D-SED circulation and sediment transport model, the WAM wind-wave model, and the WCBL marine bottom boundary layer sub-model. COMAPS takes advantage of high performance computing facilities to generate fully coupled predictions in a timely manner. The ultimate goal of this research is to improve predictive accuracy and thereby enhance the support of military, regulatory, leisure, and commercial activities.

COMPONENT MODELS AND INTER-MODEL COUPLING

The CH3D-SED model (Spasojevic and Holly, 1994) resulted from the addition of a sediment transport module (SED) to the original CH3D marine circulation code. CH3D is based on Reynolds-averaged three-dimensional conservation equations for water mass, momentum, temperature, and salinity. The basic equations are nondimensionalized and adapted for use with a curvilinear horizontal grid and a sigma-layer (terrain-following) vertical grid. A Mellor-Yamada level 2.5 turbulence closure model is used. Mixed explicit and implicit finite difference discretizations are used and solutions are reached by a mode-splitting technique, where the external mode (barotropic; depth-averaged) motions are calculated at a much smaller time-step than the internal (baroclinic; depth varying) motions. The SED module is based on three-dimensional suspended sediment transport equations and active-layer sediment mass conservation equations applied to an arbitrary number of user-defined sediment size classes. In the active layer, SED accounts for erosion, deposition, and bedload transport. SED is fully integrated with CH3D, using shared arrays, the same grids, and similar numerical schemes. COMAPS uses a parallel-processing version of CH3D-SED, based on one-dimensional domain-decomposition and the Message Passing Interface (MPI) library. A pre-processor divides the horizontal grid into a user-specified number of laterally sliced blocks. The calculations for each block are then performed on an individual processor.

The WAM model (WAMDI, 1988) is based on the conservation equation for wave action (wave energy divided by frequency) applied to a spherical (longitude-latitude) grid. Wave action is conserved for each component of a user-defined frequency-direction spectrum, with spectral source/sink terms included for wind input, nonlinear wave-wave interaction, whitecapping, and bottom friction. WAM is classed as a third-generation wave model, based on the complexity of its wave-wave interaction calculations. WAM is also a finite difference code, with upwind, explicit terms used for propagation and centered, implicit terms used for the sources and sinks. The COMAPS version of WAM was parallelized using the OpenMP library. In contrast to MPI, OpenMP offers loop-level parallelism, where high-demand sections of the code are defined as multi-threaded parallel regions. The number of WAM threads is specified at run-time.

COMAPS accounts for wave-current coupling at the air-sea boundary layer. Two effects of waves on currents are represented. Firstly, there is a surface momentum transfer due to spatial gradients of the wave field's radiation stress (Phillips, 1977). Secondly, increased surface roughness is parameterized using an enhanced CH3D drag coefficient expression that involves the ratio of wave-related surface stress to total surface stress (Janssen, 1991). Breaking waves have a major effect on nearshore currents, but the surf-zone is considered outwith the scope of COMAPS. Surface current effects on waves are accounted for in the propagation and wind input terms of the WAM wave action transport equation. In the propagation calculations, current vectors directly advect each wave component and spatial variations in the currents cause wave refraction. In the wind input source term, the wind vector relative to each wave component is vectorially shifted to reflect the current-induced propagation. Additional, minor coupling effects on waves result from the unsteady variation of depths due to storm surges, seiches, and tides. These variations modify depth-related wave refraction and are accounted for in COMAPS.

COMAPS calculates wave-current-sediment coupling in the bottom boundary layer using the WCBL model (Welsh et al., 2000). WCBL parameterizes interactions between wave and current boundary layers and related sediment transport. The code is based on the work of Grant and Madsen (1979) and Glenn and Grant (1987). Due to the oscillatory nature of wave motion, wave boundary layers are generally much thinner, but more turbulent, than current boundary layers. When near-bed wave motion is significant, this results in currents perceiving a wave-enhanced, apparent bottom roughness. WCBL calculates the maximum combined shear velocity and the time-averaged shear velocity in the wave-current boundary layer using (1) and (2), respectively:

$$U_{*cw} = \left(0.5 f_{cw} U_b^2 \alpha\right)^{0.5} \tag{1}$$

$$\overline{U}_* = \left(0.5 f_{cw} U_b^2 V_2\right)^{0.5}, \tag{2}$$

where U_b is the near-bottom wave-induced velocity; α and V_2 are dimensionless functions that account for the periodic, vectorial combination of U_b and the near-bed current, U_c; and f_{cw} is the wave-current friction factor, which depends on U_b, U_c, and the physical bed roughness, z_0. The apparent bed roughness that currents above the wave-current boundary layer perceive is calculated using

$$z_{0cw} = z_0 \left(\delta_{cw}/z_0\right)^{\left(1 - \overline{U}_*/U_{*cw}\right)}, \tag{3}$$

where δ_{cw}, the height of the combined boundary layer, is calculated from U_{*cw} and the peak radian wave frequency. This leads to the bottom drag coefficient used in CH3D-SED:

$$C_{bcw} = \left[\kappa/\ln\left(z_r/z_{0cw}\right)\right]^2, \tag{4}$$

where κ ($= 0.4$) is the von Karman constant, and z_r is the height of the lowest half-sigma layer in CH3D-SED.

When the combined bed shear stress is beyond a critical value, sediment transport will result, including bed ripples and/or sheet flow. This mobile bed behavior increases the physical bottom roughness. COMAPS calculates bed ripple and sheet flow roughness terms using the algorithms of Grant and Madsen (1982) and Nielsen (1992), respectively. WCBL uses the theory of Glenn and Grant (1987) to account for sediment-induced stratification.

This occurs if sufficient sediment is suspended to cause significant vertical gradients in the density of the water/sediment mix. In such cases, the stable stratification will partially damp out vertical turbulent mixing in the wave-current boundary layer.

At every horizontal grid point, the WAM inputs to WCBL are near-bed wave orbital velocity, near-bed wave excursion amplitude, and mean wave direction. The CH3D-SED inputs to WCBL are horizontal current components at the lowest half-sigma level, the elevation of the lowest half-sigma level, and sediment class sizes, densities, and bed composition fractions. WCBL then provides WAM with a bottom friction factor and provides CH3D-SED with apparent bottom roughness, skin-friction bed shear stress for each sediment class and reference concentration for each sediment class. CH3D-SED calls WCBL as a subroutine. This permits the domain-decomposition strategy used in CH3D-SED to be easily passed onto WCBL. Inter-model communications use MPI function calls.

ADRIATIC SEA HINDCASTS

COMAPS has been tested using an Adriatic Sea computational grid. A contour plot of the bathymetry is shown in figure 1. A logarithmic scale has been used to emphasize the shallowness of the northern part of the sea. Grid cells are 5-minutes longitude by 5-minutes latitude. The overall grid size is 97 cells by 73 cells. The grid was used directly in WAM and converted to the equivalent Cartesian coordinate grid for use in CH3D-SED. The WAM deployment uses 25 frequency bins and 24 direction bins. The CH3D-SED vertical grid uses 20 regularly spaced sigma layers. WAM and CH3D-SED each used computational time-steps of 2 minutes. One day of model time required approximately 1 hour of wallclock time using 12 CPU's (8 for WAM and 4 for CH3D-SED/WCBL) on the U.S. Army Engineer Research and Development Center's SGI Origin 3000.

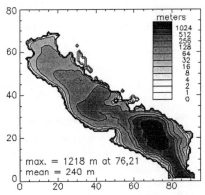

Figure 1. Depth contours and computational cell numbers for the Adriatic Sea grid

Hindcasts were performed for the period 2/1/99, 0 UTC to 2/8/99, 0 UTC. Wind stress and heat flux fields were provided by the Mediterranean Sea deployment of the U.S. Navy COAMPS atmospheric circulation model. The hindcast period was selected due to strong wind and wave activity. Wind stresses were highest during the first 36 hours of the period and tended to be largest in the northern part of the Adriatic. Throughout the period, wind directions were typically northeasterly.

Quiescent initial conditions were used for WAM and CH3D. In CH3D, initial temperature and salinity values were set to 4°C and 35 ppt. Three sediment size classes were specified in SED, corresponding to typical fine sand (150 microns), coarse silt (50 microns), and coarse clay (4 microns) grains. Initial suspended sediment concentrations were set to 15 ppm (mass fraction). The initial distribution of sediment sizes in the SED active and sub-stratum layers was set to 0.34:0.33:0.33 for the sand, silt, and clay classes, respectively. At the open boundary at the southern end of the grid, a semi-diurnal tidal boundary condition was imposed, with an amplitude of 1 m. The temperature, salinity, and concentration boundary conditions were fixed at 4°C, 35 ppt, and 15 ppm, respectively.

Hindcasts were performed for no model coupling ("0-way"); two-way atmospheric boundary layer coupling only, with WCBL switched off ("2-wayA"); and two-way atmospheric and bottom boundary layer coupling ("2-wayB"). Based on initial test runs, a coupling frequency of 10 minutes was selected. Comparisons of the predictions from the three sets of hindcasts are given below. Results are shown for hour-36 of the hindcasts. Maximum storm activity occurred around hour-12, but hour-36 offers more time for the development of basin-scale effects. The maximum wind stresses at hours 12 and 36 convert to 10-meter wind speeds of approximately 30 m/s and 20 m/s, respectively.

Figure 2 shows significant wave height contours and vectors at hour-36 of the 0-way hindcast. The vector lengths are scaled with significant wave height and are aligned with mean wave direction. The wave vectors are closely aligned with the wind stresses and wave heights are largest downwind of the highest wind stress regions. This illustrates fetch-limited growth, with wave heights then falling in shallow water as bottom friction increases.

Figure 3 shows significant wave heights at hour-36 of the 2-wayA hindcast. The contour pattern is similar to figure 2, but the basin-wide maximum height has been reduced by 5% and local maxima in the southern part of the sea have also fallen. This indicates the impact of the atmospheric boundary layer (ABL) couplings. The dominant coupling mechanism in this case is the use of effective wind vectors. Wave propagation by the surface currents (refer to figure 5) reduces the wind velocity relative to individual waves, leading to lower wave growth rates.

Significant wave heights at hour-36 of the 2-wayB hindcast (not shown) are almost identical to figure 3 with one important exception. The maximum wave height is unchanged, but there has been an increase of around 25% in wave heights in the downwind shallow water regions in the northern quarter of the sea. This was caused by the use of WCBL friction factors in the calculation of WAM bottom friction coefficients. The standard WAM bottom friction coefficient is kept constant at 0.0077 m/s. Further investigation shows that the ratio of WCBL and WAM bottom friction coefficients at hour-36 is very small in deep water and varies from approximately 0.1 to 0.6 in the shallow, northern part of the sea. The spatial variation of the WCBL bottom friction coefficient is reasonable, based on the consideration of water depths and the onset of mobile bed bottom roughness terms, but the magnitudes of the coefficient require evaluation using field data. The relatively large WAM coefficient may have been selected to compensate for the omission of depth-limited wave breaking in the model.

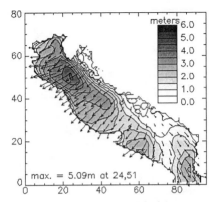

Figure 2. Significant wave height at hour-36 of the no-coupling hindcast

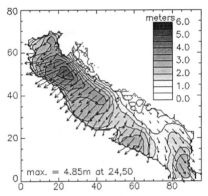

Figure 3. Significant wave height at hour-36 of the two-way coupling (WCBL off) hindcast

Figure 4 shows water elevation contours and surface current vectors at hour-36 of the 0-way hindcast. The scale of the vector lengths is shown in the figure. Momentum input from wind stresses has led to a two-gyre surface circulation pattern, with a counter-clockwise gyre in the north and a clockwise gyre in the south. The northern gyre is the stronger due to the higher wind stresses in that region. The currents in the high-stress region are aligned to the right of the wind vectors due to the Coriolis force. A return flow along the western shore completes the northern gyre. The circulation pattern has caused a 25 cm storm surge at the northwestern end of the sea.

The circulation pattern at hour-36 of the 2-wayA hindcast (not shown) is similar to figure 4, but the storm surge and surface current magnitudes in the northern part of the sea have increased by approximately 20%. The maximum surface current is 46.1 cm/s at grid point (26,32). The increases are due to a combination of the enhanced surface drag coefficient and radiation stress momentum transfer coupling mechanisms. Both mechanisms have a significant effect and the maximum current and water elevation modifications coincide with the largest waves in figure 3, as one would expect.

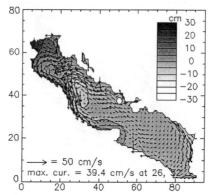

Figure 4. Water elevation contours and surface current vectors at hour-36 of the no-coupling hindcast

Figure 5 shows water elevations and surface currents at hour-36 of the 2-wayB hindcast. The use of WCBL reduces the storm surge by approximately 50% of the 0-way surge. Related reductions in surface current are on the order of 30%. The current reductions are localized in the northern part of the sea, however, in water depths less than 40 m (refer to figure 1), and the maximum current actually shows a small increase from hindcast 2-wayA. The surge and current modifications in figure 5 are due to the increased bottom roughnesses supplied by WCBL. Within the 40 m depth contour, the near-bottom wave motion is significant, and the apparent roughness of the wave-current boundary layer becomes much greater than the physical bed roughness. The resulting bottom drag coefficients are much larger than the constant value of 0.0025 used in the standard CH3D-SED model, leading to increased energy dissipation and reduced water transport.

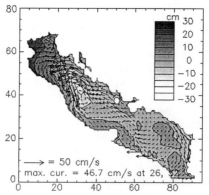

Figure 5. Water elevation contours and surface current vectors at hour-36 of the two-way coupling (WCBL on) hindcast

Figure 6 shows logarithmically scaled contours of the sand size class concentration at the lowest sigma layer at hour-36 of the 0-way hindcast. There is a limited region of low concentration that coincides with the relatively large return flow currents of the northern gyre in figure 4. Figure 4 shows surface currents, but the near-bed currents in the return flow region are also relatively high.

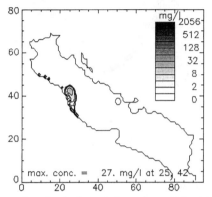

Figure 6. Sand concentrations in the lowest sigma layer at hour-36 of the no-coupling hindcast

The areal extent of sand suspension at hour-36 of the 2-wayA hindcast (not shown) is similar to figure 6, but concentrations have approximately doubled. The maximum concentration in the lowest sigma layer is 57 mg/l at grid point (25,42). The enhanced suspension is a result of the increased surface momentum input to the circulation field when the surface boundary layer couplings are switched on. This results in a general increase in surface, mid-depth, and near-bed currents.

Figure 7 shows sand concentrations for the 2-wayB hindcast. The use of WCBL greatly increases both the areal extent and magnitude of sand suspension. Widespread suspension is now predicted in the shallow northwestern part of the sea. This is quite reasonable since large waves are predicted in that region and wave forcing is known to be the dominant mechanism of sediment transport in shallow water. The standard CH3D-SED model cannot account for this, as indicated by the sediment distribution in figure 6. The 1 mg/l contour in figure 7 is similar to the 32 m depth contour in figure 1. This suggests that WCBL may have a localized effect on sediment concentrations, but concentration plots for the silt and clay size classes at upper sigma layers show that the finer particles are advected over significant distances following suspension, indicating basin-scale effects on sediment tranport. This is confirmed by the rapid removal of silt and clay size classes in shallow depths in hindcast 2-wayB. At hour-36, in depths less than 10 m, the original 33% mass fractions of silt and clay have become virtually zero. This evolution also corresponds to the natural formation of sand beaches. The concentrations in figure 6 in the region of grid point (27,45) become zero in figure 7. This is due to the reduction in CH3D-SED currents when WCBL is used. The water depth at (27,45) is approximately 60 m, meaning that currents, rather than waves, will dominate sediment erosion.

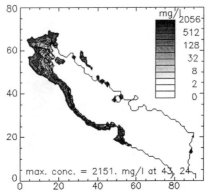

Figure 7. Sand concentrations in the lowest sigma layer at hour-36 of the two-way coupling (WCBL on) hindcast

CONCLUSIONS

This paper has addressed the impact of contemporary wave-current-sediment coupling theories on the numerical prediction of marine conditions. The COupled MArine Prediction System (COMAPS) has been built using parallel-processing versions of the CH3D-SED circulation and sediment transport model, the WAM wind-wave model, and the WCBL bottom boundary layer model. COMAPS includes inter-model coupling physics at both the air-sea and bottom boundary layers. Adriatic Sea hindcasts have been performed using different levels of coupling. This has led to the following conclusions:

- The inclusion of water surface coupling mechanisms between currents and waves increases surface currents and storm surges on the order of 20%, while reducing wave heights on the order of 5%. The increased surface momentum input, combined with vertical mixing, leads to increased near-bed currents, which can approximately double suspended sediment concentrations. The increased momentum input is due to both radiation stress and enhanced surface roughness mechanisms. The dominant cause of wave height effects is the use of the effective wind vector in wind input calculations.

- The inclusion of bottom boundary layer coupling between waves, currents, and sediment reduces coastal currents and water elevations on the order of 30% and increases coastal wave heights on the order of 25%. These effects are caused by the use of spatially and temporally variable WCBL bottom friction coefficients, as opposed to the standard, fixed values in CH3D-SED and WAM. Coastal wave heights increase because the standard bottom friction coefficient in WAM is relatively high.

- Bottom boundary layer coupling has a major effect on sediment suspension and transport. Sediment suspension becomes far more widespread and concentrations can increase by up to two orders of magnitude. The standard CH3D-SED model only considers current-related shear stress. This neglects the fact that waves dominate nearshore sediment transport. The use of WCBL parameters leads to more realistic suspended sediment distributions. In addition, the long-term suspension and advection of the finer sediments results in basin-scale transport effects and realistic sand beach formation.

In summary, this work has shown that the inclusion of coupling mechanisms between regional wind-wave, marine circulation, and sediment transport models can significantly affect predictions of wave heights, currents, water levels, sediment concentrations and net sediment transport. The nature and magnitudes of the coupling impacts in COMAPS appear to be reasonable, but further evaluation and tuning is required using field data. With the availability of high performance computing facilities, the simultaneous execution and coupling of the component models in COMAPS can be achieved in a timely manner.

ACKNOWLEDGEMENTS

This work was supported in part by a grant of HPC time from the DoD HPC Modernization Program.

REFERENCES

Glenn, S.M. and Grant, W.D., 1987. A suspended sediment stratification correction for combined wave and current flows, *Journal of Geophysical Research*, 92: 8244—8264.

Grant, W.D. and Madsen, O.S., 1979. Combined wave and current interaction with a rough bottom, *Journal of Geophysical Research*, 84: 1797-1808.

Grant, W.D. and Madsen, O.S., 1982. Moveable bed roughness in unsteady oscillatory flow, *Journal of Geophysical Research*, 87: 469-481.

Janssen, P.A.E.M., 1991. Quasi-linear theory of wind-wave generation applied to wave forecasting, *Journal of Physical Oceanography*, 21: 1631-1642.

Mastenbroek, C., Burgers, G.J.H., and Janssen, P.A.E.M., 1993. The dynamical coupling of a wave model and a storm surge model through the atmospheric boundary layer, *Journal of Physical Oceanography*, 23: 1856-1866.

Nielsen, P., 1992. Coastal bottom boundary layers and sediment transport, *Advanced Series on Ocean Engineering*, 4, World Scientific, Singapore, pp. 324.

Phillips, O.M., 1977. The Sea Surface, in *Modelling and Prediction of the Upper Layers of the Ocean*, pp. 229-237, Pergamon Press, Elmsford, NY.

Signell, R.P., Beardsley, R.C., Graber, H.C., and Capotondi, A., 1990. Effect of wave-current interaction on wind-driven circulation in narrow, shallow embayments, *Journal of Geophysical Research*, 95: 9671-9678.

Spasojevic, M. and Holly, M.J., 1994. Three dimensional numerical simulation of mobile-bed hydrodynamics, *Technical Report HL-94-2*, U.S. Army Engineer Waterways Experiment Station, Vicksburg, MS.

Tolman, H.L., 1991. Effects of tides and storm surges on North Sea wind waves, *Journal of Physical Oceanography*, 21: 766-781.

WAMDI, 1988. The WAM model – a third generation ocean wave prediction model, *Journal of Physical Oceanography*, 18: 1775-1810.

Welsh, D.J.S., Bedford, K.W., Wang, R., and Sadayappan, P., 2000. A parallel-processing coupled wave/current/sediment transport model, *Technical Report ERDC MSRC/PET TR/00-20*, U.S. Army Engineer Research and Development Center, Vicksburg, MS.

Xie, L., Wu, K., Pietrafesa, L., and Zhang, C., 2001. A numerical study of wave-current interaction through surface and bottom stresses, *Journal of Geophysical Research*, 106: 16841-16855.

TIME-DEPENDENT SEDIMENT SUSPENSION AND TRANSPORT UNDER IRREGULAR BREAKING WAVES

Nobuhisa Kobayashi[1] and Yukiko Tega[2]

Abstract: The time-dependent cross-shore sediment transport model CBREAK is compared with the irregular wave experiment on dune profile evolution due to overwash. CBREAK includes sediment suspension by turbulence generated by irregular breaking waves and bottom friction. CBREAK with the bottom friction factor calibrated to reproduce the measured wave overtopping rate is used to predict the irregular wave breaking and overtopping of a dune as well as the sediment suspension and net onshore sediment transport on the overwashed dune. The computed results are qualitatively realistic but CBRAEK in its present form can not predict fairly uniform erosion on the seaward face of the dune.

INTRODUCTION

Existing cross-shore sand transport models can not predict both beach erosion and accretion partly because sand suspension mechanisms in surf and swash zones are poorly understood. Kobayashi and Johnson (2001) developed a time-dependent cross-shore sediment transport model in the surf and swash zones on beaches under the assumptions of alongshore uniformity and normally incident waves. This model called CBREAK is based on the depth-integrated sediment continuity equation which includes sediment suspension by turbulence generated by wave breaking and bottom friction, sediment storage in the entire water column, sediment advection by waves and wave-induced return current, and sediment settling on the movable bottom. CBREAK was compared with three large-scale laboratory tests with accretional, neutral, and erosional beach profile changes under regular waves. CBREAK is compared herein with the small-scale laboratory tests by Kobayashi et al. (1996) who measured irregular wave overtopping and overwash rates as well as beach and dune profile evolutions.

CBREAK is based on the depth-integrated continuity equation for sediment per unit horizontal area expressed as

$$\frac{\partial}{\partial t}(hC) + \frac{\partial}{\partial x}(hCU_s) = S - w_f C \tag{1}$$

where t = time associated with the wave motion; x = cross-shore coordinate, positive onshore; h = instantaneous water depth; C = depth-averaged volumetric sediment concentration; U_s = horizontal sediment velocity; S = upward sediment suspension rate from the bottom; and w_f = sediment fall velocity. The sediment velocity U_s is

---

[1]Professor & Director, Ctr. for Appl. Coast. Res., Univ. of Delaware, Newark, DE 19716
[2]Ph.D. Student, Dept. of Civ. & Envir. Engrg., Univ. of Delaware, Newark, DE 19716

assumed to be given by $U_s = (U - w_f)$ where U = depth-averaged horizontal fluid velocity.

The suspension rate S per unit horizontal area is related to the wave energy dissipation rates using the equation for turbulent kinetic energy

$$S = S_B + S_f \quad ; \quad S_B = \frac{e_B D_B}{\rho g(s-1)h} \quad ; \quad S_f = \frac{e_f D_f}{\rho g(s-1)h} \quad (2)$$

where D_B = energy dissipation rate due to wave breaking; D_f = energy dissipation rate due to bottom friction estimated as $D_f = 0.5\rho f_b |U|^3$ with f_b = bottom friction factor; ρ = fluid density; g = gravitational acceleration; s = sediment specific gravity; e_B = suspension efficiency for D_B of the order of 0.005; and e_f = suspension efficiency for D_f of the order of 0.01.

The depth-integrated continuity, momentum and energy equations for finite-amplitude shallow-water waves (Kobayashi and Wurjanto 1992) are used to predict $h(t, x)$, $U(t, x)$ and $D_B(t, x)$. Use is then made of (2) to obtain $S_B(t, x)$, $S_f(t, x)$ and $S(t, x)$ and (1) to predict $C(t, x)$ as explained by Kobayashi and Johnson (2001). The rate of the bottom elevation change is found using the volume conservation of bottom sediment

$$\frac{\partial Z_b}{\partial t} = \frac{w_f C - S}{1 - n_p} \quad (3)$$

where Z_b = bottom elevation relative to the still water level (SWL); and n_p = bottom sediment porosity.

COMPARISON WITH IRREGULAR WAVE OVERWASH TEST

CBREAK is compared with the dune overwash experiment by Kobayashi et al. (1996) conducted in a wave tank that was 30 m long, 2.44 m wide, and 1.5 m high. Sand was placed in front of a 53 cm high basin used to collect both overtopped water and overwashed sand. The top of the collection basin was above SWL and the experiment was limited to subaerial dunes. The sand was well-sorted and its median diameter was 0.38 mm. The measured specific gravity, porosity and fall velocity were $s = 2.66$, $n_p = 0.41$ and $w_f = 5.29$ cm/s. The incident irregular waves generated by a wave paddle were measured using three wave gauges in still water depth of approximately 44 cm. The duration of each run was 325 s. Beach and dune profiles were measured using an ultrasound profiler. Seven (A to G) tests consisting of 72 runs were conducted for minor to major dune overwash. The incident significant wave height H_s was in the narrow range of 11.8–12.8 cm. The incident wave spectral peak period T_p was in the range of 1.2–2.0 s.

The measured time-averaged wave overtopping and overwash rates were analyzed empirically by Kobayashi et al. (1996). The cross-shore sediment transport rates were analyzed by Tega and Kobayashi (1996) using the measured profile changes and the finite-amplitude shallow-water wave model (Kobayashi and Wurjanto 1992) coupled with two sediment transport formulas. The measured overwashed dune profile evolution was predicted by Tega and Kobayashi (2000) using a time-averaged sediment transport model combined with the empirical formulas for wave overtopping and overwash rates. The time-averaged model is limited to the region below SWL and needs to be extrapolated to the dune crest using the empirical formulas developed using the same experiment.

CBREAK is compared with several representative test runs using the time series of the free surface elevation measured by the most seaward wave gauge located at $x = 0$ as input to CBREAK. The computed results turn out to be similar and the results for run 3 for test E are presented in the following. For test run E3, the significant wave height $H_s = 12.71$ cm and the spectral peak period $T_p = 2.0$ s in the still water depth of 43.2 cm at the seaward boundary $x = 0$ of the computation domain. The bottom friction factor f_b is calibrated to be $f_b = 0.01$ for which the computed time-averaged overtopping rate $Q = 8.43$ cm^2/s in comparison to the measured rate $Q = 8.38$ cm^2/s where the time averaging is performed for the entire run duration of 325 s. The suspension efficiency e_f for bottom friction is assumed to be $e_f = 0.01$ without any calibration. The suspension efficiency e_B for breaking waves is calibrated in the range $e_B = 0.02$–0.001. The computed results presented in the following are based on $e_B = e_f \exp(-h/H_s)$ so that e_B increases from about 0.03 e_f at $x = 0$ to e_f in the region where the instantaneous water depth h is very small.

For the actual computation using (1) and (2), the dimensional variables are normalized by the following representative scales: fluid time $T_p = 2$ s; vertical length $H_s = 12.71$ cm; horizontal velocity $\sqrt{gH_s} = 112$ cm/s; horizontal length $T_p\sqrt{gH_s} = 223$ cm; and reference concentration $C_r = [e_f/(s-1)] = 0.00602$. The normalized equations (1) and (2) depend on the ratio e_B/e_f and the normalized fall velocity $(w_f T_p/H_s) = 0.83$ which is large in this small-scale experiment. The morphological time scale for the profile change based on (3) is $[(1-n_p)T_p/C_r] = 196$ s which is much larger than the wave period $T_p = 2$ s. As a result, the profile $Z_b(x)$ measured before test run E3 is used for the computation of C and S using (1) and (2). Use is then made of (3) to compute the profile after E3. The normalized variables are used hereafter.

Fig. 1 shows the cross-shore variations of h, U, S and C at time $t = 161.125$ (left) and $t = 161.75$ (right). The water depth h is plotted above the bottom elevation Z_b to depict the free surface elevation $(h + Z_b)$ above SWL. The depth-averaged fluid velocity U is positive onshore. The large onshore velocity occurs on the dune crest at the time $(t = 161.125)$ of wave overtopping. The large offshore (negative) velocity occurs at the time $(t = 161.75)$ of wave downrush on the seaward slope of the dune. The sediment suspension rate S becomes large at the steep front of a breaking wave and in the region of the large onshore or offshore velocity. The depth-averaged sediment concentration C is large in the region of large S and small h.

Figs. 2–4 show the temporal variations (left) and spectra (right) of h, U, S and C at $x = 2.81$, 3.23 and 3.39, respectively. The computed time series are plotted for $100 \le t \le 120$. The spectra are computed from the entire time series for $0 \le t \le 162.5$ and the normalized frequency f is unity at the spectral peak of the incident irregular waves at $x = 0$. The spectrum of h in each figure corresponds to the spectrum of the free surface elevation at the same location. Fig. 2 shows that U and h contain appreciable low-frequency components at $x = 2.81$ corresponding to $Z_b = -1.32$ slightly seaward of the small terrace in Fig. 1. The suspension rate S becomes large intermittently when h and U increase rapidly at the steep front of each breaking wave. The depth-averaging concentration C does not respond to S due to sediment storage in the entire water column and is dominated by low frequency components.

Fig. 3 shows the computed uprush (large h and $U > 0$) and downrush (small h and $U < 0$) at the still water shoreline $Z_b = 0$ caused mostly by individual waves where the foreshore slope was approximately 1:4 and very steep. The suspension rate S remains

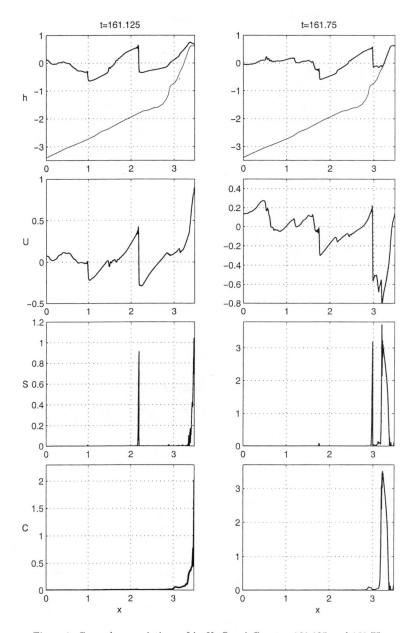

Figure 1: Cross-shore variations of h, U, S and C st $t = 161.125$ and 161.75.

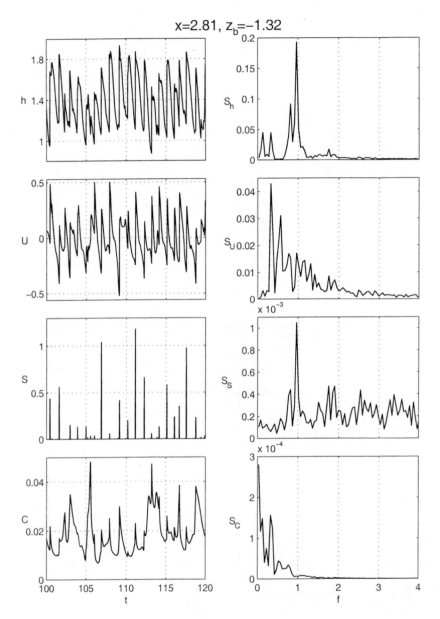

Figure 2: Temporal variations and spectra of h, U, S and C at $x = 2.81$.

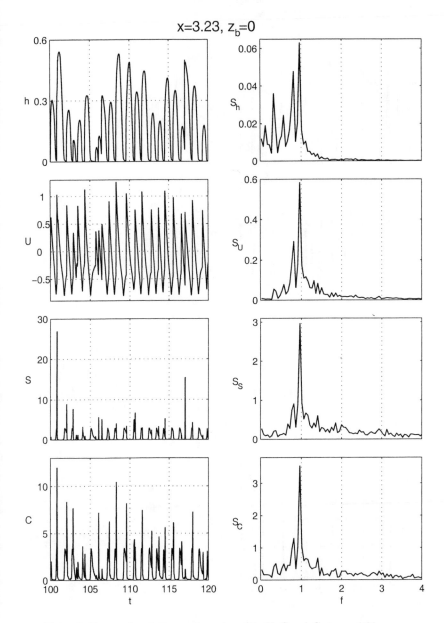

Figure 3: Temporal variations and spectra of h, U, S and C at $x = 3.23$.

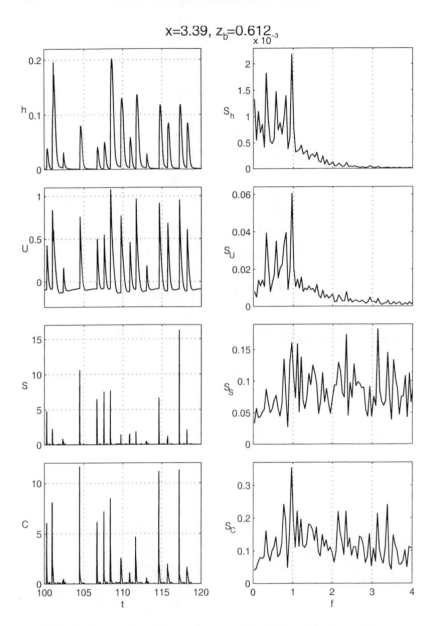

Figure 4: Temporal variations and spectra of h, U, S and C at $x = 3.39$.

large during the downrush of each wave which lasts longer than the corresponding uprush with respect to the velocity U. The concentration C tends to respond to S without delay because of the small water depth h during the wave downrush. Fig. 4 shows the overtopping flow with $h > 0$ and $U > 0$ on the dune crest located at $Z_b = 0.612$. The low frequency components of the spectra S_h and S_u indicate that a group of large waves tend to overtop the dune with the steep foreshore slope. The suspension rate S and the concentration C become large under the steep front of a overtopping bore.

Fig. 5 shows the cross-shore variations of $\overline{C}$, $\overline{S_f}$; $\overline{S_B}$ and $\overline{U}$ where the overbar denotes time averaging for the entire test run. The time-averaged concentration $\overline{C}$ is the maximum near the still water shoreline located at $x = 3.23$. The time-averaged suspension rates $\overline{S_f}$ and $\overline{S_B}$ due to bottom friction and wave breaking are small seaward of the small terrace in the region $x < 2.9$ in Fig. 1. Fig. 5 indicates that $\overline{S_f}$ is dominant above SWL but $\overline{S_B}$ is more important below SWL. The time-averaged velocity $\overline{U}$ is negative due to return current except for $\overline{U} > 0$ on the dune crest due to wave overtopping.

Fig. 6 shows the cross-shore variation of the time-averaged cross-shore sediment transport rate $\overline{hCU_s}$ which is essentially positive (onshore) and the maximum slightly below SWL. The net transport rate can be separated as

$$\overline{hCU_s} = \overline{hC}\ \overline{U_s} + \overline{(hC - \overline{hC})\ (U_s - \overline{U_s})} \quad ; \quad \overline{U_s} = \overline{U} - w_f \qquad (4)$$

The product of the time-averaged sediment volume $\overline{hC}$ per unit horizontal area and the time-averaged sediment velocity $\overline{U_s}$ causes offshore sediment transport due to return current except for the zone affected by wave overtopping of the dune. The oscillatory components $(hC - \overline{hC})$ and $(U_s - \overline{U_s})$ are positively correlated and the time-averaged product of these oscillatory components causes onshore sediment transport. The net onshore sediment transport predicted by CBREAK is consistent with that obtained from the measured profile change for E3 but the measured onshore sediment transport rate increased landward above SWL unlike the solid line shown in Fig. 6.

Fig. 7 shows the cross-shore variations of the time-averaged upward suspension rate $\overline{S}$, the time-averaged downward settling rate $(-w_f\overline{C})$, and the net suspension rate $(\overline{S} - w_f\overline{C})$ which is positive for erosion and negative for accretion. CBREAK predicts erosion below SWL and accretion above SWL where the still water shoreline is located at $x = 3.23$. Fig. 8 shows the comparison between the predicted and measured profiles after E3 relative to the initial profile for E3. The measured profiles indicate fairly uniform erosion on the seaward face of the dune. CBREAK in its present form can not predict erosion above SWL.

CONCLUSIONS

The time-dependent cross-shore sediment transport model CBREAK is compared with the irregular wave overwash experiment. CBREAK with the calibrated bottom friction factor can reproduce the measured time-averaged wave overtopping rate accurately but can not predict the measured erosion on the seaward face of the dune using the suspension efficiencies $e_f = 0.01$ and $e_B = e_f \exp(-h/H_s)$. The detailed measurements of fluid and sediment dynamics will be required to improve the relatively simple formulas for the time-dependent sediment suspension rates in (2).

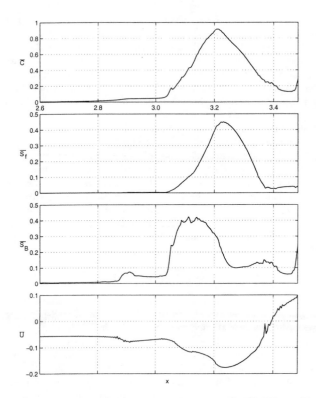

Figure 5: Cross-shore variations of time averaged $\overline{C}$, $\overline{S}_f$, $\overline{S}_B$ and $\overline{U}$.

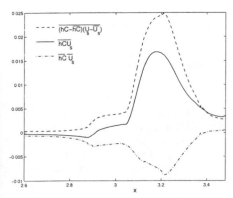

Figure 6: Cross-shore variations of sediment transport rates.

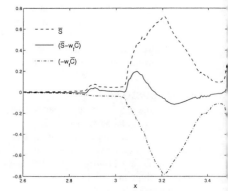

Figure 7: Cross-shore variations of suspension and settling rates.

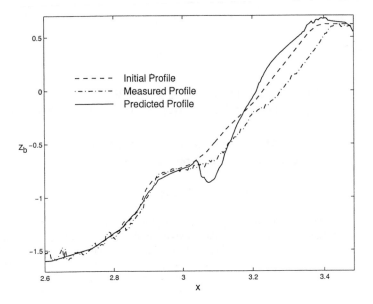

Figure 8: Measured and predicted profiles after test run E3.

ACKNOWLEDGMENTS
This study was supported by the NOAA Office of Sea Grant, Department of Commerce, under Grant No. NA85AA-D-SG033 (Project SG R/OE-29) and the National Science Foundation under Grant OCE-9901471.

REFERENCES
Kobayashi, N. and Johnson, B.D. (2001). "Sand suspension, storage, advection, and settling in surf and swash zones," *J. Geophys. Res.*, 106(C5), 9363–9376.

Kobayashi, N., Tega, Y. and Hancock, M. (1996). "Wave reflection and overwash of dunes." *J. Wtrwy. Port, Coast. and Oc. Engrg.*, ASCE, 122(3), 150–153.

Kobayashi, N. and Wurjanto, A. (1992). "Irregular wave setup and run-up on beaches." *J. Wtrwy. Port, Coast. and Oc. Engrg.*, ASCE, 118(4), 368–386.

Tega, Y. and Kobayashi, N. (1996). "Wave overwash of subaerial dunes." *Proc. 25th Coast. Engrg. Conf.*, ASCE, 4148–4160.

Tega, Y. and Kobayashi, N. (2000). "Dune profile evolution due to overwash." *Proc. 27th Coast. Engrg. Conf.*, ASCE, 2634–2647.

FLUIDIZATION MODEL FOR CROSS-SHORE SEDIMENT TRANSPORT BY WATER WAVES

Mostafa A. Foda & Chi-Ming Huang
University of California, Berkeley CA 94720

Abstract

A new sediment transport model is proposed, which uses linear wave parameters (e.g. wave height H, and period T) to predict wave-induced net bedload transport in a given water depth h, and for given seabed properties and a given seafloor grain-size distribution. Steady-state streaming and net differential shear stress on the bed yields a shear-induced sediment transport in the direction of wave propagation (i.e. onshore). On the other hand, pressure-induced dilatation effect will force sediment transport opposite to wave propagation (i.e. offshore). The net wave-induced sediment transport is the superposition of these shear and pressure effects. It is shown that under large (winter) waves, pressure effect dominates - resulting in a net *offshore* transport near the shore (e.g. sediment erosion from the beach region), and under mild (summer) waves, shear effects dominate - resulting in a net *onshore* transport near the shore (e.g. sediment accretion into the beach region).

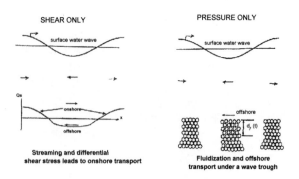

The conceptual model of pressure-and-shear induced sediment transport under a progressive water wave.

INTRODUCTION:

Loading from ocean waves propagating in shallow water will normally set some of the bottom sediment into motion, resulting in well-known and far-reaching consequences. In the nearshore zone, the net sediment transport is known to be dependent on wave conditions. A beach-building shoreward transport is commonly associated with mild incident wave heights, while stormy seas usually drive beach sediment away to the offshore. A purely linear wave is incapable of producing a wave-averaged net transport. Therefore, nonlinear effects such as mass-transport current, wave asymmetry, as well as topology are employed in the formulation of the various coastal sediment transport models (e.g. Baillard and Inman 1980, Grant and Madsen 1976). A recent review of this topic, including the performance of many of the available models vs. field and laboratory data is given by Dean (1995).

In this model study, our new contribution is in the explicit inclusion of wave pressure effect, which is conspicuously absent in previous sediment transport studies. We will demonstrate that wave pressure, and associated compressibility effect, may play a role commensurate to the role of wave shear in forcing sediment motion. Furthermore, the effect of wave pressure on sediment transport is shown to be distinctly different from that of wave shear. Therefore, a wave-induced sediment transport model needs to account for both effects: shear and pressure, *separately and explicitly.*

OUTER PROBLEM *(frictionless fluid):*

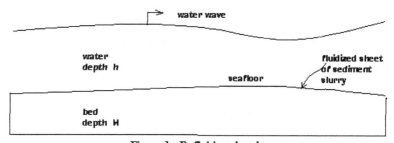

Figure 1: Definition sketch

In addition to sediment movement on the seafloor, the passage of a water wave will strain the underlying bed, as shown in the above sketch. We are particularly interested in the movement (vertical and horizontal) of the seafloor interface, which will provide the needed background for our subsequent sediment transport treatment. We employ the poroelastic solution of Mei & Foda (1981) who demonstrated that for typical wave-seabed conditions, one may ignore fluid viscosity both above and below the seafloor interface and obtain a good approximation of such seafloor movements. Their *"outer solution"* for an inviscid wave, with a water-surface deflection $\xi = A\,e^{\,i(Kx-\omega t)}$, above an undrained poroelastic bed gives the following seafloor vertical deflection η and seafloor horizontal velocity v_o :

$$\eta = -\hat{a}e^{i(Kx-\omega t)}$$

$$v_o = -\hat{v}_o e^{i(Kx-\omega t)}$$

where

$$\hat{a} = \frac{P_o \Gamma}{GK\Lambda}[(1-4\Gamma)(1-e^{-4KH})+4KHe^{-2KH}]$$

$$\hat{v}_o = \frac{P_o \omega}{2GK}\left\{1 - \frac{2\Gamma}{\Lambda}[(1-4\Gamma)(1+e^{-4KH})-2e^{-2KH}]\right\}$$

and

$$\Lambda = (1-4\Gamma)(1+4e^{-4KH}) - [1+4(KH)^2+(1-4\Gamma)^2]e^{-2KH}; \qquad \Gamma = \frac{\lambda+2G-G/(1-m)}{2(\lambda+G)}$$

$$P_o = \rho\, gA\ \mathrm{sec}\, h\ Kh; \qquad \lambda = \frac{2Gv}{1-2v}; \qquad m = \frac{nG/\beta}{1-2v}$$

and where G is the shear modulus of the surface seabed layer of depth H (v is the layer's Poisson's ratio, λ the associated Lame constant and n is the bed porosity), and P_o is the amplitude of wave pressure on the seafloor. The wave frequency ω is related to the wavenumber K through the dispersion relation: $\omega^2 = gK \tanh Kh$. For a very soft (e.g. mud) bed, where its shear velocity is small enough to be comparable to the wave's phase speed, a coupled wave-bed dispersion relation needs to be used instead (e.g. see Foda 1995). As intuitively expected, the seafloor deflection η is 180° out-of-phase with the water surface deflection ξ (the negative sign in the expression for η above. That is because, under a wave crest, water is pressing on the bed, resulting in a trough in seafloor undulation. Consequently, seafloor horizontal velocity is also 180° out-of-phase with near bottom inviscid water velocity. This implies that ignoring seafloor movements will result, among other things, in an *underestimation* of shear on the sea floor!

BOUNDARY-LAYER PROBLEM:

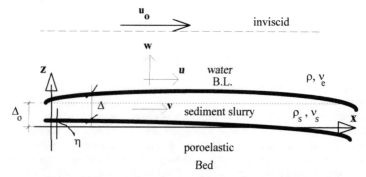

Figure 2: Boundary-layer structure at water-bed interface

Next, we focus attention on the local behavior at and near the water-bed interface (the seafloor). Let us first present our conceptual model of such a behavior. We assume, first, that fluid shear is confined within a thin water boundary layer, characterized by a constant eddy viscosity v_e. We further assume a thin sheet of fully fluidized sediment slurry, moving as a bed load on the seafloor. The thickness Δ of this sheet is allowed to be non-uniform and unsteady, in response to the non-uniform and unsteady wave forcing, but with an average thickness Δ_o. Below the slurry sheet, only partial fluidization is possible, since the hydrodynamic forcing - by definition - is not strong enough to fluidize the entire grain-size population. Immediately below the sheet, all but the largest grains will be assumed in motion. The mixture of fluidized grains in water will now represent the *pore fluid* flowing through the solid matrix made of the largest unfluidized grains. Gradually deeper into the bed more grains will be able to withstand the reduced hydrodynamic forcing and hence remain as part of the solid skeleton. Consequently, the pore-fluid suspension will have less and less solid concentration with increasing depth below the sheet. Deep enough into the bed, the entire grain population will be claimed by the solid skeleton, and the pore fluid will become just pure water.

In order to model the above conceptual behavior, we assume that below the slurry sheet, there is a porolestic bed, with a pore-fluid of variable density and viscosity (to account for the gradual change in suspension concentration). A porolestic boundary-layer may exist (Mei & Foda 1982) where seepage effect (ignored in the above outer solution) could influence pore-pressure behavior immediately below the seafloor. Mei & Foda (1981), however, demonstrated that the seepage flow itself (i.e. the magnitude of pore-fluid flux, *relative to solid, into and out of the seafloor*) is negligibly small compared to the outer-solution $\partial \eta / \partial t$. Therefore, we also ignore seepage flow and assume that η represents both the fluid and solid vertical movement at the seafloor. In the tangential (horizontal) direction, we employ continuity conditions for shear stress and horizontal velocity at the water-slurry interface, and employ a modified Saffman (1971) and Beavers & Joseph (1967) boundary condition for the fluid slip velocity at the slurry-porous bed interface. Saffman (1971) solved the tangential creeping-flow problem for fluid flow through a statistically random porous wall in order to rigorously derive the slip condition (originally proposed by Beaver & Joseph 1967 as an empirical condition). The solution, however, was for a fixed porous boundary, with a viscous fluid of a fixed viscosity. The extension to a moving porous wall is trivial, since inertia is neglected for the creeping seepage flow. Next, we re-derive Saffman's solution after allowing the pore fluid to become a solid-water suspension having a viscosity coefficient $\mu_s = \rho_s v_s$ which varies with z in a general manner. Surprisingly, this extra complication in the formulation can be easily shown to have no effect on the form of the final result, as suggested by Saffman, and still gives the following slip condition at the slurry-porous bed interface:

$$\frac{\partial v}{\partial z} = \frac{\alpha}{\sqrt{k}}(v - v_o) \qquad at \quad z = \eta$$

where v is the slurry horizontal velocity, k is the seafloor permeability coefficient, and α is a non-dimensional parameter dependent on the structure of the porous medium (approximately ≈ 0.01 for medium-size sand). According to our conceptual model, however, the relevant permeability coefficient should be associated with the largest grain-size portion of the grain-size distribution at the seafloor (since the rest is fluidized and is not actually part of the porous solid skelton at the seafloor). Here, we employ the following empirical Kozeny-carmen formula for k in terms of D_{90} grain size, and the porosity n_{90} based on D_{90} population alone:

$$k = 0.8 \, D_{90}^2 \, \frac{n_{90}^3}{(1 - n_{90})^2}$$

This is a substantially larger coefficient than the permeability corresponding to the entire grain-size distribution of the bed sediment.

Lastly, we will assume that the high-concentration slurry sheet will behave as a visco-elastic fluid. Numerous rheology studies of high-concentration suspensions (see, e.g. Davidson et. al. 1985 for review) strongly suggest that the corresponding complex viscosity v_s is a function of the frequency of the motion. It is anticipated that the slurry motion will have an oscillatory component (with a frequency equal to the driving water-wave frequency, plus super-harmonics), and a steady-state, or streaming component. Therefore, each of these components will have a different complex viscosity coefficients: v_{1s} for first-harmonic component, v_{2s} for streaming component, etc ..... (see Davidson et al 1985 for models).

Slurry Velocity

We now proceed to solve for the velocity v of the slurry sheet placed between a water layer above and a partially fluidized porous bed below. Above the slurry sheet, there is a turbulent water boundary layer. The water boundary layer is matched above into the inviscid velocity u_o and pressure p, obtained from the outer solution at the seafloor. Inside the boundary layer, we have the nonlinear boundary-layer equations:

$$\frac{\partial u}{\partial x} + \frac{\partial w}{\partial z} = 0$$

$$\frac{\partial u}{\partial t} + u \frac{\partial u}{\partial x} + w \frac{\partial u}{\partial z} = -\frac{1}{\rho} \frac{\partial p}{\partial x} + v_e \frac{\partial^2 u}{\partial z^2}$$

with the boundary conditions:

$$\frac{\partial u}{\partial z} \to 0 \qquad as \qquad z \to \infty$$

$$u \to u_o = U_o e^{i(Kx - \omega t)} \qquad as \qquad z \to \infty$$

$$u = v \qquad at \qquad z = \eta + \Delta$$

$$\rho v_e \frac{\partial u}{\partial z} = \rho_s v_s \frac{\partial v}{\partial z} \qquad at \qquad z = \eta + \Delta$$

$$\frac{D \eta}{Dt} = w \qquad at \qquad z = \eta$$

$$\frac{\partial v}{\partial z} = \frac{\alpha}{\sqrt{k}} (v - v_o) \qquad at \qquad z = \eta$$

In the above six boundary conditions; the first two guarantee matching with the inviscid wave solution at the bottom, with U_o being the amplitude of the inviscid bottom wave velocity. The next two conditions (at $z = \eta + \Delta$) are statements of continuity of tangential velocity and stress at the water-slurry interface. The last two conditions represent, respectively, the kinematic continuity condition, and Saffman's slip-velocity condition at the slurry-porous bed interface. We may slightly simplify the problem by assuming a small enough slurry-sheet thickness Δ so that slurry velocity v is constant across the sheet. Therefore, we may use the velocity and shear-continuity conditions at the top of the sheet and the slip-velocity condition at the bottom of the sheet to eliminate v and arrive at the equivalent boundary condition on u:

$$\frac{\partial u}{\partial z} = \frac{\alpha}{(\frac{\rho v_e}{\rho_s v_s})\sqrt{k}}(u - v_o) \qquad at \qquad z = \eta$$

The general three-layer problem (for arbitrary thickness Δ) will have to await a future study. Following a conventional perturbation expansion scheme, we solve for the velocity u in the water boundary layer. We assume the following series expansion for u:

$$u = u_1 e^{i(Kx - \omega t)} + \{\bar{u}_2 + u_2 e^{i2(Kx - \omega t)}\} + \ldots\ldots + *$$

where * denotes complex conjugate. The first term in the expansion represents the leading-order solution for u, and is obtained from the linear version of the above boundary-layer problem (by dropping quadratic and higher order terms). It is straightforward to show that the amplitude u_1 of this leading solution is given by

$$u_1 = U_o(1 - ae^{-(1-i)z/\delta})$$

where , $\delta = \sqrt{(2v_e/\omega)}$ is the thickness of the boundary layer, and

$$a = a_r + ia_i = \frac{1 + \hat{v}_o / U_o}{1 + (1 - i)\frac{\rho v_e}{\rho_s v_{1s}}\frac{\sqrt{k}}{\alpha\delta}}$$

with a_r, a_i being the real and imaginary parts of the complex coefficient a. Note that we used v_{1s} as the representative complex viscosity of the visco-elastic slurry for this first-harmonic motion (frequency-dependent viscosity).

At the second order, we arrive at a steady-state, or streaming, velocity $\bar{u}_2$. This streaming velocity does not vanish at the outer edge of the boundary layer, but instead is given by:

$$\bar{u}_{2\infty} = \frac{4K}{\omega}(a_r - 0.25|a|^2)U_o^2 + 4\frac{\hat{a}}{\delta}U_o(a_r + a_i) + \frac{\rho v_e}{\rho_s v_{2s}}\frac{\sqrt{k}}{\alpha\delta}U_o\left[\frac{2K}{\omega}U_o(a_r + a_i) + 8\frac{\hat{a}a_i}{\delta}\right]$$

where v_{2s} is the steady-state viscosity of the slurry, which by definition should be real (Davidson et al 1985). Notice that in the special case of a rigid, non-porous bed, $k = \eta = v_o = 0$, we recover the classical Longuet-Higgins (1953) solution:

$$\bar{u}_{2\infty} = \frac{3K}{\omega}U_o^2$$

which is peculiar in the sense that it is independent of fluid viscosity, even though this streaming velocity owes its existence to viscous effect. Our more general solution shows dependence, not only on viscosity, but also on the bed's permeability and stiffness. A more important result for our purpose here is the slip velocity u_{slip} at the seafloor interace. Again, the Longuet-Higgin's solution yields, by construction, a zero slip velocity. In our solution, the leading-order first-harmonic solution yields, at $z = 0$:

$$u_{1_{slip}} = U_o(1 - a)$$

and the second-order streaming solution yields also at $z = 0$:

$$\bar{u}_{2_{slip}} = 2\frac{\hat{a}}{\delta}U_o(a_r + a_i) + \frac{\rho v_e}{\rho v_{2s}}\frac{\sqrt{k}}{\alpha\delta}\left[\frac{2K}{\omega}U_o^2(a_r + a_i) + 8\frac{\hat{a}a_i}{\delta}U_o\right]$$

By virtue of the velocity-continuity condition, the above will also represent the slurry velocities: both the leading-order oscillatory as well as second-order streaming, respectively.

Thickness of slurry sheet

We employ a Coulomb-type criterion to define the depth Δ (x,t) of the fluidized slurry sheet. For unfluidized grains immediatly below the sheet, we assume that these grains are on the verge of fluidization, and hence satisfy the following Coulomb criterion:

$$\frac{|\tau|}{N} = \gamma$$

where γ is a Coulomb friction coefficient (~ 0.3), N is the normal stress and τ is the shear stress acting on these grains. The normal stress N is acting to press these grains against underlying grains. The situation is represented in an idealized manner in the sketch below. Two elastic spheres of diameter D are pressed against each other by a compressive force F (=N*πD^2 /4). There will be elastic strain, so that the distance between the centers of the spheres is given by $d = D(1 + \varepsilon)$, with the strain ε given by the classical Hertz's (1895) solution

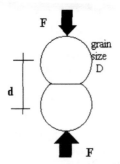

Figure 3: Definition sketch for Hertz Problem.

$$\varepsilon = -0.813\left[\pi\frac{N}{E_{grain}}\right]^{2/3}$$

where, E_{grain} is the Young's modulus of the solid grain itself (e.g. quartz), which is very different from the small elastic modulus of the loose solid skeleton of the porous bed at the sea floor. We may also express this total strain, alternatively, as a superposition of static strain ε_{static} and dynamic strain $\varepsilon_{dyn.}$ The static strain is obtained by employing Hertz solution, again, using the static overburden weight associated with a depth Δ for N.

Recall that, by definition, the weight of a bed-load of moving sediment is transmitted, effectively, to the solid phase of the underlying bed (Bagnold 1956), as opposed to a suspension-load which is supported by the pore pressure. Thus, adopting the buoyant weight of the solid phase of sediment in the fluidized sheet for N we obtain:

$$\varepsilon_{static} = -0.813 \left[\pi(1-n) \frac{(\rho_s - \rho)g\Delta}{E_{grain}} \right]^{2/3}$$

assuming the same porosity (total) as that of the underlying bed. Next, we turn to the dynamic strain. Since these particles are below the fluidized sheet, they belong to the poro-elastic bed, and hence satisfy the generalized Biot's equations (without the restriction of a linear elasticity assumption for the porous solid skeleton). Conservation of mass is expressed in terms of the so-called storage equation (e.g. Mei & Foda 1981):

$$n\nabla(v - u) + \nabla v = -\frac{n}{\beta} \frac{\partial p}{\partial t}$$

where, β is the bulk-modulus of waterfor, and u and v are, respectively, the fluid and solid velocity. For brevity n is used to refer to the partially-fluidized seafloor porosity n_{90}. Again, we follow Mei & Foda (1981) in ignoring seepage flow (u-v), and assume that vertical gradient dominates horizontal gradient for the vertical solid velocity, so that $\nabla v \sim \partial v / \partial z = \partial \varepsilon / \partial t$, where ε is the dynamic normal strain of the solid skeleton in the vertical direction. Therefore, the above storage equation is reduced to this simple 1-D compressional stress-strain relation:

$$\varepsilon_{dyn.} = -\frac{n}{\beta} p$$

where, p is the dynamic pressure (in excess of the hydro-static pressure on the seafloor, i.e. the wave pressure). Therefore, we equate $\varepsilon = \varepsilon_{static} + \varepsilon_{dyn.}$ and employ the Coulomb criterion to substitute for N in the expression for ε, and then Fourier expand $|\tau|$ into wave harmonics, to get the following series for the sheet thickness Δ:

$$\Delta = \Delta_o - s P_o e^{i(Kx-\omega t)} + \ldots\ldots + *$$

where,

$$\Delta_o = \left[2\frac{\rho v_e}{\delta}(a_r + a_i)U_o \right] / \left[\gamma(\rho_s - \rho)g(1-n) \right]$$

$$s = \frac{n}{0.542\beta} \left[\pi(1-n) \frac{(\rho_s - \rho)g}{E_{grain}\sqrt{\Delta_o}} \right]^{-2/3}$$

NET SEDIMENT TRANSPORT:
The instantanous volume flux in the slurry sheet is simply given by $Q = u_{slip} \cdot \Delta$. We substitute the various harmonics, and isolate the leading time-independent components to obtain the leading-order steady-state, or **net** sediment transport $\overline{Q}$:

$$\overline{Q} = \overline{u}_{2_{slip}} \Delta_o - s \ u_{1_{slip}} P_o + \ldots\ldots$$

The first term represents streaming flow of the slurry layer, which is normally in the direction of wave propagation (positive x direction). The second term, which is obtained at the same order-of-magnitude in our perturbation scheme, is the new pressure-induced contribution. In simple terms, compressibility at the seafloor will cause a slight expansion in the granular bed under a wave trough, where pressure is minimum and a slight compression under a wave crest where pressure is maximum (see sketch in the Abstract). Therefore, more sediment grains are fluidized and hence included into the slurry sheet under the trough than under the crest. This causes a first-harmonic perturbation (or more precisely, 180° out-of-phase with wave pressure) to the thickness of the slurry sheet around its average value Δ_o . Combing this perturbation with the oscillatory first-harmonic motion of the slurry will clearly yield a negative net transport, as represented by the second term above. Again, a negative sign means transport in the negative x direction, opposite to the direction of wave propagation.

Summer vs Winter Profiles
Figures 4 and 5 below show sample results of calculation using the obtained formula for net sediment transport under two different incident waves. Reasonable values were assigned to the various physical parameters, corresponding to expected conditions in an open-ocean sandy coastal region.

Bedload Transport

summer wave, T = 6s

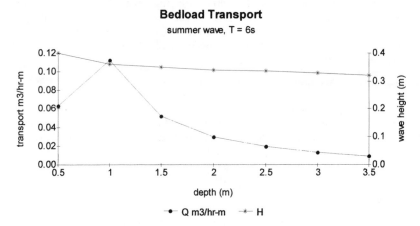

depth (m)

Figure 4: Model calculation showing a net onshore transport under a mild "summer" incident wave, with a maximum transport at depth h ~ 1 m. This implies a tendency to build a mild terrace onshore of that depth.

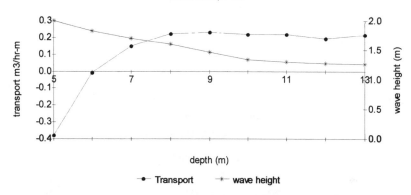

Figure 4: Model calculation showing onshore/offshore transport under a large incident wave. Sediment transport converges toawrds a depth h ~ 6 m. Sediment from shallower and deeper water are being transported to this convergence zone, suggesting a bar-building process. No equivalent convergence takes place in the water flow above.

Acknowlegement: This research was supported in part by grants from the National Science Foundation (CTS-9215889) and the California Department of Boating and Waterways (DBW-99-105-097).

REFERENCES:

Bagnold, R.A. 1956 The flow of cohesionless grains in fluids, *Phil. Trans. Royal Soc., London,* A249, 235-296.

Bailard, J.A. & Inman, D.L. 1980 An energetic bedload model for a plane sloping beach, local transport, *J. Geophys. Res.,* 86(C3), 2035-2043.

Beavers, G.S. & Joseph, D.D. 1967 *J. Fluid Mech.,* 30, 197.

Davidson, J.F., Clift, R. & Harrison, D. 1985 **Fluidization**, 2nd ed., Academic Press, NY.

Dean, R.G. 1995 Cross-Shore Sediment Transport Processes, In *Adv. Coastal & Ocean Eng.,*P. L.-F. Liu, Ed., vol 1, World Scientific, Singapore, 159-220.

Foda, M.A. 1995 Sea Floor Dynamics, In *Adv. Coastal & Ocean Eng.,*P. L.-F. Liu, Ed., vol 1, World Scientific, Singapore, 77-124.

Hertz, H. 1895 Gesammelte Werke, 1, p.155, Liepzig.

Longuet-Higgins, M.S. 1953 Mass transport in water waves, *Phil. Trans. Royal Soc., London,* A245, 535-581.

Madsen, O.S. & Grant, W.D. 1976 Sediment transport in the coastal environment, M.I.T. Ralph M. Parsons Lab. Rep. 209.

Mei, C.C. & Foda, M.A. 1981 Wave-induced response in a fluid-filled poro-elastic solid with a free surface - a boundary-layer theory, *Geophys. J. R. Astr. Soc.,* 66, 597-631.

Saffman, P.G. 1971 On the boundary condition at the surface of a porous medium, *Studies Appl. Math.,* vol 7(2), 93-101.

USING WAVE STATISTICS TO DRIVE A SIMPLE SEDIMENT TRANSPORT MODEL

Barry M. Lesht[1] and Nathan Hawley[2]

Abstract: Because both contaminant and nutrient cycles in the Laurentian Great Lakes depend on particle behavior and movement, sediment transport is a critical component of many of the water quality models being developed to understand and manage this important resource. To avoid complicated models that cannot be supported by the available field data, we have used observation-based, empirical analysis as the basis for developing methods of predicting sediment resuspension from relatively simple measurements of the surface wave field. Our modeling is based on data obtained from instrumented tripods designed to measure near-bottom hydrodynamic and sedimentological conditions for extended periods of time. Because of the long duration of the deployments, it usually is impractical to both sample and record the data at the high frequency that would be needed to resolve the effects of individual surface waves. Instead, we have used a system of burst sampling, in which we sample the sensors at high frequency during a period of time that is repeated at an interval appropriate for the deployment duration. Rather than record the individual samples during the burst, we record only statistics obtained from the individual samples. Our results show that simple representations of the surface wave field obtained from the burst statistics can be used to model sediment transport in wave-dominated environments. We also show that once the model parameters are determined, the forcing wave conditions can be derived from other sources, including wind-driven wave models, with comparable success.

---

[1] Associate Director, Environmental Research Division, Argonne National Laboratory, Argonne, Illinois 60439, barry.lesht@anl.gov
[2] Research Oceanographer, Great Lakes Environmental Research Laboratory, National Oceanic and Atmospheric Administration, Ann Arbor, Michigan 48105, hawley@glerl.noaa.gov

INTRODUCTION

Although wind-driven, large-scale circulations are important in the Great Lakes, sediment transport in the lakes is almost always initiated by surface wave action (Lesht, 1989; Hawley and Lesht, 1995). Sediment transport is included in the detailed water quality models being developed for the lakes via sub-models that can be quite complicated, with high spatial resolution, many sediment layers, several sediment size classes, and various parameterizations describing the space- and time-dependent response of the sediment bed to the imposed hydrodynamic forcing, which usually is computed by other sub-models. See Lick *et al.* (1994), Lou *et al.* (2000), Li and Amos (2001), and Harris and Wiberg (2001) for examples. Though impressive in formulation, these combined sediment transport-hydrodynamic models are generally much more detailed and complex than are the available field data (either hydrodynamic or sedimentological), and therefore the model output cannot easily be compared with, or evaluated against, field observations. This limitation makes it difficult either to quantify the uncertainty associated with the model forecasts or to have confidence in the model calibrations. Furthermore, the output of the high-resolution sediment transport models often must be aggregated spatially to match the much lower resolution of the water quality models. In an alternative approach, we have used observation-based, empirical analysis as the basis for developing simple methods of predicting sediment resuspension from relatively basic measurements of the surface wave field. The purpose of this paper is to describe our method for converting measurements of wave statistics to information that can be used to drive sediment transport models and to demonstrate the application of these models to a recent study of sediment transport in Lake Michigan.

METHODS

Our modeling is based on data obtained from instrumented tripods designed to measure near-bottom hydrodynamic and sedimentological conditions for extended (weeks to months) periods of time. The tripods (Lesht and Hawley, 1987) are equipped to measure horizontal flow velocity, wave conditions, suspended sediment concentration, and water temperature. In the configuration described here, the instruments included a Marsh-McBirney* 512 OEM two-dimensional electromagnetic current meter, two SeaTech* 25-cm-pathlength transmissometers, a Paroscientific* 8130 digital quartz pressure transducer, a solid state temperature sensor, and a compass and tilt sensors to monitor the tripod orientation on the bottom.

Data Sampling

Because of the long duration of the deployments, it usually is impractical to both sample and record data at the high frequency that would be needed to resolve the detailed

---

* Mention of trade names is for information only and does not constitute endorsement of any commercial product by Argonne National Laboratory, the National Oceanic and Atmospheric Administration, or the U.S. Department of Energy.

effects of individual surface waves. Instead, we have used a system of burst sampling, in which we sample the sensors at high frequency for a defined period of time, repeated at an interval appropriate for the deployment duration, and record only burst statistics obtained from the individual samples. These statistics include the means, standard deviations, and minima and maxima for all sensors; the covariance of the pressure deviations with each horizontal component of the horizontal flow; and the number of times the pressure deviations change sign during a burst. For the experiments described here, all the sensors were sampled at 4 Hz, the burst length was 5 minutes, and the bursts occurred every 30 minutes. Thus, rather than recording 1,200 samples per sensor per burst, we record between 4-6 statistics per sensor per burst.

Wave Pressure Analysis

Extracting information about wave processes from the burst statistics requires that we make several assumptions. First, we assume that the fluctuating pressure is Gaussian and stationary within bursts. Thus, we use only the first and second moments to characterize the wave distribution. This assumption is not terribly restrictive, because we are interested in the processes occurring near the bottom, not at the surface. By making our measurements near the bottom, we take advantage of the filtering effect of depth to reduce contributions of higher-frequency components, and the signal that remains tends to be nearly monochromatic. Second, we assume that the 5-minute burst length is sufficient to collect stable statistical values. Our choice of a 5-minute burst results from our desire to minimize power consumption. By using this value, along with a 2-minute warm-up period in each burst, the sensors are energized for only 14 minutes every hour, considerably extending the potential duration of our deployments. Finally, we assume that our near-bottom pressure measurements are sufficiently sensitive to sample the range of wave processes that will have a sedimentological effect on the bottom.   The data acquisition system allows us to measure a pressure change corresponding to about 0.7 mm of water.

Because we use an absolute pressure sensor, each individual pressure sample includes contributions from the atmospheric pressure, from the mean water depth, and from the deviation in water depth due to surface waves. We do not make direct, real-time measurements of atmospheric pressure, but we assume that the contribution from atmospheric pressure will vary slowly relative to the time scale of our measurements and can be removed in post-processing. However, we have found it useful to subtract the contribution of the mean water depth to the total pressure signal in real time to facilitate calculation of the average wave period and of the covariances between the pressure and horizontal velocity fluctuations. We estimate the mean depth by sampling the pressure during the two-minute instrument warm-up period and subtract this value from the pressure measurements made during the five-minute data burst. We also calculate the average total pressure measured during the data burst so that we can compare both the means and variances of the two estimates. The agreement between the two is excellent.

During the more than 12,000 data bursts we have collected in deployments done since 1998, the maximum absolute difference between the average water depths recorded during the warm-up and data sampling periods was 0.13 m (with fewer than 0.2% greater than 0.05 m), the average difference was 0.0003 m, and the standard deviation of the differences was 0.009 m.

Determining the average wave period from the pressure fluctuations is critical for estimating near-bottom wave orbital velocity. We estimate the average wave period, which in a monochromatic field is equivalent to the peak energy period, by dividing the data burst length in seconds by the average number of pressure fluctuation sign changes during the burst.

Current Meter Analysis

We also sample both axes of the current meter at 4 Hz and record burst statistics. For each burst, we record the mean, standard deviation, and range of the individual axis values, the mean speed, the magnitude and direction of the mean velocity vector, the standard deviation of the current direction, and the covariance of the magnitude of each flow component and the pressure fluctuations. These values are sufficient for us to estimate the mean horizontal current speed, the near-bottom wave orbital velocity, and the direction of the waves relative to the mean flow.

We assume that the near-bottom current velocity components $u(t)$ and $v(t)$ consist of steady, or slowly changing relative to the burst length, components U and V, along with fluctuating components $u'(t)$, $v'(t)$. Further assuming that the fluctuating components result from monochromatic wave of frequency ω ($\omega = 2\Pi/T$, where T is the wave period in seconds), traveling in direction Θ relative to the V axis of the current meter and having a maximum near-bottom orbital velocity of R, we have,

$$u'(t) = R\cos\theta\cos(\omega t)$$
$$v'(t) = R\sin\theta\cos(\omega t) .$$

$$(1)$$

Clearly, calculating the averages of $u(t)$ and $v(t)$ over long periods of time relative to the time scale of the fluctuations provides estimates of U and V. The magnitude of the near-bottom orbital velocity (R) is simply obtained from σ_u^2 and σ_v^2, the variances of $u(t)$ and $v(t)$, by

$$R = \sqrt{2\left(\sigma_u^2 + \sigma_v^2\right)} .$$

$$(2)$$

The relationship between the near-bottom pressure fluctuations due to surface waves and the wave orbital velocity may be obtained from linear wave theory (*e.g.*, Kinsman, 1965) and written as

$$R = P\kappa/\rho\omega ,$$

$$(3)$$

in which P is the amplitude of the pressure fluctuation measured at distance z above the bottom, κ is the wave number ($2\Pi/L$ where L is the wave length in m), and ρ is the water density. Taking $\Delta p = P\cos(\omega t)$, we can use σ_p, the measured standard deviation of Δp, to estimate P by $P = \sqrt{2} \bullet \sigma_p$. Thus, with Eqs. 2 and 3, we have two independent measurements of the near-bottom wave orbital velocity obtained from the burst statistics. The agreement between these two independent estimates (Fig. 1) is excellent.

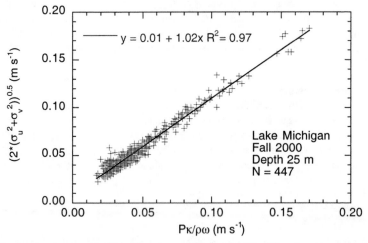

Fig. 1. Current meter [Eq. 2] and pressure sensor [Eq. 3] estimates of wave orbital velocity for fall 2000 experiment. Bursts without waves are not included.

Sediment Resuspension Model

We use a very simple model (Hawley and Lesht, 1992) that relates the suspended sediment concentration near the bottom to the local properties of the sediment and to the hydrodynamic forcing. The model, which includes the upward flux of bottom sediment due to resuspension and the downward flux due to settling, may be written

$$D\frac{dC}{dt} = \alpha \left|\frac{\tau - \tau_c}{\tau_r}\right| - w\left(C - C_{bak}\right) \quad \text{for} \quad \tau > \tau_c$$

and   (4)

$$D\frac{dC}{dt} = w\left(C - C_{bak}\right) \qquad \text{for} \quad \tau \le \tau_c,$$

where D is total water depth, C is the depth-averaged suspended sediment concentration (kg m^{-3}), C_{bak} is a background concentration, τ is the bottom shear stress (Pa), τ_c is a threshold stress value for the initiation of sediment transport, τ_r is a reference stress value used to make the excess stress term dimensionless, w represents the sediment settling velocity (m s^{-1}), and α (kg m^{-3}) represents the rate at which sediment is eroded from the bottom.

Although we have found that it is possible to express the hydrodynamic forcing directly in terms of wave orbital velocity (Lesht and Hawley, 1987), thereby eliminating the problem of estimating the bottom shear stress, we use shear stress as the forcing in the present example. Because we use our observations to estimate the parameter values, the choice of forcing flow parameter is arbitrary so long as it is used consistently in applying the model to different locations.

Field Experiments

The goals of our research are to document the frequency and intensity of sediment transport events, to establish constraints on the output of the detailed sediment transport models, and to provide the basis for developing simple empirical models that relate sediment transport to some easily measured or modeled feature of the flow. We have conducted studies in the Great Lakes using these methods since the mid 1980s (Lesht, 1989; Hawley and Lesht, 1995; Hawley and Murthy, 1995; Lee and Hawley, 1998; Hawley and Lee, 1999). A common result of this research is that although other processes such as coastal upwelling have a role, sediment transport in the Great Lakes is dominated by the effects of wind-driven surface waves. In this paper, we use data collected during the recent Episodic Events – Great Lakes Experiment (EEGLE) program (Eadie et al., 1996) to demonstrate how simple empirical models based on wave forcing can be constructed from field observations.

RESULTS

The basic data obtained from a recent (fall 2000) tripod deployment are shown in Fig. 2. A major sediment resuspension event, the only one during the 48-day deployment, occurred on day 264. At its peak, the near-bottom optical attenuation reached 5.9 m^{-1}, roughly corresponding to a suspended sediment (TSM) concentration of 11 kg m^{-3} (Hawley and Zyren, 1990). This resuspension event was clearly associated with a concurrent increase in near-bottom wave orbital velocity that reached 0.18 m s^{-1}. Although the near-bottom wave orbital velocity exceeded 0.10 m s^{-1} later in the deployment at day 280, there is only a slight increase in attenuation, suggesting that wave-driven local resuspension did not occur at this time. Although unidirectional currents near the bottom also exceeded 0.10 m s^{-1} at times, our goal here is to find a consistent set of model parameters that will allow us to reproduce the near-bottom sediment concentration time series from knowledge of the surface wave conditions alone. Having such a set of model parameters will greatly simplify the process of integrating sediment resuspension and transport into large-scale water quality models.

We determined a set of optimal model parameters by minimizing the differences between the observed and predicted sediment concentration time series through use of

Willmott's (1982) index of agreement and percent unsystematic error statistics as our criteria for evaluating the fit of the model to the data. The simplicity of the model formulation makes it easy to compare the success of different choices of the forcing variable (*e.g.*, bottom shear stress, wave orbital velocity) and to evaluate the variability in model parameters with bottom type.

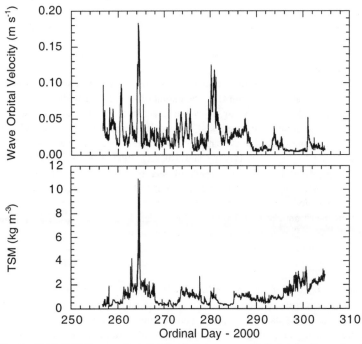

Fig. 2. Fall 2000 time series of near-bottom wave orbital velocity and sediment concentration 0.7 m above the bottom at 25-m depth in southern Lake Michigan.

Figure 3a shows the suspended sediment concentration predicted by using our model (Eq. 4) with optimized model parameters and two different estimates of the wave bottom shear stress: that estimated from the wave statistics measured by the tripod and that estimated from wave properties calculated with a simple wind-driven surface wave model (Schwab *et al.*, 1981). Because biological fouling began to affect the transmissometer late in the experiment, we limited the modeling to the 38-day period between the beginning of the deployment (day 257) and day 296.

DISCUSSION

The calibrated model forced with wave bottom shear stress estimated from the tripod observations did well in reproducing the observed near-bottom sediment concentrations.

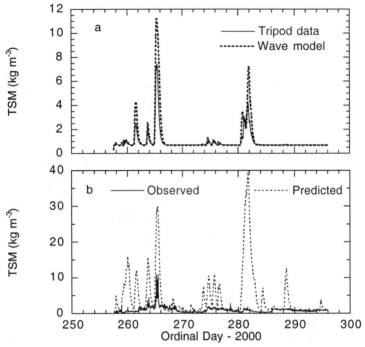

Fig. 3. Time series of suspended sediment concentrations predicted using estimates of (a) wave bottom shear stress made from the tripod data and from the wave model and (b) combined wave-current stress from the tripod data.

The same model parameter values produced a very similar result when the model was forced with shear stress estimated from the wave conditions calculated by using the wind-driven wave model. Both models successfully reproduce the major resuspension event that occurred on day 265, and both over-predict the observed concentration on day 282. The fact that the model results shown in Fig. 3a are so similar indicates that the wind-driven wave model fairly accurately simulates the observed wave conditions. We found that for this deployment, the wave model wave heights tended to be higher than those measured at the tripod, but the model's wave periods were shorter than the observations. These factors tended to offset one another when the wave stress was calculated.

A combined wave-current shear-stress model (Lou and Ridd, 1996) used with the same set of parameter values (Fig. 3b) greatly over-predicted the sediment concentration. The amount of over-prediction depended on the magnitude of the current component, which suggests that either our model assumptions are violated when currents dominate the flow field or that our point measurements of sediment concentration are insufficient to represent the flux of material off the bottom into the flow. Of course, there may also be a problem with the estimated shear stress. In any event, further analysis of this case is required. The degree to which the model results depend on the shear stress calculation is an important point. Because we do not measure shear stress directly, we must rely on values calculated from other measurements, typically current velocities, or, as in the case described here, wave orbital velocities. Although modeling the sediment response in terms of shear stress is theoretically sound, models may suffer from the uncertainty added to the calculation by converting the current or wave orbital velocities to shear stress.

CONCLUSIONS

Our simple sediment resuspension model was very successful in reproducing the major features of the observed sediment concentration when forced with either the wave properties derived from the statistics recorded by the tripod or the wave properties predicted by the wind-driven wave model. Because sediment resuspension in the Great Lakes is primarily wave driven, this result suggests that large-scale modeling of sediment transport in these waters can be simplified by limiting resuspension calculations to shallower regions near the shore and by using a parameterization of resuspension based on modeled wave properties. Further work is needed to understand how best to incorporate combined wave-current flows into the simple model formulation. We also need to better understand the sensitivities of the model parameters and how they vary with sediment type. Given the limitations of sediment transport field observations, we believe that this simple approach provides adequate accuracy and precision for most modeling applications.

ACKNOWLEDGEMENTS

Work at Argonne National Laboratory was supported by the NOAA Coastal Ocean Program though interagency agreement with the U. S. Department of Energy, through contract W-31-109-Eng-38, as part of the the Episodic Events – Great Lakes Experiment (EEGLE). This is NOAA/GLERL contribution No.1212.

REFERENCES
Eadie, B. J., Schwab, D. J., Assel, R. A., Hawley, N., Lansing, M. B., Miller, C. S., Morehead, N. R., Robbins, J. A., Van Hoof, P. L., Leshkevich, G. A., Johengen, T. H., Lavrentyev, P., and Holland, R. E. 1996. Development of a Recurrent Coastal Plume in Lake Michigan Observed for the First Time. *EOS, Transactions of the American Geophysical Union*, 77:337-338.

Harris, C. and Wiberg, P. L. 2001. A Two-Dimensional, Time-Dependent Model of Suspended Sediment Transport and Bed Reworking for Continental Shelves. *Computers and Geosciences*, 27(6):675-690.

Hawley, N. and Lee, C.-H. 1999. Sediment Resuspension and Transport in Lake Michigan During the Unstratified Period. *Sedimentology*, 46:791-805.

Hawley, N. and Lesht, B. M. 1992. Sediment Resuspension in Lake St. Clair. *Limnol and Oceanog.*, 37(8):1720-1737.

Hawley, N. and Lesht, B. M. 1995. Does Local Resuspension Maintain the Benthic Nepheloid Layer in Lake Michigan? *J. Sediment. Res.*, A65:69-76.

Hawley, N. and Murthy, C. R. 1995. The Response of the Benthic Nepheloid Layer to a Downwelling Event. *J. Great Lakes Res.*, 21:641-651.

Hawley, N. and Zyren, J. E. 1990. Transparency calibration for Lake St. Clair and Lake Michigan. *J. Great Lakes Res.*, 16:113-120.

Kinsman, B. 1965. *Wind Waves*, Prentice-Hall, Englewood Cliffs, NJ, 676 pp.

Lee, C.-H. and Hawley, N. 1998. The Response of Suspended Particulate Material to Upwelling and Downwelling Events in Southern Lake Michigan. *J. Sediment. Res.*, 68(5):819-831.

Lesht, B. M. 1989. Climatology of Sediment Transport on Indiana Shoals, Lake Michigan. *J. Great Lakes Res.*, 15:486-497.

Lesht, B. M. and Hawley, N. 1987. Near-Bottom Currents and Suspended Sediment Concentration in Southeastern Lake Michigan. *J. Great Lakes Res.*, 13:375-386.

Li, M. Z. and Amos, C. L. 2001. SEDTRANS96: The Upgraded and Better Calibrated Sediment-Transport Model for Continental Shelves. *Computers and Geosciences*, 27(6):619-646.

Lick, W., Lick, J., and Ziegler, C. K. 1994. The Resuspension and Transport of Fine-Grained Sediments in Lake Erie. *J. Great Lakes Res.*, 20(4):599-612.

Lou, J. and Ridd, P. 1996. Wave-Current Bottom Shear Stresses and Sediment Transport in Cleveland Bay, Australia. *Coastal Eng.*, 29:169-186.

Lou, J., Schwab, D. J., Beletsky, D., and Hawley, N. 2000. A Model of Sediment Transport and Dynamics in Southern Lake Michigan. *J. Geophys. Res.*, 105(C3):6591-6610.

Schwab, D. J., Bennett, J. R., and Liu, P. C. 1981. Application of a Simple Numerical Wave Prediction Model to Lake Erie. *J. Geophys. Res.*, 89:3586-3592.

Schwab, D. J., Beletsky, D., and Lou, J. 2000. The 1998 Coastal Turbidity Plume in Lake Michigan. *Estuarine, Coastal, and Shelf Science*, 50:49-58.

Willmott, C. J. 1982. Some Comments on the Evaluation of Model Performance. *Bull. Am. Meteorol. Soc.*, 63:1309-1313.

WAVE BREAKING AND SEDIMENT SUSPENSION IN SURF ZONES

Michael A. Giovannozzi[1], Nobuhisa Kobayashi[2], and Bradley D. Johnson[3]

Abstract: The time-averaged, depth-integrated suspended sediment model based on the sediment suspension and settling rates is combined with the nonlinear time-averaged irregular wave model CSHORE. In this sediment model, the suspension rate is proportional to the wave energy dissipation rate per unit volume of water. Beach profile evolution tests were conducted to calibrate CSHORE and estimate the suspension rates which were found to be much larger than the rates of the bottom elevation change. Equilibrium profile tests were also conducted to verify CSHORE for its capability in predicting the cross-shore variations of the statistics of the free surface elevation and cross-shore velocity from outside the surf zone to the inner surf zone. The sediment model combined with the calibrated CSHORE is shown to be capable at least qualitatively of predicting the cross-shore variation of the measured suspended sediment concentration above the equilibrium profile.

INTRODUCTION

Linear time-averaged models [e.g., Battjes and Stive, 1985] based on the time-averaged momentum and energy equations have been applied successfully to predict the cross-shore variations of the wave setup and root-mean-square wave height in surf zones on beaches. Kobayashi and Johnson [1998] developed the nonlinear time-averaged model CSHORE based on the momentum, energy and continuity equations to predict the cross-shore variations of the wave setup, root-mean-square wave height and return current from outside the surf zone to the swash zone on natural beaches. The numerical model CSHORE was calibrated and compared with data from three laboratory tests on a 1:16 slope and two tests with equilibrium beach profiles [Johnson and Kobayashi, 1998]. CSHORE was verified using an additional five tests on a 1:30 slope and field data collected at Duck, North Carolina [Johnson and Kobayashi, 2000]. Since CSHORE is relatively simple and efficient computationally, an attempt is made here to extend CSHORE to predict the cross-shore variations of the time-averaged sediment suspension rate, cross-shore sediment transport rate and bottom elevation change on the basis of the time-averaged, depth-integrated continuity equation for suspended sediment proposed by Kobayashi et al. [2000].

---

[1] Civil Engr., U.S. Army Corps of Engrs., Phila. Distr., Philadelphia, PA 19107

[2] Professor & Director, Ctr. for Appl. Coast. Res., Univ. of Delaware, Newark, DE 19716, USA

[3] Hydr. Engr., U.S. Army Engrs. Res. & Dev. Ctr., Coast. Hydr. Lab., Vickburg, MS 39180-6199

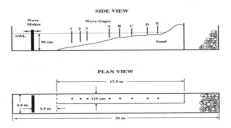

Figure 1: Experimental setup.

TIME-AVEARGED SEDIMENT MODEL

The time-averaged, depth-integrated continuity equation for suspended sediment in surf zones may be expressed as [Kobayashi et al., 2000]

$$\frac{\partial q_s}{\partial x} = S - w_f\, C \quad , \qquad S = \frac{e_B\, D_B}{\rho g(s-1)h} \tag{1}$$

where x is the cross-shore coordinate taken to be positive onshore, q_s is the cross-shore sediment transport rate (positive onshore), S is the upward sediment suspension rate from the bottom per unit horizontal area, w_f is the sediment fall velocity, C is the depth-averaged volumetric sediment concentration, e_B is the suspension efficiency on the order of 0.01 for the turbulence induced by wave breaking, D_B is the wave energy dissipation rate per unit horizontal area due to wave breaking, ρ is the fluid density, g is the gravitational acceleration, s is the specific gravity of the sediment, and h is the mean water depth including wave setup. The overbar denoting time-averaging is omitted in (1) and q_s, S, C, D_B and h are the time-averaged quantities. The effect of bottom friction on the suspension rate S may not be negligible in the swash zone but may be neglected in the surf zone [Kobayashi and Johnson, 2001].

The continuity equation of bottom sediment is expressed as [Kobayashi and Johnson, 2001]

$$(1 - n_p)\, \frac{\partial Z_b}{\partial t} = -\frac{\partial q_s}{\partial x} = w_f\, C - S \tag{2}$$

where Z_b is the bottom elevation taken to be positive upward with $Z_b = 0$ at the still water level (SWL), t is the morphological time for the beach profile evolution, and n_p is the porosity of the bottom sediment. The temporal rate of the bottom elevation change is normally related to the cross-shore gradient of the sediment transport rate in (2) but is expressed here as the difference between the sediment settling rate $w_f C$

and the sediment suspension rate S by use of (1).

To predict the temporal and cross-shore variations of Z_b using (2), it is necessary to estimate C and S as a function of t and x for the specified incident irregular waves and water level. The cross-shore variation of S at a given morphological time is estimated using the relationship in (1) where D_B and h are estimated using CSHORE for Z_b at this given time. This assumption is appropriate if the beach profile changes very little during the estimation of D_B, h and S. An additional relationship for C is required to obtain Z_b and q_s using (2). Kobayashi et al. [2000] expressed C in terms of S and S_e with S_e being the value of S for an equilibrium profile and obtained a semi-analytical solution for beach profile evolution. The analytical solution predicted both shoreline accretion and erosion but was not accurate enough to predict detailed features. An experiment was conducted in this study to examine the temporal and cross-shore variations of S and C.

PROFILE EVOLUTION TESTS

An experiment was conducted in a wave tank that was 30 m long, 2.4 m wide, and 1.5 m high as shown in Figure 1. The water depth in the tank was 90 cm. Irregular waves based on the TMA spectrum were generated with a piston-type wave paddle. A rock slope was located at the other end of the tank to absorb waves. A divider wall was constructed along the centerline in the tank to reduce the volume of sand required for the experiment. A fine sand beach was built in the 115 cm-wide flume. The sand was well-sorted and its median diameter was 0.18 mm. The measured fall velocity, specific gravity, and porosity of the sand were 2.0 cm/s, 2.6 and 0.4, respectively.

The sand beach was exposed to the incident irregular waves generated in a burst of 900 s to reduce the generation of seiche in the wave tank. Beach profiles were measured along three cross-shore transects using a vernier pointer in the swash zone and a ultrasonic depth gauge in deeper water. The measured profiles varied very little alongshore and the average profile is used in the following. Figure 2 shows the beach profiles measured at time $t = 0$ (initial), 1, 7, 11.5, 18 and 50.5 hr where this time t is the duration of the wave action generated in bursts. Erosion occurred near the shoreline below SWL. The profile at $t = 50.5$ hr was quasi-equilibrium with the bottom profile changes less than 1 cm/hr. The horizontal distance in this figure and other figures in this section is the onshore distance from the entrance of the 115-cm wide flume in Figure 1.

Eight capacitance wave gauges as shown in Figure 1 were used to measure the free surface elevation at a sampling rate of 20 Hz at eight cross-shore locations immediately after each of the six profile measurements. The eight gauges were moved somewhat as the profile evolved. These six sets of the wave and profile data are simply called tests P1–P6 in this section. The free surface measurements were made in a single burst lasting for 900 s. The initial transitional waves of 60 s were removed from the measured time series. The free surface data at wave gauges 1–3 in Figure 1 were used to obtain the incident and reflected wave spectra. The spectral peak period T_p was approximately 4.8 s. The wave reflection coefficient R based on the zero-moments of the incident and reflected wave spectra increased from 0.24 for test P1 to 0.36 for test P6 with the increase of the foreshore slope caused by the erosion below SWL as shown in Figure 2. The mean, $\bar{\eta}$, standard deviation, σ_η, and skewness, $Skew$, of the free surface elevation η were also calculated for each time series of 840 s sampled at

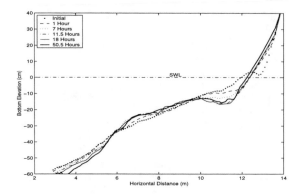

Figure 2: Measured beach profiles.

20 Hz. The root-mean-square wave height H_{rms} is defined as $H_{rms} = \sqrt{8}\, \sigma_\eta$. Table 1 lists the still water depth d and the measured values of T_p, R, $\bar{\eta}$ and H_{rms} for the six tests conducted at the given time t. The wave conditions varied somewhat partly because of the profile evolution and partly because of the repair of the wave paddle during this experiment.

Table 1: Wave Statistics at Wave Gauge 1 for Six Tests

Test	t, hr	d, cm	T_p, s	R	$\bar{\eta}$, cm	H_{rms}, cm
P1	0	88.5	4.7	0.24	-0.3	12.4
P2	1	88.5	4.8	0.29	-0.2	13.9
P3	7	88.5	4.8	0.34	-0.1	14.2
P4	11.5	88.5	4.8	0.35	-0.1	14.2
P5	18	88.5	4.8	0.34	-0.3	12.8
P6	50.5	68.5	4.4	0.36	-0.2	12.2

To assess the accuracy of CSHORE in predicting the measured irregular wave transformation, comparisons were made of the measured and computed cross-shore variations of $\bar{\eta}$, H_{rms} and $Skew$ for each test as reported by Giovannozzi and Kobayashi [2001]. The values of T_p, $\bar{\eta}$ and H_{rms} at wave gauge 1 were specified as input to CSHORE. To present the six comparisons together, the measured and computed values of H_{rms} are divided by the value of H_{rms} at wave gauge 1 for each test. The normalized wave height H^*_{rms} is unity at wave gauge 1. Figure 3 shows the measured and computed cross-shore variations of $\bar{\eta}$, H^*_{rms} and $Skew$ for the six tests for the six profiles shown in Figure 2. These wave quantities vary noticeably due to the profile changes near the shoreline. The predicted wave height H_{rms} increases landward over the scour hole for tests P3-P6 but the predicted increase may not be realistic.

To improve the agreement in Figure 3, the empirical formulas adopted in CSHORE are calibrated as follows. The local fraction Q of breaking waves introduced by Battjes and Stive [1985] is used to separate the outer zone with $0 \le Q < 1$ and the inner zone with $Q = 1$ for the beach profiles without any bar as shown in Figure 2. In the

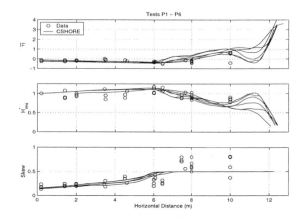

Figure 3: Measured and computed $\overline{\eta}, H^*_{\mathrm{rms}}$ and $Skew$ for six tests.

inner zone, CSHORE employs the empirical formula for the ratio $H_* = H_{\mathrm{rms}}/h$ which increases landward. The empirical parameter γ_s specifies the value of H_* at the still water shoreline. Use is made of $\gamma_s = 1$ instead of $\gamma_s = 2$ to reduce the landward increase of H_{rms} near the shoreline. The nonlinear model CSHORE requires the skewness $Skew$ in the cross-shore radiation stress and wave energy flux. The empirical relationship between $Skew$ and H_* is expressed by three straight lines with $Skew = 1.0$ at $H_* = 0.5$ which is reduced to $Skew = 0.5$ at $H_* = 0.5$ in order to improve the agreement for the skewness in Figure 3.

CSHORE with the calibrated parameters is used to compute the cross-shore variation of the suspension rate S for each test using the relationship between S and D_B in (1) where the suspension efficiency is assumed to be given by $e_B = 0.01$ without any calibration. The suspension rate for the interval from test PJ to test P(J+1) with $J = 1, 2, 3, 4$ and 5 is estimated as the average of the suspension rates at the same cross-shore location computed for the two tests. This average suspension rate S is compared with the left hand side of (2) expressing the net rate of the sand volume deposited on the bottom of unit horizontal area which is approximated by the finite difference involving the measured profiles Z_b for the two tests conducted at the known morphological time as listed in Table 1. Figure 4 shows the cross-shore variations of the average suspension rate and the net deposition (positive) or erosion (negative) rate for the five intervals of the six tests. Figure 4 indicates that the small difference between the relatively large $w_f C$ and S in (2) causes the bottom elevation change and the cross-shore gradient of the sediment transport rate q_s. However, the accuracy of the computed suspension rates is uncertain because the concentration C was not measured in these profile evolution tests.

EQUILIBRIUM PROFILE TESTS

Thirty-seven tests were conducted on the equilibrium profile sketched in Figure 1. The slopes of the foreshore, terrace and offshore zone were approximately 1:6, 1:25, and 1:10, respectively. Each test employed the same irregular waves lasting for 900 s. The offshore wave gauges 1–3 were used to ensure the repeatability of the incident and

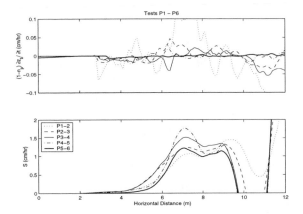

Figure 4: Cross-shore variations of measured net rates and computed suspension rates for five intervals of six tests.

reflected waves as reported by Giovannozzi and Kobayashi [2001]. The wave reflection coefficient based on the zero-moments of the incident and reflected wave spectra was 0.33. The nearshore wave gauges A–E were placed at the cross-shore locations of the velocity and concentration measurements as shown in Figure 5. The velocities and concentrations were measured at the elevations of 1, 2, $\cdots$, n cm above the local bottom where $n = 20$, 7, 4, 3 and 3 at lines A, B, C, D and E, respectively. Each test is identified by the location of the velocity and concentration measurements denoted by the letter A to E followed by the numerical $1 - n$.

For each of the 37 tests, two acoustic-Doppler velocimeters (ADV) were deployed to measure the cross-shore velocity u at the two alongshore locations 13.5 cm from the center line of the 115-cm wide flume. The measured time series of u at the two symmetric locations were in phase and indistinguishable apart from high frequency oscillations associated with vortices and turbulence caused by wave breaking. A fiber optic sediment monitor (FOBS-7) with two sensors was used to measure the sand concentration at the two symmetric locations in the same way as the two locations for u except that the alongshore distance from the center line was 4.5 cm instead of 13.5 cm due to the finite sizes of these intrusive sensors. The measured sand concentrations at the two symmetric locations were not always in phase but intermittent high concentration events occurred simultaneously in both time series.

The time series of the cross-shore velocities and concentrations as well as the time series of the free surface elevations η above SWL at the eight fixed wave gauges were measured for 900 s for each of the 37 tests. The sampling rate for all the time series was 20 Hz. The initial transition of 60 s starting from no wave action initially was removed from each time series for the subsequent statistical analysis. The bottom elevation measured before and after the 37 tests lasting 9.25 hours of wave action changed less than 1 cm. The corresponding rate of the bottom elevation change was less than 0.1 cm/hr.

Figure 6 shows the measured and computed cross-shore variations of $\overline{\eta}$, H_{rms} and

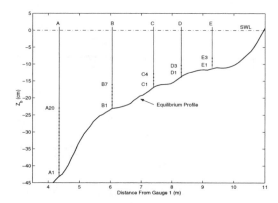

Figure 5: Locations of velocity and concentration measurements.

Skew of η where x is the onshore distance from wave gauge 1 located in the still water depth $d = 80$ cm. The input to CSHORE is the measured values of $\overline{\eta}$ = -0.1 cm, $H_{\mathrm{rms}} = 11.5$ cm and $T_p = 4.8$ s at wave gauge 1. The time series of the free surface elevation η at each of the eight wave gauges were repeatable within approximately 1% differences for the 37 tests. The spectral peak period $T_p = 4.8$ s remained the same except that the secondary peak in the low frequency range became as large as the spectral peak at $T_p = 4.8$ s at gauge E. The mean $\overline{\eta}$ decreased to -0.5 cm (wave set-down) at gauge B ($x = 6.05$ m) and increased to 0.8 cm (wave setup) at gauge E ($x = 9.30$ m). The root-mean-square wave height H_{rms} increased to 13.1 cm at gauge A ($x = 4.35$ m), decreased to 10.6 cm at gauge C ($x = 7.40$ m), and increased slightly at gauge D ($x = 8.30$ m) before the decrease at gauge E. The incident irregular waves did not break at gauge A, were breaking frequently at gauge B, broke intensely sometimes at gauges C and D, and became bores at gauge E. Gauge D was at the transition from breaking waves to bores. On the other hand, the skewness *Skew* of η increased landward except for the decrease at $x = 7.40$ m. CSHORE predicts the cross-shore variations of these wave statistics fairly well without any additional calibration.

The cross-shore velocities u measured by the two velocimeters varied very little vertically at lines A–E in Figure 5. The waves landward of line A were essentially in shallow water. The ratio between the linear wavelength L_p based on $T_p = 4.8$ s and the mean water depth $h = (d + \overline{\eta})$ was 23 at line A where $d = 43.1$ cm and $h = 42.9$ cm at line A. The mean $\overline{u}$ and the standard deviation σ_u of u measured by the two velocimeters at the different elevations are averaged vertically to obtain the averaged values of $\overline{u}$ and σ_u at lines A–E. These values are plotted as a function of x in Figure 7. The cross-shore velocity u is taken to be positive landward and the negative mean current $\overline{u}$ is undertow. The undertow was approximately -7 cm/s at lines C and D where intense wave breaking occurred. The standard deviation σ_u related to the magnitude of the time-varying velocity components increased landward outside the surf zone and was approximately constant at lines B–E inside the surf zone. The computed cross-shore variations of $\overline{u}$ and σ_u are based on the following relationships

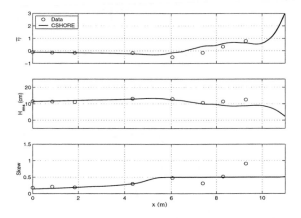

Figure 6: Measured and computed $\bar{\eta}$, H_{rms} and *Skew* for equilibrium profile tests.

derived from linear progressive long-wave theory [Kobayashi et al., 1998]:

$$\bar{u} = -(gh)^{0.5} \left(\frac{\sigma_\eta}{h}\right)^2 , \qquad \sigma_u = \left(\frac{g}{h}\right)^{0.5} \sigma_\eta \qquad (3)$$

where $h = (d + \bar{\eta})$ and $\sigma_\eta = H_{\mathrm{rms}}/\sqrt{8}$ are based on the computed $\bar{\eta}$ and H_{rms} shown in Figure 6. CSHORE with (3) predicts the measured cross-shore variations of $\bar{u}$ and σ_u well although (3) does not account for the additional water volume flux due to rollers.

The sand concentrations measured by the two sensors are expressed in terms of the sand mass in grams divided by the mixture volume in liters. The measured concentrations exhibited intermittent temporal variations in which the instantaneous concentration was intermittently much larger than the mean concentration. The intermittent suspension events are reported separately by Giovannozzi and Kobayashi [2001] and the following analysis is limited to the mean concentration. The mean concentration decreased upward rapidly at line A outside the surf zone and slowly at lines B–E inside the surf zone. Three-dimensional ripples were present in the vicinity and seaward of line A in Figure 5. The height and cross-shore wavelength of these ripples were approximately 2 cm and 11 cm, respectively. The ripples became more two-dimensional and their heights decreased landward. No ripples were visible landward of line B. At line A, the mean concentration was about 3 g/ℓ in the region affected by vortices ejected from the ripples, less than 1 g/ℓ at 4 cm from the bottom and 0.2 g/ℓ at 20 cm from the bottom. At lines B–E, the mean concentration decreased less than 40% from 1 cm to n cm above the bottom where $n = 7$, 4, 3 and 3 at lines B, C, D and E.

The mean concentrations measured by the two sensors at the different elevations above the bottom are averaged vertically to obtain the average mean concentration C at lines A–E. This concentration C is plotted as a function of x in Figure 8. The cross-shore variation of C is consistent with the visually observed intensity of wave breaking which was the maximum at line C and reduced noticeably under bores at line E. The measured average concentration C is not the same as the depth-averaged concentration C used in (1) and (2) because no concentration measurements were made

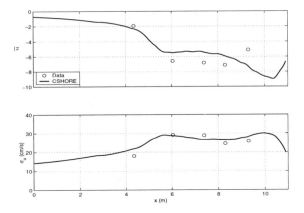

Figure 7: Measured and computed $\overline{u}$ and σ_u for equilibrium profile tests.

in the large area near the free surface as shown in Figure 5 and the sheet-flow layer of high sand concentrations on the bed. Nevertheless, these concentrations are compared in Figure 8. The computed concentration C is based on $C = S/w_f$ obtained from (2) for the equilibrium profile with $\partial Z_b/\partial t = 0$ where the corresponding suspension rate S is computed using CSHORE for the equilibrium profile. The computed volumetric concentration is converted to the concentration in grams per liter where the density of the sand was 2.6 g/cm^3. The measured cross-shore variation of the sand concentration is predicted at least qualitatively by the linear relationship between S and D_B/h in (1) with D_B and h computed by CSHORE. This indicates that the suspended sand concentration is closely related to the wave energy dissipation rate per unit volume due to wave breaking as discussed by Kobayashi et al. [2000].

CONCLUSIONS

The time-averaged, depth-integrated model for suspended sediment proposed by Kobayashi et al. [2000] is combined with the nonlinear time-averaged wave model CSHORE developed by Kobayashi and Johnson [1998]. In this relatively simple sediment model, the temporal change of the bottom elevation change is expressed as the difference between the sediment suspension and settling rates where the suspension rate is proportional to the wave energy dissipation rate per unit volume of water. Profile evolution tests were conducted to estimate the sediment suspension rates in comparison to the rates of the bottom elevation changes. CSHORE is calibrated using the measured cross-shore variations of the wave statistics. The sediment suspension rates are found to be much larger than the rates of the bottom elevation change.

Furthermore, equilibrium profile tests were conducted to verify the calibrated CSHORE and assess the capability of the sediment model in predicting the cross-shore variation of the suspended sediment concentration. The measured cross-shore variations of the wave statistics are predicted fairly well by CSHORE without any additional calibration. CSHORE combined with the simple relationships based on linear long-wave theory is also shown to predict the cross-shore variations of the mean and standard deviation of the measured cross-shore velocity. The cross-shore variation of

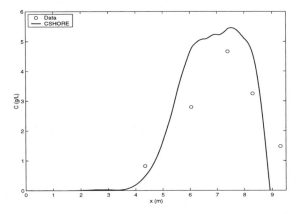

Figure 8: Measured and computed sand concentrations above equilibrium profile.

the measured suspended sediment concentration above the equilibrium profile can be explained at least qualitatively by the sediment model combined with CSHORE probably because of the importance of the irregular wave breaking and energy dissipation for the sediment suspension in the surf zone. However, an additional relationship for C is still required to predict the beach profile evolution using (1) and (2).

ACKNOWLEDGMENTS

This study was supported by the National Science Foundation under Grant OCE-9901471 and by the NOAA Office of Sea Grant, Department of Commerce, under Grant No. NA85AA-D-SG033 (Project SG R/OE-29).

REFERENCES

Battjes, J.A., and M.J.F. Stive, Calibration and verification of a dissipation model for random breaking waves, *J. Geophys. Res.*, 90, 9159–9167, 1985.

Giovannozzi, M.A., and N. Kobayashi, Sediment suspension in surf zones on eroding and equilibrium beaches, *Res. Rep. CACR-01-05*, Ctr. for Appl. Coast. Res., Univ. of Del., Newark, Delaware, 2001.

Johnson, B.D., and N. Kobayashi, Nonlinear time-averaged model in surf and swash zones, *Proc. 26th Coast. Engrg. Conf.*, ASCE, 1998.

Johnson, B.D., and N. Kobayashi, Free surface statistics and probabilities in surf zones on beaches, *Proc. 27th Coast. Engrg. Conf.*, ASCE, 2000.

Kobayashi, N., and B.D. Johnson, Computer program CSHORE for predicting cross-shore transformation of irregular breaking waves, *Res. Rep., CACR-98-04*, Ctr. for Appl. Coast. Res., Univ. of Del., Newark, Delaware, 1998.

Kobayashi, N., and B.D. Johnson, Sand suspension, storage, advection, and settling in surf and swash zones, *J. Geophys. Res.*, 106, 9363–9376, 2001.

Kobayashi, N., E.A. Karjadi, and B.D. Johnson, Cross-shore sand transport on beaches, Proc. 27th Coast. Engrg. Conf., ASCE, 2000.

Kobayashi, N., M.N. Herrman, B.D. Johnson, and M.D. Orzech, Probability distribution of surface elevation in surf and swash zones, *J. Waterw. Port Coastal Ocean Eng.*, 124, 99–107, 1998.

SIMULATION OF SEDIMENT SUSPENSION USING TWO-PHASE APPROACH

Tian-Jian Hsu[1], J. T. Jenkins [2], M.ASCE and Philip L.-F. Liu [3], F.ASCE

Abstract: A dilute sediment transport model based on the two-phase mass and momentum equations is introduced with appropriate closures on the fluid turbulence. Due to the presence of the sediment phase, an important damping mechanism in the fluid turbulent kinetic energy equation is derived and modeled. The proposed model is solve both analytically and numerically to study the sediment transport experiment in a steady uniform open channel flow (Sumer et al., 1996). In the analytical approach, we made additional approximations in order to obtain simple solutions. The analytical solution show clear improvement, which is due to a better modeling on the additional damping mechanism in the fluid turbulent kinetic energy equation, on the calculated concentration profile as compare with the solutions from the Rouse formula (Rouse, 1937). A numerical model, which solves the complete dilute two-phase equations, is also developed. The accuracy of the numerical is checked with the experimental data. With an appropriate closure on the particle stress, the numerical model can be extended to solve the sheet-flow problems in the future.

INTRODUCTION

To study sediment transport in an open channel flow, we present a model based on two-phase equations for a turbulent flow (Drew, 1975, 1983; McTigue 1981). The model is verified by the laboratory measurements of Sumer et al. (1996). Imposing boundary-layer approximations on the proposed dilute two-phase model, we seek for the semi-analytical solutions for fluid turbulence and sediment concentration in the water column. From the semi-analytical solutions, we are able to identify and quantify the contributions of the two-phase model in describing sediment transport. Those contributions are due to a better description of the sediment impact on the fluid turbulence. We demonstrate an clear improvement of the concentration profile as compared with the classic Rouse formula (Rouse, 1937).

---

[1]School of Civil and Environmental Engineering. th40@cornell.edu.

[2]Department of Theoretical and Applied Mechanics. jtj2@cornell.edu

[3]School of Civil and Environmental Engineering, Cornell University, Ithaca, NY 14853, USA. pll3@cornell.edu

To investigate the validity of the boundary-layer approximation employed in the semi-analytical approach. We further developed a numerical model that solves the complete dilute two-phase equations directly. The numerical results agree with the measurements by Sumer et al. (1996) better than the semi-analytical solutions, especially in the region away from the bed. With a further closure efforts on the particle stress, this numerical model can serve as a powerful tool to study sheet flow problem in the near future.

GOVERNING EQUATIONS

Consider a two-dimensional unidirectional free surface flow, driven by gravity, in a channel with an angle of inclination θ and uniform water depth h (figure 1). We consider only the steady state uniform flow in x-direction. Under this condition, the free surface is flat and has the same slope as the channel bed.

Drew (1983) derive the mass and momentum balance of a general two-phase system by ensemble averaging the conservation equations of each phase. In studying sediment transport, to account for the turbulence in the scale of the mean flow variations, we need to further ensemble average the two-phase mass and momentum equations. The Favre-averaging, which is an ensemble average based on concentration-weighted variables, is used in the present model. The resulting continuity equations for the fluid-phase and sediment-phase in the uniform flow condition ($\frac{\partial}{\partial x} = 0$) are

$$\frac{\partial(1 - \bar{c})}{\partial t} + \frac{\partial (1 - \bar{c}) \, \tilde{w}^f}{\partial z} = 0; \tag{1}$$

$$\frac{\partial \bar{c}}{\partial t} + \frac{\partial \bar{c} \tilde{w}^s}{\partial z} = 0 \tag{2}$$

in which x denotes the direction of flow and z is normal to the channel bottom (see Figure 1), $\bar{c}$ is the mean sediment concentration, $\tilde{w}^f$ and $\tilde{w}^s$ are the mean vertical velocity of the fluid phase and sediment phase. The dilute fluid-phase momentum equations in the x and z directions for a uniform flow are:

$$\frac{\partial \rho^f \left(1 - \bar{c}\right) \tilde{u}^f}{\partial t} = -\frac{\partial \rho^f \left(1 - \bar{c}\right) \tilde{u}^f \tilde{w}^f}{\partial z} + \frac{\partial T_{xz}^{fT}}{\partial z} - \rho^f \left(1 - \bar{c}\right) g\sin\theta \\ -\beta\bar{c}(\tilde{u}^f - \tilde{u}^s) \tag{3}$$

and

$$\frac{\partial \rho^f \left(1 - \bar{c}\right) \tilde{w}^f}{\partial t} = -\frac{\partial \rho^f \left(1 - \bar{c}\right) \tilde{w}^f \tilde{w}^f}{\partial z} - (1 - \bar{c})\frac{\partial \bar{P}^f}{\partial z} + \frac{\partial T_{zz}^{fT}}{\partial z} + \rho^f \left(1 - \bar{c}\right) g \\ +\beta\nu_{ft}\frac{\partial \bar{c}}{\partial z} - \beta\bar{c}(\tilde{w}^f - \tilde{w}^s) \tag{4}$$

in which ρ^f and ρ^s are the densities of the fluid and sediment phase, and $\bar{P}^f$ is the mean fluid pressure. The last row of both equation (3) and (4) are the linear drag force term due to the presence of sediment with β being the linear drag coefficient determined by the empirical formula for sedimenting particles (Richardson & Zaki,1954):

$$\beta = \frac{18\mu_f}{d^2(1-\bar{c})^\eta}$$

where μ_f is the dynamic viscosity of the fluid, d is the diameter of the sediment particle and η is a parameter determined from the formula based on the experimental measurements of Richardson & Zaki. The total fluid shear stress T_{xz}^{fT} is defined as the sum of viscous stress τ_{xz}^{fv} and the Reynolds stress τ_{xz}^{ft} due to the Farve-averaging:

$$T_{xz}^{fT} = \tau_{xz}^{fv} + \tau_{xz}^{ft} = \rho^f(\nu_f + \nu_{ft})\frac{\partial \tilde{u}^f}{\partial z} \tag{5}$$

and T_{zz}^{fT} is the corresponding total fluid normal stress:

$$T_{zz}^{fT} = \tau_{zz}^{fv} + \tau_{zz}^{ft} = \frac{4}{3}\rho^f(\nu_f + \nu_{ft})\frac{\partial \tilde{w}^f}{\partial z} - \rho^f\frac{2}{3}(1-\bar{c})k_f. \tag{6}$$

with ν_f the viscosity of the fluid, ν_{ft} the eddy viscosity and k_f the fluid phase turbulent kinetic energy.

The dilute sediment-phase momentum equations in the x and z directions for a uniform flow are

$$\begin{aligned}\frac{\partial \rho^s \bar{c}\tilde{u}^s}{\partial t} &= -\frac{\partial \rho^s \bar{c}\tilde{u}^s\tilde{w}^s}{\partial z} - \rho^s\bar{c}g\sin\theta \\ &\quad +\beta\bar{c}(\tilde{u}^f - \tilde{u}^s)\end{aligned} \tag{7}$$

and

$$\begin{aligned}\frac{\partial \rho^s \bar{c}\tilde{w}^s}{\partial t} &= -\frac{\partial \rho^s \bar{c}\tilde{w}^s\tilde{w}^s}{\partial z} - \bar{c}\frac{\partial \bar{P}^f}{\partial z} + \rho^s\bar{c}g \\ &\quad +\beta\bar{c}(\tilde{w}^f - \tilde{w}^s) - \beta\nu_{ft}\frac{\partial \bar{c}}{\partial z}\end{aligned} \tag{8}$$

The same linear drag force terms also appear in equations (7) and (8) but with an opposite sign. Notice that due to the dilute assumption, we neglect the particle stress terms in the sediment phase momentum equations and solve the sediment transport above the bed load region.

The fluid eddy viscosity is assumed to be calculated from the fluid turbulent kinetic energy k_f and its dissipation rate ϵ_f

$$\nu_{ft} = C_\mu \frac{k_f^2(1-\bar{c})}{\epsilon_f} \tag{9}$$

with C_μ being a numerical coefficient. The k_f and ϵ_f are calculated by their transport equation:

$$\begin{aligned}\frac{\partial \rho^f(1-\bar{c})k_f}{\partial t} &+ \frac{\partial \rho^f(1-\bar{c})k_f\tilde{w}^f}{\partial z} = \tau_{xz}^{ft}\frac{\partial \tilde{u}^f}{\partial z} + \tau_{zz}^{ft}\frac{\partial \tilde{w}^f}{\partial z} \\ &+ \frac{\partial}{\partial z}\left[\left(\nu + \frac{\nu_{ft}}{\sigma_k}\right)\frac{\partial \rho^f(1-\bar{c})k_f}{\partial z}\right] - \rho^f(1-\bar{c})\epsilon_f \\ &- \beta\nu_{ft}\frac{\partial \bar{c}}{\partial z}(\tilde{w}^f - \tilde{w}^s) - 2\beta\bar{c}k_f(1-\alpha),\end{aligned} \tag{10}$$

and

$$\frac{\partial \rho^f (1-\bar{c})\epsilon_f}{\partial t} + \frac{\partial \rho^f (1-\bar{c})\epsilon_f \tilde{w}^f}{\partial z} = C_{\epsilon 1}\frac{\epsilon_f}{k_f}\left(\tau_{xz}^{ft}\frac{\partial \tilde{u}^f}{\partial z} + \tau_{zz}^{ft}\frac{\partial \tilde{w}^f}{\partial z}\right)$$

$$\frac{\partial}{\partial z}\left[(\nu + \frac{\nu_{ft}}{\sigma_\epsilon})\frac{\partial \rho^f (1-\bar{c})\epsilon_f}{\partial z}\right] - C_{\epsilon 2}\frac{\epsilon_f}{k_f}\rho^f(1-\bar{c})\epsilon_f$$

$$-C_{\epsilon 3}\frac{\epsilon_f}{k_f}\left[\beta\nu_{ft}\frac{\partial \bar{c}}{\partial z}(\tilde{w}^f - \tilde{w}^s)\right] - C_{\epsilon 3}\frac{\epsilon_f}{k_f}\left[2\beta\bar{c}k_f(1-\alpha)\right]. \tag{11}$$

The modeled k_f and ϵ_f reduce to the clear fluid $k - \epsilon$ equations as the concentration approaches zero. However, due to the linear drag terms in the fluid phase momentum equations, additional sink (or source) terms in the third row of equation (10) and (11) appear. Specifically, the last term in both equation (10) and (11) represents important damping mechanism of fluid turbulence due to the presence of sediment with α being defined as

$$\alpha = \frac{1}{1 + T_p/T_L}$$

where $T_p = \frac{\rho^s}{\beta_1}$ is the particle response time representing a measure of the time for a single particle to be accelerated from rest to the velocity of the surrounding fluid (Drew, 1976), while the fluid turbulence time scale $T_L = 0.165\frac{k_f}{\epsilon_f}$ is an estimate of the turn-over time of the most energetic eddy (Elghobashi & Abou-Arab, 1983).

Due to the dilute assumption, the coefficients in the above equations are assumed to be the same as those used in the standard $k - \epsilon$ model for a clear fluid:

$$C_\mu = 0.09, C_{\epsilon 1} = 1.44, C_{\epsilon 1} = 1.44, \sigma_k = 1.0, \sigma_\epsilon = 1.3 \tag{12}$$

For the new coefficient $C_{\epsilon 3}$, we use 1.2 as suggested by the research result for a sediment-laden jet (Elghobashi & Abou-Arab, 1983).

We remark here that the uniform flow condition can be reached only if the flow is also steady state. We keep the time derivative terms in the above equations so that a time integration numerical scheme can be developed to find the steady-state solutions.

SEMI-ANALYTICAL APPROACH

Adopting the boundary-layer approximations in the previously proposed dilute two-phase model, we seek for the steady solution analytically. For the steady flow, the time derivative terms in the governing equations, (1) – (8), vanish. We can integrate the steady state continuity equations for both fluid and sediment phase with respect to z to get

$$(1 - \bar{c})\,\tilde{w}^f = \text{constant}, \bar{c}\tilde{w}^s = \text{constant}.$$

Since the free surface is flat and has the same slope with the bed, both $\tilde{w}^f$ and $\tilde{w}^s$ must vanish on the free surface $(z = h)$. Thus

$$\tilde{w}^f = \tilde{w}^s = 0 \tag{13}$$

in the entire water depth.

Using (13), the steady-state sediment phase vertical momentum equation, (8), is reduced to:

$$\bar{c}\frac{1}{\beta}\left[\frac{d\bar{P}^f}{dz} - \rho^s g\right] + \nu_{ft}\frac{d\bar{c}}{dz} = 0 \tag{14}$$

Now we turn our attention to the steady-state vertical momentum equation of the fluid-phase. With (13), equation (4) is reduced to:

$$\frac{d\bar{P}^f}{dz} = \rho^f g + \beta\nu_{ft}\frac{d\bar{c}}{dz}. \tag{15}$$

Notice that, we are interested in the transport of small particles ($d \approx O(0.1 \sim 1.0)mm$) above the bed load region. It is reasonable to assume $T_p \ll T_L$ and neglect the fluid Reynolds stress term in the above equation. Substituting (15) into (14), we obtain

$$\bar{c}\frac{\left(\rho^f - \rho^s\right)g}{\beta} + (1 + \bar{c})\nu_{ft}\frac{d\bar{c}}{dz} = 0. \tag{16}$$

Equation (16) serves as the governing equation describing the sediment concentration distribution, which describe a balance between the upward turbulent flux and the downward flux due to gravity.

Using (13), the convection terms in the horizontal momentum equation of both phases (equation (3) and (7)) vanish. Combining the resulting two equations, we have:

$$\frac{d\tau_{xz}^{ft}}{dz} = \left[\rho^f\left(1 - \bar{c}\right) + \rho^s\bar{c}\right]g\sin\theta \tag{17}$$

where the fluid viscous stress is also neglected. Applying $\bar{c} \ll 1$ in equation (17) and integrating the resulting equation from z to h, we obtain:

$$\tau_{xz}^{ft}(z) = \rho^f u_*^2\left(1 - \frac{z}{h}\right), \tag{18}$$

in which the boundary condition $\tau_{xz}^{ft}(h) = 0$, has been applied. In the above equation, $u_* = \sqrt{gh\sin\theta}$ and it requires $z > \delta_b$, where δ_b is the thickness of the bed load.

To solve equation (16) for the sediment concentration, we need a closure for the fluid eddy viscosity ν_{ft}. We consider a dilute region in the sediment boundary layer where $z \ll h$ (but $z > \delta_b$). Within this region, the fluid velocity profile still follows the logarithmic law. Based on (18), with $z \ll h$, we have:

$$\tau_{xz}^{ft}(z) \approx \rho^f u_*^2 = \rho^f \nu_{ft}\frac{\partial \tilde{u}^f}{\partial z}. \tag{19}$$

In a boundary layer, the fluid turbulent dissipation rate ϵ_f can be represented by the mixing length κz and the turbulence kinetic energy k_f:

$$\epsilon_f = \frac{C_\mu^{3/4}k_f^{3/2}}{\kappa z} \tag{20}$$

Therefore, ν_{ft} in (9) can be rewritten as a function of k_f and κz:

$$\nu_{ft} = C_\mu^{1/4} \kappa z k_f^{1/2} (1 - \bar{c}) \tag{21}$$

where $\kappa = 0.41$ is the von Karman constant. Substituting equation (21) into equation (19), we can rewrite the fluid velocity gradient as:

$$\frac{\partial \tilde{u}^f}{\partial z} = \frac{u_*^2}{C_\mu^{1/4} \kappa z k_f^{1/2} (1 - \bar{c})}. \tag{22}$$

We solve the fluid turbulent kinetic energy k_f using equations (10). Similar to the approximation implemented in clear fluid flow turbulent boundary layer, in the region where the velocity profile follows the logarithmic law, we neglect the convection and diffusion terms. The resulting simplified equation for k_f becomes the balance between the production and dissipations:

$$\tau_{xz}^{ft} \frac{d\tilde{u}^f}{dz} - \rho^f (1 - \bar{c}) \epsilon_f - 2\bar{c} k_f \beta (1 - \alpha) = 0. \tag{23}$$

Substituting equation (19), (20), and (22) into (23), we have:

$$(1 - \bar{c}) \frac{C_\mu^{3/4}}{\kappa z} k_f^2 + 2s\bar{c} \frac{(1 - \alpha)}{T_p} k_f^{3/2} - \frac{u_*^4}{C_\mu^{1/4} \kappa z (1 - \bar{c})} = 0. \tag{24}$$

Notice that without the sediment ($\bar{c} = 0$), equation (24) gives the familiar value for k_f in the steady uniform boundary layer:

$$k_{f0} = \frac{u_*^2}{\sqrt{C_\mu}}. \tag{25}$$

We can solve k_f by finding the positive real root of equation (24) using numerical root-finding tools. However, since $\bar{c} \ll 1$, we expect the dilute sediment can only slightly alter the magnitude of fluid turbulent kinetic energy from k_{f0}. Thus, we assume that the solution of (24) has the following form:

$$k_f = k_{f0} \left(1 - \frac{\Delta k_f}{k_{f0}} \right),$$

where Δk_f is a small deviation of k_f from k_{f0}. Substituting above relationship into equation (24), approximating $\left(1 - \frac{\Delta k_f}{k_{f0}} \right)^2 \approx \left(1 - 2\frac{\Delta k_f}{k_{f0}} \right)$, $\left(1 - \frac{\Delta k_f}{k_{f0}} \right)^{3/2} \approx \left(1 - \frac{3}{2} \frac{\Delta k_f}{k_{f0}} \right)$ and neglecting the higher order terms (e.g., $O(\bar{c} \Delta k_f)$), we can solve Δk_f and k_f becomes:

$$k_f = k_{f0} \left[1 - \bar{c} \left(\frac{s\alpha}{C_u^{3/4}} - 1 \right) \right]. \tag{26}$$

with $s = \rho^s/\rho^f$ being the specific gravity of the sediment. From (21) and (26), the fluid eddy viscosity can be expressed as:

$$\nu_{ft} = \nu_{ft0} (1 - \bar{c}) \left[1 - \frac{\bar{c}}{2} \left(\frac{s\alpha}{C_u^{3/4}} - 1 \right) \right] \tag{27}$$

$$= \nu_{ft0} \left[1 - \frac{\bar{c}}{2} \left(\frac{s\alpha}{C_u^{3/4}} + 1 \right) + O\left(\bar{c}^2 \right) \right], \tag{28}$$

where $\nu_{ft0} = C_\mu^{1/4} \kappa z k_{f0}^{1/2}$ is the fluid eddy viscosity without the presence of sediment. Substituting equation (27) into (16), we obtain an nonlinear ordinary differential equation for the sediment concentration:

$$\frac{d\bar{c}}{dz} = \frac{\bar{c} \frac{(\rho^s - \rho^f)g}{\beta}}{\kappa z u_*} \left[1 + \frac{\bar{c}}{2} \left(\frac{s\alpha}{C_u^{3/4}} - 1 \right) \right]. \tag{29}$$

Given the sediment concentration $\bar{c}_a$ at an appropriate location above the bed (i.e., $z = a$, $\delta_b < a < h$) as the initial condition, equation (29) can be integrated using any numerical package.

Sumer et al. (1996) conducted a series of experiments for studying the sediment transport in open channel flows. The detailed measurements of sediment concentration are reported for sand ($d = 0.13$mm, $s = 2.65$, $\eta = 3.25$) and acrylic ($d = 0.6$mm, $s = 1.13$, $\eta = 2.47$) at various Shields parameters. They are all compared with the present model. Only one case is shown in this paper. Figure 2 presents the results for the case of sand with Shields parameter 1.10. Two different initial conditions are tested; Figure 2(a) shows the results with the initial condition location ($z = a$) at the concentration around 5 percent and solutions in Figure 2(b) are calculated with the initial condition location at the concentration around 10 percent. The corresponding results based on the single-phase approach, i.e., the Rouse formula are also displayed. The two-phase results agree better with the measurements. Moreover, we observe that the two-phase approach results are not very sensitive to the choice of the initial condition location (i.e., the choice of a), while the results from the Rouse formula are much more sensitive. We conclude that the two-phase approach indeed provide an improvement to the Rouse formula.

NUMERICAL APPROACH

A numerical approach is also taken to solve the proposed two-phase equations (1) to (8), along with the $k_f - \epsilon_f$ equations, (10) and (11). In the two-phase equations, the sediment-phase variables are coupled with the fluid-phase variables. Therefore, we solve the sediment-phase continuity and momentum equations at the beginning of the computational cycle. A second-order predictor-corrector scheme is implemented to integrate both the sediment-phase continuity and momentum equation in time. After obtaining $\bar{c}$, $\tilde{u}^s$ and $\tilde{w}^s$ at a new time step from the sediment-phase equations, this information is substituted into the fluid-phase equations. The fluid-phase velocities and pressure are solved by a two-step projection method. The detailed implementation of the boundary conditions for the present numerical model is similar to that of Hsu et al., (2001) and is not discussed here.

In the numerical model, we solve the proposed equations with the flow field and sediment bed initially at rest. After calculating the transient part of the problem for a certain time, we obtain the steady state results. Figure 3 presents the comparison between the steady state concentration profiles from the numerical model and experimental data. Two cases in Sumer et al. (1996) (sand, Shields parameter=1.67, and 2.18) are shown; the initial condition location ($z = a$) corresponds to concentration around 10 percent. The semi-analytical solution from equation (29) is also shown. As expected, near the bed, the numerical result is very close to the solution from (29). Overall, the numerical result indeed agrees better with the experimental measurement since the numerical model involves fewer approximations.

CONCLUSION

From the analysis, we show that the additional damping term in the fluid-phase turbulent kinetic energy equation, plays an important role in modifying the fluid turbulence in the presence of sediment. This additional damping term is derived from the linear drag force due to the interaction between fluid and sediment phase. Solutions obtained from the two-phase model agree well with experimental data by Sumer et al. (1996).

A numerical model that solves the unsteady uniform version of the two-phase mass and momentum equations together with the proposed $k_f - \epsilon_f$ equations is also developed. This model can be used for modeling the sheet-flow problem and two or three-dimensional large-scale sediment transport problem in the coastal region.

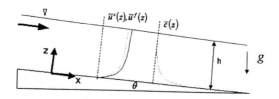

Fig. 1.   A gravity driven channel

REFERENCES

Drew, D. A. 1975 "Turbulent sediment transport over a flat bottom using momentum balance", *J. Applied Mech.*, **42**, 38–44.

Drew, D. A. 1976 "Production and dissipation of energy in the turbulent flow of a particle-fluid mixture, with some results on drag reduction", *J. Applied Mech.*, **43**, 543–547.

Drew, D. A. 1983 "Mathematical modeling of two-phase flow", *Ann. Rev. Fluid Mech.*, **15**, 261–291.

Elghobashi, S. E. & Abou-Arab, T. W. 1983 "A two-equation turbulence model for two-phase flows", *Phys. Fluids*, **26**, No. 4, 931–938.

Hsu, T-J, Jenkins, J.T., Liu, P. L.-F., 2001 "Modeling of Sediment Suspension–A Two-Phase Flow Approach", *Proc. 4th Int. Conf. on Coastal Dynamics.* , Lund, Sweden.

McTigue, D. F. 1981 "Mixture theory for suspended sediment transport", *Proc. ASCE, J. Hydr. Div.*, **107**, No. HY6, 659–673.

Richardson, J.F., and Zaki, W.N. 1954 "Sedimentation and fluidisation", Part 1, *Trans. Instn. Chem. Engrs.*, **32**, 35–53.

Rouse, H. 1937 "Modern conceptions of the mechanics of turbulence", Transactions of the ASCE, **102**, Paper No. 1965, 463–543.

Sumer, B. M., Kozakiewicz, A., Fredsoe, J. & Deigaard, R. 1996 "Velocity and concentration profiles in sheet-flow layer of movable bed", *J. Hydr. Eng.*, ASCE,**122**, NO. 10, 549–558.

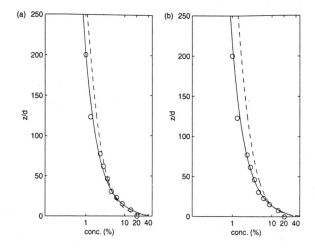

Fig. 2. Comparison of the concentration profile for sand with $\theta = 1.1$. (a) $a = 29d, \bar{c}_a = 4.99\%$, (b) $a = 8d, \bar{c}_a = 10.4\%$. ○: measurement; —: present model; - -: Rouse formula.

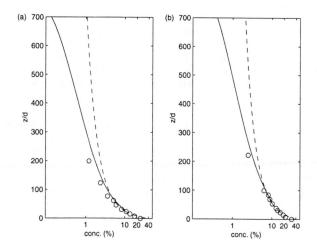

Fig. 3. Comparison of the concentration profile for sand with (a) case#2 $\theta = 1.68$, $a = 25d, \bar{c}_a = 10.24\%$, (b) case#3 $\theta = 2.18$, $a = 57d, \bar{c}_a = 10.03\%$. ○: measurement; —: numerical model; - -: approximated solution.

NUMERICAL PREDICTION OF RIP CURRENTS ON BARRED BEACHES

Okey G. Nwogu[1]

ABSTRACT: The time-dependent evolution of rip currents is investigated in this paper using a numerical model based on Boussinesq-type equations. A numerical wave basin was set up to reproduce the laboratory experiments of Haller *et al.* (1997). The bathymetry consists of a barred beach with two rip channels. Waves propagating over the bar break on the bar while those propagating through the gap break closer to the shoreline. This sets up a spatial variation of the wave-induced excess momentum flux and drives a time-varying circulation pattern. Large-scale horizontal eddies are also generated at the edges of the rip channels and advected offshore. Time histories of the measured velocities have been compared with the Boussinesq model predictions. The numerical model reproduces the overall trend in the unsteady behavior of the rip current but pronounced differences are observed in the flow exiting the rip channel.

1.0 INTRODUCTION

Rip currents are narrow, jet-like currents that flow from the surf zone towards the open ocean. Statistics from the United States Lifesaving Association (USLA) indicate that rip currents are the primary cause of over 80% of swimmer rescues on beaches. Swimmers typically get caught in the currents because they are not easily visible to an untrained eye. Rip currents are also an important conduit for transporting suspended sediments from the surf zone to deeper water.

Rip currents are typically generated when there are strong alongshore variations in the wave breaking pattern. Alongshore variations could be associated with bathymetric effects (e.g. gaps in sandbars, submarine canyons), the presence of structures (piers, jetties, etc) or edge waves (Bowen, 1969). One characteristic feature of rip currents is their unsteady nature. The currents tend to occur episodically (Smith and Largier, 1995) as well as oscillate in the horizontal plane (Haller *et al.*, 1997). The unsteady behavior of rip currents might be due the instability of jet-type flows (Haller and Dalrymple, 2001), temporal variabilities in the wave-breaking pattern (e.g. due to wave groups), low-frequency fluctuations of the mean water level, etc.

A number of numerical models have been developed to predict the mean flow field generated by breaking waves. These models can be loosely classified into models that solve

[1] Research Associate Professor, Davidson Laboratory, Stevens Institute of Technology, Hoboken, NJ 07030, onwogu@stevens-tech.edu

separately for the mean fluid and wave motions, and models that simultaneously solve for the mean fluid and wave motions. The decoupled models are based on depth and time-averaged mass and momentum equations with forcing from radiation stress gradients derived from a phase-averaged wave model (e.g. Haas *et al.*, 1998). Fully coupled models are based on the vertically integrated, time-dependent mass and momentum equations or Boussinesq equations (e.g. Chen *et al.*, 1999). The equations can implicitly describe hydrodynamic processes occurring at different time scales from tidal and infragravity fluid motions to short-period waves.

Different types of Boussinesq equations for weakly dispersive waves in shallow and intermediate water depths have been derived in the literature. Nwogu (1993) derived a set of equations for weakly nonlinear waves by vertically integrating the mass and momentum equations and retaining nonlinear terms up to the order of truncation of the dispersive terms. Wei *et al.* (1995) derived a fully nonlinear set of the equations in which the momentum equations were obtained from the dynamic boundary condition at the free surface. Nwogu (1996) derived a fully nonlinear form of the momentum equations by evaluating the Euler equations at the free surface. The different formulations have slightly different nonlinear convective terms, which are important for simulating the dynamics of weakly rotational flows in the horizontal plane.

Different semi-empirical techniques have also been used to extend the Boussinesq equations to simulate the effect of wave breaking in the surf zone. Schaffer *et al.* (1993) introduced a convective term to the momentum equation to account for the effect of a surface roller. Svendsen *et al.* (1996) modified the 1-D equations to include the effect of the vorticity generated by breaking waves. Nwogu (1996) introduced an eddy viscosity term to the momentum equation to simulate the effect of wave energy dissipation. Different formulations have also been used to characterize the onset of wave breaking. Schaffer *et al.* (1993) used a breaking criterion based on the local slope of the sea surface elevation, while Kennedy *et al.* (2000) used one based on the time-derivative of the sea surface elevation. Nwogu (1996) used the ratio of the horizontal velocity at the free surface to the phase velocity.

In this paper, a time-dependent numerical model based on Boussinesq-type equations is used to investigate the spatial and temporal evolution of rip currents on a barred beach. Numerical model results are compared with data from the laboratory experiments of Haller *et al.* (2000).

2.0 NUMERICAL MODEL

2.1 Governing Equations

The numerical model is based on a fully nonlinear set of Boussinesq-type equations, which represent vertically integrated equations for the conservation of mass and momentum for weakly dispersive waves propagating over water of variable depth. The vertical profile of the flow field is obtained from a second-order Taylor series expansion of the velocity potential about a reference elevation z_α in the water column (Nwogu, 1993):

$$u(x,z,t) = u_\alpha + (z_\alpha - z)[\nabla(u_\alpha \cdot \nabla h) + (\nabla \cdot u_\alpha)\nabla h] + \frac{1}{2}[(z_\alpha + h)^2 - (z + h)^2]\nabla(\nabla \cdot u_\alpha) \quad (1)$$

$$w(x,z,t) = -[u_\alpha \cdot \nabla h + (z + h)\nabla \cdot u_\alpha] \quad (2)$$

where u_α is the horizontal velocity at $z = z_\alpha$ and $h(x)$ is the water depth. The depth-integrated mass conservation equation can be written as:

$$\eta_t + \nabla \cdot u_f = 0 \quad (3)$$

where $\eta(x,t)$ is the free surface elevation and $u_f(x,t)$ is the volume flux density given by:

$$u_f = \int_{-h}^{\eta} u\, dz = (h+\eta)\left\{ u_\alpha + \left[(z_\alpha + h) - \frac{(h+\eta)}{2} \right][\nabla(u_\alpha \cdot \nabla h) + (\nabla \cdot u_\alpha)\nabla h] \right.$$

$$\left. + \left[\frac{(z_\alpha + h)^2}{2} - \frac{(h+\eta)^2}{6} \right]\nabla(\nabla \cdot u_\alpha) \right\} \quad (4)$$

The momentum equations are derived by evaluating the Euler equations at the free surface and are given by:

$$u_{\alpha,t} + g\nabla\eta + (u_\eta \cdot \nabla)u_\eta + w_\eta \nabla w_\eta + (z_\alpha - \eta)[\nabla(u_{\alpha,t} \cdot \nabla h) + (\nabla \cdot u_{\alpha,t})\nabla h]$$

$$+ \frac{1}{2}[(z_\alpha + h)^2 - (h+\eta)^2]\nabla(\nabla \cdot u_{\alpha,t}) - [(u_{\alpha,t} \cdot \nabla h) + (h+\eta)\nabla \cdot u_{\alpha,t}]\nabla\eta$$

$$+ [\nabla(u_{\alpha,t} \cdot \nabla h) + (\nabla \cdot u_{\alpha,t})\nabla h + (z_\alpha + h)\nabla(\nabla \cdot u_\alpha)]z_{\alpha,t}$$

$$+ \frac{1}{h+\eta}\nabla\{v_t(h+\eta)\nabla \cdot u_\alpha\} + \frac{1}{h+\eta}f_w u_b|u_b| = 0 \quad (5)$$

where g is the gravitational acceleration, $u_\eta = u(x,\eta,t)$, $w_\eta = w(x,\eta,t)$ are the horizontal and vertical velocities at the free surface, v_t is the turbulent eddy viscosity used to simulate the effect of wave energy dissipation due to breaking, f_w is the wave friction factor used to simulate effect of wave energy dissipation in the turbulent boundary layer at the seabed, and $u_b = u(x,-h,t)$ is the velocity at the seabed. The elevation of the velocity variable z_α is chosen to minimize differences between the linear dispersion characteristics of the model and the exact dispersion relation for small amplitude waves and is given by $z_\alpha + h = 0.465(h+\eta)$.

The eddy viscosity is determined from the amount of turbulent kinetic energy, k, produced by wave breaking, and a turbulence length scale, l_t, using:

$$v_t = \sqrt{k}\, l_t \quad (6)$$

A one-equation model is used to simulate the production, advection, diffusion, and dissipation of the turbulent kinetic energy produced by wave breaking (Nwogu, 1996):

$$k_t + u_\eta \cdot \nabla k = \sigma\nabla \cdot \nabla(v_t k) + B\frac{l_t^2}{\sqrt{C_D}}\left[\left(\frac{\partial u}{\partial z}\right)^2 + \left(\frac{\partial v}{\partial z}\right)^2 \right]_{z=\eta}^{3/2} - C_D\frac{k^{3/2}}{l_t} \quad (7)$$

where C_D and σ are empirical constants which have been chosen as 0.2 and 0.02 respectively. The waves are assumed to start breaking when the horizontal component of the orbital velocity at the free surface, u_η, exceeds the phase velocity of the waves, C. The parameter B is introduced to ensure that production of turbulence occurs after the waves break, i.e.,

$$B = \begin{cases} 0 & |\boldsymbol{u}_\eta| < C \\ 1 & |\boldsymbol{u}_\eta| \geq C \end{cases} \tag{8}$$

The phase velocity, C, is obtained from the linear dispersion relation of the equations using the average zero-crossing period of the incident wave train. The turbulent length scale, l_t, is a free parameter that is typically chosen to be equal to the incident wave height.

2.2 Numerical Solution

The governing equations have been solved using a time-stepping, finite difference method. The computational domain is discretized as a rectangular grid with the surface elevation η defined at the grid nodes while the velocities u_α, and v_α are defined half a grid point on either side of the elevation grid points. An implicit Crank-Nicolson scheme is used to solve the equations with a predictor-corrector method used to provide the initial estimate. The partial derivatives are approximated using a forward difference scheme for time and central difference schemes for the spatial variables. The numerical solution procedure consists of solving an algebraic expression for η at all grid points and tri-diagonal equations for u_α and v_α along lines in the x and y directions respectively.

Along external or internal wave generation boundaries, time histories of velocities u_α or v_α, and flux densities u_f or v_f corresponding to an incident wave condition are specified. Waves propagating out of the computational domain are absorbed in damping regions placed around the perimeter of the computational domain. Artificial dissipation of wave energy in damping layers is achieved through the introduction of a term proportional to the surface elevation into the right-hand side of the mass equation and a term proportional horizontal velocity into the right-hand side of the momentum equation. Details on the numerical scheme are provided in Nwogu and Demirbilek (2001).

3.0 NUMERICAL SIMULATIONS

Haller *et al.* (2000) carried out laboratory experiments to investigate wave-induced currents on a barred beach with rip channels. The experiments were performed in the Ocean Engineering Basin of the University of Delaware. The basin is 17.2m long, 18.2m wide and is equipped with a directional wavemaker. A 1:30 concrete beach was constructed in the basin. An alongshore parallel bar with two 1.8m wide gaps was placed on the beach as shown in Figure 1. The bar is 1.2m wide and has a crest elevation of 6cm. Ten capacitance-wire wave gauges and three acoustic Doppler velocimeters (ADVs) were used to measure the surface elevation and velocities at various locations in the basin.

Numerical simulations were carried out for Test Series B which corresponds to a normally incident regular wave with height, $H = 4.5$cm, and period, $T = 1$s. The simulations were carried out using grid spacing $\Delta x = \Delta y = 0.06$m and time-step size $\Delta t = 0.015$s. Two parameters that govern the nearshore circulation pattern are the turbulent length scale, l_t, which controls the rate of energy dissipation due to wave breaking, and the wave friction factor, f_w, which controls the rate of energy dissipation in the bottom boundary layer. The turbulent length scale is usually set equal to incident wave height but can be adjusted to match the surf zone wave height distribution. Estimates of the wave friction factor from similar laboratory-scale experiments range from 0.005 to 0.035 (e.g. Cox et al., 1995). The initial simulations were carried out using $l_t = 4.5$cm and $f_w = 0.01$.

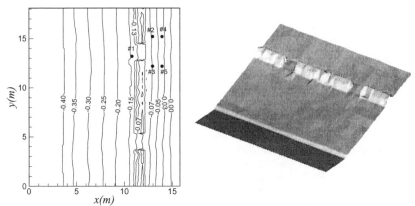

Figure 1. Two and three-dimensional views of the bathymetry for the numerical simulations.

The time-dependent evolution of the wave-induced current and vorticity[2] field is shown in Figures 2(a) to (d). A moving average low-pass filter with a cutoff frequency of 0.1Hz was used to separate the breaking-induced currents from the wave orbital velocities. Figure 2(a) shows the filtered velocity and vorticity distribution 20s into the simulation. Waves propagating over the bar break on the bar while those propagating through the gap break closer to the shoreline. Sharp gradients in wave energy dissipation rates near the gap lead to the formation of two counter-rotating eddies at the edge of the rip channel. At $t = 50s$ (Figure 2(b)), the longshore feeder currents become more prominent and the eddies are advected further seawards. Two secondary circulation cells are also observed near the shoreline. At $t = 70s$ (Figure 2(c)), the flow meanders towards the north wall of the basin. At $t = 120s$ (Figure 2(d)), the current reverses direction and flows towards the center of the basin. The oscillation in the currents is most likely due to the alongshore non-homogeneity in the bathymetry both on the bars on either side of the rip channel and shoreward of the bars. Slight differences in the bathymetry of the order of 1cm can lead to significant differences in wave breaking pattern, given the incident wave height of 4.5cm.

The evolving circulation pattern also affected the incident wave field. To illustrate the effect of wave-current interaction, we performed a zero-crossing analysis on the predicted surface elevation time histories along a transect though the rip channel ($y = 13.34m$). The average zero-crossing wave height and mean current distribution are plotted in Figure 3 for two different time periods, $t = 15$ to 25s and $t = 45$ to 55s. Early in the simulation ($t = 15$ to 25s), the breaking-induced currents are still weak and the waves propagating through the gap break close to the shoreline. However, as the simulation progresses, the magnitude of the rip currents increases. This steepens the waves propagating through the gap shifting the wave breaking location from close to the shoreline to the middle of the rip channel.

---

[2] The vertical vorticity is evaluated at z_α, i.e. $\Omega = v_{\alpha,x} - u_{\alpha,y}$

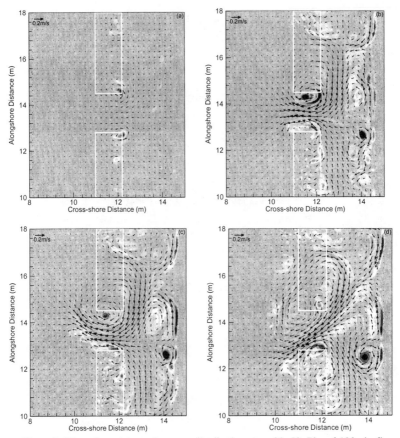

Figure 2. Filtered vorticity and current distribution at t = 20, 50, 70 and 120s (a-d).

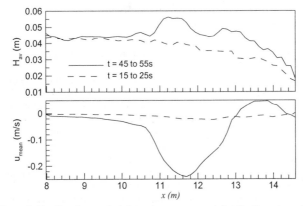

Figure 3. Predicted wave height and mean current distribution at y = 13.34m.

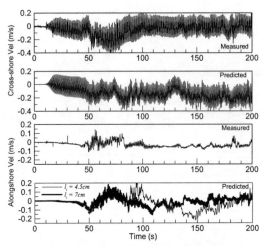

Figure 4. Measured and predicted velocity time histories at $x = 10.8$m, $y = 13.2$m.

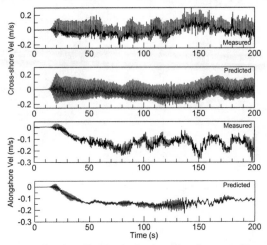

Figure 5. Measured and predicted velocity time histories at $x = 13$m, $y = 15.2$m.

The measured and predicted velocity time histories at five gauge locations (shown in Figure 1) are compared in Figures 4 to 8. The velocities were measured at 3cm above the seabed in the laboratory experiments and approximately at mid-depth for the numerical simulations. Figure 4 shows the velocity time histories at gauge location #1 ($x = 10.8$m, $y = 13.2$m) which is just seaward of the rip channel. The numerical and experimental results both show a gradual increase in the offshore-directed mean flow between $t = 0$ to $t = 50$s. For $t > 50$s, significant differences are observed between the measured and predicted velocity time histories, particularly for the cross-shore component of the flow. The measured offshore-directed mean flow increased to 0.18m/s between $t = 60$s and $t = 80$s before decreasing to 0.01m/s between $t = 100$ and $t = 200$s. The numerically predicted mean flow essentially oscillated around a mean value of 0.15m/s between $t = 50$ and $t = 200$s.

Additional numerical tests were carried out with different values of the turbulent length scale, l_t, and wave friction factor, f_w. Increasing the value of l_t to 7cm slightly improved the model-data match as shown in Figure 4. Decreasing the value of f_w to 0.005 led to a more chaotic flow while increasing the value of f_w to 0.03 led to a highly damped flow. We suspect that the oscillation in the measured cross-shore mean flow at gauge location #1 is due to a downwelling/upwelling type effect. The currents were measured at 3cm above the seabed in water 14.6cm deep. Later experimental measurements of the vertical structure of the currents suggested that the flow tends to be depth-uniform in the surf zone but becomes more of a surface current as it flows seawards.

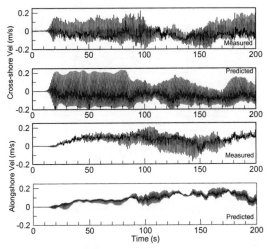

Figure 6. Measured and predicted velocity time histories at $x = 13$m, $y = 12.2$m.

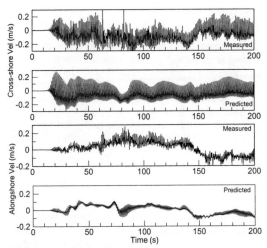

Figure 7. Measured and predicted velocity time histories at $x = 14$m, $y = 15.2$m

The velocity time histories and gauge locations #2 ($x = 13.0$m, $y = 15.2$m) and #3 ($x = 13.0$m, $y = 12.2$m) which are just shoreward of the bar are plotted in Figures 5 and 6 respectively. The mean flows at both locations are predominantly in the alongshore direction. The measured and predicted alongshore velocities are virtually identical. The velocity time histories and gauge locations #4 ($x = 14.0$m, $y = 15.2$m) and #5 ($x = 14.0$m, $y = 12.2$m) which are close to the shoreline in relatively shallow water ($h = 3.4$cm) are plotted in Figures 7 and 8 respectively. At gauge location #4, the flow is initially directed towards the north wall until t = 150s when it changes direction and flows southwards. The numerical model reproduces the observed trend.

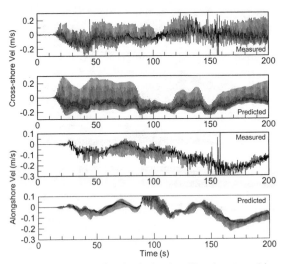

Figure 8. Measured and predicted velocity time histories at $x = 14$m, $y = 12.2$m.

CONCLUSIONS

A numerical model based on Boussinesq-type equations has been used to investigate the time-dependent evolution of rip currents on a barred beach with rip channels. The rip currents were highly unsteady, even though the incident waves were periodic. The circulation pattern is also sensitive to values of the turbulent length scale, which controls the rate of breaking-induced wave energy dissipation and the wave friction factor. Numerical model predictions have been compared with data from the laboratory experiments of Haller et al. (1997). The numerical model was able to reproduce the overall magnitude of the mean velocities and low-frequency oscillations. Pronounced differences were, however, observed in the flow exiting the rip channel probably due to three-dimensional effects.

ACKNOWLEDGEMENTS

The author would like to thank Dr. Merrick Haller for providing the experimental data. The Coastal and Hydraulics Laboratory of the U.S. Army Corps of Engineers supported this research project through contract number DACW42-01-P-0109 to Stevens Institute of Technology.

REFERENCES

Bowen, A.J. 1969. Rip Currents. 1. Theoretical Investigations. *Journal of Geophysical Research*, **74**, 5467-5478.

Chen, Q., Dalrymple, R.A., Kirby, J.T., Kennedy, A.B. and Haller, M.C. 1999. Boussinesq Modeling of a Rip Current System. *Journal of Geophysical Research*, **104**, 20617-20637.

Cox, D.T., Kobayashi, N. and Okayasu, A. 1995. Bottom Shear Stress in the Surf Zone. *Journal of Geophysical Research*, **101**, 14337-14348.

Haas, K.A., Svendsen, I.A. and Haller, M.C. 1998. Numerical Modeling of Nearshore Circulation on a Barred Beach with Rip Channels. *Proc. 26th International Conference on Coastal Engineering*, Copenhagen, Denmark, Vol. 1, 801-814.

Haller, M.C., Dalrymple, R.A. and Svendsen, I.A. 1997. Rip Channels and Nearshore Circulation. *Proc. 3rd Coastal Dynamics Conference, Coastal Dynamics '97*, Plymouth, U.K., 594-603.

Haller, M.C., Dalrymple, R.A. and Svendsen, I.A. 2000. Experiments on Rip Currents and Nearshore Circulation: Data Report. *Report No. CACR-00-04*, Center for Applied Coastal Research, University of Delaware.

Haller, M.C. and Dalrymple, R.A.. 2001. Rip Current Instabilities. *Journal of Fluid Mechanics*, **433**, 161-192.

Kennedy, A.B., Chen, Q., Kirby, J.T., and Dalrymple, R.A. 2000. Boussinesq modeling of wave transformation, breaking and runup. I: 1D. *Journal of Waterway, Port, Coastal and Ocean Engineering*, ASCE, **126**(1), 39-37.

Nwogu, O. 1993. Alternative Form of Boussinesq Equations for Nearshore Wave Propagation. *Journal of Waterway, Port, Coastal and Ocean Engineering*, ASCE, **119**(6), 618-638.

Nwogu, O.G. 1996. Numerical Prediction of Breaking Waves and Currents with a Boussinesq Model. *Proc. 25th International Conference on Coastal Engineering*, Orlando, Vol. 4, 4807-4820.

Nwogu, O.G. and Demirbilek, Z. 2001. BOUSS-2D: A Boussinesq Wave Model for Coastal Regions and Harbors. Technical Report ERDC/CHL TR-01-XX, U.S. Army Engineer Research and Development Center, Vicksburg, MS.

Schaffer, H.A., Madsen, P.A. and Deigaard, R. 1993. A Boussinesq Model for Wave Breaking in Shallow Water. *Coastal Engineering*, **20**, 185-202.

Smith, J.A. and Largier, J.L. 1995. Observations of Nearshore Circulation: Rip Currents. *Journal of Geophysical Research*, **100**, 10967-10975.

Svendsen, I.A., Yu, K. and Veeramony, J. 1996. A Boussinesq Breaking Wave Model with Vorticity. *Proc. 25th International Conference on Coastal Engineering*, Orlando, Vol. 1, 1192-1204.

EFFECTS OF CROSS-SHORE BOUNDARIES ON LONHSHORE CURRENT SIMULATIONS

Qin Chen[1] and Ib A. Svendsen[2]

Abstract: The paper analyzes how prescribed errors in the boundary conditions along both upstream and downstream cross-shore boundaries spread inside the computational domain. For simplicity we consider a longshore uniform plane beach with monochromatic, obliquely incident waves. It is found that the mismatch at the upstream cross-shore boundary tends to propagate away from the surf zone as if choosing a least-resistance path while the influence of the errors in the downstream boundary condition is limited to the adjacent area of the downstream boundary. Both the numerical and analytical solutions show that errors in the longshore current specified at the upstream boundary exponentially decay in the surf zone at a rate proportional to the bottom friction. In the case of excessive flux given at the cross-shore boundaries, a circulation cell tends to develop in the off-shore region where the errors caused by the boundary mismatch increase with the cross-shore width of the model domain.

INTRODUCTION

In nearshore modeling it is only possible to simulate the waves and currents in a small excerpt of the ocean. The boundary conditions along the free boundaries facing the ocean in the cross-shore and the longshore directions represent the effect of the flow outside the computational domain, which is not modeled and therefore by definition is unknown. Therefore, unless special information is known about the outside flow conditions, these boundary conditions will always be estimates only and hence subject to errors relative to actual flow patterns.

The problem of boundary conditions for a limited model domain has mainly been discussed in connection with laboratory experiments of longshore currents. Dalrymple et al. (1977), Visser (1991), Renier and Battjes (1997), and Hamilton and Ebersole

---

[1]Assistant Professor, Department of Civil Engineering, University of South Alabama, Mobile, AL 36688 USA. qchen@jaguar1.usouthal.edu

[2]Professor, Center for Applied Coastal Research, University of Delaware, Newark, DE 19716 USA. ias@coastal.udel.edu

(2001), among others, demonstrated that a virtually uniform longshore current could be obtained in a portion of the model basin in the laboratory by minimizing the circulation in the offshore region resulting from the errors at the upstream and downstream boundaries. Our understanding of the influence of the cross-shore boundary conditions on the longshore current simulations, however, is still very limited and consequently no tool is available to quantitatively estimate such boundary effect in the nearshore modeling.

The present paper analyzes how prescribed errors in the boundary conditions along both upstream and downstream cross-shore boundaries spread inside the computational domain. First, starting from the shallow water equations with the wave forcing, a perturbation method is used to develop a set of equations governing the spreading of the errors introduced at the cross-shore boundaries in the computational domain. Next, an analytical solution is obtained from the perturbed equations, which describes the influence of the upstream boundary mismatch on the flow field inside the domain. After that, a numerical model developed at the University of Delaware for quasi-3D nearshore circulation is utilized to conduct a series of numerical experiments with different cross-shore boundary conditions. The numerical solutions are then compared with the analytical results followed by an analysis of the momentum and mass balance of the computed flow field. Finally, the findings and conclusions are summarized.

THEORETICAL ANALYSIS

Basic Equations

The nonlinear shallow water (NSW) equations driven by the forcing of the radiation stresses have been commonly used in the modeling of breaking-generated nearshore horizontal circulation. Svendsen and Putrevu (1994) pointed out the importance of vertical variations of the nearshore currents for the momentum mixing in the nearshore circulation. The dispersive mixing term resulting from the depth variation of the currents has a similar form to the diffusion term in the depth-integrated equations. To reduce the complexity of the problem, we use the NSW equations with the forcing, bottom frcition and diffusion terms as the starting point of our analysis of the effect of errors in the cross-shore boundary conditions on the longshroe current simulations.

Using the volume flux as the variables, the depth-integrated, wave-averaged equations for depth-uniform currents can be expressed as

$$\frac{\partial \zeta}{\partial t} + \frac{\partial Q_x}{\partial x} + \frac{\partial Q_y}{\partial y} = 0 \tag{1}$$

$$\frac{\partial Q_x}{\partial t} + \frac{\partial}{\partial x}\left(\frac{Q_x^2}{h}\right) + \frac{\partial}{\partial y}\left(\frac{Q_x Q_y}{h}\right) + gh\frac{\partial \zeta}{\partial x} = -\frac{1}{\rho}\left(\frac{\partial S_{xx}}{\partial x} + \frac{\partial S_{yx}}{\partial y}\right) - \frac{f}{h}Q_x\sqrt{Q_x^2 + Q_y^2} + M_x \tag{2}$$

$$\frac{\partial Q_y}{\partial t} + \frac{\partial}{\partial x}\left(\frac{Q_x Q_y}{h}\right) + \frac{\partial}{\partial y}\left(\frac{Q_y^2}{h}\right) + gh\frac{\partial \zeta}{\partial y} = -\frac{1}{\rho}\left(\frac{\partial S_{xy}}{\partial x} + \frac{\partial S_{yy}}{\partial y}\right) - \frac{f}{h}Q_y\sqrt{Q_x^2 + Q_y^2} + M_y \tag{3}$$

where Q_x and Q_y are the cross-shore and longshore volume fluxes, ζ is the free surface elevation averaged over the short wave period, S_{ij} is the radiation stress owing to the short wave motion, h is the total water depth, f is the bottom friction coefficient, g is the gravitational acceleration, and M_x and M_y are the lateral mixing terms.

Perturbation Analysis

In order to develop insight into the physical controls of the spread of errors specified at the cross-shore boundaries, we analyze the governing equations using a perturbation expansion of the type

$$\zeta = \zeta^{(0)} + \epsilon\zeta^{(1)} + \cdots; \quad Q_x = Q_x^{(0)} + \epsilon Q_x^{(1)} + \cdots; \quad Q_y = Q_y^{(0)} + \epsilon Q_y^{(1)} + \cdots \quad (4)$$

where ϵ is the expansion parameter that is smaller than unity, and the superscripts 0 and 1 denote the background flow without boundary mismatches and the disturbances introduced at the cross-shore boundaries, respectively. The magnitude of the boundary disturbance is assumed to be weaker than that of the background flow generated by wave breaking. For simplicity, we consider a steady state and a longshore uniform plane beach. Consequently, we have $\frac{\partial}{\partial t} = 0$ in both the background and residual flows, and $\frac{\partial}{\partial y} = 0$ in the background flow only.

Substituting (4) into (1) - (3), and collecting the terms with the same ordering lead to the zero-order equations governing the wave setup and longshore uniform current generated by wave breaking, and the first-order equations

$$\frac{\partial Q_x^{(1)}}{\partial x} + \frac{\partial Q_y^{(1)}}{\partial y} = 0 \quad (5)$$

$$\frac{Q_y^{(0)}}{h}\frac{\partial Q_x^{(1)}}{\partial y} + gh\frac{\partial \zeta^{(1)}}{\partial x} = -\frac{f}{h}Q_x^{(1)}|Q_y^{(0)}| + M_x^{(1)} \quad (6)$$

$$Q_x^{(1)}\frac{\partial}{\partial x}(\frac{Q_y^{(0)}}{h}) + \frac{Q_y^{(0)}}{h}\frac{\partial Q_y^{(1)}}{\partial y} + gh\frac{\partial \zeta^{(1)}}{\partial y} = -\frac{2f}{h}Q_y^{(1)}|Q_y^{(0)}| + M_y^{(1)} \quad (7)$$

where $M_x^{(1)}$ and $M_y^{(1)}$ are the disturbances in the lateral mixing terms owing to the cross-boundary errors. Equations (5) - (7) govern the residual flow induced by the errors in the cross-shore boundary conditions.

ANALYTICAL SOLUTION

To obtain an analytical solution to the first-order equations governing the spatial distribution of the errors caused by the cross-shore boundary mismatches, simplifications are needed. First, we focus on the area where the longshore current is located. If there were no boundary mismatch, a uniform longshore current would exist in this region as a balance between the cross-shore gradient of the radiation stress S_{yx}, the bottom friction, and the lateral mixing. The longshore pressure gradient is obviously absent in the momentum balance. It may be reasonable to assume that the disturbances in the longshore pressure gradient and lateral mixing owing to errors in the cross-shore boundary conditions are small in comparison with other terms in the first-order governing equations. The justification of this assumption will be given in the later section in connection with the analysis of the momentum balance based on the numerical model solutions.

Neglecting the longshore pressure gradient and diffusion terms, (7) reduce to

$$Q_x^{(1)}\frac{\partial}{\partial x}(\frac{Q_y^{(0)}}{h}) + \frac{Q_y^{(0)}}{h}\frac{\partial Q_y^{(1)}}{\partial y} = -\frac{2f}{h}Q_y^{(1)}|Q_y^{(0)}| \quad (8)$$

The continuity equation (5) and the simplified momentum equation (8) have two unknowns $Q_x^{(1)}$ and $Q_y^{(1)}$ subject to the boundary conditions:

$$Q_x^{(1)}|_{x=0} = 0; \qquad Q_y^{(1)}|_{y=0} = Q_{yb}^{(1)} \tag{9}$$

in which $Q_{yb}^{(1)}$ is the error in the flux specified at the cross-shore boundaries.

Upon applying $\frac{\partial}{\partial x}$ on (8) and invoking the continuity equation (5) to eliminate $\frac{\partial Q_x^{(1)}}{\partial x}$, we obtain

$$a\frac{\partial^2 Q_y^{(1)}}{\partial x \partial y} + a\frac{2f}{h}\frac{\partial Q_y^{(1)}}{\partial x} - b\frac{\partial Q_y^{(1)}}{\partial y} + \frac{2f}{h}(ch - b)Q_y^{(1)} = 0 \tag{10}$$

where $a = V^{(0)}\frac{\partial V^{(0)}}{\partial x}$, $b = V^{(0)}\frac{\partial^2 V^{(0)}}{\partial x^2}$, $c = \frac{\partial V^{(0)}}{\partial x}\frac{\partial}{\partial x}\left(\frac{V^{(0)}}{h}\right)$, and $V^{(0)} = \frac{Q_y^{(0)}}{h}$. Equation (10) with the boundary condition (9) can be solved by using the method of separation of variables. The solution may be expressed as

$$Q_y^{(1)} = Q_{yb}^{(1)} e^{-\frac{2f}{h}(1+p)y} \tag{11}$$

in which

$$p = chQ_{yb}^{(1)}\left(a\frac{\partial Q_{yb}^{(1)}}{\partial x} - bQ_{yb}^{(1)}\right)^{-1} \tag{12}$$

Obviously, p depends on the cross-shore distribution of the longshore current and the error in the boundary condition. At the location of the maximum velocity, $p = 0$, as $\frac{\partial V^{(0)}}{\partial x} = 0$. Thus (11) becomes

$$Q_y^{(1)} = Q_{yb}^{(1)} e^{-\frac{2f}{h}y} \tag{13}$$

Equation (13) provides us with a tool to quickly estimate the influence of the error in the upstream boundary conditions on the flow downstream of the boundary without knowing the cross-shore distribution of the longshore current. The simple analytical solution states that errors in the upstream boundary condition decay at an exponential rate in the longshore direction. The stronger the bed shear stress is, the faster the boundary influence decays. Also the effect of the error in the upstream boundary condition decays faster in the area closer to the shoreline where the water depth is small. It is worth mentioning that the analytical solution (11) or (13) is valid only if the assumptions used to simplify the governing equations are justified. In the following section, we shall verify the analytical model using numerical experiments.

NUMERICAL MODELLING

The Total Flow

The model employed for the numerical experiments is the quasi-3D model SHORE-CIRC that includes the lateral mixing owing to the depth variations of the currents. The detailed description of the model is given in e.g. van Dongeren and Svendsen (2000).

Visser (1991) was the first to present comprehensive data sets of uniform longshore currents measured in the laboratory. We choose one of his data sets as a reference of our numerical experiments. Figure 1 shows the computed velocity field and the model/data comparison in the case of Test 4 of Visser's experiments. The numerical wave basin used here is 70 m long and 14 m wide. The slope of the longshore uniform plane beach is 1/20 and the offshore water depth is 35 cm. The obliquely incident wave train has a wave period of 1.02 s and a wave height of 7.8 cm, and the angle of incidence is 15.4 degrees. Because of the high degree of longshore uniformity of the current measured by Visser, longshore periodicity is assumed along the cross-shore boundaries.

The numerical model SHORECIRC is then driven by the gradients of the radiation stresses that are the output of the short wave model REFDIF. Wave breaking occurs about 2 m seaward of the shoreline and the obliquely incident wave generates a uniform longshore current. Sufficiently long computation is carried out to ensure that the current reaches its steady state. No shear instabilities occur in the model. As shown by the bottom panels in Figure 1, the computed wave height, wave setup, and cross-shore distribution of the velocity agree well with Visser's measurements except for a slight over-prediction of the longshore current in the offshore region. The corresponding bottom friction coefficient in (2) and (3) is 0.00325 (i.e. $f_w = 0.0065$ in SHORECIRC).

The numerical solution with the periodic cross-shore boundary condition as shown in the top panel of Figure 1 is defined as the undisturbed flow field which shall be used as the true solution, or the reference in the numerical experiments with boundary mis-matches. By replacing the longshore periodicity of the shore-normal boundaries with a specified flux boundary condition, we intentionally either increase or decrease the flux at both upstream and downstream boundaries by 10 % from the true solution. The input wave condition and all model parameters remain unchanged. We shall refer to the case with the increased flux at the shore-normal boundaries as excessive flux and to the case with the decreased flux boundary condition as reduced flux.

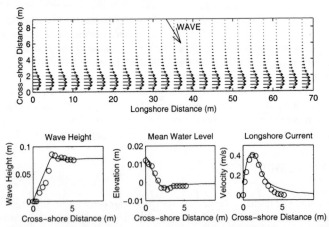

Fig. 1. Top panel: Computed velocity field of uniform longshore current. Bottom panels: Comparison of Visser's (1991) laboratory measurements (circles) and the numerical solutions (solid lines) with periodic cross-shore boundaries.

The Spreading of Boundary Errors

To analyze the spatial distribution of the error caused by the boundary mismatch, we subtract the true solution of the undisturbed flow from the solutions with the boundary errors. Figure 2 shows the resultant residual velocity field in the case of excessive flux. A similar residual velocity field that is not shown here is also obtained in the case of reduced flux. The upper panels of Figure 2 illustrate the spatial distribution of the residual flow, or the errors caused by the disturbance at the shore-normal boundaries, and the bottom panels depict the cross-shore profiles of the longshore component of the residual velocity at six different longshore locations.

We notice that the error introduced at the upstream boundary originally is concentrated near the shoreline in the area close to the upstream boundary because the assumed disturbance is proportional to the true solution. The upstream boundary error spreads into the computational domain in the manner that it tends to propagate away from the surf zone. This is illustrated more clearly by the residual velocity profiles in the bottom panels of Figure 2, where the dashed lines are the residual velocity profile at the upstream boundary as a reference. For instance, at the location of 40 m away from the upstream boundary, the residual velocity in the surf zone becomes nearly zero. In contrast to the influence of the upstream boundary error that gradually decays in the surf zone, close to the downstream boundary the error is apparently confined to the adjacent area and only spreads over a small distance. We have to blow up the area adjacent to the downstream boundary in order to examine the residual velocity field in that area as shown in the left top panel of Figure 2. We see that the area disturbed by the downstream boundary error is much smaller than that influenced by the upstream boundary error. The implication of the difference in the effects of upstream and downstream boundaries on longshore current simulations is that the downstream boundary may be placed closer to the area of interest than the upstream boundary in a computational domain to prevent the flow from being contaminated by the unavoidable boundary errors.

Our discussion so far has focused on the spatial structure of the residual flow in the area near the shoreline (i.e. $1 \sim 2$ times the surf zone width). If we take a close look at the residual flow in the offshore area in Figure 2, it is found that the residual flux may not vanish in the area far away from the cross-shore boundaries. Instead, the residual flux may increase in the offshore area owing to the spreading of the boundary error. In the case of reduced flux at the cross-shore boundaries, the requirement of mass conservation dictates a return flow near the offshore boundary. A return flow also exists in the case of excessive flux. It is found that the magnitude of the return flow in the case of excessive flux depends on the ratio of the cross-shore width to the longshore length of the computational domain. Acting as a jet, the excessive flux entrains the surrounding fluid. The larger the domain width is, the greater the return flow becomes.

The upper panels in Figure 3 show the comparison of the analytical and numerical residual fluxes normalized by the boundary mismatch dQ_b. Three longshore transects at different cross-shore locations $x/x_p = 0.5$, 1.0 and 1.5 are presented, where x is the distance from the shoreline and x_p is the location of the peak velocity. The dashed lines are the numerical results and the solid lines are the analytical solutions of Equations (13) (thick lines) and (11) (thin lines). Both cases of excessive and reduced fluxes give an identical set of dashed lines for these three transects.

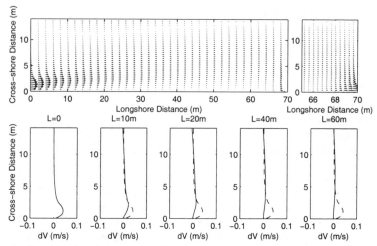

Fig. 2. Top panels: Computed residual velocity field in the case of excessive flux. Bottom panels: Cross-shore distributions of the residual velocity at six longshore locations. (−−): The residual velocity at the upstream boundary.

It is seen that both the analytical solutions given by (11) and (13) agree very well with the numerical results in the area shoreward of the peak velocity except for the region adjacent to the downstream boundary where the analytical model is invalid. In the area seaward of x_p, surprisingly, the simplified solution given by (13) is in better agreement with the numerical model than is the full solution that takes the cross-shore variation of the uniform longshore current into account. A close inspection on the contribution by the cross-shore variation of the longshore current indicates that there are considerable irregularities in the area seaward of x_p. This may be attributed to small numerical oscillations imbedded in the velocity profile that are amplified by the term containing the first and second derivatives of the velocity with respect to x. Nevertheless, the simplified model is more useful than the full solution because the longshore current is simply unknown before the numerical simulation.

The bottom panels in Figure 3 show the comparison of the residual mean water levels corresponding to the excessive (dashed lines) and reduced (solid lines) fluxes along the three longshore transects. It is seen that the mean water level is not affected by the upstream boundary mismatch. However, the excessive flux results in a considerable drawdown of the mean water surface near the downstream boundary while the reduced flux leads to an elevated mean water level in that area. Obviously, the downstream boundary mismatch causes rapid changes in the mean water level within a short distance, which leads to a large longshore pressure gradient in the area adjacent to the downstream boundary.

In the offshore area, the analytical model deviates from the numerical solution, which is not shown. The reason shall be given by the analysis of the balance of the residual momentum flux in the following subsection.

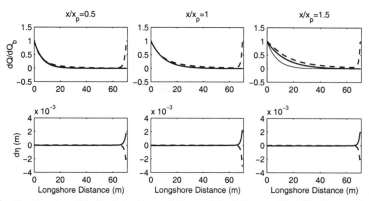

Fig. 3. Top panels: Comparison of the residual fluxes given by the analytical and numerical models. Thick solid lines: (13); thin solid lines: (11); dashed lines: numerical result. Bottom panels: The changes in the mean water level in the cases of excessive (dashed lines) and reduced (solid lines) fluxes.

The Variations of the Residual Momentum Flux

The first-order momentum equations resulting from the perturbation analysis in the preceding section governs the spreading of boundary errors. To obtain insight into the residual flow, we analyze the balance of the residual momentum flux on the basis of the numerical results given by SHORECIRC. By subtracting the true solution, or the longshore uniform flow from the flow with the boundary mismatch, we obtain the residual flux, residual velocity, and residual mean water level. The residual flow field is then used to evaluate each term in the first-order momentum equations.

Figure 4 illustrates the cross-shore variations of each term in the longshore component of the first-order momentum equation (7) at six different longshore locations. The dashed lines and thick solid lines represent the convective terms. The longshore pressure gradient is shown by the dotted lines and the bottom friction is shown by the thin solid lines. First, we notice that the convective terms play an import role in the residual flow. Second, it is seen that the longshore pressure gradient is rather weak except in the area adjacent to the downstream boundary (e.g. $L = 68$ m) where the bottom friction becomes unimportant in comparison with other terms.

On the basis of the computed first-order momentum flux, the residual flow field may be divided into three different regimes. The first one is the longshore strip of area near the shoreline, covering $1 \sim 2$ times the surf zone width. In this area starting from the upstream boundary, the longshore pressure gradient is negligible in comparison to other terms in the momentum equation. Thus the residual flow results from a balance between the convective terms and the bottom friction. This supports the assumption of the analytical model. In other words, the analytical model is applicable in this regime.

The second distinct regime of the residual flow is the area adjacent to the downstream boundary. The numerical results show that the longshore pressure gradient balances the convective term in this area where the bottom friction is negligible. A semi-analytical solution may be obtained from the simplified second-order momentum equation, but it is omitted here.

The rest of the area in the offshore region of the computational domain may be considered as a third regime of the residual flow. In this area, no term in the first-order momentum equations can be neglected because all of them are of the same order of magnitude. Thus no simple analytical solution can be obtained for this flow regime. A close inspection of the numerical results indicates that there exists a circulation in this complex flow regime where the convection and momentum mixing play an important role. Notice that the residual lateral mixing term that is negligible in the first and second flow regimes is not shown in Figure 4.

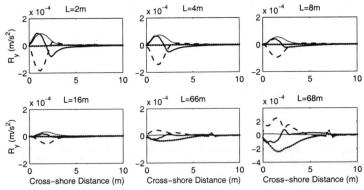

Fig. 4. Computed cross-shore variations of the terms (R_y) in (7) in the case of excessive flux. Thick solid lines: $Q_x^{(1)} \frac{\partial}{\partial x}(\frac{Q_y^{(0)}}{h})$; **dashed lines:** $\frac{Q_y^{(0)}}{h} \frac{\partial Q_y^{(1)}}{\partial y}$; **dotted lines:** $gh \frac{\partial \zeta^{(1)}}{\partial y}$; **thin solid lines:** $\frac{2f}{h} Q_y^{(1)} |Q_y^{(0)}|$.

The Entrainment and Balance of Volume Flux

From the mass conservation point of view, the total volume rate of the excessive or reduced flux introduced at the cross-shore boundaries should remain unchanged at any cross-shore section in the computational domain. If no circulation had occurred in the offshore region, the spread of the boundary errors into the offshore region would have diminished the influence of the boundary mismatch on the flow field far away from the cross-shore boundary. Unfortunately, this is not the case because the laboratory experiments by Visser (1991), Renier and Battjes (1997) and Hamilton and Ebersole (2001) as well as the numerical results in this study show that recirculation occurs if there is an error at the cross-shore boundaries. Essentially, the excessive volume flux acts as a jet that entrains additional volume flux from the surrounding fluid. At the downstream boundary where only the specified volume flux including the prescribed error is let out, the additional flux is returned and forms a recirculation. Further numerical experiments with different cross-shore width of the computational domain reveal the dependence of the strength of the circulation on the width of the domain. It is found that the entrained volume flux increases with the domain width. The growth rate seems to slow down, however, when a large basin width is used.

CONCLUSIONS

The paper analyzed how an unavoidable error in the cross-shore boundary condition influences the accuracy of the predicted longshore currents inside the computational domain. We began with a perturbation analysis of the nonlinear shallow water equation governing the spatial distribution of the boundary error, followed by a series of numerical experiments on the effects of the cross-shore boundaries on the flow field using a nearshore circulation model SHORECIRC. The numerical results suggest that the mismatch at the upstream cross-shore boundary tends to propagate away from the surf zone and the influence of the error at the downstream boundary is limited to the adjacent area. A simple formula was developed to have a quick estimate of the influence distance from the upstream boundary.

Both the analytical and numerical solutions show that errors in the longshore current specified at the upstream boundary decay exponentially in the surf zone where the bottom friction balances the convective terms in the residual momentum equation. In contrast, the longshore pressure gradient balances the convection in the area adjacent to the downstream boundary where boundary mismatches cause considerable changes in the mean water level. In the case of excessive flux given at the cross-shore boundaries, a circulation cell tends to develop in the offshore region where the errors generated by the boundary mismatch increase with the cross-shore width of the model domain. The analyses of the momentum balance and the mass balance on the basis of the numerical solutions not only confirm the hypothesis used to obtain the analytical model, but also provide insight into the complex residual flow field generated by errors in the cross-shore boundary conditions.

ACKNOWLEDGEMENTS

Funding for this study was provided by the Office of Naval Research grant N00014-99-1-0398. Discussions with Drs. Fengyan Shi and Kevin Haas are acknowledged.

REFERENCES

Dalrymple, R. A., Eubanks, R. A., and Birkemeier, W. A. 1977. Wave-induced circulation in shallow basins. *J. Waterway, Port, Coastal Ocean Div.* 103, ASCE, 117-135.

Hamilton, D. G., and Ebersole, B. A. 2001. Establishing uniform longshore currents in a large-scale sediment transport facility. *Coastal Eng.*, 42, 199-218.

Reniers, A. J. H. M., and Battjes, J. A. 1997. A laboratory study of longshore currents over barred and non-barred beaches, *Coastal Eng., 30*, 1-22.

Svendsen, I. A., and Putrevu,U. 1994. Nearshore mixing and dispersion, *Proc. Roy. Soc. Lond. A., 445*, 1-16.

van Dongeren, A. R., and Svendsen, I. A. 2000. Nonlinear and quasi 3D effects in leaky infragravity waves. *Coastal Eng., 41*, 467-496.

Visser, P. J., 1991. Laboratory measurements of uniform longshore currents, *Coastal Eng., 15*, 563-593.

WAVES AND CURRENTS ON ACCRETIONAL BARRED BEACHES

Kevin A Haas, Ib A Svendsen, member ASCE, and Qun Zhao [1]

ABSTRACT

This paper analyzes the waves and currents on beaches with rhythmic or crescentic bars. The goal is to determine under what conditions does the flow in the trough between the bar and shore show similarities with the flow patterns observed on accretional beaches. The nearshore circulation numerical model system SHORECIRC is used for this study. The crescentic shape of the bars is found to refract the waves toward the wide section of the bars driving currents toward the narrow sections of the bar generating rip currents. Channels formed on the narrow parts of the bar reinforce the rip currents within these channels. It is found that waves with an incident angle create stronger flows in the trough. In addition, larger depths over the bar are also found to generate stronger flow in the trough. The flow patterns are seen to be insensitive to the wave height and period for the particular conditions studied.

INTRODUCTION

The dynamics of the nearshore region are complex and the source of much interest. A qualitative three-dimensional model for the evolution of beach morphology has previously been established by Short (1979) and Wright and Short (1984) based on extensive observations from Narrabeen Beach and other beaches in Australia. The beaches are classified as being under either erosional or accretionary conditions. Two extreme beach stages, fully reflective and highly dissipative, are defined along with four intermediate stages. Similar beach stages have been observed on video images and classified by Lippmann and Holman (1990).

The sequence in Fig 3a in Short (1979) and similarly in Fig 2 in Wright and Short (1984) begins with a wide offshore bar on a dissipative beach. As the wave conditions decrease, the bar begins to migrate shoreward, becoming crescentic as sand is deposited in a longshore nonuniform way on the lee side of the bar. The waves break on the seaward face of the bar and bores are transmitted across the increasingly wider and shallower sections of the bar with little or no return flow. The narrower sections of the bar at the nodes provide the seaward return

[1]Center for Applied Coastal Research, Ocean Engineering Lab, University of Delaware, Newark, DE 19716, USA email: haas@coastal.udel.edu

flow as rip currents. Under certain conditions the flow in the trough is strong, similar to river flow, whereas under other conditions the flow in the trough is weak. The trough region shoreward of the bar becomes narrower until the bar eventually attaches to the shoreline creating a transverse bar and rip channel system. Eventually, the rip channels completely fill in as the entire bar becomes attached to the shore resulting in a steeper more reflective beach.

This paper analyzes the waves and the flow patterns during the early stages of this process where the inflow is concentrated over the wide, shallow bars and the outflow is in the rip channels in the narrow sections of the bar on beaches with rhythmic or crescentic bars. The goal is to to determine under what conditions the flow in the trough is increased. Particular emphasis is placed on the effect of the deepening rip channel on the overall circulation pattern. Qualitative analysis of the perceived sediment transport is also presented in order to determine if the morphological changes observed by Short (1978) can occur.

NUMERICAL MODEL

The tool for the analysis is the nearshore circulation model SHORECIRC (SC) which is used for computational experiments. It consists of a shortwave transformation component and a shortwave-averaged model, interacting simultaneously to simulate short and long wave motions in nearshore regions. The wave driver used is REF/DIF (Kirby and Dalrymple 1994) which accounts for the combined effects of bottom induced refraction-diffraction, current induced refraction and wave breaking dissipation. The basic part of the model was originally described in Van Dongeren et al. (1994). In its present version 2.0 the SC model is essentially a quasi-3D large eddy simulation model as described by Haas and Svendsen (2000). The accuracy of the SC system for modeling circulation on barred beaches was demonstrated by Haas et al. (1998), Haas et al. (2000) and Svendsen et al. (2000).

METHODOLOGY

In the setup we have for reasons of illustration, chosen absolute geometric dimensions for the test parameters, however the actual variables of the problem are of course dimensionless combinations of the test parameters. An example of the bathymetry used in the simulations is shown in Fig 1. The bathymetry consists of a mild plane beach with a slope of 1/50, a flat bar, a trough and another 1/50 plane beach. The bar is crescentic with a longshore wave length of 160 m. Channels of varying depth are placed at the narrow locations of the bars. Fig 2 shows definitions used to distinguish the varying bathymetric conditions. The depth over the crest of the bar is called h_c and the depth of the channel relative to the crest of the bar is called h_{ch}. Test are run with values of h_c equal to 10, 20, 30, 40 and 50 cm and h_{ch} equal to 0, 5, 10 and 20 cm. Most tests presented in this paper are with $h_c = 30$ cm.

Wave conditions are varied as well. Incoming wave heights used in the tests are 20, 30 and 40 cm. The incoming wave angles are 0, 5 and 10 degrees. The

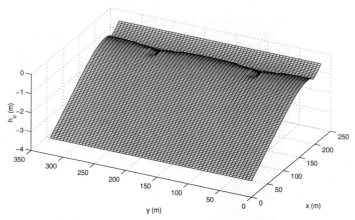

FIG. 1.   Example of the bathymetry used.

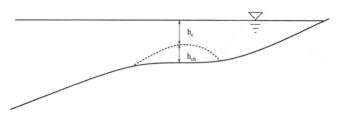

FIG. 2.   Definition sketch.

wave period is 4, 6 and 8 seconds. Cases presented in this paper have a wave height of 30 cm, period of 6 seconds and a wave angle of 0 degrees unless otherwise noted. For all the conditions, the waves break on the offshore side of the bar and continue breaking over the bar. They stop breaking in the trough and break again near the shoreline.

All simulations begin with a cold start where the mean water level (MWL) and currents are all equal to zero. The total simulation time is around 1000 seconds and the flow essentially reaches steady state. The offshore boundary condition is the absorbing generating boundary condition by Van Dongeren and Svendsen (1997), the shoreline is simulated as a wall with a shallow depth in front and the lateral boundaries are taken to be periodic.

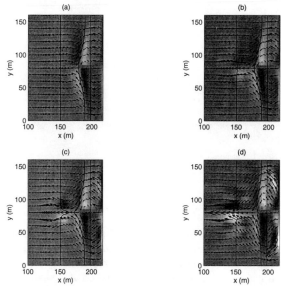

FIG. 3. Vorticity and velocity vectors for (a) $h_{ch} = 0$ m, (b) $h_{ch} = 0.05$ m, (c) $h_{ch} = 0.1$ m, (d) $h_{ch} = 0.2$ m. The solid black line is the depth contour 11 cm below the crest of the bar, the dark color is negative vorticity and the light color is positive vorticity. Only a portion of the computational domain is shown.

RESULTS

The effect of the rip channel on the circulation is established by comparing the flow patterns with different values of h_{ch} under otherwise identical conditions. Fig 3 shows excerpts of the vorticity and velocity vectors from around one of the channels for the four values of h_{ch}. The velocity vectors represent the below trough velocity, not including the short wave induced volume flux, which explains why the net flow appears to be offshore.

In Fig 3a for the case with no channel the crescentic shape of the bar still causes wave refraction focusing the waves behind the wide section of the bar driving the current toward narrow sections of the bar. Therefore a weak rip current is started even though there is no rip channel present. The effect of this rip on the sediment transport would be the creation of a channel. As the rip channel forms and increases in depth the converging currents in the trough and the associated rip current grow in strength (Fig 3b-d). Therefore, the creation of the rip channel is a self-sustaining mechanism. The circulation pattern near the shore suggests a sediment transport along the shoreline behind the narrow part toward the wide part of the bar. This could lead to the formation of cusped beaches as suggested by Short (1979).

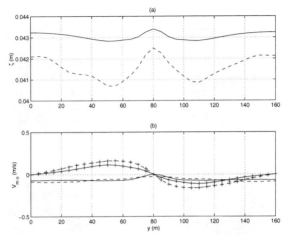

FIG. 4. Longshore section at $x = 185$ **m of (a) the MWL** ζ **for (-)** $h_{ch} = 0$ **m and (- -)** $h_{ch} = 0.2$ **m and (b) the cross-shore current** U_m **for (-)** $h_{ch} = 0$ **m and (- -)** $h_{ch} = 0.2$ **m and the longshore current** V_m **for (-+)** $h_{ch} = 0$ **m and (- + -)** $h_{ch} = 0.2$ **m. The channel is centered at** $y = 80$ **m.**

A longshore section in the trough between the bar and shore of the MWL and currents for the case without and with channels is shown in Fig 4. The currents along this section are representative of the feeder currents along other sections in the trough. The qualitative longshore variation of both the mean water level and the currents are similar whether or not channels are present. The MWL is decreasing toward the center (channel) although there is a bump in the mean water level close to the center. The longshore currents for both cases are converging toward the center.

The longshore momentum balance shows more interesting features than the cross-shore balance, therefore only the longshore momentum balance will be presented in this short paper. The longshore momentum balances with and without channels shown in Fig 5 are qualitatively similar for the the two cases. The primary difference is that the magnitude of the terms are larger for the case with the channel. Interestingly, the pressure gradient near the center has the same sign as the bottom stress, which provides a clear indication of the direction of the longshore flow. Therefore, the pressure gradient is not driving the flow but is actually opposing the flow. The radiation stress gradient has the opposite sign and is driving the converging feeder currents.

Fig 6 shows a longshore section along the offshore edge of the bar of the MWL and the currents for the case with no rip channel and the case with the deepest rip channel. In Fig 6 it is clear that the longshore variability of the MWL is

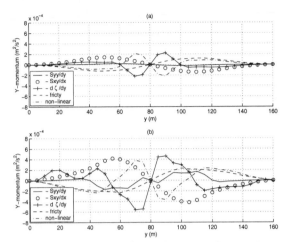

FIG. 5. Longshore section at $x = 185$ m of the longshore momentum balance for **(a)** $h_{ch} = 0$ m and **(b)** $h_{ch} = 0.2$ m. The channel is centered at $y = 80$ m.

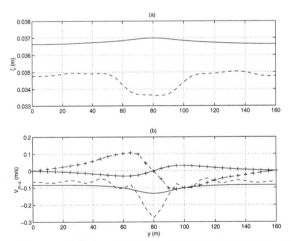

FIG. 6. Longshore section at $x = 160$ m of **(a)** the MWL ζ for (-) $h_{ch} = 0$ m and (- -) $h_{ch} = 0.2$ m and **(b)** the cross-shore current U_m for (-) $h_{ch} = 0$ m and (- -) $h_{ch} = 0.2$ m and the longshore current V_m for (-+) $h_{ch} = 0$ m and (- + -) $h_{ch} = 0.2$ m. The channel is centered at $y = 80$ m.

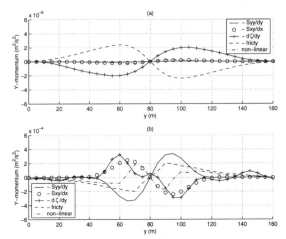

FIG. 7. Longshore section at x = 160 m of the longshore momentum balance for (a) $h_{ch} = 0$ m and (b) $h_{ch} = 0.2$ m. The channel is centered at $y = 80$ m. The scale on the vertical axis in (a) is an order of magnitude smaller than the scale in (b).

different for the two cases. Without the channel ($h_{ch} = 0$ cm) the MWL has a maximum in the center (where the channel would be located), whereas with the channel the MWL has a minimum in the channel. The longshore currents (V_m) for these two cases along this section are flowing in the opposite direction. Without the channel the longshore current is diverging away from the rip; with the channel the longshore current is converging toward the rip. This indicates that the presence of the channel creates feeder currents over a broader region, including on top of the bar. Also the cross-shore currents (U_m) for the case with the channel are much stronger in the region of the rip, confirming that the rip is much stronger with the channel present.

The longshore momentum balance for the two cases along the same section is shown in Fig 7. In (a) with no channels the terms in the longshore momentum balance are an order of magnitude smaller than the case with a channel in (b). The primary momentum balance shown in (a) is between the bottom stress and the pressure gradient. Clearly the pressure gradient is driving the flow away from the channel. For the case with the channels the pressure gradient is much larger and in the opposite directions, which drives the currents toward the rip in the center as indicated by the bottom stress term. The radiation stress components are larger for this case as well because of the increased wave refraction resulting from the presence of the channel. In addition the non-linear terms are larger because of the converging flow which then is turned offshore in the rip current.

The influence of the incoming wave conditions is examined by varying the wave height and period. Fig 8 shows the vorticity and velocity vectors for three

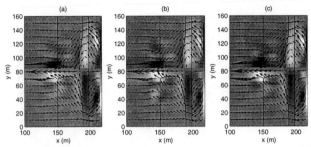

FIG. 8. Vorticity and velocity vectors for (a) H=0.3 m, T = 6 s, (b) H=0.3 m, T = 4 s, (c) H=0.4 m, T = 6 s. The solid black line is the depth contour 11 cm below the crest of the bar, the dark color is negative vorticity and the light color is positive vorticity. Only a portion of the computational domain is shown.

cases with different combinations of wave height and period. Remarkably, the overall flow patterns remain virtually identical regardless of the incoming wave conditions. At least for the range of conditions, in terms of wave height and period, studied thus far, the overall flow patterns are insensitive to the incoming wave conditions.

In the cases examined thus far, the flow in the trough does not appear to be flowing like a river. The natural next step in trying to simulate the strong flow in the trough is to do a simulation using waves with an incident angle. Fig 9 shows the vorticity and velocity vectors with an incoming wave angle of 10 degrees for the four values of h_{ch}. Clearly, the longshore current is much stronger both offshore of the bar and in the trough for these cases, although it still does not resemble the river flow. However, the effect of the channel is quite dramatic. When the channel is not present there is virtually no rip current, only a small reversal of flow on the shoreward edge of the bar. However, with the deepest channel, there are large feeder currents and a significant rip current.

The effect of the total depth over the bar is examined by varying h_c while holding all other parameters constant. Fig 10 shows the vorticity and velocity vectors for two different depths over the bar. The flow over and offshore of the bar is fairly similar for the two conditions. The biggest difference is the strength of the flow in the trough between the bar and the shore. The case with the deeper depth over the bar has stronger flow in the trough.

While the waves are still breaking over the bar, the larger depth over the bar means that more of the wave energy will be passing over the bar into the trough region leading to larger radiation stresses. Because the radiation stress gradients due to the breaking waves are driving the currents in the trough region, the larger radiation stresses drive stronger currents for the case with the deeper depth over the bar.

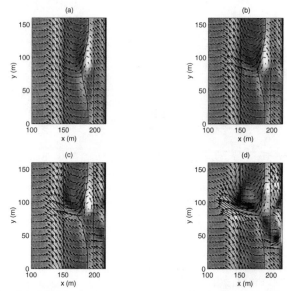

FIG. 9. Vorticity and velocity vectors for a 10 degree incident wave with (a) $h_{ch} = 0$ m, (b) $h_{ch} = 0.05$ m, (c) $h_{ch} = 0.1$ m, (d) $h_{ch} = 0.2$ m. The solid black line is the depth contour 11 cm below the crest of the bar, the dark color is negative vorticity and the light color is positive vorticity. Only a portion of the computational domain is shown.

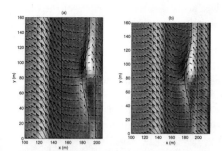

FIG. 10. Vorticity and velocity vectors for a 10 degree incident wave angle with (a) $h_c = 0.1$ m and (b) $h_c = 0.3$ m. The solid black line is the depth contour 11 cm below the crest of the bar, the dark color is negative vorticity and the light color is positive vorticity. Only a portion of the computational domain is shown.

CONCLUSIONS

Numerical simulations have been performed on mild beaches with wide, flat crescentic type bars. The crescentic shape of the bars is found to refract the waves toward the wide section of the bars. This in turn drives the flow toward the narrow section of the bar where it then turns offshore as a small rip current. These small rip currents will most likely create channels in the bar. As these channels increase in depth, the waves undergo stronger refraction and more flow is driven toward the channels, thereby creating stronger rip currents. Therefore, the mechanism of crescentic bars generating rip currents and channels appears to be self sustaining.

The goal of generating strong river-like flow in the channel was never completely obtained for the selection of parameter values studied so far. However, it was seen that waves with an incident angle do create stronger flows in the trough. In addition larger depths over the bar also generate stronger flow in the trough. The flow pattern does seem to be insensitive to the wave height and period for these particular conditions. The flow in the trough for the present conditions appear to be driven by radiation stresses rather than pressure gradients like rivers. Further work needs to be done to determine under what conditions will the flow be driven by pressure gradients strong enough to create the river-like flow.

ACKNOWLEDGMENTS

This work was sponsored by Sea Grant, under Award No NA96RG0029 and by ONR (contract no N0014-99-1-1051).

REFERENCES

Haas, K. A. and Svendsen, I. A. (2000). "Three-dimensional modeling of rip current systems." *(Ph. D. Dissertation), Res. Report CACR-00-06*, Center for Applied Coastal Research, Univ. of Delaware.

Haas, K. A., Svendsen, I. A., and Haller, M. (1998). "Numerical modeling of nearshore circulation on a barred beach with rip channels." *Proc. 26th Coastal Engineering Conference*, Vol. 1, Copenhagen. ASCE, 801–814.

Haas, K. A., Svendsen, I. A., and Zhao, Q. (2000). "3D modeling of rip currents." *Proc. 27th Coastal Engineering Conference*, Vol. 2, Sydney. ASCE, 1113–1126.

Kirby, J. and Dalrymple, R. (1994). "Combined refraction/diffraction model REF/DIF 1, version 2.5." *Report No. CACR-94-22*, Center for Applied Coastal Research, Univ. of Delaware.

Lippmann, T. C. and Holman, R. A. (1990). "The spatial and temporal variability of sand bar morphology." *J. Geophys. Res.*, 95, 11,575–11,590.

Short, A. D. (1978). "Wave power and beach-stages: A global model." *Proc. 16th Coastal Engineering Conference*, Vol. 2, Hamburg. ASCE, 1145–1162.

Short, A. D. (1979). "Three dimensional beach-stage model." *Journal of Geology*, 87, 553–571.

Svendsen, I. A., Haas, K. A., and Zhao, Q. (2000). "Analysis of rip current systems." *Proc. 27th Coastal Engineering Conference*, Vol. 2, Sydney. ASCE, 1127–1140.

Van Dongeren, A., Sancho, F., Svendsen, I. A., and Putrevu, U. (1994). "SHORECIRC: A quasi-3D nearshore model." *Proc. 24th Coastal Engineering Conference*. 2741–2754.

Van Dongeren, A. R. and Svendsen, I. A. (1997). "An absorbing-generating boundary condition for shallow water models." *J. of Waterway, Port, Coastal and Ocean Engng.*, 123(6), 303–313.

Wright, L. and Short, A. D. (1984). "Morphydynamic variability of surf zones and beaches: A synthesis." *Marine Geology*, 56, 93–118.

Wave Group Forcing of Rip Currents

Andrew B. Kennedy,[1] and Robert A. Dalrymple[2]

Abstract

Experiments have been performed in a laboratory wave basin investigating the effect of wave groups on rip currents. Results show a strong group signature in velocity time series measured in the rip neck. In some cases, mean quantities are also significantly affected. Velocity fluctuations with periods greater than the group period do not appear to be affected by the group forcing. Fundamental differences are found between flow at startup and at later times. These startup transients are extremely high, and the differences from later flow are hypothesised to be because wave-current interaction is not significant in the initial stages of flow development, and because large-scale circulation patterns have not yet developed.

Introduction

In the field, rip currents have long been observed to be unsteady on many time scales. In one of the earliest recorded observations of rip currents, Shepard and Inman (1950) write "each group of high breakers was followed in sequence by (1) a decrease in the offshore current velocity (sometimes a complete reversal to an onshore current was observed), and (2) a rise in sea level near shore. Following this rise in sea level there was a rapid development of an offshore current which built up to a maximum during the group of low waves." Similar observations of the unsteady nature of rip currents have been made by Huntley et al. (1988). In fact, it is difficult to find field observations which do not mention the time varying properties of rip currents.

Some of these variations may be due to a jet instability of the rip current, recently observed and analysed in laboratory flows by Haller and Dalrymple (2001). However, a major cause of the time-varying properties of rip currents in the field, as noted by several investigators, appears to be due to the influence of wave group forcing on the rip flow. Unfortunately, these field observations have been more qualitative than quantitative. In this paper, we will examine the results of laboratory experiments for topographically controlled rip currents, comparing results

---

[1]University of Florida, 124 Yon Hall, Gainesville, FL 32611-6580, USA

[2]Center for Applied Coastal Research, University of Delaware, Newark, DE 19716 USA.

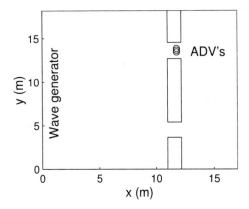

Figure 1: Plan view of the laboratory bar-channel rip current setup, showing location of bars, rip channels, and ADV's

from groupy waves and the equivalent steady forcing. A wide range of variations about two base conditions are used, giving a range of forcing with different group strengths, group periods, and wave heights.

Experimental Setup

All experiments were performed in the multidirectional basin at the University of Delaware using the setup of Haller et al. (1997,2001). Here, a bar on an otherwise planar beach is cut by two channels. The difference in wave breaking on the bar and rip channels produces differences in wave-induced setup, which drives strong offshore-directed currents through the channels. Figure 1 shows a schematic diagram of the experimental setup. Acoustic doppler velocimeters (ADVs) were used to measure current velocities simultaneously at the centerline of the right rip channel (looking offshore) and at 20 cm on either side in the longshore direction, at x=11.8 m . Incident waves were measured at eight locations in the longshore, just offshore of the bar at x=10.4 m. The water depth used was identical to that of Haller and Dalrymple (2001), test B.

Wave groups were produced using bichromatic wave inputs to the wave paddle. By varying the relative strengths and frequency separation of the two components, wave groups with independently varying groupiness factors (GF, Funke and Mansard, 1980) and group periods were created. All wave conditions used 1 s period base waves. For groupy and monchromatic tests two main wave heights, measured just offshore of the bar, were used, $H_{RMS} = 2.8$ cm, and $H_{RMS} = 4.1$ cm. Three group periods, $T_g = 16, 32, 64$ s were used along with four groupiness factors, $GF = 0, 0.47, 0.89, 1.0$. ($GF = 0.0$ is a regular wave, while $GF = 1.0$ corresponds to the strongest possible wave group.) Groupiness factors were computed theoretically by assuming the time scale of the wave groups is much longer than the time scale of the individual waves. Every possible permutation of these parameters was used, except that $GF = 1.0$ was not used with the larger wave height $H_{RMS} = 4.1$ cm, as entrained air caused by strongly breaking waves in the rip channel made measurements unreliable.

Of the three group periods used, two (32, 64 s) were longer than the longest seiching mode of the wave basin (27.4 s, Haller and Dalrymple, 2001). The shortest group period (16 s) was shorter than the longest basin scale mode, but the response in the vicinity of the rip channels appeared to have dwarfed any small basin modes.

Time Series of Rip Current Response

Figure 2 shows 1100 s of the measured time series of the offshore component of rip current response to groupy and steady wave forcing. For each subfigure, the top traces are wave motion, measured just offshore of the bar, while below are measured velocities from the three ADV's in the rip neck. Wave time series in cm have been multiplied by a factor of 5 to make them more visible. Velocities have been low-pass filtered with a 5 s cutoff to remove the incident wave frequencies. Limitations of space do not permit us to show velocity time series for all conditions of wave forcing, but the results here are typical. Two wave height base conditions were used, $H_{RMS} = 2.8$ cm and 4.1 cm. Groupy waves had group period T=32 s, and $GF = 0.47$. Other group periods and group strengths differed in detail but general features were similar, for the most part.

Even at first glance, the patterns differ strongly between the monochromatic and groupy forcing. A strong response of the rip current at the 32 s group period is immediately obvious for both groupy conditions. In contrast, the results for steady wave forcing show no such response. Velocities for the small monochromatic case show episodic, transient increases in current velocities which quickly die down. For the large monochromatic case, there are two main features: small quasi-periodic disturbances of $O(20s)$, and a strong, long period oscillation seen mainly in the first half of the time series. These 20 s disturbances are somewhat less than the fastest-growing jet instability for this condition, which from the similarities to Haller and Dalrymple (2001) case B, would have a frequency of $f < 0.01$.

For all time series, it will be noted that velocities varied between ADV's. The greatest difference appeared to be between the two measurements symmetric about the centerline. These differences were hypothesised to have two causes: time-dependent changes in circulation not related to wave groups, and a persistent bias of the rip current in one direction. The rip bias has been observed in this basin before and appears to be because one side of the wave basin is approximately 1 cm deeper than the other side in the vicinity of the bar (K. Haas, pers. commun.). The time-varying circulation may be the result of either the jet instability of Haller and Dalrymple (2001), as a result of chaotic wave-current

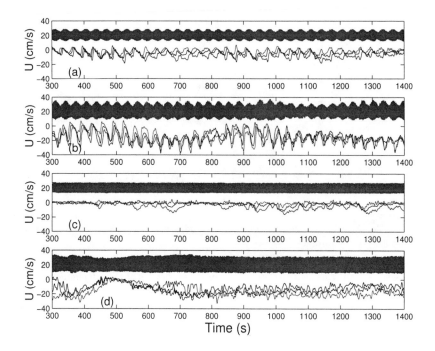

Figure 2: Low-pass filtered cross-shore velocities in rip channel. (a) $H_{RMS} = 2.8$ cm, groupy forcing; (b) $H_{RMS} = 4.1$ cm, groupy forcing; (c) $H_{RMS}=2.8$ cm, steady forcing; (d) $H_{RMS} = 4.1$ cm, steady forcing.

interaction changing the circulation patterns, or from other unidentified sources.

Evidence for chaotic wave-current interaction can be seen in Figure 2(d) at around 450 s, where a significant change in nearshore circulation appears to correspond to changes in wave height. Similarly, Figure 2(b) at 1100 s shows changes in both waves and currents.

In Figures 2(a,b,d), velocity fluctuations with very long time scales may be seen clearly. For the larger wave height (2b,d), slow variations of the mean current with time scales of $O(100$'s of s) affect velocities at all three measurement locations. For both groupy cases, (2a,b), response at the group frequency changes very slowly, in both magnitude and sometimes in the relative phase between ADV's.

Mean Currents

Although spectral analysis of velocity records can be extremely useful, stable estimates of the very long period oscillations of the rip current would require data

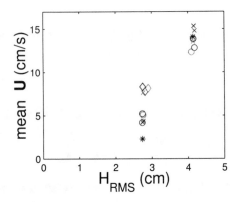

Figure 3:　Mean cross-shore velocities in rip channel. (*) Steady forcing; other symbols, groupy forcing.

in excess of that taken. Thus, we will examine properties of the rip currents over fairly wide frequency bands, which will be significantly more stable.

The most basic statistical estimate is the mean current. Figure 3 shows the mean offshore current averaged across the three ADV's. All monochromatic and groupy waves are shown. Mean current increases strongly with increasing H_{RMS}, which is not surprising. However, for the small wave condition, it is surprising that all groupy tests show much stronger mean velocities than for the regular wave case. Differences in mean velocity were in fact so large that they could be seen clearly during the experiments. This difference is particularly perplexing since, for the larger wave condition, mean velocities are very similar for monochromatic and groupy waves.

A possible explanation may be seen by considering the time-varying properties of wave forcing, after making some assumptions. First, we assume that all waves begin breaking at approximately the same location, and finish breaking a short distance later at the bar crest, where they all finish with the same height, H_0. Next, we assume that there is no breaking in the rip channel, and that waves with height $H < H_0$ do not break on the bar.

We will now consider two limiting cases. In the first (case 1), monochromatic waves are too small to break over the bar. However, the large waves of a groupy wave train with the same H_{RMS} do break over the bar. In this case, the groupy waves provide finite forcing to drive rip currents, while forcing from the equivalent monochromatic waves is essentially zero.

For case 2, we will assume that monochromatic waves and all groupy waves break on the bar. In this case, the mean forcing is essentially identical.

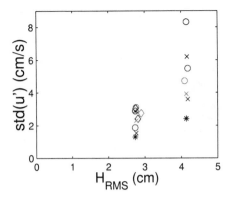

Figure 4: Cross-shore current fluctuations in wave group band, 5-70 s. (*) Steady forcing; other symbols, groupy forcing.

The larger wave height in figure 3 is an example of case 2, where groupy and regular waves provide essentially the same forcing. The small wave cases are not quite an example of case 1, as the monochromatic waves do break (weakly) over the bar, but it appears to be quite close to the limiting case and thus explains the stronger rip currents arising from groupy forcing. Further support comes from the observation that wave trains with the largest groupiness factors (GF=1.0, diamonds) have the largest mean currents, as would be predicted in this case.

Of course, these arguments are not exact. Many factors will affect them, from the fact that larger waves break in deeper depths, thus modifying the radiation stress forcing, to the effects of time-varying wave-current interaction in the groupy wave cases. However, we believe that the above arguments contain the basic explanation for why mean currents are stronger for groupy waves with the smaller wave height.

Current fluctuations in wave group band (5-70 s)

Current fluctuations in the wave group band, defined very broadly as 5-70 s to include all three wave group periods, 16,32,64 s, were investigated using zero-phase bandpass FIR filters, which only let through energy in the wave group band. The standard deviation of this filtered time series, excluding the ends of the record, was then computed to give an overall indication of the energy in the wave group band.

Figure 4 shows fluctuations in the cross shore velocity in the wave group band. As expected, groupy waves have a much larger response. For the larger wave height with any given group period, energy in this band increases monotonically with increasing groupiness factor (GF). However, different group periods with

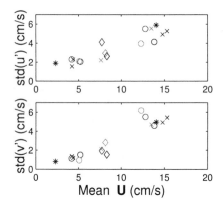

Figure 5: Very long period (> 70 s) instabilities in cross-shore and longshore flow in the rip neck

the same wave height and groupiness factor may have a significantly different response for reasons which do not appear to be simple. For the small wave height, the situation is more complex. Wave conditions with the largest GF tend to have large responses, but the progression is not as clear. This may be because of the significantly varying mean velocities, which was not the case for the larger wave height.

Very long period fluctuations (> 70 s)

To investigate very low frequency fluctuations, a low-pass zero-phase FIR filter was applied to the raw data, and the standard deviation of the resulting time series was then computed. From Haller and Dalrymple (2001), jet instabilities for the larger monochromatic case would appear to have instabilities in this range ($f < 0.01$ s). Figure 5 shows results for the cross-shore and longshore velocities, plotted against the mean cross-shore velocity. For both velocities, energy in the long period band increases strongly with increasing mean current. Little scatter is seen and fluctuations in the longshore and cross-shore are roughly comparable at this location. In fact, it appears that very long period fluctuations are only dependent on the mean cross-shore velocity, and not on any wave group parameters. The long-period oscillations (which may be seen in Figure 2 to have periods of hundreds of seconds) may thus be due to flow instabilities, chaotic wave-current interaction, or other factors completely unrelated to wave group forcing.

This, however, does not preclude the possibility that wave groups and jet instabilities may interact in some fashion when their characteristic periods are closer together, although the nature of such an interaction remains unclear.

Response to transient wave pulses

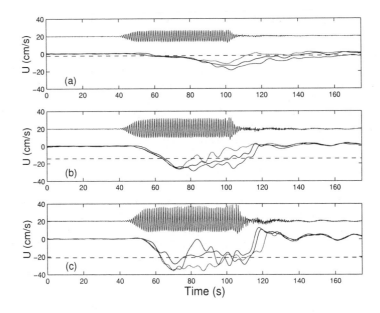

Figure 6:   Low-pass filtered cross-shore velocities in rip channel. 64 s nominal wave pulse. (a) $H_{RMS} = 2.8$ cm; (b) $H_{RMS} = 4.1$ cm; (c) $H_{RMS} = 5.5$ cm.

Many instances of swimmers caught in rip currents occur when a wave group significantly larger than the background wave climate causes a strong, transient pulsation of the rip current (e.g. Wall, 2001). Thus, tests on single, isolated wave groups were performed. In these tests, a group of waves was generated at the wavemaker with no forcing either before or after.

Figure 6 shows results for a nominal 64 s wave pulse for three incident wave heights. As might be expected, currents begin once waves start breaking and decrease once the last waves of the wave group have broken. For the larger two wave heights, this decrease appears to begin even before the waves stop breaking. Also included in the plots are mean values of the cross-shore current for monochromatic waves with the same height. The maximum current for the wave pulses is in all cases significantly over the mean current value.

This peak current just after startup was also found in tests (not shown) with several different group periods, and has also been noted by Haas and Svendsen (2000). It is highly repeatable, as it takes time for chaotic flow instabilities and circulation cells to develop. This implies that isolated large wave groups in the field will cause larger velocities than would be predicted using the equivalent steady forcing. Since it is highly deterministic, it also implies that the time-varying current response to large wave groups in the field is to a large degree predictable.

A highly noticeable feature in all tests is that the response time of the rip current decreases strongly as wave heights increase. This difference in response times is believed to be because the larger waves produce higher gradients in radiation stresses which consequently produce larger accelerations, with the associated smaller response times. If the startup flow is conceptualised as two counter-rotating vortices on the edge of the bar, then from Peregrine (1998), we see that the rate of change of circulation, Γ, is

$$\frac{\partial \Gamma}{\partial t} = E_D \tag{1}$$

for a material circuit cutting through a bore, where E_D is the rate of energy loss from breaking. The average dissipation, and thus the fluid acceleration, will be approximately proportional to $H_i^2 - H_0^2$, where H_i is the incident wave height, and H_0 is the wave height just after the bar. Qualitatively, this explains the much faster response times for larger waves. No strong quantitative conclusions may be made with only three measurements, although a preliminary analysis (not shown) seems reasonable.

Of course, this is a gross simplification and offers no explanation for the large peak velocities, which must certainly be studied in more detail. However, we believe that the core of the reason for differing response times may be found in the above explanation.

Conclusions

Even from a relatively cursory analysis of the rip current data at relatively few locations, significant conclusions may be made.

First, rip currents respond strongly to time variations in wave height. With some small lag, the time series of cross-shore velocity in the rip neck mimicked closely those of wave forcing offshore of the bar. In these time series, rip current fluctuations in a broad band were dominated by the response at the group frequency.

In contrast, the wave group forcing used appeared to have a negligible effect on very low frequency (100's of s) fluctuations of the rip current. It was hypothesised that these were due either to instabilities of very large circulation cells, jet instabilities, chaotic wave-current interaction, or some combination.

It was found that for relatively weak currents, time-varying forcing could change mean current strength, while for large currents, there was a negligible effect. An explanation for this was developed based on time-averaged values of the forcing for the rip current.

Finally, it was demonstrated that highly repeatable, large amplitude, transient currents were produced by isolated wave groups in an otherwise quiescent body of water. These always produced peak currents significantly greater than the mean currents resulting from monochromatic waves with the same height. Current response times decreased strongly as wave heights increased.

In the future, this work is to be extended using theoretical analysis, numerical modelling, and experimental surface drifter data. Future experiments with comparable wave group and jet instability periods would also be helpful to examine their possible interaction. In these ways, some general observations and preliminary conclusions made here may be examined more rigorously to give a more complete picture of unsteady rip current response.

Acknowledgements

Work for this project was performed while ABK was supported by the ONR grant N00014-97-1-0283, and by the NOPP grant N00014-99-1-0398. Their support is gratefully acknowledged.

References

Funke, E.R., and Mansard, E.P.D. (1980). "On the synthesis of realistic sea states", in *Proc. Int. Conf. Coastal Eng.*, Sydney, 2974-2991.

Haas, K.A., and Svendsen, I.A. (2000). "Three-dimensional modeling of a rip current system", *Res. Rept. CACR-00-06*, Ctr. Appl. Coastal Res., University of Delaware, 250 pp.

Haller, M.C., Dalrymple, R.A., and Svendsen, I.A. (1997). "Experimental modeling of a rip current system," *Proc. Ocean Wave Meas. Anal., Waves'97*, ASCE, 750-764.

Haller, M.C., and Dalrymple, R.A. (2001). "Rip current instabilities," *J. Fluid Mech.*, **433**, 161-192.

Haller, M.C., Dalrymple, R.A., and Svendsen, I.A. (2001). "Experimental study of nearshore dynamics on a barred beach with rip channels," *J. Geophys. Res.*, in press.

Huntley, D.A., Hendry, M.D., Haines, J., and Greenidge, B. (1988). "Waves and rip currents on a Caribbean pocket beach, Jamaica," *J. Coastal Res.* **4**(1), 69-79.

Peregrine, D.H. (1998). "Surf zone currents," *Theor. Comp. Fluid Dyn.*, **10**, 295-309.

Shepard, F.P., and Inman, D.L. (1950). "Nearshore circulation related to bottom topography and wave refraction," *Trans. AGU* **31**(2), 196-212.

Wall, R (2001). "The sea of heartache so nearly claims another family," *The Age, Feb. 1, 2001*, Melbourne, Australia.

INITIAL MOTION OF SEDIMENT UNDER WAVES: A GENERAL CRITERION

Renata Gentile[1], Laura Rebaudengo Landò[2] and Giulio Scarsi[3]

Abstract: Some of the models already available to identify the onset of general motion of sediment under wave action are considered and a new one is proposed. A laboratory investigation to analyse the phenomenon under observation is carried out by assuming a criterion related to the formation of ripples in their initial stage. The above investigation is referred to regular waves and to monogranular sand bottoms in different initial conditions synthetically described by proper geotechnical parameters. The experimental results are interpreted using the mentioned analytical models, which in any case, do not take into account the initial conditions of sediment.

INTRODUCTION

Numerous studies usually based on experimental investigations and referred to the onset of general motion of sediment under wave action have been carried out by different authors, like Bagnold (1946), Manohar (1955), Vincent (1958), Goddet (1960), Rubatta (1965), Zoccoli (1965), Horikawa and Watanabe (1967), Rance and Warren (1968), Chan et al. (1972), Komar and Miller (1973), Madsen and Grant (1976), Sleath (1978), Davies (1980), Lenhoff (1982), Rebaudengo Landò and Scarsi (1982) and Tanaka and Van To (1995).

In the present paper, the model suggested by Tanaka and Van To (1995), hereafter indicated by the notation TVT, and the one proposed by Rebaudengo Landò and Scarsi (1982) named RLS, are described, and a new model, hereafter identified by the notation

1 Research Scientist, University of Genova, Department of Environmental Engineering, Via Montallegro 1, 16145 Genova, Italy, renata@diam.unige.it
2 Full Professor, University of Genova, Department of Environmental Engineering, Via Montallegro 1, 16145 Genova, Italy, lreb@diam.unige.it
3 Full Professor, University of Genova, Department of Environmental Engineering, Via Montallegro 1, 16145 Genova, Italy, scarsi@diam.unige.it

RLSM, is formalised, which represents a modification of the RLS model and allows an average curve to be defined in the plane $[S_*, (\tau_*)_c]$, S_* being a dimensionless parameter related to properties of sediment and fluid and $(\tau_*)_c$ the dimensionless shear stress at the bottom corresponding to critical conditions represented here by the onset of general motion.

The above models are used to interpret, among others, some results deduced on the basis of a new laboratory investigation carried out by the present authors.

The experiments were performed in a wave tank located in the laboratory of the Department of Environmental Engineering of the University of Genova by adopting incoherent bottoms of substantially monogranular sand of different diameter and regular waves with proper values of the wave period and height able to give rise to the formation of ripples in their initial stage. This condition, supported by experimental evidence, is assumed as the criterion adopted here to identify the onset of general motion. It should be pointed out that the phenomenon under observation appears to be rather difficult to identify and the authors who have dealt with the matter have introduced different criteria, also based on statistical approaches formalised starting from the number of particles which have been moved due to wave motion.

In the present analysis, different initial conditions of sediment, approximately indicated as not pressed bottom and very pressed one, are alternately prepared, which may be schematically quantified starting from the values of some geotechnical parameters like the relative density index D_R or the void ratio e. It is to be borne in mind that these initial conditions seem to play a significant role as regards the onset of general motion, which is more and more delayed when the value of D_R increases.

In the following: some models to evaluate the friction coefficient are introduced, the TVT, RLS and RLSM models are described and, finally, the results of the laboratory investigation are given and compared with the analytical ones.

SOME MODELS PROPOSED TO EVALUATE THE FRICTION COEFFICIENT RELEVANT TO WAVE MOTION

The models derived to identify the flow regime within the bottom boundary layer and to represent the relevant motion due to regular waves without currents superimposed on wave motion have reached great levels of reliability and a large number of authors has dealt with their formalisation. The recent contribution given by Tanaka and Thu (1994) gives rise to a model able to provide in explicit form amplitude and phase lead of the friction coefficient with reference to the different flow regimes which may occur in the bottom boundary layer. The above model, indicated by the notation TTF, will be briefly illustrated in the following, together with the one suggested by Rebaudengo Landò and Scarsi (1981), named RLSF, on which the RLS and RLSM models are based.

The TTF Model

The TTF model provides approximate formulae of wave friction coefficient and phase lead, spanning all flow regimes (full range equations). In the following, only the relationships used in the present analysis are given, that is the ones referred to the amplitude f_w of the friction coefficient, corresponding to the flow regime under observation. It turns out that

$$f_w = f_2\{f_1 f_{w(L)} + (1-f_1) f_{w(S)}\} + (1-f_2) f_{w(R)} \tag{1}$$

where $f_{w(L)}, f_{w(S)}$ and $f_{w(R)}$ are referred, respectively, to Laminar, Smooth turbulent and Rough turbulent flow regimes and f_1 and f_2 are weight functions.
In more detail, the following expressions may be adopted

$$f_{w(L)} = 2 / (R_a)^{0.5} \tag{2}$$

$$f_{w(R)} = \exp \{-7.53 + 8.07 \zeta^{-0.100}\} \tag{3}$$

$$f_{w(S)} = \exp \{-7.94 + 7.35 R_a^{-0.0748}\} \tag{4}$$

$$f_1 = \exp \{-0.0513 (R_a / 2.5\ 10^5)^{4.65}\} \tag{5}$$

$$f_2 = \exp \{-0.0101 (R_a / R_l)^{2.06}\} \tag{6}$$

$$R_a = \hat{U}_w \hat{a}_m / v \quad ; \quad R_l = 0.501\ \zeta^{1.15} \quad ; \quad \zeta = \hat{U}_w / \sigma z_o \tag{7}$$

where $\hat{U}_w$ and $\hat{a}_m$ are, respectively, the amplitude of the wave-induced velocity and of the orbital excursion just outside the boundary layer, v is the kinematic viscosity of water, σ is the angular frequency of the wave motion and z_o the roughness length ($z_o = k_s / 30$ with k_s Nikuradse's equivalent roughness).
The flow regime which occurs in the bottom boundary layer may be deduced starting from Figure 3 in the original paper (Tanaka and Thu 1994).

The RLSF Model

The RLSF model provides formulae of wave-friction coefficient and phase lead with reference to the different flow regimes which may occur in the bottom boundary layer, included the separation ones. In the following, only the relationships used in the present analysis are given, that is the ones referred to the amplitude $\hat{C}$ of the friction coefficient, corresponding to the flow regime under observation.

Laminar flow regime

$$\hat{C} = 1/ (R_a)^{0.5} \tag{8}$$

Rough turbulent flow regime

$$1/\hat{C}^{0.5} = 7.50 \log \{ \hat{C}^{0.5} \hat{a}_m /0.30\ k_s\}; \quad \hat{a}_m / k_s > 2.67 \tag{9}$$

$$\hat{C} = 0.094; \qquad\qquad \hat{a}_m / k_s \leq 2.67 \tag{10}$$

Smooth turbulent flow regime

$$1/\hat{C}^{0.5} = 6.70 \log \{ \hat{C} R_a/ 0.93\} \tag{11}$$

Transition between laminar and smooth turbulent flow regimes

$$\log \hat{C} = 2.425 - 1.616 \log R_a + 0.127 (\log R_a)^2 \qquad (12)$$

Other transitional flow regimes

$$\log \hat{C} = \log (\hat{C})_A + [\log (\hat{C})_B - \log (\hat{C})_A] \, W \qquad (13)$$

$$W = \{\log R_a - \log (R_a)_A\} / \{\log (R_a)_B - \log (R_a)_A\} \qquad (14)$$

It is worth noting that the fields of existence of the different flow regimes and the quantities characterised by the indexes A and B are given in the original paper (Rebaudengo Landò and Scarsi 1981).

SOME MODELS PROPOSED TO DEFINE THE ONSET OF GENERAL MOTION OF SEDIMENT

In the following, the model proposed by Tanaka and Van To (TVT), referred here to waves only and the one suggested by Rebaudengo Landò and Scarsi (RLS) are briefly summarised, together with the modified Rebaudengo Landò and Scarsi (RLSM) model proposed here.

The TVT Model

In order to formalise the model, three sets of data from Manohar (1955), Goddet (1960) and Horikawa and Watanabe (1967) have been used by Tanaka and Van To to evaluate the Shields parameter $(\tau_*)_c$ under general motion. It should be pointed out that the values of bed friction relevant to the onset of general motion are greater than those required for initial motion of ~10-15%, as Figure 1, taken from the original paper, shows. The model provides an explicit formula for general motion, supplied by

$$(\tau_*)_c = 0.078 \{1 - \exp [-0.01 (S_*)^{1.01}] \} + 0.207 (S_*)^{-0.69} \qquad (15)$$

$$(\tau_*)_c = (\hat{\tau}_*)_c = \hat{\tau}_c / (\rho_s - \rho) \, g \, d \qquad (16)$$

$$S_* = d [(\rho_s / \rho - 1) \, g \, d]^{0.5} / 4 \, v \qquad (17)$$

In the previous equations ρ and ρ_s are the water and sediment density, g is the acceleration of gravity, d is the sand diameter and $\hat{\tau}_c$ is the amplitude $\hat{\tau}$ of the shear stress at the bottom evaluated with reference to critical conditions, noting that $\hat{\tau}$ is given by

$$\hat{\tau} = 0.5 \, \rho f_w \, \hat{U}_w^2 \qquad (18)$$

The curve given by Eq. 15 is shown in Figure 1 (broken line).

It is worth noting that not all the data considered in the above figure are in the laminar or turbulent regimes, but are referred also to transitional flow regimes from laminar to rough turbulent, thus the explicit full range formula for the wave friction coefficient proposed by Tanaka and Thu has been used.

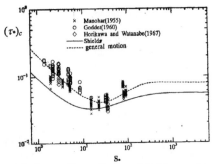

Fig. 1. Experimental observations of general motion and critical curve obtained using the TVT model (broken line, Eq. 15).

The RLS Model

In order to formalise the model, data from Manohar (1955), Goddet (1960), Rubatta (1965), Zoccoli (1965), Horikawa and Watanabe (1967), Chan et al. (1972), Sleath (1978) and Davies (1980) have been considered by Rebaudengo Landò and Scarsi. The model interprets the above data relevant to the onset of general motion directly with the original Shields curve (see Figure 2) and gives rise to the following relationships

$$(\tau_*)_c = 0.055 \{ 1 - \exp [-0.09 \, (S_*)^{0.58}] \} + 0.09 \, (S_*)^{-0.72} \tag{19}$$

$$(\tau_*)_c = (\tau_{*m})_c = (\tau_m)_c / (\rho_s - \rho) \, g \, d \tag{20}$$

In the previous equation $(\tau_m)_c$ is a mean value τ_m of the shear stress at the bottom evaluated with reference to critical conditions, noting that τ_m is given by

$$\tau_m = 0.405 \, \alpha \, \hat{\tau} \tag{21}$$

$$\hat{\tau} = \rho \, \hat{C} \, \hat{U}_w^2 \tag{22}$$

α being a proper coefficient (see Rebaudengo Landò and Scarsi 1982).

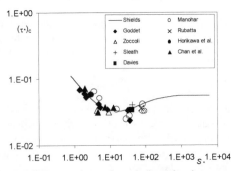

Fig. 2. Experimental observations of general motion interpreted using the RLS model and critical curve represented by the original Shields curve.

The RLSM Model

The RLSM model, proposed here, represents a modification of the original RLS model.

It consists in interpreting the data provided by the authors already considered with reference to the RLS model by using a least square regression method and starting from the amplitude of the shear stress. The following relationship is obtained

$$(\tau_*)_c = 0.083 \{1 - 0.85 \exp [-0.02 (S_*)^{0.95}] \} + 0.18 (S_*)^{-0.69} \tag{23}$$

which gives rise to the curve plotted in Figure 3, noting that $(\tau_*)_c$ is supplied by Eq. (16), related here to Eq. (22).

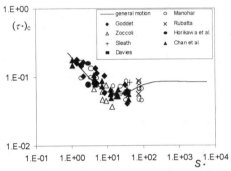

Fig. 3. Experimental observations of general motion and critical curve obtained using the RLSM model (Eq. 23).

EXPERIMENTAL INVESTIGATION

An experimental investigation is carried out by using a wave tank located in the laboratory of the Department of Environmental Engineering of the University of Genova. This investigation is based on the generation of regular waves of assigned period and proper height able to produce ripples in their initial stage, noting that the sediment is substantially made of monogranular sand in different initial conditions of compactness.

After a brief description of the experimental equipment, the tests performed are illustrated and some laboratory results are summarised.

Experimental Equipment

The wave tank, sketched in Figure 4, has a length of ~ 15 m, a width of ~ 3 m and a usable height of ~ 0.8 m. It is provided with a wave maker, wave filters, wave guide and dampers and is connected to equipment essentially consisting of an electric signal generator, a device to acquire and record the signals and a probe to monitor the wave surface.

The bottom of the tank has, starting from the horizontal section where the wave maker is located, a horizontal cemented section ~ 4 m long, followed by a section, also cemented, with a slope of ~ 6%, linked to a second horizontal section ~ 4 m long (test

section), partially cemented (0.80 m long) and partially covered by the monogranular sand, followed by a final section with a slope ~ 8% leading to the wave dampers.
The wave maker consists of an oscillating plate hinged to the bottom and operated by an oleodynamic system, controlled by a computer through the software LABview, which guarantees high precision and reliability in the reproduction of electric signals.
The resistive probe is located approximately in the middle of the test section.

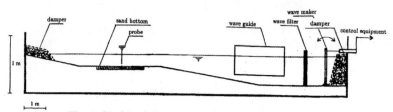

Fig. 4. Sketch of the tank used for experimental tests.

Experimental Tests

The experimental analysis is based on several tests referred to different values of the wave period T and of the sand diameter d and to different initial conditions of the sediment, synthetically described by the relative density index D_R or the void ratio e, whose values are obtained by carrying out preliminary operations and in situ standard measurements.
Wave periods in the range $T = 0.8$ s to $T = 1.3$ s are adopted, sand bottoms (characterised by $\rho_s = 2650$ kg/m^3) with $d = 0.8$ mm and $d = 0.3$ mm are used and initial conditions for the sediment ranging from a not pressed (NP) bottom to a very pressed (VP) one are considered.
In fact, at the beginning of each test, the bottom does not present a uniform spatial degree of compactness, as the values of D_R measured in different areas of the test section reveal. For example, with respect to $d = 0.3$ mm the NP bottom shows a minimum value of D_R given by $(D_R)_{min} = 0.07$, a maximum value $(D_R)_{max} = 0.55$, with a mean indicative value $(D_R)_{mean} = 0.24$; the VP bottom exhibits the values $(D_R)_{min} = 0.45$, $(D_R)_{max} = 0.93$ and $(D_R)_{mean} = 0.75$. Besides, the non uniform distribution of sediment does not represent a limitation for the purpose of the present analysis, as, with reference to each test performed, the relevant values of D_R have been evaluated in the spatial areas where ripples were generated.
In the following: only the results referred to $d = 0.3$ mm are given, taking into account that the formation of ripples was in most cases prevented with $d = 0.8$ mm, as the wave height able to cause the phenomenon could not be achieved in the tank; the results referred to NP and VP bottoms are illustrated in detail and situations in the test section characterised by the water depth $h = 0.15$ m are considered, noting that a different value (for example $h = 0.20$ m) gives rise to the same dimensionless values in the plane $[S*, (\tau*)_c]$.
Tables 1 and 2, referred, respectively, to NP and VP bottom conditions, summarise the laboratory results obtained in the test section. The above tables refer to tests lasting 30

minutes each, which are associated with the wave height H able to give rise to the formation of ripples in their initial stage, the wave period T being assigned. In the same tables, a characterisation relevant to the degree of development of ripples is given in a first approach scheme together with the values of D_R.

Table 1. Bottom in NP Conditions: Wave Period T and Wave Height H associated with Formation of Ripples* and Relative Density Index D_R .

T, s	H, m	Degree of development of ripples	D_R
0.8	0.045	❷	0.07
0.9	0.045	❷	0.07
1.0	0.046	❸	0.07
1.1	0.051	❸	0.14
1.2	0.053	❸	0.14
1.3	0.050	❷	0.14

* ❶ = few extended ripples; ❷ = more extended ripples; ❸ = extended ripples

Table 2. Bottom in VP Conditions: Wave Period T and Wave Height H associated with Formation of Ripples* and Relative Density Index D_R .

T, s	H, m	Degree of development of ripples	D_R
0.8	0.050	❶	0.45
0.9	0.053	❷	0.65
1.0	0.064	❸	0.76
1.1	0.070	❸	0.79
1.2	0.069	❸	0.80
1.3	0.067	❷	0.79

* ❶ = few extended ripples; ❷ = more extended ripples; ❸ = extended ripples

Tables 1 and 2 show that, under the same period, going from NP to VP bottoms, the values of the wave height increase (minimum difference ~10%, maximum difference ~40%). Moreover, in the different tests, it is really difficult to obtain the same degree of development of ripples, as the qualitative description given in the above tables shows. It is worth noting that the values of D_R are, as expected, lower with reference to NP conditions and higher with regard to VP ones. Furthermore, in each test and with respect to each situation considered (NP and VP bottoms), ripples began to form in spatial areas characterised by values of D_R lower than the ones associated with the other zones.

COMPARISONS BETWEEN THE RESULTS DERIVED FROM THE PRESENT EXPERIMENTAL INVESTIGATION AND THOSE OBTAINED WITH THE MODELS CONSIDERED

In order to compare the results obtained starting from the present experimental investigation and the ones deduced by using the analytical models, first of all the TVT, RLS and RLSM models are employed to derive, with respect to each wave period, the value of the wave height able to give rise to critical values of the shear stress at the bottom, corresponding respectively to Eqs. 15, 19 and 23.
Figure 5 shows the critical values of the wave height versus the values of the wave

period, obtained both from the models considered and from the laboratory investigation.

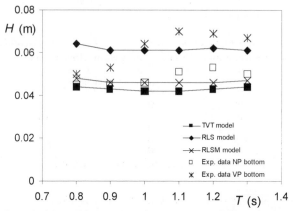

Fig. 5. Wave height versus wave period: the onset of general motion interpreted using the models considered and starting from the laboratory investigation.

The figure shows that the experimental results referred to the NP bottom are feebly underestimated by the TVT model and better interpreted by the RLSM model, whereas the ones related to the VP bottom are globally represented with sufficient approximation by the RLS model.

The values of wave height and period derived from the tests performed are therefore introduced in the analytical models in order to deduce the relevant values of the shear stress at the bottom, which are compared with the critical ones obtained with the same models, as Figure 6 shows.

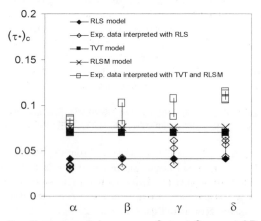

Fig. 6. Dimensionless shear stress $(\tau_*)_c$ versus α, β, γ and δ ranges of D_R: laboratory results interpreted with the theoretical models and analytical results.

In the above figure, the four ranges of D_R (indicated by the notations α, β, γ and δ) in which laboratory data have been arranged, are given by $\alpha = [0\text{-}0.25]$, $\beta = [0.25\text{-}0.50]$, $\gamma = [0.50\text{-}0.75]$ and $\delta = [0.75\text{-}1.00]$.

According to the final considerations referred to Figure 5, Figure 6 shows that the TVT model seems to feebly underestimate and the RLSM model to better interpret laboratory data in the α range of D_R, the differences becoming important in the other ranges. The RLS model seems to represent the data with sufficient accuracy in the β and γ ranges.

CONCLUDING REMARKS

The following conclusions may be drawn.

1. The criterion proposed to interpret the onset of general motion of sediment made of monogranular sand is based on the experimental evidence related to the formation of ripples in their initial stage.

2. The experimental investigation carried out shows that the phenomenon under observation is influenced, among others, by the geotechnical characteristics of the bottom itself, which are synthetically described here in terms of the relative density index D_R or void ratio e.

3. The experimental results are interpreted using some of the analytical models available (indicated by the notations TVT, RLS and RLSM), which do not take into account the above-mentioned circumstance.

4. The experimental results referred to the not pressed (NP) bottom are feebly underestimated by the TVT model and better interpreted by the RLSM model, whereas the ones related to the very pressed (VP) bottom are globally represented with sufficient approximation by the RLS model.

5. A new model should be developed, also taking into account a synthetic parameter like D_R (or e) which influences, as already stated, the phenomenon under observation. For this purpose, further laboratory investigations are needed, with particular reference to different grain sizes characterising the sand bottom. This new model could be useful, for example, to analyse the behaviour of artificial beaches prior to their final arrangement.

ACKNOWLEDGEMENTS

The authors wish to thank Dr. Stefania Remigio for the contribution given in performing the laboratory tests and in analysing the geotechnical aspects.

REFERENCES

Bagnold, R.A. 1946. Motion of Waves in Shallow Water. Interaction between Waves and Sand Bottom. *Proceedings of the Royal Society of London*, 187 (A), 1-15.

Chan, K.W., Baird, M.H.I., and Round, G.F. 1972. Behaviour of Beds of Dense Particles in a Horizontally Oscillating Liquid. *Proceedings of the Royal Society of London*, 330 (A).

Davies A.G. 1980. Field Observations of the Threshold of Sand Motion in a Transitional Wave Boundary Layer. *Coastal Engineering*, 4: 23-46.

Goddet, J. 1960. Etude du Debut d'Entrainement des Materieux Mobiles sous l'Action de la Houle. *La Houille Blanche*, 15: 122-135.

Horikawa, K. and Watanabe, A. 1967. A Study of Sand Movement due to Wave Action. *Coastal Engineering in Japan*, 10: 39-57.

Komar, P.D. and Miller, M.C. 1973. The Threshold of Sediment Movement under Oscillatory Water Waves. *Journal of Sediment. Petrol.*, 45: 362-367.

Lenhoff, L. 1982. Incipient Motion of Sediment Particles. *Proceedings of the 18th International Conference on Coastal Engineering*, Cape Town, ASCE, 1555-1568.

Madsen, O.S. and Grant, W.D. 1976. Quantitative Description of Sediment Transport by Waves. *Proceedings of the 15th International Conference on Coastal Engineering*, Honolulu, ASCE, 1093-1112.

Manohar, M. 1955. Mechanics of Bottom Sediment Movement Due to Wave Action. *Technical Memory*, US Army Corps of Eng., Beach Erosion Board, Washington DC, 75: 121 p.

Rance, P.J. and Warren, N.F. 1968. The Threshold of Movement of Coarse Material in Oscillating Flow. *Proceedings of the 11th International Conference on Coastal Engineering*, London, ASCE, 487-491.

Rebaudengo Landò, L. and Scarsi, G. 1981. Sulla Potenza dissipata dalle Onde di Gravità Progressive. *Atti dell'Accademia Ligure di Scienze e Lettere*, XXXVII, 3-24.

Rebaudengo Landò, L. and Scarsi, G. 1982. L'Inizio del Modellamento del Fondo per Azione del Moto Ondoso. *Idrotecnica*, 1: 29-40.

Rubatta, A. 1965. Inizio del Modellamento indotto dal Moto Ondoso su Fondo Incoerente a Grossa Granulometria. *Atti IX Convegno di Idraulica e Costruzioni Idrauliche*, Trieste, Italy, 106-109.

Sleath, J.F.A. 1978. Measurements of Bed Load in Oscillatory Flow. *Journal of Waterway, Port and Coastal Engineering*, 104: 291-307.

Tanaka, H. and Thu, A. 1994. Full Range Equation of Friction Coefficient and Phase Difference in a Wave Current Boundary Layer. *Coastal Engineering*, 22: 237-254.

Tanaka, H. and Van To, D. 1995. Initial Motion of Sediment under Waves and Wave-Current combined Motion. *Coastal Engineering*, 25: 153-163.

Vincent, G.E. 1958. Contribution to the Study of Sediment Transport on a Horizontal Bed due to Wave Action. *Proceedings of the 6th International Conference on Coastal Engineering*, ASCE, 326-355.

Zoccoli, F. 1965. Azione di Moti Ondosi su Fondo Mobile. Primi Risultati Sperimentali. *Atti IX Convegno di Idraulica e Costruzioni Idrauliche*, Trieste, Italy, 100-105.

ANALYSIS ON TYPHOON-INDUCED SHORELINE CHANGES USING REMOTE SENSED IMAGES

Toshiyuki Asano[1], M. ASCE,   Jun-ichi Khono[2],
Tomoaki Komaguchi[3] and Takao Sato[4]

Abstract: To study beach morphological change, applicability of remote sensed images has been investigated. Shoreline displacements analyzed by satellite imageries were comparing with field survey data in the corresponding period. The agreements are found to be satisfactory, thus the present analysis using satellite imageries are validated. Next, short term beach evolutions induced by typhoon attack are discussed. Although the analyzed shoreline displacements do not directly correspond to typhoon induced shoreline change, most of the obtained characteristics are found to be physically reasonable. Also, the succeeding shoreline change suggests that beach recovery process occurs after the typhoon-induced erosion. Furthermore, to explain the physical process of the typhoon induced shoreline change, numerical analyses on the wave hindcasting and wave transformation have been conducted. The results of typhoon hindcasting and shallow water transformation reveal that even a small change in wave incident angle results in a great difference in the wave height distributions in the alongshore direction of the study site. This implies that temporal change of wave angle associated with the typhoon-passing is an important factor on beach deformation.

INTRODUCTION

Conventional ground-based survey, covering several tens of kilometers, is limited economically to only once or twice each year. Meanwhile, the use of remote-sensed images provides frequent morphological information at lower costs. Recently, the accuracy of imagery has much advanced and the availability has also improved.

Along the Pacific coast in southern Japan islands, severe beach erosions due to

---

1. Prof., Dept. of Ocean Civil Engrg., Kagoshima Univ., 1-21-40, Korimoto, Kagoshima, 890-0065,   JAPAN, Fax: +81-99-285-8482, e-mail : asano@oce.kagoshima-u.ac.jp
2. Senior Engineer, Kagoshima Environmental Science Association, 1-15 Nanatsu-jima, Kagoshima, 891-0132, JAPAN, Fax: +81-99-262-1705, jkohno@kagoshima-env.or.jp
3. Manager, Tetra Co., Div. of Promotion for Development of Fishery Infrastructure, 6-3-1 Nishi-shinjuku, Shinjuku, Tokyo, 160-8350, JAPAN, komaguchi@tetra.co.jp
4. Former Head, Shimonoseki Investigation & Design Office, MLIT, 4-6-1, Takesaki-cho, Shimono- seki, 750-0025, JAPAN, Fax: +81-832-28-1108

typhoon attacks occur almost every year. These short-term beach deformations are little understood because field surveys are seldom conducted just before and after the storm. On the other hand, since the recurrence period of the sensing satellite is only around one month, it is possible to investigate the short-term beach morphologic changes over a wide area using remote sensed images.

This paper investigates the applicability of satellite images in beach morphology change analysis, and its verification by comparing with field survey data. In addition, to study the physical process on typhoon induced shoreline change, numerical analyses on the wave hindcasting and wave transformation, for an objective coastal area and specific typhoons, have been conducted.

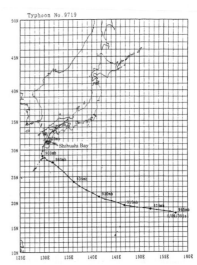

Fig. 1   Location of Shibushi Bay
and track of Typhoon T9719

METHOD AND PROCEDURE
Study area and objective typhoon

Judging from the availability of remote sensed images and field survey data, Shibushi-Bay was chosen as the study site. The bay is located in the south east coast of Kyushu mainland and opens to the Pacific Ocean.

The objective typhoon was adopted as typhoon No. 19 in 1997 (T9719). The typhoon struck the Shibushi-Bay around the 16th of September, 1997, causing severe beach erosion. Typhoon T9719 originated at the north latitude 13.6° and east longitude 179.7° on the 4th of September, 1997. It moved eastwards near Chichi-jima Island at three o'clock of the 12th, and the lowest central atmospheric pressure was recorded at 915hPa and the maximum wind speed at 50m/s. Then, it turned northbound, and landed

Table 1   Analyzed satellite imageries

	No.1	No.2	No.3	No.4	No.5
Obs. Date	11 Dec. 96	4 May 97	12 June 97	24 Sep. 97	30 Mar. 98
Sattellite	SPOT-2	IRS-1C	SPOT-2	SPOT-2	SPOT-2
Sensor	XS	PAN	PAN	PAN	PAN
Spat. Resol.	20m	5.8m	10m	10m	10m
Cloud Rate	0	3	7	3	0
Tidal Level	85cm	40cm	139cm	134cm	89cm

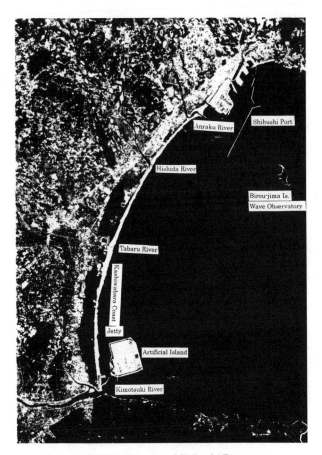

Fig. 2 SPOT/HRV-P image of Shibushi Bay

in Makurazaki city, southern tip of Kyushu mainland at eight o'clock of the 16th. It crossed Kyushu mainland, and finally landed again in Okayama Prefecture on the 17th. The typhoon weakened to 998hPa at that time. Figure 1 indicates the location of Shibushi-Bay and illustrates the track of Typhoon T9719. Figure 2 is the SPOT/HRV-P image of Shibushi-Bay taken at the 30th of March, 1998, where geographic landmarks are also shown.

For the objective area and period, proper imageries with clear visibility were searched, and five imageries taken by SPOT and IRS satellites were adopted (Table 1). Imagery No. 1 in Table-1 was taken by SPOT multi-band sensor (XS), of which terrestrial resolution was 20m. Imagery No.2 was taken by IRS satellite, of which the resolution was 5.8m. The rest were by SPOT panchromatic data with 10m resolution.

Pre-processing of Imageries

Positioning among imageries (image to image registration) with different pixel size was conducted as follows. First, a digital map of the study area scaled 1/25000 was projected into Universal Transverse Mercator (UTM) coordinate system with 2m pixel size. This was used as a map coordinate. Next, SPOT imageries were warped to match the UPM map coordinate. In this step, these imageries were re-sampled by a cubic convolution method, and this process increased their resolutions up to the accuracy of 2m. Third, the precise positioning for the

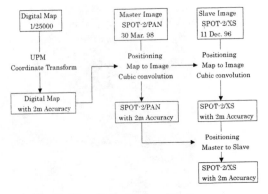

Fig. 3   Flow diagram of positioning

imageries was conducted using more than 15 Ground Control Points (GCP). These GCP were herein selected as tips of breakwater or corners of big buildings, etc, because they have clear visibility. No. 5 imagery in Table 1 was herein adopted as the base image geometry, because this imagery has the clearest visibility. Finally, all the rest imageries were corrected to match the base image geometry. The accuracy of transform using GCP was found to be within 0.3 pixel size. To summarize the above positioning procedures, the flow diagram is shown in Figure 3.

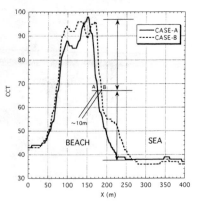

Fig. 4   CCT value profiles in cross shore direction

Reading shoreline position

Using the converted images with resolution 2m/pixel, the shoreline positions were detected. Figure 4 shows one example of the CCT value profiles in the cross-shore direction. The reflection intensity becomes high in the sand area and low in the water area. In the profile-B, a plateau like profile was observed around at x=200-250m, which would be caused by white caps induced by breaking waves. The similar disturbance will be arisen by white splash generated at run-up wave front. Here, the shoreline position was determined as the position providing the mean CCT value between the sandy beach area and sea area illustrated A-B in Fig. 4. Such determination would include some inaccuracy due to the above mentioned disturbance factors. Therefore, as the next step, the detected shoreline positions were carefully checked in the alongshore direction, and the data containing local disturbance effects were discard and replaced by the smoothed surrounding data. Finally, the obtained shoreline positions were corrected by referring the tidal level when the imagery was taken. The tidal correction was conducted by assuming the local beach slope to be 1/10. It should be noted that beach slopes near river mouths or near jetties tend to be much milder and the shoreline topographies are complex, thus, not negligible errors would be included around these areas through the tidal correction process.

RESULTS ON SHORELINE DISPLACEMENTS

The shoreline displacements between 1996/12/11 and 1998/3/30, analyzed by the satellite imageries, are shown in Figure 5. The imagery in the figure was taken on the former date. The displacements are expressed as the shoreline advances or retreats from the positions on 96/12/11 as the basis. The gray curve denotes the moving averaged shoreline position over a width of 500m. From the figure, shoreline advances are noticed in three points; the area around x=2000m, the right hand side of Hishida River, and behind the artificial island. On the other hand, shoreline retreats are observed at the left hand side of Hishida River, and over the wide area along Kashiwabara coast from x=6000m to 11000m. The detailed comparisons with the corresponding results by the field data will be given in the next chapter.

Next, the shorter-term beach evolution was examined using the imageries taken at 97/5/4 (No.2 image in Table 1) and 97/9/24(No.4 in Table 1). Note that the time interval is around 140 days, and the latter image was taken just one week after the typhoon attack. So, the shoreline displacements would be considered to show the typhoon induced shoreline change, although it contains the preceding shoreline change starting from 97/5/4. Furthermore, the succeeding shoreline changes after 97/9/24 were also investigated by analyzing No.5 imagery, which was taken at 98/3/30. The shoreline displacement during this period would be considered to represent the beach recovery

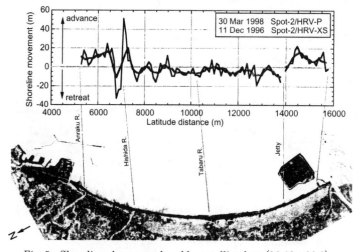

Fig. 5   Shoreline change analyzed by satellite data (96.12～98.3)

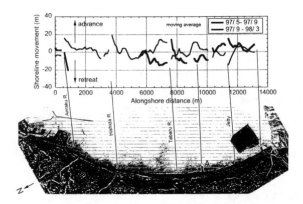

Fig. 6   Short term shoreline change analyzed by satellite data
(97.5〜97.9, 97.9〜98.3)

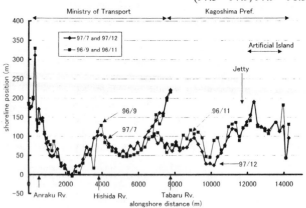

Fig. 7   Shoreline positions by field survey

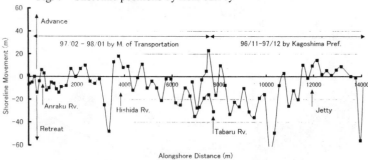

Fig. 8   Shoreline change evaluated by field survey data

process after the typhoon attack.

The obtained shoreline displacements are represented in Figure 6. It is noted that the curves are only shown partially because the image was interrupted by clouds over the shoreline. The results during 97/5/4 to 97/9/24 show around 10m retreats both north of Tabaru River (x=6000-7500m) and Kashiwabaru coast (x=8000-10000m). As Fig. 5 also illustrates erosion trends around these areas, these areas would be considered as vulnerable for erosion.   Meanwhile, the results during 97/9/24 to 98/3/30 indicate that the preceding shoreline movements during 97/5/4 to 9/24 were offset. This implies that the beach recovery against the preceding erosion can be attained.

COMPARISON WITH FIELD SURVEY DATA

The bathymetric field surveys have been conducted separately for the Shibushi Bay area. The area north of the Tabaru River has been surveyed by the Ministry of Land, Infrastructure and Transport (MLIT), while the area of south of the river has been surveyed by Kagoshima Prefecture.   The field survey data for the corresponding period to Fig. 5 were analyzed. The survey dates are slightly different from those of the imagery data, and the survey dates by MLIT and by Kagoshima Prefecture are also slightly different. The results on the shoreline position are illustrated in Figure 7. From the results of Fig. 7, the shoreline advance/retreat was evaluated as shown in Figure 8.

Comparisons between Fig. 5 and Fig. 8 reveal good agreements both for the location and magnitude of the shoreline change. Specifically, the both indicate roughly:
      * 40m retreat at the left bank of Hishida River,
      * 20m advance at the right bank,
      * 20m retreat over a wide range from x=6000m to x=10000m,
      * 20m retreat in the north region of the jetty,
      * 20m advance at the region behind the artificial island.

NUMERICAL SIMULATIONS

In order to investigate the physical process of the obtained beach morphologic changes shown above, numerical calculations have been conducted to reproduce the typhoon induced wave fields at the study site. The calculations are comprised of wave hindcasting for Typhoon T9719 and the wave deformation in the shallow area.

Wave hindcasting for offshore waves

Wave hindcasting model used here is based on spectral wave model (Komaguchi, Tsuchiya and Shiraishi: 1990).   This model is categorized as a so-called second generation spectral model, to describe two-dimensional properties of wave evolution and decay processes. The total energy is evaluated by integrating the energy balance equation as follows:

$$\frac{\partial \bar{E}(f,\theta,\bar{x},t)}{\partial t} + \nabla \bar{E}(f,\theta,\bar{x},t)\bar{C}_g = S(f,\theta,\bar{x},t),$$

where, $\bar{E}$ is the two dimensional energy density spectrum for wave frequency f and angle θ, $\bar{C}_g$ is the group velocity derived from linear wave theory. Source function S represents the three different stages of the sea state: the growing, decaying and opposing wind stages.

The calculating domain is discretized into varied mesh sizes. The coarsest mesh size is $0.5°$ (around 56km), and the finest is 4km. The southern end of the calculation domain was set at $10°$ in the north latitude. The transferred swell wave energy through the southern boundary was considered based on Ross (1976) model.

Figure 9 shows the comparisons between the hindcasted and observed results on the temporal variations of significant wave heights, wave period and wave angle at the Birou Islands. For the wave height, the agreements are satisfactory especially around the peak energy period from 9/14 to 9/16. Meanwhile, for the wave period, the hindcasted results are not able to reproduce swell waves in the early growing stage around the date of 9/13.

Figure 9 (b) illustrates the comparison on the energy averaged wave angle. The observed wave angle varied from SE $(135°)$ to SSE $(157.5°)$, meanwhile, the hindcasted wave angle yields a slightly greater angle than the observed one. Although, the position of the typhoon changed rapidly, the observed and hindcasted wave angles notably varied in small ranges. This result may be caused by the configuration of Shibushi Bay, where the bay opens south-east to the ocean, and both capes play sheltering effects.

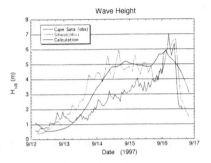

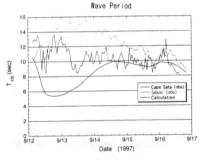

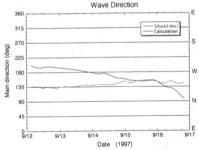

Fig. 9   Comparisons with hindcasted and observed waves induced by Typhoon T9719

Wave deformations in shallow area

Wave height distributions in Shibushi Bay have been calculated based on the energy balance equation (Karlsson: 1969). The hindcasted wave series was used as the boundary condition at the offshore side of the bay. The time history of hindcasted waves was not used directly, but instead the equivalent energy averaged wave properties were given. Based on the hindcasted and observed results shown in Fig. 9, the following representative wave properties were given as the input waves: significant wave height $H_{1/3}$=5m, wave period $T_{1/3}$=12 s, the

spreading parameter S_{max}=75, and the incident wave angle $\theta = 130°$ and $\theta = 150°$ for

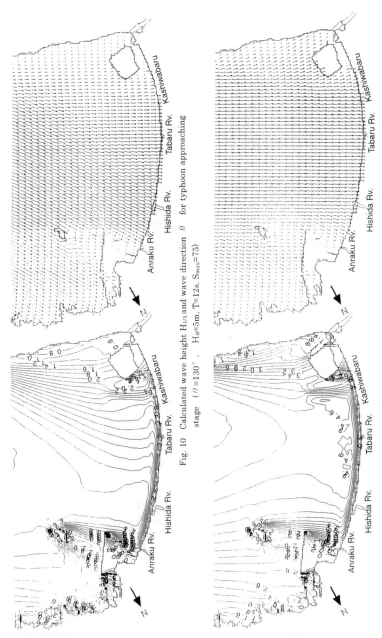

Fig. 10 Calculated wave height $H_{1/3}$ and wave direction θ for typhoon approaching stage ($\theta = 130°$, $H_0 = 5m$, $T = 12s$, $S_{max} = 75$)

Fig. 11 Calculated wave height $H_{1/3}$ and wave direction θ for typhoon most closing stage ($\theta = 150°$, $H_0 = 5m$, $T = 12s$, $S_{max} = 75$)

the typhoon- approaching and the nearest stage, respectively. The spectrum shape is assumed as Bretschneider-Mitsuyasu type. The frequency domain in the calculation covers $0.04 - 0.25$Hz in 0.01Hz intervals. The directional domain is divided into $10°$ intervals.   The calculation mesh size is 100m.

Figure 10 (a) and (b) illustrate the wave height and wave angle distributions for the typhoon-approaching stage. In the alongshore direction, the peak wave height reaches about 5m around the mouth of Hishida River. Observation of south of the Hishida River shows that the wave height decreases monotonously due to the sheltering effect of the south cape. Meanwhile, the corresponding results for the nearest stage depict different distributions in Figure 11 (a) and (b), even though the incident wave angle only differs by $20°$. The area of peak wave height in the alongshore direction is further south, around Kashiwabara coast. In particular, quite large wave heights up to 4.8m are noticed just north of the artificial island.   Compared with the corresponding wave heights of about 3.0m when $\theta = 130°$, this difference more than doubles in terms of wave energy. Concerning the wave angle, large differences are found near the entrance of Shibushi Harbor and sheltered region behind the artificial island.   This suggests that typhoon-induced shoreline change would depend on how wave angle varies from arriving stage to the leaving stage of a storm, as well as how long the storm wave lasts.

CONCLUSIONS

1) By warping SPOT and IRS imageries with different pixel sizes in the UPM map coordinate system, and by invoking the cubic convolution re-sampling, precise shoreline positioning has been conducted with the higher resolution of 2m than the original pixel size.

2) The shoreline displacements analyzed by the satellite imageries are found to be in good agreements with the field survey data in the corresponding period.

3) Using two imageries taken before and after the typhoon attack, short-term beach evolutions are discussed. The succeeding shoreline change suggests that beach recovery process occurs after the typhoon-induced erosion.

4) Numerical simulations on typhoon hindcasting and shallow water transformation reveal that even a small change in wave incident angle results in a great difference in the wave height distributions in the alongshore direction. This implies that temporal change of wave angle associated with the typhoon-passing is an important factor on beach deformation.

ACKNOWLEDGEMENTS

The authors thank the support of the satellite imageries by National Space Development Agency (NASDA) Japan, and the support of the field survey data for the southern area of Tabaru River by river engineering branch, Kagoshima Prefecture.

REFERENCES

Karlsson, T. (1969): Refraction of continuous ocean spectra, J. Waterways and Harbor Div., Vol.95, pp.437-448.

Komaguchi T., Y. Tsuchiya and N. Shiraishi (1990): Generation mechanism of abnormal waves along the Japan coast, Proc. of 22nd Intern. Conf. of Coastal Engineering, Vol. 1, pp.769-782.

Ross, D. B. (1976): A simplified model for forecasting hurricane generated waves (Abstract), Bull. Am. Meteorological Soc., 113p.

MODELING OF THE 2ND ORDER LOW FREQUENCY WAVE SPECTRUM IN A SHALLOW WATER REGION

Akira Kimura [1]   and   Kohsaku Tanaka [2]

Abstract: This study deals with the modeling of the 2nd order low frequency directional wave spectrum. The product of Bretschneider-Mitsuyasu frequency spectrum and Mitsuyasu directional function is used for the standard 1st order directional wave spectrum. The 2nd order spectrum is calculated by the convolution developed by Kimura (1984). The sea state parameters $(H_{1/3})_0$, $T_{1/3}$ and S_{max} in the 1st order spectrum are selected so that they satisfy the average relation shown by Goda et al. (1975). The frequency spectrum and the directional function of the 2nd order low frequency waves are almost independent. The frequency spectrum and directional function are approximated by simple equations individually. The results show good agreements with the theoretical values.

INTRODUCTION

The second order non-linear waves have attracted special interest as driving forces in several phenomena such as harbor resonance (Bowers, 1977, Kimura et al., 1998), slow drift oscillation of floating bodies (Remery et al., 1971, Pinkster, 1974) and formation of multiple longshore bars (Katou, 1988). Taking its effect on these phenomena into consideration, its spectral property becomes important. Sharma and Dean (1979), Sand (1982) derived the 2nd order directional low frequency wave spectrum. The spectrum, however, involves phases of the 1st order component waves. Since the phases are averaged and eliminated in the calculation of the spectrum in principle, application of their model is not possible when only the 1st order directional spectrum is given. Kimura (1984) developed a convolution to calculate the 2nd order directional low frequency wave spectrum. This can calculates the 2nd order directional wave spectrum only with the 1st order directional wave spectrum. For further application of the 2nd order directional wave spectrum in a design procedure, for example, it is convenient to have the explicit expression of the spectrum. The present study aims at modeling the 2nd order directional spectrum which is given in a simple equation.

---

[1]Prof., Tottori Univ., Koyama Minami 4-101, Tottori, Japan
[2]Kajima Road Co. Ltd. Chuo-ku Shimanouti 1-20-19, Osaka, Japan

2ND ORDER DIRECTIONAL SPECTRUM (Kimura, 1984)

The directional spectrum of the 1st order wave $E^{(1)}$ is usually expressed as,

$$E^{(1)}(f,\theta) = S^{(1)}(f)G^{(1)}(\theta|f), \tag{1}$$

in which $S^{(1)}$ and $G^{(1)}$ are the frequency spectrum and the directional function, f and θ are the frequency and the wave direction, respectively. Applying eq.(1), the 1st order wave profile is given as

$$\eta^{(1)}(x,t) = \sum_{i_1=1}^{\infty} \sum_{i_2=1}^{\infty} a_{i_1} a_{i_2} \cos(\psi_{i_1 i_2}), \tag{2}$$

where

$$a_{i_1} = \sqrt{2S^{(1)}(f_{i_1})df}, \qquad a_{i_2} = \sqrt{G^{(1)}(\theta_{i_2}|f_{i_1})d\theta}, \tag{3}$$

$$\psi_{i_1 i_2} = \boldsymbol{k}_{i_1 i_2}\boldsymbol{x} - 2\pi f_{i_1} t + \varepsilon_{i_1 i_2}, \tag{4}$$

in which $\boldsymbol{k}$ is the wave number and $\boldsymbol{x} = (x\boldsymbol{i}\ ,\ y\boldsymbol{j})$ is the distance on the water surface $(x-y$ plane), in which $\boldsymbol{i}$ and $\boldsymbol{j}$ are the unit vectors in x and y directions, respectively. t is the time, ε is the initial phase and $k=|\boldsymbol{k}|$. Subscripts i_1 and j_1 indicate the property of the component wave with frequency f_{i_1} and direction θ_{j_1}.

When the 1st order wave profile is given by eq.(2), the 2nd order wave profile is given as

$$\eta_L^{(2)} = \frac{1}{8}\sum_{i_1=1}^{\infty}\sum_{i_2=1}^{\infty}\sum_{j_1=1}^{\infty}\sum_{j_2=1}^{\infty} a_{i_1}a_{i_2}a_{j_1}a_{j_2}\alpha_{i_1 i_2 j_1 j_2}\cos(\psi_{i_1 i_2} - \psi_{j_1 j_2}), \tag{5}$$

where

$$\alpha_{i_1 i_2 j_1 j_2} = \frac{D_{i_1 i_2 j_1 j_2}^{-} - k_{i_1} k_{j_1}\cos(\theta_{i_2} - \theta_{j_2}) - R_{i_1} R_{j_1}}{2\sqrt{R_{i_1} R_{j_1}}} + R_{i_1} + R_{j_1}, \tag{6}$$

$$D_{i_1 i_2 j_1 j_2}^{-} = \left[\left(\sqrt{R_{i_1}} - \sqrt{R_{j_1}}\right)\left\{\sqrt{R_{j_1}}(k_{i_1}^2 - R_{i_1}^2) - \sqrt{R_{i_1}}(k_{j_1}^2 - R_{j_1}^2)\right\}\right.$$
$$\left. + 2\left(\sqrt{R_{i_1}} - \sqrt{R_{j_1}}\right)^2\{k_{i_1}k_{j_1}\cos(\theta_{i_2} - \theta_{j_2}) + R_{i_1}R_{j_1}\}\right]$$
$$\left/\left(\left(\sqrt{R_{i_1}} - \sqrt{R_{j_1}}\right)^2 - k_{i_1 i_2 j_1 j_2}^{-}\tanh k_{i_1 i_2 j_1 j_2}^{-}h\right)\right., \tag{7}$$

$$k_{i_1 i_2 j_1 j_2}^{-} = \sqrt{k_{i_1}^2 + k_{j_1}^2 - 2k_{i_1}k_{j_1}\cos(\theta_{i_2} - \theta_{j_2})}, \tag{8}$$

$$R_{i_1} = k_{i_1}\tanh(k_{i_1}h) = (2\pi f_{i_1})^2/g, \tag{9}$$

in which h is the water depth, g is the gravitational acceleration. From eq.(5), the 2nd order directional low frequency wave spectrum is derived as

$$R^{(2)}(f^*,\theta^*)df^* d\theta^*$$

$$= \left\{\frac{1}{2}\sum_{i_1=1}^{\infty}\sum_{i_2=1}^{\infty} a_{i_1}a_{i_2}a_{j_1}a_{j_2}\alpha_{i_1 i_2 j_1 j_2}(\cos\varepsilon_{i_1 i_2}\cos\varepsilon_{j_1 j_2} + \sin\varepsilon_{i_1 i_2}\sin\varepsilon_{j_1 j_2})\right\}^2$$

$$+ \left\{\frac{1}{2}\sum_{i_1=1}^{\infty}\sum_{i_2=1}^{\infty} a_{i_1}a_{i_2}a_{j_1}a_{j_2}\alpha_{i_1 i_2 j_1 j_2}(\sin\varepsilon_{i_1 i_2}\cos\varepsilon_{j_1 j_2} - \cos\varepsilon_{i_1 i_2}\sin\varepsilon_{j_1 j_2})\right\}^2, \tag{10}$$

in which f^* and θ^* are the frequency and the direction of the 2nd order wave, respectively. Summations with respect to j_1 and j_2 are explained in eqs.(12), (13) and (14). This spectrum involves the initial phase of the 1st order component waves (Sharma and Dean, 1979, Sand, 1982). Averaging the phases, the averaged 2nd order directional low frequency wave spectrum is derived as

$$E^{(2)}(f^*, \theta^*) = \frac{1}{4} \sum_{i_1=1}^{\infty} \sum_{i_2=1}^{\infty} a_{i_1}^2 a_{i_2}^2 a_{j_1}^2 a_{j_2}^2 \alpha_{i_1 i_2 j_1 j_2}^2, \tag{11}$$

in which summations of i_1 and i_2 are made so that f_{j_1} and θ_{j_2} satisfy the relations,

$$f_{j_1} = f_{i_1} - f^* \qquad (f_{i_1} > f_{j_1} > 0), \tag{12}$$

$$\theta_{j_2} = \theta^* - \sin^{-1}\{(k_{i_1}/k_{j_1})\sin(\theta^* - \theta_{i_2})\}, \tag{13}$$

or

$$\theta_{j_2} = \theta^* + \sin^{-1}\{(k_{i_1}/k_{j_1})\sin(\theta^* - \theta_{i_2})\}. \tag{14}$$

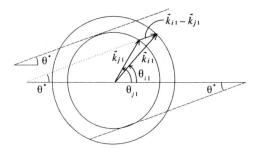

Fig.1. Relation of wave numbers k_{i_1}, k_{j_1} and $k_{i_1} - k_{j_1}$

There are infinite combinations of $\boldsymbol{k}_i$ and $\boldsymbol{k}_j$ which induce $\boldsymbol{k}_i - \boldsymbol{k}_j$ with the same f^* and θ^* (Fig.1). $\boldsymbol{k}_i - \boldsymbol{k}_j$ has, therefore, no definite dispersion relation (Kimura et al., 1997). Summations in eqs.(10), (11) are made for i_1 and i_2 so that f_{i_1} and θ_{i_2} exist within the following regions:

$$f_{i_1} > f^*, \tag{15}$$

$$\theta^* - \sin^{-1}(k_{j_1}/k_{i_1}) \le \theta_{i_2} \le \theta^* + \sin^{-1}(k_{j_1}/k_{i_1}), \tag{16}$$

and

$$\theta^* + \pi - \sin^{-1}(k_{j_1}/k_{i_1}) \le \theta_{i_2} \le \theta^* + \pi + \sin^{-1}(k_{j_1}/k_{i_1}). \tag{17}$$

Equation (11) is rewritten also as

$$E^{(2)}(f^*, \theta^*) = \frac{1}{4} \int_{f_{i_1}} \int_{\theta_{i_2}} E^{(1)}(f_{i_1}, \theta_{i_2}) E^{(1)}(f_{j_1}, \theta_{j_2}) \alpha^2(f_{i_1}, f_{j_1}, \theta_{i_2}, \theta_{j_2}) df_{i_1} d\theta_{i_2}, \tag{18}$$

where

$$\alpha(f_{i_1}, f_{j_1}, \theta_{i_2}, \theta_{j_2}) = \alpha_{i_1 i_2 j_1 j_2}. \tag{19}$$

The frequency spectrum $S^{(2)}$ and the directional function $G^{(2)}$ of the 2nd order low frequency waves are determined as

$$S^{(2)}(f^*) = \int_{-\pi}^{+\pi} E^{(2)}(f^*, \theta^*)d\theta^*, \tag{20}$$

$$G^{(2)}(\theta^*|f^*) = E^{(2)}(f^*, \theta^*)/S^{(2)}(f^*). \tag{21}$$

STANDARD DIRECTIONAL SPECTRUM OF THE 2ND ORDER LOW FREQUENCY WAVES

Directional function

The Bretschneider-Mitsuyasu spectrum is used for the 1st order frequency spectrum.

$$S_0^{(1)}(f) = 0.2573[(H_{1/3})_0^2/T_{1/3}^4]f^{-5}\exp(-1.029f^{-4}/T_{1/3}^4), \tag{22}$$

in which $H_{1/3}$ and $T_{1/3}$ are the significant wave height and period, respectively. Subscript 0 indicates the property in a deep water condition. The Mitsuyasu directional function modified by Goda et al. (1975) is used.

$$E_0^{(1)}(f, \theta) = S_0^{(1)}(f)G_0^{(1)}(\theta|f), \tag{23}$$

$$G_0^{(1)}(\theta|f) = \frac{1}{\pi}2^{2s-1}\frac{\Gamma^2(s+1)}{\Gamma(2s+1)}\cos^{2s}\left(\frac{\theta}{2}\right), \tag{24}$$

$$s = \begin{cases} S_{max}(f/f_p)^5 & f \le f_p \\ S_{max}(f/f_p)^{-2.5} & f > f_p \end{cases}, \tag{25}$$

in which $\Gamma(\ )$ is the gamma function. S_{max} is the spreading parameter of the directional function and f_p is the peak frequency of the spectrum. The 1st order directional spectrum in a shallow water condition $E^{(1)}$ is derived from the relation,

$$E^{(1)}(f, \theta) = E_0^{(1)}(f, \theta_0)\frac{\sinh(2kh)}{\tanh^2(kh)\{2kh + \sinh(2kh)\}}, \tag{26}$$

in which h is the water depth.

In the present study, 3 different wave periods for $E^{(1)}$ are use ($T_{1/3} = 8s$, $10s$ and $12s$). The incident mean wave angle is fixed normal to the shore line. $(H_{1/3})_0$ is determined so that $(H_{1/3})_0/(L_{1/3})_0$ satisfy the relation shown by Goda et al. (1975) in which $(L_{1/3})_0$ is the wave length of the significant wave. Table 1 shows the wave conditions which satisfy the relation when $T_{1/3} = 8s$.

Table 1. Wave conditions ($T_{1/3} = 8s$)

Case	S_{max}	$(H_{1/3})_0$	Case	S_{max}	$(H_{1/3})_0$	Case	S_{max}	$(H_{1/3})_0$
1.1	5	4.3m	1.5	25	2.6m	1.9	60	1.2m
1.2	10	3.7m	1.6	30	2.2m	1.10	75	1.0m
1.3	15	3.3m	1.7	40	1.7m	1.11	100	0.77m
1.4	20	3.0m	1.8	50	1.4m			

$E^{(1)}$ is calculated at $h/(L_{1/3})_0 = 0.5, 0.3, 0.2, 0.1, 0.05$.

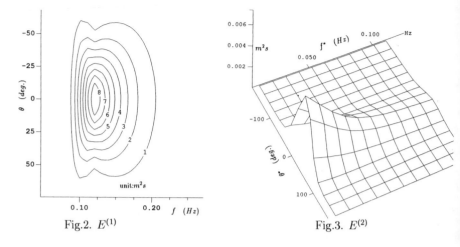

Fig.2. $E^{(1)}$

Fig.3. $E^{(2)}$

Figures 2 and 3 show the contour line of $E^{(1)}$ and the bird's-eye view of $E^{(2)}$, respectively (Case 1.2: $h/(L_{1/3})_0 = 0.2$).

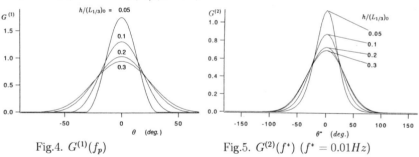

Fig.4. $G^{(1)}(f_p)$

Fig.5. $G^{(2)}(f^*)$ $(f^* = 0.01Hz)$

Figure 4 shows $G^{(1)}(\theta|f_p)$ at several water depths as shown in the figure. Figure 5 shows $G^{(2)}(\theta|f^*)$ at the same points when $f^* = 0.01Hz$ (Case 1.2). $G^{(2)}$ becomes narrower with decreasing water depth as $S^{(1)}$. Figure 6 shows $G^{(2)}(\theta|f^*)$ when f^* changes from $0.01Hz$ to $0.07Hz$ at different water depths (Case 1.2). Since the change of $G^{(2)}$ with f^* is very small, $G^{(2)}$ is assumed to be independent of f^* in this study. i.e.:

$$E^{(2)}(f^*, \theta^*) = S^{(2)}(f^*)G_m^{(2)}(\theta^*), \qquad (27)$$

$G_m^{(2)}$ is determined by fitting eq.(24) to the calculated $G^{(2)}$. s for $G_m^{(2)}(f_i^*)$ is determined so that s minimizes

$$\sum_{j=1}^{N}\left\{ G^{(2)}(\theta_j^*|f_i^*) - \frac{1}{\pi}2^{2s_i-1}\frac{\Gamma^2(s_i+1)}{\Gamma(2s_i+1)}\cos^{2s_i}\left(\frac{\theta^*}{2}\right)\right\}^2 \quad (i=1,n), \qquad (28)$$

in which $N = 36$ $(\theta_j^* = -180° \sim 180°)$, $n = 7$ $(f_i^* = 0.01 \sim 0.07Hz)$. The weighted average of s_i is calculated using $S^{(2)}(f_i^*)$.

$$s_{av} = \sum_{i=1}^{7} S^{(2)}(f_i^*)s_i \bigg/ \sum_{i=1}^{7} S^{(2)}(f_i^*), \quad (29)$$

and $G_m^{(2)}$ is given as,

$$G_m^{(2)}(\theta^*) = \frac{1}{\pi} 2^{2s_{av}-1} \frac{\Gamma^2(s_{av}+1)}{\Gamma(2s_{av}+1)} \cos^{2s_{av}}\left(\frac{\theta^*}{2}\right). \quad (30)$$

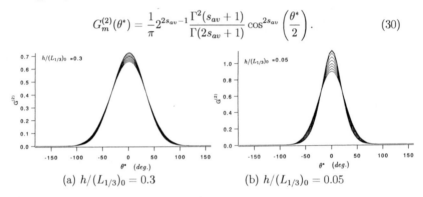

(a) $h/(L_{1/3})_0 = 0.3$ (b) $h/(L_{1/3})_0 = 0.05$

Fig.6. $G^{(2)}$ (Case 1.2)

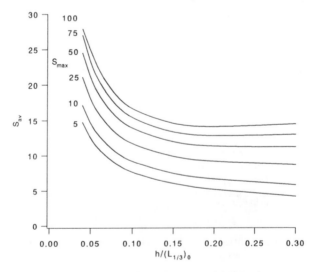

Fig.7. Relation between s_{av} and $h/(L_{1/3})_0$

Figure 7 shows the relation between s_{av} and $h/(L_{1/3})_0$. S_{max} values in the figure are those in a deep water condition. If the wave steepness $(H_{1/3})_0/(L_{1/3})_0$

and S_{max} satisfy the relation shown by Goda et al. (1975), the same results are obtained regardless of $T_{1/3}$ and h.

When the dominant direction of the incident wave spectrum is not normal to the shore line, s_{av} must be reduced slightly (Kimura et al., 1997).

Frequency spectrum

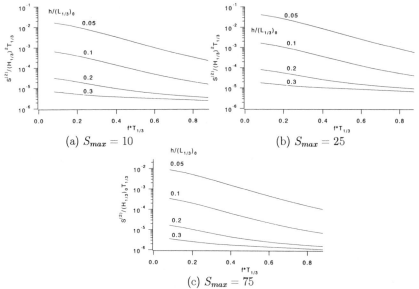

Fig.8. Relation between $S^{(2)}/[(H_{1/3})_0^2 T_{1/3}]$ and $f^* T_{1/3}$

Figure 8 shows the relation between $S^{(2)}/[(H_{1/3})_0^2 T_{1/3}]$ and $f^* T_{1/3}$ at $h/(L_{1/3})_0 = 0.05$, 0.1, 0.2 and 0.3 (solid lines) on semi-log paper when (a) $S_{max} = 10$, (b) $S_{max} = 25$ and (c) $S_{max} = 75$. Since the lines change almost linearly on the graphs, the relation is approximated with

$$\ln \left(\frac{S^{(2)}(f^*)}{(H_{1/3})_0^2 T_{1/3}} \right) = a f^* T_{1/3} + b, \tag{31}$$

in which a and b are constants. Since these constants still change slightly with $h/(L_{1/3})_0$, the next approximation is tried in addition:

$$a = c_1 \{ h/(L_{1/3})_0 \} + c_2,$$
$$b = c_3 \log \{ h/(L_{1/3})_0 \} + c_4, \tag{32}$$

in which c_1, c_2, c_3 and c_4 are constants. These constants are determined so that

$$\sum_{i=1}^{7} \left\{ \frac{S^{(2)}(f_i^*)}{(H_{1/3})_0^2 T_{1/3}} - \left[\{ c_1 h/(L_{1/3})_0 + c_2 \} f_i^* T_{1/3} + c_3 \log \{ h/(L_{1/3})_0 \} + c_4 \right] \right\}^2, \tag{33}$$

becomes minimum.

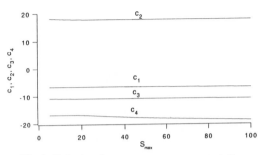

Fig.9. Relation between c_1, c_2, c_3, c_4 and S_{max}

Figure 9 shows the relations between c_1, c_2, c_3, c_4 and S_{max}. c_i $(i = 1, 2, 3)$ take constant values.

$$c_1 \simeq 18.0, \qquad c_2 \simeq -6.5, \qquad c_3 \simeq -10.8. \tag{34}$$

c_4 also takes a constant value -16.8 when $S_{max} \leq 20$, but reduces gradually when $S_{max} > 20$.

In a deep sea condition, S_{max} approximately takes the following values (Goda et al., 1975):

1) Wind waves $S_{max} \simeq 10$,
2) Swell with short decay distance $S_{max} \simeq 25$,
3) Swell with long decay distance $S_{max} \simeq 75$.

c_4 takes the values approximately -17.0 and -18.5 when $S_{max} = 25$ and 75, respectively.

Finally, $S^{(2)}$ is given as,

$$\frac{S^{(2)}(f^*)}{(H_{1/3})_0^2 T_{1/3}} \simeq \exp\left\{\left[18.0h/(L_{1/3})_0 - 6.5\right] f^* T_{1/3} - 10.8\log\{h/(L_{1/3})_0\} + c_4\right\}. \tag{35}$$

$c_4 \simeq -16.8$, -17.0 and -18.5 when $S_{max} \simeq 10$, 25 and 75, respectively.

The variance of $\eta_L^{(2)}$ is determined as

$$\overline{\left(\eta_L^{(2)}\right)^2} = \int_0^\infty S^{(2)}(f^*)df^*, \tag{36}$$

Substituting eq.(35) into eq.(36), $\overline{\left(\eta_L^{(2)}\right)^2}$ is given as,

$$\frac{\overline{\left(\eta_L^{(2)}\right)^2}}{(H_{1/3})_0^2} = \frac{\exp[-10.8\log\{h/(L_{1/3})_0\} + c_4]}{-18.0\{h/(L_{1/3})_0\} + 6.5}. \tag{37}$$

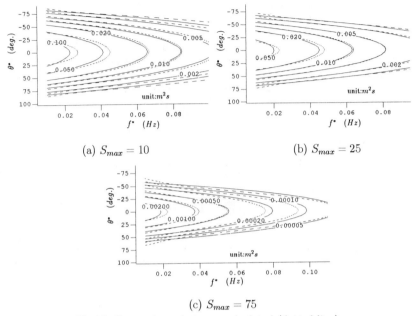

(a) $S_{max} = 10$ (b) $S_{max} = 25$

(c) $S_{max} = 75$

Fig.10. Comparisons between calculated (dotted line)
and approximated (solid line) $E^{(2)}$

VERIFICATION OF THE MODEL

From eqs.(30) and (35), $E^{(2)}$ is approximated as,

$$
\begin{aligned}
E^{(2)}\left(f^*, \theta^*\right) = {}& (H_{1/3})_0^2 T_{1/3} \exp\left\{\left[18.0h/(L_{1/3})_0 - 6.5\right] f^* T_{1/3}\right. \\
& \left. - 10.8\log\{h/(L_{1/3})_0\} + c_4\right\} \\
& \times \frac{1}{\pi} 2^{2s_{av}-1} \frac{\Gamma^2(s_{av}+1)}{\Gamma(2s_{av}+1)} \cos^{2s_{av}}\left(\frac{\theta^*}{2}\right).
\end{aligned} \tag{38}
$$

$c_4 \simeq -16.8$, -17.0 and -18.5 when $S_{max} \simeq 10$, 25 and 75, respectively.

Figure 10 shows the energy contour of the calculated (dotted line) and approximated (solid line) $E^{(2)}$ when (a) $S_{max} = 10$ (Case 1.2), (b) $S_{max} = 25$ (Case 1.5) and (c) $S_{max} = 75$ (Case 1.10) at water depth $h/(L_{1/3})_0 = 0.1$. Similar agreements are obtained in all other cases, regardless of $T_{1/3}$ and $h/(L_{1/3})_0$, if $(H_{1/3})_0/(L_{1/3})_0$ and S_{max} satisfy the relation shown by Goda et al.(1975).

CONCLUDING REMARKS

This study aims at modeling the 2nd order low frequency directional wave spectrum. Substituting the Bretschneider-Mitsuyasu frequency spectrum and Mitsuyasu directional function into the convolution developed by Kimura (1984), standard 2nd order directional spectrum is calculated. Since the frequency spectrum and the directional function of the 2nd order directional spectrum can be assumed independent, the frequency spectrum and the directional function are approximated by simple equations, respectively. The approximated equation shows wide appricability regardless of S_{max}, $T_{1/3}$ and $(H_{1/3})_0$, if these sea state parameters satisfy the average relation shown by Goda et al. (1975).

REFERENCES

[1] Bowers, E. C. 1977: Harbor resonance due to set-down beneath wave groups, Jour. Fluid Mech. Vol. 79, pp.71-92.

[2] Goda, Y. and Y. Suzuki 1975: Computation of refraction and diffraction of sea waves with Mitsuyasu's spectrum, Tech. Note of PHRI No.230, 45p.

[3] Katoh, K. 1984: Multiple longshore bars formed by long period standing waves, Rept. of PHRI, Vol.23, No.3, pp.3-46.

[4] Kimura, A. 1984: Averaged two-dimensional low frequency wave spectrum of wind waves, Comm. on Hydraulics, Rep. No. 84-3, Dept. of Civil Eng., Delft Univ. Tech., 54p.

[5] Kimura, A. 1998: Low frequency oscillation of a rectangular habar induced by multi-directional irregular waves, Proc. Coastal Eng., JSCE Vol.45, pp.226-230. (in Japanese)

[6] Pinkster, J. A. 1980: Low frequency second order wave exciting forces on floating structures, Doctral Thesis, Delft Univ. Tech., 204p.

[7] Remery G. F. M. and A. J. Hermans 1971: The slow drift osscilations of moored object in random seas, OTC, Papar No. 1500.

[8] Sand, S. E. 1982: Long waves in directional seas, Coastal Eng. 6, pp.195-208.

[9] Sharma, J. N. and R. G. Dean 1979: Development and evaluation of a procedure for simulating a random directional second order sea surface and associated wave forces, Ocean Eng., Rept. No.20, Dept. Civil Eng., Univ. of Delaware, 139p.

LATERAL MIXING AND SHEAR WAVES

Qun Zhao   Ib A. Svendsen   and Kevin Haas[1]

ABSTRACT

Study of 3D shear waves has been carried out utilizing the quasi-3D nearshore circulation model SHORECIRC. The emphasis of this study is to investigate the effect of the 3D lateral mixing has on the shear waves, and vice-versa. The kinetic energy equation for the mean flow and for the shear waves has been derived. The momentum and energy balance have been discussed using the numerical results from idealized Duck bathemetry and the wave condition on Oct.16, 1994. The effect of the 3D dispersion due to the depth varying current on shear waves has been illustrated.

INTRODUCTION

Shear waves were first identified by Oltman-Shay *et al.* (1989). Since then, they have been studied intensively. Most of these studies are either based on linear stability analysis (Bowen and Holman, 1989; Dodd *et.al.*, 1992; Falqués and Iranzo, 1994, etc.), or studied by direct simulation of finite amplitude shear waves by using nonlinear shallow water equations (Allen *et. al.* 1996; Deigaard *et. al.*, 1994; Slinn *et.al.*, 1998; Özkan-Haller and Kirby, 1999). Lateral mixing is incorporated in some of these models in a classic way by using large eddy viscosity. It is found that the general effect of including the "lateral mixing", whether a constant eddy viscosity coefficient (Deigaard *et. al.*, 1994), or a cross-shore varying eddy viscosity(Özkan-Haller and Kirby, 1999), is damping. However, the "lateral mixing" is not known a *priori*. Furthermore, Svendsen and Putrevu (1994) discovered that a major part of the lateral mixing in longshore currents is caused by the depth variation of the currents. This so called dispersive mixing is deterministic and follows from the three-dimensional flow pattern. It is also known that shear waves themselves provide tremendous lateral mixing. Therefore, the classic approach to mimic mixing is only a crude remedy. Moreover, shear waves are strong vortical motions. Vortex stretching, which is essentially a three-dimensional phenomenon, is missing in a two-dimensional model. Hence important physical mechanisms are left out in a two-dimensional simulation.

---

[1]Center for Applied Coastal Research, University of Delaware, Newark, DE 19716. Correspondence email: zhao@coastal.udel.edu

In the present work, we will utilize the quasi-3D nearshore circulation model SHORECIRC (Svendsen *et. al*, 2000) to simulate finite amplitude shear waves. Our purpose is to investigate how the three-dimensional dispersive mixing and the lateral mixing provided by the shear waves affect each other. To achieve this, 2D and quasi-3D numerical experiments are carried out simultaneously.

DESCRIPTION OF THE MODEL

Following Putrevu and Svendsen (1999), we split the total particle velocity $u_i(x, y, z, t)$ into three components

$$u_i(x, y, z, t) = u_i(x, y, z, t)' + u_{wi}(x, y, z, t) + V_i(x, y, z, t) \qquad (1)$$

$i = 1, 2$ for horizontal velocities. The x and y denote the cross-shore and long-shore directions, while z is the vertical coordinate. In the above, u_i', u_{wi}, V_i are the turbulent component, the wave component and the current component, respectively. We then further split the current velocity into the depth uniform part and the depth varying current part,

$$V_i(x, y, z, t) = \tilde{V}_i(x, y, t) + V_{1i}(x, y, z, t) \qquad (2)$$

where the depth uniform part is defined as,

$$\tilde{V}_i(x, y, t) = \frac{1}{h} \overline{\int_{-h_0}^{\zeta} u_i(x, y, z, t) dz} = \frac{\bar{Q}(x, y, t)}{h} \qquad (3)$$

in which the overbar denotes averaging over short wave period, h_0 and ζ represent the still water depth and the instantaneous water surface elevation, respectively. And $h = h_0 + \bar{\zeta}$ represent total water depth.

The second component of the short-wave-averaged velocity $V_{1i}(x, y, z, t)$ accounts for the vertical variation of the current and satisfies

$$\overline{\int_{-h_0}^{\zeta} V_{1i} dz} = -\overline{\int_{\zeta_t}^{\zeta} u_{wi} dz} = -Q_{wi} \qquad (4)$$

where Q_{wi} is the short-wave-induced volume flux, and ζ_t is the surface elevation of the wave trough level. A sketch of this split of current velocity is shown in Figure (1).

Then the depth-integrated, short-wave-averaged governing equations read (Putrevu and Svendsen, 1999),

$$\frac{\partial \bar{\zeta}}{\partial t} + \frac{\partial}{\partial x_i}(\tilde{V}_i h) = 0 \qquad (5)$$

$$\frac{\partial}{\partial t}(\tilde{V}_i h) + \frac{\partial}{\partial x_j}(\tilde{V}_i \tilde{V}_j h) + \frac{1}{\rho}\frac{\partial L_{ij}}{\partial x_j} + \frac{1}{\rho}\frac{\partial S_{ij}}{\partial x_j} + \frac{1}{\rho}\frac{\partial T_{ij}}{\partial x_j} + gh\frac{\partial \bar{\zeta}}{\partial x_j} + \frac{\tau_i^B}{\rho} = 0 \qquad (6)$$

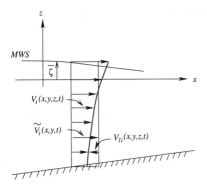

FIG. 1. Sketch showing definition of current velocities

where S_{ij}, T_{ij}, τ_i^B are short-wave induced radiation stress, depth-integrated Reynolds stress and bottom shear stress, respectively. L_{ij} is the contribution of the 3D dispersive terms,

$$L_{ij} = \rho[\overline{\int_{-h_0}^{\zeta} V_{1i}V_{1j}dz} + \overline{\int_{\zeta_t}^{\zeta} (u_{wi}V_{1j} + u_{wj}V_{1i}dz]} \tag{7}$$

and the radiation stress is defined by

$$S_{ij} = \overline{(\rho u_{wi}u_{wj} + p\delta_{ij})dz} - \frac{1}{2}\rho g h^2 \delta_{ij} \tag{8}$$

ANALYSIS OF THREE-DIMENSIONAL SHEAR WAVE SYSTEM

Momentum Balance

The flow to be discussed in this work is assumed to be long-shore uniform in average, we therefore define the instantaneous mean flow as the longshore average of the instantaneous flow.

$$V_S(x,t) = \overline{\tilde{V}(x,y,t)}^{L_y} = \frac{1}{L_y}\int_0^{L_y} \tilde{V}(x,y,t)dy \tag{9}$$

where

$$\tilde{V}(x,y,t) = V_S(x,t) + V'(x,y,t) \tag{10}$$

$V'(x,y,t)$ is the shear wave velocity, L_y is the longshore domain length. Using the longshore averaged current as the "mean" current allow us to investigate the time evolution of shear waves and their effect on the mean current at each cross-shore distance. In this paper, the longshore averaging is denoted by $\overline{\cdot}^{L_y}$, while the time averaging is denoted by $< \cdot >$.

Substitutiong (9), (10) into the momentum equation (6) and perform longshore averaging, leads to the momentum equation for the mean flow.

$$\frac{\partial V_{Si}}{\partial t} + V_{Sj}\frac{\partial V_{Si}}{\partial x_j} = -g\frac{\partial \bar{\zeta}_S}{\partial x_i} - \frac{1}{\rho h}\frac{\partial S_{ij}}{\partial x_j} - \frac{1}{\rho h}\frac{\partial T_{Sij}}{\partial x_j} - \frac{1}{\rho h}\frac{\partial L_{Sij}}{\partial x_j} - \frac{1}{\rho h}\tau_{Si}^B + \frac{V_{Si}}{h}\frac{\partial \zeta_S}{\partial t}$$
$$-\frac{\partial}{\partial x_j}\overline{V_i'V_j'}^{L_y} + \frac{\overline{V_i'\frac{\partial \zeta'}{\partial t}}^{L_y}}{h} \quad (11)$$

It's clear that the term $-\rho\overline{V_i'V_j'}^{L_y}$ has the form of a Reynolds stress, and we will call it the Reynolds stress due to shear waves. By definition, we know that the $-\frac{1}{\rho h}\frac{\partial L_{Sij}}{\partial x_j}$ term in the mean momentum equation (11) represents the 3D dispersive mixing between the depth uniform current and the depth varying current. Similarly, the $-\frac{\partial}{\partial x_j}\overline{V_i'V_j'}^{L_y}$ term provides lateral mixing to the mean flow due to the existence of shear waves. Our purpose is to investigate how the 3D dispersive mixing and the lateral mixing due to shear waves affect each other.

Energy Balance

To study the overall energy balance, we form the dot product of the mean flow momentum equation (11) and $V_S(x,t)$, integrate it over the whole calculation domain, and notice that all the convection terms go to zero upon using divergence theorem and applying the boundary conditions. The kinetic energy equation for the mean flow then becomes

$$\int_0^{L_x}\int_0^{L_y}\frac{\partial}{\partial t}\frac{V_{Si}V_{Si}}{2}dydx = \int_0^{L_x}\int_0^{L_y}\overline{V_i'V_j'}^{L_y}\frac{\partial V_{Si}}{\partial x_j}dydx$$
$$- \int_0^{L_x}\int_0^{L_y}\frac{V_{Si}}{\rho h}\frac{\partial S_{ij}}{\partial x_j}dydx - \int_0^{L_x}\int_0^{L_y}\frac{V_{Si}}{\rho h}\frac{\partial T_{Sij}}{\partial x_j}dydx$$
$$- \int_0^{L_x}\int_0^{L_y}\frac{V_{Si}}{\rho h}\frac{\partial L_{Sij}}{\partial x_j}dydx - \int_0^{L_x}\int_0^{L_y}\frac{V_{Si}}{\rho h}\tau_{Si}^B dydx$$
$$(12)$$

Equation (12) is the instant kinetic energy equation for the mean flow. The left hand side represents the change of mean kinetic energy of the mean flow, the first term at the right hand side represents the loss of mean flow kinetic energy to the fluctuations. The last four terms at the right hand side of the equation represent the work done by the radiation stress, the turbulent shear stress, the 3D lateral mixing, and the bottom friction of the mean flow, respectively.

The momentum equation for the shear waves is obtained by subtracting the mean momentum equation from the instantaneous momentum equation. Taking the dot product of the shear wave momentum equation and V_i', we obtain the local kinetic energy equation for the shear waves. Integrating it over the whole domain, we get the kinetic energy equation for shear waves equivalent to (12).

$$\int_0^{L_x}\int_0^{L_y}\frac{\partial}{\partial t}\frac{V_i'V_i'}{2}dydx = - \int_0^{L_x}\int_0^{L_y}\overline{V_i'V_j'}^{L_y}\frac{\partial V_{Si}}{\partial x_j}dydx - \int_0^{L_x}\int_0^{L_y}\frac{V_i'}{\rho h}\frac{\partial T_{ij}'}{\partial x_j}dydx$$

$$- \int_0^{L_x} \int_0^{L_y} \frac{V_i'}{\rho h} \frac{\partial L_{ij}'}{\partial x_j} dy dx - \int_0^{L_x} \int_0^{L_y} \frac{V_i'}{\rho h} \tau_i'^B dy dx$$

$$(13)$$

The first term at left hand side of (13) represents the change of kinetic energy of the shear waves ; the first term at the right hand side represents the production of shear wave kinetic energy by the mean flow; the last three terms at the right hand side represent the work done by Reynolds shear wave stress, the 3D dispersion terms and the bottom friction due to shear wave velocities.

NUMERICAL RESULTS

In this section, we will utilize SHORECIRC to simulate and analyze shear waves in a longshore uniform barred beach. The input we use here is from Oct.16, Duck 94 Experiment. Following Dodd et al. (1992) and Özkan-Haller and Kirby(1999), we form a longshore uniform topography by using measured Duck bathymetry from one transect, then extend it alongshore . The periodic boundary condition are applied at the lateral boundaries. The details of the numerical scheme and closure models are discussed in Svendsen et al. (2000).

The top panel of Figure (2) shows the measured (cycle) and simulated (line) wave heights by REF/DIF1 (Kirby and Dalrymple, 1994). The agreement is satisfactory. The lower panel in Figure (2) shows the idealized Duck bathemetry used in the present simulation.

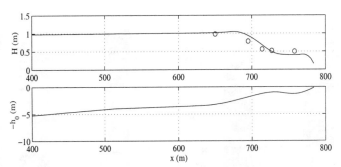

FIG. 2. Cross-shore distribution of wave height (top) and idealized Duck topography (bottom).

In order to investigate how the 3D dispersive mixing affect the shear waves and how the 3D dispersive mixing and the lateral mixing provided by the shear waves affects each other, the 2D and quasi-3D numerical experiments are carried out simultaneously by switching on/off the 3D dispersion. In the following, we consider conditions of the fully developed quasi-steady motions.

Figure (3) shows the snapshots of shear wave vorticity fields (top panels) and kinetic energy contours (bottom panel) at two time intervals for the 3D simulation (left panels) and 2D simulation (right panels). The vorticity fields show that the shear wave flow is more organized in the 3D simulation than that in the 2D simulation. The integrated kinetic energy of the whole domain at the time of the snap shots is $652.8(m^4/s^2)$ and $561.6(m^4/s^2)$ for the 3D simulation, and $1478.4(m^4/s^2)$ and $2197.7(m^4/s^2)$ for the 2D simulation, which indicates that shear waves in the 3D simulation is less energetic than that in the 2D simulation.

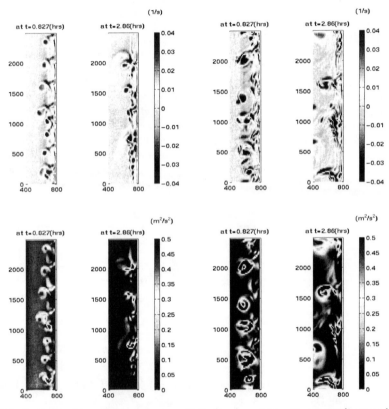

FIG. 3. Snap shot of shear wave vorticity(top) and kinetic energy(bottom) of 3D simulation(left) and 2D simulation(right). The beach is to the right and the flow is moving upward.

Comparison of the cross-shore variation of the 3D dispersive mixing, solid line in Figure (4) and the lateral mixing due to shear waves in the 3D simulation, dashed line, shows that, for the case studied here, the momentum mixing provided by shear waves is at some points larger than that by 3D dispersive mixing. The 3D dispersive mixing mainly takes place in the cross-shore direction, while the lateral mixing provided by the shear waves in the longshore direction is at the same order as that in the cross-shore direction.

Comparison of the mixing provided by shear waves in the 2D, dashed line in Figure (5), and the 3D case, solid line in Figure (5), shows that in the cross-shore direction, the lateral mixing provided by the shear waves is only slightly stronger in the 2D case than that in the 3D case. In the longshore direction, however, there is little difference. Therefore, comparison of Figure (4) and Figure (5) shows that the total mixing in the 3D simulation (the sum of the dashed and solid line in Figure (4)) is larger than in the 2D simulation (dashed line in Figure (5)), and different distributed in the cross-shore.

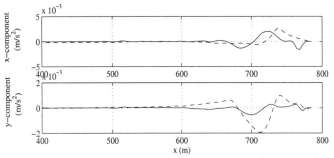

FIG. 4. Time (and longshore) averaged 3D lateral mixing terms and shear wave mixing terms. solid: mean 3D lateral mixing $< -\frac{1}{\rho h}\frac{\partial L_{Sij}}{\partial x_j} >$ **dashed: shear wave mixing** $< -\frac{\partial}{\partial x_j}\overline{V_i'V_j'}^{L_y} >$

Figure (6) shows, from top to bottom, the temporal evolution of the kinetic energy, the kinetic energy production, work done by the turbulence, the 3D dispersion terms and the bottom friction terms due to shear wave velocities, respectively. This figure shows that there is a surge at the initial stage in the shear wave developing process. After that the shear wave kinetic energy fluctuates about the mean. The contribution of the production term is always positive. The work done by the 3D dispersive terms, the turbulence and the bottom friction terms are always negative indicating they remove energy from the shear waves. Also the contribution of the 3D dispersive terms is at least one order larger than that of turbulence from the shear waves.

An alternative way to illustrate the contribution of the 3D dispersive terms on the shear waves is to look at the cross-shore distribution of time and longshore averaged work done by the 3D dispersion terms of shear waves $< -\frac{\overline{V_i'}\frac{\partial L_{ij}'}{\partial x_j}}{\rho h}^{L_y} >$,

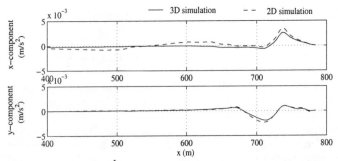

FIG. 5. $< -\frac{\partial}{\partial x_j} \overline{V_i' V_j'}^{L_y} >$: **Time and longshore averaged shear wave mixing terms.**

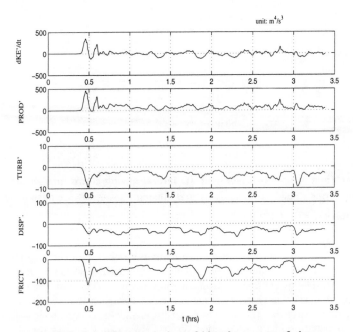

FIG. 6.   Time evolution of kinetic energy of shear waves .

as shown in the upper panel of Figure (7). Again, it is shown that the global contribution of the work done by the 3D dispersive terms is negative.

The next question we would ask is does the 3D dispersive terms always drain energy from the depth uniform current? To answer this question, we will look at the work done against the mean current V_{Si} by the mean 3D dispersive terms

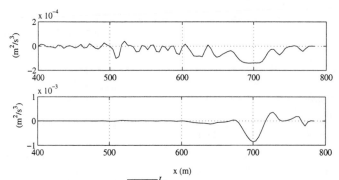

x (m)

FIG. 7. Top panel $< -\frac{V'_i}{\rho h}\frac{\partial L'_{ij}}{\partial x_j}^{L_y} >$: **Time and longshore averaged work done by the 3D lateral mixing terms of shear waves. Lower panel** $< -\frac{V_{Si}}{\rho h}\frac{\partial L_{Sij}}{\partial x_j} >$: **Time-averaged work done by the 3D lateral mixing terms due to the mean currents**

$< -\frac{V_{Si}}{\rho h}\frac{\partial L_{Sij}}{\partial x_j} >$ in (12) . The lower panel of Figure (7) shows the time-averaged cross-shore distribution of the work done by the 3D dispersive terms due to the mean currents $< -\frac{V_{Si}}{\rho h}\frac{\partial L_{Sij}}{\partial x_j} >$. It shows that unlike the term $< -\frac{V'_i}{\rho h}\frac{\partial L'_{ij}}{\partial x_j}^{L_y} >$, which always pull out kinetic energy from the depth uniform current to benefit the depth varying current, the $< -\frac{V_{Si}}{\rho h}\frac{\partial L_{Sij}}{\partial x_j} >$ term actually exchanges kinetic energy between the depth uniform current and the depth varying current in the cross-shore direction.

CONCLUSIONS

Our study on three-dimensional shear waves has showed that shear waves are less energetic in a 3D simulation than that in a 2D simulation. The three-dimensional dispersive mixing terms of the mean flow will allow energy exchange between the mean and the depth varying current. The three-dimensional dispersive mixing terms of the shear waves will pull out energy from the shear waves to benefit the depth varying current. Shear waves provide strong horizontal mixing, in certain situations, this mixing is even stronger than that provided by the three-dimensional dispersive mixing but always less than the combined mixing of the two mechanisms.

ACKNOWLEDGEMENTS

Thanks are due to Nick Dodd and Özkan-Haller for providing the bathymetry data files. Thanks are also go to the staffs at the Army Corps Field Research Facility (FRF) at Duck, North Carolina, for their contributions to the field data acquisition and distribution. This research has been sponsored by Office of Naval Rearch (ONR), Coastal Dynamics Program under grant number N00014-99-1-0291, and National Oceanographic Partnership Program by ONR (contract no

REFERENCES

Allen J. S., P. A. Newberger, and R. A. Holman (1996) : Nonlinear shear instabilities of along-shore currents on plane beaches. *J Fluid Mech.* 310: 181-213.

Bowen, A.J. and R.A. Holman,(1989) : Shear instabilities of the mean long-shore-current. 1. Theory. *J. Geophys. Res-Oceans* 94: (C12) 18023-18030.

Deigaard, R., E. D. Christensen, J.S. Damgaard, and J.Fredsoe(1994) : Numerical simulation of finite amplitude shear waves and sediment transport. In *Proc. 24th Int. Conf. Coastal.Engng.* 230-231.

Dodd, N. and E. B.Thornton (1990) : Growth and energetics of shear waves in the nearshore. *J. Geophys. Res-Oceans* 95: (C9) 16075-16083.

Dodd, N., J. Oltman-Shay , and E. B.Thornton (1992) : Shear Instabilities In The long-shore-current - A Comparison OF observation and theory. *J. Phys. Oceanogr.* 22: (1) 62-82.

Falqués, A. and V. Iranzo, (1994) : Numerical-Simulation of vorticity waves in the nearshore. *J. Geophys. Res-Oceans* 99: (C1) 825-841.

Kirby, J. and Dalrymple, R.(1994): Combined refraction/diffraction model REF/DIF 1, version 2.5. Internal Report *CACR-94-22*, Center for Applied Coastal Research, University of Delaware.

Oltman-Shay, J., P.A. Howd and W.A. Birkemeier,(1989) : Shear instabilities of the mean longshore-current. 2. Field Observations. *J. Geophys. Res-Oceans* 94: (C12) 18031-18042.

Özkan-Haller, T. and J.T. Kirby (1999) : Nonlinear evolution of shear instabilities of the long-shore current: A comparison of observations and computations. *J. Geophys. Res-Oceans* 104: (C11) 25953-25984.

Pùtrevu, U. and I. A. Svendsen, (1992) : Shear instability of longshore currents-a numerical study. *J. Geophys. Res-Oceans*97: (C5) 7283-7303.

Putrevu, U. and I. A. Svendsen (1999) : Three-dimensional dispersion of momentum in wave-induced nearshore currents. *Eur. J. Mech. B/Fluids*, 83-101.

Slinn, D.N., J.A. Allen, P. A. Newberger, and R.A. Holman (1998): Nonlinear shear instabilities of alongshore currents over barred beaches. *J. Geophys. Res-Oceans*103: 18,357-18,379.

Svendsen, I. A. and U. Putrevu, (1994) : Nearshore mixing and dispersion. *Proc. R. Soc.Lond.* A 445, 561-576.

Svendsen, I. A., K. Haas, and Q. Zhao (2000) : Quasi-3D nearshore circulation model SHORE-CIRC. CACR Report, Center for Applied Coastal Research, University of Delaware.

AN APPLICATION OF WAVELET TRANSFORM ANALYSIS TO LANDSLIDE-GENERATED IMPULSE WAVES

Hermann M. Fritz[1], and Paul C. Liu[2]

Abstract: This paper presents an analysis of applying continuous wavelet transform spectrum analysis to the results of laboratory measurement of landslide-generated impulse waves. As the measured results are understandably unsteady, nonlinear, and non-stationary, the application of time-localized wavelet transform analysis is shown to be a suitable as well as useful approach. The analysis on correlating the time-frequency wavelet spectrum configurations in connection with the ambient parameters that drive the impulse wave process, with respect to time and space, leads to interesting and stimulating insights not previously known.

INTRODUCTION

Large water waves in reservoirs, lakes, bays, and oceans are sometimes generated by the occurrence of landslides. In the oceans, landslides are responsible for the creation of dangerous "surprise tsunamis" and are known to have accounted for 10 percent of worldwide tsunamis events. For fjords, lakes, and bays, the impulse generated waves are particularly significant, as the effects tend to be amplified due to steep shore, narrow reservoir geometry, possible large slide masses, and high impact velocities. The resulting impulse waves can cause disaster due to run-up along the coastal shoreline and overtopping of dams. Because of the unpredictable nature of the phenomenon, exploration of landslide waves can best be conducted through laboratory experiments. The complexity of the landslide circumstances presents formidable challenges for a successful experiment that encompass laboratory set-up, measurement techniques, and data analysis. As the results of measurement are understandably unsteady, nonlinear and non-stationary, the application of time-localized wavelet transform analysis would be a rational approach to pursue. This paper presents such an endeavor of applying continuous wavelet transform spectrum analysis to the laboratory measurements of landslide-generated impulse waves.

1 Laboratory of Hydraulics, Hydrology and Glaciology (VAW), Swiss Federal Institute of Technology (ETH), CH-8092, Zurich , Switzerland, fritz@vaw.baug.ethz.ch
2 NOAA Great Lakes Environmental Research Laboratory, Ann Arbor, Michigan 48105-2945, U.S.A. Paul.C.Liu@noaa.gov

THE LABORATORY SET-UP

The laboratory experiment was conducted in a 1 m high, 0.5 m wide, and 11 m long rectangular prismatic water-wave channel with a pneumatic landslide generator. An overview on the implementation set-up and combination of the various systems used is shown in Fig. 1. Three different measurement techniques were built into the physical model: Laser distance sensors (LDS), particle image velocimetry PIV (Fritz, 2000), and capacitance wave gauges (CWG). Thus the system allows simultaneous acquisition of slide characteristics before impact, instantaneous whole field velocity vector fields in the impact area and wave generation area as well as water wave evolution in the near field propagation area.

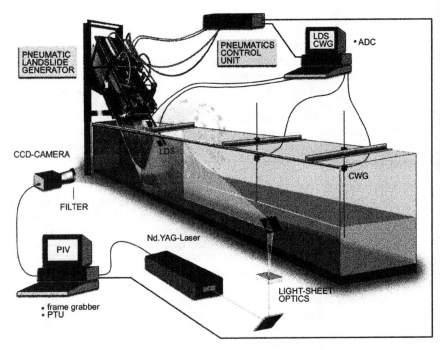

Fig. 1. Example of an experiment in progress.

The picture shown in Fig. 2 in the next page presents an example of actual experiment of impulse-generated waves in progress.

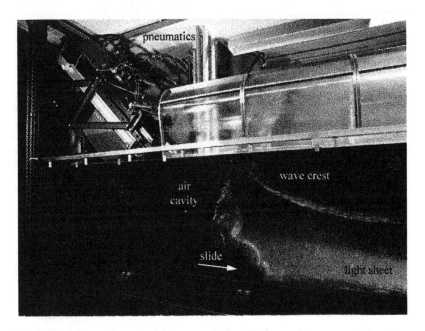

Fig. 2. Experimental setup with pneumatic installation and measurement systems

Landslides were modeled with artificial granular material that has the density and granulometry of typical natural rock formations. Impact characteristics of slide mass were controlled by means of a novel pneumatic acceleration to allow exact reproduction and independent variation of single dynamic slide parameters. Waves were recorded from 7 capacitance wave gages placed longitudinally along the channel at 1m apart as schematically shown in Fig. 3.

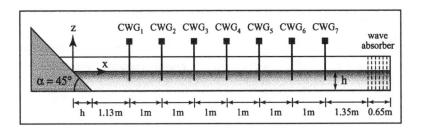

Fig. 3. Configuration of the wave gages set-up.

THE DATA

The experiment consists of creating a variety of landslide conditions from using three water depths and varying relevant governing parameters such as slide mass, slope angle, impact velocity, and impact shape. A total of 137 different runs for each of the 7 wave gages are currently available for detailed and systematic study.

The varied runs resulted in interesting and diversified outcomes of impulse-generated waves. Some start with a group of waves, some emanate a soliton kind of wave, and others evolve into transient bore-like waves. Fig. 4 shows an example, that of run no. 15, of time series plots of laboratory measured landslide-generated impulse waves. The dotted lines toward the end of the plots show the part of data that was generally subjected to the influence of the effect of wave reflections.

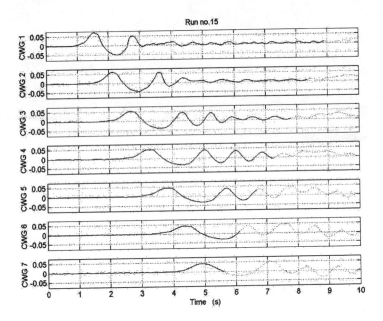

Fig. 4. An example of time series plots of a run of measured landslide-generated impulse waves. The dotted lines show the part that was generally influenced by wave reflection.

WAVELET TRANSFORM ANALYSIS

Wavelet transform analysis, developed during the last two decades, is an ideal tool for the study of the measured time series of nonstationary and transient nature. Briefly for a given time series or signal data function, $X(t)$, which is assumed to be square integrable, its conventional Fourier transform, $\hat{X}(\omega)$, is

$$\hat{X}(\omega) = \int_{-\infty}^{\infty} X(t)\exp(-i\omega t)dt,$$

which transforms the function in the time domain to the frequency domain. In order to examine the characteristics of the transformed function in the frequency domain as well as the time domain, a time windowing function has been used to modify the above equation as

$$\hat{X}(\omega,\tau) = \int_{-\infty}^{\infty} X(t)g(t-\tau)\exp(-i\omega t)dt.$$

Wavelet transform can be considered as generalization of the windowed Fourier transform by substituting the window function with a new family of functions, ψ, sometimes called the mother wavelet. By further discretizing the time, t, and frequency, ω, with a position parameter, a, and scale parameter, b, then leads to the wavelet transform, $\widetilde{X}(a,b)$, as

$$\widetilde{X}(a,b) = \int_{-\infty}^{\infty} X(t)|a|^{-1/2}\psi^{*}\left(\frac{t-b}{b}\right)dt,$$

where $a>0$, $-\infty<b<\infty$, and the asterisk superscript indicates the complex conjugate. Suffice to assert that the wavelet spectrum presented in this study, which is based on the continuous wavelet transform and Morlet wavelet (Liu, 2000), represents a natural extension of the familiar, conventional Fourier spectrum analysis.

RESULTS AND DISCUSSION
The Wavelet Spectrum

In essence, a wavelet spectrum provides a detailed glimpse of the characteristics of the data in the time-frequency domain. Fig. 5 presents the results of run no.13, gage 2, and run no.15, gage 3, as examples of exposition. Each case is represented by three panels of graphs: the top panel plots the time series data with the corresponding wavelet spectrum shown in the middle panel, where the energy density is contoured in the time-frequency domain. The lower third panel draws the total time localized energy with respect to time, which is the integration of the wavelet spectrum over frequency. The results show that one might be able to anticipate a major peak in the case of a single soliton-like impulse wave crest as in run no.13, but it is unlikely that two separate energy peaks can be preconceived from the small group of impulse generated waves in run no.15. There are even cases of three or more energy peaks shown. Sometimes they may also merge into an elongated energy peak hill. Most of the peak hills lasted temporally for 1 to 3 seconds.

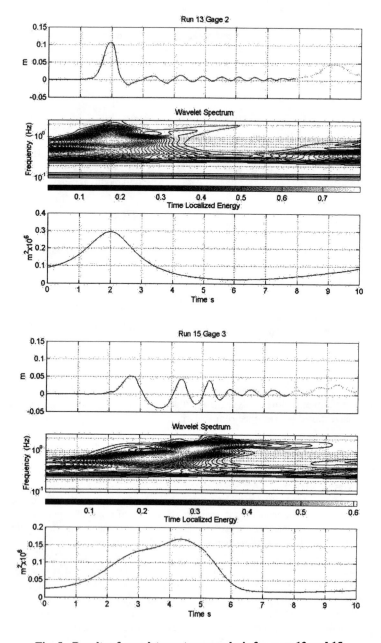

Fig. 5. Results of wavelet spectrum analysis for runs 13 and 15.

The Localized Total Energy

While wavelet spectrum generally provided ample visual interests, it is much less transparent with respect to seeking physical insights. One attempt we made is to examine the advances of the localized total energy from all seven of the wave gages in the time domain as shown in Figs. 6 and 7 for runs no. 13 and 15 respectively. The numerals near the top of the curves denote the corresponding wave gages as positioned in Fig. 3. As one may expect these figures show that there is some dissipation immediately following the impulse waves first generated by the landslide, but the dissipation tends to slow down and become stabilized. Clearly in nature these impulse waves subsequently evolve into tsunami waves.

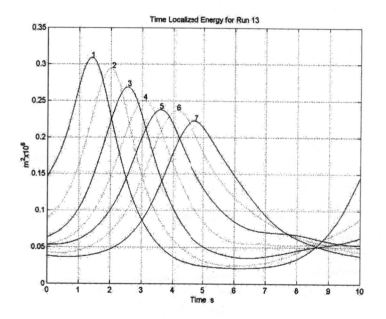

Fig. 6. The succession of localized total energy for run 13.

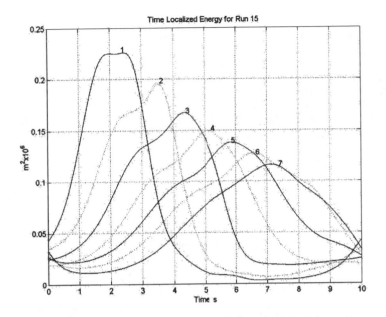

Fig. 7. The succession of localized total energy for run 15.

Impulse Wave Frequency

To examine the significance of the frequency represented by the peaks in the wavelet spectrum, we opt to compare them with the frequency derived from the speed of the impulse waves. The impulse wave speed is obtained by measuring the time of the first wave crest between the first and second wave gages. The comparison can be done readily for the deep-water case, since the wave speed c is given by $c = g/\omega$, where g is the gravity force and ω the radian frequency. In the depth dependent, intermediate water case, the linearized Airy wave theory also provides the basic relations $c^2 = (g/k)\tanh kh$ and $\omega^2 = gk \tanh kh$, where k is the radian wave number and h the water depth, for the simple task.

One would intuitively expect the laboratory wave to be clearly depth dependent. Fig. 8 shows that the correlation between the frequency from the peaks of wavelet spectrum and the frequency from the impulse wave speed display satisfactory results as shown by the upper scatter plot. However when the depth-dependent, intermediate water relations are used, the results in the lower scatter plot show almost no correlation. It appears that depth plays only a minor role here. It can be shown that there is really no difference between the frequency calculated from deep water or intermediate water formula at low wave speed cases, but as wave speed gets higher, the results of intermediate water cases start to diverge and diverge significantly at some point. However, it does not seem that the significant divergence is really an intrinsically physical matter. In a number of cases where the experiment resulted in the wave speed, c, to be actually greater than the prescribed shallow-water wave speed, $\sqrt{gh}$, which inevitably led to the zero frequencies shown in the lower panel of Fig. 8 owing to the linear Airy theory.

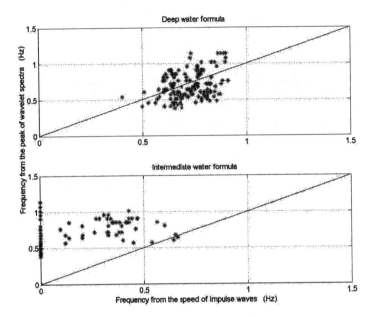

Fig. 8. Correlation of peak spectral frequency with frequency calculated from observed progression of wave crest using deep water formula (upper panel) and intermediate water formula (lower panel.)

CONCLUDING REMARKS

Landslide and landslide generated waves in nature usually and invariably signify disaster and damage that cannot be predicted. Landslide generated impulse waves conducted in the laboratory, on the other hand, represent a fascinating endeavor to discern this elusive phenomenon. The 137 runs conducted in this study provided an extensive collection of interesting as well as stimulating cases for detailed data analysis. Though this paper took only a brief glimpse of the available data, we can nevertheless surmise tenably that wavelet transform analysis will be an essential tool for the further analysis of the available data.

ACKNOWLEDGEMENTS

HMF would like to express his gratitude towards Prof. Dr. H.-E. Minor, director of VAW, and Prof. Dr.W.H. Hager for their continued advise and support. The laboratory research was supported by the Swiss National Science Foundation, grant number 2100-050586.97. This is GLERL Contribution No.1216.

REFERENCES

Fritz, H.M. 2000. PIV Applied to Landslide Generated Impulse Waves. *Proceedings of 10th International Symposium on Applications of Laser Techniques to Fluid Mechanics*.

Liu, P.C. 2000. Wavelet Transform and New Perspective on Coastal and Oceanic Engineering Data Analysis, in *Advances in Coastal and Ocean Engineering*, Vol. 6, Phillip L.F. Liu, Ed., World Scientific. 57-101.

RUN-UP HEIGHTS OF NEARSHORE TSUNAMI BASED
ON QUADTREE GRIDS

Koo-Yong Park[1], Yong-Sik Cho[2] and Byung-Ho Choi[3]

Abstract :   To investigate the run-up heights of nearshore tsunami on the lee side of a circular island, a model has been developed on quadtree grids. The model is based on the Liu *et al.* (1995) using the shallow water equations. The result has been validated experimentally and numerically. The good agreement has been achieved. It should be noted that the run-up heights are mostly dependent on the crest length of incident waves when the crest length is less than the base diameter of the island but are almost insensitive if the crest length is greater than twice the base diameter.

INTRODUCTION

Recently several devastating tsunamis have been occurred around the Pacific Ocean area (Gonzalez, 1999). These tsunamis are not only killed many human beings but also caused serious property damages. A coastal inundation map based on the maximum run-up heights is essential to mitigate coastal casualties from unexpected tsunami attacks.

In this paper, an adaptive quadtree grid system will be employed to investigate the run-up height of nearshore tsunami around a circular island in detail. Based on the numerical model developed by Liu *et al.* (1995), the maximum run-up heights of tsunami attacking a circular island will be re-calculated. The adaptive hierarchical quadtree grids will be generated around a circular island. The uniformed rectangular meshes provided by Liu et al. will be used out of the island, while a localized high resolution meshes will be adapted around boundary of the island. The newly obtained maximum run-up heights will be compared with those of Liu *et al.* and laboratory measurements done at the Coastal Engineering Research Center, U.S. Army Corps of Engineers (Liu *et al.*, 1995).

---

[1]   Senior Engineer, Hyundai Engineering and Construction Co. 140-2 Kye-dong, Chongro-gu, Seoul 110-793, Korea
      (tel) 82-2-746-2230, (fax) 82-2-746-4841, (e-mail) kypark@hdec.co.kr

[2]   Assistant Professor, Department of Civil Engineering, Hanyang University 17 Haengdang-dong, Seongdong-gu, Seoul 133-791, Korea (tel) 82-2-2290-0393, (fax) 82-2-2293-9977, (e-mail) ysc59@email.hanyang.ac.kr

[3]   Professor, Department of Civil Engineering, Sungkyunkwan University 300 Chonchon-dong, Changan-gu, Suwon, Kyonggi 440-746, Korea (tel) 82-31-296-1934, (fax) 82-31-290-7549, (e-mail) bhchoi@yurim.skku.ac.kr

A brief description of laboratory facility will firstly be introduced for completeness in the paper. The governing equations will be discretized with finite difference scheme on the adapted quadtree grids. Particular attention will be paid to validating the computational results with both Liu et al.'s numerical solutions and experimental measurements for the maximum run-up heights of nearshore tsunami around a circular island. Finally, careful discussion on the numerical solutions and concluding remarks will be described.

LABORATORY EXPERIMENT

Experiment was performed in a large scale basin at the CERC (Coastal Engineering Research Center). The basin was 30m wide and 25m long as Fig. 1. The center of the circular island was located at (x=15m, y = 13m). The surface of the island and the floor of the basin were smooth concrete. A directional spectral wave generator has 60 independent paddles. A sketch of the island geometry is shown in Fig. 1.

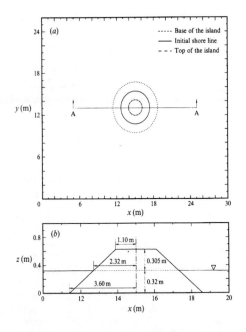

Fig. 1.   Top view of the wave basin and the island and the vertical view of the circular island on the cross-section A-A

Q-TREE GRID GENERATION

Q-tree techniques were used image processing by Samet (1982). Q-tree grids are robust to compute, cheap to produce and easy to adapt. Although they are not exactly boundary-fitting, Q-tree grids can be nearly approach the boundary by use of high resolution boundary cells.

In order to generate a Q-tree grid, the domain of interest is normalised to fit within a unit square firstly. Then the square is sub-divided into four quadrants of equal size. Next subdivision of the square into sub-panels carries out recursively about the seeding points. This involves checking each new panel in turn and subdividing if more than a prescribed minimum number of seeding points are found in the panel. A 2:1 ratio grid regularisation is carried out whereby each cell is checked in order and subdivided if adjacent to a cell that is more than one level smaller. This regulation can reduce the number of cell configurations. Q-tree adaptation is achieved straightforward using grid enrichment and coarsening by additional square subdivision.

The tree-structure of the Q-tree is very helpful for cell identification, and neighbour finding by traversal. For the applications described in this paper, information on the Q-tree grid is held using a linked list with memory pointers. Each link built with pointers from the object cell to its parent cell and its subdivided cells. A systematic search of the tree reveals each cell and its neighbours by means of the *Nearest Common Ancestor* (NCA); which is the smallest branching cell that is shared by the object cell and the neighbour under construction. Park (1999) gave further details of the cell numbering system and search procedures. Fig. 2 depicts a typical Q-tree grid generated about seeding points. The associated hierarchical tree structure is illustrated in Fig. 3.

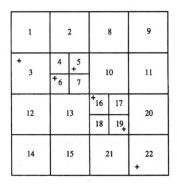

Fig. 2.   Simple Q-tree grid

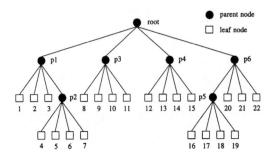

Fig. 3.   Tree structure associated with Fig. 2

GOVERNING EQUATION AND NUMERICAL MODEL

The nonlinear shallow-water equations on the staggered quadtree girds are shown as follows:

$$\frac{\partial \eta}{\partial t} + \frac{\partial P}{\partial x} + \frac{\partial Q}{\partial y} = 0$$

$$\frac{\partial P}{\partial t} + \frac{\partial}{\partial x}\left(\frac{P^2}{H}\right) + \frac{\partial}{\partial y}\left(\frac{PQ}{H}\right) + gH\frac{\partial \eta}{\partial x} = 0$$

$$\frac{\partial Q}{\partial t} + \frac{\partial}{\partial x}\left(\frac{PQ}{H}\right) + \frac{\partial}{\partial y}\left(\frac{Q^2}{H}\right) + gH\frac{\partial \eta}{\partial y} = 0$$

where

η : The free-surface displacement

P : The horizontal components of the volume flux in the x-direction

Q : The horizontal components of the volume flux in the y-direction

H : The total water depth(still water depth + free-surface displacement)

A staggered explicit finite difference leap-fog scheme is used to solve the governing equations. A moving boundary treatment was implemented along the shoreline around the circular island, and a radiation boundary condition for outer boundaries are given by Liu. *et al.* (1995)

RESULTS

The numerical model has been applied on the quadtree grids (Fig. 4). The island is located inside of rectangular domain. The Fig. 5 are shown a sequence of snapshot-type figures of the free surface displacement of a $\varepsilon = 0.1$ wave.

From surface profiles have been clearly formed according to the time which we expected, and similar to experimental data. Fig. 6 is presents comparision between experimental data and numerical results for time history of surface displacement. It is shown that good agreement has been achieved.

Lastly, velocity distributions at different time are shown in Fig. 7. The results are quite comparable with results from uniformed grids.

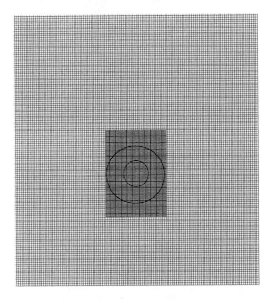

Fig. 4. Applied quadtree mesh

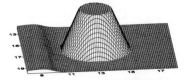 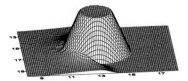

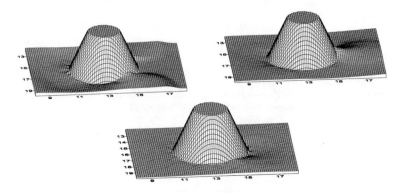

Fig. 5. Snapshots of free surface profile

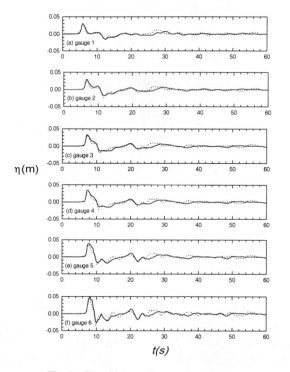

Fig. 6. Time history of free surface displacements
— numerical solution ; .... experimental data

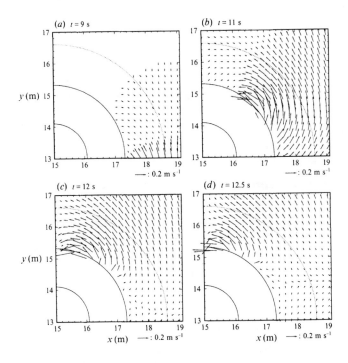

**Fig. 7.   Snapshots of velocity distribution at different times
at the lee of the island**

CONCLUSIONS

The model is based on the shallow-water equations, which developed from Liu *et al.*
(1995) has been applied on the quadtree grids for evaluation of runup around a circular
island. The results has been compared with laboratory data. It should be noted that when
the laboratory manifestation of the wave does not break on the front of the island, the
result has shown good agreement between laboratory data and numerical results. It can be
concluded that the solitary wave runup around a circular island may produce enhanced
runup of the lee side of the island, are strongly depending on the crest length of incident
waves.

However, they are almost independent when the crest length is greater than twice the
base diameter of the island. In general, a smaller island is more vulnerable than a larger
one for tsunami attack. The numerical scheme introduced in this study can be used to
forecast the maximum run-up heights on circular-shaped islands for possible tsunami

attacks for more realistic numerical results using by quadtree grids which are straight forward to generate automatically and also they can be locally refined according to flow.

REFERENCES

Gonzalez, F.I., 1999. Tsunami, *Scientific American*, May, pp. 56-65.

Liu, P.L.-F., Cho, Y.-S., Briggs, M.J., Synolakis, C.E. and Kanoglu, U., 1995. Run-up of solitary wave on a circular island, *Journal of Fluid Mechanics*, Vol. 302, pp. 259-285.

Briggs, M. J., C. E. Synolakis, and G. S. Harkins, 1994. Tsunami run-up on a conical island, in *Proceedings of an International Symposium*: Waves - Physical and Numerical Modelling, pp. 446-455, Vancouver, British Columbia.

Cho, Y.-S., 1995. *Numerical solutions of tsunami propagation and run-up*, Ph.D. thesis, 264 pp., Cornell Univ., Ithaca, N.Y.

Liu, P. L.-F., Y.-S. Cho, and K. Fujima, 1994. Numerical solutions of three-dimensional run-up on a circular island, in *Proceedings of an International Symposium*: Waves - Physical and Numerical Modelling, pp. 1031-1040, Vancouver, British Columbia.

Liu, P. L.-F., Y.-S. Cho, M. J. Briggs, U. Kanoglu, and C.E. Synolakis, 1995. Run-up of solitary wave on a circular island, *J. Fluid Mech.*, 302, pp. 259-285.

Park, K.-Y., 1999, *Quadtree grid numerical model of nearshore wave-current interaction*, D.Phil. Thesis, University of Oxford, U.K.

Titov, V. V., and C. E. Synolakis, 1998. Numerical modeling of tidal wave runup, *J Watern Port Coastal Ocean Eng.*, 124, pp. 157-171.

Yeh, H. H., P. L.-F. Liu, and C. E. Synolakis, 1996. Long-Wave Runup Models, pp. 403, *World Sci.*, River Edge, N. J.

LANDSLIDE TSUNAMI AMPLITUDE PREDICTION IN A NUMERICAL WAVE TANK

Stéphan T. Grilli[1], M.ASCE, Sylvia Vogelmann[1], and Philip Watts [2]

Abstract: Tsunami generation by underwater landslides is simulated in a three-dimensional (3D) Numerical Wave Tank (NWT) solving fully nonlinear potential flow equations. The solution is based on a higher-order Boundary Element Method (BEM). New features are added to the NWT to model underwater landslide geometry and motion and specify corresponding boundary conditions in the BEM model. A snake absorbing piston boundary condition is implemented on the onshore and offshore boundaries. Results compare well with recent laboratory experiments. Sensitivity analyses of numerical results to the width and length of the discretized domain are conducted, to determine optimal numerical parameters. The (3D) effect of landslide width on tsunami generated is then estimated. Results show that the 2D approximation is applicable when the ratio of landslide width over length is greater than three.

INTRODUCTION

Tsunamis generated by underwater landslides triggered on the continental slope appear to be one of the major coastal hazards for moderate earthquakes (e.g., Tappin et al. 2001). Such tsunamis, indeed, are only limited in height by the landslide vertical displacement, which may reach several thousand meters (Murty 1979; Watts 1998). Hence, huge coastal tsunamis, offering little time for warning, can be produced (Watts 2000).

Predicting landslide tsunamis requires complex numerical models in which both landslide and bottom geometry must be accuratedly represented. The models must also account for nonlinear interactions between landslide motion and surface wave field. Grilli and Watts (1999) implemented such of a two-dimensional (2D) numerical model, based on a higher-order Boundary Element Method (BEM), i.e., a Numerical Wave Tank (NWT). Reviews of the literature to

---

[1]Dept. of Ocean Engng., University of Rhode Island, Narragansett, RI 02882, USA. E-mail: grilli@oce.uri.edu.

[2]Applied Fluids Engng., Inc., Mail Box #237, 5710 E. 7th Street, Long Beach, CA 90803, USA. E-mail: phil.watts@appliedfluids.com

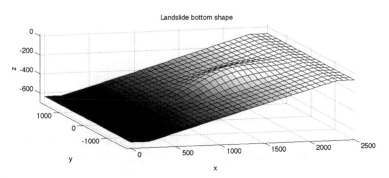

FIG. 1. Example of underwater landslide geometry over a plane 15 degree slope, modeled in the 3D-NWT.

date regarding tsunamis generated by underwater landslides and their numerical modeling can be found in the latter paper and in Watts and Grilli (2001). Here, we describe the current implementation, validation, and simulation of tsunami generation by underwater landslides in the three-dimensional (3D) NWT developed by Grilli et al. (2001). Fully nonlinear potential flow equations are solved in this NWT, based on a higher-order Boundary Element Method and an explicit time stepping scheme. Wave overturning can be modeled if it occurs in the computations. Grilli et al. validated their 3D-NWT for solitary wave shoaling and breaking over slopes, by comparing results both to experiments and to an earlier numerical solution. The agreement was excellent.

In the present work, various improvements were made to the 3D-NWT, to simulate tsunamis caused by underwater landslides. Open boundary conditions were implemented, and validated for solitary wave propagation over constant depth (Grilli and Watts 2001). These conditions extend to 3D the piston-like boundary condition developed by Clément (1996) and Grilli and Horrillo (1997). The landslide shape and kinematics were modeled on a way similar to the Grilli and Watts (1999) 2D model, by assuming a smooth initial shape for the landslide, moving down a planar slope (e.g., Figs. 1 and 2).

NWTs enable many outputs to be obtained with minimal error, and in virtually no setup time (free surface profiles, numerical wave gages, runup, etc...). Here, however, as done in earlier 2D studies, we will represent results of the 3D-NWT by a characteristic wave amplitude calculated above the initial landslide position, at the location of maximum landslide thickness (defined at horizontal location $(x_g, 0)$; Fig. 3). Our characteristic wave amplitude is thus an implicit

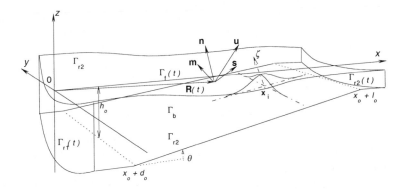

FIG. 2.   Sketch of 3D-NWT used for landslide tsunami modeling.

function of the underwater landslide shape and motion input parameters.

THE NUMERICAL WAVETANK

Governing equations and boundary conditions

Fully nonlinear potential free surface flow equations are solved in the 3D-NWT. These are summarized below. The velocity potential $\phi(\boldsymbol{x}, t)$ describes inviscid irrotational 3D flows in Cartesian coordinates $\boldsymbol{x} = (x, y, z)$, with z the vertical upward direction (and $z = 0$ at the undisturbed free surface; Fig. 2). The velocity is defined by, $\boldsymbol{u} = \boldsymbol{\nabla}\phi = (u, v, w)$.

Mass conservation in the fluid domain $\Omega(t)$, with boundary $\Gamma(t)$, is a Laplace's equation for the potential,

$$\nabla^2\phi = 0 \qquad\qquad \text{in } \Omega(t) \qquad\qquad (1)$$

Green's second identity transforms Eq. (1) into the Boundary Integral Equation (BIE),

$$\alpha_l\,\phi_l = \int_\Gamma \left\{ \frac{\partial\phi}{\partial n}(\boldsymbol{x})\,G(\boldsymbol{x}, \boldsymbol{x}_l) - \phi(\boldsymbol{x})\,\frac{\partial G}{\partial n}(\boldsymbol{x}, \boldsymbol{x}_l) \right\} \mathrm{d}\Gamma \qquad (2)$$

in which $\alpha_l = \alpha(\boldsymbol{x}_l) = \theta_l/(4\pi)$, with θ_l the exterior solid angle made by the boundary at point $\boldsymbol{x}_l$ (i.e., 2π for a smooth boundary), with the 3D free space Green's functions defined as,

$$G = \frac{1}{4\pi\,r} \quad \text{with} \quad \frac{\partial G}{\partial n} = -\frac{1}{4\pi}\,\frac{\boldsymbol{r}\cdot\boldsymbol{n}}{r^3} \qquad (3)$$

where $\boldsymbol{r} = \boldsymbol{x} - \boldsymbol{x}_l$, $r = |\,\boldsymbol{r}\,|$, $\boldsymbol{x}$ and $\boldsymbol{x}_l = (x_l, y_l, z_l)$ are points on boundary Γ, and $\boldsymbol{n}$ is the outward unit vector normal to the boundary at point $\boldsymbol{x}$.

The boundary is divided into various sections, with different boundary conditions (Fig. 2). On the free surface $\Gamma_f(t)$, ϕ satisfies the nonlinear kinematic and dynamic boundary conditions,

$$\frac{D\,\mathbf{R}}{D\,t} = \mathbf{u} = \boldsymbol{\nabla}\phi \qquad\qquad \text{on } \Gamma_f(t) \qquad (4)$$

$$\frac{D\,\phi}{D\,t} = -gz + \frac{1}{2}\,\boldsymbol{\nabla}\phi \cdot \boldsymbol{\nabla}\phi - \frac{p_a}{\rho_w} \qquad \text{on } \Gamma_f(t) \qquad (5)$$

respectively, with $\mathbf{R}$ the position vector of a free surface fluid particle, g the acceleration due to gravity, p_a the atmospheric pressure, ρ_w the fluid density, and D/Dt the material derivative.

Various methods can be used for wave generation in the NWT. Here, tsunamis are generated on the free surface due to a landslide motion $\mathbf{x}_\ell(t)$ specified on the bottom boundary Γ_b (Figs. 1 and 2). We have, for landslide geometry and kinematics,

$$\overline{\mathbf{x}} = \mathbf{x}_\ell \; ; \; \overline{\frac{\partial\phi}{\partial n}} = \mathbf{u}_\ell \cdot \mathbf{n} = \frac{\mathrm{d}\mathbf{x}_\ell}{\mathrm{d}\,t} \cdot \mathbf{n} \qquad \text{on } \Gamma_b(t) \qquad (6)$$

where overlines denote specified values, and the time derivative follows the landslide motion. See below for details.

Along stationary parts of the boundary, such as part of the bottom and some lateral parts of Γ_{r2}, a no-flow condition is prescribed as,

$$\overline{\frac{\partial\phi}{\partial n}} = 0 \qquad\qquad \text{on } (\Gamma_{r2}), (\Gamma_b). \qquad (7)$$

Assuming the landslide motion is in the negative x direction, actively absorbing boundary conditions are specified at one or both extremities of the NWT in the x direction, initially at $x = x_o$ and $x_o + L_0 + l_1$ (Figs. 2 and 3). These are modeled as pressure sensitive "snake" absorbing piston wavemakers. The piston normal velocity is specified as,

$$\overline{\frac{\partial\phi}{\partial n}} = u_{ap}(\sigma, t) \qquad\qquad \text{on } \Gamma_{r2}(t), \text{ with,} \qquad (8)$$

$$u_{ap}(\sigma, t) = \frac{1}{\rho_w h_0 \sqrt{gh_0}} \int_{-h_0}^{\eta_{ap}(\sigma,t)} p_D(\sigma, z, t)\,\mathrm{d}z \qquad (9)$$

calculated at curvilinear abscissa σ, horizontally measured along the piston boundary, where η_{ap} is the surface elevation at the piston and $p_D = -\rho_w\{\frac{\partial\phi}{\partial t} + \frac{1}{2}\boldsymbol{\nabla}\phi \cdot \boldsymbol{\nabla}\phi\}$ denotes the dynamic pressure. The integral in Eq. (9) represents the horizontal hydrodynamic force $F_D(\sigma, t)$ acting on the piston at time t, as a function of σ.

For well-posed problems, we have, $\Gamma \equiv \Gamma_f \cup \Gamma_b \cup \Gamma_{r1} \cup \Gamma_{r2}$.

Time integration

Second-order explicit Taylor series expansions are used to express both the new position $\boldsymbol{R}(t + \Delta t)$ and the potential $\phi(\boldsymbol{R}(t + \Delta t))$ on the free surface, in a mixed Eulerian-Lagrangian formulation. The adaptive time step Δt in the Taylor series is calculated at each time, from the minimum distance between nodes on the free surface, ΔR_o, and a constant mesh Courant number $\mathcal{C}_o = \Delta t \sqrt{g h_o}/\Delta R_o \simeq 0.5$ (see Grilli et al. 2001 for details).

First-order coefficients in the Taylor series are given by Eqs. (4) and (5), which requires calculating $(\phi, \frac{\partial \phi}{\partial n})$ on the free surface. This is done by solving Eq. (2) at time t, with boundary conditions (6) to (9). Second-order coefficients are obtained from the material derivative of Eqs. (4) and (5), which requires calculating $(\frac{\partial \phi}{\partial t}, \frac{\partial^2 \phi}{\partial t \partial n})$ at time t. This is done by solving a BIE similar to Eq. (2) for the $\frac{\partial \phi}{\partial t}$ field. The free surface boundary condition for this second BIE is obtained from Bernoulli Eq. (4), after solution of the first BIE for ϕ.

BEM discretization

The spatial discretization in the NWT follows that of the Grilli et al. (2001) model. All details can be found in the latter reference.

In short, the BIEs for ϕ and $\frac{\partial \phi}{\partial t}$ are solved by a BEM. The boundary is discretized into collocation nodes and 2D cubic sliding boundary elements, based on polynomial shape functions. These are expressed over 4 by 4 node reference elements, of which only one 4-node quadrilateral is used as the actual boundary element. Curvilinear changes of variables are used for expressing boundary integrals over reference elements and deriving discretized equations. Discretized boundary integrals, both regular and singular, are calculated for each collocation node by numerical Gauss quadrature (special methods are used to regularize singular integrals, based on polar coordinate transformations in the reference element). Double and triple nodes and edges are used at intersecting parts of the boundary. The BEM algebraic system of equations is solved in the present applications using a direct elimination method.

Tangential derivatives, e.g., needed in the Taylor series, are calculated on the boundary in a local curvilinear coordinate system $(\boldsymbol{s}, \boldsymbol{m}, \boldsymbol{n})$ defined at each boundary node (Fig. 2), with $\boldsymbol{s} = \boldsymbol{x}_s$, $\boldsymbol{m} = \boldsymbol{x}_m$, and $\boldsymbol{n} = \boldsymbol{s} \times \boldsymbol{m}$ (subscripts indicate partial derivatives). Derivatives of the geometry and field variables in tangential directions $\boldsymbol{s}$ and $\boldsymbol{m}$ are computed, by defining, around each node, a local 5 node by 5 node, 4th-order, sliding element.

Landslide geometry and discretization

In the Grilli and Watts (1999) model, semi-ellipses were used to represent the geometry of 2D vertical landslide cross-sections, which were then moved downslope according to a specified landslide kinematics. Sharp corners occurred at the intersections between the landslide and the planar slope, causing singularities in the BEM solution. This required refining the discretization near corners.

Here, 3D underwater landslides are represented by a fully submerged smooth sediment mound of density ρ_ℓ, sitting over a plane slope of angle θ (Figs. 1 to

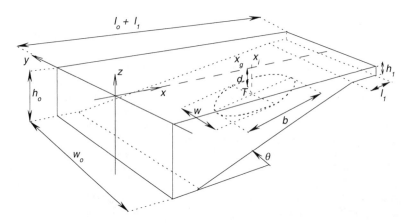

FIG. 3.   Definition of 3D-NWT and landslide geometrical parameters.

3). The landslide has maximum thickness T (measured perpendicularly to the slope). The middle of the landslide surface is located in depth d, at a distance x_g along the x-axis. The geometry is represented by $T \operatorname{sech}^2(k\, r_\ell)$ curves, in polar coordinates (r_ℓ, φ_ℓ) defined within the slope and centered on the landslide axis intersection with the slope, at point $\boldsymbol{x}_i = (x_i, y_i, z_i)$. These curves are truncated at points where they reach an elevation εT above the slope. This geometric model provides for a smoother bottom geometry than ellipses. The landslide, in fact, is simply treated as a "wave" of bottom elevation moving downslope. The landslide footprint on the slope is defined as an ellipse of major axis b (in the x direction) and minor axis w (in the y direction) (Fig. 3). These dimensions are expressed as functions of specified characteristic dimensions (B, W) which, with T, define the landslide volume. Coefficients $k(\varepsilon, \varphi_l, b, w)$ in the definition of landslide geometry are defined based on these parameters. For the sake of simplicity and comparison with other work, it is assumed that the landslide volume is identical to that of the semi-ellipsoid defined by (B, W, T).

In the NWT, the maximum number of nodes/elements that can be used in the domain discretization is limited by the available computer memory. To select an optimal size and resolution of the discretization, a tsunami characteristic wave length was defined by Grilli and Watts (1999), based on theoretical scaling considerations, as,

$$\lambda_o \simeq t_o \sqrt{g\, d} \qquad (10)$$

where t_o is a characteristic time defined below. In all cases, the domain length (x direction) was selected close to that value, and at least $M_x = 20$ elements were used to discretize one theoretical wavelength.

Landslide motion

We follow the wavemaker formalism developed by Watts (1998). Dimensional analysis shows that, within a family of similar landslide geometry, landslide motion and tsunami characteristics are functions of the five nondimensional independent parameters : $\gamma = \rho_\ell/\rho_w$, θ, d/b, T/b, and w/b. Accordingly, one can derive an approximate equation describing the center of mass motion $S(t)$ (parallel to the planar slope in the negative x direction), for rigid underwater landslides, starting at rest at $t = 0$ (see Watts and Grilli (2001) for details). We find,

$$S(t) = S_o \ln\left(\cosh\frac{t}{t_o}\right) \tag{11}$$

with,

$$S_o = \frac{u_t^2}{a_o} \; ; \quad t_o = \frac{u_t}{a_o} \tag{12}$$

where S_o and t_o are characteristic length and time of motion, respectively, and a_o and u_t denote landslide initial acceleration and terminal velocity, respectively, given by,

$$a_o = g\,\frac{\gamma - 1}{\gamma + C_m}\,\sin\theta \tag{13}$$

where C_m is an approximate added mass coefficient, and,

$$u_t = \sqrt{gB}\,\sqrt{\frac{\pi(\gamma - 1)}{2\,C_d}}\,\sin\theta \tag{14}$$

where C_d is an approximate drag coefficient. Watts (1998) and Watts (2000) found added mass and drag coefficients to be of $\mathcal{O}(1)$ for 2D and quasi-2D landslides (for which $W \gg B$). [Note that Coulomb friction was neglected in this analysis.]

APPLICATIONS

To compare results with an earlier 2D model and with recent laboratory experiments, a quasi-2D landslide case was run by Grilli and Watts (2001), for which $W \gg B$ and, hence, there is no lateral (y) variation in landslide geometry. Therefore, the 3D-NWT problem becomes equivalent to a 2D slice along a uniform landslide, similar to cases solved in the 2D model by Grilli and Watts (1999). The agreement with 2D results and with the experiments was found to be quite good for both the characteristic tsunami amplitude and the time evolution of free surface elevation at numerical gages.

Here, we consider a landslide of similar parameters as in this quasi-2D case, i.e. : slope angle $\theta = 15°$, rigid landslide with average density $\rho_\ell = 1,900$ kg/m^3 (and thus $\gamma = 1.845$ for $\rho_w = 1,030$ kg/m^3), length $B = 1,000$ m (and thus $b = 1,299$ m for $\varepsilon = 0.5$) and maximum thickness $T = 50$ m, and initial submergence $d = 260$ m. Now, however, there is a lateral y variation in landslide geometry and a full 3D landslide geometry is modeled as discussed above (Fig. 1). The

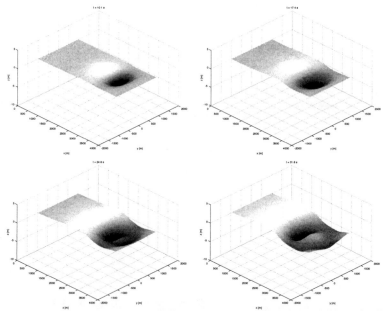

FIG. 4. Surface elevation at different times for a 3D landslide tsunami simulation with : $\theta = 15$ **deg.,** $B = 1000$ **m,** $W = 1000$ **m,** $T = 50$ **m,** $d = 260$ **m,** $L_0 = l_0 + l_1 = \lambda_0 = 3888$ **m,** $w_0 = 2000$ **m,** $d_0 = 400$ **m,** $h_0 = 1,031$ **m,** $l_1 = 300$ **m,** $h_1 = 70$ **m.**

computational domain is such as sketched in Fig. 3. The offshore boundary is represented by an absorbing piston and all other lateral boundaries are assumed impermeable. The shallow shelf at the onshore extremity has depth $h_1 = 70$ m and length $l_1 = 300$ m. The constant depth offshore region has a length $d_o = 400$ m (with $x_o = 0$), and its depth $h_o = h_1 + L_0 \tan \theta$ depends on the selected domain length. The landslide kinematics is computed based on Eqs. (11) to (14). The hydrodynamic coefficients are set to the standard values $C_m = 1$, $C_d = 1$. With this data, the landslide kinematics is defined by : $S_o = 4,469$ m, $u_t = 58.1$ m/s, $a_o = 0.754$ m/s², and $t_o = 77.0$ s, and, based on this, we find an approximate tsunami wavelength $\lambda_o = 3,888$ m.

To assess the effects of discretization and boundary conditions on 3D-NWT results, we study the sensitivity of tsunami characteristic amplitude η_c to the length $L_0 = L_0 + l_1$ and width w_0 of the computational domain. Sensitivity of results to slope angle θ, landslide length B, thickness T, initial submergence d, and density, was studied by Watts and Grilli (2001), using the 2D model by Grilli and Watts (1999). Empirical relationships were derived, through curve fitting of numerical results, between the dimensionless tsunami characteristic amplitude

η_c/S_o and dimensionless parameters : $\sin \theta, B/d, T/B$, and γ. These relationships are essentially applicable to 3D results. Here, to assess 3D effects on tsunami generation, we study the sensitivity of η_c to the dimensionless landslide width W/B.

Only rigid landslides are simulated in this study. Watts and Grilli (2001) investigated the influence of landslide deformation, in the form of a linear landslide extension in the downslope direction during failure. They found that the impact of this deformation on the characteristic amplitude was negligible as a first approximation.

For a domain length $L_0 = \lambda_o$, with $h_o = 1031$ m, $W = B$, and $w_0 = 2W$, we specify 20 BEM elements in the x direction, of initial length $\Delta x_o = 194.4$ m, 10 elements in the width y direction, of length $\Delta y_o = 200$ m, and 4 elements over depth. With this data, the initial time step is set to $\Delta t_o = 0.45 \Delta x_o/\sqrt{g\,h_o} = 0.87$ s. Fig. 4 shows free surface elevations computed in this case, at various times $t = 0$ to 31.6 s. We see the appearance of an initial dipole-like surface elevation, with a wave of depression forming above the landslide initial location and a wave of elevation propagating offshore of the landslide. Later on ($t > 21$ s), a smaller wave of elevation is seen to also propagate onshore, which would eventually induce coastal flooding and runup. For this case, the maximum error on mass conservation was 0.82% of the landslide volume.

We maintain the same density of discretization and initial time step for other cases studied in the following sensitivity analysis, using different domain length and width, and landslide width.

Effect of domain length

For simplicity, cases with no variation in the y-direction ($W = \infty$), are modeled. The domain width is set to $w_0 = B$ for all calculations. The total domain length L_0 is set to multiples of the theoretical wavelength λ_o (0.6 to 3). For fully developed quasi-2D tsunami generation we find $\eta_{c2D} \simeq 6.0$ m. For $L_0 > \lambda_o$, little change is observed in η_{c2D}. Hence, this value is used for domain length in the following.

Effect of domain width

Here, we verify at which distance, $y = \pm w_0/2$, the impermeable lateral boundaries have to be located to avoid perturbing the tsunami generation process, i.e., changing η_c through reflection. To do so, landslides of different widths $W = B$, $2B$ and $3B$ are modeled and, for each width, calculations are performed using different domain widths $w_0 = W$ to $3 - 6W$. Results show that, for a ratio $w_0/W \geq 2$, η_c only varies very slightly. Hence, this ratio is used in the following.

Effect of landslide width

Landslides of various widths $W = B/2$ to $3B$ are modeled and the resulting η_c compared. Results show an increase in η_c with increasing landslide aspect ratio W/B. The large initial increase with W/B, however, becomes slower for $W/B > 2$, where the amplitude eventually approaches its 2D value asymptotically. For the current results, we find $\eta_c \simeq \eta_{c2D} - 2.36(W/B)^{-1.35}$ (with $R^2 = 0.95$).

CONCLUSIONS

Landslide tsunami generation mechanisms were explored in a 3D-NWT solving fully nonlinear potential flow equations, using a higher-order BEM. Accurate modeling of underwater landslide geometry and motion, and snake absorbing piston boundaries and their efficiency, were illustrated in the applications. Experimental validation of results was reported elsewhere for a quasi-2D landslide.

For one realistic set of landslide parameter values, we conclude that the characteristic tsunami amplitude η_c is not significantly affected when the width of the computational domain $w_0 \geq 2W$ and the length of the computational domain $L_0 \geq \lambda_o$, respectively. We also find that the 3D η_c converges towards the 2D value for $W \geq 2B$. The latter result is of importance as 2D analyses (such as described in Grilli and Watts (1999) and Watts and Grilli (2001)) are much easier and quicker to carry out than 3D studies, particularly in a realtime forecasting situation. It is therefore of interest to assess whether 2D conditions can be assumed in a given case and the corresponding simplified model applied.

REFERENCES

Clément, A. (1996). "Coupling of two absorbing boundary conditions for 2d time-domain simulations of free surface gravity waves." *J. Comp. Phys.*, 26, 139–151.

Grilli, S., Guyenne, P., and Dias, F. (2001). "A fully nonlinear model for three-dimensional overturning waves over arbitrary bottom." *Intl. J. Numer. Methods in Fluids*, 35(7), 829–867.

Grilli, S. and Horrillo, J. (1997). "Numerical generation and absorption of fully nonlinear periodic waves." *J. Engng. Mech.*, 123(10), 1060–1069.

Grilli, S. and Watts, P. (1999). "Modeling of waves generated by a moving submerged body. Application to underwater landslides." *Engng. Analysis with Boundary Elts.*, 23, 645–656.

Grilli, S. and Watts, P. (2001). "Modeling of tsunami generation by an underwater landslide in a 3D-NWT." *Proc. 11th Offshore and Polar Engng. Conf. (ISOPE01, Stavanger, Norway)*, Vol. 3. 132–139.

Murty, T. (1979). "Submarine slide-generated water waves in kitimat inlet, british columbia." *J. Geophys. Res.*, 84(C12), 7777–7779.

Tappin, D., Watts, P., McMurtry, G., Lafoy, Y., and Matsumoto, T. (2001). "The Sissano, Papua New Guinea tsunami of July 1998 – offshore evidence on the source mechanism." *Marine Geology*, 175, 1–23.

Watts, P. (1998). "Wavemaker curves for tsunamis generated bu underwater landslides." *J. Waterway, Port, Coastal, and Ocean Engng.*, 124(3), 127–137.

Watts, P. (2000). "Tsunami features of solid block underwater landslides." *J. Waterway, Port, Coastal, and Ocean Engng.*, 126(3), 144–152.

Watts, P. and Grilli, S. (2001). "Tsunami generation by submarine mass failure. Part I : Wavemaker models." *To be submitted to J. Waterway, Port, Coastal, and Ocean Engng.*

BENCHMARK CASES FOR TSUNAMIS GENERATED
BY UNDERWATER LANDSLIDES

Philip Watts[1], Fumihiko Imamura[2], Aaron Bengston[3], and Stephan T. Grilli[3], M.ASCE

Abstract: Three benchmark cases are proposed to study tsunamis generated by underwater landslides. Two distinct numerical models are applied to each benchmark case. Each model involves distinct center of mass motions and rates of landslide deformation. Computed tsunami amplitudes agree reasonably well for both models, although there are differences that remain to be explained. One of the benchmark cases is compared to laboratory experiments. The agreement is quite good with the models. Other researchers are encouraged to employ these benchmark cases in future experimental or numerical work.

INTRODUCTION

Tsunamis generated by underwater landslides are receiving more attention following analyses demonstrating that the surprisingly large local tsunami documented during the 1998 Papua New Guinea catastrophe was generated by submarine mass failure (Kawata *et al.*, 1999; Tappin *et al.*, 1999, 2001; Watts *et al.*, 2001). In response to these and other studies, recent work by marine geologists now considers the tsunamigenic potential of submarine mass failure scars (Goldfinger *et al.*, 2000). Despite these advances in the observational science, there remain few instances of validation of the numerical models currently in use (Heinrich, 1992; Grilli and Watts, 2001). Consequently, the ability of scientists to simulate tsunami generation remains in doubt. We note that underwater landslides are also called submarine mass failures and display a wide range of morphological features (Prior and Coleman, 1979; Edgers and Karlsrud, 1982; Hampton *et al.*, 1996).

Researchers have tackled tsunami generation by underwater landslides with a wide variety of numerical methods incorporating many different assumptions. Iwasaki (1987, 1997) and

1 President, Applied Fluids Engineering, Inc., Private Mail Box #237, 5710 E. 7th Street, Long Beach, CA 90803 USA. phil.watts@appliedfluids.com
2 Professor, Disaster Control Research Center, School of Engineering, Tohoku University, Aoba 06, Sendai 980-8579, Japan. imamura@tsunami2.civil.tohoku.ac.jp
3 Professor, Department of Ocean Engineering University of Rhode Island, Narragansett, RI 02882 USA. grilli@oce.uri.edu

Verriere and Lenoir (1992) utilized linear potential theory to simulate wave generation by moving the domain boundary. Depth-averaged Nonlinear Shallow Water (NSW) wave equations were solved by Fine *et al.* (1998), Harbitz (1992), Imamura and Gica (1996), and Jiang and LeBlond (1992, 1993, 1994), in combination with disparate landslide models. Fully nonlinear fluid dynamic field equations were solved by Assier Rzadkiewicz *et al.* (1997), Grilli and Watts (1999) (in an irrotational and inviscid approximation), and Heinrich (1992), in concert with assorted landslide models. For the most part, scientists have studied vastly different landslide geometries, motions, and constitutive behaviors.

To further advance research in tsunami generation by underwater landslides, benchmark cases are needed to validate numerical models and to help explain the origins of any discrepancies that may exist, both between numerical models, and with comparisons to experimental results. Benchmark cases are already available for tsunami propagation and inundation (Yeh *et al.*, 1996). Our goal in this work is to establish three benchmark cases for future reference by researchers interested in tsunami generation by underwater landslides, and to compare simulations, for one of these cases, to recently performed laboratory experiments. Each case is two-dimensional in order to reduce computational or experimental effort. We hope that this work will promote future numerical comparisons and experimental realizations of these benchmark cases. The comparisons made here are not an end of this effort. We expect to repeat this experiment at a larger laboratory scale.

BENCHMARK CASES

To facilitate their experimental realization, the benchmark cases chosen for this work are based in part on the sliding block experiments of previous researchers (Heinrich, 1992; Iwasaki, 1982; Watts, 1997; Wiegel, 1955). A straight incline forms a planar beach with the coordinate origin at the undisturbed beach and the positive x-axis oriented horizontally away from the shoreline (Fig. 1). A semi-ellipse approximates the initial landslide geometry with a maximum thickness T along half of the minor axis that is perpendicular to a major axis of total length b. The nominal underwater landslide length measured along the incline is $b = 1000$ m for all three benchmark cases given here. The semi-ellipse has an initial vertical submergence d at the middle of the landslide surface. All underwater landslides are assumed to have a bulk density $\rho_b = 1900$ kg/m^3 and fail in sea water of density $\rho_o = 1030$ kg/m^3. The size and density of a frictionless landslide suffices to describe center of mass motion following failure given the slope. Landslide deformation is permitted following incipient motion of the semi-ellipse so that we may document the influence of deformation on tsunami generation. A solid block landslide would undergo no deformation. Landslide deformation, if any is present, is assumed to occur about the center of mass motion.

The geometrical parameters for each benchmark case are given in Table 1. The key to establishing these benchmark cases is to specify only the initial condition of the frictionless landslides. Each benchmark case therefore depends on the model assumptions or realization of landslide mechanics. The initial submergence at the middle of the landslide, $x = x_g$, was obtained from a scaled reference equation $d = b \sin\theta$, while the initial landslide thickness was calculated from another scaled reference equation, $T = 0.2\ b \sin\theta$ (Watts and Grilli, 2001). A wave gage was situated above the middle of the initial landslide position at $x_g = (d + T/\cos\theta)/\tan\theta$, and recorded tsunami amplitude $\eta(t)$. Dimensional quantities are presented throughout because different numerical techniques employ different non-dimensional schemes. Watts (1998) provides the correct Froude scaling to perform these benchmark experiments at laboratory scale.

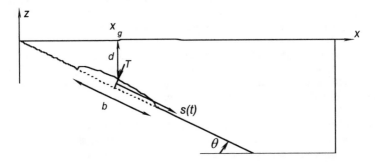

Fig. 1. Schematic diagram of underwater landslide parameters

A laboratory experiment was conducted in the University of Rhode Island wave tank (length 30 m, width 3.6 m, depth 1.8 m). This tank is equipped with a modular beach made of 8 independently adjustable panels (3.6 m by 2.4 m) whose difference in slope can be up to 15°. Benchmark case 2 was tested in the wave tank at 1:1000 scale. Two beach panels were set to an angle $\theta = 15°$ and covered by a smooth aluminum plate. A quasi two-dimensional experiment was realized by building vertical (plywood) side walls at a small distance (about 15 cm) from each other. A semi-elliptical wood and plastic landslide model was built and installed between the walls. The model was equipped with low-friction wheels and a lead ballast was added to achieve the correct bulk density. An accelerometer was attached to the model center of gravity to measure landslide kinematics. Experiments were repeated at least five times and the repeatability of results was very good.

Table 1. Underwater Landslide and Wave Gauge Parameters

Benchmark Case	Incline Angle, °	Landslide Length, m	Landslide Thickness, m	Mean Depth, m
Case 1	30	1000	100	500
Case 2	15	1000	51.8	259
Case 3	5	1000	17.4	87.2

NUMERICAL MODEL DESCRIPTIONS

Imamura and Imteaz (1995) developed a mathematical model for a two-layer flow along a non-horizontal bottom. Conservation of mass and momentum equations were depth-integrated in each layer, and nonlinear kinematic and dynamic conditions were specified at the free surface and at the interface between fluids. Both fluids had uniform densities and were immiscible. Vertical velocity distributions were assumed within each fluid layer. The landslide fluid was ascribed a uniform viscosity, which sensitivity analyses show has very little effect on wave records over a range of viscosities 1-100 times that of water. A staggered leap-frog finite difference scheme, with a second-order truncation error was used to solve the governing equations. Landslides were thus modeled as immiscible fluid flows comprising a second layer, as in the work of Jiang and LeBlond (1992, 1993, 1994). An instantaneous local force balance governed landslide motion. Hence, this motion resulted from the solution of the problem itself and was not externally specified as a boundary condition. We will refer to this numerical model as the II Model below.

Grilli *et al.* (1989, 1996) developed and validated a two-dimensional Boundary Element Model (BEM) of inviscid, irrotational free surface flows (i.e., potential flow theory). Cubic boundary elements were used for the discretization of boundary geometry, combined with fully nonlinear boundary conditions and second-order accurate time updating of free surface position. The model was experimentally validated for long wave propagation and runup or breaking over slopes by Grilli *et al.* (1994, 1998). Model predictions are surprisingly accurate; for instance, the maximum discrepancy for solitary waves shoaling over slopes is 2% at the breaking point, between computed and measured wave shapes. Grilli and Watts (1999) applied this BEM model to water wave generation by underwater landslides and performed a sensitivity analysis for one underwater landslide scenario. The landslide center of mass motion along the incline was prescribed by the analytical solutions of Watts (1998, 2000). In these computations, the landslide retained its semi-elliptic shape while translating along the incline. Grilli and Watts (2001) recently validated the model experimentally. We will refer to this numerical model as the GW Model below.

Both the II and GW Models are used in the following to simulate tsunamis generated by underwater landslides of identical initial characteristics corresponding to the three benchmark cases in Table 1. For discretization techniques and numerical parameters used in both models, please refer to Imamura and Imteaz (1995) and Grilli and Watts (1999).

UNDERWATER LANDSLIDE CHARACTERISTICS

Water wave amplitudes above an underwater landslide scale with characteristics of center of mass motion (Hammack, 1973; Watts, 1998, 2000). We contrast the center of mass motions of submarine slides and slumps by the dominant retardive force providing convenient end members for these two types of motion. Underwater slides are identified by translational failure and distant center of mass displacement, while underwater slumps are defined to undergo rotational failure and center of mass displacement typically less than the slump length (Schwab *et al.*, 1993). Center of mass motion is parametrized by the characteristic time of motion t_o and the characteristic distance of motion s_o (Watts, 1998, 2000). These quantities are dependent on the sediment type, sediment density, sediment shear strength, and landslide shape. The center of mass motion of a slide (retarded primarily by fluid dynamic drag) is significantly different from that of a slump (retarded primarily by basal friction). Likewise, the tsunamis generated by these two forms of submarine mass failure can differ substantially. To make this point clearly, we compare tsunami generation by a slide with that of a slump given identical landslide shape and density. Fig. 2 shows computed surface elevations above the initial landslide position as shown on Fig. 1. The maximum tsunami trough (or "characteristic") amplitude generated by the underwater slide is four times that of the underwater slump, despite having similar initial accelerations. This is not surprising because: i) the duration of acceleration lasts about twice as long for this slide, ii) the landslide displacement near $t = 0$ goes as time squared, and iii) the tsunami amplitude scales foremost with center of mass displacement (Watts, 2000). These three facts result in a four-fold increase in tsunami amplitude in favor of the slide. The slide wavelength appears to be about double that of the slump. This wavelength difference is consistent with the different durations of acceleration between the slide and the slump (Watts, 2000). Our benchmark cases represent underwater slides because these are easier to reproduce at laboratory scale.

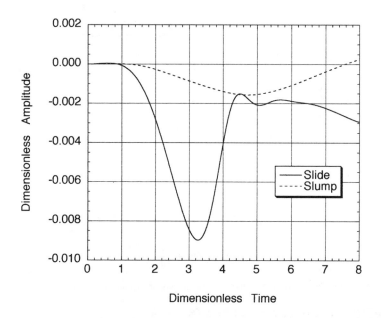

Fig. 2. Comparison of underwater slide and slump tsunami generation

By the time an underwater slide comes to rest within some oceanic deep, considerable deformation and spreading may have occurred. Therefore, we need to consider landslide deformation in addition to center of mass motion. Watts (1997) found experimentally that the primary mode of underwater slide deformation is through extension parallel to the incline. Watts *et al.* (2000) found the same result for numerical simulations of deformable slides performed with the model of Imamura and Imteaz (1995). For both experimental and numerical results, the rate of extension was constant (i.e., slide length grew linearly in time) following an initial transient. The maximum landslide thickness T remained more or less constant over time. Watts and Grilli (2001) derive a maximum rate of landslide extension based on observations of actual landslide deposits. Fig. 3 demonstrates that the maximum rate of extension changes the tsunami amplitude by less than 10% and has no significant effect on tsunami wavelength at a particular wave gauge situated above the landslide. Center of mass motion appears to affect tsunami features much more than landslide deformation (Watts *et al.*, 2000; Watts and Grilli, 2001). We surmise that center of mass motion proceeds much more rapidly than deformation at those early times applicable to tsunami generation.

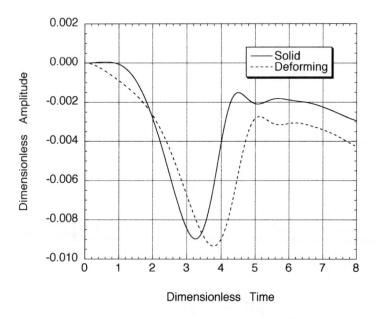

Fig. 3. Comparison of solid and deforming underwater slide tsunami generation

SIMULATION AND EXPERIMENTAL RESULTS

Descriptions of tsunami generation by underwater landslides should begin by documenting landslide center of mass motion and rates of deformation. Since both motion and deformation were prescribed in the GW Model, we proceed to describe the results obtained from the II Model and compare these results with the GW Model. We also relate the experimental initial acceleration obtained for case 2. Assuming the center of mass motion $s(t)$ is parallel to the incline (Fig. 1), we verified that the simple equation

$$s(t) = \frac{a_0 t^2}{2} \qquad (1)$$

provides an accurate fit of these motions (Watts *et al.*, 2000). Eq. (1) is the first term in a Taylor series expansion of landslide motion beginning at rest (Watts and Grilli, 2001). Values of initial landslide accelerations a_0 for the II Model obtained by curve fitting Eq. (1) can be found in Table 2. The experimental initial acceleration was $a_0 = 0.73$ m/s^2 for case 2. This compares favorably with the value from the GW Model given in Table 2.

Table 2. Landslide Initial Accelerations and Rates of Deformation

Benchmark Case	II Initial Acceleration, m/s^2	GW Initial Acceleration, m/s^2	II Rate of Deformation, s^{-1}	GW Rate of Deformation, s^{-1}
Case 1	3.11	1.47	0.062	0.0
Case 2	1.29	0.76	0.035	0.0
Case 3	0.40	0.26	0.017	0.0

Landslide deformation in the II Model was manifested foremost as an extension in time, $b(t)$ of the initial landslide length b_0 (Watts *et al.*, 2000). The non-dimensional ratio b/b_0 varied almost linearly with time, following an initial transient, similar to the experimental observations made by Watts (1997) for a submerged granular mass. A semi-empirical expression that describes landslide extension is

$$b(t) \ = \ b_0 \, \{1 + \varGamma t \ [1 - \exp(-K t)]\} \tag{2}$$

where $\varGamma$ is the eventual linear rate of extension and the exponential term describes an initial transient, with $K = a_0/b_0 \, \varGamma$ (Watts and Grilli, 2001). The parameter K is chosen to fix the uppermost landslide corner in place as the center of mass begins to accelerate. Table 2 gives values of $\varGamma$ for the II Model found from curve fits of Eq. (2).

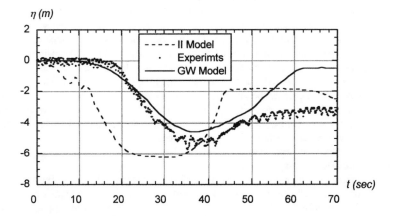

Fig. 4. Comparison of model and experimental results for benchmark case 2

Fig. 4 shows the tsunami simulation results of both numerical models and the laboratory experiment all performed for benchmark case 2. The GW and II Model results agree qualitatively for all three cases, although the GW Model produces slightly smaller wave amplitudes (Watts *et al.*, 2000). The II Model produces more acute free surface curvature near $t = 0$ as well as longer tsunami periods. We note that the II Model has water wave disturbances in the first 5-20 s of each simulation brought on by a Kelvin-Helmholtz type instability along the landslide-water interface.

DISCUSSION

The results presented here, as well as the more detailed results in Watts *et al.* (2000), represent a first attempt to provide benchmark cases for underwater landslide tsunami generation. Both the numerical and experimental results suffer from either incomplete knowledge or noticeable errors, as the case may be. However, the existence of benchmark cases enables further (numerical and experimental) realizations to improve on the accuracy given here. Some of us are in the process of planning a second round of benchmark experiments. The more vexing problem is to address the many assumptions built into each numerical model in an attempt to identify their respective ranges of applicability. The two numerical models compared here are far too different to draw substantial conclusions regarding tsunami generation (see Watts *et al.*, 2000). We believe that many more benchmark cases and numerical simulations are required to accomplish this task. As a first step to encouraging further numerical work, we have placed the experimental results provided above on the web site *www.tsunamicommunity.org* for downloading.

CONCLUSIONS

Case studies of actual tsunami events cannot be expected to provide sufficient data of sufficient accuracy with which to validate numerical models (Watts, 2001). To overcome this difficulty, three benchmark cases for tsunamis generated by underwater landslides are proposed in this paper. The underwater landslide initial acceleration and rate of deformation are both needed to compare benchmark simulations or experiments. Underwater landslide center of mass motion during tsunami generation can be described by the initial acceleration. Larger initial accelerations produce larger tsunami amplitudes. The characteristic tsunami amplitudes differed by up to 13% for the two numerical models compared here. Experimental results and numerical simulations to date indicate that the primary mode of landslide deformation consists of a linear rate of extension. Experimental results available for benchmark case 2 showed a agreement with the GW Model results, in part because landslide deformation changes the shape of the wave gage record. Interpretation of existing numerical model differences awaits more detailed model comparisons. We have endeavored to begin a process of comparing numerical simulations and experimental realizations for three benchmark cases. We hope the process will continue.

ACKNOWLEDGEMENTS

This work arose from a discussion at the 1999 IUGG General Meeting in Birmingham, U. K. at which time all tsunami scientists were invited to participate in landslide tsunami benchmark simulations. The authors are grateful for support from Applied Fluids Engineering, Inc. and the Disaster Control Research Center at Tohoku University.

REFERENCES

Assier Rzadkiewicz, S., Mariotti, C., and Heinrich, P. (1997). Numerical simulation of submarine landslides and their hydraulic effects. *J. Wtrwy, Port, Coast, and Oc. Engrg.*, ASCE, 123(4), 149-157.

Edgers, L., and Karlsrud, K. (1982). Soil flows generated by submarine slides: Case studies and consequences. *Nor. Geotech. Inst. Bull.*, 143, 1-11.

Fine, I. V., Rabinovich, A. B., Kulikov, E. A., Thomson, R. E., and Bornhold, B. D. (1998). Numerical modelling of landslide-generated tsunamis with application to the Skagway Harbor tsunami of November 3, 1994. *Proc. Tsunami Symp.*, Paris.

Goldfinger, C., Kulm, L. D., McNeill, L. C., Watts, P. (2000). Super-scale failure of the Southern Oregon Cascadia Margin. *PAGEOPH*, 157, 1189-1226.

Grilli, S. T., Skourup, J., and Svendsen, I. A. (1989). An efficient boundary element method for nonlinear water waves. *Engrg. Analysis with Boundary Elements*, 6(2), 97-107.

Grilli, S. T., Subramanya, R., Svendsen, I. A. and Veeramony, J. (1994). Shoaling of Solitary Waves on Plane Beaches. *J. Wtrwy, Port, Coast, and Oc. Engrg.*, ASCE, 120(6), 609-628.

Grilli, S. T., and Subramanya, R. (1996). Numerical modeling of wave breaking induced by fixed or moving boundaries. *Comp. Mech.*, 17, 374-391.

Grilli, S. T., Svendsen, I. A., and Subramanya, R. (1998). Breaking criterion and characteristics for solitary waves on slopes -- Closure. *J. Wtrwy, Port, Coast, and Oc. Engrg.*, ASCE, 124(6), 333-335.

Grilli, S. T., and Watts, P. (1999). Modeling of waves generated by a moving submerged body: Applications to underwater landslides. *Engrg. Analysis with Boundary Elements*, 23(8), 645-656.

Grilli, S. T., and Watts, P. (2001). Modeling of tsunami generation by an underwater landslide in a 3D-Numerical Wave Tank. *Proc. 11th Offshore and Polar Engrg. Conf., ISOPE01, Stavanger, Norway*, 3, 132-139.

Hammack, J. L. (1973). "A note on tsunamis: Their generation and propagation in an ocean of uniform depth." *J. Fluid Mech.*, 60, 769-799.

Hampton, M. A., Lee, H. J., and Locat, J. (1996). Submarine landslides. *Rev. Geophys.*, 34(1), 33-59.

Harbitz, C. B. (1992). Model simulations of tsunamis generated by the Storegga slides. *Marine Geology*, 105, 1-21.

Heinrich, P. (1992). Nonlinear water waves generated by submarine and aerial landslides. *J. Wtrwy, Port, Coast, and Oc. Engrg.*, ASCE, 118(3), 249-266.

Imamura, F., and Imteaz, M. M. A. (1995). Long waves in two-layers: Governing equations and numerical model. *J. Sci. Tsunami Hazards*, 13, 3-24.

Imamura, F., and Gica, E. C. (1996). Numerical model for tsunami generation due to subaqueous landslide along a coast. *Sci. Tsunami Hazards*, 14, 13-28.

Iwasaki, S. (1982). Experimental study of a tsunami generated by a horizontal motion of a sloping bottom. *Bull. Earth. Res. Inst.*, 57, 239-262.

Iwasaki, S. (1987). On the estimation of a tsunami generated by a submarine landslide. *Proc., Int. Tsunami Symp.*, Vancouver, B.C., 134-138.

Iwasaki, S. (1997). The wave forms and directivity of a tsunami generated by an earthquake and a landslide. *Sci. Tsunami Hazards*, 15, 23-40.

Jiang, L., and LeBlond, P. H. (1992). The coupling of a submarine slide and the surface waves which it generates. *J. Geoph. Res.*, 97(C8), 12731-12744.

Jiang, L., and LeBlond, P. H. (1993). Numerical modeling of an underwater Bingham plastic mudslide and the waves which it generates. *J. Geoph. Res.*, 98(C6), 10303-10317.

Jiang, L., and LeBlond, P. H. (1994). Three-dimensional modeling of tsunami generation due to a submarine mudslide. *J. Phys. Ocean.*, 24, 559-573.

Kawata, Y. and International Tsunami Survey Team members. (1999). Tsunami in papua New guinea was intense as first thought. *Eos, Trans. Am. Geophys. Union*, 80(9), 101.

Prior, D. B., and Coleman, J. M. (1979). Submarine landslides: Geometry and nomenclature. *Z. Geomorph. N. F.*, 23(4), 415-426.

Schwab, W. C., Lee, H. J., Twichell, D. C. (1993). *Submarine landslides: Selected studies in the U.S. exclusive economic zone*. Bull. 2002, U.S. Geol. Surv., U.S., Dept. of Interior, Washington, DC.

Tappin, D. R. , Matsumoto, T., and shipboard scientists. (1999). Offshore surveys identify sediment slump as likely cause of devastating Papua New Guinea tsunami 1998. *Eos, Trans. Am. Geophys. Union*, 80(30), 329.

Tappin, D. R., Watts, P., McMurtry, G. M., Lafoy, Y., Matsumoto, T. (2000). The Sissano Papua New Guinea tsunami of July 1998 – Offshore evidence on the source mechanism. *Marine Geol.*, 175, 1-23.

Verriere, M., and Lenoir, M. (1992). Computation of waves generated by submarine landslides. *Int. J. Num. Methods Fluids*, 14, 403-421.

Watts, P. (1997). Water waves generated by underwater landslides. PhD thesis, California Inst. of Technol., Pasadena, CA.

Watts, P. (1998). Wavemaker curves for tsunamis generated by underwater landslides. *J. Wtrwy, Port, Coast, and Oc. Engrg.*, ASCE, 124(3), 127-137.

Watts, P. (2000). Tsunami features of solid block underwater landslides. *J. Wtrwy, Port, Coast, and Oc. Engrg.*, ASCE, 126(3), 144-152.

Watts, P., Imamura, F., and Grilli, S. T. (2000). Comparing model simulations of three benchmark tsunami generation cases. *Sci. Tsunami Hazards*, 18(2), 107-124.

Watts, P., Borrero, J. C., Tappin, D. R., Bardet, J.-P., Grilli, S. T., Synolakis, C. E. (2001). Novel simulation technique employed on the 1998 Papua New Guinea tsunami. In: G. T. Hebenstreit (ed), *Tsunami research at the end of a critical decade*, Kluwer, Dordrecht, The Netherlands.

Watts, P. (2001). Some opportunities of the landslide tsunami hypothesis. *Sci. Tsunami Hazards*, in press.

Watts, P., and Grilli, S. T. (2001). Tsunami generation by submarine mass failure part I: Wavemaker models. *J. Wtrwy, Port, Coast, and Oc. Engrg.*, ASCE, to be submitted.

Wiegel, R. L. (1955). Laboratory studies of gravity waves generated by the movement of a submarine body. *Trans. Am. Geophys. Union*, 36(5), 759-774.

Yeh, H. H., Liu, P.L.-F., Synolakis, C. E. (1996). *Long-wave runup models: Friday Harbor, USA, 12-17 Sept. 1995*. World Scientific, Singapore.

NUMERICAL MODELING OF TSUNAMI PROPAGATION OVER VARYING WATER DEPTH

Sung B. Yoon[1], Chul-Sun Choi[2], and Sung-Myeon Yi[3]

Abstract: A dispersion-correction finite difference numerical scheme is employed to simulate the propagation of distant tsunamis over slowly varying topography with improved dispersion effect of waves. The scheme solves the shallow-water equations on a uniform grid system. However, the actual computation is made on an imaginary grid system whose grid size is adjustable according to the condition required to satisfy local dispersion relationships. The present model is applied to the simulation of a historical tsunami event. Numerical results are compared with the tide gage records, and reasonably good agreements are obtained.

INTRODUCTION

Most of the numerical models developed so far for the simulation of tsunamis are based on the shallow-water equations and have been successfully applied to the near-field tsunamis. However, when the tsunamis are generated far from the region of interest, the propagation of tsunamis should be modeled using Boussinesq equations to take into consideration the dispersion effect.

The Boussinesq numerical model, however, requires a small mesh size to suppress the numerical dispersion and consumes huge computer resources due to the implicit nature of the solution technique to deal with dispersion terms. Thus, the Boussinesq model is not preferred for the simulation of distant tsunamis. Imamura *et al.* (1988)

1 Assoc. Prof., Department of Civil and Environmental Engineering, Hanyang University, Ansan, Kyunggi 425-791, Korea. e-mail : sbyoon@hanyang.ac.kr
2 Senior Coastal Engineer, Civil and Architectural Engineering Department, Korea Power Engineering Co., Inc., Yongin, Kyunggi 449-713, Korea. e-mail : cschoi@ns.kopec.co.kr
3 Senior Coastal Engineer, Civil and Architectural Engineering Department, Korea Power Engineering Co., Inc., Yongin, Kyunggi 449-713, Korea. e-mail : hydroyi@ns.kopec.co.kr

presented a finite difference model for the simulation of transoceanic tsunamis which solves the shallow-water equations using the leap-frog scheme. The dispersion effect neglected in the shallow-water equations is taken into account in the simulation by utilizing the numerical dispersion inherent in the leap-frog scheme. This can be done if the grid size is selected for the given water depth and the time step according to the criterion proposed by Imamura *et al.* The use of numerical models developed by Imamura *et al.*, however, is restricted to the case of constant water depth if a uniform finite difference grid is employed.

For the accurate simulation of tsunamis propagating in a varying topography, the dispersion effect of waves should be carefully considered in the whole computational domain. Since the mesh size, which is determined by the criterion proposed by Imamura *et al.*, changes according to the given water depth and time step, an adjustable grid system should be employed for varying topography. Although the finite difference models of Imamura *et al.* (1988) is computationally efficient, the mesh size of the models can not be adjusted for two-dimensional cases.

In the present study we employed a dispersion-correction finite difference scheme developed by Yoon (2001) to simulate a historical tsunami event. The scheme uses an imaginary grid size determined from the condition of Imamura *et al.* instead of actual grid size to satisfy local dispersion relationships of waves for the case of slowly varying topography. The variables on the imaginary grid points are obtained by the interpolation of the variables assigned on the actual uniform grid points.

GOVERNING EQUATIONS

The wave lengths of tsunamis are much longer than those of wind generated waves and shorter than those of tides. Thus, the dispersion effects of tsunamis are relatively strong and should be properly considered in the numerical simulation of tsunami propagation for accuracy. Since the free surface displacements are small compared to water depth, the nonlinearity of waves can be neglected for the case of transoceanic propagation.

As a result, the linear Boussinesq equations can be used as governing equations. However, the Boussinesq equations include dispersion terms which introduce numerical difficulties in practice due to the presence of mixed type of differentiations with respect to time and space at the same time. Thus, the following shallow-water equations are generally employed instead, and the physical dispersion of waves is replaced by the numerical dispersion arising from the explicit scheme.

$$\frac{\partial \eta}{\partial t} + \frac{\partial P}{\partial x} + \frac{\partial Q}{\partial y} = 0, \quad \frac{\partial P}{\partial t} + gh\frac{\partial \eta}{\partial x} = 0, \quad \frac{\partial Q}{\partial t} + gh\frac{\partial \eta}{\partial y} = 0 \qquad (1)$$

where η represents the free surface displacement from still water level, P and Q are the depth-averaged volume fluxes in the x and y directions, respectively. g is the acceleration of gravity and h is the water depth.

DISPERSION-CORRECTION FINITE DIFFERENCE SCHEME

A leap-frog scheme is frequently employed to solve the shallow-water equations on the staggered mesh as shown in Fig. 1 (Imamura et al., 1988; Liu et al., 1995; Cho, 1995). This scheme gives numerical dispersion which is a kind of numerical error resulting from the discretization of the governing equations.

Imamura et al. (1988) is the first who analyzed the numerical dispersion produced by the leap-frog scheme and used it intentionally to compensate for the physical dispersion effect neglected in the shallow-water equations. For this purpose, Imamura et al. assumed uniform waves propagating in one direction. The finite difference approximation using the leap-frog scheme gives

$$\frac{\eta_i^{n+1/2} - \eta_i^{n-1/2}}{\Delta t} + \frac{P_{i+1/2}^n - P_{i-1/2}^n}{\Delta x} = 0 \tag{2}$$

$$\frac{P_{i+1/2}^{n+1} - P_{i+1/2}^n}{\Delta t} + gh\frac{\eta_{i+1}^{n+1/2} - \eta_i^{n+1/2}}{\Delta x} = 0 \tag{3}$$

As shown in Fig. 1, indices i and n represent spatial grid point and time level, respectively. Δx is the spatial grid size, and Δt is the time step.

If the grid size Δx in (2) and (3) is chosen to satisfy the following Imamura condition,

$$\Delta x = \sqrt{4h^2 + gh(\Delta t)^2} \tag{4}$$

then the physical dispersion of Boussinesq equations can be replaced by the numerical dispersion produced by the leap-frog finite difference approximation of the shallow-water equations. The numerical scheme developed by Imamura et al. is relatively simple and computationally more economical than numerical models employing the Boussinesq equations. When the water depth changes in a computational domain, the grid size should be adjusted to satisfy the Imamura condition given by (4). However, it is not an easy task to construct an adjustable grid system for the computational domain where the water depth varies in both directions. Thus, we seek a new scheme which solves the shallow-water equations on a uniform grid system, while the dispersion effect of waves is properly considered in the simulation.

The dispersion-correction scheme developed by Yoon (2001) uses an imaginary grid size Δx_* determined from (4) instead of actual Δx to satisfy local dispersion relationships of waves for the case of slowly varying topography. The variables on the imaginary grid points are obtained by the interpolation of the variables assigned on the actual uniform grid points.

In order to clarify the idea, the new scheme is presented in some details for the case of one-dimensional propagation of tsunamis over slowly varying topography. Using a uniform grid system with grid size Δx and time step Δt, the modified leap-frog scheme can be written as

$$\frac{\eta_i^{n+1/2} - \eta_i^{n-1/2}}{\Delta t} + \frac{P_{*F}^n - P_{*B}^n}{\Delta x_*} = 0 \tag{5}$$

$$\frac{P_{i+1/2}^{n+1} - P_{i+1/2}^n}{\Delta t} + gh_{i+1/2}\frac{\eta_{*F}^{n+1/2} - \eta_{*B}^{n+1/2}}{\Delta x_*} = 0 \tag{6}$$

where Δx_* denotes the local grid size of imaginary grid evaluated using (4) for given $h_{i+1/2}$ and Δt. The variables with subscripts *F and *B represent the interpolated values on the imaginary forward and backward grid points, respectively, using the values on the neighboring actual uniform grid points as shown in Fig. 1.

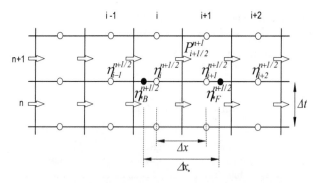

Fig. 1 Assignment of variables on the one-dimensional staggered mesh;
○, actual grid point; ●, imaginary grid point.

The extension of this dispersion-correction scheme to two-dimensional problems is straightforward, and the details can be found in Yoon (2001).

TEST OF NUMERICAL SCHEME

The accuracy of the dispersion-correction finite difference scheme is tested for the propagation of a tsunami whose initial free surface profile is given by a Gaussian hump. The computed free surface displacements are compared with the analytical solution of the linear Boussinesq equations (Carrier, 1991). The numerical tests show that the numerical model of Imamura *et al.* (1988) gives accurate solutions only when the actual grid size Δx is identical with the grid size given by (4). However, the present model is free from the choice of Δx.

APPLICATION TO ACTUAL TSUNAMI EVENT

On July 12, 1993, a destructive tsunami (M = 7.8 in JMA scale) named as Hokkaido-Nansei Tsunami (hereafter 1993 Tsunami) was generated from the deep sea area near Okushiri Island of Hokkaido, Japan. This tsunami attacked the Japanese coast and killed 202 lives. This tsunami propagated across the East Sea and caused some damages along the east coast of Korea. The 1993 Tsunami is selected for the verification of the numerical model employed in this study.

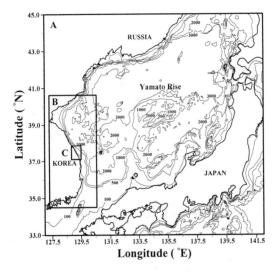

Fig. 2 Computational domains and bathymetry of East Sea (depth unit: m).

Fig. 2 shows the bathymetry of East Sea. The computational domain is divided into 900 × 931 finite difference mesh. Since the present numerical model is developed based on square grids (i.e., $\Delta x = \Delta y$), the grid system is constructed to form locally a square grid. The angular grid size in E-direction, $\Delta \psi$, is first set to 1 min, and the angular grid size in N-direction, $\Delta\phi$, is determined by

$$\Delta\phi = \cos\phi\Delta\psi \qquad (7)$$

where ψ and ϕ represent the longitude and latitude of grid point, respectively. Thus, the actual grid size is given by

$$\Delta x = \Delta y = R\cos\phi\Delta\psi \qquad (8)$$

where R (= 6.378×10^{6} m) is the radius of the earth. The grid size changes slowly according to the latitude of grid point, but is approximately 1.5 km around the central part of computational domain. Using this grid system the present numerical model developed for Cartesian coordinates can be employed with no modification to consider the curvature of the earth. Moreover, the dispersion effect of tsunamis can be correctly modeled for slowly varying topography as in the case of East Sea.

The time step Δt is chosen as 3 s. For the coastal area where the water depth is small, the wave length of tsunamis becomes short, and the resolution of grid degenerates. Thus, a series of finer grid system, B, C, D and E, is nested dynamically to the coarse grid region denoted by A. As shown in Fig. 2, Mukho Port (subregion C) is selected for near-field computations with finer grid system, because the tide gage record is available for the 1993 tsunami at this port. Table 1 presents the computational information for grid subregions. The bathymetry of each subregion is presented in Fig. 3. For the finest grid system denoted by E, nonlinear shallow-water equations (IOC, 1997) are solved, because the nonlinearity increases as the water depth decreases. Manning-type energy dissipation for bottom friction is implemented in the nonlinear model. The Manning's roughness coefficient is set to be equal to 0.015 s/m$^{1/3}$.

Table 1 Computational information for each subregion.

Region	Grid size $\Delta\psi$	Time step Δt (s)	Remark
A	1 min	3.0	Linear SWE
B	20 s	1.0	Linear SWE
C	6.67 s	0.33	Linear SWE
D	2.22 s	0.11	Linear SWE
E	0.74 s	0.11	Nonlinear SWE

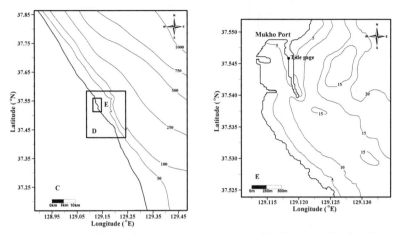

Fig. 3 Computational subregions and bathymetry around Mukho Port (depth unit: m).

The initial free surface displacement generated by the under-sea earthquake of 1993 Tsunami is obtained using the model of Mansinha & Smylie (1971) based on the fault parameters proposed by Takahashi et al. (1994) as presented in Table 2. In the table H is the focal depth, θ the orientation of fault, δ the dip angle, λ the slip angle, L the length of fault, W the width of fault, and D the dislocation of fault. The initial free surface profile is shown in Fig. 4.

The numerical simulation is performed for 250 min after the tsunami is generated. Fig. 5 shows the propagation map with 5 min time interval. The arrival time at the east coast of Korea is approximately 100 min.

Table 2 Fault parameters for 1993 Tsunami (Takahashi et al., 1994).

Date	Location		H(km)	θ(°)	δ(°)	λ(°)	L(km)	W(km)	D(m)
	Lat. (°N)	Long. (°E)							
July	40.10	139.30	5	163	60	105	24.5	25	12.00
12,	42.34	139.25	5	175	60	105	30	25	2.50
1933	43.13	139.40	10	188	35	80	90	25	5.71

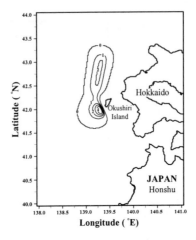

Fig. 4 Initial free surface profile of 1993 Tsunami (unit: m).

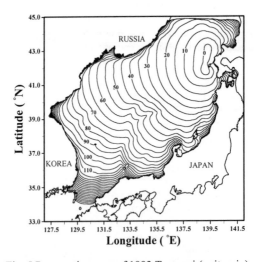

Fig. 5 Propagation map of 1993 Tsunami (unit: min).

Fig. 6 shows the comparison of calculated and measured time histories of free surface displacement at Mukho Port. The natural period of resonance at Mukho Port can be estimated theoretically as 8 min based on the averaged water depth ($h = 7$ m) and the length of harbor ($l = 1000$ m). Both the observed and the calculated wave periods agree well with the theoretical one. The calculated first wave arrives 4 min ahead of the observed one. And the calculated crest height of the first several waves is larger than the observed ones. This can be attributed to the orifice effect of the well-type tide gage. A good agreement between calculated and observed free surface displacements is also achieved for the first five waves. However, some discrepancies are evident after the sixth wave, because the waves are contaminated by the waves reflected from the coasts of Russia and Japan where the grid size is too coarse to resolve the short waves in the shallow water region.

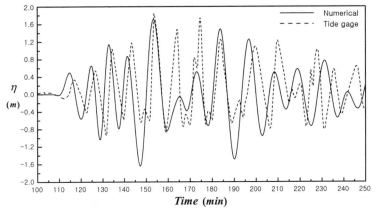

Fig. 6 Comparison of calculated and observed time histories of water level
at Mukho Port for 1993 Tsunami.

CONCLUSIONS

A dispersion-correction finite difference numerical scheme is employed for the simulation of distant tsunamis propagating over slowly varying topography. This scheme solves the shallow-water equations on a uniform grid system. The dispersion effect of waves is considered in the computation by utilizing the numerical dispersion error arising from the leap-frog scheme. The dispersion effect can be properly captured by using an imaginary grid system whose grid size is determined from the Imamura condition according to the given time step and local water depth. An application of the present model to the simulation of a historical tsunami is performed. Numerical results are compared with the tide gage records, and reasonably good agreements are obtained.

ACKNOWLEDGEMENTS

This work was supported by grant No. R01-1999-00314 from the Basic Research Program of the Korea Science & Engineering Foundation.

REFERENCES

Carrier, G.F. 1991. Tsunami propagation from a finite source. *Proc. of 2nd UJNR Tsunami Workshop*, NGDC, Hawaii, 101-115.

Cho, Y.S. 1995. *Numerical simulations of tsunami propagation and run-up*. Ph.D. Thesis, School of Civil and Environmental Engineering, Cornell University, Ithaca, NY.

Imamura, F., Shuto, N., and Goto, C. 1988. Numerical simulation of the transoceanic propagation of tsunamis. *Proc. of 6th Congress Asian and Pacific Regional Division*, IAHR, Japan, 265-271.

IOC (Intergovernmental Oceanographic Commission) 1997. *Numerical method of tsunami simulation with the leap-frog scheme*. IUGG/IOC Time Project, Manuals and Guides, UNESCO.

Liu, P.L-F., Cho, Y.-S., Yoon, S.B., and Seo, S.N. 1995. Numerical simulations of the 1960 Chilean tsunami propagation and inundation at Hilo, Hawaii. *Tsunami : Progress in Prediction, Disaster Prevention and Warning*, edited by Tsuchiya and Shuto, Kluwer Academic Publishers, 99-115.

Mansinha, L. and Smylie, D.E. 1971. The displacement fields of inclined faults. *Bull. Seismol. Soc. Amer.*, 61(5), 1433-1440.

Takahashi, T., Shuto, N., Imamura, F., and Ortis, M. 1994. Fault model to describe Hokkaido Nansei offshore earthquake for tsunami. *Jour. Coastal Eng.*, JSCE, 41, 251-255 (in Japanese).

Yoon, S.B. 2001. Propagation of distant tsunamis over slowly varying topography. *Jour. Geophys. Res.* (submitted).

CHARACTERISTIC MODES OF THE ANDREANOV TSUNAMI BASED ON THE HILBERT-HUANG TRANSFORMATION

Torsten Schlurmann[1], Torsten Dose[2] and Stefan Schimmels[3]

Abstract: This paper concerns the identification of the characteristic modes of the „Andreanov tsunami" based on the recently developed time-dependent Hilbert-Huang Transformation (Huang et al. 1998 & 1999). Results of this technique are evaluated with conventional analysing methods based on the Fourier Transformation, e. g. restraining high and band pass filters which attempt to classify a signal into predefined frequency bounded subcategories. The Hilbert-Huang Transformation (HHT) disintegrates two (three) accurate monochromatic components from the collected data of the „Andreanov tsunami" in the near-field of the epicenters in shallow water environment which are assumed to characterize resonant effects as a result of tsunami-shoreline interaction processes. Furthermore the HHT determines a distinct time-frequency representation of the tsunami in the far-field of the epicenters in deep waters.

INTRODUCTION

Aleutian Islands Earthquakes 10-Jun-1996

Awareness of tsunami hazards has been heightened by numerous earthquake generated waves around the Pacific Rim in the last decades (Bernard 1997). A major earthquake occurred in the Andreanov Islands, Aleutian Islands, on 10-Jun-1996, 04:04 UTC. The earthquake epicenter was located at 50.6°N latitude, 177.7°W longitude, in a depth of around 33 km and was strongly felt at Adak Island, which is situated approximately 160 km north-northeast of the epicenter. Earthquakes in the Adak area occur on the convergent boundary between the Pacific and North American crustal plates. This region, where the two plates are being forced directly into one another, is one of the world's most active

[1] Postdoctoral Researcher, Hydr. Eng. Section, Civil Eng. Dep., University of Wuppertal, Pauluskirchstr. 7, 42285 Wuppertal, Germany, ph.: +49-202-4394197, fax: +49-202-4394196, schlurma@uni-wuppertal.de
[2] Res. Assistant, Hydr. Eng. Section, Civil Eng. Dep., University of Wuppertal, Pauluskirchstr. 7, 42285 Wuppertal, Germany, ph.: +49-202-4394194, fax: +49-202-4394196, dose@uni-wuppertal.de
[3] Res. Assistant, Inst. for Fluid Mech. and Comp. Appl. in Civ. Eng., Univ. of Hannover, Appelstr. 9a, 30167 Hannover, Germany, ph.: +49-511-7624786, fax: +49-511-7623777, schimmels@hydromech.uni-hannover.de

seismic zones. Over one hundred earthquakes of magnitude $M_w \geq 7$ have occurred along this boundary since the last one hundred years. In particular, a magnitude $M_w = 7.9$ was measured on 10-Jun-1996 that turned out to be the most severe earthquake in North America in the past decade (1986-1996). It was preceded by several foreshocks the previous day, the largest of which had a magnitude $M_w = 6.0$. Eighty-eight aftershocks, magnitudes $M_w \geq 4.0$ have occurred, the largest of which occurred on 10-Jun-1996, 15:24 UTC, and had a magnitude $M_w = 7.2$. Some minor damages have been reported from these earthquakes, but no reports of injuries have been received (AEIC 1996). Tanioka and Gonzalez (1998) profoundly investigate these particular earthquakes and attempt to compute numerically the rupture area from the measured tsunami waveforms. Specifically the fault geometry of both earthquakes are characterized to be pure thrust type faulting with a definite geometry: strike 260°, dip 20° and rake 90° (Tanioka and Gonzalez 1998). The source time function (STF) of the first rupture (10-Jun-1996, 04:04 UTC) as the dominating generation mechanism for any subsequent tsunami along the adjacent coastlines was determined by Tanioka and Ruff (1997). It was indicated that the total duration was only 54 seconds and a seismic moment was of approximately $7.3*10^{20}$ Nm. More detailed information from the 10-Jun-1996 Aleutian Islands earthquakes, e.g. seismograms of body and surface waves, can be obtained from the MichSeis group (*http://www.geo.lsa.umich.edu/~MichSeis*).

Earthquake generated water waves (tsunamis)

Both earthquakes generated tsunamis causing shoaling wavegroups of wave heights from about 5 cm up to 105 cm along the adjacent coastlines in the Northern Pacific, while the maximum wave height was recorded off the coast of Adak (#104) very close to the epicenters (160 km) in shallow water. Measurements of the „Andreanov tsunami" were made at 127 separate locations around the entire Pacific Rim at deep-ocean, offshore and coastal stations (Eble et al. 1997) after the major earthquake occurred. Figure 1 summarizes all tsunami stations and the location of the 10-Jun-1996 Andreanov Islands earthquake epicenter and each station for which a record was obtained. Eble et al. (1997) further report that this particular tsunami is evident at 46 of these stations, not present at another 46 stations, and uncertain at 32 stations, while the remaining three stations are characterized by bad or missing data. Their results are exclusively based on restraining high and band pass filters. Public access and maintenance of this online database (*ftp.pmel.noaa.gov*) has been made through a joint effort by National Oceanic and Atmospheric Administration (NOAA) and the Pacific Marine Environmental Laboratory (PMEL).

The main objectives of this paper is to analyse the water surface elevations at four considerable stations with conventional highpass and lowpass filtering methods which attempt to sort a signal into predefined frequency bounded subcategories. It is further shown that the recently developed Hilbert-Huang Transformation (HHT) disintegrates two (three) accurate monochromatic components from the collected data of the „Andreanov tsunami" in the near-field of the epicenters in shallow water environment which are assumed to characterize resonant effects as a result of tsunami-shoreline interaction processes. Furthermore, the HHT determines a distinct time-frequency representation of the tsunami in the far-field of the epicenters in deep water conditions.

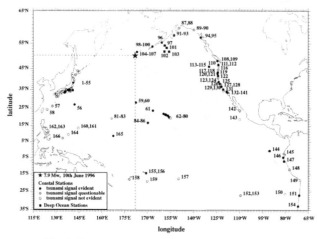

Fig. 1. Tsunami station summary and earthquake epicenter (taken from Eble et al. 1997)

MEASUREMENT DEVICES AND DATA STRUCTURE

Conventional tide gauge data are provided by the Japanese Meteorological Agency (JMA), the University of Hawaii (UH), the Pacific Tsunami Warning Center (PTWC), and the National Ocean Service (NOS). Sampling rates for JMA, PTWC and UH are 1, 2 and 4 minutes, respectively. Only, NOS data show a higher temporal resolution at either 15-second or 30-second intervals. The NOAA/NOS deep ocean "Bottom Pressure Recorder" (BPR) signals obtained from this tsunami are of particular interest in this investigation. Four of these BPR constitute an array located south of the Shumagin Islands in the Aleutian Island chain, approximately 1350 km from the earthquake epicenter (#97, #101, #102 and #103). Another group of three BPR is installed off the Washington-Oregon coast, about 3500 km from the epicenter (#113, #114 and #115). All the BPR records are of high temporal resolution at 15-seconds intervals (Eble et al. 1991 and Okada 1995). All data are edited to remove values that exceeded reasonable bounds. Outliers are replaced with linearly interpolated values from adjacent data points. Duplicate data are eliminated and gaps filled with a flag. All data are dominated by local tidal fluctuations, mainly by semi-diurnal and diurnal components. In particular, four data records of NOAA/NOS stations are considered in this investigation: *i*) from the tide gage station #104 located south of Adak, AK, USA, very close to the epicenters (160 km) in shallow water regions (depth 2.5 m), *ii*) the data series obtained at the bottom pressure recorder #102 from the Shumagin Islands almost 1350 km from the earthquake epicenter in a water depth of about 4800 m, *iii*) from the tide gage station #94 located near the coast of Sitka, AK, USA, at a distance of about 2800 km from the epicenters in a water depth of about 3800 m, and *iv*) the data series obtained from the bottom pressure recorder #113 off the Washington-Oregon coast located 3500 km from the epicenters in a depth of 1500 m. Time series of these four station are used to evaluate the earthquake generated water waves. Figure 2 presents both the records of the two tide gauges and of the two bottom pressure recorder within a time period of 24 hours (10 to 11-

Jun-1996). The dashed vertical lines indicate both earthquake events on 10-Jun-1996, 04:04:00 UTC, and 10-Jun-1996, 15:25:00 UTC. The solid vertical lines demonstrate the estimated arrival time of the tsunami at the specific station derived from numerical calculations by means of a modified PTWC travel time code (Eble et al. 1997). Obviously, the time series show that each of the four data records is dominated by a semi-diurnal and a diurnal tidal component. The tsunami is only clearly evident at the tide gauge station #104 about 40 minutes after the main shocks. The record of the BPR from station #102 indicates only minor irregular water surface elevation precisely after the main earthquake. Therefore, it can be concluded that this signal is superimposed with seismic surface waves (Rayleigh) as the BPR is directly mounted on the seafloor. Numerical calculations predict the tsunami to arrive at station #102 after 110 minutes. Both signals from #094 and #113 do not show any obvious deviations from the regular water surface elevation. Thus, a simple inspection by eye does not provide any further information about the Andreanov tsunami in the far field from the epicenters. Numerical calculations estimate the tsunami arrival time at station #094 after 260 minutes and at station #113 after 300 minutes.

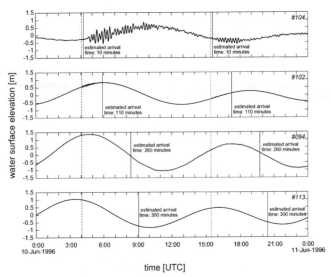

Fig. 2. Water surface elevation records within 10 to 11-Jun-1996 from #104, #102, #094 and #113 [Δt = 15 s]

ANALYSING METHODS

Digital filters are used in a wide variety of signal processing applications, such as spectrum analysis, digital image processing, and pattern recognition. They belong to the class of discrete-time LTI (linear time invariant) systems, which are characterized by the properties of causality, recursibility, and stability (Mitra & Kaiser 1993). In the following the principles of high- and bandpass filtering in discrete-time LTI systems are briefly summarized. All digital filters within this analysis were designed with the Signal Processing Toolbox in MATLAB 6® (Release 12.1). Moreover, the Hilbert-Huang Transformation

(HHT) is used in this investigation. The HHT (Huang et al. 1998 & 1999) is a new method for analyzing nonlinear and non-stationary data series. The central idea behind the HHT is the so-called Empirical Mode Decomposition (EMD) that numerically decomposes any time-dependent signal into its own underlying characteristic modes. Applying the Hilbert Transformation (HT) to each of these disintegrated Intrinsic Mode Functions (IMF) subsequently provides the Hilbert amplitude or energy spectrum. The principles are briefly discussed below. However, it is referred to Huang et al. (1998 & 1999) or Schlurmann (2000 & 2001) for a more comprehensive summary on the EMD and the HHT.

High- and bandpass filtering

In general the z-transform $Y(z)$ of a digital filter's output $y(n)$ is related to the z-transform $X(z)$ of the input $x(n)$:

$$Y(z) = H(z)X(z) = \frac{b(1) + b(2)z^{-1} + \ldots + b(nb+1)z^{-nb}}{a(1) + a(2)z^{-1} + \ldots + b(na+1)z^{-na}} X(z) \qquad (1)$$

where $H(z)$ is the filter's transfer function. It is the frequency-domain descriptor of the LTI system. Herein, the constants $b(i)$ and $a(i)$ are the filter coefficients and the order of the digital filter is the maximum of na and nb. After rearranging this equation and assuming that $a(1) = 1$ and $y(1) = 1$ (MathWorks 2000), the final structure of the filtered signal $y(n)$ is (eq. 2):

$$y(n) = b_1 x(n) + b_2 x(n-1) + \ldots b_{nb+1} x(n-nb) - a_2 y(n-1) - \ldots - a_{na+1} y(n-na) \qquad (2)$$

This is the standard time-domain representation of a digital filter, computed starting with and assuming unity initial conditions. This representation's progression is (eq. 3):

$$\begin{aligned}
y(1) &= b_1 x(1) \\
y(2) &= b_1 x(2) + b_2 x(1) - a_2 y(1) \\
y(3) &= b_1 x(3) + b_2 x(2) + b_3 x(1) - a_2 y(3) - a_3 y(1) \\
\vdots\ &=\ \vdots
\end{aligned} \qquad (3)$$

In this investigation, a Kaiser windowed FIR high- and bandpassfilter of exceptional high order ($nb = 10000$, $na = 0$) have been designed to steepen the sideband slopes in the transitation bands of both filters. According to Eble et al. (1997) the cutoff frequency for a highpassfilter is set to $f_c = 2$ hours ($\approx 1.39 * 10^{-4}$ Hz). Spectral components with lower frequencies are effectively diminished, and consequently, the diurnal and semi-diurnal tidal components in the signals are removed. The highpassfiltered data records are presented in fig. 3. The filtered record from #104 in the near-field of the epicenters shows almost steady water waves of amplitude 0.5 m for the first and 0.2 m for the second tsunami. The numerically expected tsunami arrival time is underestimated by 40 minutes for both events. Similar steady water waves are measured in other coastal areas from the Andreanov tsunami even in the far-field of the epicenters, e. g. Hilo (#78), Crescent City (#120) and Caldera (#151). Tsunami-shoreline interaction processes cause these effects. In addition to the shoaling of waves on the nearshore slope, the tsunami interacts with the shoreline generating standing wave resonances at the shore, edge waves by the impulse of the incident waves, and, trapped reflected waves by refraction. However, the highpassfiltered signal from station #102 nonetheless shows the superimposed high-frequency seismic Rayleigh waves. No tsunami evidence is apparent in the highpassfiltered data at station #102. In addition, no tsunami evidence can be confirmed from tide gauge station #094. Despite this fact the signal in the far-field of the epicenters from station #113 shows a significant deviation from the regular water surface record. Evidently, a water wave disturbance of amplitude 0.01 m and approximate duration of 12 minutes is measured 30 minutes earlier than the estimated arrival time. Eble et al. (1997) deduce that this small disturbance is the tsunami generated by the earthquake registered 300 minutes before.

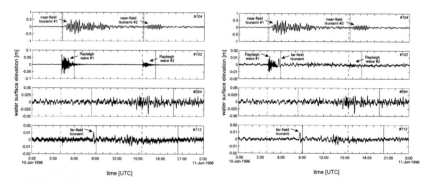

**Fig. 3. Highpassfiltered water surface   Fig. 4. Bandpasssfiltered water surface
elevation records   elevation records**

The lower cutoff frequency for the bandpassfilter is set to f_{c1} = 2 hours ($\approx 1.39 * 10^{-4}$ Hz) - while the higher cutoff frequency is defined at f_{c2} = 4 minutes (≈ 0.04167 Hz) to remove wind generated waves (infragravity waves) and seismic Rayleigh waves from the record. The bandpassfiltered data records are shown in fig. 4. No significant differences from stations #104, #94 and #113 are detectable compared with the highpassfiltered data from fig. 3. But, in contrast, the bandpassfiltering process from the BPR station #102 eliminates the superimposed seismic Rayleigh waves. A similar small water wave disturbance as from station #113 is revealed at this particular location. Eble et al. (1997) previously described that this solitary wave is only evident from BPR stations, e. g. in the southern Pacific (Hawaii and Chile) and Eastern Pacific (Japan).

Hilbert-Huang Transformation (HHT)

The HHT was developed for physical interpretation of any non-stationary and nonlinear data as a substitute for the commonly used Fourier analysis which is strictly limited to linear and stationary (periodic) time series (e.g. Titchmarsh 1948). The HHT is based on the Empirical Mode Decomposition (EMD) to disintegrate time-dependent records into their individual characteristic oscillations. Schlurmann et al. (2000) apply the HHT to decompose laboratory generated extreme waves - transient waves with exceptional large wave heights - into their characteristic oscillations revealing entirely new physical interpretation of nonlinear transient water waves. Schlurmann (2000 and 2001) introduces this technique in the analysis of extreme waves from real sea states during storm events in the Sea of Japan and summarizes in this context the fundamental algorhythm of the EMD. The central idea of HHT is the sifting process of the EMD to decompose a time series into its own characteristic Intrinsic Mode Functions (IMF). The EMD is an adaptive technique that is derived from the assumption that any signal consists of numerous IMF. Each of them ideally represents an embedded characteristic oscillation. An IMF is defined by two criteria: *i*) the number of extrema and of zero crossings must either equal or differ at most by one, and, *ii*) at any instant in time, the mean value of the envelope defined by the local maxima and the envelope of the local minima is zero. The first criterion is almost similar to the narrow band requirement of a Gaussian process, while the latter condition modifies a global requirement to a local one, and, is necessary to ensure that the instantaneous frequency, which is the rate of change of the phase of the analytical signal derived from the Hilbert transformation will not have unwanted fluctuations. Therefore, the EMD is based on the direct extraction of energy associated with various intrinsic time scales. Applying the Hilbert Transformation (HT) to each of these disintegrated functions leads to the Hilbert amplitude spectra, generating amplitude- and frequency-varying spectra in the time-domain.

The EMD numerically disintegrates the time series from station #104 into five IMF in total. The results of this decomposition are summarized in figures 5 and 6. The first three IMF are jointly shown in the middle panel of fig. 5 which characterize relative high frequency components including the earthquake generated water waves in the near-field of the epicenter. Obviously, the summation of these three IMF is very similar to the 4 min to 2 hrs bandpassfiltered signal. In addition, the characteristic modes four and five are jointly shown in the bottom panel of fig. 5, and define the low frequency components including the semi-diurnal and diurnal tidal components. The EMD automatically disintegrates frequency subcategories from the data record without any information about essential numerical restrictions, e .g. cutoff frequencies, transition, pass- or stopbands which are strongly required for high- or bandpass filtering processes.

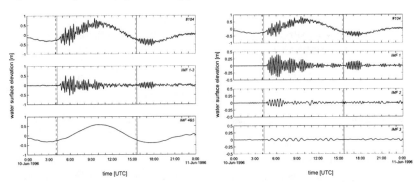

Fig. 5. Frequency subcategories of EMD (station #104)   **Fig. 6. First three IMF of EMD (station #104)**

Figure 6 shows the first three IMF of the earthquake generated wave from station #104 in detail. Mode one and two carry most of the embedded energy. The third IMF is of insignificant effect. Most interestingly, the tsunami is mainly composed of two individual stable components caused by resonant tsunami-shoreline interaction processes. In particular, after the first earthquake (04:04 UTC) from 05:00 to 07:00 UTC the tsunami is characterized by two monochromatic waves. Each of them is almost of constant frequency with slightly varying amplitudes. Similar conditions are applicable after the second earthquake from 16:00 to 18:00 UTC, where the first IMF derived from the EMD is dominating the tsunami record.

Similarly, the EMD disintegrates the record from station #102 into five IMF. The first three IMF are shown collectively in the middle panel of fig. 7, while the modes four and five are jointly added in the bottom panel. Again, the summation of IMF 1 to 3 is closely identical to the bandpassfiltered signal. Fig. 8 presents the first three IMF of the tsunami measured at station #102 in detail. The EMD disintegrates the seismic Rayleigh wave from the real tsunami. The first two IMF contain high-frequency noise and the seismic waves and the third IMF reveals the tsunami generated by the first rupture and is almost identical to the bandpassfiltered signal derived before. Nevertheless, the data from stations #094 and #113 have been processed and evaluated, but are not presented here as similar results were obtained in general.

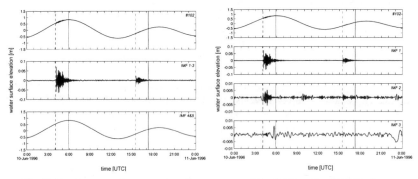

Fig. 7. Frequency subcategories of   **Fig. 8. First three IMF of**
EMD (station #102)   **EMD (station #102)**

Figures 9 and 10 present the corresponding Hilbert amplitude spectra of the decomposed data from station #104 and #102. Fig. 9 shows the first two IMF of the resonant modes in shallow water (#104) in the near-field. The Hilbert spectra are plotted time versus instantaneous frequency, here, instantaneous period, respectively. Accordingly, time-dependent frequency and amplitude modulations of the waves can be identified. The grey shaded colorbars beneath the Hilbert spectra matches the amplitudes of the corresponding IMF. It is proven that both IMF are of almost constant period. The Hilbert spectrum of IMF 1 in the second panel of fig. 9 validates an instantaneous period $T(t) = 13$ minutes. Slight frequency and amplitude modulations occur in time-domain, but on the whole this characteristic mode is of constant shape for about 24 hours. The Hilbert spectra of IMF 2 reveals that this component is characterized by an instantaneous period $T(t) = 23$ min. Nevertheless, strong frequency and amplitude modulations occur. If these disturbances are related to real physical processes of tsunamis or appear due to numerical instabilities of the EMD cannot be concluded at this time. IMF 2 and 3 of the decomposed record in the far-field (#102) in deep waters are shown in fig. 10. Both the IMF and the corresponding Hilbert spectra are presented in the same manner than previously done. IMF 1 of this record is not exposed here as it only contains irrelevant high frequency noise. Obviously, IMF 2 represents the seismic Rayleigh wave that describes a frequency subcategory in between 4 to 10 minutes. The third panel presents IMF 3. It exposes the Hilbert spectrum of the tsunami in far-field and reveals that the earthquake generated wave has only one characteristic mode with an approximate period $T(t) = 23$ to 30 minutes. Evidently, this solitary wave is characterized by a transient nature as the period varies significantly in time.

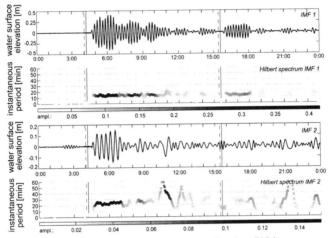

Fig. 9. IMF and Hilbert spectra (station #104)

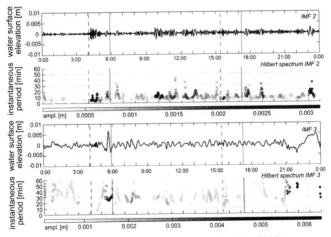

Fig. 10. IMF and Hilbert spectra (station #102)

CONCLUSION AND DISCUSSION

From these examples of tsunami records in the Northern Pacific, it seems as if the Hilbert-Huang Transformation is an extremely powerful tool to assist analysing nonlinear and non-stationary data series in the time-frequency plane. This new technique effectively supports the conventional high- and bandpass filtering methods in analysing tsunamis. Particularly, this investigation demonstrates how the EMD succeeds in detecting nearshore resonant modes in the near-field of the epicenters (#104), and how the tsunami is characterized in the far-field in deep water (#102). In this context it has been exposed that the earthquake generated wave has only one characteristic mode with time- and amplitude varying periods $T(t) = 23$ to 30 minutes. It is further shown that this solitary wave is evidently characterized by a transient nature as the period varies significantly in time.

However, the Hilbert-Huang Transformation is not an undisputed technique in respect of certain points that are heavily debated at the moment (Schlurmann 2000) which definitely need further investigations.

REFERENCES

AEIS, 1996. 10 June 1996 Adak Earthquakes. *Technical Report*, AEIS, Alaska Earthquake Information Center, Univ. of Alaska Geophysical Institute & U.S. Geological Survey

Bernard, E.N., 1997. Reducing tsunami hazards along U.S. coastlines. *In Persp. On Tsunami Hazard Reduction*, Kluwer Academics Publ., pp. 189-203

Eble, M.C. and Gonzalez, F.I., 1991. Deep-ocean bottom pressure measuremants in the northeast Pacific. *Journal of Atmos. Oceanic Tech.*, **8**, 2, pp. 221-233

Eble, M.C., Newman, J., Wendland, B., Kilonsky, D., Luther, Y., Tanioka, M., Okada and González, F.I., 1997. The 10 June 1996 Andreanov tsunami database. *Technical Report NOAA DR ERL PMEL-64* (PB98-130495)

Huang, N.E., Shen, Z., Long, S., Wu, M.C., Shih, H.H., Zheng, Q., Yen, N.-C., Tung, C.C. and Liu, H.H., 1998: The Empirical Mode Decomposition and Hilbert spectrum for non-linear and non-stationary time series analysis: *Proc. R. Soc. London A*, **454**, pp. 903-995

Huang, N.E., Shen, Z. and Long, S., 1999. A new view of non-linear water waves: The Hilbert Spectrum. *Ann. Review of Fluid Mechechanic*, **31**, pp. 417-457

MathWorks, 2000. Signal Processing Toolbox User's Guide, The MathWorks, Inc.

Mitra, S.K. and Kaiser, J., 1993. Handbook for Digital Signal Processing, J. Wiley & Sons

Okada, M., 1995. Tsunami observation by ocean bottom pressure gauge. *In Tsunami: Progress in Pred., Disaster Prev. and Warning*, eds.: Tsuchuya & Shuto, pp. 287-303

Schlurmann, T., Schimmels, S. and Dose, T., 2000. Spectral Frequency Analysis of Freak Waves using Wavelet Analysis (Morlet) and Empirical Mode Decomposition (Hilbert)", *Proc. 4th Int. Conf. on Hydroscience & Eng*, Seoul, S.-Korea, (in press)

Schlurmann, T., 2000. The Empirical Mode Decomposition and the Hilbert Spectra to analyze embedded characteristic oscilations of extreme waves. *Proc. Rogue Waves Workshop*, IFREMER, Brest, France, (in press).

Schlurmann, T., 2001. Spectral Frequency Analysis of Nonlinear Water Waves derived from the Hilbert-Huang Transformation. Proc. 20th Offshore Mechanics and Artic Engineering Conference (OMAE 2001), Rio de Janeiro, Brazil

Tanioka, Y. and Gonzalez, F.I., 1998. The Aleutian earthquake of June 10, 1996 (Mw 7.9) ruptured parts of both the Andreanof and Delarof segments. *Geophysical Research Letters*, AGU, **25**, 12, pp. 2245-2248

Tanioka, Y. and Ruff, L.J., 1997. Source Time Functions. *Seismo. Res. Let.*, **68**, pp. 386

Titchmarsh, E.C., 1948. Introduction to the theory of Fourier integrals. *Oxford Univ. Press*

NEURAL NETWORK FORECASTING OF STORM SURGES ALONG THE GULF OF MEXICO

Philippe E. Tissot[1], Daniel T. Cox[2], M.ASCE, and Patrick Michaud[3]

Abstract: Accurate water level forecasts are of vital importance along the Gulf of Mexico as its waterways play a critical economic role for a number of industries including shipping, oil and gas, tourism, and fisheries. While astronomical forcing (tides) is well tabulated, water level changes along the Gulf Coast are frequently dominated by meteorological factors. Their impact is often larger than the tidal range itself and unaccounted for in present forecasts. We have taken advantage of the increasing availability of real time data for the Texas Gulf Coast and have developed neural network models to forecast future water levels. The selected inputs to the model include water levels, wind stress, barometric pressures as well as tidal forecasts and wind forecasts. A very simple neural network structure is found to be optimal for this problem. The performance of the model is computed for forecasting times between 1 and 48 hours and compared with the tide tables. The model is alternatively trained and tested using three-month data sets from the 1997, 1998 and 1999 records of the Pleasure Pier Station located on Galveston Island near Houston, Texas. Models including wind forecasts outperform other models and are considerably more accurate than the tide tables for the forecasting time range tested, demonstrating the viability of neural network based models for the forecasting of water levels along the Gulf Coast.

INTRODUCTION

Accurate water level forecasts along the Gulf of Mexico coast, estuaries, and intracoastal waterways are of great importance to federal, state, and local agencies, industries such as ports, fisheries, construction and coastal communities. The overall economic importance of the Gulf of Mexico for the U.S. economy is high: nine out of the twelve largest U.S. ports with tonnage greater than 50 million tons are located along

1 Assistant Professor of Physics, Physical and Life Sciences Department, and Research Faculty, Conrad Blucher Institute, Texas A&M University-Corpus Christi, Corpus Christi, Texas 78411, ptissot@cbi.tamucc.edu
2 Associate Professor, Coastal and Ocean Engineering Division, Civil Engineering Department, Texas A&M University, College Station, Texas 77843-3136, dtc@tamu.edu
3 Professor of Computer Sciences, Computing and Mathematical Sciences Department, and Director, Division of Near Shore Research, Conrad Blucher Institute, Texas A&M University-Corpus Christi, Corpus Christi, Texas 78411, pmichaud@cbi.tamucc.edu

the Gulf coast and account for 52% of the U.S. tonnage (NOAA, 1999). For ports and waterways along the Atlantic and Pacific coasts water level forecasts are obtained by consulting the tide tables. In the Gulf of Mexico, water level changes are often dominated by meteorological factors the impact of which are often larger than the tidal range itself and unaccounted for in present forecasts (Cox et al., 2002). A one-month comparison between measured water levels and tidal forecasts is presented in figure 1 for the Pleasure Pier tide station located on Galveston Island near Houston, Texas, for the spring of 1997. As can be observed the difference between tidal forecasts and actual water levels can be larger than one foot for several consecutive days corresponding to the passage of frontal systems. The passage of frontal systems does in fact represent one of the major forcing for water levels and one of the main reasons for the inadequacy of the tide tables along the Gulf Coast. The passage of frontal systems across Texas and the Gulf of Mexico takes place approximately with a weekly frequency between early October and late May. In 1994 the National Oceanic and Atmospheric Administration (NOAA) conducted a current assessment program in Aransas Pass, Texas, and Corpus Christi Bay, Texas, and indicated that for typical weather conditions and for current predictions (a closely related parameter) the "presently published predictions do not meet working standards" (NOAA, 1991; NOAA 1994). Differences between observed and predicted current velocities were up to 100%, and wind was identified as the main cause with density variations, morphology and fresh water runoff playing a secondary role. To remedy the problem, both reports recommended forecasting based on real-time data including wind data. As present models are based on harmonic analysis they are fundamentally unable to account for these aperiodic forcing functions and new modeling techniques relying on real-time data must be introduced.

Sophisticated models based on finite elements and finite differences are ideal to understand the physical processes of coastal and estuarine dynamics and for simulating storm surges during hurricanes events. These models provide highly accurate solutions

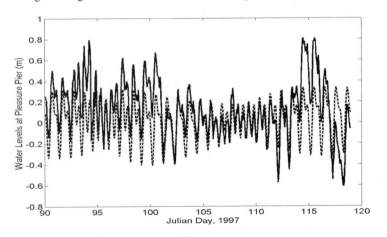

Fig. 1. Comparison between measured water levels (-) and tidal forecasts (...) at Galveston Pleasure Pier during the spring of 1997

to the governing equations of motion when the boundary conditions and time histories of the forcing functions are well prescribed. Their ability as forecasting tools, however, is limited to the accuracy of the forecasted time histories of the forcing conditions such as wind stress. On the other hand, empirical models such as those constructed using neural network techniques can yield accurate forecasts of water levels by incorporating data for established weather patterns. The prediction of water level by a neural network model at an entrance to a harbor channel, for example, can be used either directly as an aid to navigation or indirectly as a seaward boundary for a finite element model of the detailed flow in the harbor channel. In the second case, the neural network model provides two advantages: it reduces the requirements of having a large computation grid outside the entrance channel, and it can easily incorporate information from land-based stations (e.g., information about frontal passages) which may be difficult to incorporate into a finite element model discretized over only the water. The purpose of this paper is to show that a relatively simple neural network model can be constructed and trained using a fairly modest data set (three months) to provide accurate forecasts of water levels on a time scale of 1 to 48 hours. Direct comparisons or integration with finite element models is beyond the scope of the present work and will be pursued in the future.

The modeling philosophy applied in this work is to include and scale data streams such as observational data and forecasts to account for the main forcing of a problem and train a neural network to establish relationships between forcing functions and future water level changes. Of course, large amounts of real-time observational data are required to apply this modeling technique. Over the past ten years Texas has seen a dramatic increase in the availability of real-time observational data along the Gulf coast including parameters such as water levels, wind speeds, wind directions, barometric pressures, water temperatures and air temperatures. The Texas Coastal Ocean Observation Network (TCOON) is one of the main sources of such data and consists of 60 platforms from Brownsville to the Louisiana border (Michaud et al., 2001). TCOON station locations are illustrated in figure 2. The location of the Pleasure Pier tide station located on Galveston Island near Houston is highlighted as data from this station is used to test the model. Increases in the performance and decreases in the cost of sensors, telecommunication and overall information processing equipment should continue for the foreseeable future and make the deployment of data intensive models possible for most coastlines. The present work takes advantage of the real-time data available through the TCOON network and the modeling capabilities of neural networks to predict water levels in real-time and alleviates the present limitations of the tide tables.

NEURAL NETWORK MODELING OF WATER LEVEL CHANGES

The concept of neural networks emerged in the sixties as scientists aimed at emulating the functioning of the brain. The main advantages and key characteristics of neural networks for water level forecasting are their non-linear modeling capability, their generic modeling capacity, their robustness to noisy data, and their ability to deal with high dimensional data (Rumelhart et al., 1995). At the heart of a neural network is the assignment of judicious weights and biases to the elemental neurons of the network. This learning process must be based on a large set of prerecorded observations such as the TCOON database. Rumelheart et al. (1986) developed a type of learning algorithm to assign such a set of optimum weights and biases called backpropagation.

Fig. 2. Gulf of Mexico with the locations of the TCOON stations, the Pleasure Pier Station on Galveston Island and schematic of a typical frontal boundary

Backpropagation neural networks use the repeated comparison between the output of a neural network for a given input and an associated set of target vectors and optimize the neuronal weights and biases by backpropagating a function of this error through the network. The use of error backpropagation has been a key for the application of neural network to a growing number of practical cases including environmental, financial, and engineering problems (Swingler, 1996, Zirili, 1997). During the past five to ten years, neural networks have also been applied successfully to a growing number of coastal and riverine cases such as the forecasting of physical or water quality parameters (Mase et al., 1995, Recknagel et al., 1997, Mase and Kitano, 1999, Moatar et al., 1999, Tsai and Lee, 1999). Neural networks are increasingly tested for the forecasting of flooding along rivers (Campolo et al., 1997, Kim and Barros, 2001) a related application. The application of neural networks to water level forecasting consists in designing and training a network that, given a time series of water levels weather observations and forecasts (wind and tidal forecasts), accurately predicts the next water levels for a period of one, six, twelve, twenty four hours or more. The typical structure of the neural networks used in this work is illustrated in figure 3 and is relatively simple with one hidden layer and one output layer. Neural networks with additional hidden layers were also tested but did not improve on the performance of the two layer models. The elements of the input decks are chosen to track the variation of the main forcing

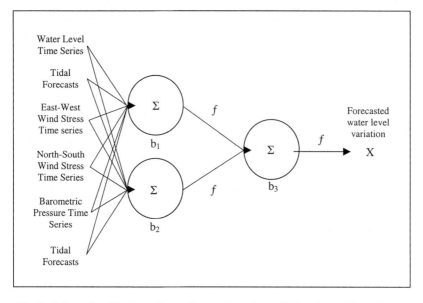

Fig. 3. Schematic of the type of neural network model applied to the problem of water level forecasting including outputs, inputs, and neural network

functions to the problem. They consist of time series of previous water levels, barometric pressures, wind speeds, and wind directions. Also included as part of the input decks are tidal forecasts computed using Xtide 2 (Hopper, 2000). The tidal forcing is included in the model by using water level differences between measured and forecasted water levels and the water levels predicted by the tide tables. The changes in the resulting water level differences are then a direct function of the meteorological forcing. Finally the model forecasts changes in water level differences rather than absolute water level differences. Focusing the model on changes in water level differences allows a more direct relationship between short-term forcing and changes in water levels. Also this allows inclusion of long-term effects such as steric effects as part of the input to each short-term forecasts. The models were tested with and without wind forecasts. When tested with wind forecasts the models initially included exact wind predictions with actual future measurements used as the forecasts. The influence of the accuracy of the forecasts was then evaluated by adding an error to the wind forecasts proportional to the wind and forecasting time. A discussion of the influence of the wind forecasts on the model accuracy as well as discussions on the characteristics of the optimum input deck and neural network structure are included as part of the results and discussion section. 'Tansig' transfer functions are used for the hidden and output layers while the input decks are scaled to a [-1,1] range. The neural network models are trained using a backpropagation algorithm and all computations are performed within the MATLAB 5.3/version 3 of the Neural Network Toolbox (The Math Works Inc., 1998) computational environment running on a Pentium III PC.

RESULTS AND DISCUSSION

The model was tested for the Pleasure Pier tide station on the Gulf Coast side of Galveston Island near Houston, Texas, (http://dnr.cbi.tamucc.edu/overview/022). The station is located near the ship channel leading to the port of Houston--one of the largest ports in United States. To test and optimize the models, the neural networks were successively trained over three data sets, spring 1997, spring 1998, and spring 1999. The approximately three-month data sets were chosen as they are representative of the type of conditions that can be encountered during frontal passages. The model was trained over one of the data sets and then applied to the two other data sets to assess the model performance. Figure 4 compares the performance of the model when trained over the 1999 data set and applied to make 24-hour forecasts for the 1997 data set. As can be observed the neural network outperforms significantly the tide tables during the frontal passages when wind forcing becomes the primary forcing driving water level changes. Figure 5 compares the performance of the model trained during 1997 and applied to the 1999 data set. The 24-hour water level forecasts displayed in figure 5 improve considerably on the tide tables matching closely the measured water levels and demonstrate the ability of the neural network to predict water levels during frontal passages. The input deck of the neural network used for the examples in figures 4 and 5 consists of 5-hour time series of previous water levels, tidal forecasts, wind speeds and wind directions, 20-hour times series of barometric pressures and an exact wind forecast for the time of forecast, i.e. a 24-hour wind forecast. The structure of the neural network is very simple with only one neuron in both the hidden and output layers. The optimization of the model is discussed later in this section. The fact that a very simple neural network is capable of making relatively accurate water level predictions

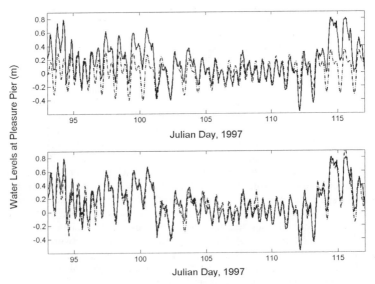

Fig. 4. Comparison of water levels measured at Galveston's Pleasure (-) with tide tables (top, ...) and neural-network forecasts (bottom chart, ...)

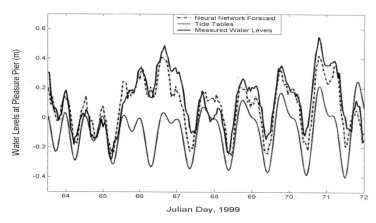

Fig. 5. Performance of the neural network model when trained over the 1997 data set and applied to a strong frontal passage during the spring of 1999

should not be surprising as Cox et al. (2002) have shown that a simple linear model based on the wind stress and a time delay could already lead to significant improvements of water level forecasts as compared to the tide tables.

To compare different versions of the model quantitatively over the complete data sets, models were evaluated by computing an error index including all forecasted water levels. The error index is detailed below and is the relative standard deviation of the water level differences or the ratio of the root mean square of the differences between forecasted and measured water levels and the root mean square of the measured water levels (Cox et al., 2002). The error index is zero for a perfect forecast and approaches one when the accuracy of the forecasts is of the same order as the water level variation.

$$E = \frac{\left[\dfrac{1}{N} \sum_{i=1}^{N} (H_i - X_i)^2 \right]^{1/2}}{\left[\dfrac{1}{N} \sum_{i=1}^{N} (H_i)^2 \right]^{1/2}} \tag{1}$$

where N = number of water level measurements and forecasts; H_i = observed water levels (m); and X_i = forecasted water levels (m). To assess the model performance over the three data sets the models are first trained over one of the data sets and then applied to the two other data sets. The operation is repeated by rotating the training data sets and the data sets over which the models are tested. An average model performance is then computed by averaging the performance of the models for the six resulting time series of forecasted water levels. The performance variability of the models is estimated by computing the standard deviation over the six different cases.

The error index was used to optimize the network structure and the input deck. The network parameters were optimized for 24-hour forecasts and verified by comparing with the optimum parameters obtained for 9-hour forecasts as well. There were no substantial differences between the optimized neural network parameters optimized for 9 and 24-hour forecasts. The extent of the past wind speeds and wind directions time series was varied between 0 and 30 hours. The optimal time series length was found to be 5 hours with relatively small changes in performance when increasing or decreasing the length of the time series in the 1 to 20-hour range. Input time series longer than 20 hours led to decrease in performance. It should be noted that including longer time series in the input deck does not necessarily mean that the additional input data will be taken into account by the model. As the weights of the neural network are optimized during the training procedure the weights of the additional inputs can be zero. A more in-depth study of the variation of neural network weights and biases when optimizing the model is ongoing. Although the neural network should optimize itself and not take into account unnecessary inputs, larger input decks will affect the training times and possibly the final weights and biases. The length of the input deck was therefore chosen as the smallest time series leading to the best model performance. The optimum past water level measurements and tidal levels time series was 5 hours while the optimum barometric pressure time series length was 20 hours. The barometric pressure time series had a relatively small impact on the model performance on the order of 5%. This is likely due to the fact that winds are in large part a result of the pressure differences across the frontal boundaries and that therefore the barometric pressure effect is in large part already included through the wind inputs. As will be discussed later in this section, the addition of a wind prediction for the time of the forecast has a significant effect on the performance of the model for prediction times longer than 6 hours. The optimized model for this study includes 1 hidden and 1 output neuron, 5 hour time series of previous wind speeds, wind directions, water level measurements and tidal predictions, 20 hour time series of barometric pressures and a wind forecast for the time of the forecast. The performance of the model is displayed for a range of 1 to 48 hours forecasts in figure 6. As can be observed, the model outperforms significantly the tide tables indicating that neural network models can indeed factor in meteorological forcing and lead to more accurate water level forecasts along the Gulf Coast.

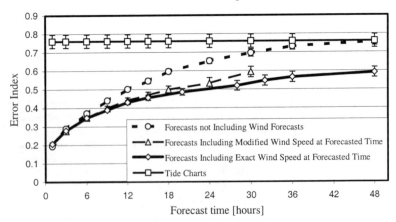

Fig. 6. Comparison of the performance of neural network models forecasting future water levels at Galveston, Texas, Pleasure Pier Station.

To assess the sensitivity of the model to the accuracy of the wind forecasts, an error proportional to the magnitude of the wind forecast and the forecasting time was added to the exact wind forecasts (Cox et al., 2002). The model was then trained on the exact wind forecasts and tested with the modified wind forecast. As can be observed in figure 6, the model performance is not affected for forecasting times below 15 hours and not affected significantly for longer forecasting times. The model was also tested while excluding wind forecasts. The results are displayed in figure 6 and show that a model without wind forecasts performs well for up to 6-hour forecasts and does not provide significant improvements over the tide tables for forecasting times longer than 24 hours. The accuracy of the model without wind forecasts will likely vary depend on the location as it depends on the natural time lags between the onset of frontal driven winds and the water level response (Cox et al., 2002). As the wind forecasts are the primary factor for accuracy of the model past 12 to 15 hours, the model is presently being modified to include wind forecasts for up to 60-hour predictions extracted in real time from the National Center for Environmental Predictions Meso Eta model. The real time inclusion of the wind forecasts is conducted as part of a collaboration with the National Weather Service. The Meso Eta forecasts will allow access not only to accurate wind forecasts for the station for which the model is trained but also wind forecasts for large portions of the Gulf of Mexico allowing for the development of more sophisticated models.

Present plans include the application of the model to other locations along the coast of Texas and its real time access through the World Wide Web. Depending on the location the models could require the addition of other inputs such as precipitation and riverine inflows. As the availability of real-time data is constantly improving along the Texas Gulf Coast this should not be a significant limiting factor. This technique will be more difficult to apply for the case of predicting storm surges during strong tropical storms and hurricanes for several reasons. First, compared to frontal events which occur almost weekly in the fall and spring months, tropical storms are more episodic and the available data base is small. Second, frontal events cover a large spatial area with conditions (wind speed, direction) relatively constant over that area compared to tropical storms which are more localized. Different types of neural network modeling techniques may have to be adapted to address the dynamic and localized nature of tropical storms.

CONCLUSIONS

A new data intensive forecasting method based on neural network modeling was developed to predict water levels along the Gulf Coast. The neural networks are trained to forecast water levels by establishing a relationship between time series of previous water levels, wind stress, and barometric pressures and future water levels. The technique was tested over a period of three years for a tide station along Galveston Island near Houston, Texas. Models were tested with and without including wind forecasts. Models including wind forecasts significantly outperformed the tide tables for the tested forecasting range of 1 hour to 48 hours. The models not including wind forecasts performed well up to 12 to 18 hours. Further work includes the addition of more sophisticated wind forecasts extracted in real time from the National Center for

Environmental Predictions Meso Eta model and expanding the model to other locations along the Texas coast.

REFERENCES

Campolo, M., Andreussi, P., and Soldati, A. 1997. River Flood Forecasting with a Neural Network Model. *Water Resources Research*, 35 (4), 1191-1197.

Cox, D., Tissot, P, and Michaud, P. 2002. Water Level Observations and Short Term Predictions Including Meteorological Events for the Entrance of Galveston Bay, Texas. *Journal of Waterways, Port, Coastal and Ocean Engineering*, accepted for publication, January 2002, ASCE.

Hopper, M. 2000. WXTide32 version 2.6, January 20, 2000, Copyright © 1998-2000 Michael Hopper, http:www.GeoCities.com/SiliconValley/Horizon/1195.html.

Kim, G., and Barros, A. 2001. Quantitative Flood Forecasting Using Multisensor Data and Neural Networks. *Journal of Hydrology*, 246, 45-62.

Mase, H. and Kitano, T. 1999. Prediction Model for Occurrence of Impact Wave Force. *Ocean Engineering*, 26 (10), 949-961.

Mase, H., Sakamoto, M., and Sakai, T. 1995. Neural Network For Stability Analysis of Rubble-Mound Breakwaters. *Journal of Waterway, Port, Coastal, and Ocean Engineering*, 121 (6), ASCE, 294-299.

Michaud, P., Jeffress, G., Dannelly, R., and Steidly, C. 2001. Real-Time Data Collection and the Texas Coastal Ocean Observation Network. *Proceedings of INTERMAC '01 Joint Technical Conference*, Tokyo, Japan, accepted.

Moatar F., Fessant, F., and Poirel, A. 1999. pH Modelling by Neural Networks. Application of Control and Validation Data Series in the Middle Loire River. *Ecological Modeling*, 120, 141-156.

NOAA 1999. Assessment of the National Ocean Service's tidal current program. NOAA Tech. Rpt. NOS Co-OPS 022, United States Department of Commerce, 71p.

NOAA 1994. Special 1994 tidal current predictions for Aransas Pass, Corpus Christi, Texas. NOAA Technical Memorandum NOS OES 8, United States Department of Commerce, 26p.

NOAA 1991. Corpus Christi Bay Current Prediction Quality Assurance Miniproject. *NOAA Technical Memorandum NOS OMA 60*, United States Department of Commerce, 46p.

Recknagel, F., French, M., Harkonen, P., and Yabunaka, K-I. 1997. Artificial Neural Network Approach for Modeling and Prediction of Algal Blooms. *Ecological Modeling*, 96, 11-28.

Rumelhart, D. E., Durbin, R., Golden, R., and Chauvin, Y. 1995. Backpropagation: The Basic Theory. *Backpropagation: Theory, Architectures, and Applications*, Rumelhart, D. E., Chauvin, Y., eds, Lawrence Erlbaum Associates, Publishers, Hillsdale, 1-34.

Rumelhart, D. E., Hinton, G. E., and Williams, R.J. 1986. Learning Representations by Back-Propagating Errors. *Nature*, 323, 533-534.

Swingler, K. 1996. *Applying Neural Networks, A Practical Guide*, Academic Press Limited, London, 303 p.

The MathWorks, Inc. 1998. *Neural Network Toolbox for use with Matlab 5.3/version 3*, The MathWorks, Natick, MA.

Tsai, C-P., and Lee, T-L. 1999. Back-Propagation Neural Network in Tidal-Level Forecasting. *Journal of Waterway, Port, Coastal, and Ocean Engineering*, 125 (4), ASCE, 195-202.

Zirili, S.J. 1997. *Financial Predictions Using Neural Networks*, International Thomson Computer Press, London, 135 p.

EFFECTS OF NONLINEARITY AND BOTTOM FRICTION ON HURRICANE-GENERATED STORM SURGE IN CENTRAL PACIFIC OCEAN

Shi-Jun Liao[1], Y. Wei[2], Michelle H. Teng[3], Philip L. F. Liu[4], Kwok Fai Cheung[5], C. S. Wu[6]

Abstract: A storm surge model based on the shallow water equations and solved by a finite difference scheme has been developed. This model is adapted from the original COMCOT tsunami simulation model developed by Liu's group at Cornell. The modified COMCOT model was first applied to simulate the storm surge generated by both Hurricane Iniki and Hurricane Iwa for model validation. Then the linear and nonlinear versions of the model along with or without bottom friction were tested, and the simulated results are compared in order to examine the nonlinear effects on the generation and evolution of storm surge in the Pacific Ocean. Our results show that the nonlinear effects including convective acceleration and the bottom friction terms are not significant in storm surge modeling for the Pacific insular environment.

INTRODUCTION

Hurricanes are a frequent natural hazard in the tropical regions in the world. The Hawaii Island chain experienced several devastating hurricanes in the past including the 1992 Hurricane Iniki and the 1982 Hurricane Iwa. While research on storm surge has been very active on the US mainland, in Europe and Asia, most of the existing studies have been carried out for regions with continental shelves. Studies on hurricane-generated storm surge in the Pacific insular environment, such as the Hawaiian coastal waters, have been very limited.

1 Visiting scientist, Dept. of Ocean and Resources Engineering, University of Hawaii at Manoa, Honolulu, Hawaii 96822 USA. Currently: professor, School of Naval Architecture and Ocean Engineering, Shanghai Jiao Tong University, Shanghai, China. sjliao@mail.sjtu.edu.cn
2 Graduate Research Assistant, Department of Ocean and Resources Engineering, University of Hawaii at Manoa, Honolulu, Hawaii 96822 USA. yongwei@oe.eng.hawaii.edu
3 Associate Professor, Dept. of Civil Engineering, University of Hawaii at Manoa, Honolulu, Hawaii 96822 USA. teng@wiliki.eng.hawaii.edu
4 Professor, School of Civil and Environmental Engineering, Cornell University, Ithaca, NY 14850 USA. P113@cornell.edu
5 Associate Professor, Dept. of Ocean and Resources Engineering, University of Hawaii at Manoa, Honolulu, Hawaii 96822 USA. cheung@oe.eng.hawaii.edu
6 Senior Scientist, Marine Techniques Branch, National Weather Service, NOAA, Silver Spring, Maryland 20910 USA. Chung-sheng.Wu@noaa.gov

It was reported in several papers (Kolar *et al* 1994, Luettich *et al* 1992, Westerink *et al* 1992) that the nonlinear effects including both the convective acceleration and the bottom friction may be important in storm surge generation and evolution along the east coast and in the Gulf of Mexico, where large continental shelves exist. For the Pacific insular environment, the slopes of the islands are quite steep and water depth increases rapidly from the coast to the open ocean. Whether the nonlinear effects such as bottom friction and convective acceleration are important in storm surge generation and evolution in the Pacific Ocean remains to be investigated.

In the present study, a storm surge model based on the shallow water equations and solved by a finite difference scheme has been developed. This model is adapted from the original COMCOT tsunami simulation model developed by Liu's group at Cornell (Liu *et al* 1995, 1998). The modified COMCOT model was first applied to simulate the storm surge generated by both Hurricane Iniki and Hurricane Iwa, and the simulated results are compared with the field data for model validation. Then the linear and nonlinear versions of the model along with or without bottom friction were tested, and the simulated results are compared in order to examine the nonlinear effects on the generation and evolution of storm surge in the Pacific Ocean.

GOVERNING EQUATIONS

Storm surges are long waves and therefore can be described by the depth integrated continuity and momentum equations

$$\frac{\partial \zeta}{\partial t} + \frac{\partial (hu)}{\partial x} + \frac{\partial (hv)}{\partial y} = 0, \tag{1}$$

$$\frac{\partial u}{\partial t} + u\frac{\partial u}{\partial x} + v\frac{\partial u}{\partial y} - \gamma\, v = -g\frac{\partial \zeta}{\partial x} - \frac{1}{\rho}\frac{\partial P_a}{\partial x} + \frac{1}{\rho h}(\tau^x - \tau_b^x), \tag{2}$$

$$\frac{\partial v}{\partial t} + u\frac{\partial v}{\partial x} + v\frac{\partial v}{\partial y} + \gamma\, u = -g\frac{\partial \zeta}{\partial y} - \frac{1}{\rho}\frac{\partial P_a}{\partial y} + \frac{1}{\rho h}(\tau^y - \tau_b^y), \tag{3}$$

where u, v are fluid velocities, ζ the elevation of water surface, $h = \zeta + D$ the total fluid depth, D the bathymetric depth, P_a the air pressure at the sea surface, γ the Coriolis coefficient, ρ the density of water, g the gravity acceleration, (τ^x, τ^y) the wind stress on the surface, (τ_b^x, τ_b^y) the bottom friction stress, t the temporal variable and (x, y) the horizontal spatial coordinates, respectively.

The bottom friction stress is given by

$$\bar{\tau}_b = \frac{gn^2 |\bar{u}|\bar{u}}{H^{10/3}}, \tag{4}$$

where n is the Manning coefficient. The shallow water equations are solved by the finite difference method and a related code is developed, which is based on the original COMCOT tsunami simulation code provided by Liu's group at Cornell. In the present study, the total surface stress (τ^x, τ^y) is computed by applying the well-known WAM model (The WAMDI Group 1988, Gunther et al 1992, Koman et al 1996). The WAM model is based on the energy or action balance equation that can be used to predict surface waves generated by wind fields. Here, the surface shear stress calculated from the WAM output includes both wind stress and wave stress which are input into the storm surge model as forcing (Li and Zhang 1997). The coupling between the WAM model and the storm surge model is performed at each time step.

The parameterized wind and pressure model used in the present study is the modified Rankine model:

$$W(r) = W_{max}\left(\frac{r}{R}\right)^X \quad \text{for } r < R; \quad \text{and} \quad W(r) = W_{max}\left(\frac{R}{r}\right)^X \quad \text{for } r > R, \quad (5)$$

$$P_a(r) = P_c + (P_\infty - P_c)e^{-(R/r)^B} \quad (6)$$

where W is the wind speed, W_{max} the maximum wind speed, R the radius to the maximum winds, and P, P_c and P_∞ are the local, central and unperturbed pressures, respectively. The empirical parameter X typically has a value between 0.4 and 0.6, and B ranges between 1.0 and 2.5. In the present study, $X = 0.5$ and $B = 1$.

To investigate the nonlinear effects of the bottom friction and the convective acceleration, we will compare the simulated results based on the fully nonlinear model (1)-(3) with that based on the following linearized equations

$$\frac{\partial \zeta}{\partial t} + \frac{\partial (Du)}{\partial x} + \frac{\partial (Dv)}{\partial y} = 0, \quad (7)$$

$$\frac{\partial u}{\partial t} - \gamma\, v = -g\frac{\partial \zeta}{\partial x} - \frac{1}{\rho}\frac{\partial P_a}{\partial x} + \frac{\tau^x}{\rho h}, \quad (8)$$

$$\frac{\partial v}{\partial t} + \gamma\, u = -g\frac{\partial \zeta}{\partial y} - \frac{1}{\rho}\frac{\partial P_a}{\partial y} + \frac{\tau^y}{\rho h}, \quad (9)$$

where the bottom friction and the convective acceleration are neglected.

NUMERICAL SIMULATIONS OF HURRICANE INIKI AND IWA

Hurricane Iniki (1992) and Hurricane Iwa (1982) are two hurricanes that caused the most severe damage in Hawaii in recent decades. The track and history of the two hurricanes are shown in Fig.1 and Fig.5.

Simulation of Storm Surge Generated by Hurricane Iniki

The computational domain used in the present simulation is a square domain with longitude varying from $-165°$ to $-150°$ and latitude from $10°$ to $25°$ (Part of the computational domain along with water depth contour around Kauai Island are shown in Fig.2). The spatial resolution is $0.1°$ which is about 12 km. (A spatial resolution of 1.2 km was also tested, and the difference in surge height between the two different resolutions was found to be about 5 cm.) The time step is 2 sec and the Manning coefficient is 0.04.

For model validation, we compare the simulated surge height with the measured sea surface elevation recorded by a tidal gage at Port Allen ($-159.59°$, $21.90°$). Figure 3 shows that the simulated surge height reaches its peak at the same time as the recorded sea surface elevation. However, the recorded surface elevation is higher than the simulated surge. The reason is that the recorded sea surface elevation includes tides, storm surge and wave set-up while the simulated result represents only the storm surge. It was reported that, during Hurricane Iniki's landfall on Kauai, the maximum surge coincided with the high tide which is estimated to be 0.64 m (Sea Engineering 1993). Based on the NOS buoy data, the significant wave height near Port Allen is about 11 m with a wave period of about 14 sec. By applying the empirical formulas developed by the US Army Corps of Engineers, the wave set-up near Port Allen was estimated to be about 1.0 meter by Fletcher et al (1995). In the present study, wave set-up was re-calculated based on simulated wave height from the WAM model and the simulated wave set-up is calculated to be about 0.60 m. Since the buoy location is quite far (about 400 km) from Port Allen, while the simulated wave height is for Port Allen specifically, we will use the simulated wave set-up in our discussions below. Table 1 presents the comparison between the simulated surge height and the measured data by a tidal gage inside Port Allen. The results show that the simulated and recorded maximum surge heights agree with each other reasonably well. This verifies the storm surge simulation performed in the present study.

Table 1. Comparison between Simulated and Recorded Maximum Surge Height at Port Allen in Kauai

Simulated maximum surge height (m)	Recorded maximum sea level (m)	Predicted maximum tide (m)	Estimated wave set-up (m)	Corrected recording for the max. surge (m)
0.51	1.89	0.64	0.60	0.65

To investigate the nonlinear effects of both the bottom friction and the convective acceleration, we compare the simulated storm surge generated by Hurricane Iniki based on the aforementioned nonlinear (1)-(3) and linear models (7)-(9) for two locations in Kauai, i.e. Port Allen and Nawiliwili ($-159.36°$, $21.96°$). The simulated results are presented in Fig.4 and Fig.5 showing that the nonlinear and linear results are almost identical. In other words, the nonlinear effects are insignificant in the generation of storm surge in this case. (Please note that the time in the figures is UTC time which is 10 hours ahead of local Hawaiian time.)

Simulation of Storm Surge Generated by Hurricane Iwa

The track of Hurricane Iwa is shown in Fig. 6. In this case, the computational domain is a square domain with longitude varying from $-168°$ to $-153°$ and altitude from $13°$ to $28°$. The spatial resolution, time step and Manning's coefficient are the same as in the simulation for Hurricane Iniki. The simulated storm surge at Nawiliwili is shown in Fig. 7. Again, we can see that the nonlinear and linear results are almost identical. This further verifies that the nonlinear effects of both convective acceleration and the bottom friction terms are not significant in storm surge generation in the Pacific insular environment.

This result is mainly due to the bathymetric characteristics in the Pacific insular environment. When a storm surge evolves along the east coast or in the Gulf of Mexico, where large continental shelves exist, it moves in considerably long time on the continental shelf in rather shallow water such that the bottom friction and convective acceleration terms become important. However, in the Pacific insular environment, the water depth near the islands increases rather quickly and the storms often pass the islands fast. Thus, the bottom friction and convective acceleration do not have enough time to affect the evolution of storm surge.

CONCLUSIONS

In this paper, a storm surge model based on the shallow water equations and solved by a finite difference scheme has been developed. This model is adapted from the original COMCOT tsunami simulation model developed by Liu's group at Cornell and is coupled with the well-known WAM model. The nonlinear effects of the bottom friction and convective acceleration in storm surge generation and evolution in the Pacific insular environment are investigated. Our results show that, due to the bathymetric characteristics around the Pacific islands, both the bottom friction and the convective acceleration are *not* important in the Pacific insular environment.

ACKNOWLEDGEMENTS

This work is funded by NASA under grant #NAG5-8748.

REFERENCES

Fletcher, C.H., Richmond, B.M., Barnes, G.M. and Schroeder, T.A. 1995. Marine flooding on the coast of Kauai during Hurricane Iniki: hindcasting inundation components and delineating washoever. *J. of Coastal Research*, 11(1), 188-204.

Gunther, H., *et al* 1992. The WAM Model, Technical Report No. 4, GKSS, Hamburg.

Kolar, R.L., Westerink, M.E., Cantekin, M.E., and Blain, C.A. 1994. Aspects of Nonlinear Simulations Using Shallow Water Models Based on the Wave Continuity Equation, *Computers in Fluids*, 23(3), 523-538.

Koman, G. J., et al, 1996. *Dynamics and modeling of Ocean Waves*, Cambridge University Press, New York.

Li, Y.S. and Zhang, M.Y. 1997. Dynamics coupling of wave and surge models by Eulerian-Langangian method, *Journal of Water Way, Port, Coast and Ocean Engineering*, 1 –7.

Liu, P.L.F., Cho, Y-S, Briggs, M.J., Kanoglu, U. and Synolakis, C.E. 1995. Run-up of solitary waves on a circular island. *JFM*, 302, 259-285.

Liu, P. L. F., Woo, S.B. and Cho, Y-S. 1998. Computer Programs for Tsunami Progagation and
    Inundation, Report, School of Civil and Environmental Engineering, Cornell University, Ithaca,
    USA.
Luettich, R.A., Westerink, J.J. and Scheffner, N.W. 1992 . ADCIRC: An Advanced
    Three-Dimensional Circulation Model for Shelves, Coasts, and Estuaries, Report 1: Theory
    and Methodology of ASCIRC-2DDI and ADCIRC-3DL, US Army Corps of Engineers.
Sea Engineering, Inc. 1993. Hurricane Iniki coastal inundation modeling. Technical Report
    prepared for the U.S. Army Corps of Engineers, Pacific Ocean Division.
The WAMDI Group, 1988. The WAM Model: a third generation ocean wave prediction mode,
    Journal of Phys. Oceanography, 1775-1810.
Westerink, J.J., Luettich, R.A., Baptista, N.W., Scheffner, N.W. and Farrar, P. 1992. Tide and
    Storm Surge Prediction Using Finite Element Model, *Journal of Hydraulic Engineering*, 118
    (10), 1373-1390.

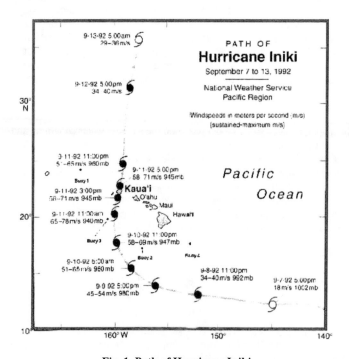

Fig. 1. Path of Hurricane Iniki

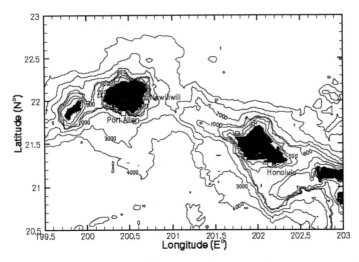

Fig.2. Part of the computational domain and water depth contour around Kauai Island in Hawaii

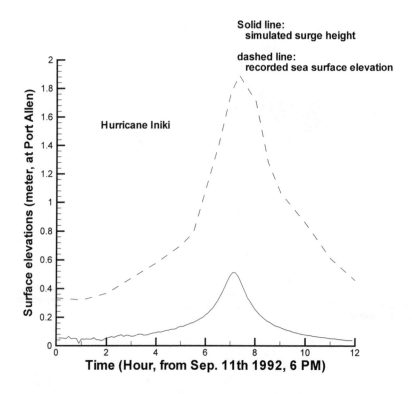

Figure 3.   Comparison between simulated surge height with the recorded sea surface elevation during Hurricane Iniki at Port Allen in Kauai

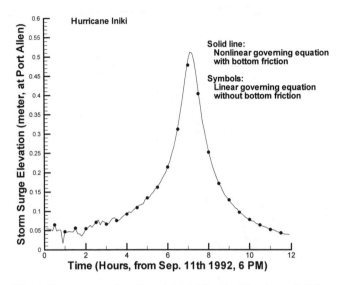

Fig. 4. Storm surge elevation at Port Allen for Hurricane Iniki

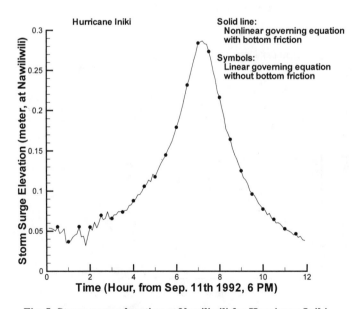

Fig. 5. Storm surge elevation at Nawiliwili for Hurricane Iniki

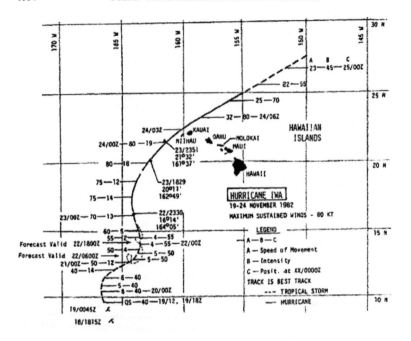

Fig. 6. Path of the Hurricane Iwa

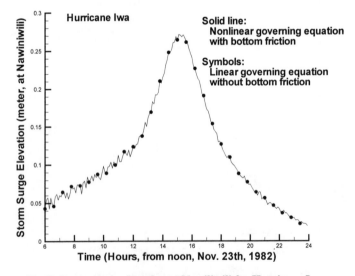

Fig. 7. Storm surge elevation at Nawiliwili for Hurricane Iwa

NUMERICAL MODELING OF STORM SURGE GENERATED BY HURRICANE INIKI IN HAWAII

Edison Gica[1], Michelle H. Teng[2], M.ASCE, Richard A. Luettich, Jr.[3], Kwok Fai Cheung[4], Cheryl Ann Blain[5], Chung-Sheng Wu[6] and Norman W. Scheffner[7]

Abstract: The U.S. Army Corps of Engineers' ADvanced CIRCulation (ADCIRC) model (Luettich, et. al. 1992) is applied to simulate the storm surge generated by Hurricane Iniki (1992) in Hawaii. The purpose of this study is to examine the validity of the ADCIRC model in predicting hurricane-generated storm surge in the Pacific insular environment and to investigate the sensitivity of storm surge generation to different wind models for this particular region. The simulated results showed good agreement with the recorded storm surge near the southwest shore of Kauai Island where Iniki made landfall. Our present numerical results revealed that the dominant forcing factor for the storm surge generation by Hurricane Iniki in Hawaii is the drop in atmospheric pressure near the eye of the hurricane, while the wind-generated surface stress has little direct effect in generating the surge.

[1] Research Assistant, Department of Civil Engineering, University of Hawaii at Manoa, Honolulu, Hawaii 96822, gica@wiliki.eng.hawaii.edu
[2] Associate Professor, Department of Civil Engineering, University of Hawaii at Manoa, Honolulu, Hawaii 96822
[3] Professor, Insititute of Marine Science, University of North Carolina at Chapel Hill, Morehead City, North Carolina, NC 28557
[4] Associate Professor, Department of Ocean and Resources Engineering, University of Hawaii at Manoa, Honolulu, Hawaii 96822
[5] Oceanographer, Oceanography Division, Naval Research Laboratory, Stennis Space Center, Mississippi 39529
[6] Physical Scientist, Office of Science and Technology, National Weather Service, NOAA, Silver Spring, Maryland 20910
[7] Senior Research Hydraulic Engineer, US Army Engineering Waterways Experiment Station, Coastal Engineering Research Center, Vicksburg, Mississippi 39180

INTRODUCTION

Hurricanes can cause severe damage to coastal structures through both direct wind impact and coastal flooding due to hurricane-generated high waves and surge. For civil and coastal engineers, accurate prediction of the hurricane waves and storm surge is critical in their structural design and mitigation plan. The present paper focuses on studying storm surge generation by hurricanes in the Pacific insular environment and by examining Hurricane Iniki as a case study.

Developing accurate numerical models for predicting hurricane-generated storm surge has been an active research topic for many decades. One of the earlier widely used storm surge models was the SLOSH model developed by scientists at the National Weather Service of NOAA in the US. The SLOSH model is relatively simple, and has been applied to modeling storm surge in many regions. The SLOSH model is based on the linearized shallow water equations and does not consider surge-tide interaction. In recent years, more advanced nonlinear storm surge models have been developed. One of these new models is the ADvanced CIRCulation Model, or the ADCIRC model, which was developed by Luettich et al (1992). This model is based on a modified version of the depth-averaged shallow water equations. In the modified version, the original continuity equation is replaced by the generalized wave-continuity equation which was obtained by combining the continuity and momentum equations (Westerink et al. 1992, Kolar et al. 1994). Although the equations are depth-averaged, an extended algorithm has been developed to calculate the 3-D flow components (Luettich and Westerink 1991, Grenier et al. 1995). The forcing terms in this model include surface pressure, surface shear stress due to wind, and tide potential. Nonlinear advection, bottom friction and Coriolis effect are included. The model is computed by using a finite element scheme with triangular elements. One of the advantages of the ADCIRC model is that the grids can be variable in size, e.g., coarse in the open ocean and fine in the coastal region.

We should mention that although research on storm surge has been very active on the US mainland, in Europe and Asia, most of the existing studies have been carried out for oceanic regions with gradually sloped continental shelves such as the US east coast and the Gulf of Mexico. Studies have been very limited on hurricane-generated storm surge in the Pacific insular environment, such as the Hawaiian coastal waters, where continental shelves are absent and water depth increases rapidly from shallow coastal waters to deep open ocean. To study the coastal impact due to Hurricane Iniki in Hawaii, Fletcher et al. (1995) presented a very detailed and useful post-event analysis for each inundation component based on empirical formulas. However, so far very few dynamic simulations of storm surge generated by hurricanes in Hawaiian waters have been reported.

One of the objectives of the present study is to simulate the storm surge generated by Hurricane Iniki (1992) in the Hawaiian coastal waters by applying the ADCIRC model and to examine the validity of the ADCIRC model for simulating storm surge in the Pacific insular environment by comparing the simulated results with the recorded field data.

The other objective is to investigate the sensitivity of storm surge generation to different wind field models. As we know, wind stress acting on the sea surface is one of the forcing factors in generating surges during a storm event. For any storm surge simulation model, the wind field of a hurricane must be prescribed as an input function of space and time. Since the detailed wind field of a hurricane is very difficult to measure during a hurricane event, many different parametric wind models have been developed to approximate the wind field based on several hurricane parameters including the central pressure and the radius to the maximum wind. The effect of wind forcing on storm surge generation and how sensitive the simulated surge is to different wind models for the Pacific insular environment have not been fully investigated thus far.

In the following sections, the history of Hurricane Iniki, the parametric wind models tested in the present study and the simulated results will be presented and discussed.

HURRICANE INIKI

Hurricane Iniki was initially formed as a tropical depression near (12°N, 135°W) on September 5, 1992. After that, it moved towards the Hawaiian island chain along a west-northwest trajectory as shown in figure 1. With its sustained maximum wind continuing to increase, Iniki was upgraded to tropical storm and hurricane on September 7 and 8, 1992, respectively. On the afternoon (around 3 pm local time) of September 11, 1992, Hurricane Iniki made landfall on the southwest coast of Kauai Island and caused severe property damage due to its high winds and the associated coastal flooding. Causing total losses of $1.6 billion, Iniki has been ranked as one of the most costly hurricanes in the U.S. history.

The path and strength (including central pressure and maximum wind) of Hurricane Iniki during the period of September 7-13, 1992 are shown in figure 1 (National Weather Service 1993, Fletcher et al. 1995). Compared with earlier hurricanes in Hawaii, Hurricane Iniki is relatively well documented. The available field data include wave height recorded by three buoys and surge height by several tidal gauges near different islands. These data are valuable for validation of hurricane wave and storm surge simulation models.

PARAMETRIC WIND AND PRESSURE FIELD MODELS

For storm surge generation, the forcing sources include the atmospheric pressure and the surface shear stress generated by wind. During a hurricane event, usually only a finite number of parameters at a limited number of locations can be measured in the field. Therefore, a mathematical model for describing the entire wind and pressure field based on the limited field data is necessary. These mathematical models are often based on the following parameters: the central pressure, maximum wind, radius to the maximum wind as well as other parameters. In the present study, we will test on two commonly used models for describing the wind and pressure field, namely, the Holland model (Holland 1980, Bode and Hardy 1997) and the modified Rankine model (Fletcher et al. 1995, Martino 2000).

Holland Model

This model is given as

$$P(r) = P_c + (P_\infty - P_c)e^{-(R/r)^B} \tag{1}$$

$$W(r) = \left[\frac{B}{\rho_a} \left(\frac{R}{r}\right)^B (P_\infty - P_c)e^{-(R/r)^B} + \left(\frac{rf}{2}\right)^2 \right]^{1/2} - \frac{rf}{2}, \quad 1.0 < B < 2.5 \tag{2}$$

$$U_{10} = KW(r), \quad K \approx 0.8 \tag{3}$$

$$\tau_s = (\tau_{sx}, \tau_{sy}) = C_{10}\rho_a|U_{10}|U_{10} \tag{4}$$

where P, P_c and P_∞ are the local, central and unperturbed pressures, respectively, R is the radius to maximum winds, W the wind speed, f the Coriolis coefficient, U_{10} the wind speed at a height of 10 m above the mean sea level, τ_s the shear stress at the water surface due to wind, ρ_a the air density, and C_{10} the drag coefficient which is given by $10^3 C_{10} = 0.8 + 0.065 U_{10}$ (Wu 1982). In the shallow water equations, the forcing term directly related to wind is the shear stress τ_s generated by wind, which is calculated through equations (2)-(4).

Modified Rankine Model

The modified Rankine model is a simpler model with its wind field given as

$$W(r) = W_{max}\left(\frac{r}{R}\right)^X \quad \text{for } r < R; \quad \text{and} \quad W(r) = W_{max}\left(\frac{R}{r}\right)^X \quad \text{for } r > R, \tag{5}$$

where W_{max} is the maximum wind speed, and the empirical parameter X typically has a value between 0.4 and 0.6. The pressure field $P(r)$, wind speed U_{10} and shear stress τ_s are given by the same equations (1), (3), (4), respectively.

NUMERICAL SIMULATION

In the present study, the ADCIRC finite element model is applied to simulate the storm surge generated by Hurricane Iniki. As shown in a recent separate study (Liao et al 2001), the nonlinear effect due to convective acceleration is insignificant in storm surge generation in the Hawaiian waters where the water depth increases rapidly from the shoreline to the deep ocean. Based on this result, the nonlinear convective term in the ADCIRC model is neglected in the present simulation. In addition, surge-tide interactions are not considered in the present study in order to focus on examining the effects of wind models on storm surge generation.

The computational domain is shown in figure 2(a). It is noted that the finite element grids are made finer not only near shore but along the entire track of Hurricane Iniki in order to provide sufficient resolution for the simulation of storm surge generated by the hurricane. A close-up picture of the variable finite element grid system near the Kauai Island is shown in figure 2(b). In the present study, the grid size ranges from 100 m in the coastal waters near the southwest side of Kauai where the hurricane made the landfall, 2 km along the track of the hurricane, to about 100 km in the unperturbed regions of the open ocean.

The measured data for hurricane trajectory, central pressure P_c, and maximum wind W_{max} were obtained from the survey report on Hurricane Iniki prepared by the National Weather Service, NOAA (1993). These data were given for 6-hour intervals while typical time step used in our simulation is 2.5 second. Therefore, between each two given data points, linear interpolation was applied to obtain the values of different input parameters for each time instant in our simulation. Similar to other parameters such as hurricane trajectory, central pressure and maximum wind speed, the radius to the maximum winds R should also be a variable changing in time as the hurricane evolves. However, due to the limitations in current technology for hurricane monitoring, it is difficult to measure the radius to the maximum winds at every time instant during the hurricane event. Based on the available field data and analysis by Fletcher et al 1995, in the present study, a constant value of 12 km is assumed for R. For the wind and pressure models, the parameters B and X are given the values of 1.25 and 0.5, respectively.

RESULTS AND DISCUSSIONS

Our simulation results show that the maximum surge occurred near the hurricane center at about 7 hours before the landfall. At that hour, the central pressure dropped to its lowest value of 93800 Pa. The simulated maximum surge height near the hurricane center at this hour is 0.77 m by the Holland wind model and 0.80 m by the modified Rankine wind model.

Table 1 presents the comparison between the simulated maximum surge height (0.38 m by Holland wind model and 0.43 m by the modified Rankine wind model) and the recorded maximum sea level of 1.89 m by a tidal gauge at Port Allen (see figure 2) on Kauai about the time Iniki made the landfall on Kauai's southwest shore. We should point out that the recorded water level by a tidal gauge consists of several components of water surface rise including the tide, storm surge, and wave set-up. In order to compare the simulated storm surge with the recorded surge, the total water level recorded by the tidal gage must be decomposed in order to isolate the storm surge component.

In table 1, the data for the recorded maximum sea level and the predicted maximum tide were obtained from the technical report prepared by Sea Engineering, Inc. (1993). The wave set-up of 0.80 m is estimated by applying the empirical formulas given in the US Army Corps of Engineer's Shore Protection Manual (Fletcher et al 1995). The calculations are based on the recorded hurricane wave height of 6 m and period of 10 sec by Buoy No. 51002 offshore Kauai Island. (This estimated wave set-up is slightly higher than the estimated wave set-up of 0.60 m based on the SWAN model in Liao et al 2001). After subtracting tide and wave set-up from the recorded sea surface elevation, the corrected recording for storm surge at Port Allen is obtained as 0.45 m, which is very close to the simulated results. A more detailed comparison between the recorded sea surface elevation and the simulated surge height at Port Allen at different time instants (UTC time is 10 hours ahead of the local time in Hawaii) is presented in figure 3. This figure shows that the recorded and predicted time instants when the maximum surge height occurred at Port Allen agree with each other very well.

In addition, the numerical results show that the simulated surge heights based on the two different wind models (Holland and the modified Rankine wind models) are almost the same. When Hurricane Iniki made landfall near Port Allen on Kauai Island, the center of the hurricane was about 14 km away from Port Allen. Based on the pressure model (1), the atmospheric pressure at Port Allen was calculated to be 97660 Pa which should generate a static surge height of 0.37 m. This result indicates that the pressure drop induced by Hurricane Iniki is the dominant forcing factor in generating storm surge in Hawaii's coastal waters while the effect of wind as forcing is negligible here. One reason for the negligible wind effect can be that Hawaii does not have a continental shelf and water depth increases rapidly offshore. Another reason may be that Hurricane Iniki passed the Kauai Island relatively fast. If the hurricane had remained in the coastal waters of Kauai for a much longer time period, the wind effect could have been more significant. This issue as well as surge-tide interaction are currently being further studied and the results will be presented in a separate paper.

Table 1. Comparison between the Simulated and Recorded Maximum Storm Surge at Port Allen of Kauai Island in Hawaii, September 11, 1992.

Wind Models	Simulated maximum surge height (m)	Recorded maximum sea level (m)	Predicted maximum tide (m) (Sea Engineering, Inc. 1993)	Estimated wave set-up (m)	Corrected recording for storm surge (m)
Holland	0.38	1.89	0.64	0.80	0.45
Modified Rankine	0.43	1.89	0.64	0.80	0.45

CONCLUSION

The ADCIRC model was applied to simulate the storm surge generated by Hurricane Iniki (1992) in Hawaii. The simulated surge height was compared with the recorded water level at Port Allen in Kauai. The two were found to be in good agreement after the effects of tides and wave set-up were subtracted from the recorded data at Port Allen. These results show that the ADCIRC model can predict accurately storm surge generated by hurricanes in the central Pacific insular environment.

Two different wind input models, namely, the modified Rankine model and the Holland model were tested in the present study. The simulated surge heights by Hurricane Iniki based on these two wind models were found to agree with each other closely. Our results also show that the dominant forcing factor in generating storm surge during Hurricane Iniki is the drop in the atmospheric pressure while the wind has little effect. In addition, our study revealed that the hurricane-generated waves are much higher than the hurricane-generated storm surge during Iniki. These results are specifically applicable to the Pacific insular environment where coastal bathymetry is usually very steep. For storm surge generated by hurricanes over gradually sloped continental shelves, where hurricanes may remain for a much longer time period, the storm surge can be significantly higher.

ACKNOWLEDGEMENT

This study is funded by a grant from the NASA Office of Earth Science Award No. NAG5-8748 to the University of Hawaii at Manoa. Helpful discussions with the following colleagues and project team members are greatly appreciated: Professor Shi Jun Liao from the Shanghai Jiaotong University, Professor Philip Liu from the Cornell University, Yong Wei, Dr. Amal Phadke, and Raymond Rojas from the University of Hawaii at Manoa, Dr. Eiji Nakazaki from Sea Engineering, Inc. in Hawaii, and Chris Martino from the U.S. Navy Pacific Missile Range Facility in Hawaii. CPU allocation on Cray T90 from the San Diego Supercomputer Center (SDSC) is also gratefully acknowledged.

REFERENCES

Bode, L. and Hardy, T.A. 1997. Progress and Recent Development in Storm Surge Modeling. *Journal of Hydraulic Engineering*, 123(4), 315-331.

Fletcher, C.H., Richmond, B.M., Barnes, G.M. and Schroeder, T.A. 1995. Marine Flooding on the Coast of Kauai during Hurricane Iniki: Hindcasting Inundation Components and Delineating Washover. *Journal of Coastal Research*, 11(1), 188-204.

Grenier, R.R., Jr., Luettich, Jr., R.A. and Westerink, J.J. 1995. A Comparison of the Nonlinear Frictional Characteristics of Two-Dimensional and Three-Dimensional Models of a Shallow Tidal Embayment. *Journal of Geophysical Research*, 100(C7), 13719-13736.

Holland, G.J. 1980. An Analytical Model of the Wind and Pressure Profiles in Hurricanes. *Monthly Weather Review*, 108, 1212-1218.

Kolar, R.L., Westerink, J.J. Cantekin, M.E. and Blain, C.A. 1994. Aspects of Nonlinear Simulations Using Shallow Water Models Based on the Wave Continuity Equation. *Computers and Fluids*, 23(3), 523-538.

Liao, S., Wei, Y., Teng, M.H., Liu, P.L.F., Cheung, K.F., and Wu, C.S. 2001. Effects of Nonlinearity and Bottom Friction on Hurricane-Generated Storm Surge in Central Pacific Ocean. *Proceedings of WAVES2001*, ASCE, San Francisco, California, accepted.

Luettich, R.A. and Westerink, J.J. 1991. A Solution for the Vertical Variation of Stress, Rather than Velocity, in a Three-Dimensional Circulation Model. *International Journal for Numerical Methods in Fluids*, 12, 911-928.

Luettich, R.A., Jr., Westerink, J.J. and Scheffner, N.W. 1992. ADCIRC: An Advanced Three-Dimensional Circulation Model for Shelves Coasts and Estuaries, Report 1: Theory and Methodology of ADCIRC-2DDI and ADCIRC-3DL. *Dredging Research Program Technical Report DRP-92-6*, U.S. Army Engineers Waterways Experiment Station, Vicksburg, MS, 137p.

Martino, C.D. 2000. Modeling of Hurricane Waves in Hawaiian Waters. M.S. Thesis, University of Hawaii at Manoa, Honolulu, USA.

National Weather Service 1993. Hurricane Iniki, September 6-13, 1992. *Natural Disaster Survey Report*, US National Oceanic and Atmospheric Administration (NOAA).

Sea Engineering, Inc. 1993. Hurricane Iniki Coastal Inundation Modeling. *Technical Report prepared for the U.S. Army Corps of Engineers*, Pacific Ocean Division.

Westerink, J.J., Luettich, R.A., Baptista, A.M., Scheffner, N.W. and Farrar, P. 1992. Tide and Storm Surge Predictions Using Finite Element Model. *ASCE Journal of Hydraulic Engineering*, 118(10), 1373-1390.

Wu, J. 1982. Wind Stress Coefficients over Sea Surface from Sea Breeze to Hurricane. *J. Geophys. Res.*, 87, 9704-9706.

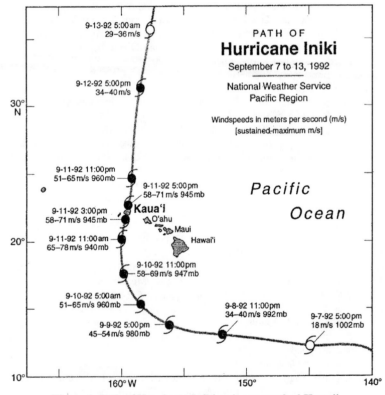

Figure 1. Path of Hurricane Iniki as it approached Hawaii.

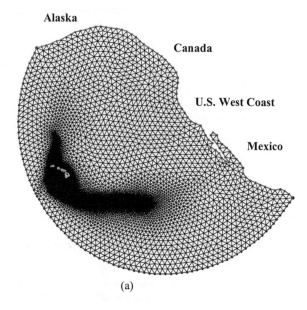

(a)

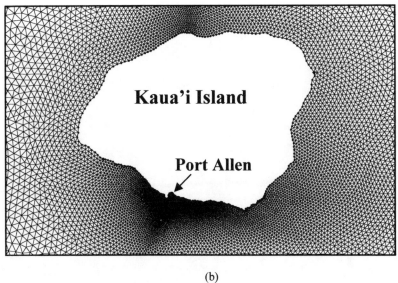

(b)

Figure 2. Computational domain (a) and finite element grids (b).

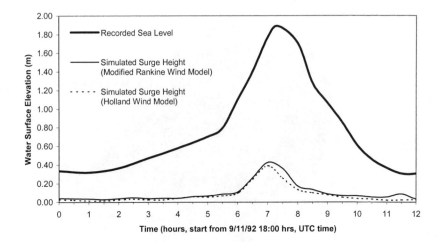

Figure 3. Comparison between the simulated surge height and the recorded sea water level at Port Allen, Kauai Island, Hawaii.

Effects of Natural Sea States on
Wave Overtopping of Seadikes

from

H. Schüttrumpf[1]
J. Möller[1]
H. Oumeraci[1]
J. Grüne[2]
R. Weissmann[3]

1. INTRODUCTION

The influence of natural sea states on wave overtopping constitutes one of the investigated aspects within the German research project „Loading of the Inner Slope of Seadikes by Wave Overtopping" (BMBF KIS 009). In fact, the incident wave spectrum is influenced by the very complex morphology in shallow coastal zones like wadden seas and shallow forelands. Therefore, the wave spectra in these areas are quite different from the commonly used theoretical single peak wave spectra. Wave run-up and wave overtopping are influenced by the shape of these wave spectra and can not be simply described on the basis of theoretical single peak wave spectra like PM, JONSWAP or TMA wave spectra.

The objective of this paper is to study experimentally the influence of different wave spectra on wave overtopping for seadikes by means of small and large scale model tests. For this purpose, model tests were run for the following boundary conditions: (i) Regular waves, (ii) TMA-spectra (single peak), (iii) JONSWAP-spectra (single peak and double

[1] Leichtweiss-Institute for Hydraulics; Beethovenstr. 51a; 38106 Braunschweig;
Fax: (+49)-5313918217; E-Mail: H.Schuettrumpf@tu-bs.de ; Ja.Moeller@tu-bs.de ;
H.Oumeraci@tu-bs.de
[2] Coastal Research Centre; Merkurstr. 11; 30419 Hannover Fax: (+49)-7629219;
E-Mail: gruene@fzk.uni-hannover.de
[3] Institute for Soil Mechanics and Foundation Engineering; Universitätsstr. 15; 45117
Essen; Fax: (+49)-2011832870; E-Mail: Roland.Weissmann@uni-essen.de

peak spectra composed of two superposed JONSWAP-spectra) and (iv) Measured single, double and multi peak spectra from different locations at the German and Dutch coast.

The investigated wave spectra were provided by Dipl.-Ing. H.D. Niemeyer (Coastal Research Station, Norderney) and Dipl.-Ing. J. Grüne (Coastal Research Centre, Hannover) for several locations at the German North sea coast, by Dipl.-Ing. H. Bergmann (LWI) for Probstei / Baltic Sea, by Dipl.-Ing. T. Trampenau (LWI) for Warnemünde / Baltic Sea, by ir. John G. de Ronde (Rijkswaterstaat) for several locations in the Netherlands. Their contributions are gratefully acknowledged. The support of the german ministery of education and research (BMBF) within the project BMBF-KIS 009 is also gratefully acknowledged.

2.   OVERTOPPING FORMULAE

The crest height of seadikes and other coastal structures is a function of the design water level and the relevant wave run-up height $R_{u2\%}$ which is exceeded by two percent of the incoming waves. If the crest height of a seadike is not high enough, wave overtopping occurs. Wave overtopping can not be avoided due to uncertainties in the prediction of the design water level and the incoming wave parameters. Therefore, wave overtopping is considered as an important input for the design of coastal structures and flood defences.

Various types of formulae to calculate the average overtopping rate q exist (see Table 1). Most dimensionless equations use an exponential function $Q_* = Q_0 \exp(-b\,R_*)$ or a relationship of the type $Q_* = Q_0 (1 - R_*)^b$ for the calculation of the average overtopping rate q. Within these formulae R_* and Q_* are dimensionless factors describing the dimensionless overtopping rate and the dimensionless freeboard, respectively. For a freeboard $R_C = 0$ ($\Rightarrow R_* = 0$) the coefficient Q_0 is equal to the dimensionless overtopping rate Q_* and for a large freeboard there is no overtopping ($R_* \to \infty \Rightarrow Q_*=0$).

For the design of seadikes in Germany, EAK A2 (2001) (guidelines for the design of coastal structures in Schüttrumpf & Oumeraci, 2000) recommend the formulae by Van Der Meer et al. (1995, 1998) with the restriction, that this method can only be applied for single peak wave spectra:

For $\xi_d < 2.0$
$$\frac{q}{\sqrt{g\,H_S^3}} \sqrt{\frac{H_S/L_0}{\tan\alpha}} = 0.06 \exp\left(-5.2 \frac{R_C}{H_S} \frac{\sqrt{H_S/L_0}}{\tan\alpha}\right) \qquad (1)$$

For $\xi_d \geq 2.0$
$$\frac{q}{\sqrt{g\,H_S^3}} = 0.2 \exp\left(-2.6 \frac{R_C}{H_S}\right) \qquad (2)$$

with: L_0 $= (g\, T_p{}^2)/(2\pi)$, T_P = peak period of the spectrum
        H_S = significant wave height; R_C = freeboard
        $\tan \alpha$ = slope; ξ_d = surf similarity parameter

$$\xi_d = \frac{\tan \alpha}{\sqrt{H_S/L_0}} \tag{3}$$

Schüttrumpf (2001) reported that the existing overtopping models for average overtopping rates by Van Der Meer et al. and also Van Gent (1999) are not valid for the boundary conditions $R_C=0$ and $R_C=R_{u,\,max}$.

Therefore, Schüttrumpf (2001) conducted model tests with no freeboard ($R_C=0$) and without overtopping ($R_C>R_{max}$) and derived the following formulae:

$$\frac{q}{\sqrt{2\, g\, H_S^3}} = 0.038 \cdot \xi_d \cdot \exp\left(-5.5\, \frac{R_C}{R_{u,2\%}} \right) \quad \text{for } \xi_d < \xi_{gr} \tag{4}$$

$$\frac{q}{\sqrt{2\, g\, H_S^3}} = \left(0.096 - \frac{0.160}{\xi_d^3} \right) \cdot \exp\left(-5.5\, \frac{R_C}{R_{u,2\%}} \right) \quad \text{for } \xi_d \geq \xi_{gr} \tag{5}$$

with: ξ_{gr} = 2.0
        $R_{u,\,2\%}$ = run-up height exceeded by 2% of the incident waves ($=1,5\cdot\xi_d\cdot H_S$)
        ξ_d = surf similarity parameter (here calculated with T_m)

A comparison of both overtopping models to the measured average overtopping rates will be performed in the following.

3.   MODEL-SET-UP

Model tests have been performed in model and prototype scale. Small scale model tests were performed in the small wave flume of the LEICHTWEISS-INSTITUTE (LWI) for Hydraulics of the Technical University of Braunschweig. The flume is about 100m long, 2m wide and 1.25m deep. The flap type wave paddle is capable to generate regular and irregular waves with wave heights up to 0.25m and wave periods between 1.5s and 6.0s in 0.60m to 0.80m water depth. A smooth and impermeable 1 on 6 dike slope was tested. The water depth was kept constant (d=0.70m) for all model tests resulting in a constant freeboard of $R_C=0.10m$ (Fig. 1). The width of the crest was kept constant B = 0.30m.

The incoming wave was measured by 10 wave gauges. The measured wave field was separated in incident and reflected wave field by using a 3 gauge procedure (Mansard and Funke, 1980). The overtopping volume was measured by an overtopping tank mounted on three weighing cells. This system is capable to measure individual overtopping volumes as well as average overtopping rates. A more detailed description of the small scale model

Fig. 1: Model Set-up for the 1:6 Dike Slope in the LWI-wave flume
- View from the seaward slope.

tests is given by Oumeraci et al. (1999).Prototype tests were performed in the Large Wave Channel (GWK) of the Coastal Research Centre FZK, Hannover. The wave flume has a length of 324m, a width of 5.0m and a depth of 7.0m. Regular waves up to a height of 2.0m and wave spectra up to a significant wave height of 1.4m were generated.

Within these tests a dike with an impermeable 1 on 6 slope on the seaward side and a 1 on 3 slope on the landward side was investigated. The crest of the dike was 2.0m wide. The dike height was 6.0m. These dimensions correspond to a model scale of about $1:N_L = 1:1$ to 1:2.5. The investigated dike geometry is typical for seadikes at the German North Sea coast. Figure 2 shows a cross-section of the dike in the wave flume and photos of the incident waves, the up-rushing waves on the seaward slope and the subsequent overtopping flow on the inner slope.

The objective of these tests was to obtain detailed information about the hydrodynamic processes on the seaward slope, the crest and the inner slope of the dike profile. In order to analyse these processes the following parameters were measured:

- wave parameters in the flume and on the dike,
- pressures on the dike surface (seaward slope, crest and inner slope),
- velocities on the dike surface (seaward slope, crest and inner slope),
- wave run-up heights,
- layer thicknesses (seaward slope, crest and inner slope) and
- overtopping volumes.

A detailed description of the prototype tests is given by Oumeraci et al. (2001). The incident wave field and the characteristic overtopping parameters have been determined in exactly the same way as for the small scale model tests. Therefore, no differences in the methodology for data analysis exist.

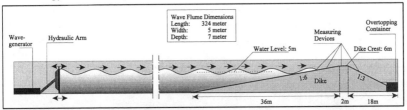

Large Wave Flume, Hannover

incident waves   wave run-up (seaward slope)   wave overtopping (inner slope)

<u>Fig. 2:</u> Model Set-up in the Large Wave Flume (GWK)

4. TEST PROGRAMME

Existing design concepts for wave overtopping are usually based on investigations with standard wave spectra (TMA, JONSWAP, PM, etc.). As already mentioned, the incident wave spectra at the German coasts are generally influenced by the complex morphology of shallow zones like wadden seas and shallow forelands. In previous publications the effect of different wave characteristics on wave run-up already was pointed out (Grüne, 1982). Thus, the influence of these mostly double or multi peak wave spectra on wave overtopping at seadikes should be investigated. Regular waves and different wave spectra (standard wave spectra and wave spectra from the field) are used for the model investigations.

Theoretical (single peak) spectra:
TMA spectra are the most frequently used wave spectra for shallow waters. In the past, many model tests on wave overtopping were performed with TMA spectra. For comparison of the results with existing design formulae and in order to analyse the differences between standard and natural wave spectra, JONSWAP and TMA spectra were applied for these model tests.

Natural sea spectra:

To analyse the influence of the shape of a wave spectrum on wave overtopping, different wave spectra, recorded at several locations in the field, were included in the test programme. The natural wave spectra are generally multi-peaked in comparison to the theoretical wave spectra (TMA) with a single peak. Multi-peak spectra are typical for the German North Sea coast (Fig. 3). Some typical spectra and the corresponding field locations along the coasts are given in Figure 4.

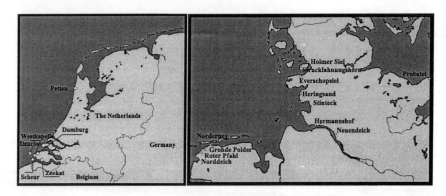

Fig. 3: Overview of Field Locations where Wave Spectra were measured.

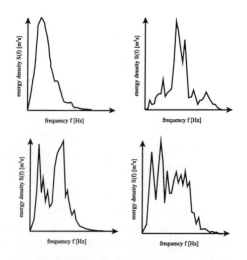

Fig. 4: Typical measured wave spectra in wadden
seas and over shallow forelands

5.   OVERTOPPING ANALYSIS AND RESULTS

The overtopping analysis in this paper will be limited to the the model by Van Der Meer et al. (1995, 1998) and the model by Schüttrumpf (2001) because these two models were found to be the most appropiate in former analyses (Oumeraci et al. 2001).

a) Overtopping Model by Van Der Meer et al. (1995, 1998)

The results from the overtopping tests at LWI (Oumeraci et al., 1999) with natural sea spectra have shown that the overtopping model by Van Der Meer fits well for theoretical single peak wave spectra, but not for natural wave spectra with more than one peak. Nevertheless, this formula is still the most commonly used method for the calculation of the average overtopping rate q on seadikes. Therefore, this method is checked first for the overtopping data from the large scale model tests and also for the data from the small scale model tests at LWI. The comparison for the large scale experiments is drawn in Fig. 5. It shows the formulae by Van Der Meer et al. and the measured data from the model tests and also the confidence intervals of the Van der Meer formula. For TMA spectra the model fits the data relatively well, but for natural wave spectra with more than one peak there is a large scatter. Former analysis has shown that this scatter is mainly due to the use of the peak period T_p, which is not appropriate to describe natural wave spectra having more than one peak. Therefore, other wave parameters will be used for further analysis.

b) Overtopping model by Schüttrumpf (2001)

Analysing the small scale model tests at LWI for natural wave spectra and also for theoretical wave spectra it was found, that the model by Schüttrumpf fits the data better than the model by Van Der Meer. Since T_p was not suitable for the description of wave overtopping in the case of multi-peak wave spectra, Schüttrumpf (2001) suggested to use the mean wave period T_m (as already recommended for wave run-up by Grüne & Wang, 2000 from results of field and large-scale investigations) and the wave height H_{m0}. These parameters are more appropriate for the analysis of natural wave spectra. Figure 6 shows the overtopping data calculated with the overtopping model by Schüttrumpf for TMA spectra and natural wave spectra. The reliability of the overtopping function is given by taking the b-coefficient (b=-5.5/1.5=-3.67) as a normally distributed stochastic variable with an average of -3.67 and a standard deviation of 0.55. Considering this standard deviation the confidence bands are calculated and also plotted in Figure 6.

The dimensionless overtopping parameters show some scattering for natural wave spectra. Nevertheless, the overtopping model by Schüttrumpf fits much better than the model by Van Der Meer et al. and will be used for further analysis. All data will be analysed in more detail in the following for H_{m0} and T_m.

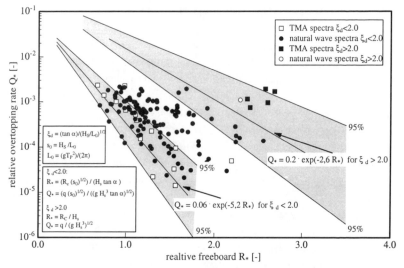

<u>Fig. 5</u>: Comparison of the measured Data (Large Scale Model Tests, GWK) with the Overtopping Model by Van der Meer et al. (1995 and 1998)

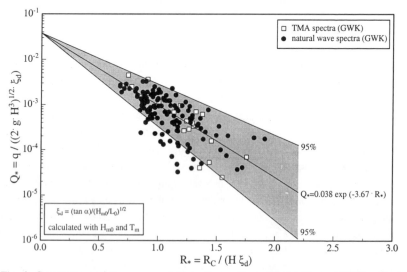

<u>Fig. 6</u>: Comparison of the measured Data (Large Scale Model Tests, GWK) with the Overtopping Model by Schüttrumpf (2001)

Detailed Analysis of natural wave spectra

As mentioned above, the natural wave spectra collected at the German and Dutch coast are generally multi peak wave spectra with a complex shape. In a first step, the wave spectra will be divided into wave spectra that are measured in wadden seas (e.g. Norddeich) and in spectra from open coasts (e.g. Norderney) (Fig. 3).

In Figure 7 the dimensionless overtopping rates for the natural wave spectra are shown. For comparison the results from the large scale model tests and from the small scale model tests in the wave flume of LWI are plotted. In this figure it is shown that there is an influence of the field location of the wave spectrum in nature and therefore of the shape of the wave spectrum. Wave spectra from the open coasts result in higher overtopping rates than wave spectra measured in wadden seas. Therefore, different overtopping functions were found for wave spectra from open coasts and wave spectra from wadden seas. The more generic overtopping function developed by Schüttrumpf lies between that for open coasts and that for wadden seas. The generic formula by Schüttrumpf can be used for all wave conditions. The standard deviation σ is quite equal for all formulas. The overtopping formulae only differ in the b-coefficient For high relative freeboards the formula for open coasts leads to a higher relative overtopping rate than the generic formula and the formula for wadden seas leads to a smaller relative overtopping rate than the generic overtopping function. This fact can be explained by the shape of the wave spectra in wadden seas which are strongly influenced by wave breaking. The generic function for wave overtopping can be used for all natural wave spectra because it fits relatively well for all data.

6.   SUMMARY AND CONCLUSIONS

Wave run-up and wave overtopping investigations have yet been based on theoretical single peak wave spectra. Field measurements in wadden seas and foreland areas show that wave spectra are composed by the very complex processes due to wind, swell, wave transformation and wave breaking and can not be simply described by single peak wave spectra. Therefore, the peak period T_p is of no use anymore for the description of wave run-up and wave overtopping. A new overtopping model has been derived in this paper based on measured wave spectra and the mean period T_m which provides overtopping formulas for open coasts as well as for wadden seas. By this way, an overtopping formula has been derived which takes into account the morphological conditions in front of the dike.

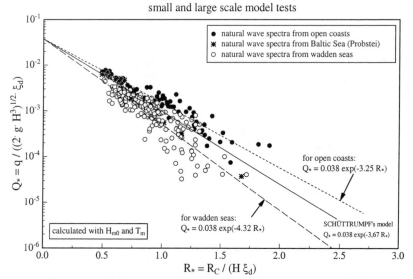

Fig. 7: Dimensionless Overtopping Parameter splitted in Natural Wave Spectra from Wadden Seas and Estuaries (e.g. Everschopsiel), from the Baltic Sea (e.g. Probstei) and from open Coasts (e.g. Norderney) calculated with the Model by Schüttrumpf (2001)

References

GRÜNE, J. (1982): Wave run-up caused by natural storm surge waves. Proc. 18th Intern. Conf. on Coastal Engineering (ICCE´82), Cape Town, South Africa..

GRÜNE, J.; WANG, Z. (2000): Wave run-up on sloping seadykes and revetments. Proc. 27th Intern. Conf. on Coastal Engineering, (ICCE´2000), Sydney, Australia.

MANSARD, E.P.D.; FUNKE, E.R.: The measurement of incident and reflected spectra using a least squares method. Proceedings International Conference Coastal Engineering (ICCE); Vol. 17, No. 1; pp. 154-172, 1980

OUMERACI, H.; SCHÜTTRUMPF, H.; SAUER, W.; MÖLLER, J.; DROSTE, T. (1999) Physical Model Tests on Wave Overtopping with natural Sea States - 2D Model Tests with single, double and multi-peaked Wave Energy Spectra. LWI-Report No. 852

OUMERACI, H.; SCHÜTTRUMPF, H.; MÖLLER, J.; KUDELLA, M. (2001) Large Scale Model Tests on Wave Overtopping. LWI-Report No. 858

SCHÜTTRUMPF, H.; OUMERACI, H.: EAK-Empfehlungen A2 - Wellenauflauf und Wellenüberlauf (Kurzfassung) Hansa - Schiffahrt - Schiffbau - Hafen. 2000 (in German)

VAN DER MEER, J.W. and JANSSEN, J.P.F.M.: Wave Run-up and Wave Overtopping at Dikes. In: Wave Forces inclined and Vertical Structures. Ed. Z.DEMIRBILEK. pp. 1-27, 1995

VAN DER MEER, J.W., TÖNJES, P. and DE WAAL, J.P.: A code for dike height design and examination. Proceedings Int. Conf. on Coastlines, Structures and Breakwaters. (Ed. N.W.H. Allsop) Thomas Telford, London, 1998

VAN GENT, M.R.A.: Physical Model Investigations on Coastal Structures with Shallow Foreshores. 2D model tests with single and double-peaked wave energy spectra. Delft Hydraulics. Report H3608, 1999

WAVE RUN-UP AND OVERTOPPING OF SEA DIKES: RESULTS FROM NEW MODEL STUDIES

Jimmy Murphy[1], Holger Schüttrumpf[2] and Tony Lewis[3]

Abstract: Sea dikes are used as coastal defence structures in many countries worldwide and have been subject to rigorous study. However most of the previous work and indeed the resultant design formulae for wave run-up and overtopping have been based on 2D testing and field measurements. This current study combined both 2D and 3D tests at different scales such that the influence of such parameters as wave direction and directional spreading could be examined. The results show that these parameters have a significant influence on both the run-up heights and overtopping volumes.

INTRODUCTION

Knowledge of wave run-up and overtopping magnitudes is important to the design of many coastal structures. Very often a structure's crest level is designed by use of formulae that calculate run-up exceedence levels or expected overtopping volumes for specified wave conditions. Such formulae are empirically based with their derivation relying on the results of physical model tests. Normal model tests for this application are carried out in wave flumes where the test section is subject to direct long crested wave attack. The influence of such parameters as wave directionality and directional spreading which are often representative of real sea conditions cannot be examined in wave flumes.

There has not been a great deal of research on the influence that these two parameters have on wave run-up and overtopping. In TAW (1974), wave obliqueness was included as

1 Senior Engineer, Hydraulics and Maritime Research Centre, Department of Civil and Environmental Engineering, University College Cork, Cork, Ireland, jm.hmrc@ucc.ie
2 Senior Engineer, Federal Waterways Engineering and Research Institute, Wedeler Landstr. 157, 22559 Hamburg, Germany, schuettrumpf@hamburg.baw.de
3 Senior Lecturer, Department of Civil and Environmental Engineering, University College Cork, Cork, Ireland, tlewis@ucc.ie.

one of a number of run-up reduction factors to be applied when conditions differ from direct attack on a smooth impermeable slope. A general equation was thus proposed for determining run-up heights and is given below as,

$$\frac{Z_{u2\%}}{H_s} = 1.6\gamma\xi_p \text{ with a maximum of } 3.2\gamma \tag{1}$$

where $Z_{u2\%}$ is the run-up level exceeded by 2% of waves, H_s is the significant wave height at the toe of the structure, γ is the reduction factor and ξ_p is the Iribarren number given by $tan\alpha/\sqrt{2\pi H_s/gT_p^2}$, α is the slope angle, g is the acceleration due to gravity and T_p is the peak wave period.

Other research has shown an increase in wave run-up and wave overtopping for low angles of wave attack (up to 20 degrees). The model tests by Owen (1980), Tautenhain (1982) and Van der Meer (1995) have shown an increase in the wave run-up height of about 10% as compared to normal wave attack for wave directions between 10 and 20 degree. It was unclear from these studies, whether this increase was real or influenced by model effects.

The objective of the tests carried out in this study was to investigate the influence of the wave directionality and directional spreading on wave run-up and overtopping. The tests were performed as part of the EU-MAST funded OPTICREST (OPTImisation of CREST level design of sloping coastal structures through prototype monitoring and modelling) project. The 3D tests were undertaken in a wave basin at the Hydraulics and Maritime Research Centre (HMRC), Cork, Ireland whilst 2D tests were carried out in a wave flume at the Leichtweis Institute (LWI) in Braunschweig, Germany. This paper will concentrate on the 3D tests but the 2D results will be referenced.

TEST FACILITY/MODEL SET-UP

The run-up and overtopping tests were undertaken in the Ocean Wave Basin at the HMRC. This basin is shown in Figure 1 and has dimensions 25m x 18m x 1m. Waves are generated by means of a multi segment wave generator consisting of 40 individually controlled wave paddles with active wave absorption. A wide variety of sea states can be simulated ranging from simple monochromatic waves to complex wave spectra which can be either short crested of long crested.

In the OPTICREST test programme three different sets of model tests were undertaken in the Ocean wave basin corresponding to,

a. Petten dike

This model was used to compare laboratory and prototype run-up on the Petten Sea Defence in the Netherlands. See Van Gent et al. (2001) for a detailed description of the work undertaken.

b. 1 in 6 low crested dike

A low crested dike with a slope of 1 in 6 was constructed after completion of the Petten model tests. This dike had a freeboard of 0.05m and was used for overtopping tests. It corresponded to a 1:2 scale of a model tested in a wave flume in Braunschweig University. The toe of the structure was placed in 1m water depth at a distance of 5.5m from the paddles and it extended over the 1:10 slope of the Petten model. The width of the model at toe level was 13m and it reduced uniformly to 5m at the crest level. Tapering the model in this manner was considered necessary in order to ensure a smooth transformation of oblique waves as they propagated across the slope. The base width was designed taking the directionality of the incoming waves into account. Wave absorbing material was placed at either side of the model to reduce the effects of wave reflections. Two collecting containers were placed directly behind the structures crest. These containers were designed to collect and store all the water that overtopped the crest at two different locations along the crest.

c. 1 in 6 high crested dike

This model was constructed by merely extending the low crested dike such that it had a freeboard of 0.35m. This enabled a series of run-up tests to be carried out. The extended section had had a 5m width from a vertical distance of 0.05m above the still water level to the structure crest. As for the overtopping tests portable wave absorbers were placed along the sides of the 1 in 6 slope in order to minimize wave reflections. A step gauge was designed and fabricated in house at the HMRC and was used to measure run-up. The gauge consists of 24 pairs of parallel conductivity probes.

To measure wave conditions a multi probe array was located at the toe of the structure as shown in figure 1. This multi probe array and analysis software were designed and developed by Aalborg University to measure three-dimensional waves in laboratory basins. It is suitable for use in tests with shortcrested, directional sea states and can also separate incident and reflected waves.

Figure 1 below shows the high crested dike model layout but details of the other two model set-ups can be seen from this diagram.

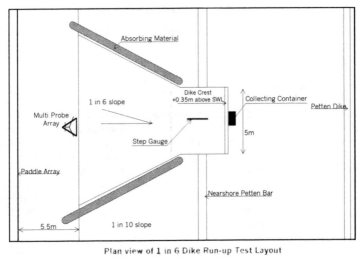

Plan view of 1 in 6 Dike Run-up Test Layout

Sectional View

Figure 1 Dike (1 in 6) Model Layout with Instrumentation Locations

TEST CONDITIONS

The test matrix consisted of a comprehensive set of experiments incorporating a range of parameters including prototype storms at Petten and tests conditions used by Braunschweig University at the 1:1 scale. A summary of the various test series is provided below.

- Series 1 Petten Storms
- Series 2 Direct Long Crested Waves
- Series 2 Oblique Long Crested Waves
- Series 4 Direct Short crested waves
- Series 5 Oblique Short-Crested waves

Apart from the Petten storms the spectral shape used to define the input wave conditions corresponded to that of the JONSWAP spectrum. The table below shows the range of variation in the test wave parameters.

Wave Parameter	Range	Step
H_s	0.04 – 0.1m	0.02m
T_p	1.06 – 2.36sec	approx 0.35sec
Direction	0 – 25 degrees	5 degrees
Spreading Index (n)	1, 5, 7, 10, 15, 25	

Table 1 List of Test Wave Parameters

The application of a spreading function to a specified wave spectrum results in the spectral energy being distributed over a range of directions. The particular spreading function applied in the case of this study was, $\cos^{2n}(\theta/2)$. The factor n (or s as used in other studies) defines the range of angles over which the incident energy is spread. As the value of n increases the amount of spreading decreases until the waves eventually become longcrested.

Figure 2 Wave Run-up Test

RUN-UP TEST RESULTS

Figure 3 and 4 presents the data from test series 2-4 as were described above. The dimensionless run-up is plotted on the y-axis and the Iribarren Number (surf similarity parameter) on the x-axis. If the results from longcrested waves incident normal to the structure are first considered (N/S in figure 4 and 0 deg in figure 5) it can be seen that the points can be described by the standard design formula for calculating run-up on smooth slopes as given in equation 1. It is important that these results fit existing formulae data as it gives greater confidence in the reliability of subsequent results.

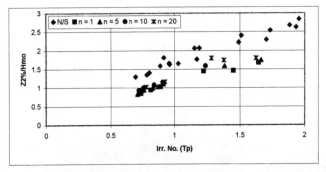

Figure 3 Test Results showing influence of Directional Spreading

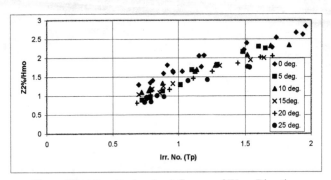

Figure 4 Test Results showing influence of Wave Direction

From figure 3 it can be seen that the application of a spreading factor causes a significant reduction in the run-up levels as compared with the longcrested situation. In general the trend is for lower spreading index values to give higher reductions in wave run-up, although over the range of values tested the differences are not great. Run-up reduction factors were calculated for each *n-value* by dividing the dimensionless run-up values by constants such that they were brought in line with the longcrested results The results show that the maximum reduction in wave run-up for waves spread about a direction normal to the structure is about 31%. It should be noted that tests carried out on the Petten model showed only minor reductions in run-up as a result of the application of spreading (see Van Gent et al. 2001). One reason why this parameter was not considered to be as influential in the Petten case was due to the presence of a shallow foreshore, which had the effect of reducing the level of spreading in front of the structure.

Figure 4 shows that run-up decreases with increasing wave obliquity for similar values of the Iribarren number. Run-up reduction factors were determined for each angle of incidence in the same manner as for the directional spreading tests and the results of this analysis are shown in figure 5.

Figure 5 Variation of Run-up Reduction factor with Wave Angle

Over the range of values tested, which are relatively low levels of obliquity, the following simple linear relationship can be used to define the magnitude of the run-up reduction factor,

$$\gamma_{dir} = 1 - 0.0095(\theta) \qquad \text{where } \theta \text{ is the angle of incidence} \qquad (2)$$

What this formula is stating is that a 1% reduction in run-up occurs per degree of wave obliquity. Obviously this relationship is not sustainable over a wide range of angles and it would be expected that the lower limit for run-up would approximate to the wave height at the toe of the structure.

The final part of the run-up analysis concerns test series 5 which is the combined application of wave directionality and directional spreading. Run-up reduction factors were calculated in a manner similar to that described previously. The results show that there is a greater reduction to run-up due to the combined effects of wave obliqueness and spreading than due to either individually. A maximum reduction in the dimensionless run-up of about 45% was obtained. It was found that by multiplying the reduction factors determined for the individual directional spreading and wave directionality run-up tests that the result would approximately correspond to the combined factor for the same conditions. The following example illustrates this point,

Direction only: $\theta = 10^0 \ \gamma_{dir} = 0.885,$
Spreading only: $n = 10 \quad \gamma_{spr} = 0.769$ $\qquad$ $\gamma_{dir}\gamma_{spr} = 0.68$

Combined: $\theta = 10^0, n = 10 \quad \gamma_{com} = 0.654$

OVERTOPPING TEST RESULTS

The results from the wave overtopping tests showed the same general trend as the wave run-up results. Both directional waves and shortcrested seas caused a reduction in the overtopping volumes as compared with the longcrested direct wave situation. Only results from oblique wave tests are presented as similar trends were obtained when directional spreading was applied. Figures 5 and 6 below plot all measurements in dimensionless form. The dimensionless factors Q_* (dimensionless discharge volume) and R_* (dimensionless freeboard parameter) were taken from the Van der Meer formula:

$$Q_* = \frac{q}{\sqrt{gH_S^3}} \sqrt{\frac{H_S/L_0}{\tan\alpha}} = Q_0 \exp\left(-\frac{b}{\gamma_\theta} R_*\right) = Q_0 \exp\left(-\frac{b}{\gamma_\theta} \frac{R_C}{H_S} \frac{\sqrt{H_S/L_0}}{\tan\alpha}\right) \qquad (3)$$

where q is the average overtopping rate, α is the slope angle, L_0 is the deep water wave length, b and Q_0 are dimensionless coefficients, R_C is the freeboard and γ_θ is a correction coefficient for oblique wave attack. Figure 6 shows that there is relatively good agreement between longcrested direct wave test results. The HMRC values are in general slightly higher than the 2D test results but are within the range of normal variations about the main trend. Figure 7 shows that in general that overtopping volumes reduce as the angle of wave incidence increases.

To determine the correction factor, Q_0 was determined for all model tests with $0°$. Than, the factor b was calculated for each wave direction with $Q_0=0.0296=$const. The correction factor γ_θ was calculated in the next step by,

$$\gamma_\theta = \frac{b(\theta = 0°)}{b(\theta)}$$

By this way, the influence of wave direction on wave run-up was determined for long-crested waves. The results are given in table 2 and indicate the magnitude of reduction in the overtopping volumes for the range of conditions tested. In addition, the standard deviations and coefficients of determinations are also given in table 2 for the coefficient b.

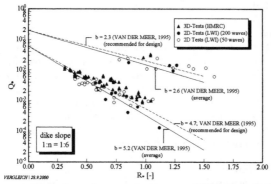

Figure 6 Comparison of test results from two different test facilities

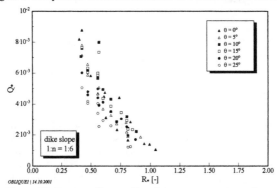

Figure 7 Influence of wave direction on overtopping volumes

θ	b	σ	r^2	γ_θ
[°]	[-]	[-]	[-]	[-]
0	2.79	±0.071	0.91	1.000
5	2.80	±0.062	0.97	0.996
10	2.80	±0.051	0.78	0.996
15	2.78	±0.049	0.95	1.004
20	3.24	±0.055	0.91	0.861
25	3.85	±0.069	0.91	0.725

Table 2: Influence of wave direction on wave overtopping for long crested waves

It can thus be concluded, that no increase in wave overtopping occurs for long crested waves. This conclusion was also found by an independent investigation at NRC / Canada (Möller et al., 2001).

CONCLUSIONS

This study has shown that wave directionality and directional spreading can have a significant influence on wave run-up heights and overtopping volumes. The general indication is that the magnitudes of these parameters are reduced which is contrary to some of the previous research in this area. However the results of this study would not be regarded as being the definitive statement on this topic but just another step to achieving a better understanding as to the nature of wave interactions with coastal structures. Needless to say much more research is required on this subject particularly in relation to 3D model tests and field monitoring of prototype structures.

ACKNOWLEDGEMENTS

The research presented in this paper was undertaken within the European Union funded research project MAST-OPTICREST (contract MAS3-CT97-0116)

REFERENCES:

Möller, J.; Ohle, N.; Schüttrumpf, H.; Daemrich, K.F.; Oumeraci, H.; Zimmermann, C. 2001. Einfluss der Wellenangriffsrichtung auf Wellenauflauf und Wellenüberlauf (Influence of the wave direction on wave run-up and wave overtopping) *Proc. 3rd FZK-Kolloquium*. Hannover. Germany

Owen, M.W. 1980. Design of seawalls allowing for wave overtopping, *Report No. EX924*, Hydraulics Research, Wallingford.

TAW, 1974. Technical Advisory Committee on Protection against Inundation. Wave Run-up and Overtopping. Government Publishing House, The Hague, The Netherlands

Tautenhain, E. et al, 1982. Wave run-up at Sea dikes under oblique Wave Approach. *Proceedings Coastal Engineering*. pp 804-810

Van der Meer, J.W. et al. 1990. Waterbeweging op taluds. Invloed van berm, rewheid, ondiep voorland en scheve lang-en kortkammige golfaanval *Delft Hydraulics H. 1256*

Van der Meer, J.W. 1993. Conceptual Design of Rubble Mound Breakwaters, *Delft Hydraulics, publication number 483*

Van der Meer, J.W. et al 1995. Wave Run-up and Overtopping at Dikes. *ASCE book on "Wave Forces on inclined and vertical wall structures"* Ed. Z. Demirbilek

Van Gent et al. 2001. Field Measurements and Laboratory Investigations on Wave Propagation and Run-up. *Proceedings of Waves 2001*, San Francisco, USA.

Estimating Overtopping Impacts in Los Angeles/Long Beach Harbors with a Distorted-Scale Physical Model

David D. McGehee, P.E., M.ASCE,[1] Chuck Mesa, P.E.[2], and Robert D. Carver, M.ASCE[3]

ABSTRACT: Recent topographic surveys of the breakwaters for Los Angeles/Long Beach Harbors revealed an average crest elevation of the San Pedro Breakwater of approximately 1.5 ft below the design crest of + 14 ft MLLW. A physical model study was conducted to estimate future impacts on harbor operations if crest loss continues or if the crest is restored to the design elevation. A 3-dimensional distorted-scale physical model (1:100 vertical, 1:400 horizontal) was used to investigate the change in total energy and low frequency wave energy (periods > ~85 sec) at numerous locations within the harbor as a function of the breakwater crest elevation. Constraints of the distorted scale model were overcome to generate incident overtopping waves that induced low frequency oscillations in the harbor. Sensitivity of these oscillations to crest elevation is presented.

INTRODUCTION

The San Pedro breakwater is the westernmost component of the three breakwaters that protect the Los Angeles/Long Beach Harbor (LA/LB) complex. A survey of the entire three-breakwater system was recently performed (USACE 1996) that revealed the San Pedro Breakwater is between 1 to 3.5 ft below the + 14 ft MLLW design elevation. The causes are not known at this time, nor is it known whether settlement will be progressive. However, the protection level afforded to the harbor is reduced, relative to the design elevation. Of particular concern is the potential for impacts on Port operations from increased long wave energy levels at various terminals inside the harbor. The present study was designed to address two questions:

Principal, Emerald Ocean Engineering, 107 Ariola Drive, Pensacola Beach, Fl 32561,

Coastal Engineer, US Army Engineer District, Los Angeles, 911 Wilshire Blvd, Los Angeles, CA 90017-3401 Attn: CESPL-ED-D

Research Hydraulic Engineer, US Army Engineer Research and Development Center, 3909 Halls Ferry Rd., Vicksburg, MS 39180,

1. Does the present condition of the breakwater result in significantly more low frequency energy in the harbor, relative to a structure at the design crest elevation?
2. What are potential future impacts on low frequency energy in the harbor if crest elevation loss continues?

The amplitude of incident long waves at the prototype scale is negligible relative to the breakwater dimensions, so it is expected that the crest elevation will have little or no effect on direct transmission of low frequency wave energy until the crest approaches the still water level. Overtopping of the breakwater by wind waves is common during high energy sea states, and the amount of wave energy transmitted into the harbor by overtopping is a function of crest elevation. Low frequency wave energy in the harbor has been strongly correlated with wind wave energy offshore (Seabergh, 1999), so it is reasonable to question if increased overtopping wind waves will affect low frequency wave energy in the harbor

A 3-d physical model of the LA/LB has been in use at the US Army Engineer Research and Development Center (ERDC) since 1973 to investigate many effects of harbor modification on tidal circulation and harbor resonance (Seabergh, 1993). Previous harbor resonance studies utilizing this model produced low-amplitude long waves with the wave generator that propagate throughout the harbor. In the present study it was desired to generate energetic wind wave spectra capable of overtopping the breakwater, and measure the resulting low frequency wave energy in the harbor as a function of various crest elevations. Due to model scale and wave generator limitations, this study required developing a testing program that could adapt the model to a purpose not intended in its original design.

METHODOLOGY
Model Description
The Los Angeles and Long Beach Harbors model is molded in concrete grout at a vertical scale of 1:100 and a horizontal scale of 1:400 and reproduces San Pedro Bay and the Pacific Ocean seaward of the harbor out to the -91.5-m (-300-ft) MLLW contour. Model wave data were collected at two sites between the wave paddle (Gages 21 and 24) and the harbor, and at 28 locations throughout the harbor (Figure 1).

The LA/LB model breakwater was originally constructed with an impermeable crest at an elevation of +14 ft MLLW (prototype), made from concrete cap segments, placed on top of a permeable rubble base. A survey conducted at the beginning of the study showed the model crest elevation ranged between +13 to +14 ft MLLW, indicating some settlement had occurred over the years. The first two test series were run using this construction method. To test a wide range of crest elevations more efficiently, the concrete segments were removed and replaced with segments made from high-density, water resistant clay. The clay crest segments were easy to shape for any desired crest elevation and could be placed rapidly by hand.

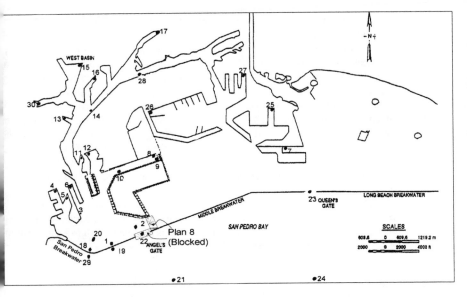

Figure 1. Model layout with wave gage locations

Constraints

In all previous studies of harbor resonance in the LA/LB model, non-overtopping long waves were generated by the wave generator. The ideal plan would generate scaled wind waves energetic enough to overtop the breakwater, and measure the resulting long wave energy in the harbor for various crest elevations.

Seabergh (1993) describes how similitude is obtained in a distorted model by assuming the shallow water approximation for celerity. Using this assumption the effects of the distortion on refraction for waves with periods over 85 sec (prototype) can be neglected, and the time scale can be shown to be 1:40 for long waves, with an error of under 1 per cent. The error in this assumption increases with increasing depth or decreasing wavelength, and would be quite large for the wind wave portion of the spectrum. Physical limitations of the wave generator rendered this potential distortion error moot. Stroke and command signal limitations made the generator incapable of generating high frequency waves with sufficient height to overtop the structure. Preliminary tests determined that the lower limit of wave periods for waves high enough to overtop (O 10 ft prototype scale wave height) was 62.8 sec at prototype scale. Such high energy-long period overtopping waves are extremely unlikely to occur naturally. The decision to proceed with the tests within these constraints was based on the following considerations and assumptions:

1. The incident waves will introduce wave energy into the harbor that varies with breakwater crest elevation.

2. Initial tests confirmed that these incident waves, just as wind waves in the prototype, produced oscillations in harbor basins at resonant frequencies well below the incident wave frequencies - frequencies consistent with those measured in previous model studies

3. Measured harbor energy will be normalized by incident energy. Trends in amplification of incident energy with crest elevation and incident wave energy at each site will be similar as for wind waves. By normalizing the amplifications again by the amplification at the existing crest elevation, the relative sensitivity of the amplification at any site to future changes in crest elevation can be obtained. Conversion of incident or harbor energy to prototype scale is neither valid nor required.

4. Overtopping low frequency waves in the model transmit much more energy than overtopping wind waves of the same height, so they provide an upper limit for the prototype situation. If the effect of crest elevation is relatively small at a site for this study, then it will not likely be significant for wind waves in the prototype.

Testing Protocol

Incident Waves   Incident waves were generated with a JONSWAP spectrum having a peak model period of 1.57 sec and a gamma of 3.3. Five intensities of the spectra, corresponded to electronic gains of 10, 20, 30, 40, and 50 percent of full gain, were used which produced wave heights at Gage 21, in 2 ft of water directly in front of the generator, of 0.02, 0.04, 0.06, 0.08, and 0.10 ft. This produced conditions which range from relatively calm to major overtopping. Two repeat tests were conducted for each combination of crest elevation and gain level.

Breakwater Plans   A total of 9 plans were tested, Table 1. Plan 0 was the existing condition of the model. In Plan 1the model crest was manually corrected to an average elevation +12.5 ft. Plans 2-7 were conducted with the clay crest set from 0 ft MLLW (overtopping occurs for all gain levels) to + 24 ft MLLW (no overtopping for any gain level). Plan 8 used the same crest elevation as Plan 7, but also eliminated wave transmission through the entrance channel by blocking the opening to Angel's Gate and the passage between the Middle Breakwater and Pier 400 (see Figure 1), so the only energy entering the harbor came through the porous breakwater.

Table 1. Summary of Breakwater Plans

Plan	0	1	2	3	4	5	6	7	8
Crest (ft) MLLW	+ 13.5*	+12.5*	0.0	+7.5	+10.0	+12.5	+14.0	+24.0	+24.0
Crest Material	concrete	concrete	clay	clay	clay	clay	clay	clay	clay

* Original construction   ** Entrances blocked

Uncertainty   The two major sources of uncertainty in the results are random errors in the generating/measurement system and model distortion effects in the scaling relationships. Generally, wave height measurements repeat within $\pm$ 5 percent and the wave generator reproduces the desired command signal within $\pm$ 3 percent. Since measured wave height at each gage linearly related to the command signal (see Figure 2, below) the maximum uncertainty in wave height measurements can be calculated to be about 3 per cent. Seabergh, 1993, estimated the distortion error in the calculated time scale based upon the shallow water approximation for wave celerity. The time scale ratio of 1:40 obtained from the approximation overpredicts the prototype period by 0.5 % for a 2.125-sec (85-sec prototype) waves in 1 m of water, compared to the exact, intermediate water wave solution. By the same criterion, the maximum error for a 62.8-sec wave is 1.2 per cent.

RESULTS[4]

Total Energy   The total energy measured at a site is characterized by the H_{m0}. Figure 2 (l) is an example plot of model H_{m0} as a function of gain at Gage 5 for all repeat tests of Plan 0. Most of the repeat tests at each gain are within the 3 percent error bars of each other. Generally, all of the gages and all of the plans show a similar clustering of repeat test data and the linear model response as the incident wave heights are increased. For visual clarity, the remaining analysis will plot just one test for each crest/gain combination, and will focus on Plans 2-8 only.

Gages located behind the breakwater in the outer harbor away from basins – Gages 1, 2, 18 and 20 – best demonstrate the expected trend with crest elevation and gain. Gage number 20, (Figure 2r) is typical of this group: energy decreases with increasing crest height and decreasing wave height, and cessation of major overtopping is evidenced by the flattening of the slope.

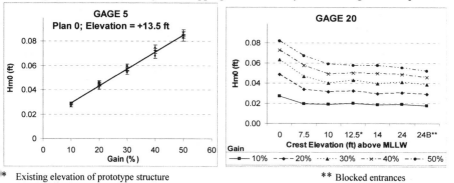

*   Existing elevation of prototype structure   ** Blocked entrances

Figure 2 - (l) H_{m0} vs. gain, Plan 0, Gage 5; (r) H_{m0} vs. crest elevation, all plans, Gage 20

---

4 Complete study results, including plots of all measurements, can be found in the McGhee, 2001

H_{m0} amplification (*Hamp*) is the ratio of H_{m0} at each gage to the measured incident H_{m0} at Gage 21 for that run. It allows nondimensional comparison of total energy between sites. To compare the sensitivity of the energy level at a site to changes in crest elevation, amplification was normalized (*Hnrml*) by dividing by the amplification for the existing condition (Plan 5) for that run. A value of *Hnrml* less (greater) than 1 means the total energy at that site is decreased (increased) relative to the existing conditions. Figure 3 (l) plots *Hamp* and Figure 3(r) plots Hnrml for Gage 20. The trend is identical to Figure 2, but the relative order of the gain curves is reversed because of increasing losses associated with breaking and friction at higher gains. There is significant energy reduction as the crest elevation increases from 0 to + 7.5 ft, moderate improvement as the crest increases to the existing elevation of +12.5 ft, little additional benefit in increasing the crest beyond this elevation except for at the highest gain level, and only slightly more benefit in closing off the entrances altogether (Elev. 24B). This is because a significant amount of energy is transmitted through the porous structure.

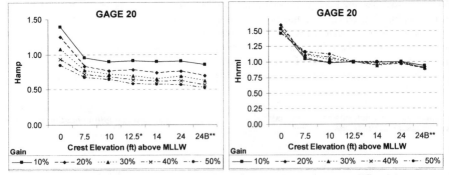

Figure 3 - (l) Gage 20 H_{m0} amplification; (r) normalized amplification

Note that *Hamp* > 1 for three cases, but transmitted energy does not actually exceed incident energy at these sites. Though the formula for *Hamp* looks similar to that for the transmission coefficient, *Kt*, they are not identical because H_{m0} at gages near harbor structures includes a significant amount of reflected, in addition to the transmitted, energy.

Almost all gages show the same general trends seen in Figures 2 and 3: H_{m0}, *Hamp* and *Hnrml* curves are nearly parallel for the five gain levels. This allows use of the site-averaged *Hamp* and *Hnrml* to succinctly characterize the harbor's overall sensitivity to changes in the crest elevation. Figure 4 is the average *Hamp* and *Hnrml* for all interior harbor gages (3-17, 25-28 and 30). Interior harbor gages exhibit the trends seen above, but with one interesting difference: all show an increase in relative energy at the higher crest elevations, and some even continue to increase when the entrances are blocked. While the observation of this "harbor paradox" is not without precedent (Miles and Munk, 1961), it is curious is that the breakwater appears to have "settled" into a low energy state that is only slightly more energetic than an enclosed harbor.

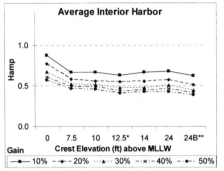
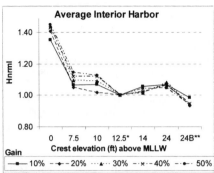

Figure 4 - (l) Average interior harbor H_{m0} amplification; (r) normalized amplification

Low Frequency Energy The spectrum at any gage in the harbor is dominated by the energy at the peak frequency of the incident waves – 0.64 Hz – but includes energy above and below the peak. The main objective of the study is to examine the response of the harbor basins at frequencies below the (unrealistically) high-energy peak. The low frequency portion of the energy spectrum is defined as all energy at or below 0.47 Hz (lower than 85-sec period, prototype). For most of the gages, the sum of the low frequency energy follows the same general trends seen in Figures 2. Low frequency amplification (*Lamp*) is the square root of the ratio of measured low frequency energy at a site to the measured low frequency energy at Gage 21, and *Lnrml* is *Lamp* normalized by Plan 5 *Lamp*. Plots of *Lamp* and *Lnrml* at each gage are not always as well behaved as Gage 20 (Figures 3), but the general trend is preserved. Figure 5 shows the average *Lamp* for all gains and the average *Lnrml* for Plans 3 – 7 for the interior harbor gages. (Plans 2 and 8 are omitted for simplicity, as they are unlikely to ever occur in nature).

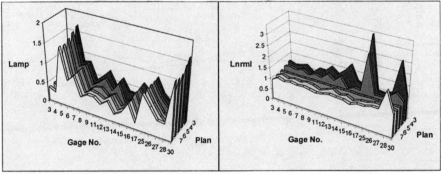

Figure 5 – (l) Average low frequency amplification; (r) average normalized low frequency amplification for interior harbor gages

Sites 5, 7, 11, 13, 16, 25 and 30 are particularly responsive compared to other sites, while Sites 3, 4, 14, 17, and 28 exhibits relatively low excitation. The normalized amplification highlights the sensitivity of the site to changes in the crest elevation. Sites 17 and 28 would react strongly to additional crest loss; Sites 11 and 13 react to a much less degree. Most of the sites are relatively insensitive to positive or negative changes on the order of a few ft.

Figure 6 plots the *Lamp* for Plan5, the existing condition, together with the normalized amplification for Plans 3, 4, 6, and 7. (*Lnrml* for Plan 5 is 1.0, by definition, so it is omitted for clarity). The gages are reordered by decreasing Plan 5 amplification. This allows identification of the few sites that showed changes of a significant magnitude with changes in crest height.

Sites 5 and 30, while energetic, would experience changes no more than 10 percent. Sites 11 and 13 are only moderately energetic, but loss of crest elevation increases their response by 30 to 50 percent. Site 30 and, particularly, Site 17, are very sensitive to crest elevation. Loss of an additional 2 ½ ft of crest would make Sites 30 and 17 about as energetic as Site 13 and 11, respectively, are now with the existing crest elevation. Note the increase in energy at the two higher crest elevations by some gages, especially at Gage 28.

Only Sites 16 and 11 show a significant reduction in energy when the crest is raised to its design elevation, +14 ft MLLW. The rest show reduction, no change, or even a slight increase.

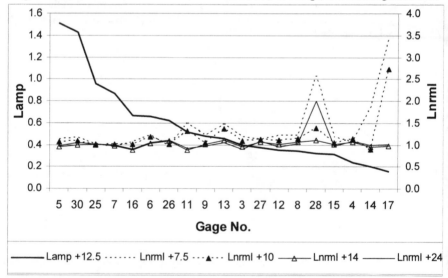

Figure 6. *Lamp* (Plan 5) and *Lnrml* (Plans 3, 4, 6, and 7)

CONCLUSIONS

1. The LA/LB model is able to generate low frequency waves that overtop the San Pedro Breakwater at its current crest elevation.

2. The overtopping waves produce low frequency oscillations in the various harbor basins. The energy in the basins increases linearly with incident energy and is sensitive to crest elevation.

3. The average total energy inside the harbor decreases as crest elevation is raised, reaches a minimum near a crest elevation of +12.5 ft MLLW, and then increases for higher elevations. Average measured energy inside the harbor for the existing crest elevation is no more than 10 per cent higher than the configuration with no overtopping allowed and the harbor entrances blocked.

4. Sites with the highest low frequency amplification are relatively less sensitive to crest elevation. Sites with the most sensitivity have relatively less low frequency amplification.

5. Sites 11, 13, 28 and 17 would experience significant increase in low frequency energy if the crest loss continued to an elevation of +10 ft MLLW.

6. Sites 5 and 11 would experience modest reductions in low frequency energy if the crest elevation was raised to +14 ft MLLW.

ACKNOWLEDGEMENTS

Funds for this study were provided by the San Pedro Breakwater Operations and Maintenance Study of the US Army Engineer District, Los Angeles. Messrs. John Heggins, Tony Brogdon and Tim Nicely of ERDC operated the LA/LB model and collected and reduced model data. Assistance with report preparation was provided by Ms. Allison Dunford of Alatar Enterprises.

Conversion Factors, non-SI to SI Units of Measurement

Multiply	By	To Obtain
feet	0.3048	meters
miles	1.852	kilometers
degrees (angular)	0.0174	radians

REFERENCES

McGehee, D.D., 2001, "3-Dimensional Hydraulic Physical Model Evaluation for the Los Angeles/Long Beach Harbor Operation and Maintenance Study," Final Technical Report, Emerald Ocean Engineering, Pensacola, FL. 26 p, plus Appendices.

Miles, J.W. and Munk, Walter, 1961, "Harbor Paradox," Journal, Waterways and Harbors Division, Proc. ASCE, Vol. 87, pp 111-130.

Seabergh, William C. and Thomas, Leonete J., 1999, *Los Angeles and Long Beach Harbors Model Enhancement Program: Long Waves and Harbor Resonance Analysis*, Technical Report CERC-99-20, US Army Engineer Waterways Experiment Station, Vicksburg, MS. 56 p.

Seabergh, William C. and Thomas, Leonetee J., 1993 *Los Angeles and Long Beach Harbors Model Enhancement Program, Improved Physical Model Harbor Resonance Methodology*, Technical Report CERC-93-17, US Army Engineer Waterways Experiment Station, Vicksburg, MS. 32 p., plus Appendices.

U.S. Army Corps of Engineers, Los Angeles District, 1996, *Comprehensive Condition Survey, San Pedro Breakwater, Los Angeles Harbor, Los Angeles County, CA.*

GENERATION OF SEICHES BY TWO
TYPES OF COLD FRONTS

M.P.C. de Jong[1], J.A. Battjes[2] and L.H. Holthuijsen[3]

Abstract: The origin of seiches in the Port of Rotterdam is investigated with observations at sea and in the harbour and with numerical simulations. A wavelet analysis of the observations shows that low-frequency energy (0.1–2.0 mHz) occurs at sea prior to each seiche in the Port of Rotterdam. An analysis of weather charts (from 1995 till 2000) indicates that all seiches coincided with the passage of a storm. Some of these storms included a sharp cold front, whereas others included a more gradual cold front. Numerical simulations with a hydrodynamic model driven by meteorological observations reproduced seiches for the sharp cold fronts but not for the gradual cold fronts.

1 INTRODUCTION

1.1 Seiches

Occasionally standing waves occur inside harbour basins, apparently in response to low-frequency energy at sea. These standing waves are called seiches. The response of harbour basins to such excitation is relatively well known (see e.g. Mei 1989). However, the origin of the low frequency energy at sea is usually not known.

Such low-frequency energy at sea can be generated by a number of mechanisms, e.g. surf beat, tsunamis, internal waves, and atmospheric disturbances (Wilson 1972, Korgen 1995 and Giese & Chapman 1993). The dominating source may differ from harbour to harbour depending on the occurrence of such mechanisms and the specific

[1]Ph.D. student, Fluid Mechanics Section, Delft University of Technology, The Netherlands, e-mail: m.p.c.dejong@ct.tudelft.nl

[2]Professor, Fluid Mechanics Section, Delft University of Technology, The Netherlands, e-mail: j.battjes@ct.tudelft.nl

[3]Associate Professor, Fluid Mechanics Section, Delft University of Technology, The Netherlands, e-mail: l.h.holthuijsen@ct.tudelft.nl

geographic location of the harbour. The present study is aimed at the source of the seiche energy at sea for the western section of the Port of Rotterdam, The Netherlands (see Figure 1). In this area seiches mainly occur in the *Calandkanaal* (length $\approx$ 20 km, depth $\approx$ 20 m) with measured crest-to-trough heights of up to 1.7 m and in most cases a period of 90 minutes.

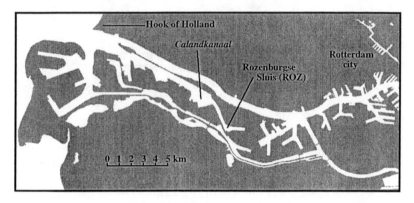

Figure 1: The western section of the Port of Rotterdam.

In general, previous studies of the seiches in the Port of Rotterdam have focussed on the phenomenon inside the harbour (see e.g. Looff & Veldman 1994), without considering the situation at sea, although in some cases suggesting that these seiches could be caused by cold fronts. This suggestion is supported by an analysis of weather maps (1995-2000) made for this study for the days on which seiches have occurred. This paper, therefore, analyses a number of these events to establish whether or not the seiches are generated by cold fronts.

1.2 Earlier studies

The generation of seiches by atmospheric disturbances has been studied earlier by Hibiya & Kajiura (1982), who describe the analysis of a seiche at Nagasaki Bay, Japan (period 30 minutes). They showed with numerical simulations that a sudden increase ("jump") of 3 hPa (mbar) followed by a gradual decrease in pressure can generate long waves at sea, which subsequently causes a seiche.

Monserrat et al. (1991a), Monserrat et al. (1991b), Rabinovich & Monserrat (1996) and Rabinovich & Monserrat (1998) studied a number of seiches at the *Ciutadella* inlet of the Balearic Islands (period 10 minutes). They too find a high correlation between atmospheric pressure fluctuations and seiches inside the inlet. Vidal & Matín (2001) showed with numerical simulations forced by meteorological observations that seiches in this inlet are generated by a standing wave at sea between two of the Balearic Islands.

To address the problem for the Port of Rotterdam, measurements of water levels, atmospheric pressure and wind speed have been obtained from a number of locations at sea and inside the harbour. These measurements are analysed with a wavelet technique to identify the occurrence of low-frequency energy (0.1–2.0 mHz) at sea and inside the harbour. The generation of low-frequency energy at sea has been numerically simulated for a number of seiche events with a shallow water model (Gerritsen et al. 2000) driven by artificial pressure fields.

2 Measurements

2.1 Data acquisition

Water level, wind speed and atmospheric pressure at sea were measured at *Europlatform* (EUR), *Lichteiland Goeree* (GOE) and *Meetpost Noordwijk* (MPN). Figure 2 shows these offshore platforms, located approximately 20 km and 40 km from the harbour mouth.

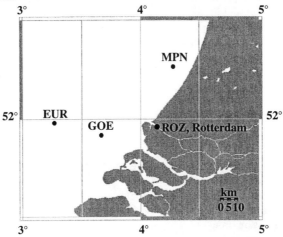

Figure 2: Overview of the Dutch coast near the harbour mouth of the Port of Rotterdam and the measurement locations at sea.

Water level (from step gauges and radar altimeters), atmopsheric pressure and wind speed measurements are available at a number of these platforms for one period of five months and a number of periods of a few days. The sample rate is 4 Hz (prior to 2000) or 2.56 Hz (2000 and later). The atmospheric pressure data consisted of ten minute averages and the wind speed data of one minute averages. For events without meteorological observations, observations from an onshore location near the coast 80 km north of in the Port of Rotterdam were used to asses the meteorological situation of that day.

Inside the harbour water level measurements are available from *Rozenburgse Sluis* (ROZ), located near the closed end of the *Calandkanaal* (see Figure 1). These measurements are taken with a floater, sampled at a 60 second interval. These data have been stored continuously since 1995.

2.2 Data analysis

A wavelet analysis based on the *Morlet* wavelet (Morlet et al. 1982) has been used to determine wavelet power spectra (WPS) or scalograms of the water level, wind and atmospheric pressure measurements. These WPS can be interpreted as wave energy density as functions of time and period (e.g. Farge 1992, Torrence & Compo 1998). From these spectra the occurrence of seiches or of low-frequency energy at sea can be identified, even if visual inspection of the water level measurements at sea does not clearly indicate such occurrence.

To reduce the computer capacity needed for the wavelet computations, the water level signals were diluted to one value per minute (after the wind waves were removed from the signals to avoid aliasing). The low frequency tide components have also been removed from the measurements to isolate the seiche frequency band (0.1–2.0 mHz). This filtering of the water level measurements does not influence the wavelet results for the seiche frequency band; it merely enables one to make a visual inspection of the time signals in the seiche frequency band.

3 ATMOSPHERIC DISTURBANCES

3.1 Storms

A visual inspection of the weather charts for the days that seiches have occurred between 1995 and 2000 shows that all events coincided with a storm passing over the North Sea region. In some cases the seiche coincided with the passage of a cold front, whereas in other cases the seiche coincided with the passage of a local minimum of atmospheric pressure (trough).

These findings are confirmed by the measurements. During a number of the seiches, a cold front passed that showed up as one or more sharp changes ("jumps") in atmospheric pressure or wind direction, whereas in other cases the passing low pressure system included a cold front that was accompanied by gradual changes in atmospheric pressure.

These two situations correspond to the two types of cold fronts that are distinguished in the meteorological literature (e.g. Bader et al. 1995 and Browning & Monk 1982). The first is the classical cold front (also called Ana cold front or type 1 cold front), which is characterised by sharp changes in atmospheric pressure, wind direction, and temperature. During heavy events it is accompanied by a narrow line of intense precipitation, called line convection. The second type of cold front is the split cold front (Kata cold front, type 2 cold front), which is characterised by more gradual changes in meteorological parameters. This second type of cold fronts results in widespread, mild precipitation.

3.2 Cold fronts (observed)

The most recent seiche in the Port of Rotterdam in the available measurements coincided with a classical cold front that passed over the North Sea on 30 October 2000 in easterly direction (see Figure 3). It showed up as a line of showers (line convection) on precipitation radar images (not shown here) and as a small jump in atmospheric pressure of 2 hPa at the platforms at sea. Figure 4 shows the seiche at ROZ and the jump in atmospheric pressure at EUR.

More gradual changes in atmospheric pressure were found on 5 - 6 April 1997 (Figure 5). Figure 6 shows the seiche that occurred inside the *Calandkanaal* and the gradual changes in the atmospheric pressure measurements from Hook of Holland (HOH).

4 NUMERICAL SIMULATIONS

4.1 Numerical model

The hydrodynamic model (Delft 3D, see e.g. Gerritsen et al. 2000) that was used for the simulations of the low-frequency energy (at sea and in the harbour), solves the continuity equation and the non-linear momentum balance equation on orthogonal, curvilinear boundary-fitted co-ordinates. The model includes the effects of atmospheric pressure gradients, wind stress, bottom friction, Coriolis and horizontal eddy viscosity.

These simulations were run in a vertically integrated (2D) mode and used the model set-up of the PROMISE project (Gerritsen et al. 2000), with the resolution getting finer towards the harbour mouth. Weakly reflective boundary conditions were applied at the open boundaries of the computational domain, located at the Norwegian Sea and the English Channel. These computations were driven by synthetic fields of atmospheric pressure and wind speed, which were obtained by converting the measured time series of these parameters into frozen spatial fields advected with the inferred velocity of the front inferred from a number of point measurements of atmospheric pressure and a sequence of precipitation radar images.

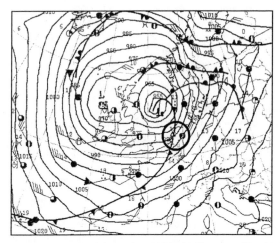

Figure 3: Weather analysis for 30 October 2000, 12:00 GMT. The Netherlands are indicated by the circle. Source image: Royal Dutch Meteorological Institute (KNMI).

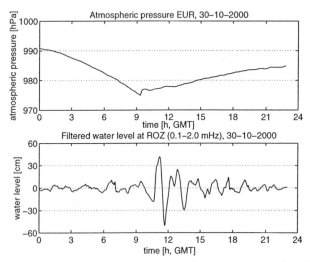

Figure 4: Atmospheric pressure at EUR and water level at ROZ on 30 October 2000.

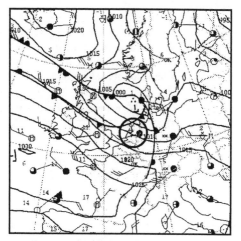

Figure 5: Weather analysis for five April 1997, 12:00 GMT. The Netherlands are indicated by the circle. Source image: Royal Dutch Meteorological Institute (KNMI).

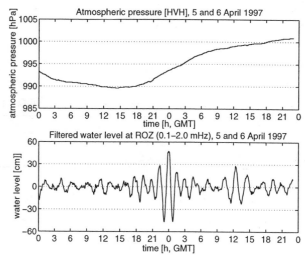

Figure 6: Atmospheric pressure at HOH and water level at ROZ on 5 and 6 April 1997.

4.2 Classical cold front

The wavelet power spectrum of the water level measured at location EUR (see Figure 7) and at location ROZ (see Figure 7) shows that energy at the eigen period of the *Calandkanaal* (5400 s) occurred at sea just prior to the seiche. The generation of this low-frequency energy at sea is reproduced in the simulations as shown by the wavelet power spectrum of the simulation at location EUR.

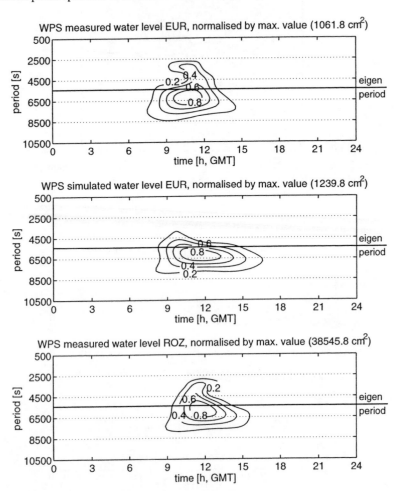

Figure 7: Normalised wavelet power spectra of measured water level at EUR (top panel), simulated water level at EUR (middle panel) and the measured water level at ROZ (bottom panel) for 30 October 2000.

4.3 Split cold front

The seiche that occurred with the split cold front of 5-6 April 1997 (see Figure 6) could not be reproduced numerically. Other simulations of conditions with a split cold front failed similarly.

5 CONCLUSIONS

The seiches which have been analysed for this study all coincided with a storm, more specifically: with the passage of a cold front. A number of these seiches coincided with a classical cold front with sharp changes in atmospheric pressure and wind direction. Other seiches coincided with cold fronts with gradual changes of these meteorological parameters. The numerical simulations reproduce the generation of seiches by the classical cold fronts: the wavelet power spectra of the simulations indicate that low-frequency energy is generated at sea near the eigen frequency of the *Calandkanaal*. On the other hand, the numerical simulations could not reproduce the generation of the low-frequency energy at sea just prior of the seiche for a split cold front.

To investigate the failure of the numerical simulations for split cold fronts, more information regarding the front type, velocity and direction for future events will be obtained together with precipitation images. In adddition, more detailed observations at Hook of Holland are planned. Moreover, future numerical simulations will use atmospheric pressure and wind fields from a high resolution (2 km) numerical meteorological model instead of the synthetic fields described in this paper.

ACKNOWLEDGMENTS

The authors acknowledge the financial support of the Dr. Ir. Cornelis Lely Foundation and of the Directorate-General for Public Works and Water Management (RWS). Furthermore, the authors thank J. Starke and J. Rozema of the North Sea Directorate of RWS for the measurement data from sea. In addition, we thank C. Mooiman, R. Kamphuis and N. Neerven of the Municipal Harbour Organisation Rotterdam for supplying the measurement data from the Port of Rotterdam. We thank WL | Delft Hydraulics for placing the Delft 3D simulation package at our disposal and for permission to use its curvilinear EU/PROMISE model set-up. Wavelet software was provided by C. Torrence and G. Compo (Torrence & Compo 1998), and is available through URL: 'http://paos.colorado.edu/research/wavelets/'.

REFERENCES

BADER, M.J., FORBES, G. S., GRANT, J. R., RILLEY, R. B. E., & WATERS, A. J., 1995, *Images in weather forecasting - A practical guide for interpreting satellite and radar imagery*, Cambridge University Press, 499 pp., ISBN: 0-521-45111-6

BROWNING, K. A., & MONK, G. A., 1982, *A simple model for the synoptic analysis of cold fronts*, Quart. J. R. Met. Soc., **108**, 435–452

C. VIDAL, R. MEDINA, S. MONSERRAT, & MATÍN, F. L., 2001, Harbor resonance induced by pressure-forced surface waves, *Pages 3615–3628 of: Coastal Engineering 2000 (ICCE 2000)*, vol. 4

FARGE, M., 1992, *Wavelet transforms and their applications to turbulence*, Annual reviews, **1992**, 395–457

GERRITSEN, H., VOS, R. J., VAN DER KAAIJ, TH., LANE, A., & BOON, J. G., 2000, *Suspended sediment modelling in a shelf sea (North Sea)*, Coastal Engineering, **41**(1-3), 317–352

GIESE, G. S., & CHAPMAN, D. C., 1993, *Coastal seiches*, Oceanus, **36**(1), 38–46

HIBIYA, T., & KAJIURA, K., 1982, *Origin of the Abiki phenomenon (a kind of seiche) in Nagasaki Bay*, Journal of the Oceaonographical Society of Japan, **38**, 172–182

KORGEN, B. J., 1995, *Seiches, transient standing-wave oscillations in water bodies can create hazards to navigation and unexpected changes in water conditions*, American Scientist, **83**, 330–341

LOOFF, H. DE, & VELDMAN, J. J., 1994, *Seiches in the port of Rotterdam during storm surges*, Proceedings of the International Symposium: Waves-Physical and Numerical Modelling, Vancouver, 287–286

MEI, C. C., 1989, *The applied dynamics of ocean surface waves*, New York: Wiley, pp 183–252, ISBN: 0-471-06407-6

MONSERRAT, S., IBBETSON, A., & THORPE, A. J., 1991a, *Atmospheric gravity waves and the 'Rissaga' phenomenon*, Q. J. R. Meteorol. Soc., **117**, 553–570

MONSERRAT, S., RAMIS, C., & THORPE, A. J., 1991b, *Large-amplitude pressure oscillations in the western Mediterranean*, Geophysical research letters, **18**(2), 183–186

MORLET, J., ARENS, G., FURGEAU, E., & GIARD, D., 1982, *Wave propagation and sampling theory - Part 2: Sampling theory and complex waves*, Geophysics, **47**(2), 222 – 236

RABINOVICH, A. B., & MONSERRAT, S., 1996, *Meteorological Tsunamis near the Balearic and Kuril Islands: Descriptive and statistical analysis*, Natural Hazards, **13**(1), 55–90

RABINOVICH, A. B., & MONSERRAT, S., 1998, *Generation of meteorological tsunamis (Large amplitude seiches) near the Balearic and Kuril Islands*, Natural hazards, **18**, 27–55

TORRENCE, C., & COMPO, G. P., 1998, *A practical guide to wavelet analysis*, Bulletin of the American Meteorological society, **79**, 61–78

WILSON, B. W., 1972, *Seiches*, Adv. Hydroscience, **8**, 1–94

A Tidal Constituent Database for the East Coast of Florida

D. Michael Parrish[1], ASCE, and Scott C. Hagen[2], M.ASCE

Abstract: Florida's St. Johns River Water Management District requires tidal constituents (amplitudes and phases) at open water boundaries on the continental shelf from the Florida/Georgia border to Jupiter, Florida. These tidal constituents are used as boundary conditions for hydrodynamic studies on the continental shelf and within estuaries. Since the historical tidal record is limited, this research (the production of the necessary tidal constituents) will result in the improvement of the accuracy of future coastal and estuarine simulations.

A significant part of the overall effort was to refine an existing unstructured mesh (Eastcoast2001) of the Western North Atlantic Tidal (WNAT) model domain. The WNAT model domain encompasses the Gulf of Mexico, the Caribbean Sea, and the North Atlantic Ocean west of the 60° W meridian. Eastcoast2001 has been refined in three distinct areas: near the Florida shoreline, along the eastern Floridan Plateau-edge, and across the continental slope. Refinement is on the order of three in the shallow regions and 10 across the continental rise. The new mesh retains the bathymetric data of Eastcoast2001 and incorporates more detailed shoreline data.

Simulations were performed with a two-dimensional, finite element code for coastal and ocean circulation (ADCIRC-2DDI), which was forced with M_2, M_4, M_6, O_1, N_2, S_2, K_1, and steady constituents. Model verification is demonstrated by comparison of simulation results to historical tidal data.

---

1 Graduate Student, Water Resources Engineering Program, Civil and Environmental Engineering Department, University of Central Florida, Orlando, Florida 32816-2450. dmp24688@ucf.edu
2 Assistant Professor, Civil and Environmental Engineering Department, University of Central Florida, Orlando, Florida 32816-2450. shagen@mail.ucf.edu

INTRODUCTION

The St. Johns River Water Management District (District) is employing hydrodynamics models as management tools for monitoring and protecting each of the estuarine systems within its boundaries, which extend from about Fernandina Beach at the north to St. Lucie Inlet at the south. These models will be used for assessing and quantifying tidal effects on the health of the estuaries and, eventually, for the establishment of total maximum daily load for major bodies of water, including the Lower St. Johns River, Indian River Lagoon, and Tolomato / Guana / Matanzas Rivers. The performance and accuracy of these models are greatly dependent upon the accuracy and resolution of the data used to calibrate, verify, and apply the model.

Tidal constituents at open water boundaries on the continental shelf along the coast of Florida, and within estuaries are essential for hydrodynamics assessments. However, the historical tidal record is limited and site-dependent, which restricts the number and accuracy of tidal constituents practicable for model boundary forcing. At present, tidal constituents (amplitudes and phases of the tidal harmonic coefficients) are only available at a few dozen coastal locations. Therefore the District lacks the capability to define boundary conditions at many different locations beyond the shore.

In this paper we present results from simulations that will provide for the District's need for boundary condition specification. We compare our model results to historical data from 11 tide gages that are located along the eastern Florida coastline. As a direct result of the modeling process, we have produced the SJR mesh, described herein.

MODEL DESCRIPTION

ADCIRC-2DDI (Advanced Circulation Model for Oceanic, Coastal, and Estuarine Waters—Two Dimensional Depth Integrated option) solves the shallow water equations over the Western North Atlantic Tidal (WANT) model domain to generate a tidal response that is harmonically analyzed to produce tidal constituents. ADCIRC-2DDI assumes incompressibility, hydrostatic pressure, and the Boussinesq approximation, and neglects baroclinic terms as well as lateral diffusion and dispersion terms. Given the assumptions just stated, the simplest form of the shallow water equations, in spherical coordinates is (Luettich, et al. 1991):

$$\frac{\partial \zeta}{\partial t} + \frac{1}{R\cos\phi}\left[\frac{\partial UH}{\partial \lambda} + \frac{\partial (VH\cos\phi)}{\partial \phi}\right] = 0 \tag{1}$$

$$\frac{\partial U}{\partial t} + \frac{1}{R\cos\phi}U\frac{\partial U}{\partial \lambda} + \frac{1}{R}V\frac{\partial U}{\partial \phi} - \left(\frac{\tan\phi}{R}U + f\right)V = -\frac{1}{R\cos\phi}\frac{\partial}{\partial \lambda}\left[\frac{p_s}{\rho_0} + g(\zeta - \eta)\right] + \frac{\tau_{s\lambda}}{\rho_0 H} - \tau.U \tag{2}$$

$$\frac{\partial V}{\partial t} + \frac{1}{R\cos\phi}U\frac{\partial V}{\partial \lambda} + \frac{1}{R}V\frac{\partial V}{\partial \phi} + \left(\frac{\tan\phi}{R}U + f\right)U = -\frac{1}{R}\frac{\partial}{\partial \phi}\left[\frac{p_s}{\rho_0} + g(\zeta - \eta)\right] + \frac{\tau_{s\phi}}{\rho_0 H} - \tau.V \tag{3}$$

where λ = longitude (°); ϕ = latitude (°); ζ = free surface elevation relative to the geoid; U = depth averaged horizontal velocity, λ direction; V = depth averaged horizontal velocity, ϕ direction; R = radius of the Earth; H = total water column, h + ζ; h = bathymetric depth, relative to geoid; $f = 2\Omega\sin\phi$ = Coriolis parameter; Ω = angular speed of the Earth; p_s = atmospheric pressure at the free surface; g = acceleration due to gravity; ρ_0 = reference density of water; η = effective Newtonian tide potential; $\tau_{s\lambda}$ = applied free surface stress, λ direction; $\tau_{s\phi}$ = applied free surface stress, ϕ direction;

$$\tau_* = C_f \left(U^2 + V^2\right)^{1/2} / H \tag{4}$$

C_f = bottom friction coefficient;

The finite element (FE) solution is more readily obtained when the equations are first converted to the Carte Parallelogramatique (CP) projection:

$$x' = R(\lambda - \lambda_0)\cos\phi_0 \tag{5}$$

$$y' = R\phi \tag{6}$$

Therefore the shallow water equations are, by the CP projection:

$$\frac{\partial\zeta}{\partial t} + \frac{\cos\phi_0}{\cos\phi}\frac{\partial(UH)}{\partial x'} + \frac{1}{\cos\phi}\frac{\partial(VH\cos\phi)}{\partial y'} = 0 \tag{7}$$

$$\frac{\partial U}{\partial t} + \frac{\cos\phi_0}{\cos\phi}U\frac{\partial U}{\partial x'} + V\frac{\partial U}{\partial y'} - \left(\frac{\tan\phi}{R}U + f\right)V = -\frac{\cos\phi_0}{\cos\phi}\frac{\partial}{\partial x'}\left[\frac{p_s}{\rho_0} + g(\zeta - \eta)\right] + \frac{\tau_{s\lambda}}{\rho_0 H} - \tau_* U \tag{8}$$

$$\frac{\partial V}{\partial t} + \frac{\cos\phi_0}{\cos\phi}U\frac{\partial V}{\partial x'} + V\frac{\partial V}{\partial y'} - \left(\frac{\tan\phi}{R}U + f\right)U = -\frac{\partial}{\partial y'}\left[\frac{p_s}{\rho_0} + g(\zeta - \eta)\right] + \frac{\tau_{s\phi}}{\rho_0 H} - \tau_* V \tag{9}$$

Solving the shallow water equations in this form gives rise to numerical instabilities. Therefore, it is necessary to reformulate the equations in order to provide for a stable solution. Thus ADCIRC-2DDI solves the Generalized Wave Continuity Equation (GWCE) together with the equations of momentum conservation. The GWCE is obtained by applying the following steps: First write the primitive non-conservative momentum equations into conservative form; then take spatial derivatives of these results. Substitute the spatial derivatives into a time-differentiated primitive continuity equation. To this result, add a weighting parameter multiplied by the primitive continuity equation; this yields the GWCE. In CP form, the GWCE is:

$$\frac{\partial^2 \zeta}{\partial t^2} + \tau_0 \frac{\partial \zeta}{\partial t} + \frac{\cos\phi_0}{\cos\phi} \frac{\partial}{\partial x'} \frac{\partial \zeta}{\partial t} \left\{ U - \frac{\cos\phi_0}{\cos\phi} UH \frac{\partial U}{\partial x'} - VH \frac{\partial U}{\partial y'} + \left(\frac{\tan\phi}{R} U + f \right) VH \right\} \qquad (10)$$

$$- H \frac{\cos\phi_0}{\cos\phi} \frac{\partial}{\partial x'} \left[\frac{P_s}{\rho_0} + g(\zeta - \eta) \right] - (\tau_* - \tau_0) UH + \frac{\tau_{s\lambda}}{\rho_0} \right\} + \frac{\partial}{\partial y'} \left\{ V \frac{\partial \zeta}{\partial t} - \frac{\cos\phi_0}{\cos\phi} UH \frac{\partial V}{\partial x'} - VH \frac{\partial V}{\partial y'} \right.$$

$$- \left(\frac{\tan\phi}{R} U + f \right) UH - H \frac{\partial}{\partial y'} \left[\frac{P_s}{\rho_0} + g(\zeta - \eta) \right] - (\tau_* - \tau_0) VH + \frac{\tau_{s\phi}}{\rho_0} \right\} - \frac{\partial}{\partial t} \left(\frac{\tan\phi}{R} VH \right) + \tau_0 \left(\frac{\tan\phi}{R} VH \right) = 0$$

The GWCE is solved in conjunction with the primitive momentum equations (8) and (9) using a FE technique that allows for arbitrary bathymetry, coastlines, and open boundaries (Luettich et al. 1991, Scheffner and Carson 2001).

DESCRIPTION OF THE COMPUTATIONAL DOMAIN

The utility of large computational domains has been demonstrated by previous research (Westerink, et al., 1993, 1995). Coastal models with large computational domains allow accurate specification of boundary conditions, since the open boundaries may be placed in the deep ocean where flow behavior is linear, and tidal constituents may be specified accurately. We see deep-water placement of the open boundary as advantageous compared to placement on the continental shelf or at the shelf break, where bathymetry--and therefore the flow field--changes relatively rapidly.

The open ocean boundary of the WNAT model domain coincides with the 60° west meridian, and lies predominantly in the deep ocean. The coastal boundaries are composed of the South, Central, and North American coastlines. Because of its great size (8.347×10^6 km²) and since high resolution is required in coastal regions to adequately represent geometry and tidal flow, it is infeasible to discretize the domain with a uniform grid. Rather, we apply to the present model application an unstructured mesh so that we may provide high resolution in areas of shallow water, steep bathymetry, and rapidly changing bathymetric gradient, while providing lower, though still adequate, resolution in the deep ocean.

DEVELOPMENT OF THE COMPUTATIONAL MESH
Design of Eastcoast2001

In 1994, Westerink et al. conducted a WNAT model domain convergence study on a series of meshes in a successful effort to produce a FE mesh that ADCIRC-2DDI may use to solve the shallow water equations, thus providing for tidal constituents at any location specified within the domain.

Since then, a series of further studies have provided solutions of greater accuracy with meshes of higher resolution. In 1998, Roe produced an unstructured mesh from scratch, making use of the conclusions of Westerink's 1994 study and other studies. Roe's mesh contains 32,947 nodes and 61,705 elements.

In 2000, Mukai developed a new mesh for the WNAT model domain (see Figure 1a), substantially increasing mesh resolution and bathymetric resolution beyond that of Roe's mesh. After numerous simulations, several adjustments were made to the mesh to produce Eastcoast2001, which was employed in successful simulations. Eastcoast2001 has 254,565 nodes, 492,182 elements. Along the Florida coastline and at the eastern Florida Plateau-edge, resolution is approximately 3 km × 3 km. Resolution in the vicinity of the continental shelf break is about 10 km × 10 km (Figure 2).

Design of the SJR mesh

We have designed the SJR mesh (Figure 1b) to provide greatest solution accuracy along the east coast of Florida. Rather than create a new FE mesh, we elected to refine an existing FE mesh, namely Eastcoast2001.

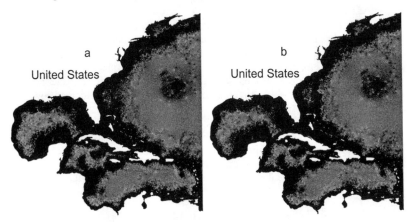

Fig. 1. a) Eastcoast2001; b) the SJR mesh

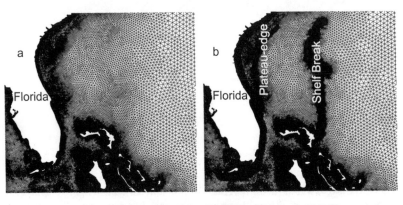

Fig. 2. a) Eastcoast2001 (detail); b) the SJR mesh (detail)

Our refinements are located in three areas. 1) Along the shore of North America, from southern South Carolina to Key West, we have provided a band of 1.0 km × 1.0 km resolution that is 30 km wide. 2) We have placed a narrow (3 km) band of 1.0 km × 1.0 km resolution along the eastern Florida Plateau-edge. 3) We have placed another region of 1.0 km × 1.0 km resolution across the continental shelf break, from Great Abaco north to 32.8° N latitude (Figure 2). The modifications of Eastcoast2001 amount to refinements on the orders of 3, 3, and 10 for the respective regions.

The justification for refining these specific areas is as follows: 1) long-period waves traveling from deeper to shallower water undergo dispersion, producing numerous constituent waves; some of the constituent waves are of shorter period, and therefore require greater resolution if they are to be captured and propagated to the shore. 2) Recent research (Hagen, et al. 1998, 2000, 2001) indicates that refinement of areas with steep bathymetric gradients and sudden changes in the gradients (e.g., continental shelf break) may be as important or more important than near shore refinement. 3) The continental slope is the steepest (about 6°) large-scale feature in the entire domain— only slopes near some islands are steeper—and research conducted by Hagen, et al. (1998, 2000, 2001) has revealed that steep regions require the same or greater level of resolution than the shallow areas in order to produce accurate simulation of flow.

The SJR mesh differs from Eastcoast2001 in two additional ways: it contains more inlets, and therefore possesses a slightly different coastal boundary and has slightly different bathymetry in the vicinity (within about 3.0 km) of these inlets. See Figure 3 for an example of these differences. These inlets have been added in order to accommodate additional tidal stations beyond the number provided in the simulations with Eastcoast2001.

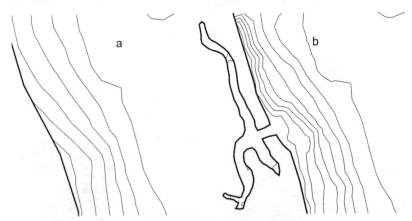

Fig. 3. a) Eastcoast2001 (detail); b) the SJR mesh (detail). Each figure displays the same range in the two different meshes. The dark line represents the shoreline. Contours are every 2.0 m, the contour furthest from the shore being 18.0 m.

Bathymetry for the SJR mesh was largely obtained by interpolation from Eastcoast2001. Only in areas where the boundary was altered were bathymetries not obtained in this manner. In these regions, we interpolated bathymetry from the National Geophysical Data Center onto the SJR mesh; the resolution of this bathymetric data is greater than that of the SJR mesh.

BOUNDARY FORCING

The tidal model is depth-forced on the open-ocean boundary by eight constituents, (M_2, M_4, M_6, O_1, N_2, S_2, K_1, steady). The global ocean model by Le Provost et al. (1998) defines boundary conditions for the deep-ocean nodes. For areas near the open boundary not in deep water, tide records exist from which boundary conditions are interpreted. All land boundary nodes employ a no-flow boundary condition.

PARAMETERS

The model parameters follow; they are common to both the Eastcoast2001 and the SJR mesh simulations. Seven tidal potential forcings (M_2, M_4, M_6, O_1, N_2, S_2, K_1) are applied over the entire domain. The bottom friction coefficient, C_f, is set at 0.003. The simulations are begun from a cold start. Boundary forcings are ramped over a period of 20 days to promote solution stability. The ramp function does not significantly affect the simulated tides of the last 45 days of simulation (from a total of 90), to which harmonic analysis is applied. The time step of 5.0 seconds ensures a Courant number less than 1.0 everywhere in the mesh. Wetting and drying elements are employed in some coastal areas; also, the minimum depth is set to 0.25 m; this means, for example, that even if the mesh possesses bathymetric points more shallow than 0.25 m, ADCIRC-2DDI sets the bathymetry to 0.25 m at the corresponding points.

COMPARISON BETWEEN MODEL RESULTS AND HISTORICAL DATA
Discussion and Conclusions

Table 1 and Figure 4 present 16 tidal stations that were used in the analysis of model results. In Table 2, we compare model M_2 amplitudes and phases (from the harmonically analyzed results of the last 45 days of the 90-day simulation) to historical amplitudes and phases. Amplitude errors are presented as a percentage and are normalized by the historical value. Phase errors are presented as the absolute difference between historical and model phases. Good model agreement with the historical can be seen upon close examination of Table 2. The only amplitude error above 10% is located at Station No. 2 and it is 11.5%. However, it should be noted that the corresponding absolute error is merely eight centimeters, a depth that for practical applications would be negligible. The majority of the phase errors are 5° or less, with the exceptions at Stations 2 and 11 being near 20° error and Station 13 near 10° error. Six additional constituents (Q_1, O_1, K_1, N_2, S_2, K_2) were analyzed and show the same general level of performance. Those results are not included in the interests of the brevity of this paper.

Table 1. Tidal stations

No.	Gage ID	Description	Latitude	Longitude
1	NOS 8661070	Springmaid Pier, S.C.	33.655	-78.918
2	NOS 8665530	Charleston, Cooper R. Entrance, S.C.	32.783	-79.917
3	NOS 8720587	St. Augustine Beach, Atlantic Ocean	29.857	-81.263
4	NOS 8721020	Daytona Beach (Ocean), Fla.	29.147	-80.963
5	NOS 8722670	Lake Worth Pier, N. Miami Bch., Fla.	26.612	-80.033
6	NOS 8723080	Haulover Pier, N. Miami Bch., Fla.	25.903	-80.120
7	NOS 8723170	Miami Beach (City Pier), Fla.	25.768	-80.130
8	NOS 8723962	Key Colony Beach, Fla.	24.718	-81.018
9	NOS 9710441	Settlement Point, Grand Bahamas	26.710	-78.997
10	IHO 315	Nassau	25.083	-77.350
11	IHO 313	Eleuthera	24.767	-76.150
12	IHO 338	IAPSO #30-1.2.10	26.467	-69.333
13	IHO 355	IAPSO #30-1.2.14	28.017	-76.783
14	IHO 357	IAPSO #30-1.2.4	28.133	-69.750
15	IHO 360	IAPSO #30-1.2.15	28.450	-76.800
16	IHO 41	IAPSO #30-1.2.11	30.433	-76.417

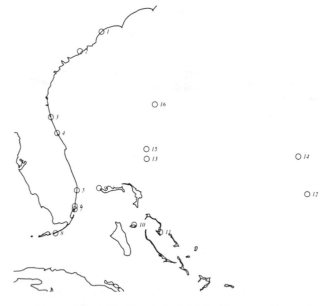

Fig. 4. The location of tidal stations (see Table 1).

Finally, time series plots are presented in Figures 5 and 6. These plots are re-synthesized model and historical time series using the seven constituents (Q_1, O_1, K_1, N_2, M_2, S_2, K_2) from the harmonic analysis of model output and historical data. Model amplitude and phasing closely matches that of the historically-generated time series.

Table 2. M_2 amplitude and phase errors: SJR mesh simulations compared to historical amplitudes and phases.

Station No.	Amplitude Historical (m)	Amplitude Model (m)	Error (%)	Phase Historical (°)	Phase Model (°)	Error (°)
1	0.751	0.697	7.2	357.20	353.47	3.73
2	0.758	0.671	11.5	12.30	349.56	22.74
3	0.662	0.661	0.2	14.30	8.83	5.47
4	0.595	0.571	4.0	10.90	6.64	4.26
5	0.419	0.392	6.4	13.40	9.19	4.21
6	0.373	0.353	5.4	19.60	14.23	5.37
7	0.373	0.345	7.6	20.40	15.34	5.06
8	0.243	0.220	9.6	41.00	35.39	5.61
9	0.405	0.389	4.1	10.60	6.41	4.19
10	0.379	0.369	2.7	7.72	3.68	4.04
11	0.321	0.331	3.2	20.32	2.18	18.14
12	0.327	0.300	8.2	3.00	358.97	4.03
13	0.402	0.379	6.0	9.00	359.17	9.83
14	0.351	0.326	7.1	1.00	357.58	3.42
15	0.411	0.384	6.7	2.00	358.59	3.41
16	0.441	0.404	8.5	358.00	355.66	2.34

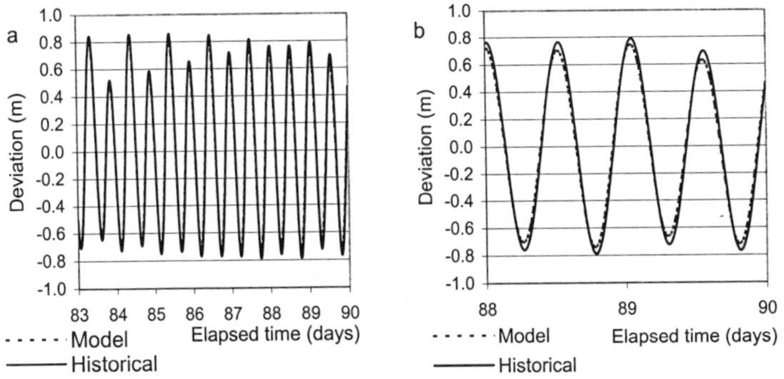

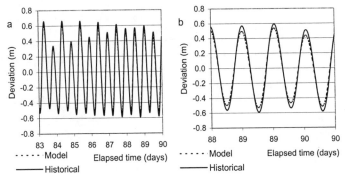

Fig. 5. Time series at Station 4 for the last 7 (a) and 2 (b) days of model time.

Fig. 6. Time series at Station 16 for the last 7 (a) and 2 (b) days of simulated time.

ACKNOWLEDGEMENTS
This research was funded by the St. Johns River Water Management District, Florida.
We especially thank Ann Mukai and Dr. Joannes J. Westerink, who supplied us with the
Eastcoast2001 mesh and simulation results. We also thank Dr. A. M. Baptista of the
Oregon Graduate Institute for allowing us to use the software package Gredit.

REFERENCES
Hagen, S. C. 1998. *Finite element grids based on a localized truncation error analysis*,
    Ph.D. Dissertation, Department of Civil Engineering and Geological Sciences,
    University of Notre Dame, Ind.
Hagen, S. C., Westerink, J. J., and Kolar, R. L. 2000. One-dimensional finite element
    meshes based on a localized truncation error analysis. *International Journal of
    Numerical Methods in Fluids,* **32,** 241-261.
Hagen, S.C., Westerink, J. J., and Kolar, R. L. and Horstman, O. 2001. Two-
    dimensional, Unstructured Mesh Generation for Shallow Water Models: Comparing
    a Wavelength-based Approach to a Localized Truncation Error Analysis.
    International Journal of Numerical Methods in Fluids, **35**, 669-686.
Le Provost, C., Lyard, F., Molines, J. M., Genco, M. L., and Rabilloud, F. 1998. A
    hydrodynamic ocean tide model improved by assimilating a satellite altimeter-
    derived data set. *Journal of Geophysical Research*, **103**, 5513-5529.
Luettich, R.A., Westerink, J. J., and Scheffner, N. W. 1991. *ADCIRC: An Advanced
    Three-Dimensional Circulation Model for Shelves, Coasts and Esturaries, Report 1:
    Theory and Methodology of ADCIRC-2DDI and ADCIRC-3DL*, Technical Report
    DRP-92-6, Department of the Army, USACE, Washington D.C.
Mukai, A. 2001. *A finite element mesh for the Western North Atlantic*, Masters Thesis,
    Department of Civil Engineering and Geological Sciences, University of Notre
    Dame, Ind.
National Geophysical Data Center. 2001. http://rimmer.ngdc.noaa.gov/coast/
    getcoast.html. On-line database. Boulder, Col.
Roe, M. J. 1998. *Achieving a Dynamic Steady State in the Western North Atlantic/Gulf
    of Mexico/Caribbean Using Graded Finite Element Grids*, Master's Thesis,
    Department of Civil Engineering and Geological Sciences, University of Notre
    Dame, Ind.
Scheffner, N. W. and Fulton C. C. 2001. *Coast of South Carolina Storm Surge Study*,
    U.S. Army Corps of Engineers Engineer Research and Development Center. Coastal
    and Hydraulics Laboratory. Report ERDC/CHL TR-01-11.
Westerink, J. J., Luettich, R. A., and Muccino, J. C. 1994. Modeling tides in the
    western North Atlantic using unstructured graded grids. *Tellus*, **46A**, 178-199.
Westerink, J. J., Luettich, R. A., Blain, C. A. and Hagen, S. C. ed. Carey, G. F. 1995.
    Surface Elevation and Circulation in Continental Margin Waters. in *Finite Element
    Modeling of Environmental Problems*, Wiley.

Evidence of Near-Surface Density Stratification as a Factor in Extreme Seiche Events at Ciutadella Harbor, Menorca Island

Graham S. Giese and David C. Chapman [1]

Abstract

Large-amplitude harbor seiches at Ciutadella on Menorca Island in the western Mediterranean are related to local atmospheric conditions and it has been suggested that the seiche forcing mechanism involves the transfer of atmospheric energy to baroclinic modes in offshore waters. To explore this concept we examined the relationship between seasonal patterns of harbor seiche activity and water temperature. Results indicate that during the season of harbor water cooling, the probability of large harbor seiche activity varies linearly with water temperature. During the season of water warming, however, the probability of large seiche activity varies exponentially with the rate of water warming which is maximum at the summer solstice. These patterns in large seiche activity are in accord with published accounts and may result from seasonal changes in near-surface density stratification of the water offshore of Ciutadella.

Introduction

Many studies of the "rissaga" phenomenon, extreme resonant harbor oscillations at Ciutadella on Menorca Island in the western Mediterranean (Figure 1), have been reviewed by Rabinovich and Monserrat (1996). All evidence supports the view that these destructive short period sea-level oscillations are related to local atmospheric conditions. Previously, we suggested that the atmospheric forcing involves wind-driven flows of stratified water through the shallow passage separating Menorca and Mallorca Islands (Giese and Chapman 1998). We speculated that such flows produce highly nonlinear internal waves that transfer energy to a wide range of barotropic modes in the passage (cf. Chapman and Giese 1990), and that the resulting high-frequency barotropic modes excite resonant oscillations in nearby bays and harbors. This hypothesis differs from other

---

[1] Woods Hole Oceanographic Institution, Woods Hole, MA 02543
email: ggiese@whoi.ed, dchapman@whoi.edu

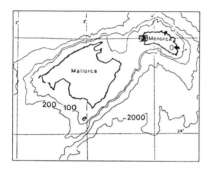 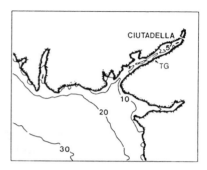

Figure 1: (left) Location of Ciutadella Harbor area (shaded box) on west coast of Menorca Island. Note the shallow water between Menorca and Mallorca. (right) Ciutadella Harbor, showing location of tide and water temperature recorder (TG). [Modified from Gomis et al. (1993) Fig. 1].

mechanisms proposed to explain the phenomenon (e.g., Tintoré et al. 1988) in that it assigns a key role to the formation of, and transfer of energy from, internal waves.

Methods

    The propagation of internal waves in the sea depends upon the density stratification of the sea water, and in the interior of the western Mediterranean, where there is little influence of rivers, variations in near-surface density stratification depend primarily on water temperature. In this study we examine the relationship between water stratification and seasonal harbor seiche patterns, using the temperature of near-surface water in Ciutadella Harbor as a proxy for that of near-surface water in the passage outside the harbor. Such a relationship is expected, indeed necessary, if internal waves are the key to the transfer of atmospheric energy to the harbor seiches.

    Our sea level and temperature recorder, with sensors at a depth of 3 m, operated in Ciutadella Harbor (TG, Figure 1) between 12 June 1992, and 25 December 1997, under the care of our colleagues at the University of the Ballearic Islands (see Acknowledgements). Over that period we obtained 1-minute sea level and temperature data for 1313 days.

    The sea-level record (Figure 2) is dominated by astronomical tides and resonant seiche oscillations with a spectral peak period of 10.6 minutes. Daily sea-level variance at the fundamental seiche frequency (9-12 minute band) provided the initial measure of seiche activity for this study, while the daily maximum sea-level variation over a single seiche period was used for later analyses. The

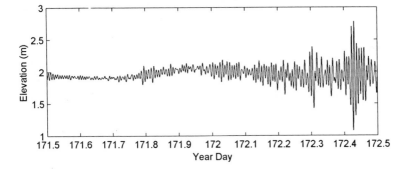

Figure 2: One-minute sea-level data from Ciutadella Harbor, 20 June (12:00) - 21 June (12:00) 1992.

temperature data show a pronounced seasonal variation as well as smaller daily variations resulting from diurnal solar heating and less-regular changes due primarily to wind-driven upwelling and downwelling. The daily water temperature data used in this study are the daily means of the 1-minute measurements.

Seasonal Patterns

We reduced these data by averaging the temperature for each year-day and found that the minimum daily mean temperature, 13.3°C, occurred on year-day 60 (YD60) and the maximum, 27.4°C, occurred on YD235. Figure 3a shows the mean year-day temperatures in a plot that begins and ends with the coolest day, YD60. The modeled year-day temperatures (solid curves) were determined by fitting third-order polynomials separately to the warming and cooling segments of the mean year-day temperature series. The equations for the modeled temperature are:

$$T_w = -2.32 \times 10^{-6} Y_D^3 + 1.18 \times 10^{-3} Y_D^2 - 9.94 \times 10^{-2} Y_D + 15.8 \qquad (1)$$

and

$$T_c = 1.61 \times 10^{-6} Y_D^3 - 1.29 \times 10^{-3} Y_D^2 + 0.246 Y_D + 19.9 \qquad (2)$$

where T_w is modeled temperature during the warming period, T_c is modeled temperature during cooling and Y_D is the year-day. Year-day for the months of January and February of the cooling period is the actual year-day plus 366.

Figure 3b shows the time derivatives of the modeled temperatures, i.e., dT_w/dt and dT_c/dt. During the warming period, temperature increases at an increasing rate until a maximum rate-of-warming (dT_w/dt) is reached at YD171, after which temperature increases at a decreasing rate until the maximum temperature is reached (YD235). YD171 corresponds to 21 June, i.e., the maximum rate-of-warming occurs about the same time as the summer solstice. The summer and winter solstices are indicated by vertical dotted lines.

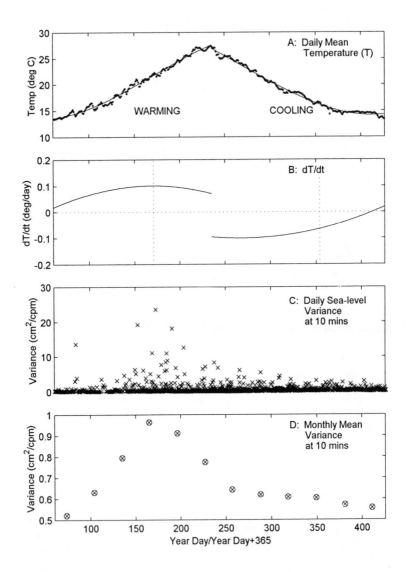

Figure 3: (a) Mean water temperature (dots) for each year-day together with third-order least squares fits (solid lines). (b) Time derivatives of temperature vs. year-day curves. (c) Daily seiche activity. (d)
Smoothed monthly mean seiche activity.

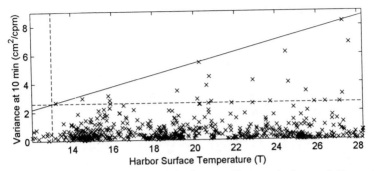

Figure 4: Daily seiche activity during cooling period vs. daily water temperature (see text for details).

Pronounced seasonal variations in activity of the 10.6-minute fundamental harbor seiche mode are apparent in Figure 3c which shows the daily sea-level variance at 10 minutes as a function of year-day. Seiching reaches a maximum during the warming period. During cooling seiche activity decreases steadily in a pattern similar to that of the harbor water temperature, but during warming the pattern of activity more closely resembles that of the rate-of-warming (Figure 3b). The smoothed monthly means (Figure 3d) show the pattern more clearly - seiche activity increases rapidly from a minimum in March to a maximum in June; thereafter the monthly activity decreases. Because activity patterns differ so markedly in the warming and cooling periods, we examine seiche behavior in the two seasons separately.

Activity During the Cooling Period

Figure 4 presents daily seiche activity during cooling plotted with respect to observed daily mean harbor water temperature. The solid line was drawn through the maximum activity values so as to provide a linear envelope to the distribution. The equation of the line is:

$$A_{max} = 0.4T - 2.6 \qquad (3)$$

in which A_{max} is the activity maximum and T is water temperature.

We do not know the extent to which factors not involving water temperature contribute to the seiching, but we estimate the maximum value of the background seiching (i.e., independent of water temperature) to be 2.6 cm²/cpm (dashed horizontal line in Figure 4). According to (3), maximum activity reaches the background level at a temperature of 13°C (dashed vertical line in Figure 4), so we rewrite (3) to reflect that fact:

$$A_{max} = 0.4(T - 13) + 2.6 \qquad (4)$$

The minimum temperature of the western Mediterranean is close to 13°C and is typically found within 50-60 m of the surface in the study area, so we designate

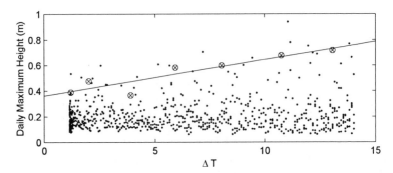

Figure 5: Daily maximum seiche heights during cooling period (dots)
vs. modeled water temperature minus 13°C together with estimates
of expected maximum seiche height for all but 2.3% of observed days
($\times$ filled circles and solid line).

the temperature difference between this minimum and the harbor surface water
temperature as ΔT. Summarizing, our observations suggest that during the 6-
month cooling period the maximum level of resonant harbor seiching at Ciutadella
above the background level is proportional to the difference between the harbor
water temperature and the minimum offshore water temperature, i.e., $A_{max}(T) \sim$
ΔT.

Useful though this relationship is in suggesting processes that may control the
harbor seiche activity during the cooling period, it is of limited use to the harbor
manager who wants to know the probability of extreme activity at any time. For
that purpose we make use of daily maximum seiche height (D_{max}) rather than
daily sea-level variance at the seiche frequency. While the latter measure is a
more robust statistic (being based on 1440 data points), maximum height is of
greatest concern to harbor managers. The search for probablility estimates is
much simplified by the fact that D_{max} during both cooling and warming have
approximately lognormal distributions. Thus the quantity $\log_{10}(D_{max} - C)$ is
examined, where C is the minimum value of D_{max} for the chosen period.

We sorted the values of $\log_{10}(D_{max} - C)$ during cooling by temperature, and
then divided the result into 7 subgroups for which we calculated their mean values
and their standard deviations. The results are presented in Figure 5 (converted
back to D_{max}) in the form of circles filled with $\times$ to represent those means plus
two standard deviations of each subgroup, together with the linear least squares
best fit:

$$D_{max}(97.7\%) = (0.0283\Delta T) + 0.364 \qquad (5)$$

in which $D_{max}(97.7\%)$ is the expected maximum seiche height for all but 2.3% of
observed days. During the cooling period, maximum seiche height varies linearly
with water temperature minus 13°C.

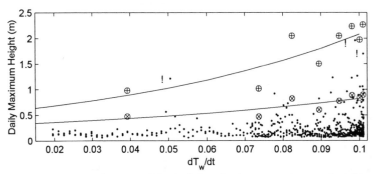

Figure 6: Daily maximum seiche heights during warming period (dots) vs. dT_w/dt together with estimates of expected maximum seiche height for all but 2.3% of observed days ($\times$ filled circles and lower solid line) and for all but 0.1% of observed days ($+$ filled circles and upper solid line).

Activity During the Warming Period

Earlier we noted that during the warming period both seiche activity and the harbor water rate-of-warming, dT_w/dt, reach a maximum in June. Following the methodology used for the cooling period, we explore this relationship by plotting D_{max} during warming against dT_w/dt (Figure 6). The obvious disparity in data density results in part from the sparcity of measurements from March through May, in part from the decreasing rate of temperature change as the maximum (summer solstice) is approached, and in part because the data "fold back" into decreasing rate-of-warming values following the solstice.

We calculated $\log_{10}(D_{max} - C)$ for each day, divided the series into 8 subgroups and calculated the mean and standard deviation for each. We calculated the linear least squares best fit for the expected $\log_{10}(D_{max} - C)$ for all but 2.3% (mean plus two standard deviations) and 0.1% (mean plus three standard deviations) of observed days. Figure 6 shows these approximations converted back to the basic variable, D_{max}. The circles filled with $\times$ show the means plus two standard deviations, and those with $+$ show the means plus three standard deviations. The equations of the lines are:

$$D_{max}(97.7\%) = 10^{[(5.09dT_w/dt)-0.63]} + 0.061 \tag{6}$$

$$D_{max}(99.9\%) = 10^{[(6.38dT_w/dt)-0.33]} + 0.061 \tag{7}$$

Thus we see that maximum seiche height varies exponentially with temporal gradient of water temperature during the warming period.

Annual Maximum Seiche Height Probabilities

Maximum seiche height probabilities for both the cooling and warming periods, expressed in terms of year-day, are presented in Figure 7. First the

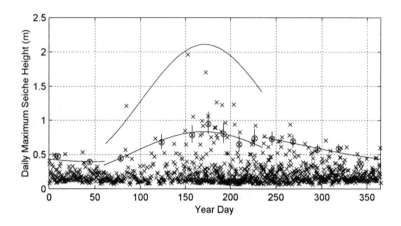

Figure 7: Daily maximum seiche heights for all days ($\times$'s) vs. year-day together with estimates of expected maximum seiche height for all but 2.3% of observed days ($\times$ filled circles) and 80% confidence intervals for the estimates (short vertical lines). The two low curves represent (5) and (6), and the high curve (7), evaluated in terms of year-day.

$\log_{10}(D_{max} - C)$ for each period were ordered by time and divided into 7 subgroups for which the means and standard deviations were calculated. The means plus two standard deviations of each subgroup, expressed in terms of daily maximum seiche height, are indicted in the figure by circles filled with $\times$. Also shown are the 80% confidence limits of the estimates. The lower two solid lines represent equations (5) and (6), predicted maximum seiche heights for the cooling and warming periods respectively at the 97.7% level, evaluated in terms of year-day. For the cooling period this was accomplished using (2) minus 13°C to obtain ΔT; while for the warming period we used the time derivative of (1) to obtain dT_w/dt. The highest solid line represents equation (7) evaluated in terms of year-day and indicates that, at the 99.9% level, daily maximum seiche heights larger than 2 m can be expected during the 6-week period centered at the summer solstice.

Tidal Effect on Seiche Activity

Giese and Chapman (1997, figure 4) noted that the days with greatest seiche activity occurred during periods with very small tidal range. The data points (events) coded '!' in Figure 6 were used as illustrations of that association in the earlier report. The fact that those large events also occurred during the warming period suggests a seasonal pattern in the relationship between seiche activity and tidal range. To explore that possibility, the daily seiche activity data were divided into two groups, one including all days with predicted tidal activity less than the median value, and the other having the days when tidal activity was larger than

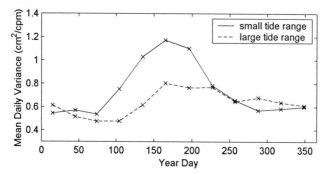

Figure 8: Monthly seiche activity for days with small tidal range (solid line) and days with large tidal range (dashed line).

(or equal to) the median. Figure 8 shows monthly mean seiche activity (mean daily sea-level variance at 10 minutes) for days with small tides ($\times$'s) and with large tides (+'s). Seiche activity on days with small tides is markedly greater during the warming months of April through July than on days with large tides. There is little difference during the other months.

Discussion and Conclusions

The seasonal pattern of extreme seiche activity at Ciutadella, with a maximum near the summer solstice, is one of the most widely recognized characteristics of this phenomenon. In Ciutadella itself rissaga events are sometimes known as "Rissagues de Sant Joan" (Pons 1987) referring (in Catalonian) to St. John's day, the 24th of June. Rabinovich and Monserrat (1996, Table 1) include 10 events at Ciutadella in a listing of large seiche oscillations in the general area. Of the 10 events, five occurred in the 6-week period centered at the summer solstice, nine occurred within June or July, and one, the largest event with a maximum height of 3 m, occurred on the 21st of June.

Results of the present study show that this characteristic seasonal maximum can be explained in terms of the rate of surface water warming. We suggest that this relationship results from increased density stratification within the upper layer of the shallow water in the passage between Menorca and Mallorca Islands. Wave guides produced by such stratification could trap internal wave energy and focus it into shallow waters thereby exciting barotropic normal modes (Chapman and Giese 1990). Similarly, we suggest that the enhancement of seiche activity that occurs when tides are extremely small during the period of maximum warming results from enhanced stratification due to reduced tidal mixing.

Similar reasoning can be applied to our finding that during the cooling period Ciutadella harbor seiche activity can be explained in terms of the temperature difference between the surface water and the minimum offshore water temperature, the 13°C water that lies within approximately 60 m of the surface. This relationship may result from the fact that during the cooling period the upper layer is not stratified but is well mixed by the convective sinking of cooled surface water. Thus the shallowest internal wave guides are at the base of the mixed layer and they increase in depth and decrease in strength as the season prgresses. Both factors would decrease the energy available to barotropic normal modes from internal waves focussed into shallow waters.

Acknowledgements

We thank our colleagues and friends whose contributions made these field results possible: J. Tintoré, D. Gomis, S. Monserrat and S. Alonso at the Universitat de les Illes Balears (UIB) in Palma de Mallorca; A. Jansa at the Centro Meteorologico Zonal in Palma de Mallorca; M. Garcies at the Universitat Politecnica de Catalunya in Barcelona; and M. Mora of Club Nautic Ciutadella. Major funding was provided by the Office of Naval Research under grant No. N00014-92-J-1200. UIB research was supported through DGICYT project PB89-0428.

References

Chapman, D.C., and G.S. Giese, 1990. A model for the generation of coastal seiches by deep-sea internal waves. *J. Phys. Oceanogr.*, 20, 1459-1467.

Giese, G.S., and D.C. Chapman, 1998. Hazardous harbor seiches, tides, wind and baroclinicity, in *Ocean Wave Measurement and Analysis*, Edge, B.L. and J.M. Hemsley eds., 1, 208-218, American Society of Civil Engineers, Reston, VA.

Gomis, D., S. Monserrat and J. Tintore, 1993. Pressure-forced seiches of large amplitude in inlets of the Balearic Islands. *J. Geophys. Res.*, 98, 14437-14445.

Pons, P.M., 1990. Ses rissagues de Sant Joan, in *Les Rissagues de Ciutadella i Altres Oscillacions de Nivell de la Mar, de Gran Amplitud a la Mediterrania*, Institut Menorqui d'Estudis, Ciutadella de Menorca, Spain.

Rabinovich, A.B. and S. Monserrat, 1996. Meteorological tsunamis near the Balearic and Kuril Islands: Descriptive and statistical analysis. *Natural Hazards*, 13, 55-90.

Tintoré, J., D. Gomis, S. Alonso and D.P. Wang, 1988. A theoretical study of large sea level oscillations in the western Mediterranean. *J. Geophys. Res.*, 93, 10797-10803.

Observations and predictions of tides and storm surges along the Gulf of Mexico

Y.J. Nam[1], D. T. Cox[1], P. Tissot[2], P. Michaud[2]

ABSTRACT: This paper shows that the subtidal energy on the Texas Coast and in Corpus Christi Bay is large due to meteorological events and that the water level anomaly can be larger than the astronomical tide itself. The relative importance of remote and local forcing on the subtidal response in Corpus Christi Bay was studied using water level and wind data observed during the winter and spring months from 1998 to 2000. The multiple coherence squared between the water level response inside the bay and the local and remote forcing was high ($MCS > 0.8$ for most locations), indicating that the local wind stress and water level on the coast are the primary forcing mechanisms inside the bay over the range of frequencies studied. The study further confirmed the importance of remote forcing for the water level response as predicted by the analytical model of Garvine (1985).

1. INTRODUCTION

The need for reliable water level forecasting is increasing with the trend toward deep-draft vessels, particularly for shallow water ports along the Gulf of Mexico (NOAA, 1999). Nine of the twelve largest U.S. ports are located along the Gulf of Mexico, and ports served by the Mobile Bay Entrance and Galveston Bay Entrance account for 46% of the total U.S. tonnage (NOAA, 1999). Although the astronomical tides in the Northern Gulf of Mexico are easily predicted by conventional harmonic analysis, it is difficult to accurately predict the total water level fluctuations because of frequent meteorological events, such as the passage of strong cold fronts. Our inability to accurately predict water level anomalies (difference between the observed water level and the tide prediction) can have severe consequences. In Galveston Bay there were over 1,240 ship groundings between 1986 and 1991, with a significant number of incidents involving petrochemicals.

[1]Div. of Coastal and Ocean Engrg., Dept. of Civil Engrg., Texas A&M Univ., College Station, TX 77843-3136 USA; dtc@tamu.edu

[2]Div. of Nearshore Research, Conrad Blucher Institute for Surveying and Science, Texas A&M Univ.–Corpus Christi, Corpus Christi, TX 78412 USA; PTissot@envcc00.cbi.tamucc.edu

To improve navigation and safety in these waterways, NOAA has established the Physical Oceanographic Real-Time System (PORTS) which includes the near real-time monitoring and reporting of water levels and meteorological conditions via telephone or Internet (www.co-ops.nos.noaa.gov/). Other agencies are developing real-time forecasting models for estuarine hydrodynamics of oil spill response and for search and rescue operations along the Texas coast (hyper20.twdb.state.tx.us/bhydpage.html). Although both systems greatly reduce navigational and environmental hazards along the northern coast of the Gulf of Mexico, they rely on harmonic analysis for either the level prediction in the estuary itself or as a seaward boundary condition for an estuarine hydrodynamic model. Presently, they do not incorporate meteorological effects. This raises two questions:

1. What is the relative importance of the remote forcing (water level at the mouth of the estuary) to the local forcing (wind stress over the estuary) for subtidal water level, setup, and current response in the estuary?

2. To what extent can the remote forcing be predicted using a simple empirical model relating meteorological forecasts to water level anomaly?

Previously, Guannel et al. (2001) studied the entrance to Galveston Bay and confirmed the importance of the remote forcing on water levels inside the bay and the importance of the local forcing on the surface slope, consistent with the analytical model of Garvine (1985). In this paper, we look at observations in Corpus Christi Bay at one location on the open coast (014 Bob Hall Pier), and at four locations in Corpus Christi Bay (009 Port Aransas; 001 Naval Air Station; 008 Texas State Aquarium; and 011 White Point) as shown in Fig. 1. Data for these stations are provided by the Conrad Blucher Institute for Surveying and Science at Texas A&M University-Corpus Christi as part of the Texas Coastal Ocean Observation Network (TCOON) (Michaud et al., 1994). TCOON consists of over 40 stations with real-time access made available through the Internet and other media and has been in operation for over 10 years. All stations report water level, and many others report wind speed, direction, gust, air temperature, water temperature and barometric pressure. A subset of the archived data were used for this study from early December to the end of March for years 1998 to 2000, for a total of 318 days of data at hourly intervals. The choice of data was determined by data availability and by the intention to restrict the study to winter and spring months when cold front frequently passed over the study area. Tropical events and sea breeze activity associated with summer and fall months were excluded.

Fig. 2 shows observation for Port Aransas for $60 < Jd < 80$, 1998, during the passage of a strong cold front. The top panel shows the measured water level (solid) and that predicted by harmonic analysis (dashed). The figure indicates that the water level anomaly caused by the meteorological event is as large as the tide range itself.

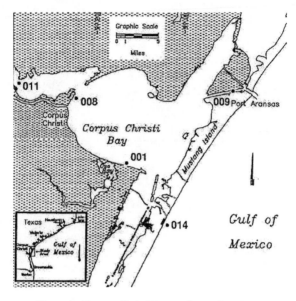

Figure 1: Corpus Christi Bay and gage locations.

Wind direction plays an important role in determining the magnitude and sign of the anomaly as was shown for Galveston Bay (Cox et al., 2002). The figure also indicates that the response of the system is on the order of only a few hours after the passage of the front.

Fig. 3 shows the power spectrum for the water levels in Corpus Christi Bay. The figure indicates the diurnal and semidiurnal tide components are damped from outside the bay (014) to the upper reaches of the bay (011). The figure also indicates that the subtidal energy is large and is not damped. Fig. 4 shows a portion of the filtered water level, η, water level setup , and wind stress. Data were filtered using a Lanczos filter with a 36 hour cutoff to remove the tidal variability and high frequency wind fluctuations, and the wind stress was estimated from the wind speed and direction following Wu (1980). The wind stress was considered rectilinear, either North-South and East-West or Shore Normal and Shore Parallel (Guannel, 2001). For this paper, winds recorded at 014 were used and assumed to be representative of the wind over the bay. The water level setup is simply calculated as the difference of the gages inside the bay, $\eta_{009} - \eta_{008}$ (solid) and $\eta_{009} - \eta_{001}$ (dash-dot). The top panel shows that the anomaly fluctuation can be as large as ± 0.2 m which is large compared to the rms of the meteorological tide at Port Aransas of 0.16 m. The figure also shows that there is a small lag on the order of several hours between the anomaly on the coast (014) and inside the bay (009). The three large positive anomalies in the upper panel ($Jd \simeq 15$,

$Jd \simeq 22$; and $Jd \simeq 28$) generally occur when there is a large component of wind stress from the east (dash-dot line in lower figure).

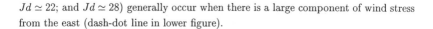

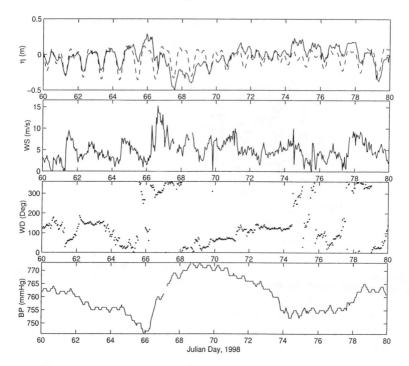

Figure 2: Observations for 009 Port Aransas Station in 1998. Top panel shows observed water level (solid) and predicted using harmonic analysis (dashed); second, third and fourth panels show observed wind speed, direction and barometric pressure. Wind and barometric pressure were measured at 014 and assumed constant over the study area.

2. Local and Remote Forcing

A number of remote mechanisms can cause water level fluctuations at the mouth of an estuary, including winds blowing parallel to the coast and the associated Eckman transport. Local winds act directly on the the bay through the surface wind stress. Garvine (1985) used scaling arguments and a simple analytical model to show that subtidal variations of water levels inside the estuary are dominated by the remote effects because the length scale of the estuary is short compared to the subtidal wavelength. The depth-averaged current is a response to the conservation of mass. The surface slope was shown to be dominated by the local wind. In considering the orientation of the estuary to the coast, Garvine (1985) demonstrated that the local and remote forcing

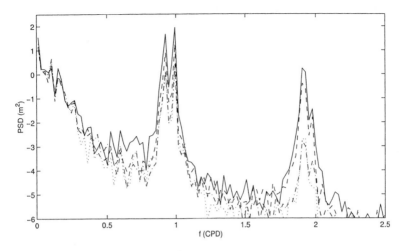

Figure 3: Power spectrum of water level for η_0 (014, solid), η_1 (009,dashed), η_2 (001, dash-dot), η_3 (008, dotted).

should have either a combined or opposite effect if the estuary is aligned parallel to the coast, and the two mechanisms should be independent if the estuary is perpendicular to the coast. Smith (1977) found that for Corpus Christi Bay there was evidence of local forcing dominance at shorter time scales (60 to 100 hours) and remote forcing dominance at lower frequencies on the bay volumes. Using a month-long set of water level and current observations, Wong and Moses-Hall (1998) confirmed the importance of the remote forcing on water levels for Delaware Bay, but found that the local wind effect dominates the current fluctuations, particularly the current structure.

The analysis method used by Wong and Moses-Hall considers the multiple and partial coherence of a two input, one output system. This method is applied here to Corpus Christi Bay which is a lagoonal estuary. In the frequency domain, the water level response η at any location in the estuary can be written

$$\eta_j = H_{1j}\,\eta_0 \;+\; H_{2j}\,\tau_{wj} \;+\; \epsilon_{\eta j} \quad j = 1, 2, ..., n \tag{1}$$

where η_j represents the water level at the j-th estuary station with $j = 1, 2, ..., n$ representing the $n = 4$ TCOON stations used in Corpus Christi Bay; η_0 is the observed coastal water level (Station 014) representing the remote effects; τ_{wj} is the local wind stress; and $\epsilon_{\eta j}$ represents the noise that is not coherent with either η_0 or τ_w. H_{1j} and H_{2j} are complex quantities representing the transfer functions between the remote (H_1) and local (H_2) forcing and response of the estuary. A similar equation can be written for the water level gradient (setup) or currents in the bay.

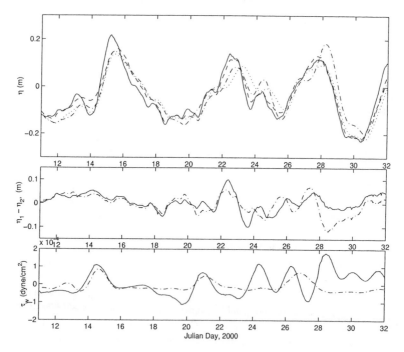

Figure 4: Observations of filtered (36 hr cut off) water level (top), water level setup (middle), and wind stress (lower) for Corpus Christi Bay. Top: water level η_0 (014, solid), η_1 (009, dash), η_2 (001, dash-dot), η_3 (008, dotted). Middle: water level setup, $\eta_1 - \eta_3$ (solid), $\eta_1 - \eta_2$ (dash-dot). Bottom: wind stress, North-South (solid), East-West (dash-dot). $14 < Jd15$ corresponds to storm with winds from the northeast.

Fig. 5a,b shows the multiple coherence squared (MCS) for the two input (η_0, τ_w), one output (η_j) system where j represents the four locations considered in the bay and where τ_w is computed using either the shore normal or shore parallel winds. For Fig. 5a,b the MCS is high (> 0.8) for most of the locations indicating that the estuary response is primarily a function of these two mechanisms. Other mechanisms such as river discharge are less important for the data considered here, with the possible exception of the Station 011. Overall, the MCS is lower in Corpus Christi Bay, however, compared to the previous study at Galveston Bay (Guannel et al, 2001), indicating that the assumption of rectilinear winds may not be as suitable as in the earlier location.

Fig. 5c shows the partial coherence squared (PCS) between η_0 (remote forcing) and η_j (response) with the local forcing shut down (Wong and Moses-Hall, 1998). The figure indicates that the PCS is high near the mouth (009) and decreases slightly the head (008, 001) and more so at the far reaches of the bay (011). Overall, the PCS is high (> 0.6) indicating that the remote effect is primarily responsible for the water level. Fig. 5e shows that the PCS is low and only slightly above the 95% significant level (0.2).

Fig. 6a-f shows similar results considering the local forcing as either North-South or East-West directed. The only curve which is significantly different than the corresponding curve in Fig. 5 is that for the station at the furthest reach of the bay (011). For this station, the MCS appears to be highest considering the East-West directed wind stress which is consistent with the analytical model of Garvine (1985).

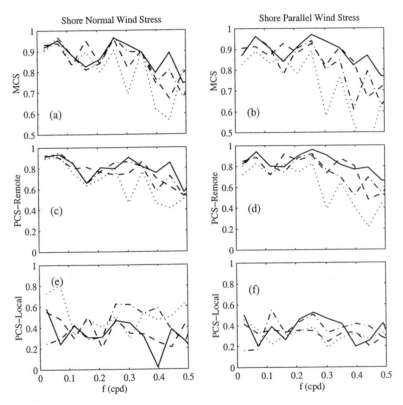

Figure 5: Water level response for shore normal wind stress (a, c, e) and shore parallel (b, d, f) for 009 (solid), 001 (dash-dot), 008 (dash), and 011 (dotted). The 95% significance level is 0.20 computed with 30 DOF.

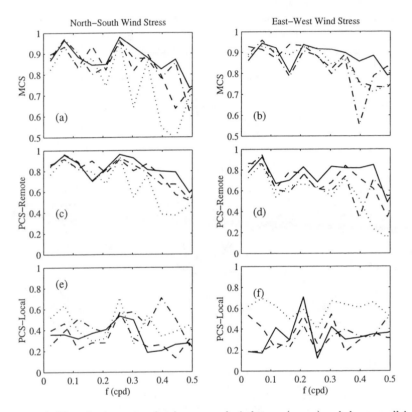

Figure 6: Water level response for shore normal wind stress (a, c, e) and shore parallel (b, d, f) for 009 (solid), 001 (dash-dot), 008 (dash), and 011 (dotted). The 95% significance level is 0.20 computed with 30 DOF.

4. Summary and Conclusions

This paper shows that the subtidal energy on the Texas Coast and in Corpus Christi Bay is large due to meteorological events and that the water level anomaly can be larger than the astronomical tide itself. The relative importance of remote and local forcing on the subtidal response in Corpus Christi Bay was studied using water level and wind data observed during the winter and spring months from 1998 to 2000. The multiple coherence squared between the water level response inside the bay and the local and remote forcing was high ($MCS > 0.8$ for most locations), indicating that the local wind stress and water level on the coast are the primary forcing mechanisms inside the bay over the range of frequencies studied. The study further confirmed the importance of remote forcing for the water level response as predicted by the analytical model of Garvine (1985).

References

Cox, D.T., Tissot, P., and Michaud, P., (2002) "Water level observations and short-term predictions including meteorological events for the entrance of Galveston Bay, Texas," *J. of Wtwy., Port, Coast., and Oc. Engrg.* In press, January issue.

Garvine, R.W. (1985) "A simple model of estuarine subtidal fluctuations forced by local and remote wind stress," *J. Geophys. Res.* 90, C6, 11945–11948.

Guannel, G., Tissot, P., Cox, D., Michaud, P.(2001) "Local and Remote Forcing of Subtidal Water Level and Setup Fluctuations in Coastal and Estuarine Environments," *Proc. Coastal Dynamics '01*, ASCE, 443-452.

Guannel, G. (2001) "Observations of local and remote forcing on subtidal variability in Galveston Bay, Texas," *M.Sc. Thesis*, Texas A&M University, College Station, TX.

Michaud, P.R., Thurlow, C.I., and Jeffress, G.A. (1994) "Collection and dissemination of marine information from the Texas Coastal Ocean Observation Network," *Proc. U.S. Hydr. Conf.*, Spec. Publ. 32, Hydrogr. Soc., Norfolk, VA, 168–173.

NOAA (1999) "Assessment of the National Ocean Service's tidal current program," *NOAA Tech. Rpt. NOS Co-OPS 022*, United States Department of Commerce.

Smith, N.P. (1977) "Meteorological and tidal exchanges between Corpus Christi Bay, Texas, and the northwestern Gulf of Mexico," *Estuarine and Coastal Marine Sci.*, 5, 511–520.

Tissot, P., Cox, D., and Michaud, P. (2001) "Neural network forecasting of storm surges along the Gulf of Mexico," *Proc. Waves 2001*, ASCE, in press.

Wu, J. (1980) "Wind stress coefficients over sea surface near neutral conditions – A revisit," *Journal of Physical Oceanography*, 10, 727–740.

Wong, K.-C. and Moses-Hall, J.E. (1998) "On the relative importance of the remote and local wind effects to the subtidal variability in a coastal plain estuary," *J. Geophys. Res.* 103, C9, 225–232.

Vessel-Generated Long-Wave Measurement and Prediction in Corpus Christi Ship Channel, TX

S. Fenical[1], H. Bermudez[2], Dr. V. Shepsis[3], D. Krams[4], Paul Carangelo[5]

Abstract: This paper discusses the analysis and results of an investigation into pressure field impacts caused by deep-draft vessel navigation on the beaches and bluffs along the Corpus Christi Ship Channel (CCSC), Corpus Christi, TX.

A steady-state, 3-D hydrodynamic numerical model was modified to analyze the pressure fields generated by deep-draft vessels in the channel. The modified Vessel Generated Pressure Field (VGPF) simulates pressure field distributions and calculates water surface elevations in the channel during deep-draft vessel passage. The model incorporates the channel geometry (flat, shallow banks and deep center channel) and vessel geometry (length, beam, draft and position in channel). The vessel-generated pressure fields cause long-period water surface elevation fluctuations as large as one meter in height and 120 seconds in period.

To validate the model, several field experiments were conducted that consisted of measuring the pressure waves during deep-draft vessel passage and tracking the positions and speeds of these vessels during a two-week period in August 2000. Data were collected continually during this period using a non-directional pressure gage deployed along the channel bank at elevation –2 meters (MSL). Since the channel banks are relatively flat (~50H:1V), the water recedes far from the bluff when the drawdown passes, then rushes up in the form of long-wave runup with a bore on the leading edge.

The results of the study have been used to identify the impacts of deepening the channel on shoreline stability and develop shoreline protection measures along the CCSC where pressure field effects have caused erosion of the beach and bluffs.

1. INTRODUCTION

[1] Project Manager, Pacific International Engineering, 2101 Webster St. 12th Floor, Oakland, CA 94612 Phone (510) 663-4208, Fax (707) 221-7228, Email scottf@piengr.com
[2] Project Manager, Pacific International Engineering, 3415 Greystone Drive, Suite 100, Austin, TX 78731 Phone (512) 420-0716, Fax (512) 420-0609, Email: hugob@piengr.com
[3] Principal Coastal Engineer, Pacific International Engineering, P.O. Box 1599, Suite 106-A, Edmonds, WA 98020, Phone (425) 921-1703, Fax (425) 744-1440, Email vladimir@piengr.com
[4] Senior Project Engineer, Port of Corpus Christi Authority, 222 Power St., Corpus Christi, TX 78401 Phone (361) 882-5633, Fax (361) 882-3079, Email: krams@pocca.com
[5] Coastal Environmental Planner, Port of Corpus Christi Authority, 222 Power St., Corpus Christi, TX 78401 Phone (361) 882-5633, Fax (361) 882-3079, Email: carangelo@pocca.com

The CCSC extends from the Gulf of Mexico through Corpus Christi Bay and into the Port of Corpus Christi, a distance of approximately 39 km. The current channel depth and width for most of its length inside the bay are 13.7 m (MLT) and 152 m, respectively.

Deep-draft vessels transiting the CCSC (see Figure 1) routinely travel at speeds of 3.1 to 5.1 m/sec (6-10 knots) and occasionally greater that results in hydrodynamic effects known as drawdown or pressure fields. The shallow and flat submerged channel banks (see Figure 2) allow a head differential to form between the passing vessels and the shoreline. The water on the banks travels offshore when the drawdown arrives, leaving the shoreline dry for up to 30-50 meters seaward of the shoreline (actual width of land that becomes dry depends on the beach slope and the drawdown height and period). When the water level increases again in the channel, large cross-shore and longshore velocities are created on the bank, causing shoreline erosion and other related problems. The objective of the study was to first determine the impacts of the pressure fields on shoreline erosion, then determine the incremental increase of these effects that may result from deepening of the CCSC.

Pressure field effects are well known phenomena in coastal engineering and deep-draft vessel navigation. However, until recently there has not been a well-established, simple engineering methodology for calculating the pressure field parameters and their impacts on shoreline stability and berthed vessels along the narrow waterways.

Figure 1. Corpus Christi Ship Channel Near Port Aransas, Texas

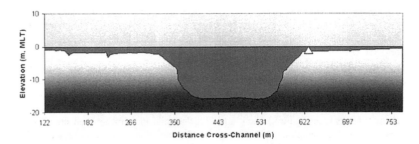

Figure 2. Corpus Christi Ship Channel Cross-Section Near Port Aransas, Texas

Pacific International (PI) Engineering, in collaboration with scientists from the Ukraine Academy of Science, has developed the Vessel-Generated Pressure Field (VGPF) numerical model. The model predicts pressure field effect parameters that can be applied toward solutions for various practical engineering problems occurring along the narrow waterways where pressure field effects exist.

An example VGPF model pressure field computation is presented in Figure 3. PI Engineering has previously and successfully applied the VGPF model on projects located in the Inner & Outer Harbor Waterways, Port of Oakland, CA, Karnafuli River, Bangladesh, Port of Tacoma, WA, and the Blair Waterway, WA. The VGPF model was designed to simulate forces and moments induced on the berthed vessel by passing vessels in restricted waterways. The model employs the method of matched asymptotic expansion and slender body theory. The model background and theoretical basis are described in detail in Shepsis et al (1998).

To develop confidence in the VGPF model predictions, verification efforts were conducted in two locations: the Port of Oakland Inner Harbor Waterway and the Corpus Christi Ship Channel.

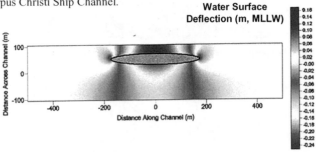

Figure 3.   VGPF Model Results Showing Pressure Field-Induced Water Surface Deflection in Restricted Waterway

3. MODEL VALIDATION

3.1 Model Validation Program at the Port of Oakland Inner Harbor Waterway

The first VGPF model validation program was conducted between May 2 and June 15, 1999 in the Inner Harbor Waterway, Port of Oakland, CA. One non-directional wave pressure gage was deployed along the channel bank during this period, and collected pressure data continuously at a frequency of 4 Hz. The data were processed and the pressure field event data were extracted.

A total of 91 vessel passing events (and the corresponding vessel-generated pressure field waves) were recorded in the Inner Harbor Waterway during the field data collection program. For each of the measured pressure field events, the corresponding passing vessel parameters were collected from the Lloyd's Register of Ships. Current speeds in the channel were analyzed based on the results of hydrodynamic modeling (RMA2 model), and tidal elevation data were estimated from NOAA predictions.

Passing vessel speed and position were collected from radar log files compiled by the Vessel Traffic Service (VTS) of the U.S. Coast Guard at Alameda, CA. Using this data, the model was validated for maximum drawdown and zero-crossing half-period. Detailed analyses of the field data and validation results based on the Port of Oakland Inner Harbor Waterway field experiments were presented in Shepsis et al (2001).

3.2 Model Validation Program at the CCSC

The VGPF model was modified for application to the CCSC (due to different channel shape) to analyze the behavior of pressure field effects over the extensive shallow shelf adjacent to the channel. Therefore, model validation was required. The field data collection program, conducted between August 3 to August 29, 2000, consisted of collecting non-directional wave pressure data along the CCSC at a frequency of 4 Hz. Statistical distribution of vessel-generated drawdowns at the CCSC is shown in Figure 4.

No vessel radar data is collected by the VTS in the CCSC. Therefore, to determine the speeds and locations of passing vessels, the field data collection program also included an experiment consisting of tracking deep-draft vessels passing through the channel in the vicinity of the wave gage (following the vessels in a small boat). This experiment, which included coordination with the pilots, provided information on vessel parameters (length, beam, draft), vessel speed, and vessel location relative to the channel centerline. A total of seven deep-draft vessel-tracking experiments were conducted. The results of the tracking experiments are presented in Table 1.

Table 1 Vessel Tracking Experiment Data

Ship Name	Date	Tidal Elevation (m)	Length (m)	Beam (m)	Draft (m)	Speed (m/sec)	Direction	Tracking Position (m)
Maritrans	8/4/00 12:20	Ebb: 0.24	160.0	25.6	5.5	2.8	Inbound	30 astern
Molda (Tanker)	8/4/00 16:00	Flood: 0.27	232.0	42.0	11.7	6.0	Inbound	30 astern, 3 stbd
Monarch	8/4/00 18:05	Flood: 0.34	188.4	31.0	8.0	4.1	Inbound	30 astern, 3 stbd
Teseo (Tanker)	8/4/00 16:40	Slack: 0.34	243.8	45.7	8.5	4.4	Outbound	30 astern, 3 stbd
CEM Princess (Light Chemical)	8/4/00 19:15	Slack: 0.34	122.7	18.0	7.1	3.7	Inbound	15-20 astern, 3 stbd
Henning Maersk (Light Chemical)	8/4/00 19:45	Ebb: 0.34	160.0	25.6	7.3	3.8	Inbound	30 astern, 3 stbd
Texas	8/4/00 21:45	Ebb: 0.27	152.4	23.8	8.0	4.1	Inbound	15-20 astern, 3 stbd

Vessel tracking data were used in combination with the measured pressure field wave data to validate the numerical model. Three of the pressure field wave data series obtained during the vessel tracking experiments were able to be compared with the predicted pressure field wave based on availability of vessel data and coincidence of consistent wave measurements. The vessel passing events used to validate the model were the Monarch, the Teseo and the Henning Maersk. Figure 5 shows a comparison of predicted and measured pressure field waves for the Henning Maersk on August 4, 2000. Figure 6 shows a comparison of predicted and measured maximum drawdowns for the three valid comparison cases.

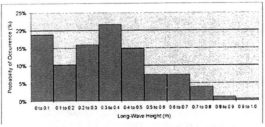

Figure 4. Distribution of Measured Pressure Field Wave Drawdowns at CCSC, August 3 to 29, 2000.

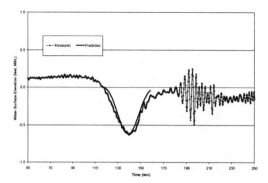

Figure 5. Comparison of Measured and Computed Pressure Field Wave for Henning Maersk, August 4, 2000 (note high-frequency vessel wakes arriving at T = 180 sec).

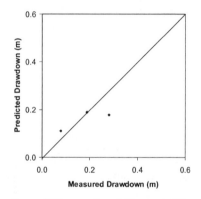

Figure 6. Comparison of Measured and Computed Pressure Field Wave Maximum Drawdowns in CCSC

Although the comparison between measured and predicted pressure field waves is limited due to lack of data, previous VGPF model verification efforts in the Inner Harbor Waterway, Port of Oakland, were considered. It is suggested that in general, the model reproduces both the height and zero-crossing half periods of the pressure field waves reasonably well.

4. PRESSURE FIELD MODEL APPLICATION

As was previously mentioned, the objective of the CCSC study was to identify impacts from the pressure field waves on shoreline/bluff erosion in conjunction with proposed channel deepening. For this purpose, the VGPF model was used to predict pressure field impacts in conjunction with the RBREAK2 model.

RBREAK2 (Kobayashi and Poff, 1994) is a 1-D time domain numerical model for simulation of wave transformation, runup and overtopping.

Pressure field waves propagate from the relatively deep water of the CCSC (depths of 13.7-15.8 m) toward the shallow bank (depths less than about 2 m) as long-period wave energy that moves cross-shore and along-shore with a significant mass of water. The RBREAK2 model was used to simulate long-period wave transformation from the deep channel (passing vessel location) toward the bluff, and predict the impact from the cross-shore bank velocities on shoreline erosion and bluff retreat.

The water motions created deep-draft vessels on the channel bank and flats may consist of several phases:
- Phase 1 – original slight movement onshore;
- Phase 2 – significant outflow toward the channel, where the bottom slope can be exposed far from the shoreline;
- Phase 3 – shoreward and along-shore water flow in a form of a bore (long-wave runup) that results in a rapid return in water level; and,
- Phase 4 – propagation of short-period waves (vessel wakes) on top of the shoreward water flow.

It should be noted that occurrence of the above four phases of motion depends on the pressure field parameters. For small pressure field waves, some phases may be so insignificant that they may not be noticed by observers or field measurements.

The pressure field effects and forces created by motions of the pressure field wave on the shallow bank are investigated for Phases 1-3 (above) in terms of long-period wave runup and bank flow velocities. Using the RBREAK2 model, the approximate long-term equilibrium bluff position was estimated using the following assumptions:
- The bluff is not armored;
- Sediment characteristics are uniform for the entire stretch of shoreline; and,
- The bluff will migrate landward with a relatively constant beach slope until the predicted bottom velocities in the long-wave runup are less than or equal to velocities required for initiation of sediment movement.

The above assumptions imply that there is some equilibrium bluff retreat distance. Specifically, if the shoreline and bluff erode to a certain point, then no further significant bluff erosion will occur due to the pressure field effects. Figures 7-9 show long-wave runup at simulation time 40.5 seconds, 102.0 seconds and 123.0 seconds from the moment the pressure field reaches the channel bank, respectively.

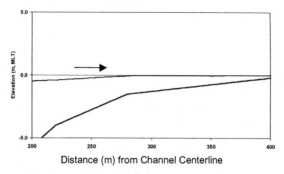

Figure 7. Pressure Field Drawdown Wave at Time = 40.5 seconds

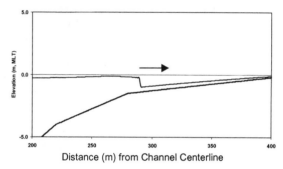

Figure 8. Pressure Field Drawdown Wave at Time = 102.0 seconds

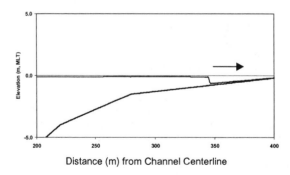

Figure 9. Pressure Field Drawdown Wave at Time = 123.0 seconds

The distance from the existing bluff position to the equilibrium bluff location is referred to as the Ultimate Bluff Recession Point (UBRP). Assuming that the shoreline and bluff along the CCSC consists of similar sediment, the UBRP and the time period when the shoreline would recede to the UBRP are functions of pressure field parameters (height and period), shoreline configuration and frequency of occurrence of these pressure field effects.

The following example demonstrates this functional relationship: For the largest vessels navigating the channel near Port Arnasas, the UBRP is estimated to be approximately 64 meters landward of the existing bluff location (the vessel "Algarrobo" was used to generate the input water surface elevation time series - length 281.3 meters, beam 54 meters, draft 14 meters). However, vessels of this size constitute only about 1% of the deep-draft vessel traffic. If only these larger vessels were in the channel, the time period between now and when the shoreline reaches UBRP position would be very long because the events are infrequent (even though each event causes more erosion than smaller craft). Medium-size and small craft, however, constitute much more of the overall vessel traffic, hence may contribute the greater percentage of erosion (even though each of these events cause less erosion than a large vessel). Therefore, a design vessel is required that can be used to describe the statistically equivalent erosion rate based on the severity of the events it causes and the frequency of these events.

To develop an objective mechanism for estimating future impacts from pressure fields and determine reasonable vessel modeling input parameters, the following approach was developed: Statistical vessel traffic data (beam, length and draft) for existing conditions in the CCSC (1997-1999) were obtained from the U.S. Army Corps of Engineers. Pressure field effects were simulated for deep-draft vessels ranging in draft from 8.8 m to 14.3 m. A total of seven final deep-draft vessels were modeled using the VGPF and RBREAK models. Passing vessel speed was assumed to be constant and equal to 5.1 m/sec (10 knots) in all modeling scenarios. The calculation results were the UBRPs corresponding to each type of vessel. UBRP values were weighted with passing frequency for each type of vessel in the channel. The statistical median UBRP for all vessels, based on severity of the event (UBRPs) and the passing frequency, were estimated for several channel areas.

5. CONCLUSIONS
- Methodology was developed to analyze existing causes of erosion, including pressure field waves, for the CCSC.
- Numerical modeling tools were utilized to both understand existing hydrodynamics in the channel, as well as predict future impacts of increased vessel traffic following widening and deepening of the navigation channel. Numerical models were validated based on the field data collected at the project sites.
- Study has provided estimates of impacts on shoreline erosion for the existing CCSC project and the existing fleet distribution. The study also provided impacts to shoreline erosion associated with the projected fleet distribution for the 52-foot project. Specifically, the study allowed quantification of the impacts that may result from the pressure field effects generated by deep-draft vessel traffic in the future channel.

6. REFERENCES

Bangash, 1993, "Impact and Explosion, Analysis and Design."

Clough and Penzien, "Dynamics of Structures."

Dean, 1974, "Evaluation and Development of Water Wave Theories For Engineering Application."

Gerald and Wheatley, "Applied Numerical Analysis."

Mast, "Design Manual, Floating Concrete Structures." (BERGER/ABAM internal Memo)

Nachlinger, "Reference Manual for MOSES, Multi-Operational Structural Engineering Simulator"

Oortmerssen, 1976, "The Motions of a Moored Ship In Waves."

Optimoor User's Guide, Table E-1 "Rope Strength Data Use in OPTIMOOR"

Pacific International Engineering,[PLLC], "Docking Facility Feasibility Study", July 27, 1999

Pacific International Engineering,[PLLC], "Hydrodynamic Study of Inner Harbor and Outer Harbor Channels, Port of Oakland, CA", May 20, 2000

Pacific International Engineering,[PLLC], "Middle Harbor Wave Data: Preliminary Results of Processing, Analysis and Recommendations", June 7, 1999

Sarpkaya and Isaacson, 1981, "Mechanics of Wave Forces on Offshore Structures."

Sea Guard Fenders, Manufacturer's Loading Specifications, p. 12-13

R. Sorensen, Investigation of Ship Generated Waves, Journal of the Waterways, Harbors and Coastal Engineering Division, ASCE, February 1967

R. Sorensen, Water Waves Produced by Ships, Journal of the Waterways, Harbors and Coastal Engineering Division, ASCE, May 1973

R. Weggel and R. Sorensen, Ship Wave Prediction for Port and Channel Design, Proceedings, Ports 86, 1986, USA

Saunders, 1957, "Hydrodynamics In Ship Design."

Texas General Land Office / PI Engineering Corpus Christi Beach Study, October 12, 2000, November 3, 2000

The Society of Naval Architects and Marine Engineers, 1988. "Principles of Naval Architecture." 2nd Rev.

U.S. Coast Guard, "Transview" Software Documentation, 1999

V. Shepsis, S. Fenical, B. Hawkins, E. Dohm, and F. Yang, Deep-Draft Vessels in Narrow Waterway, Port of Oakland 50-ft Deepening Project, Proceedings Port 2001, ASCE, Norfolk, VA 2001

V. Shepsis, R. Andrews and C. Holland, Port of Oakland Inner Harbor Waterway Design Considerations, Proceedings of Ports 98, ASCE, Long Beach, CA, 1998

Wehausen and Salvesen, 1977, "Numerical Ship Hydrodynamics."

ANALYTICAL SOLUTION OF ONE-DIMENSIONAL BAY FORCED BY SEA BREEZE

Nicholas C. Kraus[1], M.ASCE, and Adele Militello[2]

ABSTRACT: This paper introduces a closed-form analytical solution of the one-dimensional (1D), depth-averaged linearized momentum and continuity equations that incorporates linear bottom friction and the non-linear wind stress. The solution describes wind-forced motion in a 1D basin with horizontal bottom as governed by water depth, basin length, bottom friction coefficient, wind speed, and fundamental frequency of an oscillatory wind. The solution displays in compact form general behavior and dependencies of the physical processes, including generation of wind-induced harmonics of the forcing motion, damping, and resonance. The solution can serve as a benchmark test for numerical models of the shallow-water equations, as well as provide estimates of wind-induced motion in enclosed water bodies.

INTRODUCTION

The water of many bays, estuaries, harbors, lakes, and reservoirs is subjected to forcing associated with a periodic or quasi-periodic wind. Such forcing can be the sea breeze, which has solar diurnal periodicity, or the wind associated with passage of seasonal weather fronts with periods typically varying between 3 and 5 days. The leading-order responses of an enclosed body to a steady wind – set-up, set-down, mean current, and recirculating current – are well known (e.g., Ippen and Harleman 1966). Lesser known are the harmonics induced to the water body by a periodic wind. Militello and Kraus (2001) classified these as *forced harmonics* for motion generated directly by the non-linear wind stress $|W|W$ (W = component of wind speed), in contrast to *response harmonics* generated within the water body through interactions contained in the various other non-linear terms in the equations of motion.

---

1) U.S. Army Engineer Research and Development Center, Coastal and Hydraulics Laboratory, 3909 Halls Ferry Road, Vicksburg, MS 39180-6199. Nicholas.C.Kraus@erdc.usace.army.mil.

2) Coastal Analysis, LLC, 4886 Herron Road, Eureka, CA 95503. CoastalAnalysis@home.com.

In warm climates in particular, sea breeze can induce substantial diurnal motion in water bodies. Because wind forcing is a quadratic function of its speed, response harmonics generated by sea breeze are present in the water level and current, in addition to the fundamental forcing frequency (Zetler 1971). Nonlinear interactions within the water body also transfer energy into harmonic frequencies, as shown in numerous studies of tidal motion. In two-dimensional, depth-averaged horizontal flow, the quadratic bottom stress, advection, and nonlinear continuity terms generate response harmonics because they are nonlinear with respect to the current velocity, water-surface elevation, or both.

A central consideration in understanding wind-induced water motion and its harmonics is that a water body is locally forced over its entire surface. In contrast, the tide must propagate from a connection to the ocean and is damped by friction as it traverses the bay or estuary. Thus, a distinction between wind and tide is that wind is a *local* forcing whereas the tide is a *boundary* forcing. The relative strength of terms in the equations of motion is, therefore, different.

The sea breeze fluctuates with a frequency of 1 cpd (cycle per day) that is close to frequencies of the diurnal tidal constituents (K1 O1, S1, and others). Similarly, higher harmonics of the water motion induced by sea breeze (wind harmonics) lie at frequencies near the higher harmonics of the diurnal tidal frequencies. Thus, wind harmonics can be obscured by tidal motion and not easily detected. Conversely, tidal constituents must be calculated carefully if wind harmonics are present because they introduce similar motion not of gravitational origin. In embayments where the tidal amplitude is small, the sea breeze can contribute significantly to the diurnal variance of the water surface and current. This situation is common along the coast of Texas, where the strong predominant southeast wind and sea breeze can dominate the tide in producing setup and setdown in its numerous shallow estuaries and bays (Collier and Hedgpeth 1950). Militello (2000) and Militello and Kraus (2001) examined sea-breeze-induced motion at Baffin Bay, Texas, a large, non-tidal water body. Kraus and Militello (1999) document along-axis oscillations in water level exceeding 0.6 m in response to periodic fronts passing East Matagorda Bay, Texas.

This paper introduces a new closed-form analytical solution of the one-dimensional (1D), depth-averaged linearized momentum and continuity equations that incorporates linear bottom friction and the non-linear wind stress. The analytic solution describes linearized wind-forced motion in a 1D basin with horizontal bottom as governed by water depth, basin length, bottom friction coefficient, wind speed, and fundamental frequency of the oscillatory wind.

ORIGIN OF WIND HARMONICS

For focus of discussion and development of the analytic solution, a spatially uniform oscillatory wind blowing parallel to the x-axis is specified. The wind speed is then given as

$$W = w_0 + w\sin(\sigma t) \tag{1}$$

where w_0 = speed of the steady wind, w = amplitude of the oscillatory wind, and $\sigma = 2\pi/T$, in which T = period of the oscillatory wind. A sinusoidal representation for the wind with T = 24 hr is a reasonable description of sea breeze and is implemented below.

To demonstrate how harmonics are generated through wind forcing, for the special case $W \geq 0$, the quadratic wind velocity is

$$W|W| = W^2 = w_0^2 + \frac{1}{2}w^2 + 2w_0 w \sin(\sigma t) - \frac{1}{2}w^2 \cos(2\sigma t) \tag{2}$$

Eq. 2 contains three forcing components as a steady part, a fundamental diurnal frequency σ, and the first even harmonic (semi-diurnal frequency) 2σ of the fundamental. For a pure oscillatory wind, $w_0 = 0$, and the Fourier expansion of the quadratic wind velocity produced by Eq. (2) is

$$W|W| = w^2 \sum_{j=1}^{\infty} A_j \sin[(2j-1)\sigma t] \tag{3}$$

in which

$$A_j = \frac{-8}{\pi(2j-3)(2j-1)(2j+1)} \tag{4}$$

Eq. 4 shows that harmonic frequencies generated by a pure oscillatory wind are odd multiples of the fundamental frequency. Relative magnitudes of A_2 and A_3 to A_1 are 1/5 and 1/35.

ANALYTICAL SOLUTION

Equations of Motion

For a basin of uniform width and water depth $h \gg \eta$ (deviation of water surface from still water), the continuity and momentum equations of depth-averaged motion are

$$\frac{\partial \eta}{\partial t} + h\frac{\partial u}{\partial x} = 0 \tag{5}$$

$$\frac{\partial u}{\partial t} + u\frac{\partial u}{\partial x} + g\frac{\partial \eta}{\partial x} + \frac{C_f u|u|}{h} + \frac{\rho_a}{\rho}\frac{C_D W|W|}{h} = 0 \tag{6}$$

where t = time, u = horizontal water velocity, g = acceleration due to gravity, C_f = coefficient of bottom friction, ρ_a and ρ are the densities of air and water, respectively, C_D = wind-drag coefficient, and W = wind velocity. Militello and Kraus (2001) showed by scaling analysis of Eqs. 5 and 6 that the pressure gradient term is of the same order as the wind forcing and bottom friction terms, whereas the inertia and advective terms are 2-3 orders of magnitude smaller than the wind forcing term for the stated conditions.

A 1D basin of length L with vertical walls and uniform still-water depth is considered (Fig. 1), over which an along-axis sinusoidal wind blows with spatial uniformity. The governing equations are linearized, including omission of the advective term (which was shown to be small in the scaling analysis), to allow closed-form solution and to eliminate generation of response harmonics by nonlinear terms. Although it is not the intent to compare the linear and non-linear models, the Lorentz approximation for estimating the value of the linear bottom friction coefficient C_{fL} by the principle of equivalent work (Ippen and Harleman 1966) gives $C_{fL} = (8/(3\pi)) u_m C_f$, where u_m is a representative value of the magnitude of the current. The quantity C_{fL} has dimensions of velocity.

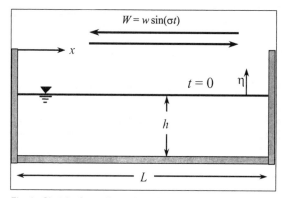

Fig. 1. Sketch of one-dimensional basin with wind forcing, $t = 0$

The continuity and momentum equations (Eqs. 5 and 6) then become

$$\frac{\partial \eta}{\partial t} = -h \frac{\partial u}{\partial x} \tag{7}$$

and

$$\frac{\partial u}{\partial t} = -g \frac{\partial \eta}{\partial x} - \frac{C_{fl}}{h} u + F \tag{8}$$

where the wind forcing is represented by the function

$$F = F(t) = C_D \frac{\rho_a}{\rho} \frac{W|W|}{h} \tag{9}$$

for pure oscillatory wind specified by Eq. 1 with $w_0 = 0$. Although the wind-drag coefficient varies with the wind speed in some formulations, it is taken to be constant for this derivation, as is C_{fl}.

From Lamb (1945), Ippen and Harleman (1966), and others, linear equation systems such as Eqs. 7 and 8 can be solved by differentiating Eq. 7 with respect to x and Eq. 8 with respect to t, then adding the resultant equations to eliminate η. The one-dimensional inhomogeneous wave equation for u is obtained,

$$u_{tt} + 2d u_t - c^2 u_{xx} = F_t \tag{10}$$

in which notation was simplified by defining $d = C_{fl}/(2h)$, and where $c^2 = gh$. The subscripts denote partial differentiation with respect to t and x. The quantity d has the dimensions of frequency, and shows that the friction term in Eq. 10 decreases inversely with the depth.

For the idealized basin, the initial and boundary conditions on u are, respectively, $u(x, 0) = u_t(x, 0) = 0$, and $u(0, t) = u(L, t) = 0$. The water surface is specified to be initially horizontal, and the wind begins blowing at $t = 0$. Symmetry indicates that the problem can be solved over half the basin, for example, on $[0, L/2]$. In the solution procedure that follows, the full interval $[0, L]$ is chosen as the spatial domain, with symmetry about $L/2$ for u and anti-symmetry for η serving as checks of the solution.

A solution is sought of the form of a Fourier expansion

$$u(x,t) = \sum_{n=1}^{N} u_n(t) \sin\left[\frac{(2n-1)\pi x}{L}\right] \tag{11}$$

which is a normal-mode equation satisfying the lateral boundary conditions. Substitution of Eq. 11 into Eq. 10 shows that the u_n satisfy the equation describing forced motion with damping,

$$(u_n)_{tt} + 2d(u_n)_t + \sigma_n^2 u_n = \sum_{n=1}^{N} F_n \tag{12}$$

with $\sigma_n = (2n-1)\pi c/L$ corresponding to odd normal modes. Eqs. 3, 9, and 10 give

$$F_n = \sum_{j=1}^{N} D_{nj} \cos(\sigma_j t) \tag{13}$$

where $\sigma_j = (2j-1)\sigma$ are the frequencies of harmonics forced by the quadratic wind stress and

$$D_{nj} = \frac{4}{(2n-1)\pi} C_D \frac{\rho_a}{\rho} \frac{\sigma}{h} w^2 A_j \tag{14}$$

The solution of Eq. 12 with the initial conditions depends on the relative values of σ_n and d, by which either underdamping ($d < \sigma_n$) or overdamping ($d > \sigma$) can occur. Note that d contains the water depth and that the σ_n will have a wide range if a reasonable number of components (e.g., $N = 7$) is assigned. Critical damping ($d = \sigma_n$) cannot occur in a practical situation for input values specified to one or two significant figures. The formal solution given below for overdamping describes both the under- and overdamping situations for complex arguments of the exponential functions appearing in it.

The solution of the linearized shallow-water wave equations for the basin with an impressed wind blowing as $W = w \sin(\sigma t)$ and with initial conditions of a flat water surface and boundary conditions of zero velocity is found to be, for the depth-averaged velocity,

$$u(x,t) = \sum_{n=1}^{N}\left(\sum_{j=1}^{J} u_{nj}\right) \sin\left[\frac{(2n-1)\pi x}{L}\right] \tag{15}$$

with the real part of

$$u_{nj} = C_{1nj} e^{\lambda_1 t} + C_{2nj} e^{\lambda_2 t} + C_{3nj} \cos\sigma_j t + C_{4nj} \sin\sigma_j t \tag{16}$$

where again, with $d = C_{fL}/(2h)$ and C_{fL} a friction coefficient for linearized bottom stress,

$$\lambda_1 = -d + \sqrt{d^2 - \sigma_n^2}$$
$$\lambda_2 = -d - \sqrt{d^2 - \sigma_n^2} \quad , \tag{17}$$

$$C_{1nj} = -\frac{\hat{D}_{nj}}{2\sqrt{d^2 - \sigma_n^2}}\left[(\sigma_n^2 - \sigma_j^2)\sqrt{d^2 - \sigma_n^2} + d(\sigma_n^2 + \sigma_j^2)\right]$$

$$C_{2nj} = -C_{1nj} - \hat{D}_{nj}(\sigma_n^2 - \sigma_j^2)$$

$$C_{3nj} = \hat{D}_{nj}(\sigma_n^2 - \sigma_j^2)$$

$$C_{4nj} = 2d\sigma_j\hat{D}_{nj}$$

(18)

$$\hat{D}_{nj} = \frac{D_{nj}}{(\sigma_n^2 - \sigma_j^2)^2 + 4d^2\sigma_j^2}$$

(19)

The water-surface elevation is given by integrating Eq. 7 with Eq. 15 for u to give

$$\eta(x,t) = \sum_{n=1}^{N}(2n-1)\left(\sum_{j=1}^{J}\eta_{nj}\right)\cos\left[\frac{(2n-1)\pi x}{L}\right]$$

(20)

with the real part of the following equation taken:

$$\eta_{nj} = \frac{C_{1nj}}{\lambda_1}(e^{\lambda_1 t} - 1) + \frac{C_{2nj}}{\lambda_2}(e^{\lambda_2 t} - 1) + \frac{C_{3nj}}{\sigma_j}\sin\sigma_j t - \frac{C_{4nj}}{\sigma_j}(\cos\sigma_j t - 1)$$

(21)

This solution describes linearized wind-forced motion in a 1D basin as governed by five parameters: water depth, basin length, bottom friction coefficient, wind speed, and fundamental frequency of the oscillatory wind. The solution includes the initial transients and possible mixed under-damping ($d < \sigma_n$) and over-damping ($d > \sigma_n$), depending on the normal modes, as can occur according to the values of λ_1 and λ_2.

RESULTS

Example Dynamics of Analytical Solution

For examining properties of the analytic solution, the geometry of an idealized basin was established that approximated Baffin Bay, Texas (Militello and Kraus 2001), as $L = 29$ km, $h = 1$ m, and $C_{fL} = 0.009$, upon which a spatially uniform sinusoidal wind was imposed with $w = 10$ m s^{-1} and $C_D = 0.0016$. Through trial runs, 3-digit reproducibility was obtained with four wind harmonics ($J = 4$) and nine normal modes ($N = 9$).

The time series of η from Eq. 20 at $x = L - 500$ m and u from Eq. 15 at the middle of the basin are plotted in Fig. 2 for 3 days. Day 0 was omitted to allow transients to disappear. The greatest variation in water level and velocity are experienced at the basin ends and middle, respectively.

The water-surface elevation and current velocity contain complex structure through the presence of both wind-generated harmonics and normal-mode frequencies. The spectra of η and u shown in Fig. 3 indicate strong motion at the fundamental frequency of 1 cpd and energy at the odd forced harmonics associated with the quadratic wind stress. The peak at 4.65 cpd is the first normal (seiching) mode of the basin. The amplitudes of the harmonics can also be obtained from the solutions, and Eqs. 15 and 20.

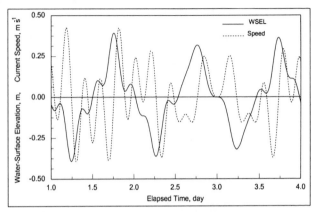

Fig. 2. Calculated time series of water-surface elevation (WSEL) and current, 1-m depth

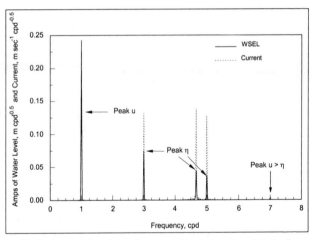

Fig.3. Spectra of water-surface elevation and current, 1-m depth

Figures 4 and 5 respectively plot the water level and velocity along the basin at hourly intervals for the first 12 hr after the start of Day 1. Because the motion of the water level and current are complicated (Fig. 2), changes at a fixed time interval are not regular. The water level fluctuates between about ±0.4 m, and the velocity fluctuates between about 0.4 and -0.3 m s^{-1} for this particular 12-hr interval. At the location $L/2$, η is anti-symmetric and u is symmetric. Also, the water surface along the basin exhibits some curvature, departing from a straight line that might be expected intuitively.

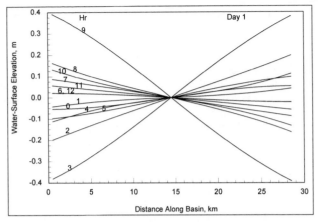

Fig. 4. Selected water levels along the basin, 1-m depth

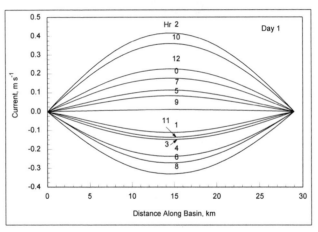

Fig. 5. Selected current velocities along the basin, 1-m depth

To demonstrate other properties of the analytic solution, the depth in the previous example was changed to $h = 2$ m. Time series of the water-surface elevation and current are plotted in Fig. 6 with corresponding spectra in Fig. 7. These are analogous to Figs. 2 and 3 for the situation of $h = 1$ m. The responses (amplitudes) of both the water level and current speed are smaller for the greater depth because the wind must move more water. Also, the first normal mode is now located at 6.60 cpd. As a consequence of the different normal modes, the time series differ for the two ambient depths.

Action of Bottom Friction in Presence of Sea Breeze

With bottom friction acting, higher-mode frequencies damp more than lower modes. Damping of water-surface elevation amplitudes for the friction coefficient C_{fL} ranging from

0 (no friction) to 0.02 (strong damping, as over a porous reef) are shown in Fig. 8 for the idealized basin. Amplitudes are normalized by the no-friction value of the corresponding frequency. Motions on the fundamental forcing frequency (1 cpd), the 1st and 2nd harmonics (3 and 5 cpd), and the first resonant mode (4.7 cpd) are present. The curves for the fundamental and harmonic frequencies indicate steep damping for smaller values of friction coefficients, tapering to mild slopes with greater values of friction. Curves for the harmonic frequencies approach near-zero slope with increased friction coefficient. Motion at these frequencies is present, even at large friction values, because it is forced over the entire surface of the water body. In contrast, the resonant frequency damps to near zero for even small values of friction, being a boundary-induced phenomenon.

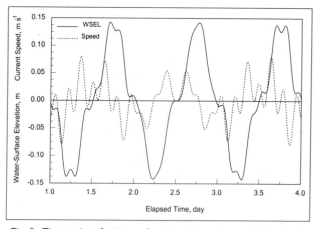

Fig. 6. Time series of water-surface elevation and current, 2-m depth

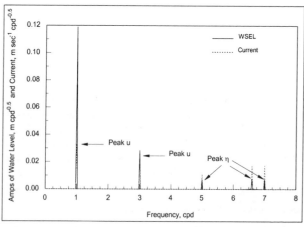

Fig. 7. Spectra of water-surface elevation and current, 2-m depth

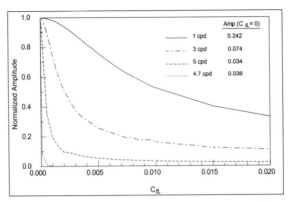

Fig. 8. Normalized amplitudes for oscillatory wind-forced water level for friction coefficient ranging from 0 to 0.020.

CONCLUSION

An analytical solution for an idealized 1D basin was developed to study the response of initially quiescent water to oscillatory wind as governed by the linearized equations of motion with quadratic wind stress. The solution displays in compact form general behavior and dependencies of the physical processes, including generation of harmonics of the motion, damping, and resonance. The solution can serve as a benchmark test for numerical models of the shallow-water equations to examine properties such as numerical damping, generation of spurious motions, symmetry, and accuracy. It also provides a convenient procedure for making first-order estimates of wind-induced motion in enclosed water bodies such as bays, estuaries, lakes, and reservoirs.

ACKNOWLEDGEMENTS

This work was conducted as part of activities of the Inlet Geomorphology and Channels Work Unit of the Coastal Inlets Research Program, U.S. Army Corps of Engineers (USACE). Permission was granted by Headquarters, USACE, to publish this information.

REFERENCES

Collier, A., and Hedgpeth, J.W. 1950. An Introduction to the Hydrography of Tidal Waters of Texas. *Pubs. Inst. Mar. Sci.*, 1(2), 125-194.

Ippen, A.T., and Harleman, D.R.F., 1966. Tidal Dynamics in Estuaries. *Estuary and Coastline Hydrodynamics*, Ippen, A.T. (Ed), McGraw-Hill, New York, 493-545.

Kraus, N.C., and Militello, A., 1999: Hydraulic Study of Multiple Inlet System: East Matagorda Bay, Texas. *J. Hydraulic Eng.* 25(3), 224-232.

Lamb, H. 1945. Hydrodynamics. 6th edition. Dover Publications, New York, 738 pp.

Militello, A. 2000: Hydrodynamic Modeling of a Sea-Breeze Dominated Embayment, Baffin Bay, Texas. *Proc. Sixth International Conf. on Estuarine and Coastal Modeling*, ASCE, 795-810

Militello, A., and Kraus, N.C. 2001. Generation of Harmonics by Sea Breeze in Nontidal Water Bodies. *Journal of Physical Oceanography*, 31(6): 1,639-1,647.

Parker, B.B., 1991. The Relative Importance of the Various Nonlinear Mechanisms in a Wide Range of Tidal Interactions (Review). *Tidal Hydrodynamics,* Parker, B. B., Ed., John Wiley & Sons, New York, 237-268.

Zetler, B.D., and Hansen, D.V. 1970. Tides in the Gulf of Mexico – A Review and Proposed Program. *Bull. Mar. Sci.*, 20(1), 57-69.

GROWTH RATES AND EQUILIBRIUM AMPLITUDES OF FORCED SEICHES IN CLOSED BASINS

David F. Hill [1]

Abstract: A weakly nonlinear analysis of two-dimensional, standing waves in a rectangular basin of arbitrary depth is presented. The waves are resonated by periodic oscillation along an axis aligned with the wavenumber vector.

First, linear analysis is pursued in order to determine the growth rate of the resonated wave as a function of the basin dimensions, mode number, forcing amplitude, and fluid parameters. The effects of viscosity and frequency mismatch (detuning) are considered. It is found that waves in transitional depth have the largest growth rates. Experiments are found to agree extremely well with the theoretical results.

Second, cubic nonlinearity is considered in order to describe limits on the maximum amplitude of the resonated wave that can not be accounted for by viscosity or initial detuning. Results indicate that, for the case of perfect resonance, the maximum amplitude can be two orders of magnitude larger than the forcing amplitude. Theoretical predictions of amplitude response curves confirm the nonlinear 'frequency reversal' that has been demonstrated previously. Experiments again agree very well with the theory.

INTRODUCTION

That harbors and closed basins such as lakes and water supply reservoirs are susceptible to seiching is not a new concept. It is widely accepted that seiches can develop either in response to wind stress or, in the case of harbors, to incident wave energy. In the latter case, which is forced, the response of the basin depends upon the proximity of the forcing frequency to one of the natural frequencies of oscillation of the basin. If the conditions are right, large amplitude waves can be excited through a simple linear resonance. In the past, therefore, significant attention has been paid to computing the natural frequencies of both simple and complex domains and to determining the so-called amplification factor of basins. In other words, given the forcing frequency and amplitude of incident waves, how large will the resonated waves in a harbor grow to become?

Less attention has been paid to the problem of wave resonance in closed basins due to horizontal oscillation. In this case, the forcing could originate from

---

[1] Dept. of Civil and Env. Engr., Pennsylvania State University, 212 Sackett Building, University Park, PA 16802. E-mail: dfhill@engr.psu.edu.

seismic oscillations. It is of clear and practical interest to be able to compute both the growth rates as well as the equilibrium amplitude of waves forced in this manner; such knowledge would allow for predictions of overtopping potential. In the current study, weakly nonlinear resonant interaction theory is used to quantify both of these quantities.

Initial experiments confirm both of the theoretical results and show remarkable quantitative agreement with the predictions. Of paramount importance, the results indicate that growth can occur very quickly and that the amplitude of the resonated wave can be two orders of magnitude larger than the amplitude of the forcing.

Horizontal Resonance

If a basin of fluid is oscillated horizontally, forced waves can be resonated. Generally speaking, an amplitude evolution equation of the form

$$\dot{a} \sim \beta + i\lambda a^2 a^*,$$

where β and λ are instability and interaction coefficients, is obtained. The initial growth is linear in time and the effects of viscosity and nonlinearity are to place a upper bound on the wave amplitude.

There have been a number of studies in the past that have dealt with horizontal resonance. Verhagen and Wijngaarden (1965) used an inviscid, shallow-water theory to address the two-dimensional oscillations induced by rotational (about a horizontal axis) forcing. While this is technically different from pure horizontal forcing, many of the concepts are the same. They found that, near perfect resonance, a bore was generated that would travel back and forth between the end walls of the tank. Borrowing from acoustics and using the method of characteristics, the authors were able to arrive at satisfactory predictions of the water surface elevation.

Chester (1968) and Chester and Bones (1968) added the effects of weak dispersion and weak viscosity in their theoretical and experimental studies of resonant waves. Of particular interest in their studies was the fact that the theoretical amplitude response curves were multi-valued; they demonstrated multiple equilibrium amplitudes at certain values of forcing frequency. Both the theoretical and experimental amplitude response curves leaned to the right, since the water depth was shallow, and demonstrated multiple 'bifurcation' points. Ockendon et al. (1986) later used asymptotic methods to further explore these multiple solutions, again for the case of shallow water. Byatt-Smith (1988) also used asymptotics to show the existence of multiple solutions, but made no attempt to determine the preferred solution.

An important work by Lepelletier and Raichlen (1988) also used long-wave theory, with dispersive and dissipative terms, and paid particular attention to the transients associated with commencement and cessation of the basin motion. Their study gave analytic equations for the initial linear growth rate of the resonated wave and noted a linear 'beating' when the forcing frequency was slightly detuned from the basin natural frequency. Solving the problem numerically, the authors produced theoretical response diagrams that showed the same lean to the right as the studies listed above. Maximum amplitudes were found to be one to two orders of magnitude greater than the forcing amplitude and experiments were found to agree very well with the theory.

Some very recent work on horizontal wave resonance has been motivated by the efficient design of tuned liquid dampers, for use in structural control of buildings. These studies have tended to focus on the experimental quantification of the rate of energy dissipation by resonated bores, but some have shed light on the details of the wave evolution. Feng (1997) reported on a purely experimental study of resonance of short waves and noted a curious transition from steady to unsteady wave motion near the frequency of the third mode. Reed et al. (1998) confirmed the finding of Lepelletier and Raichlen (1988) that the maximum amplitude response occurs at a frequency $O(10\%)$ above the basin resonant frequency. Their results, on the other hand, revealed equilibrium amplitudes only on the same order as the forcing amplitude. The shallow-water equations were solved numerically and demonstrated good agreement with the experiments.

Current Study

The current paper details a theoretical and experimental investigation of waves resonated in a rectangular basin by horizontal oscillation. In contrast to many of the previous studies discussed above, the problem is formulated in water of general depth, rather than in the shallow-water limit. Additionally, multiple-scales and perturbation methods are used in order to derive analytic evolution equations for the amplitude and phase of the wave. These coupled equations may then be integrated in order to determine the maximum transient and steady-state amplitudes in the basin. Only two-dimensional waves, with the forcing aligned with the wavenumber vector, will be considered.

First, the linear limit is considered and the effects of fluid viscosity and initial frequency detuning are included. If either damping or detuning is present, a limit is placed on the wave amplitude, although this limit may be physically meaningless, i.e. greater than the water depth.

Next, the weakly-nonlinear problem is considered in order to more completely and more accurately describe the full evolution of the wave amplitude. The effect of cubic self-interactions is to introduce additional detuning which, even in the limits of no viscosity and no initial detuning, limits the growth of the wave. Experiments are performed, using a small horizontal shaking table, and demonstrate excellent agreement with both the linear and nonlinear theoretical predictions.

THEORETICAL ANALYSIS

Formulation

As illustrated in figure 1, two-dimensional waves in a basin of depth h and length L are considered. The breadth of the basin is D. Periodic forcing of the basin in the x direction is facilitated by prescribing the velocity of the $x = 0$ and $x = L$ vertical walls to be

$$U_0 = U_L = \frac{b\omega}{2} e^{-i(\omega+\Delta)t} + c.c., \tag{1}$$

where b is a real-valued displacement amplitude, ω is a linear resonant frequency of the basin, Δ is some small detuning from this resonant frequency, and $c.c.$ denotes the complex conjugate.

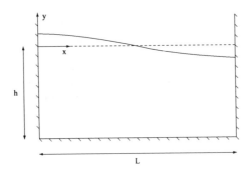

Fig. 1.   Schematic of standing wave geometry.

The infinitesimally small departure of the free surface from its equilibrium location is described by $\xi(x,t)$. If the fluid is assumed to be weakly-viscous, the velocity vector, $\underline{u} = (u, v, w)$, is given by the sum of the gradient of a potential function, $\Phi(x, y, t)$, which satisfies Laplace's equation,

$$\nabla^2 \Phi = 0 \qquad -h \leq y \leq 0, \tag{2}$$

and a rotational velocity vector $\underline{U} = (U, V, W)$. By definition, therefore, $\nabla \cdot \underline{U} = 0$.

The velocity potential and displacement variable are subject to the usual nonlinear free-surface kinematic and dynamic boundary conditions, as well as no-slip and no-flow conditions at the boundaries. By virtue of the weakly-viscous assumption, the rotational velocity vector is only of significance in the vicinity of solid boundaries. Solution for these boundary layer corrections and their incorporation into the boundary value problem are discussed at length by Mei and Liu (1973) and Mei (1989).

In order to solve this boundary value problem at successive orders, the wave steepness, $\epsilon \equiv ka$, where k is the wavenumber and a the wave amplitude, is taken to be much less than one. The forcing amplitude, b, is taken to be $O(\epsilon a)$. The viscosity, in a scaled sense, is taken to be $O(\epsilon^2)$ The detuning is assumed to be $O(\epsilon\omega)$ and, finally, a is taken to be a function of the slow time variable $\tau = \epsilon t$.

The proper ordering of the perturbation expansions can be arrived at by simple reasoning. Assume that the wave amplitude and forcing amplitude, in some scaled sense, are given by A and B respectively. The forcing terms are therefore $O(B)$ and the cubic terms are $O(A^3)$. This implies that, at equilibrium, $A \sim B^{1/3}$. The free-surface displacement is therefore taken to be

$$\xi = B^{1/3}\eta_{01}\cos(kx)e^{-i\omega t} + B^{2/3}\eta_{10} + B^{2/3}\eta_{12}\cos(2kx)e^{-2i\omega t}$$
$$+ B\eta_{21}\cos(kx)e^{-i\omega t} + c.c. \quad (3)$$

As indicated by this expansion, both a bound superhamonic and a set-down of the water surface, due to quadratic nonlinearity, are expected. The expansion for the velocity potential is similar, with the exception that there is no equivalent set-down term.

At the leading $O(B^{1/3})$ order, the effects of nonlinearity, detuning, viscosity, and forcing are all absent, leaving only the well-known solution for a linear standing wave. In order to satisfy the condition of no flow through the wall at $x = L$, the wavenumber k is given by $k = n\pi/L$, where n is an integer mode number.

At the next, $O(B^{2/3})$, order, the familiar Stokes wave solution for the superharmonic is found (Tadjbakhsh and Keller, 1960):

$$\eta_{12} = \frac{a^2 k \cosh(kh)}{16\sinh^3(kh)}(2\cosh^2(kh) + 1) \quad (4)$$

$$\phi_{12} = \frac{-3i\omega a^2}{32\sinh^4(kh)}\cosh[2k(y + h)]. \quad (5)$$

The 'zeroth' harmonic, i.e. steady-state setdown of the water surface, is given by:

$$\eta_{10} = \frac{1}{8g}\omega^2|a|^2[1 + \coth^2(kh)]\cos(2kx) - \frac{k|a|^2}{4\sinh(2kh)}. \quad (6)$$

Tadjbakhsh and Keller (1960) derived the first term, which applies to standing waves only, but omitted the second term, which is well-known (Mei, 1989; Dean and Dalrymple, 1991) and which applies to both progressive and standing waves.

Finally, at the third order of expansion, an inhomogeneous problem for the fundamental harmonic, $e^{-i(\omega+\Delta)t}$, is obtained. Due to the existence of a nontrivial solution for the homogeneous (linear) problem, an orthogonality condition must be applied in order to guarantee solvability. This leads to the following nonlinear evolution equation for a:

$$\dot{a} = i\Delta a - (1 - i)\alpha a + \beta - i\lambda a^2 a^*. \quad (7)$$

In this equation, α is a damping coefficient, given by Keulegan (1959) as

$$\alpha = \sqrt{\frac{\nu\omega}{2}}\left[\frac{1}{D} + \frac{1}{L} + \frac{k(1 - 2h/L)}{\sinh(2kh)}\right], \quad (8)$$

β is a forcing coefficient, given by

$$\beta = [1 + (-1)^{n-1}]\sqrt{\frac{g}{n\pi L}}\left[\tanh\left(\frac{n\pi h}{L}\right)\right]^{3/2}b, \tag{9}$$

and λ is a nonlinear interaction coefficient, given by

$$\lambda = \frac{\omega k^2}{256\sinh^4(kh)\cosh^2(kh)}[\cosh(6kh) - 6\cosh(4kh) - 24 - 7\cosh(2kh)]. \tag{10}$$

EXPERIMENTAL INVESTIGATION

To verify the linear and nonlinear theories, a small horizontal shaking facility (figure 2) was constructed. The facility essentially consists of an aluminum baseplate mounted via pillow block bearings onto Thomson stainless-steel linear rails. A glass tank, having a length of 61.0 cm, a width of 45.7 cm, and a height of 38.1 cm was then affixed to this baseplate.

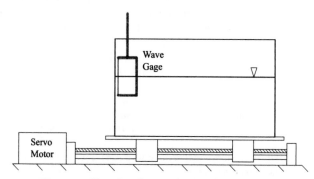

Fig. 2.   Sketch of experimental apparatus.

Harmonic motion of the tank was achieved by using a Thomson Omni-Drive brushless DC servomotor, which was connected to the baseplate by way of a precision lead-screw. The drive was operated in velocity control mode, which means that the motor delivered a velocity that was linearly proportional to a DC voltage supplied to the drive by a function generator.

To measure the free surface displacement, a custom-made, surface-piercing capacitance wave gage was used. Power was supplied to the gage by a ± 12 VDC source and the output signal was directed to the A/D card mentioned above. Calibrations of the gage, performed at the start of each experiment, were automated by mounting the gage on a Velmex Unislide assembly. Over the range of displacements reported here, the gage was found to have a very linear response. Finally, the forcing amplitude b was measured using a dial indicator ($0.01mm$ resolution).

Experiments were performed for two different water depths, $h = 15.2cm$ and $h = 22.9cm$. In both cases, first-mode waves were studied and the forcing amplitude was $b = 0.35mm$. Finally, for each depth a number of forcing frequencies, covering the range $0.9 < (\Delta + \omega)/\omega < 1.1$, were considered.

With regards to the initial growth rate, β should be independent of Δ, provided that $\Delta \ll \omega$. However, it can be difficult to obtain a reliable value of β, in the presence of seemingly small detuning, due to the fact that the wave amplitude begins to level off very quickly. Considering, therefore, only the case of zero detuning, the experiments reveal initial linear growth rates of $0.0847cm/s$ and $0.119cm/s$ for the $h = 15.2cm$ and $h = 22.9cm$ cases. The theoretical predictions, as given by 9, for these cases are $0.0841cm/s$ and $0.106cm/s$, indicating very good agreement.

Turning to the nonlinear results, figure 3 details the experimentally measured amplitude (transient) response curves for the two depths, along with the theoretical predictions. The vertical axis describes the maximum crest elevation at the side-wall, scaled by the tank depth. Note that the *relative* depths are 0.125 and 0.188 respectively and therefore bracket the critical value determined by Tadjbakhsh and Keller (1960). The frequency shift, though slight, is evident in the theoretical predictions and confirmed in the measurements. Overall, the agreement between theory and experiment is quite good, although there is some slight systematic offset between the experimental and theoretical profiles. In both cases, the experimental profiles sit slightly to the left of the theoretical profiles. The lack of perfect agreement is not altogether unexpected, however, since the theory is based upon the assumption of weak nonlinearity ($\epsilon \ll 1$). In the experiments with the maximum response, $\epsilon \sim 0.5$, so it is rather remarkable to see such good agreement.

Finally, a brief discussion of experimental uncertainty is warranted. The uncertainty in determining the water surface elevation from the wave gage is quite low. Given the 12 bit resolution of the A/D card, and the $\pm 10V$ span of the wave gage output, the uncertainty in ξ is $\sim 0.05mm$. Another slight systematic error arises from the fact that the wave gage was positioned at a distance of approximately $1cm$ from the end wall of the tank, while all of the theoretical results have been computed for $x = 0$. Given the length of the tank, however, this amounts to a difference of only 0.1%.

Of somewhat greater significance is the uncertainty in the forcing amplitude b. While the dial indicator allowed measurement of b accurate to $\pm 0.01mm$, it was observed that the amplitude of oscillation became a bit erratic when very large waves were present in the tank. This was most likely due to the inability of the small servomotor used in this study to handle the additional inertial loads induced by these waves. Mitigating this uncertainty, however, is the fact that, as discussed in §3, the equilibrium wave amplitude scales roughly by the forcing amplitude to the 1/3 power. As such, the observed maximum amplitude is not expected to be strongly sensitive to uncertainties in b.

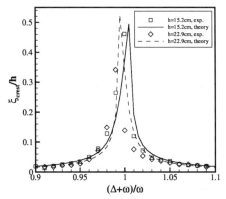

Fig. 3. Comparison of theoretical and experimental transient amplitude response curves for $h = 15.2cm$, $h = 22.9cm$. $L = 61cm$, $D = 45.7cm$, $b = 0.35mm$, $n = 1$, $\nu = 1 \times 10^{-6}m^2/s$.

CONCLUDING REMARKS

A theoretical and experimental investigation of the resonance of standing waves in closed rectangular basins has been presented. The forcing was provided by harmonic horizontal oscillation of the basin along one of its axes and the effects of weak viscosity and weak initial detuning were included. In contrast to related previous studies, no assumptions regarding the fluid depth were made. Additionally, easily integrable evolution equations for the amplitude and phase of the resonated wave were derived.

The results indicate that steady-state wave amplitudes up to two orders of magnitude greater than the forcing amplitude are possible. Additionally, it is shown that waves in water of transitional depth have the largest temporal growth rates and steady-state amplitudes. Experiments were performed and were found to agree well with both the linear and nonlinear theories.

ACKNOWLEDGEMENTS

The author would like to thank Drs. Joe Hammack and Diane Henderson for their assistance with the instrumentation used in this study. He would also like to thank Drs. Chris Duffy and Mostafa Foda for several helpful discussions.

REFERENCES

Byatt-Smith, J. (1988). "Resonant oscillations in shallow water with small mean-square disturbances." *Journal of Fluid Mechanics*, 193, 369–390.

Chester, W. (1968). "Resonant oscillations of water waves. I. Theory." *Proceedings of the Royal Society of London (A)*, 306, 5–22.

Chester, W. and Bones, J. (1968). "Resonant oscillations of water waves. II. Experiment." *Proceedings of the Royal Society of London (A)*, 306, 23–39.

Dean, R. and Dalrymple, R. (1991). *Water wave mechanics for engineers and scientists*, Vol. 2 of *Advanced Series on Ocean Engineering*. World Scientific.

Feng, Z. (1997). "Transition to traveling waves from standing waves in a rectangular container subjected to horizontal excitations." *Physical Review Letters*, 79(3), 415–418.

Keulegan, G. (1959). "Energy dissipation in standing waves in rectangular basins." *Journal of Fluid Mechanics*, 6, 33–50.

Lepelletier, T. and Raichlen, F. (1988). "Nonlinear oscillations in rectangular tanks." *Journal of Engineering Mechanics*, 114(1), 1–23.

Mei, C. (1989). *The applied dynamics of ocean surface waves*, Vol. 1 of *Advanced Series on Ocean Engineering*. World Scientific.

Mei, C. and Liu, P.-F. (1973). "The damping of surface gravity waves in a bounded liquid." *Journal of Fluid Mechanics*, 59(2), 239–256.

Ockendon, H., Ockendon, J., and Johnson, A. (1986). "Resonant sloshing in shallow water." *Journal of Fluid Mechanics*, 167, 465–479.

Reed, D., Yu, J., Yeh, H., and Gardarsson, S. (1998). "Investigation of tuned liquid dampers under large amplitude excitation." *Journal of Engineering Mechanics*, 124(4), 405–413.

Tadjbakhsh, I. and Keller, J. (1960). "Standing surface waves of finite amplitude." *Journal of Fluid Mechanics*, 8, 442–451.

Verhagen, J. and Wijngaarden, L. (1965). "Non-linear oscillations of fluid in a container." *Journal of Fluid Mechanics*, 22(4), 737–751.

Uncertainties in the Prediction of Design Waves on Shallow Foreshores of Coastal Structures

Markus Muttray[1], Hocine Oumeraci[2] and Matthias Bleck[3]

Abstract: Based on extensive hydraulic model testing, the uncertainties of available „standard" models used to predict wave heights on shallow foreshores of coastal structures are quantified. Both wave transformation on the foreshore and at the structure are taken into account. Finally, improvements of the prediction models are proposed and recommendations on the selection of the models to be used are provided.

INTRODUCTION

One of the prerequisites to help moving sustainable design of coastal protection from an academic debate into the realm of concrete work, performance and return is the development and use of probabilistic design tools (OUMERACI; 2000). However, the implementation of a reliability based design implies among others that the uncertainties associated with the prediction of wave conditions at the design site should be reliably assessed. In fact, small errors in estimating design waves may result in much larger errors for the predicted wave loads, wave overtopping, structure stability, etc.. One of the main results of a large European research project on reliability based design of coastal structures was that most uncertainties still originate from the errors in predicting wave transformation from deep water towards shallow foreshores (OUMERACI et al.; 2001). Therefore, a basic research project was initiated by the authors to determine the uncertainties associated with the prediction of wave transformation on shallow foreshores and in presence of coastal structures with different reflection properties. The project has just been completed and the results of

---

[1]   Senior Coastal Engineer; Delta Marine Consultants bv; P.O. box 63; 2800 AB Gouda, The Netherlands; mmuttray@dmc.nl

[2]   Professor and Head of Leichtweiß-Institute for Hydraulic Engineering; Beethovenstr. 51a; 38106 Braunschweig, Germany; h.oumeraci@tu-bs.de

[3]   Teaching and Research Assistant; Leichtweiß-Institute; m.bleck@tu-bs.de

the whole project have been summarized in a final report (OUMERACI and MUTTRAY, 2001).

This paper is intended to principally discuss some of the results by focusing on the uncertainties associated with the prediction of wave heights on shallow foreshores and in the presence of coastal structures.

EXPERIMENTAL SET-UP AND PROCEDURE

Hydraulic model tests using both regular and irregular waves have been conducted in the wave flume (100 m x 2 m x 1.25 m) of the Leichtweiss-Institute (LWI) in Braunschweig, Germany. A detail of the tested foreshore geometry is shown in Fig. 1. A total of 14 waves gauges were installed to record the waves in the far field and in the foreshore. A run-up gauge was also installed to record the water surface elevation directly at the sloping structures (not shown in Fig. 1). For the irregular wave tests trains of 100 waves were used for analysis. The water depth at the upper end of the foreshore was varied from h = 10 cm to h = 30 cm, the wave height from H_s = 5 cm to H_s = 25 cm and the wave period from T_p = 1.25 s to T_p = 3.55 s.

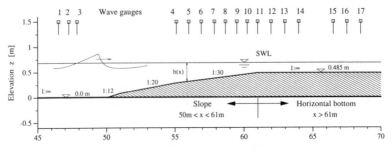

Fig. 1. Experimental Set-up and location of the wave gauges (without sloping structures)

The main objective of the hydraulic model tests is to quantify the uncertainties associated with the prediction of wave shoaling, wave height at breaking inception and after breaking, and then to draw tentative recommendations on the prediction models to be used. In addition, the effect of wave set-up and wave reflection on wave shoaling and wave breaking is also addressed.

Experimental Results and Discussion
Prediction Models, Calculations Procedure and Definition of the Uncertainties

The prediction models used to calculate the wave transformation on shallow foreshores (engineering approach) are described by MUTTRAY and OUMERACI (2000) and in more detail by OUMERACI and MUTTRAY (2001). A brief summary of these models is given in Plate 1. Regarding the wave damping due to friction at the side walls and at the bottom of the flume it should be stressed that this does not affect the local wave height in the surf zone, because the wave energy dissipation due to wave breaking already covers these losses.

1. Wave Set-Up

$$\frac{d\overline{\eta}}{dx} = \frac{1}{\overline{\eta}+h} \cdot \frac{d}{dx}\left[\frac{1}{8}\overline{H^2}\left(\frac{1}{2} + \frac{2kh}{\sinh 2kh}\right)\right] \tag{1}$$

$\overline{\eta}$ = wave set-up; $\overline{H}$ = mean wave height; h = water depth; k = $(2\pi)/L$ = wave number

2. Wave Shoaling

Non-linear shoaling koefficient k_s according to explicit approximation by OUMERACI & MUTTRAY (2001) of exact solution by SHUTTO (1974) using cnoidal theory:

$\dfrac{K_s}{K_{sA}} = 1$ (linear) for $\Pi < \dfrac{2}{15}$ with non-linearity parameter:

$\dfrac{K_s}{K_{sA}} = 1 + 0.4\left[\Pi - \dfrac{2}{15}\right]^{5/4}$ for $\dfrac{2}{15} < \Pi < 0.5$ (2)   and

$$\Pi = \frac{H}{L}\coth^3 kh \tag{3}$$

$\dfrac{K_s}{K_{sA}} = 0.69 + 0.6\,\Pi^{1/2}$ for $\Pi > 0.5$

linear shoaling coefficient:

$$K_{sA} = \left[\left(1 + \frac{2kh}{\sinh 2kh}\right)\tanh kh\right]^{-1/2} \tag{4}$$

Remark: Eq.(2) deviates by less than 2% from exact solution by SHUTTO (1974)

3. Wave Breaking

3.1 Theoretical Breaking Criterion of MICHE (1944) derived for regular waves (horizontal bottom):

$$\frac{H_b}{L} = M\,\frac{L}{L_0} = M\tanh kh \tag{5}$$

with: H_b = Breaker Height
L, L_0 = local and deep water wave length
M = 0.142

3.2 Empirical breaking criterion of WEGGEL (1972) for regular waves over sloping bottom:

$$\frac{H_b}{h_b} = b - a\frac{H_b}{gT^2} \tag{6}$$

with: $a = 43.8(1 - \exp(-19\tan\alpha))$

$b = \dfrac{1.56}{1 + \exp(-19.5\tan\alpha)}$

3.3 Semi-empirical breaking-criterion of GODA (2000) for regular and irregular waves over sloping bottom:

$$\frac{H_b}{L_0} = A\left[1 - \exp\left(1 - \exp\left(-1.5\frac{\pi h}{L_0}(1 + 15\tan^{4/3}\alpha)\right)\right)\right] \tag{7}$$

with: A = 0.17 for regular waves and A = 0.12 to 0.18 for irregular waves

4. Wave Damping due to Friction:

Formula by IWAGAKI and TSUCHIYA (1966) using linear wave theory and a laminar boundary layer at the bottom and on the side walls of the wave flume:

$$K_d = \frac{H(x)}{H_0} = \exp\left[-\sqrt{\frac{\nu T}{\pi}}\left(1 + \frac{1}{\kappa_{bw}}\right)\frac{k^2 x}{2kh + \sinh 2kh}\right] \tag{8}$$

**Plate 1. Prediction Models for Wave Transformation on Shallow Foreshores
(Engineering Approach)**

As already reported by MUTTRAY and OUMERACI (2000) two types of calculation procedures have been used to predict the local wave height in the surf zone (see Plate 1).

- *Type 1:* a simplified procedure

  taking only wave shoaling (linear or non-linear, Eq.(4) od Eq.(2)) and wave breaking (Eq.(5), Eq.(6) or Eq.(7)) into account ($H_{(x)} = f(K_s, H_{crit})$);

- *Type 2:* a more complete procedure

  taking wave shoaling, wave damping (Eq.(8)), wave set-up (Eq.(1)) and wave breaking into account ($H_{(x)} = f(K_s, K_d, \bar{\eta}, H_{crit})$).

The uncertainties of the predicted local wave parameters are described by the standard deviation, the coefficient of variation (CoV) and the systematic deviation between calculated values x_i (prediction) and expected values E_x (measurements) which are defined in MUTTRAY and OUMERACI (2000).

Wave Shoaling, Wave Breaking and Post-Breaking Wave Heights

The main findings on the uncertainties associated with the prediction of local wave heights which have recently been reported by MUTTRAY an OUMERACI (2000) are first discussed before embarking into a more detailed discussion on the effect of wave set-up and wave reflection.

Uncertainties in Wave Shoaling Prediction (Before Breaking)

The main results are summarized in Table 1 in terms of the coefficient of variation σ' of the shoaling coefficient associated with the significant (H_s) and the maximum (H_{max}) wave height of an irregular wave train as well as with the height H_m of regular waves. Focus is put on the results obtained by using the simplified procedure. The figures in parentheses are related to the results by using the complete calculation procedure.

Table 1. Uncertainties in shoaling coefficient predicted by linear and non-linear shoaling models (see Plate 1)

Type of Waves	Wave Height	Coefficient of Variation σ' for:	
		Linear Model (Eq. 4)	Non-linear Model (Eq. 2)
Irregular Waves	H_s	$\sigma'_{Hs} = 3.9\ \%\ (3.8)^*$	$\sigma'_{Hs} = 10.4\ \%\ (9.4)^*$
	H_{max}	$\sigma'_{Hmax} = 12.7\ \%\ (13.5)^*$	$\sigma'_{Hmax} = 6.1\ \%\ (6.5)^*$
Regular Waves	H_m	$\sigma'_{Hm} = 14.7\ \%\ (15.6)^*$	$\sigma'_{Hm} = 8.7\ \%\ (9.3)^*$

*Figures in ( ) are related to complete calculation procedure (type 1).

First, it is important to stress that the uncertainties are not or not significantly reduced by considering the additional effects which are accounted for by the complete calculation procedure. Therefore, the discussion is focused only on the results obtained by using the simplified calculation procedure.

Local significant wave heights H_s are better predicted by the linear shoaling model (Eq. 4) with an uncertainty of $\sigma'_{Hs} \approx 4\%$. Using a non-linear model (Eq. 2) would dramatically increase the uncertainty ($\sigma'_{Hs} \approx 10\%$).

Considering the maximum wave height H_{max} in an irregular wave train a better prediction is provided by the non-linear model (Eq. 2) with an uncertainty of $\sigma'_{Hmax} \approx 6\%$. Using the linear model would result in a much higher uncertainty ($\sigma'_{Hmax} \approx 13\%$). Obviously the transformation of the maximum waves in an irregular wave train is very similar to that of regular waves (see below).

A possible implication of the different shoaling behaviour of H_s and H_{max} is that the statistical distribution of wave heights in a spectrum will be affected when the waves enter shallow water, i.e. even before any wave breaking starts.

The uncertainties associated with the predicted height H_m of regular waves are much less when using the non-linear shoaling model ($\sigma'_{Hm} \approx 9\%$) than the linear model ($\sigma'_{Hm} \approx 15\%$).

Uncertainties in the Predicted Wave Height at Breaking

Considering wave breaking on shallow foreshores, it was found that the local wave height H_b at the breaking point is essentially governed by the shoaling process. Accounting for further processes as this is done in the complete calculation procedure (type 2) generally leads to an uncertainty reduction of less than $\Delta\sigma' = 1\%$. Overall, the GODA-model (Eq. 7) proved to be more appropriate for the prediction of maximum wave height H_{max} at the breaking point by using a coefficient $A = 0.15$. The associated uncertainty is $\sigma' \approx 7\%$, while it is $\sigma' = 32\%$ for the WEGGEL-model and $\sigma = 11\%$ (with $M = 0.14$) for the MICHE-model.

Surprisingly, for the prediction of the significant wave height at the breaking point the GODA-model appears to be less appropriate ($A = 0.10$ $\sigma' \approx 17.5\%$) than the simple breaking criterion $\gamma_b = H_s/h_b = 0.5$ ($\sigma' \approx 12\%$) which does not account for the foreshore slope.

For the prediction of the breaking of regular waves the GODA-model with $A = 0.17$ is associated with much less uncertainty ($\sigma' \approx 9\%$) than the WEGGEL-model ($\sigma' \approx 16\%$) (MUTTRAY and OUMERACI, 2000).

Uncertainties in the Predicted Post-Breaking Wave Height

For the regular waves the local wave height $H(x)$ after breaking is expressed as a function of the critical wave height H_{crit} which is predicted by the GODA-model (Eq.7):

$$H(x) = H(x) = H_{crit}\left[0.4 \, \exp\left(-0.15 \, \frac{\Delta x}{h(x)}\right) + 0.6\right] \qquad (9)$$

where x = distance from the breaking point; $h(x)$ = water depth at distance x and 0.6 $H_{crit} = H_{stable}$ is the stable wave height which prevails after a certain distance from the breaking point. Eq. (9) which is associated with uncertainties in the order of $\sigma' = 15\%$ exhibits a similar experimental decay as the empirical relationship obtained by

KWEON and GODA (1996) by re-analysing available data from model tests with regular waves. In contrast to the findings of KWEON and GODA (1996) no distinct influence of the wave period could be found (MUTTRAY and OUMERACI, 2000). Moreover, much simpler models than that suggested by KWEON and GODA (1996) have been derived for random waves (MUTTRAY and OUMERACI, 2000):

·   Maximum wave height: $H_{max}(x) = H_{crit}$ (10)
    with $\sigma' = 9\%$ and H_{crit} according to Eq. (7)

·   Significant wave height: $H_s(x) = 0.5\ h(x)$ (11)
    with $\sigma = 13\%$ and $h_{(x)}$ = local water depth.

Effect of Wave Set-Up

The comparison between the directly measured wave set-up $\bar{\eta}(x)$ and the wave set-up $\bar{\eta}\ (H_m(x))$ calculated by using Eq. (1) and the locally measured wave height ($H_m(x)$ in Fig.2) shows that the spatial distribution of the wave set-up is qualitatively well-described by the calculation while quantitatively some differences occur which are particularly significant immediately behind the breaking point.

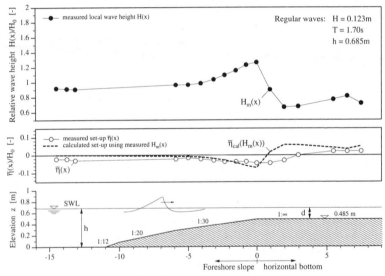

Fig. 2 Cross-shore distribution of wave height and wave set-up

By considering all the data obtained from regular wave tests and relating the wave set-up $\bar{\eta}_{meas}(x)$ by the local water depth $h(x)$ at the same location x across shore a more detailed comparison with the calculated set-up values $\bar{\eta}_{cal}(x)$ is drawn in Fig.3, showing that the calculation (Eq.1) by using measured local wave heights $H(x)$ slightly underestimates the set-down before breaking while the set-up after breaking is overestimated.

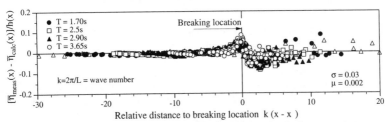

Fig.3 Comparison between measured and calculated relative wave set-up

A direct comparison of the calculated and measured wave set-up for both regular and irregular waves has led to a coefficient of variation of $\sigma' \approx 50\%$ for regular waves and $\sigma' \approx 60\%$ for irregular waves.

Overall, the tests with irregular waves resulted in much smaller set-up values $\bar{\eta}(x)$ than those with regular waves. The reasons for this difference may be seen from Fig.4, showing the set-up $\eta(t)$ recorded at wave gauge 15 (see Fig. 1) located at $x = 5.49$ m after the breaking point for a regular and an irregular wave test with approximately similar conditions. While the set-up for regular waves remains at a high and relatively constant level over the test duration, it is very variable for irregular waves. Instantaneous values $\eta(t)$ can even be higher than for regular waves.

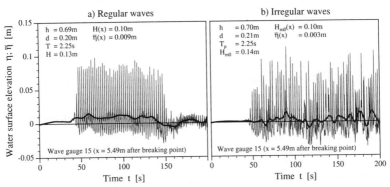

Fig.4 Recorded wave set-up for regular and irregular waves

As previously mentioned the wave attenuation due to friction does not affect the local wave height after breaking. If we assume that friction has no significant effect up to the point of breaking, then the effect of wave set-up on the local wave height along the entire cross-shore profile can be found through comparison of the results obtained by calculation procedure type 1 (simplified) and calculation procedure type 2 (complete). Based on such results and for the range of the set-up values measured in this study the following tentative conclusions can be drawn on the effect of the wave set-up on the prediction of local wave heights:

· *before and at breaking*: no significant effect could be identified,
· *after breaking*: no significant effect could be identified for the predicted regular
wave heights and for significant wave heights, but considering wave set-up reduces
the uncertainty in the prediction of maximum heights from $\sigma' = 14$ % to $\sigma' = 9$ %.

Effect of wave reflection

The effect of wave reflection on wave shoaling and wave breaking, and thus on the
local wave height in front of structures, was examined by comparative analysis of the
same model tests performed with the foreshore alone (see Fig. 1) and with different
reflecting structures added at various locations along the horizontal bottom behind the
foreshore slope. First results are discussed below examplarily for an impermeable
vertical non-overtopped wall located at the upper end of the foreshore slope.

The comparison in Fig.5 of the incident wave height obtained through (i) reflection
analysis at the wall ($H_{i,wall\ (with\ wall)} = H(x)/(1+Kr)$) and (ii) direct measurement at the
same location without wall ($H_{i,wall\ (without\ wall)}$) shows that wave reflection does not affect
the local incident wave height and that the shoaling models (Eqs. (2) & (4)) can be used
for progressive waves as well as for partially and totally standing waves.

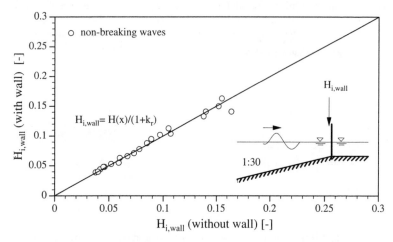

Fig.5 Incident wave height at wall location for models with and without wall

A further comparison in Fig.6 of the location of the breaking point Δx_b (relative to
wall location) with and without wall surprisingly shows that wave reflection does not
affect the location of the breaking point systematically. This is contrary to the results of
RUGGIERO and McDOUGAL (2001) suggesting that the reflected waves should
cause the incident waves to begin breaking further offshore. Analysis of further test
series will be undertaken in order to confirm this finding. As expected, however, wave
reflection is found to substantially affect the breaker types.

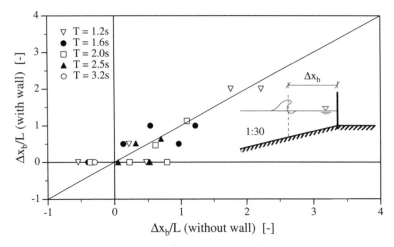

Fig.6 Effect of wave reflection on the location of the breaking point

CONCLUDING REMARKS

Prior to breaking, regular wave heights and maximum wave heights in an irregular wave train are better predicted by a non-linear shoaling model (Eq.2) resulting in uncertainties of $\sigma'_{Hm} \approx 9\%$ and $\sigma'_{Hmax} \approx 6\ \%$, respectively. For the prediction of significant wave heights a linear shoaling model (Eq.4) is more appropriate ($\sigma'_{Hs} \approx 4\ \%$).

At breaking, the GODA-model (Eq. 7) was found to be appropriate for the prediction of regular wave heights with $A = 0.17$ ($\sigma' = 9\ \%$) and the maximum wave height in an irregular wave train with $A = 0.15$ ($\sigma' \approx 7\ \%$). For the prediction of significant wave heights, the use of a simpler breaking criterion leads to less uncertainty than the GODA-model.

After breaking, the maximum wave height H_{max} and the significant wave height H_s can be predicted by simple relationships (Eqs. 10 & 11) resulting in a coefficient of variation of $\sigma'_{Hmax} = 9\ \%$ and $\sigma'_{Hs} = 13\ \%$, respectively. For regular waves a more complicated relationship (Eq. 9) with more uncertainty ($\sigma'_{Hm} \approx 15\ \%$) has been derived.

Unexpectedly, no significant effect of the wave set-up could be identified for the prediction of local wave heights before, at and after breaking. A slight reduction of the uncertainty is achieved by considering wave set-up only for the prediction of maximum wave height at the breaking.

Even more unexpected are the results related to the effect of wave reflection on the local wave height. Contrary to the results reported in the literature, the reflected waves do not cause systematically the incident waves to break further offshore, although wave reflection was observed to strongly affect the breaker type. Moreover, the shoaling models and the breaking criteria for progressive waves remain valid in the presence of wave reflection.

ACKNOWLEDGEMENTS

This work is part of the basic research project "Design Wave Parameters in front of Structures with different Reflection Properties" (OU1/3-1, 2 & 3) which is supported by the German Research Council (DFG). This support is gratefully acknowledged.

REFERENCES

Goda; Y. 1983. A unified nonlinearity parameter of water waves. *Report of the Port and Harbour Research Institute*, 22(3), 3-30.

Goda, Y. 2000. Random Seas and Design of Maritime Structures. *Advanced Series on Ocean Engineering;* Vol. 15; World Scientific.

Iwagaki, Y. and Tsuchiya, Y. 1966. Laminar damping of oscillatory waves due to bottom friction. *Proceedings International Conference Coastal Engineering* (ICCE); Vol. 10; pp. 149-174; Tokyo.

Kweon, H.-M. and Goda, Y. 1996. A parametric model for random wave deformation by breaking on arbitrary beach profiles. *Proc. 25th Intern. Conf. Coastal Eng.*, pp. 261-274; Orlando/Florida..

Miche, M. 1944. Mouvements ondulatoires de la mer en profondeur constante ou décroissante. Annales des Ponts et Chaussées; Vol. 114; pp. 25, 131, 270, 369 ff..

Muttray, M. and Oumeraci, H. 2000. Wave Transformation on the Foreshore of Coastal Structures. ASCE, Proc. ICCE 2000.

Oumeraci, H. 2000. The Sustainability Challenge in Coastal Engineering. Proc. 4th Int. Conf. of Hydrodynamics.

Oumeraci, H., Kortenhaus, A., Allsop, N.W.H., DeGroot, M.B., Crouch, R.S., Vrijling, J.K., Koortman, H.G. 2001. Probabilistic Design Tools for Vertical Breakwaters. Balkema, pp. 373.

Oumeraci, H. and Muttray, M. 2001. Bemessungswellenparameter vor Strukturen mit verschiedenen Reflexionseigenschaften. *Leichtweiss-Institute, TU Braunschweig, unpublished, Final Report,* OU 1/3-3.

Shuto, N. 1974. Nonlinear long waves in a channel of variable section. Coastal Engineering in Japan, Vol. 17, pp. 1-12, Tokyo, Japan.

Ruggiero, P. and McDougal, W.G. 2001. An analytic model for the prediction of wave wave set-up, long shore currents and sediment transport on beaches with sea walls. *Coastal Engineering*, 43 (3-4), 161-162, Elsevier.

Weggel, J.R. 1972. Maximum breaker height. *Journal of Waterways, Habours and Coastal Engineering Division*, 98(WW4), 529-548.

DESIGN WAVE EVALUATION
FOR COASTAL PROTECTION STRUCTURES IN THE WADDEN SEA

Hanz D. Niemeyer[1]& Ralf Kaiser[1]

Abstract: The evaluation of design waves for coastal protection structures is enormously important for both the safety of the hinterland and economical use of public money. The recently gained advancements in mathematical modeling of waves allows its successful application even in areas with complex morphological structures like the Wadden Sea.

A remaining major problem of coastal design wave modeling is still the evaluation of suitable boundary conditions for design conditions, particularly for offshore design waves. The following paper highlights procedures applied for the solutions of these problems in the Frisian Wadden Sea at the south North Sea coast which are partly also applicable at coastlines of any type.

INTRODUCTION

Primary aim of designing coastal protection structures is to guarantee safety for the protected area and particulary for the human beings living there. The secondary one is to achieve an acceptable economical benefit-cost ratio. Meeting both requirements depends highly on a proper evaluation of design waves. Any uncertainty in evaluation them endangers human life and safety of goods or leads to an inappropriate spending of tax payer's money. Nevertheless in many regions of the world and not only in developing countries the basis for the evaluation of design waves is rather poor.

The length of time series of wave measurements is generally insufficient with respect to a proper evaluation by probabilistic methods since the therefore remaining ranges of uncertainty are enormously high. Additionally the density of wave measuring nets is in most cases insufficient in order to cover the local variability of wave climate.
The classical wave prediction methods basing on wind input incorporate also in-accuracies, which grow the more the area of interest is characterized by significant changes of morphological features, which is particularly valid for Wadden Sea coasts.

---

[1] Coastal Research Station of the Lower Saxon Central State Board for Ecology, Fledderweg 25, D-26506 Norddeich/East Frisia, Germany

As well the complex morphology as the superposition of a number of distinct wave systems entering from offshore aggravate the application of conventional forecasting methods or make it even useless in these areas.

Mathematical wave models have proven their capability even in these morphologically complex coastal areas but the implementation of reliable boundary conditions like e. g. the offshore design wind or wave conditions remains a problem if the evaluation of design waves is required.Therefore the design of coastal protection structures at the Wadden Sea coasts of the southern North Sea is still semi-empirical, particularly for the determination of design wave run-up on sea dykes. Major reason is usually the lack of reliable design wave parameters (Niemeyer et al. 1995).

TASKS AND PROBLEMS

The CRS Hydrodynamics Section is generally involved in the design of protection structures at the Lower Saxon coast at the southern North Sea (fig. 1). Its particular task is the evaluation of water level heights, wave parameters and wave run-up being appropriate for design conditions with respect to both technical state of the art and legal boundary conditions. In the recent past local authorities asked for assistance by the design of coastal protection structure which implied specific requirements for the evaluation of coastal design waves.

In the first case the ultimate aim was the design height of dyke at a stretch of about 10 kilometers at the Frisian mainland coast (fig. 1) which shall be strengthened during forthcoming years. In order to keep the maintenance cost of the new dyke as low as possible beforehand to stretches of 250 m each were planned with pilot constructions of dyke revetments. The responsible dyke community asked therefore additionally to determination of design wave run-up

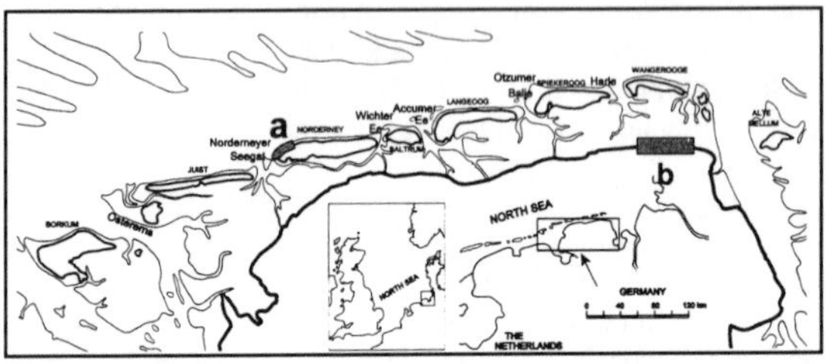

Fig. 1. Location of the investigation areas at the risia North Sea coast: a) Island of Norderney; b) Elisabethgroden-dyke

by the extrapolation-method (Niemeyer et al. 1995) for an estimation of local design wave conditions in order to get the necessary basics for the revetment design.

In the second case parts of the seawall on the island of Norderney (fig. 1) had to be investigated with respect to a probably needed redesign. The core of the construction was erected at about 1860; afterward numerous extensions and improvements have been carried out due to increasing wave loads or due to damages during storm surges (Niemeyer et al. 1996). Therefore model tests in the large wave flume at Hannover University (Führböter 1982) were planned in order to evaluate the wave loads the seawall will experience if design conditions occur. As necessary boundary conditions for the model tests wave spectra on the island's shoreface were required from CRS Hydrodynamics Section.

In the first case the spectral shape is of lower importance since the formulas for the determination of wave run-up and the evaluation on revetments usually do not consider multipeak spectra. For the second one the spectral shape for design conditions was regarded as rather important, since the steering of the wave paddle is controlled either by wave trains or by spectra. Furthermore the input of spectral information is expected to deliver an improved basis for the evaluation of wave loads on structures.

The evaluation of design waves in the inner part of estuaries by mathematical modeling has been carried out earlier with success (Niemeyer 1997; Niemeyer & Kaiser 1999, 2001) by applying the third-generation model SWAN (Ris et al. 1995; Holthuijsen et al. 1998; Booij et al. 1999). This has proven its applicability in Wadden Sea areas delivering reliable results both for normal and storm conditions (fig. 2) if driven by measured spectra at its offshore boundary. But there are no wave measuring data available matching design conditions in the southern North Sea.

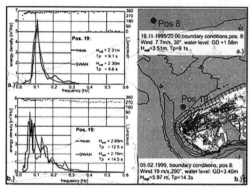

It is possible to create in SWAN a parametric spectra for chosen offshore wave parameters. Therefore chosen offshore design wave parameters can be transferred into such a parametric spectrum as boundary condition for a model run. Major disadvantage of this procedure is that specific regional features of wave spectra like e. g. multipeaks get lost and any of their possible impacts on local design waves also.

Fig. 2. Comparison of measurements and SWAN model data, island shoreface of Norderney (a: ordinary tide; b: storm surge)

PROBLEM APPROACH
Mainland coast, Feasibility study

Field measurements highlighted that waves in Wadden Sea areas are on the one hand strongly depth-controlled (fig. 3). On the other hand local water depths are also determined by local wave climate (Niemeyer 1991): The relation of wave parameters and water depths are in a dynamical equilibrium. If the water depth is larger than the waves need, sedimentation takes place. Vice versa an unsteady non-continuing increase of wave energy input beyond the level of established dynamical equilibrium will only create a higher rate of dissipation. If this holds also stand for design conditions the use of measured spectra as a seaward boundary condition for model runs could lead to reliable results. It must be stressed that such an approach would only be valid for wave parameters but not for the spectral shape. It is therefore only then successfully applicable, if the transfer of spectral shape into wave loads is not required.

Therefore tests have been carried out comparing the results of runs with SWAN for the design water level with distinct seaward boundary conditions:

1. Measured directional wave spectra from storm surges (fig. 4),

2. Parametric wave spectra with higher waves and longer periods for matching design conditions (fig. 4).

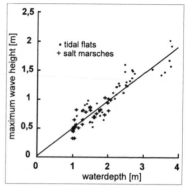

Fig.3. Wave height / water depth- relation in Wadden Sea areas (Niemeyer 1983)

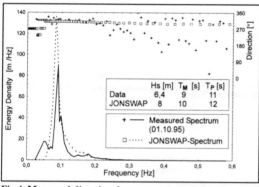

Fig.4. Measured directional wave spectrum (storm surge of January 10th, 1995) and chosen JONSWAP-Spectrum)

The comparison of the results of the two model runs is here presented as well for significant wave heights (fig. 5) as for mean wave periods (fig. 6) and for mean directions (fig. 7) for the whole model area. The results highlight that particularly in the areas of interest i. e. on the tidal flats and salt marshes in front of the sea dyke at the mainland coast the differences for the wave parameters are very small (fig. 5 - 7). They are always within the range of the model accuracy to be expected. In any case the margin of 5 % is not exceeded in the vicinity of the sea dyke.

The closer the mainland coast the more the differences of local wave parameters diminish: In the course of wave propagation to the mainland coast gradually an increasing tendency towards a uniform wave climate develops independently from the offshore wave conditions.The interactions of waves and tidal flat morphology determine dominantly local wave conditions, as already highlighted by the strong linear relationship between wave heights and water depths (fig. 3). The results of this comparison give emphasis to the effect of morphological filtering of storm waves traveling from offshore across the tidal flats. The water depth on tidal flats is determined by the midterm dynamical equilibrium of strong waves and morphology; but during singular events it acts as a limiter. This effect facilitates enormously the choice of suitable off-

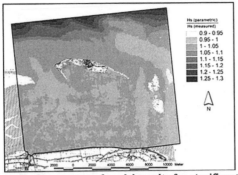

Fig.5. Comparison of model results for significant wave heights for measured and parametric design spectra as offshore boundary conditions

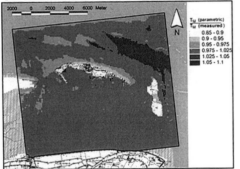

Fig. 6. Comparision of model results for mean wave period for measured and parametric spectra as offshore boundary conditions

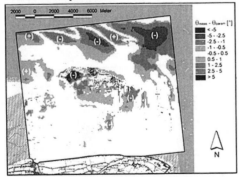

Fig. 7. Comparison of model results for mean wave direction for measured and parametric spectra as offshore boundary conditions

shore boundary conditions for mathematical modeling of design waves for dykes at the mainland coast of Wadden Sea areas by implementing measured storm surge spectra.

For the modeling of design waves an overall model representing an area of 14 x 20 km with a grid size of 100 x 100 m was established; in the vicinity of the dykes a smaller nested model for an area of 4,25 x 12,5 km and a grid size of 50 x 50 m was additionally implemented (fig. 8).

Beside the choice of the off-shore wave spectrum driving the model, water level, wind velocity and direction must be introduced as governing boundary conditions. The design water level is about

$$SWL = G. D. + 5,6 m$$

and the equivalent wind velocity and direction are:

$$u = 30 m/s, \qquad D = 315°$$

Though it is a well known fact, that the wind field is significantly affected in the coastal area (Kaiser & Niemeyer 1999) the reduction of velocity is not taken into consideration. On the one hand this assumption is on the safe side. On the other hand local wind generation is much lesser important for wave climate on tidal flats than the interaction of waves entering from offshore with local morphology.

The results of the overall model for the significant wave heights and mean directions make

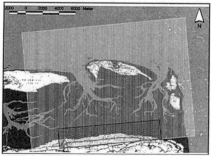

Fig. 8. Model areas and grid size- East Frisian coast seaward of the Elisabethgroden Dyke

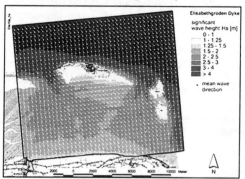

Fig. 9. Significant wave heights H_s and mean wave direction θ_m (SWL: G. D. + 5,6 m; u = 30 m/s from 315°) in the Wadden Sea area seaward of the Elisabethgroden Dyke

evident the tremendous changes waves experience propagating from offshore coastward in barrier island sheltered Wadden Sea areas (fig. 9): An enormous wave energy dissipation takes place on the ebb deltas of the tidal inlets and on the islands' shorefaces. Furthermore the barrier islands shadow large areas of the tidal flats against direct wave intrusion. Waves from offshore penetrate only towards the tidal flat area and the mainland coast via the tidal inlets. From there they are spreading out across the flats. The wave heights decrease there the more the farther from the inlet.

The evaluation of the design wave parameters in front of the dyke was carried out with the nested model (fig. 10 + 11) in order to consider sufficiently any variation for dimensioning the dyke and the revetments properly. The uniformity in local variation of the wave parameters features again the filtering effects due to interactions of waves and tidal flat morphology.

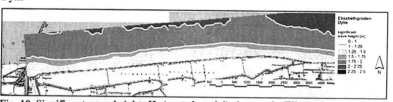

Fig. 10. Significant wave heights H_s (nested model) close to the Elisabethgroden Dyke

Fig. 11. Peak periods T_p (nested model) close to the Elisabethgroden Dyke

Island shoreface, Offshore design wave evaluation

The modeling of the design waves for the Elisabethgroden Dyke taught already the lesson that the wave-morphology-interactions act not everywhere in the coastal area as a limiter. Particularly seaward of the barrier islands an increase of wave energy has an impact on local wave climate. This is in accordance with experience gained earlier from field measurements (Niemeyer 1983).

In line with this perception as a preliminary for the modeling of design waves the effect of distinct offshore wave conditions on the wave climate in the area of interest was investigated. As offshore boundary conditions for the modeling three distinct wave spectra were chosen:

1.   Measured spectrum with $H_S = 6$ m; $T_P = 14$ s; $T_M = 10$ s;

2.   JONSWAP-spectrum with $H_S = 8$ m; $T_P = 12$ s; $T_M = 10$ s;

3.   JONSWAP-spectrum with $H_S = 10$ m; $T_P = 12$ s; $T_M = 10$ s;

The measured spectrum (fig. 12) was chosen since for this storm surge a verification of the SWAN-model is available (fig. 2). Additionally to the first case another JONSWAP-Spectrum with higher wave energy (fig. 12) was introduced which represents the wave parameters evaluated by statistical relationships for design wind conditions (Niemeyer 1983).

The morphology of the investigation area was reproduced by a model bathymetry covering an area of approximately 6,7 x 12,6 km with a grid size of 80 x 80 m. In order to avoid boundary effect the output area was reduced to approximately 5,9 x 7,4 km. For the computation of the design wave spectra a nested model with a grid size of 20 x 20 m was embedded in the shoreface area (fig. 13).

For the data transfer for the steering of the large wave flume tests in the nested model computation points were chosen being representative for stretches of the island's northwestern shore (fig. 14).

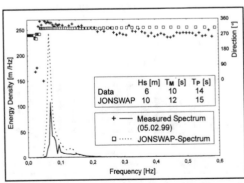

Fig. 12. Measured wave spectrum (storm surge, February 5 th 1999) and JONSWAP-spectrum ($H_S=10$m; $T_P = 15$s) offshore area

Comparing the results at computation points of the model runs with distinct boundary conditions highlights that an increase of offshore wave energy has significant effects on the wave climate on the island's shoreface for the design water level which could not be neglected

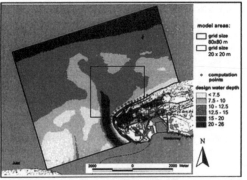

Fig. 13. Dimensions and bathmetry of the wave models Norderney

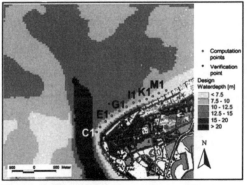

Fig. 14. Bathymetry in the model area with computation points for data transfer

(fig . 15 + 16): The significant wave heights increase there for the assumed extreme offshore wave condition up to approximately 19% (fig. 15). For the mean periods the figure is even higher with a maximum of about 24% (fig. 16). The consideration of these extreme offshore wave conditions for the redesign of the seawall at the northwestern shore of the island of Norderney is therefore indispensable.

With respect to required input signal for the large wave flume tests beside the wave parameters the spectral shape is also highly important. Since the measured spectra in the offshore area del area have a more complex structure than the JONS-WAP-spectrum it is urgently necessary to check if such boundary conditions would create unacceptable simplifications of the design spectra on the island's shoreface. In order to solve this question

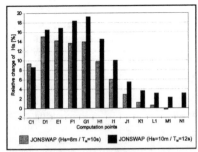

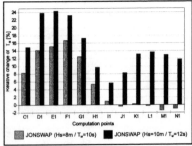

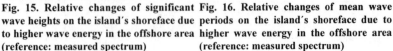

Fig. 15. Relative changes of significant wave heights on the island's shoreface due to higher wave energy in the offshore area (reference: measured spectrum)

Fig. 16. Relative changes of mean wave periods on the island's shoreface due to higher wave energy in the offshore area (reference: measured spectrum)

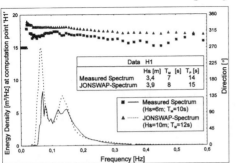

Fig. 17. Comparison of modeled wave spectra at computation point H₁ with distinct offshore boundary conditions

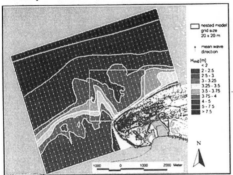

Fig. 18. Significant wave heights for design conditions in the offshore area, in the ebb delta region and on the island's shoreface

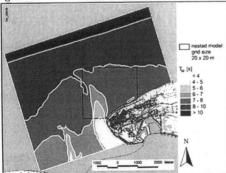

Fig. 19. Mean wave periods for design conditions in the offshore area, in the ebb delta region and on the island's shoreface

the spectra modeled with SWAN at the computation point H₁ (fig. 14) by applying the measured spectra on the one hand and the JONSWAP-spectrum for extreme conditions on the other hand as offshore boundary condition were compared (fig. 17).

The spectra on the island's shoreface have the same characteristic features: the main energy is bound in two narrow frequency ranges of the same order of magnitude. The measured spectrum has furthermore multipeaks, but they are on a considerably lower scale than the major ones and are therefore of low importance with respect to loads on structures. Summing up the use of a parametric spectrum for the representation of extreme offshore wave conditions is acceptable for the modeling of design waves in this environment; neglected effects are of minor importance. Major reason are the strong interactions between waves and morphology between the offshore area and the island's shoreface, particularly the intense wave breaking on the ebb delta shoals of the tidal inlet (Niemeyer 1987).

DESIGN WAVE EVALUATION

The model run for the chosen offshore boundary conditions produced the following results for design significant wave heights (fig. 18): From offshore in landward direction first a slightly gradual decrease takes place which is characterized by long stretched contours of wave height isobaths. They get a highly irregular shape after meeting

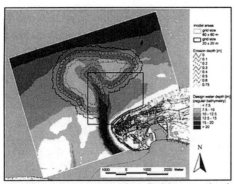

Fig. 20. Implementation of an fictitious erosion in the control partof the ebb delta

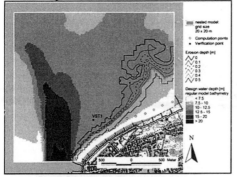

Fig. 21. Implementation of an fictitious erosion on the shoreface and on the beach

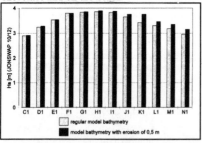

Fig. 22. Effect of a fictitious erosion of shoreface and beach (Fig. 21) on significant wave height

the ebb delta area where they reflect the differences in local wave breaking. Obvious is the penetration of relatively high waves onto the island's north-western shoreface. This is a resulting effect of the erosion of the ebb delta shoals which has taken place in that area during the last 25 years leading to a significant increase of local wave energy (Niemeyer et al. 1997; Kaiser & Niemeyer 1999). In consequence the seawall experiences in that part of the island's shoreface the relatively highest wave loads.

The change of mean periods due to decreasing water depths in the investigation area is much lesser accentuated than for wave heights (fig. 19): The relative changes from offshore to the island's shoreface are smaller and the pattern of isobaths has a less pronounced irregularity. Therefore the variabilty of the mean periods in the area of interest is rather small.

STUDY OF FURTHER EBB DELTA AND SHORE EROSION

With respect to experienced changes of local wave climate due to ebb delta erosion additional tests were carried out by implementation of further erosion of the model topography in the ebb delta area (fig. 20) on the one hand and of the shoreface and beach (fig. 21) on the other hand.

In the ebb delta area a fictitious erosion was step by step introduced: first up to 0,5 m and then up to 0,75 m (fig. 20). After both steps no significant increase of wave parameters on the island's shoreface took place: Obviously the decrease of the ebb delta shoals had already exceeded the ultimate treshhold beyond which no further impact on the landward wave climate will happen.

The fictitious beach erosion (fig. 21) had no remarkable effect in the areas with the relatively high wave loads and only a limited one on the significant wave heights in the eastern part of the investigation area (fig. 22).

CONCLUSIONS

The evaluation of design waves for coastal protection structures in Wadden Sea areas requires specific feasibility studies in order to overcome the lack of sufficiently long time series of measurements. The application of reliable mathematical wave models is very beneficial; crucial for their successful practical use is the introduction of reliable boundary conditions. Their derivation and introduction for to regions with distinct morphological features was displayed.
The morphology of tidal flats which levels are determined by mid-term interactions with from offshore penetrating waves act during extreme storm surges as a limiter for waves. Therefore wave spectra measured during storm surges can be used to drive mathematical wave models for the evaluation of design waves for dykes at the Wadden Sea mainland coast.

The ebb deltas of tidal inlets act as well as a limiter on waves penetrating from off-shore into the coastal areas as tidal flats do. But wave energy dissipation in the ebb delta region is not everywhere sufficiently high for absorbing the effects of extreme offshore wave conditions in total. This is particularly the case if the ebb delta effect on attenuating waves is weakened by severe erosion as happened in the area of the island of Norderney. But obviously there is even for continuous erosion a threshold beyond which a further increase of the from offshore penetrating wave energy does not take place.

The investigations highlighted the enormous progress in the evaluation of design waves for coastal protection structures being achievable by application of third generation wave models with the capabilities of SWAN.

The successful application of SWAN in the Wadden Sea area was possible because

a. verification data were available,

b. the knowledge gained from process-oriented analysis of field measurements was successfully introduced into the set-up of modeling.

ACKNOWLEDGEMENTS

These studies were carried out for the 3rd Dyke Community of Oldenburg and the Lower Saxon Agency for Water Management and Coastal Protection-Division Norden. Additionally basic investigations were undertaken in the framework of the research project "Design Waves for Coastal Protection Structures and Dunes" of the GERMAN COMMITTEE FOR COASTAL ENGINEERING RESEARCH (Sponsor: GERMAN FEDERAL MINISTRY FOR EDUCATION AND RESEARCH via the carrier BEO; project-code KIS 004). The authors are greatly indebted to their colleagues Günther Brandt, Detlef Glaser, Anja van Hettinga, Holger Karow and Georg Münkewarf from the CRS Hydrodynamics Section for their assistance and support.

REFERENCES

Booij, N., Ris, R.C. and Holthuijsen, L.H. 1999. A Third-Generation Wave Model for Coastal Regions, Part I, Model description and validation. J. Geoph. Research, 104, C4

Führböter, A. 1982. The Research Facility 'Large Wave Flume' and its tasks. Intermaritec ,82, Hamburg (in German)

Holthuijsen, L.H., Ris, R.C. and Booij, N. 1998. A Verification of the Third-Generation Wave Model "SWAN". Proc. 5th Int. Worksh. Wave Hindcast. a. Forecast., Melbourne, Fl./USA

Kaiser, R. and Niemeyer, H. D. 1999. Change of Local Wave Climate due to Ebb Delta Migration. Proc. 26th Int. Conf. Coast. Engrg., Copenhagen/Denmark, ASCE, New York

Kaiser, R., Weiler, B. and Niemeyer, H. D. 2001. Evaluation of Design Waves for Coastal Protection Structures in the Wadden Sea. Proc. 27th Int. Conf. Coast. Engrg., Sydney/Australia, ASCE, New York (in preparation)

Niemeyer, H.D. 1983. Wave Climate at an Island Sheltered Wadden Sea Coast. Res. Rep. MF 0203 Germ. Fed. Min. f. Res. & Techn. (in German)

Niemeyer, H.D. 1987. Changing of Wave Climate due to Breaking on a Tidal Inlet Bar. Proc. 20th Int. Conf. Coast. Engrg., Taipei/R.o.C. Taiwan, ASCE, New York

Niemeyer, H.D. 1991. Case Study Ley Bay: an Alternative to Traditional Enclosure. Proc. 3rd Conf. Coast. a. Port Engrg. i. Devel. Countr., Mombasa/Kenia

Niemeyer, H.D. 1997. Check of the Design Height of Dykes at the Lower Ems Estuary. CRS-Rep. 05/97 (in German)

Niemeyer, H.D., Gärtner, J., Kaiser, R., Peters, K.H. and Schneider, O. 1995. Estimation of Design Wave Run-up on Sea Dykes under Consideration of Overtopping Security by Using Benchmarks of Flotsam. Proc. 4th Conf. Coast. a. Port Engrg. i. Devel. Countr., Rio de Janeiro/Brazil

Niemeyer, H.D. and Kaiser, R. 1998. Investigations on the design of test slopes of the Elisabethgroden Dyke, Wangerland. CRS-Rep. 10/98 (in German)

Niemeyer, H.D. and Kaiser, R. 1999. Investigations on the Safety of Dykes at the Lower Ems estuary. Arb. Forsch.-Stelle Küste, 13 (in German)

Niemeyer, H.D. and Kaiser, R. 2001. Evaluation of Design Water Levels and Design Wave Run-up for an Estuarine Coastal Protection Masterplan. Proc. 27th Int. Conf. Coast. Engrg., Sydney/Australia, ASCE, New York (in preparation)

Niemeyer, H.D., Kaiser, R. and Knaack, H. 1997. Effectiveness of a Combined Beach and Shoreface Nourishment on the Island of Norderney/ East Frisia, Germany. Proc. 25th Int. Conf. Coast. Engrg., Orlando,Fl./USA, ASCE, New York

Ris, R., Holthuijsen, L. H. and Booij, N. 1995. A Spectral Model for Waves in the Near Shore Zone. Proc. 24th Int. Conf. Coast. Engrg., Kobe/Japan, ASCE, New York

LABORATORY INVESTIGATION OF NONLINEAR IRREGULAR WAVE KINEMATICS

Hae-jin Choi[1], Daniel T. Cox[2], M.H. Kim[3] and Sangsoo Ryu[4]

ABSTRACT: A comprehensive experimental study was conducted to produce benchmark wave kinematics data sets for moderate/steep regular and irregular waves. The experiments were conducted in a narrow glass-walled wave tank at Texas A&M University and velocities were measured using an LDV. The time series of surface elevation and wave kinematics were obtained at nine vertical positions including two points above still water level (SWL). The experimental data were compared with linear theory and fully nonlinear numerical wave tank (NWT) simulations based on potential theory, mixed Eulerian-Lagrangian approach, and boundary element method. For both regular and irregular waves, the linear theory tends to overestimate the measured kinematics especially above the SWL, and the fully nonlinear NWT simulations correlate better with experimental data. It is also shown that the exceedance probability for irregular wave crest heights is underestimated by the Rayleigh distribution, while the opposite is true for crest velocities. The joint distribution plots of wave crest/wave period and particle velocity/wave period reveal that there exists strong correlation between them.

1. INTRODUCTION

Accurate predictions of extreme wave kinematics particularly above the still water level (SWL) are crucial when estimating nonlinear wave forces on slender members. There have been many theoretical and experimental studies on predicting and measuring wave kinematics for regular and irregular waves. For example, the kinematics of regular, bichromatic, and irregular waves were investigated by Gudmestad (1993) and Longridge et al. (1996). They compared laboratory data with predictions from linear theory and widely used stretching methods, such as Wheeler stretching. They concluded that Wheeler stretching better predicts wave kinematics for a broad bandwidth spectrum, while linear extrapolation is the better predictor for a narrow banded

[1] Graduate student, hae-jin@tamu.edu
[2] Associate Professor, dtc@tamu.edu
[3] Associate Professor, m-kim3@tamu.edu
[4] Graduate student, sruy1@hotmail.com
All at Ocean Engineering Program, Civil Engineering Department, Texas A&M University, College Station, Texas, 77843-3136,

spectrum. Swan (1990) addressed the possible influence of the viscosity on the kinematics of regular waves. Gudmestad (1993) summarized many research works related to the prediction and measurement of deep water wave kinematics for regular and irregular waves.

Though there exist many papers showing wave kinematics comparisons between laboratory and theoretical prediction based on linear and high-order wave theory, no one has yet shown the comparison against fully nonlinear NWT simulations especially extreme velocities above the SWL which is important in estimating wave impact forces on slender members near free surface (Bea et al., 1999). In this paper, a comprehensive experimental study was conducted using an LDV system to produce benchmark wave kinematics data sets for both moderate/steep regular and irregular waves. The experimental data are compared with linear theory and fully nonlinear NWT simulations (Koo and Kim, 2001) based on potential theory, mixed Eulerian-Lagrangian approach, and boundary element method.

In Sec.2, the experimental setup and procedure are presented. In Sec.3, the kinematics of two moderate/steep regular waves are investigated and the experimental results are compared with linear theory and fully nonlinear NWT simulations. The kinematics of moderate/steep irregular waves are presented in Sec.4. Finally, various statistics of irregular waves and their kinematics including exceedance probability and joint probability distributions are presented. The results are summarized and concluded in Sec.5.

2. EXPERIMENT

To measure the particle velocities of regular and irregular waves, we conducted a series of experiments using the narrow wave tank at Texas A&M University. The glass walled flume was 36.1 m long, 0.91 m wide, 1.22 m high, and equipped with a permeable wave absorbing beach down-stream. Wave generation was provided by a dry-back, hinged flap wavemaker capable of producing regular and irregular waves with period ranging from 0.25 to 4 s and maximum height of 25.4 cm. The setup is shown schematically in Fig.1, where x is the horizontal coordinate positive in the direction of wave propagation with $x = 0$ at the wave maker and z is positive upward. As shown in Fig.1, we measured the free surface elevation at 7 m, 8 m and 9 m from wavemaker with two different steepness regular waves and irregular waves. The water depth in the tank was constant with h= 0.80 m.

Table 1 shows the experimental conditions for each regular wave and irregular wave test. Cases 1 and 2 are representative of linear and nonlinear test cases for regular wave, respectively, where the wave steepness is H/L =0.0298 (Case1) and H/L =0.0734 (Case 2). Cases 3 and 4 are representative of linear and nonlinear test cases for irregular waves where the wave steepness based on the significant wave height and wavelength computed using the peak period is H/L =0.0310 (Case3) and H/L =0.0750 (Case 4). Therefore, Case 1 and Case 3 can be compared to see the difference between regular and irregular waves with essentially the same characteristics. Similarly, Case 2 and Case 4 can be compared for the nonlinear conditions. Table 1 also lists the number of waves (N_W) used in each record and U which is the horizontal velocity at the SWL calculated by linear wave theory. U is used for normalization in subsequent figures.

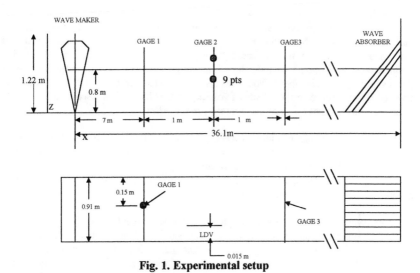

Fig. 1. Experimental setup

The free-surface elevation was recorded using resistant-type surface-piercing wave gages. The wave kinematics were measured using a LDV system at 8 m from wavemaker for seven vertical positions below SWL and two vertical positions above.

The wave crest horizontal velocity of regular waves was measured with duration of 70 s each. The time series of steady state portion (40 waves) was selected to analyze wave elevation and wave crest/trough, horizontal/vertical particle velocities.

For irregular waves (Case 3 and 4), a JONSWAP spectrum with $\gamma = 1$ (=Pierson-Moskowitz spectrum) was used. For the irregular wave data analysis, we collected time series at 8 m for 250 s. Each time series was truncated at the beginning to eliminate transient effects. About 250 waves were used for the subsequent statistical analyses.

Table 1. The conditions of experiment

CASE	TYPE	H (cm)	T (s)	N_W	H/L	ka	U (cm/s)
Case 1	Regular	4.03	0.931	40	0.0298	0.094	13.63
Case 2	Regular	9.00	0.887	40	0.0734	0.231	31.86
Case 3	Irregular	4.04	0.931	254	0.0310	0.097	13.63
Case 4	Irregular	9.18	0.887	234	0.0750	0.236	31.86

* Water depth h=80 cm

3. ANAYSIS OF REGULAR WAVES

In this section, the horizontal velocities of regular waves are studied under the wave crest. The results are presented with normalized values i.e. measured vertical position Z is normalized by wave height H and horizontal velocity u_c is normalized by linear horizontal velocity U at the SWL.

The horizontal velocities of Case 1 (H/L=0.0298) at 16 vertical positions are shown in F: 2. The measured data are presented with the results of linear theory and the fully nonlinear NW simulations. In general, the nonlinear NWT results correlate better with experimental data than t linear theory. Compared to the linear theory, appreciable reductions are observed for t measured crest horizontal velocities above the trough level and the nonlinear NWT simulatio show a similar trend. The phenomenon was also observed in Swan's (1990) experiment, where i attributed the difference to the effects of vorticity caused by vertical viscous diffusion. This ki of viscous effect can not be reproduced by the present NWT.

Fig. 3 presents similar results for Case 2 (H/L=0.0734). Compared to Case 1, the measure crest values are smaller than those of linear wave theory. The results of the nonlinear NW correlate well with the experimental data. In this case, we do not observe any abrupt reduction maximum horizontal velocities above wave trough level, possibly because viscous effects are le significant as the wave steepness increases.

In Fig. 4 and 5, both crest and trough horizontal velocities for Case 1 are presented at fo vertical positions under the wave trough. From these figures, it is seen that the measured negativ minimum values are greater than the positive maximum values, which results in small negativ mean flow. The actual peak-to-trough magnitudes of the measured data are larger (Case 1) about the same (Case 2) compared to those of linear theory. Therefore, if the amplitude is define as half of the peak-to-trough distance and it is compared against linear theory assuming zero ma: transport flow, the measured data are in better agreement with the linear theory, as pointed out b Gudmestad (1993). It is well known from the second-order Stokes wave theory that there exist some positive mean drift proportional to wave amplitude squared near the mean water level. Th cannot be proved from the fully nonlinear simulation or measurements since only the crest value are obtained near the SWL.

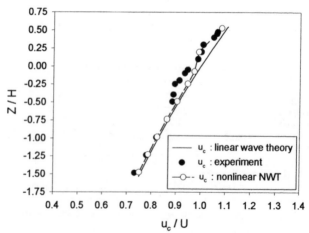

Fig. 2. Comparison with wave crest horizontal velocities for Case 1

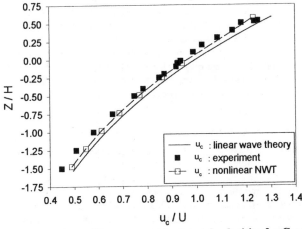

Fig. 3. Comparison with wave crest horizontal velocities for Case 2

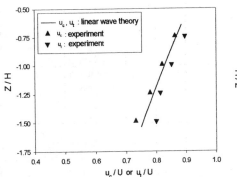

Fig. 4. Comparison of regular wave crest and trough horizontal velocities for Case 1

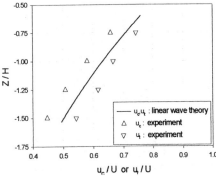

Fig. 5. Comparison of regular wave crest and trough horizontal velocities for Case 2

4. ANALYSIS OF IRREGULAR WAVES

Below trough level of the largest wave, the velocity signal was continuous, and the definition of the significant horizontal crest velocity $(u_c)_S$ follow the standard definition used to report significant wave height: $(u_c)_S$ is equal to the average of the one third highest crest velocities. For Case 3 and 4, the number of waves were $N_W = 254$ and 234, respectively, so $N_W/3 = 85$ and 78.

Above trough level, the velocity signal was not continuous because the signal would drop out if the instantaneous water level was below the measuring elevation of the LDV as would be the case for the passing of smaller waves. Therefore, the number of crest horizontal velocities measured above trough level was significantly lower than the number of waves in the record. For these elevations, $(u_c)_S$ was computed by averaging all ranked crest horizontal velocities up to $N_W/3$. If, for example, there were only 100 crest velocities measured at a particular elevation for Case 3, then $(u_c)_S$ is estimated by averaging the highest 85 crest velocities at that elevation. If however, there were only 64 measured crest velocities, then $(u_c)_S$ is based on the average of all 64 measurements.

This is a somewhat unsatisfactory method for computing $(u_c)_S$ since it uses a different number of points at elevations above trough level. An alternative method is suggested in which the vertical coordinate is defined with respect to the moving water level. This definition, however is difficult to apply for the case of most fixed probe instruments and would require either a probe which moves with the free surface or one which provides a high spatial resolution in the vertical (e.g., PIV, pulse-coherent acoustic Doppler). Nevertheless, $(u_c)_S$ is computed as defined above and elevations above trough level for which $N < N_W$, the number of velocities used to compute $(u_c)_S$ are marked on the figure.

The results of Case 3 for small steepness ($H/L = 0.031$) irregular waves are presented in Fig. 6. The linear irregular wave simulations tend to overpredict the measured data by about 10%. This trend was also illustrated by Longridge et al. (1996). The result seems consistent with Gudmestad's observation. It is also seen that the measured significant crest horizontal velocity above MWL can be greatly increased by the similar rate as the linear irregular waves. The equivalent linear regular wave model tends to overestimate below trough level and underestimate above it. In particular, the horizontal velocity near the wave crest is greatly underestimated by equivalent regular waves. Like the regular wave cases, the measured significant trough velocities tend to be greater than the crest values, which again results in small negative mean transport flow. The same phenomenon was also observed by Skjelbreia et al. (1991). The same trend can also be seen in Case 4 of steeper irregular waves. It seems that the negative mean transport flow is less significant compared to the regular wave case, which was also pointed out by Gudmestad (1993).

Fig. 7 shows similar plots for Case 4 ($H/L = 0.075$). In this case, the linear irregular-wave simulations greatly overpredict the measured horizontal velocities near wave crest. Surprisingly, the equivalent linear regular wave model gives better prediction for wave kinematics than the linear irregular wave model.

Fig. 8 shows the exceedance probability for the irregular wave crests compared with Rayleigh distribution. The measured exceedance probability is given as $P = i/(N+1)$ where i is the ith rank of the crest height and N is the number of waves listed in Table 1. The solid curve shows the extreme crest height for a Rayleigh distribution given by $P = \exp[-8(\eta_c/H_S)^2]$ as discussed in Forristall (1978). Case 3 with smaller steepness correlates better with the measured data than Case 4 except to the two extreme points

Fig. 9 shows the exceedance probability for the crest horizontal velocity measured at $Z = -1.5H_S$ below the SWL. Again, Case 4 (steeper waves) tends to deviate more from the

Rayleigh distribution. It is very interesting to note that the Rayleigh distribution underpredicts the wave crest height but overpredicts the crest horizontal velocity. We also have similar trend at other locations.

Fig. 10 presents the joint probability distributions between wave crest height, crest horizontal velocity and wave period for Case 3 and Case 4. Although 250 waves are not enough to give smooth lined joint probability distributions, it is found from Figure 10 that the wave crest height and wave crest velocity have strong correlation through the similar shape of joint probability distribution.

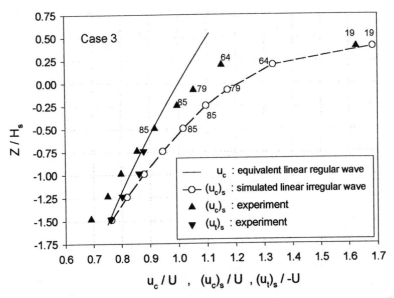

Fig. 6. Comparison with wave crest horizontal velocities for Case 3 with number of wave crest velocities used to compute $(u_c)_S$ presented near open circles.

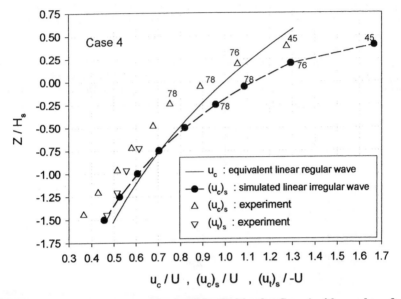

Fig. 7. Comparison with wave crest horizontal velocities for Case 4 with number of wave crest velocities used to compute $(u_c)_S$ presented near closed circles.

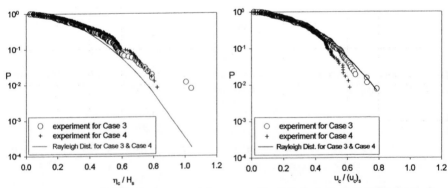

Fig. 8. Exceedance probability for wave crest & $(u_c)_S$ significant wave

Fig. 9. Exceedance probability for u_c & at Z= -1.5Hs

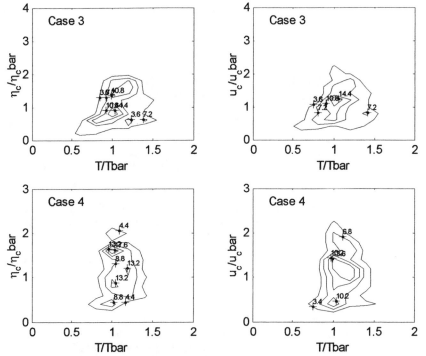

Fig. 10. Joint probability distributions of wave crest height and wave crest horizontal velocity vs. wave period for Case 3 and Case 4.

5. SUMMARY AND CONCLUSIONS

This paper reports the detailed laboratory measurements of free surface elevation and wave kinematics in regular and irregular waves. The measured data are compared with linear theory and fully nonlinear NWT simulations. The comparison shows that the fully nonlinear NWT simulations generally agree much better with measured data than linear theory. In particular, the nonlinear NWT simulations predict some negative mean transport flow below SWL, which is also observed in the present experiment. It is also seen that the linear irregular wave simulation can greatly overestimate the crest horizontal velocities above SWL. The exceedance probability for irregular wave crest heights is underestimated by the Rayleigh distribution, while the opposite is true for crest velocities. The joint distribution plots of wave crest/wave period and particle velocity/wave period reveal that there exists strong correlation between them.

ACKNOWLEDGEMENTS
The authors would like to thank Jose Alberto Ortega, undergraduate student, and Sungwo
Shin, graduate student at Texas A&M University for helping the experiment, and also John Ree
for technical support in the laboratory.

REFERENCES
Bea, R.G, Xu, T., Stear, J., Ramos, R., 1999.   Wave Forces on Decks of Offshore Platforms
   Journal of Waterway, Port, Coastal, and Ocean Engineering. 125, 3, 136-144.

Forristall, G.Z., 1978.   On the Statistical Distributions of Wave Heights in a Storm. In: *Journal
   of Geophysical Research 83 (C5),* 2353-2358

Gudmestad, O.T., 1993.   Measured and Predicted Deep Water Wave Kinematics in Regular and
   Irregular seas. In: *Proceedings of Marine Structures 6*, 1-73

Koo, W.C., Kim, M.H. 2001. Fully Nonlinear Waves and Their Kinematics: NWT Simulation VS
   Experiment, *4th International Symposium on Ocean Wave Measurement and Analysis*, WA-
   VES 2001. (Submitted)

Longridge, J.K., Randall, R.E., Zhang, J., 1996.   Comparison of Experimental Irregular Water
   Wave Elevation and Kinematic Data with New Hybrid Wave Model Predictions. *Ocean
   Engineering , Vol. 23, N o. 4*, 277-307

Skjelbreia, J., Berek, E., Bolen, J.K., Gudmestad, O.T., Heideman, J., Ohmart, R.D., Spidsoe, N.
   Torum, A., 1991.   Wave Kinematics in Irregular Waves. In: *Proceedings Offshore Mechan
   ics and Artic Engineering,* OMAE Stavanger, ASME, New York, 223-228

Swan, C., 1990.   A Viscous Modification to the Oscillatory Motion Beneath a Series of Progressive Gravity Waves. In: *Water Wave Kinematics*, 313-329

DEPTH-LIMITED WAVE BREAKING
FOR THE DESIGN OF NEARSHORE STRUCTURES

Luc Hamm [1]

Abstract: Four laboratory datasets acquired in a directional wave basin and in a large-scale flume are re-analyzed to provide practical advice to coastal engineers in charge of estimating significant wave heights in the nearshore zone as part of a coastal structure design. An estimation of the total water depth including wave set-up and long wave activity is proposed. The underestimation of the significant wave height by using a spectral approach is stressed and rules of thumb are suggested to correct it.

INTRODUCTION

A lot of confusion is still visible in engineering studies in the use of various empirical relationships relating wave heights to water depth at breakpoint (the so-called breaking index or γ parameter). These relationships published in the last 30 years, can be classified into four categories: individual waves at break point, surf zone saturation, wave height decay in the surf zone using a spectral approach and a temporal approach.

Recently, Hamm and Peronnard(1997) have stressed again the differences in the definition of the nearshore wave parameters between the temporal and the spectral approaches. As design coastal engineers are not directly interested by the saturation of the energy of waves estimated from the spectral significant wave height (H_{mo}) but rather for a prediction of the heights of individual waves propagating in the surf zone, a temporal approach is definitely needed with the use of a regular wave breaking index or a wave-by-wave transformation model.

1 Dr., SOGREAH Maritime, 6, rue de Lorraine, F-38130 Echirolles, France, +33(0)4 7633 4188, luc.hamm@sogreah.fr

In the present paper, these different concepts are illustrated from a detailed analysis of two laboratory experiments where high-quality data have been recorded, enabling an accurate analysis of the different processes involved in the surf zone. The first section of the paper explains the methodology and is followed by the description of the two experiments. The next sections are devoted respectively to the analysis of the ratio $H_{mo}/H_{1/3}$ and to breaking indexes. The paper concludes with practical recommendations to experimentalists and to design engineers.

DATA ANALYSIS METHODOLOGY

Raw data from free surface elevations recorded by capacitive wave gauges were analyzed following the principles presented in Hamm and Peronnard (1997). More precisely, the oscillatory part $\eta(t)$ of the surface elevation was first decomposed into a low-frequency part $\eta_{low}(t)$ and a high frequency part $\eta_{hi}(t)$. $\eta_{low}(t)$ was obtained by applying a low-pass filter at half of the peak frequency of the spectrum and $\eta_{hi}(t)$ by the difference to $\eta(t)$.

The spectral significant wave height H_{mo} was computed as four times the standard deviation of the signal $\eta(t)$ together with its high ($H_{mo,hi}$) and low ($H_{mo,lo}$) frequency parts obtained from $\eta_{hi}(t)$ and $\eta_{low}(t)$ respectively. In a second step, the high frequency part of the signal was analyzed in the temporal domain by a zero down-crossing analysis to get the distribution of the individual waves. This distribution was filtered in the temporal domain to eliminate the very small waves induced by the turbulent fluctuations in the surf zone. This filtered distribution was then used to compute the associated wave heights ($H_{1/3}$, H_{max}). Hamm and Peronnard (1997) demonstrated that this separation between motions is a valid way to improve the quality of experimental data published in the literature and to enhance the calibration and validation of numerical wave transformation models in the surf zone.

A further step is described here to also improve the estimation of the local water depth which is used in most breaking index formulas. For that purpose, it is recommended here to follow Goda(1975)'s conclusions to estimate the total water depth (h_{tot}) at each position of the wave gauges by adding the contributions from the still water depth (h_o), the set-up (b); and the root-mean-square amplitude of the low frequency signal ($A_{lfw} = H_{mo,lo}/\sqrt{8}$) as shown in equation (1):

$$h_{tot} = h_o + b + A_{lfw} \qquad (1)$$

It was found from previous experiments that the mean value of records obtained from capacitive wave gauges is not an accurate estimation of the set-up. In these two experiments, set-up measurements were obtained from bottom pressure sensors or by using tappings (Mory and Hamm, 1997).

SMALL-SCALE BASIN MEASUREMENTS

The first dataset selected for this analysis comes from laboratory model tests performed in 1994 in the 3D wave basin of the Laboratoire d'Hydraulique de France (Grenoble, France). The basin is 30m by 30 m and equipped on one side with a multidirectional wave generator made of 60 paddles working in a translation mode (shallow water generation). These tests have been reported in details by Mory and Hamm (1997). The sea bed was a concrete bottom consisting of a zone of constant depth of 0.33 m closest to the wave generator, an underwater plane beach sloping at 1 in 50 and an emerged plane beach sloping at 1 in 20. An offshore breakwater 6.66 m long was built in the left side of the basin while the other side was used as a reference plane profile.

For the present analysis two series of measurements were selected performed along this reference profile including unidirectional (UNI) and multidirectional (DIR) waves with a cosine square angular distribution. Both incident wave conditions were normally incident to the beach and had a Jonswap spectrum with a peak period of 1.69s and a spectral significant wave height of 0.115m. Some 600 waves were generated for each test. The paddles were steered with a linear signal without accounting for the generation or absorption of second-order low-frequency waves. As a consequence, the low-frequency waves measured in the basin were mostly parasitic free waves induced by uncontrolled processes including set-down compensation at the wave generator, surf beat induced by breaking waves, multiple reflections of these waves at the boundaries of the basin and, possibly, oscillations at the natural frequencies of the basin. This situation is not too far from that found in field observations where it is common to measure a free long-wave climate not directly linked with the wind wave sea state observed simultaneously.

The evolution of the different parameters described in the previous section along the beach profile is shown in figures 1 and 2 for the two tests. It is noted that the amplitude of the low frequency waves is much higher with unidirectional waves in very shallow water than with directional waves. Such a difference was also previously noted by Briggs and Smith (1990) in similar 3D wave basin tests. The measurements of H_{max} also show that the generation of directional waves seems to induce the highest waves. This point has not received any explanation yet.

LARGE-SCALE FLUME MEASUREMENTS

The second group of datasets selected for this analysis come from laboratory model tests performed in 1993 at Delft Hydraulics Laboratory in the Delta-flume (de Voorst, The Netherlands) as part of the LIP11D European project. This flume is 250 m long and 5 m deep and enables full scale wave generation. These tests are described in detail in Arcilla et al. (1994). The raw data from two sub-tests were analyzed in the present study: test 2A (H_{mo} of 0.85m; T_p of 5.0s; 750 waves generated on a monotonic beach profile) and test 1C (H_{mo} of 0.60m; T_p of 8.0s; 475 waves generated on a barred beach profile). A standard linear steering signal generated from a Jonswap spectrum was used to drive the piston-type wave-maker. Reflected long-waves were absorbed at the wave-maker.

Figures 3 and 4 show the evolution of the different parameters investigated together with the bottom profiles. It is noted that the amplitude of the low frequency waves is comparatively lower than in the previous small-scale basin tests.

Unidirectional waves on a 1 in 50 slope

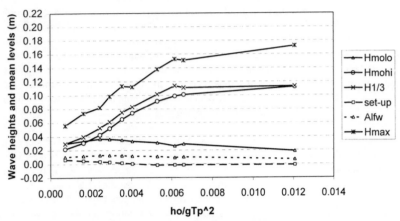

**Fig. 1. Basin tests: Wave heights and mean water levels
with the unidirectional wave condition**

Directional waves on a 1 in 50 slope

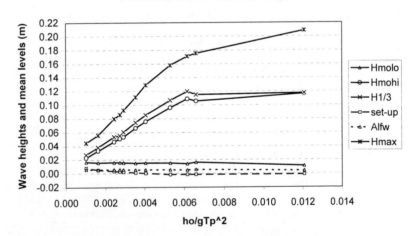

**Fig. 2. Basin tests: Wave heights and mean water levels
with the directional wave condition**

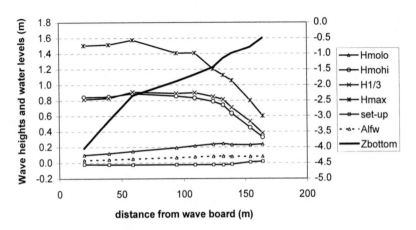

Fig. 3. LIP11D test 2A: Bottom elevation, wave heights and mean water levels

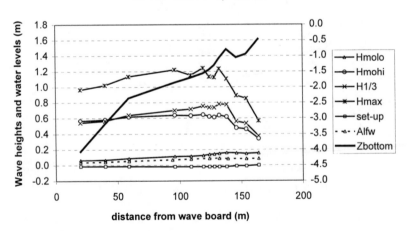

Fig. 4. LIP11D test 1C: Bottom elevation, wave heights and mean water levels

RATIO $H_{1/3}$ OVER H_{MO}

This ratio was first studied by Thomson and Vincent (1985) from laboratory tests. They found an increase from the wave generator to the breaking zone and then a decrease further inshore. They parameterized the envelope of the recorded maxima with the help of the stream function theory as:

$$\left(H_{1/3} / H_{mo}\right)_{max} = \exp\left(0.02289(h_o / gT_p^2)^{-0.43642}\right) \qquad (2)$$

The still water depth is used in this expression and the low frequency part of the energy is included into H_{mo}. The same methodology was applied for the present study and the evolution of the ratio along the beach profile was plotted to find the maximum ratio value as shown on figure 5. Thanks to a clean separation between short and long waves, it was possible to use both expressions of the spectral significant wave height namely H_{mo} and $H_{mo,hi}$ to plot this figure. A clearly visible maximum in the first case with H_{mo}. On the other hand the use of $H_{mo,hi}$ which is more realistic from a physical point of view leads to a monotonic increase of the ratio in the inner surf zone in the case of small-scale experiments. It also leads locally to higher values of the ratio. It means that small-scale flume tests are not suitable to accurately define such a ratio for short waves.

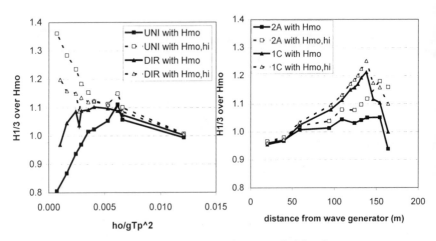

Fig. 5: Ratio $H_{1/3}$ over H_{mo}. left: basin tests and right: flume tests

To go further, the ratio $H_{1/3}/H_{mo,hi}$ was plotted in figure 6, measured at the position of the maximum value of the ratio ($H_{1/3}/H_{mo}$). Additional field data from DUCK85 reported by Ebersole and Hughes (1987) where a proper signal analysis was also performed, was then added. Figure 6 shows that equation (2) is still valid presenting an envelope to the measurement cloud with a slight overestimation of about 5%.

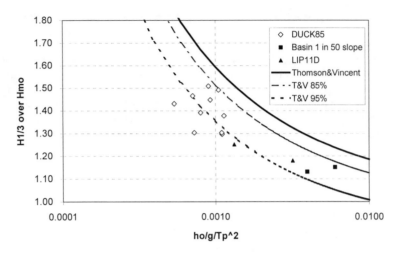

Fig. 6: Comparison of the measured ratio against equation (2)

BREAKING INDEXES

The two main breaking indexes $H_{1/3}/h$ and H_{max}/h used in design where h is the water depth were revisited next in this study. Figure 7 shows a graph of the latter index using three possible definitions of the water depth: the still water depth (h_o), the mean water depth (h_o+b) and the total water depth (h_{tot}) defined by equation (1). The Weggel (1972) breaking index computed at each position of the wave gauges is also plotted for reference.

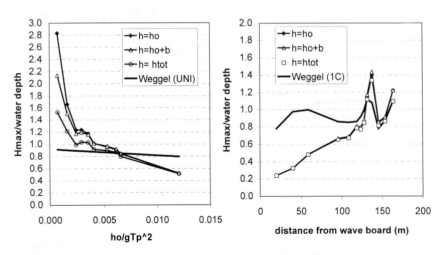

Fig. 7: Evolution of H_{max}/h. left: test UNI (basin) and right: test 1C (flume)

In the basin (small-scale tests), a large sensitivity of the breaking index to the definition of the water depth is observed. The proposed definition of the total water depth in equation (1) decreases significantly the gap observed with the Weggel formula. The influence of the long waves is clearly visible for that case at the last two measurement points nearshore where they are dominant (remember figure 1). In the flume (large-scale tests), the influence of long waves is still visible with less intensity. Figure 8 shows the measured ratios H_{max}/h_{tot} plotted against Weggel breaking index (the points where the long waves were dominant have not been plotted). This plot leads to the conclusion that a safety factor of 1.25 shall be applied to the Weggel estimation to be able to fit the data.

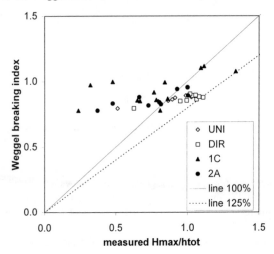

Fig. 8: Comparison between measured H_{max}/h_{tot} and Weggel breaking index

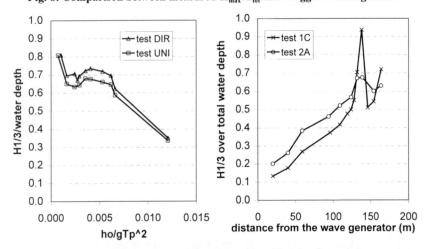

Fig. 9: Ratio $H_{1/3}$ over h_{tot}. left: basin tests and right: flume tests

Finally, figure 9 shows the evolution of $H_{1/3}/h_{tot}$. This index peaks between 0.6 and 0.8 for the different tests with a highest value of 0.95 at the top of the bar. Such values are higher than the 0.55-0.6 values commonly used by designers in shallow water.

CONCLUSIONS

The surfzone is a place where different non-linear motions are active. Providing the designer of a coastal structure with adequate information is therefore not an easy task. Goda (1975) published a key publication on this subject whose details don't seem very well known to practitioners. In this paper, advantage is taken of the availability of high quality raw data acquired during two major laboratory experiments to re-analyze the different processes which are of interest for design. A careful analysis of these data confirms that attention should be paid on the estimation of both the total water depth and the short wave heights.

The total water depth differs from the still water depth by the addition of the wave set-up and the long wave activity. We recommend a meaningful measurements of these quantities by all experimentalists involved in the field or in the lab. A procedure is shown by separating the low and high frequency parts of measured sea surface elevations in order to get a realistic view of the long wave energy. Set-up cannot be estimated accurately from capacitive wave gauges and additional instruments like pressure sensors or tappings must be used.

Looking at it from a designer's point of view, it is now established that wave set-up is a predictable quantity with an adequate wave transformation module working in the spectral or in the temporal domain. In complex cases where set-up gradients are likely to be dominant, it may be necessary to improve the prediction with a 2D hydrodynamic shallow water model driven by radiation stresses. On the other hand, long wave activity is site specific and its correlation with short-wave sea states has been estimated only for research purposes on very few places. Kamphuis (2000) recently suggested an empirical formula based on small-scale flume tests.

In what concerns short waves, published literature (Goda 1975, Dally 1992, Hamm and Roelvink 1994, van Rijn and Wijnberg 1996) has demonstrated that wave-by-wave models are adequate tools to get an estimate of wave heights like $H_{1/3}$ or $H_{1/10}$. Available models are yet limited to the case of nearly-prismatic beaches. Spectral transformation models are thus recommended in the case of complex bathymetries. They provide a reasonable estimation of the spectral significant wave height. From these computations, the Thomson and Vincent (1985) formula can be used to get the significant wave heights needed in design computations. The case of the maximum wave heights is less clear. The present analysis shows that a classical breaking index formula (Weggel formula) underestimates by up to 20% the measured maximum wave heights even when the total depth is considered. A safety factor of 1.25 is recommended as a rule of thumb to the maximum wave heights predicted by the Weggel formula for safety reason in preliminary design.

ACKNOWLEDGEMENTS
This study has been performed as part of the research project ECOMAC within the framework of the Club pour les Actions de Recherche sue les Ouvrages en Mer (CLAROM group) with the financial support of the Fond de Soutien aux Hydrocarbures (FSH) under convention n° CEP&M-M0610199 and FSH-994DM06101.

REFERENCES

Arcilla, A.S., Roelvink, J.A., O'Connor, B.A., Reniers, A. and Jimenez, J.A., 1994. The Delta Flume '93 experiment. Proc. Coastal Dynamics Conf., Barcelona, ASCE, 488-502

Briggs, M.J. and Smith, J.M., 1990. The effect of wave directionality on nearshore waves. Proc. 22nd Int. Conf. On Coastal Engng, ASCE, 267-280.

Dally, W.R. (1992). Random breaking waves: field verification of a wave-by-wave algorithm for engineering application. Coastal Engineering, 16, 369-397.

Ebersole, B.A. and Hughes, S.A., 1987. DUCK85 photopole experiment. Report M.P. 87-18, CERC, Waterways Experiments Station, US Army Corps of Engineers, Vicksburg, MS, USA.

Goda, Y. (1975). Irregular wave deformation in the surf zone. Coastal Eng. in Japan, Vol. 18, 13-26.

Hamm, L. and J.A. Roelvink (1994). Short and long wave transformation modelling in a large scale flume. Proc. Int. Conf. on the role of large scale experiments in coastal research (Coastal Dynamics'94), ASCE, 503-517

Hamm, L. and C. Peronnard (1997). Wave parameters in the nearshore: A clarification. Coastal Engineering, 32, 119-135.

Kamphuis, J.W. (2000). Designing for low frequency waves. Proc. 27th Int. Conf. On Coastal Engng, ASCE.

Mory,M. and Hamm,L., 1997. Wave heights, set-up and currents around a detached breakwater submitted to regular or random wave forcing. Coastal Engineering, 31, 77-96.

Thomson, E.F. and C.L. Vincent (1985). Significant wave height for shallow water design. J. of Waterway, Port, Coastal and Ocean Engng, 111(5), ASCE, 828-842.

Van Rijn, L.C. and Wijnberg, K.M. (1996). One-dimensional modelling of individual waves and wave-induced longshore currents in the surf zone. Coastal Engineering, 28, 121-145.

Weggel, J.R.(1972). Maximum breaker height for design. Proc. Int. Conf. On Coastal Engng., ASCE, 419-432

WAVE MODELING FOR DOS BOCAS HARBOR, MEXICO

Rajesh Srinivas, M. ASCE[1]

Abstract: The wave propagation model REFDIF was used to simulate wind-generated wave conditions in Dos Bocas Harbor to define breakwater modifications necessary to keep the harbor operational at least 85% of the time and to define the extreme wave climate in the harbor vicinity for one-berth and two-berth configurations.

BACKGROUND AND PURPOSE

Dos Bocas, located at approximately 18o 26' N and 93o 10' E, lies on the Gulf of Mexico coast of Mexico in the state of Tabasco (Figure 1). The site has been selected to develop an ambitious oil, industrial, and commercial port. Plans to develop Dos Bocas Harbor began in the late 1970s. The design involved two breakwaters — an east and a west breakwater with lengths of 3.5 km and 1.2 km, respectively. The breakwaters and supporting entrance and access channels have been partially completed, relative to original expectations, over the years owing to cyclicity of oil prices. This paper reports a study to evaluate the possibility of harbor operations through one- and two-exportation berths for 100,000 DWT crude oil-carrying ships with limited addition to existing breakwaters. These new alternatives would prove feasible if the harbor were operational at least 85% of the time every year. Study objectives are to (1) define the breakwater extensions necessary to keep the harbor operational at least 85% of the time, and (2) define the extreme wave climate in the harbor and its vicinity.

HARBOR OPERATIONAL CRITERIA

Port operational time corresponds to the amount of time that wave conditions within the port do not exceed the maximum permissible values that limit the safe

---

[1] 9000 Cypress Green Drive, Jacksonville, FL 32256, USA. rsrinivas@taylorengineering.com

Figure 1. Location Map, Dos Bocas Harbor, Mexico

operation of vessels. Several factors influence the wave conditions' limiting values including but not limited to the type and size of ship, ship operations, tug assistance, mooring facilities, type of cargo, and safe loading and unloading of cargo. Following a literature search and discussions with PEMEX and port operators, a limiting wave height of 1.25 m and a limiting wave period of 10 sec at the ship berth location were adopted as the limiting criteria for safe port operation. These criteria must be satisfied at least 85% of the time. Note that the present study solely looked at wind-generated waves; long waves generated by harbor resonance or short-wave interactions were not investigated.

OFFSHORE WAVE DATA ANALYSIS

Lin (1998) applied the STWAVE wave generation-propagation model to hindcast three-hourly wave climate at nine Wave Information Study (WIS) stations offshore Dos Bocas for a period of 20 years from 1976 to 1995. WIS station ST004, located at 18.475 °N, 93.225 °W in 20 m (MLW) water depth, provided the offshore boundary conditions for the nearshore wave modeling discussed here. The three-hourly wave climate time series at ST004 was analyzed statistically to determine the representative wave climate parameters for nearshore wave modeling. The joint marginal probability of wave height and direction revealed that the 22.5°N direction band contains the most waves (36.61%), followed in order by the 45.0N° direction band (28.96%), the 0.0N° direction band (16.61%), and the 337.5°N direction band (9.56%). Waves are much less frequent from all other directional bands. Analyzing the cumulative marginal probabilities p(H) and p(T), by considering all directions, the wave height at ST004 is less than or equal to 1.1 m for 85% of the time and the wave period is less than or equal to 5.5 sec for 85% of the time.

The design wave height and wave period at ST004 are conservatively assumed to be 1.1 m and 6 sec, respectively, for purposes of nearshore wave modeling to determine the wave conditions occurring 85% of the time inside Dos Bocas harbor. Note that wave period is conserved for propagating waves. Thus, the port operational criteria requiring wave periods less than 10 sec for 85% of the time is automatically satisfied for the harbor. However, wave heights in the harbor must be determined through a combination of nearshore wave propagation modeling and statistical techniques, as discussed next.

NEARSHORE WAVE MODELING

The REFDIF Wave Propagation Model
The propagation of waves over complex bathymetry, around offshore islands, and around offshore and coastal structures involves the interactions of many processes including shoaling, refraction, reflection, diffraction and energy dissipation (breaking, cross-frequency energy transfer, bottom and surface energy dissipation). REFDIF (Refraction Diffraction), a weakly nonlinear monochromatic wave propagation model first developed by Kirby and Dalrymple in 1983, accounts for all the above processes except cross frequency energy transfer (wave-wave interactions) and backward reflection. Note that forward or grazing reflection is simulated. The impacts of processes REFDIF does not simulate are believed negligible given the Dos Bocas Harbor geometry.

Model Methodology and Test Conditions
The model's bathymetric grid describes the specific physiographic characteristics of Dos Bocas Harbor. The area selected for the model grid scheme covers the region between the east and west breakwaters and extends sufficiently from the harbor in all directions such that the effects of the complex bathymetry and structures of the harbor vicinity on waves are fully modeled. The rectangular model grid measures 5,840 m in the shore-perpendicular (north-south) direction and 6,980 m in the shore-parallel (east-west) direction. The shore-parallel offshore boundary lies approximately 3,400 m away from the harbor shoreline. The eastern and western boundaries lies approximately 1,800 m east of the east breakwater and 2,500 m west of the west breakwater. The model grid scheme consists of 292 × 349 grid elements, each of dimensions 20×20 m. Given a particular harbor configuration (bathymetry and breakwaters), REFDIF requires the offshore wave height, wave period, and wave approach direction to be specified. Consequently, the wave statistics were further analyzed to determine the required wave boundary conditions and modeling scenarios/runs required to satisfy project objectives.

The first key objective of the present study is to determine the breakwater extensions required to keep the port operational for at least 85% of the time for the one-berth and two-berth harbor configurations. The analysis presented earlier showed offshore waves less than or equal to 1.1 m occurring 85% of the time. The joint cumulative probability of wave height and direction was calculated to determine the dominant direction bands for nearshore wave propagation modeling. This analysis revealed that wave heights less than or equal to 1.1 m occur 33.2% of the time from NNE, 28.91% of the time from NE, 9.18% of the time from N, 5.29% of the time from NNW, and 0.50% of the time from NW. Cumulatively, these five direction-height conditions occur offshore 77.09% of the time. Offshore wave heights less than 1.1 m from other directions should not affect the design wave conditions inside the harbor as they either propagate offshore or are sheltered off from the harbor by the presence of the existing breakwaters. Based on these offshore wave characteristics, the cases documented in Table 1 were modeled to determine the maximum wave height experienced at least 85% of the time inside the harbor.

For each harbor configuration, REFDIF was applied for each of the five cases to simulate the wave field at every location in the model grid. Wave heights from each of the five cases were compared to calculate the maximum wave height experienced at every grid location. Given the statistics of the offshore hindcast data and the port geometry, modeling the five cases of Table 1 to determine the maximum wave heights among the five cases at every location in turn determines the effective maximum wave height experienced at each location at least 85% of the time.

The second objective is to determine the extreme waves expected inside the harbor. Knowledge of these waves is essential for the design of harbor facilities and coastal structures. Further, knowledge of these waves are also useful to the ship, boat, and tug operators using the harbor since it quantifies the worst conditions expected in the harbor. Lin (1998) conducted an extreme wave analysis for ST004. The level of confidence in the extreme wave analysis diminishes rapidly for return periods in excess of 50 years. Consequently, extreme waves expected in 50 years were chosen to represent the worst case offshore conditions; the corresponding extreme waves in Dos Bocas Harbor were then modeled with REFDIF. Based on the offshore extreme wave height data, three additional cases shown in Table 2, representing extreme storm effects, were modeled for each harbor configuration.

Wave propagation modeling considered four harbor configurations (Table 3). Digital terrain models created with data on existing and proposed harbor bathymetries and structures derived the computational bathymetric grids for the four harbor configurations. Harbor configuration 2 describes the navigation project of the final one-berth port design which includes the existing east and west breakwaters; dredged approach and access channels and turning basins; and, both the Terminal de Usos Multiples and Puerto de Abastecimiento (Figure 2). Harbor configurations 3 and 4 describe the navigation project of the two-berth port designs which have larger dredged channels and turning basins compared to Harbor Configuration 2. Due to space constraints, the remainder of this paper will present some results for Harbor Configuration 2 only; detailed analyses on all cases and harbor configurations are presented in Srinivas et al. (1999).

Table 1. Wave Conditions Modeled to Determine Maximum Wave Conditions Experienced At Least 85% of the Time

Case	Offshore Wave Conditions		
	Height (m)	Period (sec)	Approach Direction (°N)
1	1.1	6	315
2	1.1	6	337.5
3	1.1	6	0
4	1.1	6	22.5
5	1.1	6	45

Table 2. Extreme Wave Conditions Modeled

Case	Offshore Wave Conditions		
	Height (m)	Period (sec)	Approach Direction (°N)
6	6.74	12	345
7	6.74	12	350
8	6.74	12	355

Table 3. Bathymetric and Breakwater Conditions Associated with Various Harbor Configurations

Harbor Configuration	Bathymetry	Breakwater	
		East	West
1	Existing	Existing	Existing
2	One-berth	Existing	Existing
3	Two-berth	Existing	Existing
4	Two-berth	300 m extension	Existing

Model Simulations for Harbor Configuration 2

This harbor configuration comprised the dredged one-berth bathymetry and existing breakwaters at Dos Bocas Harbor. Cases 1 through 8 were run to define the maximum waves experienced at least 85% of the time and the extreme wave climate.

Maximum Waves Experienced 85% of the Time

Model simulations were conducted for Cases 1 through 5. Figure 3a presents wave vectors in the nearshore region of Dos Bocas Harbor for Case 1. The offshore wave height and period are 1.1 m and 6 sec, respectively; the offshore wave approach angle is 315°N. These vectors are superimposed on a gray-scale image of the harbor bathymetry to show the effects of the bathymetry and the breakwaters on wave propagation. The length of each wave vector reflects the wave height at the location while the direction of each wave vector directly corresponds to the wave direction at the location. Figure 3b presents wave height contours superimposed on a gray-scale image of the harbor bathymetry for Case 1. As expected, the results (including Cases 2 through 5 which are not shown here) show that the complex bathymetry and the east and west breakwaters greatly affect waves. Wave rays tend

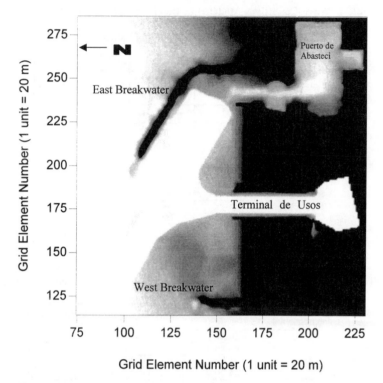

Figure 2. Inset View of Harbor Configuration 2 Model Grid Coverage

to become more and more shore perpendicular as they approach the shoreline. Diffraction effects around the breakwaters leak waves into the geometric shadow regions. The berth location on the leeward side of the east breakwater remains successfully sheltered in all cases; in fact, the east breakwater is quite effective in sheltering a fairly large area on its lee. Waves penetrate into the multiple use terminal via the access channel. Negligible wave activity occurs in the supply terminal.

Figure 4 presents contours of maximum wave height superimposed on the harbor bathymetry. These wave heights represent the maximum wave height experienced at least 85% of the time (including waves from all directions) at all locations in the harbor and its vicinity. The immediate vicinity of the west breakwater experiences wave heights of the order of 2 m with maximum wave height about 2.5 m immediately east of the landward end of the west breakwater. Wave heights of about 2 m occur east of the landward side of the east breakwater.

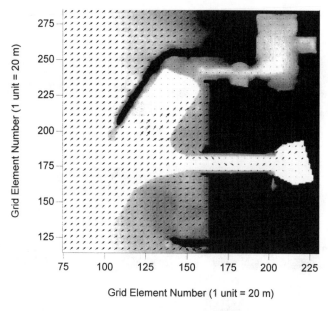

Grid Element Number (1 unit = 20 m)

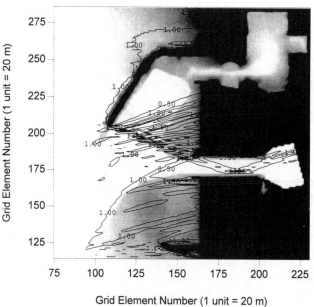

Grid Element Number (1 unit = 20 m)

Figure 3. Wave (a) Vectors and (b) Height Contours for Case 1

The protection the east breakwater affords becomes evident since the immediate lee of the breakwater, including the berth location and most of the turning basin,

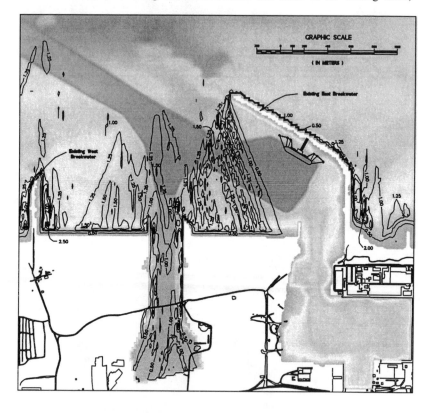

Figure 4. Maximum Wave Heights Experienced 85% of the Time

experiences wave heights less than 0.5 m. Wave heights are typically between 1.25 to 1.5 m away from the immediate lee of the east breakwater. Controlling wave heights in the access channel to the multiple use terminal and the multiple use terminal itself vary between 1.0 to 1.5 m. Waves in the approach channel to the turning basin are typically less than 0.5 m; isolated locations experience wave heights of 1.25 m. In conclusion, the berth location and its vicinity experience wave heights less than the limiting 1.25 m wave height for safe operation at least 85% of the time. Thus, given the limiting wave conditions for port operation, no extension of the breakwater is necessary for the one-berth port design. In fact, additional model

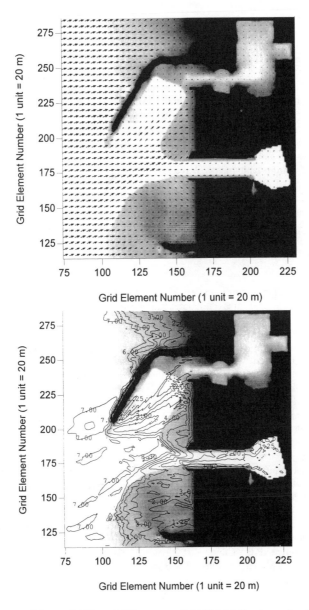

Figure 5. Wave (a)Vectors and (b)Height Contours for Case 6

runs with more severe waves indicate that the berth location should remain operable about 95% of the time.

Extreme Waves

Cases 6 through 8 were simulated to examine the effects of extreme waves. Only some of the results for Case 6 are presented here. Figures 5a and 5b present wave vectors and wave height contours in the nearshore region of Dos Bocas Harbor for Case 6. Note that the offshore wave height and period are 6.74 m and 12 s, respectively, and the offshore wave approach angle is 345°N. Again, these figures show that the local, complex bathymetry and the east and west breakwaters greatly affect waves in the harbor vicinity. These longer period waves are very sensitive to changes in the bottom bathymetry. Wave rays bend as they approach the shoreline; multiple regions of local wave energy convergence and divergence are evident. Diffraction effects around the breakwaters leak waves into the geometric shadow regions; diffraction effects of the uneven bottom bathymetry attenuate wave amplification effects. Focussing effects result in high waves at various locations, especially at the mouth of the access channel to the multiple use terminal where maximum wave heights are in the range 7.9 to 8.3 m.

CONCLUSIONS

(1) The one-berth harbor configuration with the existing east and west breakwaters proves adequate and thus eliminates the need to extend the east or west breakwaters

(2) The two-berth harbor configuration with the existing east and west breakwaters, presented in Srinivas et al. (1999), proves marginally adequate. The two-berth harbor configuration with a relatively short 300 m extension of the east breakwater along its existing axis, together with the existing west breakwater, proves adequate.

(3) Extreme wave height modeling indicates that wave heights may reach as high as 7 to 9 m in localized areas. These highest waves typically occur at the junction of the approach channel and the access channel to the multiple use terminal.

REFERENCES

Kirby, J. T., and Dalrymple, R. A. 1983. A Parabolic Equation for the Combined Refraction-Diffraction of Stokes Waves by Mildly Varying Topography. *Journal of Fluid Mechanics*, 136, 543—566.

Lin, L. 1998. *Evaluation of Wave Climate for Dos Bocas Oil Port.* Coastal and Hydraulics Laboratory. U.S. Army Corps of Engineers, Waterways Experiment Station, Vicksburg, MS.

Srinivas, R., Dompe, P.E., Craig, K.R., Trudnak, M.E., and Taylor, R.B. 1999. Nearshore Wave Modeling Study, Dos Bocas Port, Mexico. Taylor Engineering, Inc., Jacksonville, FL.

MODELLING THE EFFECTS OF STRUCTURES ON NEARSHORE FLOWS

Alessandro Mancinelli[1], Luciano Soldini[2], Maurizio Brocchini[3], Roberto Bernetti[4] and Patrizia Scalas[5].

Abstract: The effects of defense structure on nearshore water flows are analyzed by means of an efficient and robust flow solver. This is both flexible enough to incorporate bathymetric data typically collected in field surveys and accurate enough to allow for a detailed description of important flow features. Such a tool is believed to be capable of providing useful information for design activities of coastal structures like submerged breakwaters.

INTRODUCTION

The dynamics of nearshore flows is characterized by motions occurring at different time/space scales. They may range from the small scale turbulence to currents and tides. However, the energetically-dominant motion is that of wind waves (often referred to as "waves") whose typical period is of the order of ten seconds. In their approach to the coastline these waves undergo a number of transformations (mainly forced by topographic variations) like refraction, diffraction, shoaling and breaking. The latter is one of the main mechanism by which momentum and energy are transferred to both higher and lower frequency motions. In the high frequency range we find the turbulent eddies which are responsible for much energy dissipation while low frequency motions are, for example, the longshore currents, the rip currents, etc. This already complex situation is made even more complicated by the presence of artificial structures like ports, piers, emerged and submerged breakwaters, groins, sea-walls, etc. These alter the natural flow evolution hence modifying the occurrence and manifestation of the wave transformation mechanisms.

1 Prof. Eng., Istituto di Idraulica, University of Ancona, Via Brecce Bianche, 60131 Ancona, Italy, istidra@popcsi.unian.it
2 Eng., Istituto di Idraulica, University of Ancona, Via Brecce Bianche, 60131 Ancona, Italy, lsoldini@idra.unian.it
3 Dr., Dipartimento di Ingegneria Ambientale, University of Genova, Via Montallegro 1, 16145 Genova, Italy, brocchin@diam.unige.it
4 Eng., Via Strada Vecchia del Pinocchio 1/A, 60131 Ancona, Italy, bernetti@inwind.it
5 Eng., Dipartimento di Ingegneria Ambientale, University of Genova, Via Montallegro 1, 16145 Genova, Italy

When a structure is placed on the seabed made of incoherent material (silt, sand or gravel) not only the flow pattern is altered but also the sediment transport is changed usually causing a modification in the local sediment transport capacity. The most typical case concerns the shoreline changes behind detached breakwaters (e.g. Hsu and Silvester, 1990). Also crosshore structures like jetties, piers or groins induce very significant shoreline changes due to persistent gradients in longshore currents and sediment transport.

Coastal structures are built for different reasons: to provide protection to infrastructures against coastal disasters, to develop maritime activities, to protect or to increase recreational and touristic facilities, to defend the coast from erosion, etc.

Many of these problems are becoming of growing importance. In particular erosion problems are very significant along the Italian coasts (over 30% of the coast is prone to erosion). If we consider the coast of the "Marche Region" on the central Adriatic Sea, over 60% of the 172 kilometers long coast is protected by coastal structures due to erosion problems. The main causes of erosion are both the construction of many harbor structures along the coast and the decrease in sediment supply by rivers. An historic analysis of Mancinelli et al. (1999) shows that all beaches at the river mouths have been eroded in last century up to 300 meters.

Fig. 1. Left: The Marche Region. Right: Example of detached breakwaters.

The first structures used to protect eroding coasts and beaches were detached breakwater of emerging type and rubble mound seawalls. At present we can find over 40 kilometers of emerged breakwaters along the Marche coast, built from the 50's to the 80's. During these years the utilization of these defense structures has not been completely satisfactory. It is clear that emerged breakwater induce some important modifications on nearshore water flows and sediment transport. The primary effect of the breakwater is to induce two circulation cells behind the structure, due to diffraction and setup of the mean water level. Moreover, there is a zig-zag movement of water masses in the swash zone even when waves propagate normally to the beach. These flows have the net effect of conveying the eroded sediment to the sheltered area behind the breakwater and determining the generation of salients or tombolos whose size increases with the breakwater length and decreases with the distance from the shore. The other important phenomenon induced by emerged breakwater is the reduction of sediment supply to the downdrift beaches and consequently new erosion areas appear. Moreover, other problems like degradation of sand and water quality, irregularity of both the emerged beach and depth contours, degradation of the landscape may be observed. To avoid all these problems, starting from about the second half of the 80's, submerged structures were used for the realization of new coastal defense projects. Submerged breakwaters have a better impact on the landscape, reduce the waves impact on the beach, but many aspects of their efficiency are still under investigation.

At present the coast of the Marche Region is protected by submerged breakwaters for over 15% of their length. These structures have been used either to protect eroded beaches or to substitute existing breakwater of emerging type. In the first case salients or tombolos have been removed, sand and water quality have been improved, but erosion of the downdrift beaches has increased. In the second case the substitution of emerged breakwaters with submerged ones has been realized lowering the crest elevation below the mean water level and enlarging the berm width up to 10 – 14 meters. The results are not completely satisfactory because in many cases, particularly along sandy beaches, the shoreline has undergone a considerable withdrawal.

Analysis of different realizations shows that the efficiency of a submerged breakwater depends on many parameters. The distance of the structures from the shoreline and the gap width depend on both the submerged beach slope and the granulometry of the sea bed material. The breakwater submergence, the berm width and the characteristics of the incoming waves all have a fundamental role in the wave energy dissipation over the submerged structure, the short waves being less affected than the longer waves. Also the breakwater geometry can induce local scouring phenomena because of a reduced berm width or a too steep slope at the offshore side. Finally the mean water level setup due to the combined effects of tide, wind, pressure and runup, reduces the efficiency of submerged breakwaters.

As a consequence of these uncertainties the planimetric configuration of a coastal defense system is determined with empirical methods or with simple one-line models of the shoreline evolution. In this context, the mathematical model described in the following could represent a fundamental instrument to understand with higher accuracy some features of the flow interaction with coastal structures and to support design activities.

In the following the implementation of a mathematical model based on the Nonlinear Shallow Water Equations (NSWE) and some preliminary results are illustrated. The next section is devoted both to a brief introduction of the NSWE and to illustrate the numerical solution of the equations by means of the WAF method. Subsequently we show both some tests used to validate the solver and some preliminary results of nearshore circulation. Finally the obtained results are discussed and conclusions are given.

THE MODEL

Among the most favored models for nearshore circulation we find both those based on the NSWE and those based on Boussinesq-type equations. Both these types of models have strengths and weaknesses. Boussinesq-type models are clearly superior to NSWE in correctly reproducing the wave evolution in intermediate to deep waters (e.g. Madsen and Schäffer, 1998) hence allowing for a correct positioning of the breaking point. On the contrary this is not possible when using NSWE, which do not account for wave frequency dispersion. However, the latter equations are more suitable than most currently available Boussinesq-type models for representing the flow in the vicinity of the shoreline. In practice an almost-optimal solver for nearshore hydrodynamics could be achieved either by extending Boussinesq models to better represent swash zone dynamics (see Bellotti and Brocchini, 2001) or by extending the NSWE to include dispersive effects. In attempt of following this second line of action we investigate a robust and efficient solver which integrates the NSWE and which will be the basis for a more complex solver. As it stands the solver we describe is one of the most advanced for the solution of the 2DH version of the NSWE over a beach of arbitrary topography, which, on the basis of a WAF method, solves the "local Riemann problem". In the following we briefly describe the structure of the solver.

The Mesher

This is the module, which builds the computational grid from the available information on both the bathymetry and the geometry of artificial structures enclosed in the chosen domain. The aim we want to pursue is that of having a representation of the undisturbed water depth as $h = h(x, y)$ where h is the depth of the point $P(x,y)$ on the (x,y)-plane. The description of the function $h = h(x, y)$ is obtained by using spline functions starting from a rectangular mesh generated on the (x,y)-plane. A fundamental assumption of the model is that the information regarding the bathymetry are given through surveying the depths along a number of sections of the same length L_{sec}, parallel to the x-axis i.e. orthogonal to the coast. On each section a coordinate s is defined, thus $h_{ji} = h_{ji}(s_{ji})$ where s is the x-coordinate along the j-section for the point i ($j=1,2,..M$ and $i=1,2,..N_j$).

In general the number of measurement points is different from section to section, thus a first analysis is required to achieve a homogeneous subdivision. The new points, which do not give more information on the bathymetry, are generated using the following representation for the depth h on the section j:

$$Z_j(s) = \sum_k^{N_j} L_{j,k}(s) \, s_{j,k}$$

(1)

where $L_{i,j}(s)$ is a spline function C_1^0 defined as

$$L_{i,j}(s) = \begin{cases} \dfrac{s - s_{k-1}}{s_s - s_{k-1}}, & s_{k-1} < s < s_k \\ \dfrac{s - s_{k+1}}{s_s - s_{k+1}}, & s_k < s < s_{k+1}. \end{cases}$$

(2)

The new points defining the j-section are generated with a new step Δs, chosen in order to be lower than the minimum $[s_{i+1,j} - s_{i,j}]$ in all the available sections (M). Hence, the new depths are obtained through the expression $Z_{jk} = Z_j(k\Delta s)$. The new M sections have the same spacing and number of points [i.e. $k = \text{int}(L_{sec} / \Delta s)$]. Now it is possible to generate the function $h = h(x, y)$ through the definition of the following spline function C_2^0 in the (x,y)-domain

$$N_{ij}(x,y) = \begin{cases} \dfrac{x-x_{i-1}}{x_i-x_{i-1}} \cdot \dfrac{y-y_{j-1}}{y_j-y_{j-1}}, & x_{i-1}<x<x_i, \quad y_{j-1}<y<y_j \\[2mm] \dfrac{x-x_{i+1}}{x_i-x_{i+1}} \cdot \dfrac{y-y_{j-1}}{y_j-y_{j-1}}, & x_i<x<x_{i+1}, \quad y_{j-1}<y<y_j \\[2mm] \dfrac{x-x_{i-1}}{x_i-x_{i-1}} \cdot \dfrac{y-y_{j+1}}{y_j-y_{j+1}}, & x_{i-1}<x<x_i, \quad y_j<y<y_{j+1} \\[2mm] \dfrac{x-x_{i+1}}{x_i-x_{i+1}} \cdot \dfrac{y-y_{j+1}}{y_j-y_{j+1}}, & x_i<x<x_{i+1}, \quad y_j<y<y_{j+1} \end{cases} \tag{3}$$

where (x_j, y_j) are the coordinate in the global reference frame which define the M sections. Finally, the depth at the generic point $P(x,y)$ is derived from the expression

$$h(x,y) = \sum_{i=1}^{K} \sum_{j=1}^{M} N_{ij}(x,y) x_i y_j. \tag{4}$$

Once chosen the mesh size step in the x-direction (i.e. Δx) and the one in the y-direction (i.e. Δy), the generation of the computational mesh is complete. In fact, the coordinate in the global reference frame of mesh node P_{km} is (in the hypothesis of regular mesh) $P_{km}\left[k\Delta x; m\Delta y; h(k\Delta x, m\Delta y)\right]$.

On this bathymetry it is possible to set discontinuities (e.g. slumps) and defense structures (e.g. breakwater). The projection of the structure to be modeled on the (x,y)-plane can be defined according to the following inequality $f_{st}(x,y) \leq 0$. Different structures require a different relation f_{st}. Furthermore the structure is characterized by a particular distribution of the depth $h_{st} = h_{st}(x,y)$.

Thus, the coordinates of each point P_{km} for the mesh nodes are modified as $P_{km}\left[k\Delta x; m\Delta y; h(k\Delta x, m\Delta y)\right]$ where $k\Delta x$ and $m\Delta y$ are such that $f_{st}(k\Delta x; m\Delta y) < 0$.

The NSWE Solver

The solver under analysis is described in large detail in Brocchini et al. (2001) to which we refer the interested reader. The classic NSWE are put in the suitable conservative form:

$$d_t + (ud)_x + (vd)_y = 0, \tag{5a}$$

$$(ud)_t + (u^2 d + \tfrac{1}{2}gd^2)_x + (uvd)_y = gdh_x - C_\tau |\mathbf{v}|u, \tag{5b}$$

$$(vd)_t + (uvd)_x + (v^2 d + \tfrac{1}{2}gd^2)_y = gdh_y - C_\tau |\mathbf{v}|v \tag{5c}$$

in which d is the total water depth, u and v the onshore and longshore components of the depth-averaged flow velocity, g the acceleration of gravity and C_τ a Chezy-type friction coefficient.

These are cast in vectorial form suitable for numerical integration and through a splitting technique we can reduce the problem to that of solving a homogeneous system representing a pure advection 1DH hyperbolic problem. Subsequently, in order to take into account discontinuous solutions of the problem (shock waves), we use the integral form of the homogeneous equation over a rectangular volume in the space-time domain. This integral form can be written in a discrete form on the basis of the WAF method (Toro, 1997) for which a spatially piecewise constant distribution characterizes all flow variables. Such a distribution gives rise to a number of local "Riemann problems" at each cell interface. An exact "Riemann solver" is chosen to compute the flow structure in the range of influence of each discontinuity.

VALIDATION AND PRELIMINARY RESULTS

The model has been extensively tested and validated on the basis of both available analytical solutions and experimental data. Brocchini et al. (2001) reports on the model validation in the case of simple (uniform plane beach) and complex (bay-type beach) bathymetries while in the work of Scalas (2001) an extensive analysis is described of the model performances in the case of wave-structures interaction. In the latter work it is shown the model well reproduces the diffraction patterns predicted by available analytical solutions like that of Penney and Price (1952).

In order to better highlight the model capabilities we here illustrate results based on the experimental data both of Briggs et al. (1995) and of Haller et al. (1997). The former are particularly useful to represent field conditions of diffraction around a semi-infinite breakwater while the latter are very useful for investigating the onset and development of rip currents.

The diffraction case of Briggs et al. (1995)

The computational domain shown in Fig. 2 is basically identical to the experimental area and is 12.25m long in the onshore direction and 20m wide in the longshore direction. We choose a discretization such that the mesh size in the x-direction is of 0.05m, while the size in y-direction is 0.1m. A flat bottom is used with still water level depth of 0.4m. The breakwater is defined as an impermeable structure 10m long and 0.2m wide. To match the experimental conditions the shoreward side of the breakwater and the lateral boundary of the domain in the "illuminated region" are taken as reflecting boundaries, while all the other boundaries as absorbing boundaries.

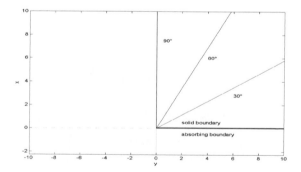

Fig. 2. Solution domain and sections position

The only difference with the experimental conditions concerns the seaward boundary which, to reduce the computational effort, is placed at a distance of one wavelength from the structure. From there two different sinusoidal signals are propagated orthogonally to the breakwater: they both have a 1.33s period (2.25m length), but wave heights are of H=0.0550m and H=0.0775m respectively.

The diffraction coefficient along three sections (α = 30°, 60°, 90°) is computed and compared with the physical model test data. In Fig. 3 comparison of computed and experimental diffraction coefficients is shown for both cases. In these figures the abscissa represents a normalized distance (radius/wavelength) taken from the breakwater tip.

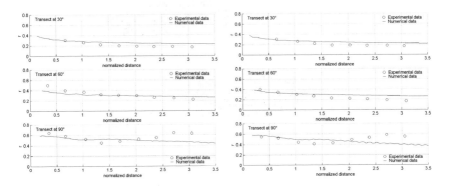

Fig. 3. Comparison of computed and experimental diffraction coefficients.
Left: case of H=0.0550m. Right: case of H = 0.0775m.

Both figures show that the numerical results are in good agreement with the experimental ones for the α = 30° and 60° sections. However, a worse comparison is found for data collected along the section α = 90°. Even if the comparison is reasonable for a distance smaller than about 2 wavelengths, the model seems inadequate to reproduce the increase which r undergoes for larger distances. Although, from one hand we cannot explain this increase in the experimental data we observe that results similar to ours have been obtained by Li et al. (1999) by means of a numerical model based on Boussinesq-type equations.

The generation of rip currents: test conditions of Haller et al. (1997)

Recently both experimental (Haller et al., 1997) and numerical (Chen et al., 1999) studies have been devoted to analyze both details of the generation mechanisms of topography-induced rip currents and their evolution. All these investigations revealed that such rip currents are not steady rather they oscillate both in time and space.

In the attempt of reproducing this important phenomenon, typical of the flow due to waves interacting with arrays of submerged obstacles, we have run a few tests by using the bottom configuration of Haller et al., (1997): the considered numerical wave basin (see Fig. 4) is 19m long and 9m wide, representing almost half the physical basin. A 1.8m wide rip channel interrupts a submerged bar (submergence of 0.05m) placed on a beach with a 1:30 slope. The water depth of the offshore flat bottom is h=0.35 m. The bars are considered as submerged structures and are fully integrated in the bathymetry (i.e. they are defined through their coordinates in the longshore sections which the Mesher uses to define the bathymetry). The lateral boundaries are taken as reflecting boundaries. This choice follows the need of reproducing the physical test in a numerical basin, which is half of the experimental one. In fact the lateral boundary at y=0m represents the physical (i.e. reflecting) boundary of the experimental domain, while the one at y=9m is the middle section of the real basin and acts as a reflecting one in the numerical domain for symmetry consideration. Similar boundary conditions were used in the numerical comparison carried out by Chen et al. (1998).

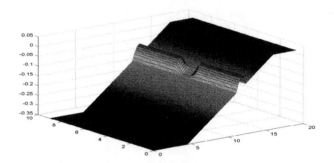

Fig. 4. Three-dimensional view of the numerical bathymetry (in meters).

For all runs the boundary signal is a sinusoidal wave of 1s period (1.42m offshore length L) and a=0.024m amplitude which propagates orthogonally to the beach. These conditions give a nonlinearity parameter a/h=0.07 and a frequency dispersion parameter kh=1.5 hence showing that the NSWE solver is used out of its range of validity for which it should be $O(a/h)=O[(kh)^2]=1$.

The results of a first, relatively short run (20s of simulation) show that the model reproduces fairly well the patterns of important wave-averaged properties. For example, the setup is found to be highest shoreward of the bar due to the shoaling and breaking processes and, in the longshore direction, it is greatest away from the channel. We also find that the onshore component of the short-wave mass flux decreases moving onshoreward due to energy dissipation caused by breaking which gives an almost monotonic decay of the wave amplitude. However, the presence of submerged bars gives a local (over the bars) increase of the mass flux due to the short waves because in shallow depths the relative importance of waves over currents is high i.e. the total mass flux is almost entirely associated with the wave motion. The opposite occurs in the deeper region of the rip channel where the mass flux is almost entirely associated with the currents.

To illustrate the spatial and temporal evolution ("meandering") of the rip current and nearshore circulation, a second test has been run on the same bathymetry for a longer period of time (180s). We here report only results concerning the flow velocity. In Fig. 5, from left to right and from top to bottom, we show the velocity patterns at some representative times. Vortex generation starts (t=10s) in correspondence of the edges of the rip channel, both at the seaward and at the shoreward sides of the channel. When vortex pairs combine they can have either a resulting offshore motion or an onshore one (t=50s). In the first case they interact with each other and move offshorewards, giving as a result the so-called "rip current". They can also interact with the bars, i.e. separate and move along the bars, until they reach the lateral boundary and then move offshorewards (t=80s). In the second case vortices separate at the seaward edge of the bars and interact with the complex current field which results in the generation of two main circulation cells (secondary circulation cells).

The motion just described is not stationary and we notice that the complexity of the current field increases as the time goes on (starting from t = 130s).

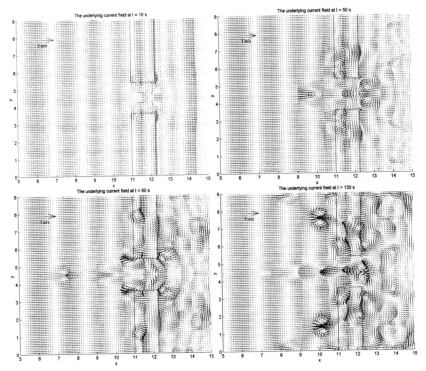

Fig. 5. Velocity field at t=10, 50, 80, 130 seconds.

These results show that the qualitative characteristics of generation and instability of the rip current can be adequately modeled by the solver, which is suitable for reproducing the essential mechanism for rip currents formation i.e. a spatial disuniformity of the radiation stresses. Physical phenomena, which are not accounted for in the model like the seabed shear stress and frequency dispersion only quantitatively, alter the final results.

DISCUSSION

The mathematical model above illustrated has shown good performances during all the tests we performed. The NSWE numerical solver (WAF) has proven efficient and robust in reproducing complex nearshore hydrodynamic flows.

Furthermore, taking into account the intrinsic limits of the model (absence of both frequency dispersion and bottom friction), which mainly result in wave breaking to start too far offshore, very good results (in comparison with both analytical and experimental tests) have been obtained in reproducing phenomena like wave shoaling, diffraction, refraction, breaking (and induced manifestations like free surface setup and breaking-generated currents). The model allows for easy and effective description of nearshore flows characterized by complex topographies and structures (both absorbing and reflective), which interact with shallow water motions.

It is evident that the solver implemented may prove very useful in designing new coastal defense structures or guide rehabilitation activities of coastal stretches. One clear example is the substitution of many coast-parallel emerged breakwaters with submerged breakwaters, in order to improve the environmental conditions of the area (tombolo dismantlement, renewal of hydrodynamic circulation, etc.). Such activity has already brought to large erosion problems. These drawbacks are due to the wrong attitude of only reducing in height the already existing emerged breakwaters without considering the distance from the shoreline.

In these cases the model can show the importance of the hydrodynamic flow which could be the fundamental factor of shore modifications behind the structures. This can be easily achieved by means of a number of sample computations which allow for analysis of the different hydrodynamic patterns associated with different design strategies (breakwater position, berm width, etc.) hence evidencing the best design layout.

ACKNOWLEDGEMENTS
Thanks are extended to the "Regione Marche" for partially supporting the research. Financial support from the European Union through the contract EVK3-2000-22038 (DELOS) is also acknowledged.

REFERENCES
Bellotti, G. and Brocchini, M. 2001. On the shoreline boundary conditions for Boussinesq-type models. *Int. J. Num. Meth. in Fluids*, 37(4), 479-500.

Briggs, M.J., Thompson, E.F. and Vincent, C.L. 1995. Wave diffraction around breakwater. *J. Wtrwy., Port,Coast., and Oc. Engrg.*, 121(1), ASCE 23-35.

Brocchini, M., Bernetti, R., Mancinelli, A. and Albertini, G. 2001. An efficient solver for nearshore flows based on the WAF method. *Coast. Engng.*, 43(2), 105-133.

Chen, Q., Dalrymple, R.A., Kirby, J.T., Kennedy, A. and Haller, M.C. 1999. Boussinesq modeling of a rip current system. *J. Geophys. Res. - Oceans*, 104, 20617-20637.

Haller, M.C., Dalrymple, R.A. and Svendsen, I.A. 1997. Rip channels and nearshore circulation. *Proc. Coastal Dynamics '97*, ASCE, 594-603.

Hsu, J.R.C. and Silvester, R. 1990. Accretion behind single offshore breakwater. *J. Wtrwy., Port, Coast., and Oc. Engrg.*, 116(3), ASCE, 362-380.

Li, Y.S., Liu, S.X., Yu, Y.X. and Lai, G.-Z. 2000. Numerical modelling of multi-directional irregular waves through breakwaters. *Appl. Math. Model.*, 24, 551-574.

Madsen, P.A. and Schäffer, H.A. 1998. Higher-order Boussinesq-type equations for surface gravity waves: derivation and analysis. *Proc. Roy. Soc. Lond. A*, 356, 3123-3184.

Mancinelli, A., Soldini, L. and Lorenzoni, C. 1999. Dinamica delle foci fluviali nelle Marche. Proposte di sistemazione e gestione. *Proc. Giornate di studio su "La Difesa Idraulica Del Territorio"*, 463-482 (in Italian).

Penney, W.G. and Price, A.T. 1952. The diffraction theory of sea waves and the shelter afforded by breakwaters. *Phil. Trans. Roy. Soc. A*, 244, 236-253.

Scalas, P. 2001. *La modellazione delle interazioni onda-struttura e onda-batimetria tramite un solutore tipo WAF*, Degree dissertation, University of Genova (in English).

Toro, E.F. 1997. *Riemann solvers and numerical methods for fluid dynamics*, Springer, 594p.

WAVE TRANSMISSION AT SUBMERGED BREAKWATERS

Karl-Friedrich Daemrich[1], Stephan Mai[1] and Nino Ohle[1]

Abstract: Measurements of wave transmission at two trapezoidal submerged structures are analysed and discussed with respect to the design formula of d'Angremond et al.. Transmission coefficients agreed well within the given range of validity, however, an appropriate crest height and crest width from the rubble mound surface has to be used. Special interest has been put on results beyond the upper limit of the formula, e.g. relatively high water levels, the variation in the mean transmitted periods, and on some results from numerical modelling.

TEST SET-UP AND HYDRAULIC BOUNDARY CONDITIONS
Data of the following two test series were used:

The first series was investigated in a side channel of a wave basin. The idea was to perform tests without the increase of water level in the transmission area which mostly will occur in channel tests but not in the field case. The structure was completely from rubble 35 to 55 mm diameter, with slope 1 over 2. The height of the structure was 0.5 m, the crest width 0.2 m. Water levels were between 0.45 and 0.7 m, significant wave heights between 2.5 and 17.5 cm with peak periods from 1 to 1.75 sec.

The second series was investigated in the *Large Wave Channel* in Hannover. The structure was formed like a summer dike sloped 1 over 7, with a height of 1.5 m above horizontal sand beach and crest width of 3 m. The structure was impermeable, constructed as sand core covered with a mattress and filled with *Colcrete Solidur*, a mixture of fine sand and cement. Water levels were from crown height to 1.5 m above the crown. Significant wave heights from 0.6 to 1.2 m with peak periods of 3.5 to 8 sec.

THE DESIGN FORMULA OF D'ANGREMOND, VAN DER MEER AND DE JONG
In the design formula of d'Angremond, van der Meer and de Jong (1996) the transmission coefficient K_t is calculated as a function of

1 Research Scientist, University of Hannover, Franzius-Institut for Hydraulic, Waterways & Coastal Engineering, Nienburger Str. 4, 30167 Hannover, Germany, Tel.: +49 / 511 / 762-3739 or -4295, Fax. +49 / 511 / 762-3737, e-mail: daekf@fi.uni-hannover.de, smai@fi.uni-hannover.de, Nino.Ohle@fi.uni-hannover.de

relative freeboard R_c/H_s,
relative crest width B/H_s,

and the Iribarren parameter $\xi = \tan\alpha / \sqrt{s_{0p}}$ $(s_{0p} = \dfrac{g}{2\pi} \cdot \dfrac{H_s}{T_p^2})$ $\qquad$ (1)

$$K_t = -a \cdot \frac{R_c}{H_s} + \left(\frac{B}{H_s}\right)^{-b} \cdot \left(1 - exp(-c \cdot \xi)\right) \cdot d \qquad (2)$$

For permeable structures the formula is given as

$$K_t = -0.4 \cdot \frac{R_c}{H_s} + \left(\frac{B}{H_s}\right)^{-0.31} \cdot \left(1 - exp(-0.5 \cdot \xi)\right) \cdot 0{,}64 \qquad (3)$$

for impermeable structures

$$K_t = -0.4 \cdot \frac{R_c}{H_s} + \left(\frac{B}{H_s}\right)^{-0.31} \cdot \left(1 - exp(-0.5 \cdot \xi)\right) \cdot 0{,}80 \qquad (4)$$

The formulae are limited to values of K_t between 0.075 and 0.80.

The formulae deliver transmission coefficients K_t for the relative freeboard $R_c/H_s = 0$ dependent on the relative crest width and the breaker number. The variation with R_c/H_s is then linear with slope −0.4 within the given limits. The general trend is sketched in Fig. 1.

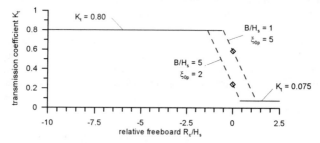

**Fig. 1: Principle of the design formula of d'Angremond et al.
with examples of results for given parameters B/H_s and ξ**

DATA OF THE INVESTIGATIONS IN THE SIDE CHANNEL OF A WAVE BASIN (SERIES 1) IN COMPARISON TO THE DESIGN FORMULA OF D'ANGREMOND ET AL.

In Fig. 2 the data from the first series are plotted according to the above mentioned scheme (transmission coefficient as a function of the relative freeboard).

The tendency, compared to the design formula, is reasonable around $R_c/H_s = 0$, but the deviation from the straight line for $R_c/H_s > 1$ or transmission coefficients higher than about 0.7 can be clearly stated.

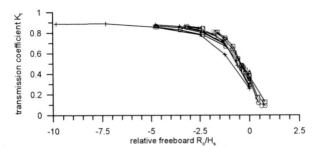

Fig. 2: Data of the investigations in the side channel of a wave basin (series 1)

Fig. 3 gives a direct comparison of measured and calculated transmission coefficients, however, without considering the range of validity, to highlight the trend near and beyond the upper limit of validity.

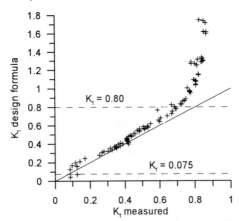

**Fig. 3: Comparison of measured transmission coefficients
with results from the design formula**

The results can be characterised as follows:

1. the scatter is relatively low,
2. within the range of validity of the design formula there is a clear trend with nearly constant too high theoretical values,
3. outside the range of validity the deviation between measured and calculated transmission coefficients is continuously increasing.

Discussing in detail the deviation of the data within the range of validity, the definition of the crest height in rubble was found as source of possible uncertainties with a strong effect on R_c as the most important parameter. Some calculations with slightly changed crest heights were performed and it was found, that with a calculated increase of the structure height of only 1 cm the overall agreement was much better, however, with slightly increased scatter, as to be seen in Fig. 4.

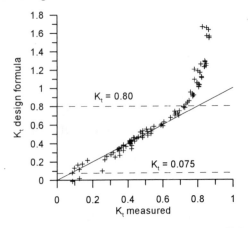

Fig. 4: Comparison of measured transmission coefficients with results from the design formula (crest height + 1 cm)

If the crest height is under discussion, the same has to hold for the crest width. And of course slightly different coefficients in the design formula could be expected for different data sets. With non-linear regression calculations the possible deviations of crest height and width as well as the coefficients of the design formula were determined. For this calculations only data from measurements with water levels from 50 to 55 cm, where the design formula should give best results (within the range of validity), were used. Results from calculations with modified coefficients and corrected crest height and crest width are shown in Fig. 5.

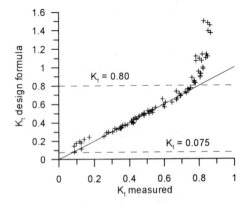

Fig. 5: Comparison of measured transmission coefficients with results from the design formula with modified coefficients and corrected crest height and width

It came out from this calculations that the crest height should be selected some 4 mm higher, the width some 8 mm wider. The differences of the coefficients are not too big:

Design formula with coefficients of d'Angremond et al.:

$$K_t = -0.4 \cdot \frac{R_c}{H_s} + \left(\frac{B}{H_s}\right)^{-0.31} \cdot \left(1 - exp(-0.5 \cdot \xi)\right) \cdot 0{,}64 \qquad (5)$$

Design formula with coefficients calculated for this data set:

$$K_t = -0.33 \cdot \frac{R_c}{H_s} + \left(\frac{B}{H_s}\right)^{-0.225} \cdot \left(1 - exp(-0.44 \cdot \xi)\right) \cdot 0{,}632 \qquad (6)$$

To examine, how good the expressions for the influence of the Iribarren number and the relative crest width fit to the data, the design formula was rearranged and the influences extracted. The result, which confirms that the used function for the influence of the relative crest width is reasonable for the range of the data, is given on the left hand side of Fig. 6. The same holds for the influence of the Iribarren number (right hand side of Fig. 6).

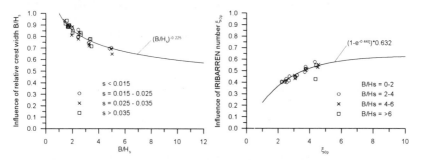

Fig. 6: Function for the influence of the relative crest width B/H$_s$ (left hand side) and of the Iribarren number (right hand side)

DATA FROM HIGH WATER LEVELS BEYOND THE UPPER LIMIT OF VALIDITY OF THE DESIGN FORMULA
There is still the problem that the design formula does not hold for the high water levels and transmission coefficients (Fig. 3, 4 and 5).

For the range of data in this series it was not too difficult to include hyperbolic terms in the R_c/H_s term. Using hyperbolic tangent and hyperbolic arc sine in the following combination (determined by non linear regression)

$$-0.33 \cdot \left(0.99 \tanh \frac{R_c}{H_s} + 0.28 \arcsin \frac{R_c}{H_s}\right) \qquad (7)$$

resulted in the scatter plot shown in Fig. 7 when using all data.

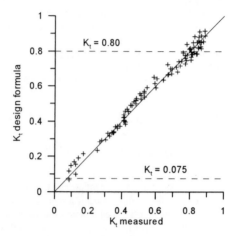

**Fig. 7: Comparison of measured transmission coefficients
with results from a design formula with hyperbolic term**

However, we are aware of the fact that such a fit is very much dependent on the range of wave parameters investigated and should be seen as a first step only to incorporate transmission coefficients beyond $K_t = 0.8$ in a design formula. As a first theoretical approximation to the upper range of data Wiegel's Power Transmission Theory (Wiegel 1964) with the transfer function method was used. For the structure investigated, the results from this calculations can be approximated by

$$K_t = tanh\left(2\pi \cdot R_c / L_{0p}\right)^{0.262} \tag{8}$$

In Fig. 8 this function is shown together with the data as a function of R_c/L_{0p}. For our range of data the application of the Power Transmission Theory was not really successful, but we still think that a possibly modified Power Transmission Theory could be of some value in selecting physical more conclusive fits of the hyperbolic terms mentioned above. For comparison the numerical model *Odiflocs* (van Gent 1992) has been used for the high water levels 0.6 m to 0.7 m. The results are shown in Fig. 9.

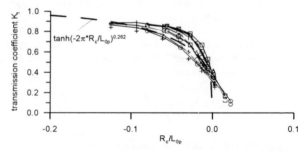

Fig. 8: Data of series 1 in comparison to Power Transmission Theory

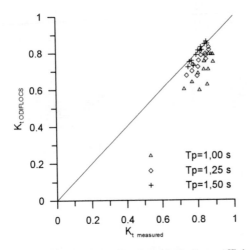

**Fig. 9: Comparison of measured transmission coefficients
with numerical calculations from *Odiflocs***

For our calculations there was the trend that the longer periods fitted quite well. With decreasing periods the results become too low, but have in principle a reasonable trend. The testing with *Odiflocs* and search for the reasons for these tendencies will go on.

VARIATION OF MEAN TRANSMITTED PERIODS

As a last point of the analysis of this data set, the change in the transmitted wave periods is treated. Plotting the relation of the mean periods of transmitted and incident waves as a function of the relative freeboard it can be seen that the reduction is strongest when the still water level is close to the crest, with a rapid increase with increasing crest height (Fig. 10).

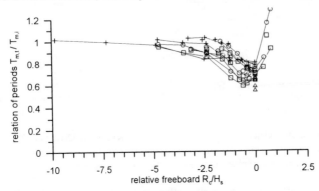

Fig. 10: Variation of periods T_m with R_c/H_s

Plotting the same data as a function of the freeboard related to the peak wave length gives an idea of a function for this data set for negative freeboards (Fig. 11). A rough estimate for the range $R_c/L_{0p} < 0$ can be taken from the formula in Fig. 12, but this is not seen as a general design recommendation without further tests and more detailed analysis.

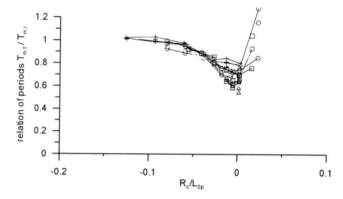

Fig. 11: Variation of periods T_m with R_c/L_{0p}

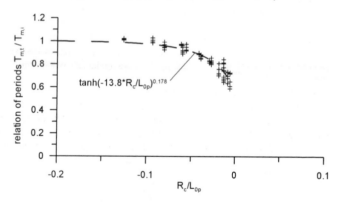

Fig. 12: Fitting function for $R_c/L_{0p} < 0$

It has to be mentioned that this relationship is based on an average from 3 wave gauges in different distances (3 m, 6 m, and 9 m) from the structure, and that there is also a trend, that the reduction of periods is stronger closer to the structure.

DATA FROM MEASUREMENTS IN THE LARGE WAVE CHANNEL (TEST SERIES 2)

In these tests the submerged structure was situated on a foreland. A sketch of the test set-up and an example of measured wave data along the channel are given in Fig. 13.

For the calculation of transmission coefficients incident waves were taken from a wave gauge 52 m in front of the structure. Transmitted waves were from a wave gauge 78 m behind the structure. Incoming significant wave heights were corrected for the shoaling influence from the deeper water to the water depth at the structure.

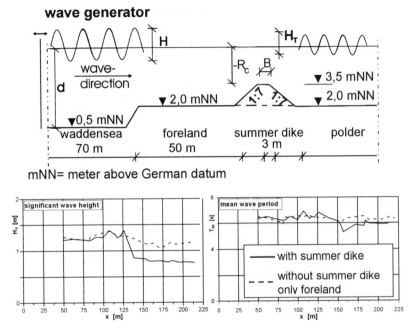

**Fig. 13: Experimental set-up and example of measurements
in the *Large Wave Channel***

Results from the design formula of d'Angremond et al. in comparison to calculated
transmission coefficients are shown in Fig. 14.

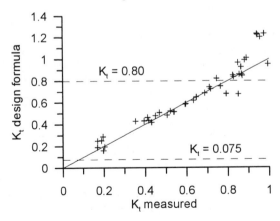

**Fig. 14: Comparison of measured and calculated transmission coefficients
at the Summer dike measured in the Large Wave Channel**

There is relative good agreement in the usual range of validity of the design formula and also the typical trend in the high transmission coefficients. As this is from tests with a very flat sloped structure, this can be seen as one prove for the quality of the term with the Iribarren parameter.

For the investigations various numerical models have been used for comparison. These models with breaker terms according to Battjes / Janssen came up with really good results. These results can be found in the literature (Mai et al., 1998, 1999a, 1999b).

CONCLUDING REMARKS

From the two test series it can be stated that the design formula of d'Angremond et al. is a good basis for analysis and control of measurements on wave transmission at submerged structures within the given range of validity. However, an appropriate crest height and width has to be used. To enable the determination of appropriate values for the effective crest height (and therewith R_c) it is strongly recommended to perform enough measurements around $R_c = 0$ with small steps of variations in the water level.

Some methods are discussed in the paper to deal with the range of high water levels beyond the up to now range of validity of the design formula, however, there is still a need for better theoretical or empirical description.

Concerning the test set-up of the experiments in the side channel of the wave basin, it is believed that investigations without model dependent increase of water level in the transmission area may be more realistic than channel tests, where the increase often is influenced by the channel dimensions.

REFERENCES

D'Angremond, K., van der Meer, J.W., de Jong, R.J. 1996. Wave Transmission at Low-crested Structures. *Proceedings of the 25th International Conference on Coastal Engineering (ICCE)*, Kobe, Japan.

Mai, S., Daemrich, K.-F., Zimmermann, C. 1998. Wellentransmission an Sommerdeichen. *Wasser und Boden*, Heft 50/11 (in German)

Mai, S., Ohle, N., Daemrich, K.-F. 1999a. Numerical Simulation of Wave Propagation Compared to Physical Modelling. *Proceedings of the HYDRALAB-Workshop on Experimental Research and Synergy Effects with Mathematical Models*, Forschungszentrum Küste (FZK), Hannover, Germany

Mai, S., Ohle, N., Zimmermann, C. 1999b. Applicability of Wave Models in Shallow Coastal Waters. *Proceedings of the 5th International Conference on Coastal and Port Engineering in Developing Countries (COPEDEC)*, Cape Town, South Africa.

Van Gent, M.R.A. 1992. Numerical Model for Wave Action on and in Coastal Structures. *Communications on Hydraulic and Geotechnical Engineering*, Report No. 92-6, Delft, The Netherlands

Wiegel, R.L. 1964. Oceanographical Engineering. Prentice Hall International Series in *Theoretical and Applied Mechanics*, Englewood Cliffs, N.J.

APPLICATION OF PERMEABLE GROINS ON TOURIST SHORE PROTECTION

Sherif Abdel-Mawla, [1] and Mootaz Khaled, [2]

ABSTRACT: The Northwestern region of Egypt is characterized by its virgin nature with its all-original rural conditions without human interference. During the last few decades, the beach strips that overlook the waterfront have witnessed an urbanization revolution. The majority of the carried out installations fall within one single sector of investment, which locally called "Tourist Village". Human intervention in the coastal balance, by installation of different kinds of measures in the near shore zone, causes more diverse side effects and environmental deterioration. The unlucky shores subjected to erosion; one of these shores is located at about 140 km to the west of Alexandria and extends to about 1100 m. The shoreline change simulation for this shore is calibrated and modeled by GENESIS. The suggested protection measure was permeable groins. These groins, with its minimal effect on the adjacent shores, verify the requirements of Northwestern coastal protection policy. The proposed permeable groin is composed mainly of supported piles as main elements. In between each two piles, a wave screen is mounted. It is characterized by less water blocking, so it permits water flow exchange, and exerting slightly less total hydrodynamic force compared to a conventional pile groin. A construction system is designed to effectively reduce the installation time and in turn low cost. The modeling results and estimated cost show that the deterioration in the shoreline could be controlled with minimal side effects on the adjacent shores and with valued low cost.

INTRODUCTION

The randomness of constructing marinas, jetties, and breakwater in the tourist projects play the major role in the morphology change in some places along the Northwestern coast. The implementation of these projects was carried out in a hasty and spontaneous manner. Consequently, most of theses projects are suffering from one or more environmental problems due to the insufficient integration between the development processes and the prevailing environmental conditions.

1 Dr. Eng., Assistant Prof., Civil Eng. Dept., Faculty of Engineering, Suez Canal University, Port-Fouad, 42523, Port-Said, EGYPT. sherif-mawla@maktoob.com
2 Dr. Eng., Assistant Prof., Irrigation & Hydraulic Eng. Dept., Faculty of Eng., Ain-Shams University, Cairo, EGYPT. mizoamk@hotmail.com

To overcome the erosion problems, it is recommended to use groins system with some constraints on the design so that it cause no more side effect on the environment and adjacent shores. Groins in a multitude of forms have been used as coastal protection structures for a long time. Yet, their expected effects are still only marginally quantifiable. From the literature it appears that less than half of the groins have performed as expected. Outlines of the practice with use of groins are given be the U.S. Army Corps of Engineers (1992). Groins can be classified as short or long, as high or low, and by the type of construction and material used. According to the material used, the groin could be permeable or pervious. We discuss here the pile groin as a special form of permeable groin.

The pile groins are generally in the form of a single row of piles with different permeability from location to location as well as along the length of the groin. The shoreline in a pile-groin field is a continuous with less saw-tooth typical shape (Raudkivi 1996). The pile groin is developed in form, as discussed in the next sections, to overcome the installation cost and the probable bed erosion at the gaps. The purpose of the present work is to discuss the problems in the Northwestern coast, focus on a case study, suggested development on pile groin, and introduces the advantageous features of the proposed system.

FEATURES OF THE NORTHWESTERN COAST OF EGYPT

The arid and the semi-arid environment of Northwestern region, scarcity of population and the unpolluted natural wildlife have been always the well-known characteristics of its terrestrial areas. Undoubtedly, the coastal zone of this domain was unique and invaluable.

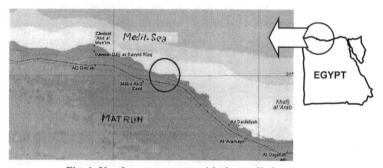

Fig. 1. Northwestern coast with the studied area

The Northwestern coast of Egypt is extending about 600 km from Alexandria to El-Salloum (Figure 1). The coastal zone beaches are distinct by its white oolitic carbonate sand, its clear blue water, its mild weather, and the sun prevailing most of the year. Before the construction of tourist projects the shoreline was in equilibrium. Recent human intervention with the sea, caused many coastal problems such as erosion in some places, accretion in others, and coastal pollution. The northwestern coast is situated in an open active area exposed to winter stormy sea waves, to the swells of summer season. The near-shore and beach face slope are relatively steep, but the slope becomes flatter in the offshore zone and they are found to be ranging between 1:3 – 1:10 and 1:20 – 1:90 respectively. The steep slope of the beach face causes the waves to be broken near the shoreline without permitting sufficient surf zone for swimming and/or other water recreation activities. The strong long-shore and rip currents created several troubles and serious hazards that threatened the safely and life of the swimmers.

PRPBLEM STATEMENT
The studied shore area is located at about 140 km to the west of Alexandria and extends to about 1100 m (marked area in fig.1). The measured change in shoreline in 1991 and 1999 is given in fig. 6. The eroded zone is noticeable and could be estimated in field by the distances between the residual sandy limestone ridges in sea and the shoreline (fig.2). Agitation and turbulent conditions at the bay cause sand loss and crakes and corrosion in the ridges.

Fig. 2. Studied shore with noticeable erosion features

Waves
In general the Egyptian Northern Coast lacks a detailed measured wave data due to financial problems. Thus, no measured data are found at the area. To overcome this problem; data gathered by Suez Canal Authority, at 40 Km to the east of the area, is used in this paper. The data could be concluded as; The predominant wave direction is from NNW, significant wave height = 1.0 m, significant wave period = 6.5 sec, only 3.5% exceed 4.0 m in height, and 0.8% exceed 11 sec in period.

Tide
The tidal range in the Egyptian Northern coast is about 20 cm. The wave setup can cause a level change of about 30 cm. Therefore; a maximum of 50-cm variation in levels is expected (CRI 1998).

Current
No exact measurements are found for the area, but in general the currents are from west to east. The main cause for littoral currents is waves and from the previous studied, it is concluded that the sediment transport is wave dominated.

Sediment
Sediment sampling analysis could be concluded as follow; surface bed layer of about 3.0 m thickness is fine sand (0.06-01mm), followed by a

layer   of about 4.0 m thickness of medium sand (0.2-0.6 mm), and
underneath layer of Coarse/ calcareous sand (0.7-2 mm).

PHILOSOPHY OF SOLUTION
The Proposed erosion control measure should meet the following requirements:
- The system has minimal effect on the adjacent shores.
- The possibility of using the shore protection structure as foot-passage in sea.
- Low cost structure.
- The design should preserve the natural environment as much as possible and
  in particular; the natural color of the near-shore water that reflects the seabed white
  sand and no sight pollution to the extended panorama of the natural sea front would
  take place, as that caused by the outcrop of a breakwater part above the sea level.

Permeable groins, especially pile groins, are   found convenient to meet the above-
mentioned requirements.

Review and Synthesis of Permeable Groins
Shoreline response = F [groin (s); beach; waves; wind, & tide). Groin length,
spacing, and   permeability (elevation, porosity) could be controlled as structure
parameters. Groins should be permeable to allow water and sand to move
alongshore, and   reduce rip current formation and cell circulation (Kraus et al.1994).
Permeable groins   act as a hydraulic roughness on the long-shore current. According
to its position with respect to shoreline, their effects on waves could be negligible.
By reducing, but not blocking, the littoral current, the velocity differential between
the velocity seaward and in the pile-groin fields is smaller than with impervious
groin (Raudikivi 1996).   The reduced littoral current velocity in the pile-groin fields
leads to a reduction of turbulence produced at the bed by the wave-current
interaction, to a reduced amount of sediment suspended. .Field observations and
model studies indicate that the pile groins approximately halve the velocity of the
littoral current through the groin field compared to the same groinless conditions
(Raudikivi 1996). A schematic longshore velocity distribution along the coast with
and without pile-groin fields is shown in fig. 3.

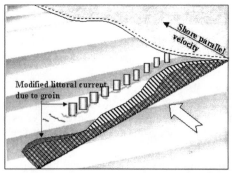

Fig. 3. Velocity distribution with and without pile-groin fields

The wave transmission coefficient K_{to} for incident waves perpendicular to a row of piles could be expressed as Wiegel (1961):

$$K_{to} = H_t / H_i = [b/(D+b)]^{1/2} = (b/c)^{1/2} \qquad (1)$$

$$K_{to} = H_t / H_i = [1 - (D/c)^2]^{1/2} \qquad (2)$$

From Gruene and Kohlhase (1975):

$$K_{t\beta} = 0.5[1 - (D/c)^2]^{1/2}[1 + cos^2 \beta] \qquad (3)$$

Where H_i = incident wave height (m); H_t = transmitted wave height (m); D = Width pile facing the waves (m); b = width of gap between the piles (m); and c = center–to center distance (m); and β = angle to the normal to the slotted wall.
For = 8.2%-43.9% (average= 25.3%); wave transmission = 40%-88% (average= 70%). For $\beta = 60°$ this leads to $K_{t\beta} = 45\%$ (Weiss 1991).

Proposed Permeable Groin

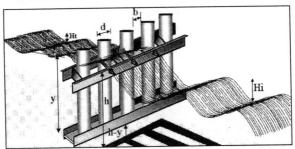

Fig. 4. Suspended pipe frame groin

Observations of pile groin frequently show a scour, of about half a meter deep and a few meters wide, along both sides of the pile groin that inversely proportional to the pile groin permeability. Bed erosion between the piles further complicates the situation. To overcome erosion problem and cost of installation many piles to form a groin, the system shown in fig.4, primarily produced by Mani and Jayakumar (1995), is introduced to build the groin with adequate draft ratio for our purpose. The pipe frame is mounted on supported piles. The foot-passage body is rested on the supported piles. Wave transmission coefficient could be expressed as (Mani 1996):

$$K_t = \frac{-1 + \sqrt{1 + 4T_t}}{2T_t} \qquad (4)$$

$$T_t = \frac{H_i}{6 L Sinh^2 kh}[K_{loss}(S_a)] \qquad (5)$$

$$S_a = 2ky + sinh\, 2kh - sinh\, 2k(h-y) \qquad (6)$$

$$K_{loss} = \left[\frac{1}{C_c \, P} - 1 \right]^2 \qquad (7)$$

$$C_c = coefficient \ of \ contraction = 0.48 + 0.4 \, P^3 \qquad (8)$$

$$P = b / b + d \qquad (9)$$

Where k= wave number ($2\pi/L$); L= wave length (m); h= water depth (m); and y= draft (m).

Laboratory Tests

Usually, the current depth-averaged flow velocity does occur at y/h = 0.63 and the groin porosity ratios recommended in literature are 8.2%-43.9% (Raudkivi, 1996). The behavior of the proposed system with pre-defined parameters is checked experimentally in a wave flume of water depth; h=30 cm, regular wave of heights H= 1.5-9.0 cm and periods 0.70 –1.19 sec., pipe diameter= 2.15 cm, gap ratio; b/d= 0.36 (permeability= 0.27), and draft ratio; y/h= 0.65. The result is compared to Mani's results and equation as shown in fig. 5.

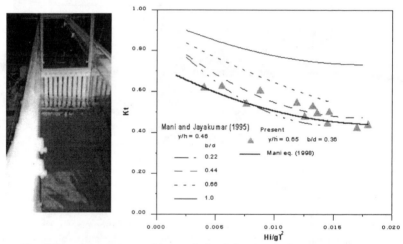

Fig. 5 Experiment view and transmission behavior against wave steepness

SIMULATION OF SHORELINE CHANGE

GENESIS (ver.3) is applied to simulate the action of single and multiple groins. To facilitate the operation of the GENESIS model, a synthetic wave data file with time step of 4 hours is constructed to simulate the conditions at the site. The modeled site consists of a stretch of beach of length of about 1750 m in length along the beach. The measured shorelines for 1991 and 1999 are used for model calibration. In order to get a reasonable calibration, the shoreline is divided into five parts and each is adjusted separately taking into account the model parameters. The following table shows the parameters used in the model.

Table 1. Model Parameters

DX =25.0	DT = 4.00	ISSTART=1	N = 56	NTS = 17520
NWAVES = 1	DCLOS = 8.0	ABH = 1.3	DZ = 40.0	D50 = 0.08
HCNGF = 0.8	ZCNGF = 1.0	ZCNGA = 8.5	K1 = 0.70	K2 = 0.35

Where:

DX : Grid Spacing in (m)   ABH : Berm height in (M)
DT : Time step in hours   DZ : Offshore depth of wave (m)
ISSTART : Grid Start of Shoreline   D50 : mean diameter of beach material in (mm)
N : Number of grid cells   HCNGF: Wave height change factor
NTS : No. Of time steps   ZCNGF: Wave Period Change factor
NWAVES : No of waves per time step   ZCNGF: Wave direction change factor
DCLOS : Closure depth in (m)   K1, K2: Calibration Coefficients.

The above parameters comply with the conditions at site after adjusting the waves according to the local north of the site, perpendicular to the baseline. Fig. 6 shows the results of the calibration runs and the calculated shoreline for 1999. It is clear from the figure that the model under predicts the shoreline advance in the eastern part. In order to check the calibration accuracy, the volumetric change of measured and calculated shoreline are compared and the following is the result:

Volume change for measured shoreline (erosion)= -317323 m3 (8 year)
Volume change for calculated shoreline (erosion)= -322252 m3 (8 year)

This indicates that the model over predicts erosion in the whole area. But as a whole the calculated shoreline is a good measure for model calibration. It is worth mentioning, that the shoreline in this area exhibits some interesting behavior as the area shaped as a bay is eroding and the area shaped as a headland is accreting. This may be due to reversed currents or a submerged rocky zone at the headland zone. This may lead to some troubles when simulating the shoreline to study the effects of the structures. The simulation work is divided into some steps. The first step is to predict the shoreline change condition without any interference. This result is shown in Fig. 7. It is clear form the figure that the trend of the shoreline is still the same with maximum advance of 70.0 m and maximum retreat of about 97.5 m. The average rate of erosion is 26 m. The second phase of work is to check the effect of groins on the shoreline. Three Groin positions are tested. The first is at a distance of 975 from the origin of the baseline, the second is 200 m to the first groin, and the third is at about 425 m from the origin. The initial length of each groin is chosen to be 130 m (1/3 width of the surf zone. These lengths of the groins caused the runs to be unstable. Thus, the lengths of the groins are changed until a stable condition is reached. The chosen lengths of the groins are 80 m each. The water depths at the tips of the groins are 3.20 m, 2.85 and 2.75 m starting form east to west.

For each case the groin is tested in case of impermeable and permeable (perm=0.27; b/d= 0.36). Table (2) concludes the most important runs and Fig. (8), Fig. (9), and Fig. (10) are examples for their results. It is clear that the permeable groins give a better shape for the beach (less sinuosity). Although permeable groins give a better shape, but increasing the number of groins increased the maximum retreat of the shore for the same simulation period. Volumetric change is decreased significantly by using the three groins. In order to reach a more convenient solution, nourishment scheme is proposed. The scheme is to nourish each run with an average berm equals the average change in the shoreline. The Nourishment is applied to the area suffering from shoreline retreat (erosion). Due to the instability of the runs, nourishment scheme is applied to the case of only one groin. The permeable groin is tested against two berm values (17.0 m and 21 m). It is clear form table (2) that no significant change is found.

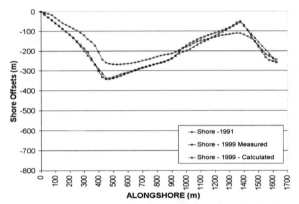

Fig. 6 Measured and calibrated shoreline change

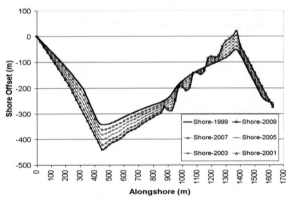

Fig. 7 Measured and Predicted shoreline change

DESIGN OF GROIN BUILDING SYSTEM

The marine equipment used usually in groin construction is the main factor in its cost. As the depth at the groin tips is relatively shallow, it is desired to use inland manual technique in installing the supported piles. Driven the piles manually is done on a special designed extendable shoring beginning from shoreline toward seaward direction. The pipe frame is prefabricated and mounted afterward on the supported piles. This technique minimizes the estimated cost for the three-groin system (fig. 10) to about 50% of the traditional groins (rubble mound) or 60% if supported piles is constructed using marine equipment.

Table 2. Results of Some Relevant Runs

	R-01	R-02	R-03	R-04	R-05	R-06	R-07	R-8	R-9	Unit
No. of Groins	1	1	1	1	1	2	2	3	3	
Permeability	0.00	0.27	0.00	0.27	0.27	0.00	0.27	0.00	0.27	
Nourishment	N.A.	N.A.	YES	YES	YES	N.A.	N.A.	N.A.	N.A.	
Berm	N.A.	N.A.	17	21	17	N.A.	N.A.	N.A.	N.A.	
Advance Max.	72.24	70.65	70.65	70.65	70.65	72.24	72.24	72.24	72.24	[m]
Retreat Max.	97.37	97.37	80.00	75.91	80.00	110.22	108.50	108.50	110.22	[m]
Aver. Change	-16.83	-20.92	-11.49	-11.93	-13.57	-15.84	-19.99	-19.99	-12.68	[m]
STDEV	52.33	50.75	47.02	45.51	46.52	54.20	54.69	54.69	51.53	[m]
Vol. Change	-252928	-314433	-172729	-179249	-203996	-238122	-300433	-300433	-190620	[m³]
Vol. Rate	-25292	-31443	-17272	-17924	-20399	-23812	-30043	-30043	-19062	[m³/y]
Vol. Gain	139323	77818	219522	213001	188255	154129	91818	91818	201631	[m³]
Gain Rate	13932	7781	21952	21300	18825	15412	9181	9181	20163	[m³/y]
Nourish. Vol.	N.A.	N.A.	157250	194250	157250	N.A.	N.A.	N.A.	N.A.	[m³]
Net Gain	N.A.	N.A.	62272	18751	31005	N.A.	N.A.	N.A.	N.A.	[m³]

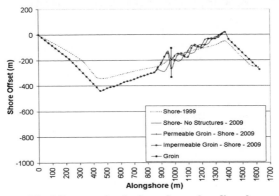

Fig. 8 Impact of a single groin on shoreline change

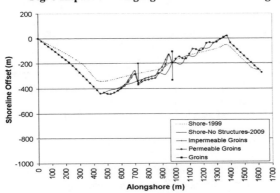

Fig. 9 Impact of two- groins on shoreline change

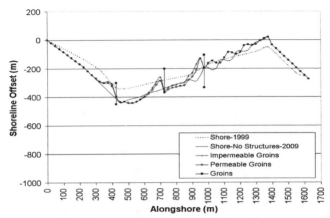

Fig. 10 Impact of three-groin on shoreline change

CONCLUSION

The area exhibits interesting behavior, which contradicts with the expected behavior of a shoreline with such a configuration. This can be attributed to currents or certain submerged morphological feature. This behavior caused instability of several runs and hindered the possibility of applying a larger number of groins. The performance of permeable groins does not differ a lot from the behavior of impermeable groins, in case of using a One-Line model. The nourishment applied to impermeable groins can be applied to permeable groins without fear of losing much benefit. Finally, aforementioned pile groin is valued low cost measure.

REFERENCES

Coastal Research Institute; CRI 1998. Wave climate along the Mediterranean Egyptian Coast, Report 2, Alexandria, EGYPT.

Gruene, G. and Kohlhase, S. 1975. Wellentransmission an Schlitzwaenden., die Kueste, Kiel, Germany, 27, 74-82 (in German).

Kraus, N, Hanson, H and Blomgren, H. 1994. Modern Functional Design of Groin System, Coastal Engineering Conf. Proc., ASCE, Kobe, Japan,1327-1342.

Mani, J S.and Jayakumar, S.1995. Wave Transmission by Suspended Pipe Breakwater, J. Wtrwy., Port Coast. And Ocean Engrg., ASCE, Vol.121,335-338.

Mani, J.S. 1998. Wave Forces on Partially Submerged Pipe Breakwater, Inter. OTRC Symposium, Houston, Texas, ASCE, 505-512.

Raudkivi, A. 1996. Permeable Pile Groin J. Wtrwy., Port Coast. And Ocean Engrg., ASCE, Vol. 122. No. 6, 267-272..

U.S. Army Corps of Engineers. 1992. Coastal Groins and Nearshore Breakwaters. Engineer Maual, No. 1110-2-1617. Washington, D.C.

Weiss, D. 1991. Einreihige Holzpfahlbulnen im Technischen Kuesten-Schutz vom Mecklenburg-vorpommern., die Kueste, Kiel, Germany, 52, 205-224 (in German).

Wiegel, R. 1961.Crossely Spaced Piles as a Breakwater., doc. and Harb. Autho., London, UK, 42 (491, 150).

BERM FOR REDUCING WAVE OVERTOPPING AT COMMERCIAL PORT ROAD, GUAM

Edward F. Thompson[1], M.ASCE, and Lincoln C. Gayagas[2]

Abstract: This paper describes the procedures and results of a typhoon overtopping-frequency analysis for a vulnerable section of the commercial port road along Cabras Island, Apra Harbor, U.S. Territory of Guam. The study was a challenging application of present numerical modeling technology to an innovative project design.

INTRODUCTION

Apra Harbor, Guam's commercial port, is located on the west side of the island. The harbor is well-protected by a combination of natural features and Glass Breakwater, a long man-made breakwater connecting into Cabras Island on the shoreward end (Figure 1). Cabras Island is a narrow, east-west oriented island which not only affords protection to the harbor, but also accommodates many of the commercial port facilities.

The port access road runs along the north side of Cabras Island. The container yard occupies most of the west-central part of Cabras Island. In this area, the road is protected from the sea by a relatively low, recurved concrete seawall fronted by a rubble-strewn beach. Subaerial beach width ranges from 30 m (100 ft) to 70 m (230 ft) along the vulnerable area. A coral reef extends seaward a distance of about 100 m (300 ft). During storms, waves can run up the beach, overtop the seawall, and cause disruption and damage to the road and port facilities.

The U.S. Army Honolulu Engineer District (HED) and U.S. Army Corps of Engineers, Pacific Ocean Division, have developed a project design to reduce vulnerability of the road and container yard to overtopping and flooding (POD 1995).

1 Hydraulic Engineer, U.S. Army Engineer Res. & Dev. Center, Coastal & Hydraulics Lab., 3909 Halls Ferry Road, Vicksburg, MS 39180-6199 USA. Edward.F.Thompson@erdc.usace.army.mil
2 Hydraulic Engineer, U.S. Army Engineer Dist., Japan; formerly with Honolulu Engineer Dist., U.S. Army Corps of Engr., Honolulu, HI 96858-5440 USA. Lincoln.C.Gayagas@poj02.usace.army.mil

OCEAN WAVE MEASUREMENT AND ANALYSIS

The project involves construction of an armored, low-crested berm in the ocean beach profile along a 957-m (3140-ft) stretch of coastline (Figure 2). The U.S. Army Engineer Research and Development Center, Coastal and Hydraulics Laboratory (CHL), recently conducted numerical model studies to assist

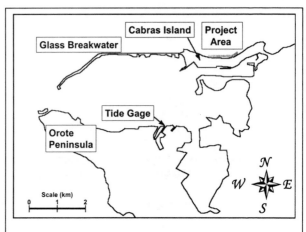

Fig. 1. Location map, Apra Harbor, Guam

HED in evaluating the expected performance of this innovative project design. Studies included modeling selected historical storms and several hypothetical variations of actual storms, calculating overtopping rates during each modeled storm, and evaluating the protection from overtopping and flooding afforded by both existing and with-project conditions.

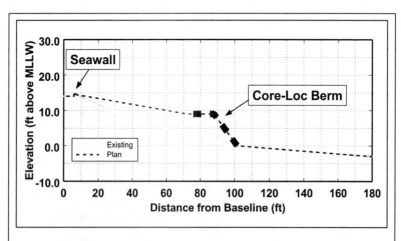

Fig. 2. Example of existing and plan profiles

STORM SELECTION

Guam's low-latitude location is favorable for tropical storm and typhoon formation and passage. The island often experiences typhoon impacts and occasionally a typhoon passes directly over the island. Typhoons usually approach Guam from the east or southeast and turn more toward the north, typically after passing the island.

Typhoon track data covering the years 1945-97 were obtained from the U.S. Navy's Joint Typhoon Warning Center (JTWC). Track data are given at 6-hour intervals, including latitude and longitude of the storm eye (with 0.1-deg precision) and maximum sustained 1-minute mean surface wind. Available information about storm impacts on Guam was also gathered and reviewed to insure the storm selection process included all important historical storm events.

Only typhoons which passed within a 322-km (200-mile) square box centered on the islands of Rota and Guam and had wind speeds of 64 knots (typhoon strength) or greater *within* the box were considered. From these typhoons, a representative storm set was selected for modeling. The set included all historical storms with eye passing within the immediate vicinity of Guam and a representative sample of storms for other typical travel paths relative to Guam. For example, Typhoon Omar (Aug 92) approached Guam from the east southeast and passed directly over the island (Figure 3).

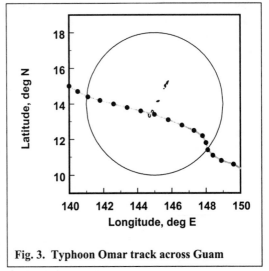

Fig. 3. Typhoon Omar track across Guam

The impact of a typhoon on the study area can be strongly affected by typhoon track. Historical data provide a valuable record, but storms with small variations in the historical tracks would have been equally likely. For analysis of extremes, it is important to capture small variations in the most damaging storms which would have caused them to be more damaging to the study area. After analysis of the effect of small track shifts in extreme historical typhoons, two hypothetical storms were added to the model storm set, giving a total of 30 storms.

Typhoons selected for modeling should be fairly representative of storm track statistics for the full set of typhoons passing into the box around Guam. Typhoons were classified according to their travel direction (Table 1). Storms selected for modeling are considered sufficiently representative of the full set of storms.

Table 1. Statistics of Typhoon Travel Direction				
	Full Set of Storms		Storms Selected for Modeling	
Travel Direction	Number of Storms	Percent	Number of Storms	Percent
Moving toward west	75	65	18	60
Moving toward north	27	23	8	27
Moving toward west & then north	11	9	3	10
Moving toward east	3	3	1	3
Total	116	100	30	100

MODELING APPROACH, OFFSHORE

Calculation of typhoon stage-frequency and overtopping relationships for Cabras Island requires application of several standard CHL numerical models and many additional processing steps. First, a Planetary Boundary Layer (PBL) wind model simulates the time history of typhoon-induced surface wind and atmospheric pressure fields for each selected storm during its general proximity to the study area (Cardone, Greenwood, and Greenwood 1992). The PBL model operates on a nested grid system centered on the storm eye. Storm tracks and maximum sustained 1-minute mean surface winds were obtained from the JTWC database. Central pressure was calculated from maximum wind speed using the relationship developed by Atkinson and Holliday (1977), based on data from Guam. Radius to maximum winds was approximated by application of relationships developed in a generalized numerical model study of storm characteristics (Jelesnianski and Taylor 1973). Wind velocities produced by the PBL model represent an averaging time of 30-60 minutes, which is appropriate for wave and storm surge modeling (Thompson and Cardone 1996).

The time history of wind information serves as input to both a long-wave hydrodynamic model ADCIRC and a wind-wave model WISWAVE. The ADCIRC model (Luettich et al. 1992, Westerink et al. 1992) provides a refined time history of typhoon-induced water levels at the study location for each storm. The computational grid developed for this study is circular with 8-deg (900-km) diameter and center at 145 deg E longitude and 14 deg N latitude. The islands of Guam and Rota are located in the central region of the grid. The grid boundary is shown as a circle in Fig. 3. Grid resolution is coarser in the open regions with increasing resolution toward the shore, reaching node spacing of about 50 m in the project area. The grid contains 15301 elements and 8410 nodes. Reefs, shallow areas, and embayments are finely resolved in and near the study area so that the hydrodynamics can be accurately calculated in these regions. Although astronomical tides were simulated for calibration, they were not included in routine storm simulations, as discussed later.

The WISWAVE model (Hubertz 1992, Resio and Perrie 1989) provides a time history of deep-water wave parameters in the general vicinity of Apra Harbor. This model is a second-generation directional spectral wave model in which spectral wave computations are based on the integration of energy over the discrete frequency spectrum. The grid was an 8-deg square with constant spacing of 0.083 deg. The islands of Guam and Rota were specified as land in the grid for accurate calculation of wave sheltering and refraction. Wind forcing for the wave model was calculated by application of the PBL model, as discussed previously. Wind speed and direction were calculated for each point on the wave grid at 1-hr intervals.

Deep-water waves produced by WISWAVE were transformed to the study area by application of the nearshore wave transformation model WAVTRAN (Jensen 1983, Gravens et al. 1991). The WAVTRAN model calculates transformation of directional wave spectra during propagation from one depth to another shallower depth, taking into account bottom contour orientation and wave sheltering. Bottom contours are assumed to be straight and parallel. Waves were transformed to 10-m depth or, in cases where waves would be breaking in that depth, the approximate nearshore depth at which breaking would begin.

MODELING APPROACH, NEARSHORE
The time history of transformed wave parameters is subsequently matched with nearshore water level information from ADCIRC and used to calculate a time history of wave ponding over the reef and nearshore setup, runup, and overtopping. Storm surge levels were typically quite small, never greater than 1 m, and did not need to be included in wave processes outside the reef.

Astronomical tide range is also relatively small in the study area, 0.7 m (2.4 ft) between mhhw and mllw. The shape of the tidal time series is asymmetrical, such that high tides rise little above mean sea level, while lower low tides can drop precipitously below the msl (Figure 4). Thus, tide levels are characteristically in a narrow 0.4-m (1.3-ft) range between msl and mhw most of the time. Astronomical tide was included in the study as a single level, mhw, a representative, but not extreme, high tide level.

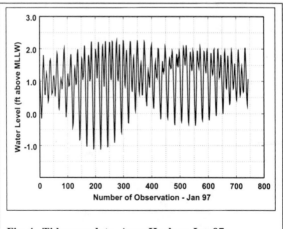

Fig. 4. Tide gage data, Apra Harbor, Jan 97

Water level in the reef lagoon was estimated as a combination of storm surge, tide, and wave-induced ponding. The increase in water level due to ponding was estimated from empirical relationships developed by Seelig (1983) in laboratory experiments with a fringing reef configuration typical of Guam. In this formulation, ponding is related to deepwater significant wave height, wave period, and water depth over the reef crest. For given incident wave and water level conditions, the estimated ponding is a fixed value with no time-varying surf beat behavior.

Significant wave height in the reef lagoon was estimated as 0.4 times the local water depth, as indicated in previous fringing reef investigations (e.g. Smith 1993). During an intense local typhoon, depth over the reef can exceed 3.0 m (10 ft), giving nearshore significant wave heights of over 1.2 m (4 ft).

Waves which have propagated across the reef lagoon encounter the nearshore slope approaching the seawall. Again, they break and cause a local increase in water level. This contribution to water level, referred to as wave *setup*, is not included in the ponding calculation. It is calculated with traditional relationships for wave setup on a sloping beach, with significant wave height and water depth in the reef lagoon serving as incident wave height and depth of wave breaking (*Shore Protection Manual* 1984).

Breaking waves at the shore intermittantly push water up the beach, creating wave *runup*. For both existing and plan nearshore profiles in the project area, runup during an intense typhoon can reach the seawall crest and continue over the top of the seawall. This wave *overtopping* can create problems along the commercial port road due to flooding, debris, and damage to the road surface. It can also cause flooding in the Apra Harbor container yard. Wave overtopping rate was calculated with the methodolgy developed by van der Meer and Janssen (1994). Reduction factors used in the calculation were determined to fit this application and produce overtopping rates consistent with qualitative observations of storm damage in the project area, as discussed in the following section.

IMPLEMENTATION OF OVERTOPPING METHOD
The methodology used to estimate overtopping rates includes four reduction factors to represent a variety of physical factors which can reduce overtopping. Implementation of these reduction factors requires a calibration/validation process to insure that the methodology is giving a reasonable representation of the project area.

Three historical typhoons which caused damage to the commercial port road in the study area are considered in the Feasibility Report (POD 1995). Typhoon Roy (Jan 88) and Typhoon Koryn (Jan 90) were reported to cause significant overtopping of the seawall and washing of rubble and debris onto the road. Typhoon Omar (Aug 92) caused similar damage but to a lesser extent due to rapidly changing conditions as the eye passed almost directly over the study area.

The same three storms were used in this study to help calibrate overtopping rate calculations to be consistent with documented experience. An overtopping rate time series was computed for each of the three calibration storms acting on the existing profiles. The berm reduction factor was determined as recommended by van der Meer and Janssen (1994). Since most existing profiles do not have a berm, this factor affected only a small number of profiles. The reduction factor for influence of a shallow foreshore was initially set equal to one. The reef presence suggests that a value less than one could be more appropriate, but behavior of waves over a reef during intense typhoon winds is not well documented. The reduction factor for influence of roughness was initially set equal to one. The presence of rubble on the shore suggests that a value less than one may be applicable, but the overall roughness impact of the rubble is unknown. The reduction factor for influence of angle of wave attack was set equal to one. This value is appropriate since the long-period waves characteristic of intense typhoons can be expected to approach nearly perpendicular to shore.

Calculated maximum overtopping rates for each storm were compared with qualitative damage reports along the commercial port road and published information about dangerous overtopping rates on roadways (e.g., CIRIA/CUR 1991). It was concluded that reduction factors for influence of shallow foreshore and roughness should be set equal to one for all applications with existing profiles.

For plan profiles, the reduction factor for influence of shallow foreshore was set equal to one, as with existing profiles. However, the reduction factor for influence of roughness will be affected by the planned addition of Core-Loc armor units to the nearshore profile. A roughness factor of 0.6 was taken for the Core-Loc portion of the profile. The section of nearshore profile one significant wave height above and below the swl was used to determine the reduction factor for influence of roughness. Using a linear weighting, factors of 0.6 for Core-Loc slope and 1.0 for other parts of the profile were combined to give the overall reduction factor, which varied with swl and significant wave height during the course of each typhoon.

DEVELOPMENT OF OVERTOPPING RELATIONSHIPS

Wave setup and overtopping rates were computed along 15 transects within the study area. Transects were specified by elevation profiles surveyed and provided by HED. Plan profiles were also provided. The project stations modeled are at 61-m (200-ft) intervals, beginning with Station 00+00 and ending with Station 28+00. Maximum overtopping rate is extracted for each nearshore profile in each storm.

The Empirical Simulation Technique (Scheffner et al. 1999) was applied to get overtopping-frequency relationships based on historical storm parameters and calculated maximum overtopping rates. Overtopping-frequency values and their standard deviations were derived for 5, 10, 25, 50, and 100-year return periods. Maximum overtopping rates with 100-yr return period illustrate variability along the coast (Figure 5). Most profiles have overtopping rates of about 0.065 m^3/sec/m (0.7 cfs per ft) for existing conditions and 0.009 m^3/m (0.1 cfs per ft) for plan profiles. Existing profiles 6 and 28 have reduced overtopping rates which are more like the plan overtopping rates, a consequence of the natural berm present on these existing profiles.

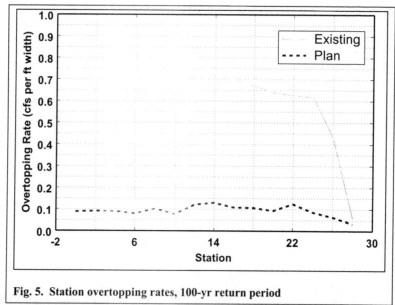

Fig. 5. Station overtopping rates, 100-yr return period

Maximum overtopping rates for the full project length can be obtained from the profile results. Profile overtopping rates are given as m^3/m (cfs per ft) of width. Since profiles are at 61-m (200-ft) intervals, each profile overtopping rate can be multiplied by that spacing to give total overtopping rate along the section of coast represented by the profile. The first and last profiles are considered to represent an 82-m (270-ft) width so that the full project length is included. Total overtopping rates along the project area are summarized in Table 2 and Figure 6. The proposed project has a strong impact on reducing overtopping rates.

Table 2. Maximum Overtopping Rates along Project Length		
Return Period, yr	Maximum Overtopping Rate, m^3 per sec (cfs)	
	Existing	Plan
2	0.0 (0.0)	0.0 (0.0)
5	1.03 (36.4)	0.0 (0.0)
10	8.15 (287.7)	0.86 (30.5)
25	25.19 (889.4)	3.11 (110.0)
50	36.68 (1295.4)	5.17 (182.4)
75	45.65 (1612.1)	7.15 (252.6)
100	50.19 (1772.3)	8.18 (288.7)

CONCLUSIONS

Numerical modeling of typhoon winds, waves, storm surge, and nearshore processes can be used to estimate wave overtopping rates along the Commercial Port Road, Guam, due to historical storm events. The EST methodology can be used to predict overtopping rate versus return period for extreme events. By modeling both existing

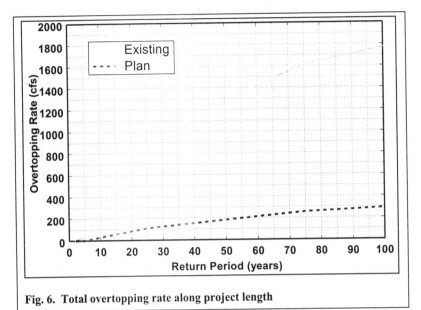

Fig. 6. Total overtopping rate along project length

nearshore profiles and proposed changes, impacts of the proposed project on protection of the road and commercial port facilities can be evaluated. Model results indicate that the proposed low-crested berm to be built in the nearshore profile will significantly reduce the vulnerability of the road and port facilities to damage due to wave overtopping.

ACKNOWLEDGEMENTS

The tests described and the resulting data presented herein, unless otherwise noted, were obtained from work at the U.S. Army Engineer Research and Development Center, supported by the Honolulu Engineer District, United States Army Corps of Engineers. Permission was granted by the Chief of Engineers to publish this information.

REFERENCES

Atkinson, G. D., and Holliday, C. R. 1977. Tropical Cyclone Minimum Sea Level Pressure/Maximum Sustained Wind Relationship for the Western North Pacific. *Monthly Weather Review*, 105, 421-427.

Cardone, V. J., Greenwood, C. V., and Greenwood, J. A. 1992. Unified Program for the Specification of Hurricane Boundary Layer Winds over Surfaces of Specified Roughness. Contract Rep. CERC-92-1, U.S. Army Engineer Waterways Experiment Station, Vicksburg, MS.

CIRIA/CUR. 1991. *Manual on the Use of Rock in Coastal and Shoreline Engineering*, CIRIA Special Pub. 83, Construction Industry Research and Information Association, London, UK; and CUR Rep. 154, Centre for Civil Engineering Research and Codes,

Gouda, The Netherlands.

Gravens, M. B., Kraus, N. C., and Hanson, H. 1991. GENESIS: Generalized Model for Simulating Shoreline Change. Report 2, Workbook and System User's Manual. Tech. Rep. CERC-89-18, U.S. Army Engineer Waterways Experiment Station, Vicksburg, MS.

Hubertz, J. M. 1992. User's Guide to the Wave Information Studies (WIS) Wave Model, Version 2.0. WIS Rep. 27, U.S. Army Engineer Waterways Experiment Station, Vicksburg, MS.

Jelesnianski, C. P., and Taylor, A. D. 1973. A Preliminary View of Storm Surges Before and After Storm Modifications. NOAA Tech. Memorandum ERL WMPO-3, Weather Modification Program Office, Boulder, CO.

Jensen, R. E. 1983. Methodology for the Calculation of a Shallow-Water Wave Climate. WIS Rep. 8, U.S. Army Engineer Waterways Experiment Station, Vicksburg, MS.

Luettich, R. A., Jr., Westerink, J. J., and Scheffner, N. W. 1992. ADCIRC: An Advanced Three-Dimensional Circulation Model for Shelves, Coasts, and Estuaries. Tech. Rep. DRP-92-6, U.S. Army Engineer Waterways Experiment Station, Vicksburg, MS.

POD. 1995. Commercial Port Road, Territory of Guam, Storm Damage Reduction Study, Feasibility Report & Environmental Assessment, Final Report. U.S. Army Corps of Engineers, Pacific Ocean Division, Honolulu, HI.

Resio, D. T., and Perrie, W. 1989. Implications of an f^4 Equilibrium Range for Wind-Generated Waves. *Journal of Physical Oceanography,* 19, 193-204.

Scheffner, N. W., Clausner, J. E., Militello, A., Borgman, L. E., Edge, B. L., and Grace, P. J. 1999. Use and Application of the Empirical Simulation Technique: User's Guide. Tech. Rep. CHL-99-21, U.S. Army Engineer Waterways Experiment Station, Vicksburg, MS.

Seelig, W. N. 1983. Laboratory Study of Reef-Lagoon System Hydraulics. *Journal of Waterway, Port, Coastal and Ocean Engineering,* ASCE, 109(4), 380-391.

Shore Protection Manual. 1984. U.S. Army Engineer Waterways Experiment Station, U.S. Government Printing Office, Washington, DC.

Smith, J. M. 1993. Nearshore Wave Breaking and Decay. Tech. Rep. CERC-93-11, U.S. Army Engineer Waterways Experiment Station, Vicksburg, MS.

Thompson, E. F., and Cardone, V. J. 1996. Practical Modeling of Hurricane Surface Wind Fields. *Journal of Waterway, Port, Coastal and Ocean Engineering*, ASCE, 122(4), 195-205.

Van der Meer, J. W., and Janssen, J.P.F.M. 1994. Wave Run-up and Wave Overtopping at Dikes and Revetments. Rep. No. 485, Delft Hydraulics, Delft, The Netherlands.

Westerink, J. J., Luettich, R. A., Jr., Baptista, A. M., Scheffner, N. W., and Farrar, P. 1992. Tide and Storm Surge Predictions Using Finite Element Model. *Journal of Hydraulic Engineering*, ASCE, 118(10), 1373-1390.

STABILITY OF ARMOR UNITS FOR A LEEWARD MOUND OF COMPOSITE BREAKWATERS

Tetsuya HAYAKAWA[1], Yasuji YAMAMOTO[2], Katsutoshi KIMURA[3]
and Yasunori WATANABE[4]

Abstract: : The armor units on the rubble mound foundation of a composite breakwater should be of sufficient mass to ensure stability against storm waves. Toward this, the stability mass of seaward units has been calculated by an equation developed from hydraulic model tests and the mass of leeward units has been designed empirically as half mass of seaward units.

This study uses numerical simulation to show that large eddies generated by wave overtopping results in a rapid caissonward flow above the leeward mound crown and that the velocity of this flow is a very important factor in the stability of armor units. Hydraulic model tests were used to formulate a function of wave conditions and structural geometry. From the balance of forces acting on an armor unit, the stability number N_S in the Hudson formula was formulated to allow design of armor units on the leeward mound.

INTRODUCTION

The rubble mound foundation of a composite breakwater is covered with armor units of concrete or stone whose mass is sufficient to prevent wave action from displacing rubble stones. Tanimoto et al. (1982) and Takahashi et al. (1990) developed a calculation method for stable mass of armor unit using hydraulic model tests.

The mass for leeward armor units has been set empirically at half that for seaward units. However, as there are few examples of damage to leeward armor units, it might be

---

1 Port Planning Section, Hokkaido Regional Development Bureau, Ministry of Land Infrastructure and Transport, Kita 8 Nishi 2, Kita-ku, 060-8511 JAPAN
2 Civil Engineering Research Institute, Independent Administrative Institution, 1-3 Hiragishi Sapporo, 062-0931 JAPAN
3 Division of Environmental Engineering, Muroran Institute of Technology,27-1,Mizumoto-cho, Muroran, 050-8585 JAPAN
4 Division of Environmental Resources Engineering, Hokkaido University, Kita 13 Nishi 8, Kita-ku, 060-8628 JAPAN

possible to omit their use under certain conditions, which would reduce construction costs.

More ever, a project is under way to create a shallow leeward mound for a composite breakwater, using dredged materials (Figure 1). Such mounds provide a favorable environment for seaweed aquaculture. However, the water

Fig. 1. Structure for environmental enhancement

depth at the mound crown must be between 4 and 5m, shallow enough to allow for sufficient photosynthesis. In such a case, the stability of armor units against wave overtopping is crucial.

This study used direct numerical simulation to clarify the fluid motion behind the breakwater that results from wave overtopping. In addition, hydraulic model experiments clarified the stability of armor unit at a leeward mound, toward development of a calculation method of stable mass for leeward armor units.

STUDY APPROACH

1) Numerical Calculation Method

To reproduce the complicated fluid motion that results from wave overtopping, direct numerical simulation was conducted applying cubic polynomial interpolation (CIP) to the continuity equation and the N-S equation, after Watanabe et al. (1996). At the movement boundary condition, the density function method was used, also applying CIP. Figure 2 shows the cross section of breakwater for calculation. Water depth h is 45 cm, caisson width B is 50 cm and height D is 55 cm. Water depth at the crown of the leeward mound d_l is 15 cm. For the wave generation boundary, two-dimensional cnoidal wave theory was applied. Wave height H is 20 cm and period T is 1.7 s, and the Reynolds number is about 10^6, using water depth as the representative length. The structure and wave conditions are the same as for the hydraulic test using regular waves.

2) Experiment method

The experiment was made in a two-dimensional wave flume 27 m long, 1.2 m deep and 0.6 m wide. A wooden model (Figure 2) was set on a horizontal mortar bed. With the caisson width B set at 50 cm and the height D set at 55 cm, the water depth h and the water depth at the crown of the leeward mound d_l were varied between 40, 45 and 50 cm and between 5, 10 and 15 cm, respectively. Six propeller-type current meters were installed on the leeward side of the caisson at intervals of 5 cm, to measure the maximum flow velocity above the mound crown. The flow velocity of overtopping wave at the leeward edge of the caisson crown also was measured. Regular waves were used, and the wave height H

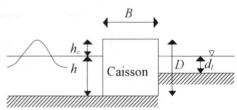

Fig. 2. Cross section of breakwater

and the period T were changed between 10, 15, 20 and 25 cm and between 1.2, 1.7 and 2.2 s, respectively.

In addition, an experiment on the stability of armor units on the leeward mound was made, using irregular waves with a significant wave height $H_{1/3}$ of 10 and 14 cm and a significant wave period $T_{1/3}$ of 1.7 s. For structural conditions, the water depth h was fixed at 45 cm, the crown height of caisson h_c was changed between 8.0, 8.4 and 11.2 cm and the water depth at the leeward mound crown d_l was changed from 5 to 30 cm. For armor units, crushed stone (two layers) of nine different masses (from 2.0 g to 60.0 g) were used. We applied 200 irregular waves in one wave group and measured the number of displaced armor units. This was repeated 15 times (3000 waves in total). The damage ratio was calculated as the percentage of displaced armor units to the total number of armor units. The observation area for displaced units were all within 60cm ($ \fallingdotseq$ 1.3 h) leeward of the caisson.

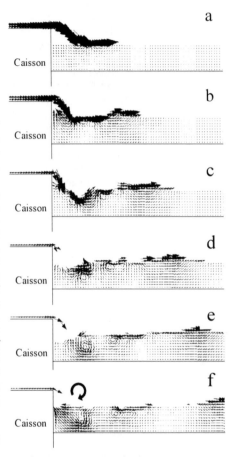

Fig. 3. Flow velocity vectors on the leeward side of the caisson

FLUID MOTION ON THE LEEWARD SIDE OF THE CAISSON

Figure 3 shows the distributions of flow velocity vectors on the leeward side of the caisson obtained by numerical simulation. The simulation results are shown in Figures 3(a) to 3(f) at an interval of $T/16$s. As shown in Figures 3(a) and 3(b), overtopping waves plunge into the leeward water surface, where the shear resistance between fluids generates eddies (Figure 3(c)). As momentum is supplied by overtopping wave, the water level drops at the plunge point and rises around that point. These eddies then interact, as shown in Figure 3(d), and ultimately two large eddies are generated (Figure 3(e)). These eddies are adverted caissonward or seaward in Figure 3(f). The negative rotation eddy that moves caissonward (left in the figure) exerts a particularly strong influence on the leeward mound crown, generating a rapid caissonward flow.

WAVE OVERTOPPING MODEL

1) Flow velocity above the caisson crown

The previous section clarified factors affecting the stability of mound armor units on the leeward side of the breakwater. Here, in order to estimate the fluid force acting on armor units, the maximum flow velocity near the leeward side of the mound crown is formulated from the results of the model experiment using regular waves.

Overtopped waves free-fall from the crown of the caisson and plunge against the water surface, affecting the leeward mound. Therefore, parameters representing the maximum flow velocity are these: flow velocity at the leeward edge of the crown U_2, crown height of the caisson h_c, and water depth at the leeward mound crown d_l.

Fujii et al. (1994) modeled the motion of overtopping waves to examine the stability of railings constructed on the caisson crown against waves and the safety of people. It is possible to calculate the water level and the flow velocity according to certain conditions. The flow velocity at the leeward edge of the crown (U_2) is expressed by the following equation.

$$U_2 = \begin{cases} (0.68 + 1.10\,H/h_m)\dfrac{C_1 \eta_1^{3/2}}{0.4\eta_1} & (H/h_m < 0.4) \\[2ex] \left(0.8 + \dfrac{0.32}{(10\,H/h_m - 4)^2 + 1}\right)\dfrac{C_1 \eta_1^{3/2}}{0.4\eta_1} & (H/h_m \geq 0.4) \end{cases} \tag{1}$$

where H = wave height; h_m = converted water depth in which consideration is given to the influence of the mound; η_1 = maximum seaward water level above the crown of the caisson; and C_1 = flow coefficient $(1.61\,\mathrm{m}^{0.5}/\mathrm{s})$.

h_m is calculated by the following equation.

$$h_m = \begin{cases} d & (B_M/L \geq 0.16) \\[1ex] d + (h-d)\dfrac{0.16 - B_M/L}{0.05} & (0.11 < B_M/L \leq 0.16) \\[1ex] h & (B_M/L < 0.11) \end{cases} \tag{2}$$

where B_M = width of the seaward mound shoulder; L = wave length; and d = water depth at the mound crown;

η_1 is calculated by the following equation.

$$\eta_1 = KH - h_c \tag{3}$$

where K = wave crest height ratio, which is expressed by the following equation:

$$\left. \begin{aligned} K &= \frac{1 + \sqrt{1 + 4h_c^*/h_m}}{2} \\[2ex] h_c^* &= h_c\,\frac{H/h_m}{2H/h_m - \dfrac{-1 + \sqrt{1 + 4h_c/h_m}}{2}} \end{aligned} \right\} \tag{4}$$

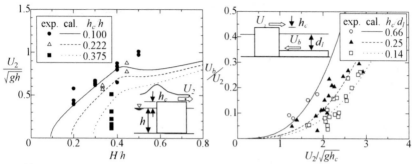

Fig. 4. Flow velocity at the crown of the caisson

Fig. 5. Flow velocity near the leeward mound crown

Figure 4 shows the experimental results of the peak value of flow velocity above the leeward edge of the caisson crown (U_2) and the results of the calculations by Equations (1) to (4). The x-axis shows the ratio of wave height to water depth (H/h), and the y-axis shows U_2 which was made dimensionless by gravitational acceleration g and water depth h. U_2 increases when H is large and h_c is small, because the wave overtopping quantity increases. Although there is some dispersion, calculated values and experimental values tend to agree roughly.

2) Flow velocity above the mound crown on the leeward side

Figure 5 shows the peak value of the horizontal flow velocity near the mound crown U_b, which was made dimensionless by U_2. The x-axis shows U_2, which was made dimensionless by the crown height h_c and the gravitational acceleration g. U_b increases as U_2 increases. This is thought to be because the velocity when overtopping waves plunge against the leeward water surface increases and the influence of eddies increases, according to numerical simulation. In addition, U_b tends to decrease when h_c/d_l decreases. This is attributed to the decreasing influence of eddies on the mound crown resulting from increases in the water depth at the leeward side of the mound crown d_l. Using the method of least squares, this tendency is formulated as follows.

$$\left.\begin{array}{l} \dfrac{U_b}{U_2} = a\left(\dfrac{U_2}{\sqrt{gh_c}}\right)^3 \\[2mm] a = 0.0399\left(\dfrac{h_c}{d_l}\right) + 0.00455 \end{array}\right\} \tag{5}$$

STABILITY OF LEEWARD MOUND ARMOR UNITS
1) Kinetic characteristics of armor units

Figure 6 shows the relationship between the number of waves N and the damage ratio D for dimensionless crown height $h_c/H_{1/3}$ of 0.8 and ratio of crown height to water depth on the leeward mound crown h_c/d_l of 0.47. If the mass of armor unit is insufficient, the damage ratio D significantly increases at the early stage of wave action, and then it plateaus. This study focused on the damage ratio D when the damage became almost constant at $N=3000$. Regarding the relationship between the mass of armor unit M and the damage ratio D, we defined the situation where D equaled 1% as the stability limit of armor unit and its corresponding mass as the critical mass for stability of armor unit.

Photo 1 shows the displacement conditions of armor units when the numbers of waves N were 1000, 2000 and 3000 under the same conditions as those described in Figure 6. The damage ratios D were 1.3%, 2.7%, and 3.3% respectively, and the movement of armor units was caissonward, showing agreement with the caissonward flow in the simulation results.

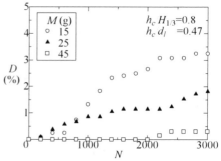

Fig. 6. Influence of wave number

2) Stable mass of armor units

Tanimoto et al. (1982) proposed that the critical mass for stability M is proportional to the 6th power of U_b (Equation (6)), on the basis of the balance among the drag force, the lift force and the the friction force against the sea floor, which can be calculated by Morison's equation.

$$M = \frac{\rho_r U_b^{\,6}}{\{Cg(S_r - 1)\}^3} \tag{6}$$

where ρ_r = density of armor unit; g = gravitational acceleration; S_r = ratio of armor unit density to fluid density; and C = a parameter determined by the shape and the coefficients of drag force and lift force of armor units.

The following is the equation using the stability number N_S, proposed by Hudson et al. (1959).

Photo 1. Damage pattern

$$M = \frac{\rho_r}{N_s^{3}\left(S_r -1\right)^{3}} H^3 \tag{7}$$

If M is eliminated from Equations (6) and (7), the stability number N_S can be described as follows:

$$N_S = \frac{C}{\left(U_b / \sqrt{gH}\right)^{2}} \tag{8}$$

By using the critical mass for stability of armor unit and the wave height H that were obtained experimentally, N_S was calculated from Equation (7). However, because the stability of leeward mound armor units at the composite breakwater tended to depend heavily on the maximum wave in the irregular wave group, the maximum wave data were used as the wave heights.

Figure 7 shows the relationship between the stability number N_S which was obtained experimentally and the horizontal flow velocity U_b near the mound crown which was made dimensionless by the maximum wave height H_{max} and the gravitational acceleration g. The values of U_b were not directly measured on site but were calculated from Equations (1) to (4) by using the maximum wave data. As U_b increased, the drag force and the lift force affecting armor units became larger and the stability number N_S decreased.

The solid line reflects the calculation result when C took the value of 0.06 in Equation (8). Although the line expresses the trend of N_S obtained from the experiment, when C is calculated using the actual drag force and lift force coefficients, the value is ten times the magnitude of 0.06.

In Equation (8), the influences of the drag force and the lift force were taken into account on the basis of Morison's equation, which used for substances under steady flow. For this reason, the drag force and the lift force were not precisely estimated, including the occurrence of large eddy behind the composite breakwater, which resulted in a

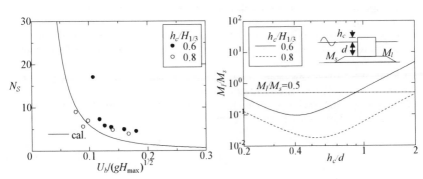

Fig. 7. Stability Number   **Fig. 8. Comparison between seaward and leeward armor weight**

discrepancy between the experimental results and the calculation results.

Figure 8 shows the unit mass ratio of the leeward armor M_l to the seaward armor M_s. The values of M_s were calculated by Tanimoto (1982). M_l/M_s in the case of $h_c/H_{1/3}=0.8$ is less than that in the case of $h_c/H_{1/3}=0.6$, because the quantity of waves overtopping is smaller in the former case. The figure also shows $M_l/M_s=0.5$ used in the empirical design. In the region where h_c/d is relatively large, the empirical design method may underestimate the mass of leeward units.

CONCLUSION

The main results of our study follow.

1) Complicated fluid motion generated by the plunge of overtopping wave was reproduced, using direct numerical simulation applying CIP. We found that when the water at the leeward mound is shallow, the plunge of overtopping wave generates large-scale eddies that result in caissionward rapid flow.

2) The influence of wave conditions and structure on the flow velocity at the leeward mound was clarified by hydraulic model tests.

3) A calculation method of stable mass of armor unit for leeward mound was proposed and was compared with conventional design methods.

In the hydraulic experiment of this study, the ratio of wave height to water depth $H_{1/3}/h$ was a relatively low 0.31, and the mound condition was relatively shallow water depth. To improve the calculation method of stable mass for armor units of the leeward mound of a composite breakwater, we need to consider diffracted waves. Also, the validity of our calculation method requires examination under a wide range of experimental conditions.

REFERENCES

Tanimoto, K., Yagyu, T., Muranaga, T., Shibata, K. and Goda, Y. 1982. Stability of Armor Units for Foundation Mounds of Composite Breakwaters by Irregular Waves Tests, *18th International Conference on Coastal Engineering*, pp. 2144-2163

Kimura, K., Takahashi, S., and Tanimoto, K., 1994. Stability of Rubble Mound Foundations of Composite Breakwaters under Oblique Wave Attack, *Proc of ICCE '94, ASCE*, pp. 1227-1240

Fujii, A., Takahashi, S. and Endo, K., 1994. An Investigation of the Wave Forces Acting on Breakwater Handrails, *24th International Conference on Coastal Engineering*, pp. 1046-1060

Hudson, R.Y. 1959. Laboratory Investigation of Rubble-Mound Breakwaters, *Proc. ASCE, Waterways and Harbors Division*, Vol. 85

Watanabe, Y. and Saeki, H. 1996. Fluid motion with wave overtopping behind a breakwater, *Proc of ICCE '99, ASCE*, pp. 183-191

DIRECTIONAL WAVE TRANSFORMATION INDUCED BY A CYLINDRICAL PERMEABLE PILE

Georges Govaere[1] and Rodolfo Silva[2], M.ASCE

Abstract: This paper presents an analytical solution for the evaluation of linear wave interaction with a porous pile. The extension to the cases of unidirectional and multidirectional waves is obtained by means of a transfer function. The results show a clear reduction in wave heights when a permeable, rather than impermeable, structure is used. It is also seen that irregular waves generate smoother variations in the wave field than those predicted by the regular wave model.

INTRODUCTION

Cylindrical piles are commonly used in ocean engineering in a variety of structures, such as lighthouses, docks, bridge piles and research or oil production platforms. A cylindrical form is used since the wave-structure interaction is lower than that induced by other forms, principally in the distribution of pressure forces on the pile, scour problems and free water oscillations.

In recent years wave interaction on cylindrical piles has been studied by a large number of authors. Among others, Roldan (1992) derived a solution of wave kinematics inside and outside a permeable vertical cylinder. Darwiche, Williams and Wang (1994) studied wave interaction with a semiporous cylindrical breakwater. Silva (1995) presented a model of the wave kinematics around a submerged permeable structure and, finally, Govaere and Silva (1999) presented a solution for the wave kinematics around a protected cylindrical pile.

---

[1] Ph.D. Student, Engineering Institute, National University of Mexico, Apto 70-472 Ciudad Universitaria, Coyoacan, C.P. 04510, México, D.F. e-mail georges@litoral.iingen.unam.mx
[2] Research Scientist, Engineering Institute, National University of Mexico, Apto 70-472 Ciudad Universitaria, Coyoacan, C.P. 04510, México, D.F. e-mail rodo@litoral.iingen.unam.mx

The main goal of this paper can be summarized as the development of a mathematical model to represent wave transformation caused by a permeable cylindrical pile.

ASSUMPTIONS
The main assumptions made in this work are:
- The flow is incompressible and irrotational, therefore the Laplace equation can be used.
- The Linear Theory is used, so linear superposition of components is valid and wave breaking is not considered.
- The porous media is homogeneous and isotropic, thus, the flow follows Darcy's law. Hence, the water velocity within the media is linearly proportional to the pressure difference across the pile.

TEORETICAL BACKGROUND
Figure 1 shows a sketch of the problem with a permeable pile of radius R_1, height $a=h-h_p$ and constant depth h, a region around the pile (region 1), a region above the pile (region 2), and a region inside the porous media (region 3).

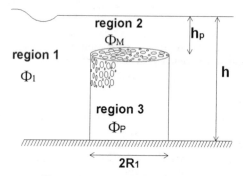

Figure 1. Sketch of the problem.

It is assumed that the flow is irrotational and incompressible and thus a velocity potential to describe the flow in the water and in the porous media can be derived. Following the methodology suggested by Roldan (1992), Silva (1995) and Govaere (1999), the velocity potentials can be described in the following way:

We express the velocity potential in the water (region 1) as the sum of incident, ϕ_I and scattered wave emanating from the cylinder ϕ_s:

$$\Phi_1(x,y,z,t) = \text{Re}\left\{\left(\phi_I(x,y) + \phi_s(x,y)\right)I(z)e^{-i\sigma t}\right\} \tag{1}$$

where the depth dependence function is given by:

$$I(z) = -\frac{igA}{\sigma}\frac{\cosh k(h+z)}{\cosh kh} \tag{2}$$

and the wave number in the water, k, is obtained through the normal dispersion relationship.

$$\frac{\sigma^2 h}{g} = kh \tanh kh \tag{3}$$

The water free surface level is given by:

$$\eta(x,y) = \text{Re}\left\{ A\phi(x,y)e^{-i\sigma t} \right\} \tag{4}$$

Using a cylindrical coordinate system (r,θ), an incident monochromatic wave train with an angle β, impinging on the pile is considered:

$$\phi_1 = e^{ikr\cos(\theta-\beta)} \tag{5}$$

In cylindrical coordinates, the incident velocity potential can be expressed as:

$$\phi_1 = \sum_{n=-\infty}^{\infty} J_n(kr)e^{in\left(\frac{\pi}{2}-\theta+\beta\right)} \tag{6}$$

The interaction of the incident wave train in the pile generates a scattered wave pattern. The general form of the scattered wave can be represented as:

$$\phi_s = \sum_{n=-\infty}^{\infty} A_n Z_n H_n^{(1)}(kr)e^{in\theta} \tag{7}$$

where

$$Z_n = \frac{J_n{}'(kR_1)}{H_n^{(1)}{}'(kR_1)} \tag{8}$$

So, in region 1, the total potential is:

$$\phi_1 = \phi_I + \phi_s = \sum_{n=-\infty}^{\infty} J_n(kr)e^{in\left(\frac{\pi}{2}-\theta+\beta\right)} + \sum_{n=-\infty}^{\infty} A_n Z_n H_n^{(1)}(kr)e^{in\theta} \tag{9}$$

In turn, in regions 2 and 3, over and inside the porous media, respectively, the solutions are of the form:

$$\Phi_M = M \sum_{n=-\infty}^{\infty} B_n J_n(Kr)e^{in\theta} \qquad\qquad -h_p \le z \le 0 \qquad\qquad (10)$$

$$\Phi_p = P \sum_{n=-\infty}^{\infty} B_n J_n(Kr)e^{in\theta} \qquad\qquad -h \le z \le -h_p \qquad\qquad (11)$$

where the depth dependence functions are (Losada *et al.* (1996)):

$$M(z) = \frac{ig}{\sigma} \frac{\cosh K(h+z) - F \sinh K(h+z)}{\cosh Kh - F \sinh Kh} \qquad\qquad (12)$$

$$P(z) = \frac{ig}{\sigma} \left(\frac{1 - F \tanh Ka}{(s-if)} \right) \frac{\cosh K(h+z)}{\cosh Kh - F \sinh Kh} \qquad\qquad (13)$$

where

$$F = \left(1 - \frac{\varepsilon}{(s-if)}\right) \frac{\tanh Ka}{1 - \frac{\varepsilon}{(s-if)} \tanh^2 Ka} \qquad\qquad (14)$$

J_m are the Bessel functions and $H^{(1)}_m$ are the Hankel functions, both of which can be resolved as described by Abramowitz (1964). $I_0(z)$, $M_0(z)$ and $P_0(z)$ are the depth dependency functions (Silva, 1995), ε is the porous media porosity, f is the friction coefficient and s is the added mass coefficient. The complex wave number K is given by the complex dispersion equation:

$$\sigma^2 - gK \tanh Kh = F \left[\sigma^2 \tanh Kh - gK \right] \qquad\qquad (15)$$

Using the boundary condition of continuity of mass and pressure:

$$\frac{\partial \Phi_1}{\partial r} = \frac{\partial \Phi_M}{\partial r} \qquad\qquad r = R_1 \ \text{ and } \ -h_p \le z \le 0 \qquad\qquad (16)$$

$$\frac{\partial \Phi_1}{\partial r} = \varepsilon \frac{\partial \Phi_p}{\partial r} \qquad\qquad r = R_1 \ \text{ and } \ -h \le z \le -h_p \qquad\qquad (17)$$

$$\Phi_1 = \Phi_M \qquad\qquad r = R_1 \ \text{ and } \ -h_p \le z \le 0 \qquad\qquad (18)$$

$$\Phi_1 = (s-if)\Phi_p \qquad\qquad r = R_1 \ \text{ and } \ -h \le z \le -h_p \qquad\qquad (19)$$

ortogonalizing, and solving the equation system, the solution for the unknown parameters A_m and B_m are:

$$A_m = \frac{(C_2 - C_1)}{(C_1 - C_2 C_3)} e^{im\left(\frac{\pi}{2} - \beta\right)} \qquad\qquad (20)$$

and

$$B_m = \frac{C_1 C_2 - C_1 C_2 C_3}{C_1 - C_2 C_3} e^{in\left(\frac{\pi}{2}-\beta\right)} \tag{21}$$

where

$$C_1 = \frac{VkJ_m{}'(kR_1)}{KJ_m{}'(KR_1)(W + \varepsilon U)}, \quad C_2 = \frac{J_m(kR_1)V}{J_m(KR_1)\left[W + (s - if)U\right]}, \quad C_3 = \frac{Z_m H_m(kR_1)}{J_m(kR_1)} \tag{22}$$

$$V = \int_{-h}^{0} [I(z)]^2 dz, \qquad W = \int_{-h_p}^{0} M(z)I(z)\ dz, \qquad U = \int_{-h}^{-h_p} P(z)I(Z)\ dz \tag{23}$$

The velocity potential in the water around the pile (region 1), is given by:

$$\Phi_1(r,\theta,z) = I(z) \sum_{n=-\infty}^{\infty} \left[J_n(kr) + Z_n H_n(kr) \frac{(C_2 - C_1)}{(C_1 - C_2 C_3)} \right] e^{in\left(\frac{\pi}{2}-\beta+\theta\right)} \tag{24}$$

The velocity potential over the porous media (region 2):

$$\Phi_M(r,\theta,z) = M(z) \sum_{n=-\infty}^{n=\infty} J_n(Kr) \frac{C_1 C_2 - C_1 C_2 C_3}{C_1 - C_2 C_3} e^{in\left(\frac{\pi}{2}-\beta+\theta\right)} \tag{25}$$

The velocity potential in the porous media (region 3):

$$\Phi_p(r,\theta,z) = P(z) \sum_{n=-\infty}^{n=\infty} J_n(Kr) \frac{C_1 C_2 - C_1 C_2 C_3}{C_1 - C_2 C_3} e^{in\left(\frac{\pi}{2}-\beta+\theta\right)} \tag{26}$$

The potentials presented above are the solutions for a monochromatic wave field. The extension of these results to the spectral multidirectional wave case was made using a transfer function, following Silva (1995). The incident frequency spectrum is divided into discrete wave components characterized by a certain frequency f_j. The local spectrum at any location can be expressed as:

$$S_p(f_j,r,\theta) = |TF|^2 S_i(f_j,r,\theta)$$

where S_p and S_i are the propagated and incident spectra, respectively, and TF is the transfer function.

The friction coefficient, f, is evaluated by an iterative method as shown in Losada et al. (1996).

$$f = \frac{\varepsilon}{\sigma} \frac{\int_V \varepsilon^2 \left\{ \frac{\nu \; \bar{q}_{rms}^2}{K_p} + \frac{\varepsilon C_f}{\sqrt{K_p}} | \vec{q}_{rms} |^3 \right\} dV}{\int_V dV \int_t^{t+T} \varepsilon \; \bar{q}_{rms}^2 \; dt} \qquad (27)$$

where

$$\bar{q}_{rms} = \sqrt{u_{r\_rms}^2 + u_{\theta\_rms}^2 + u_{z\_rms}^2} \qquad \text{and} \qquad U_{rms} = \sqrt{8 mo_U} \qquad (28)$$

mo_u is the zero-order moment of the total velocity spectrum.

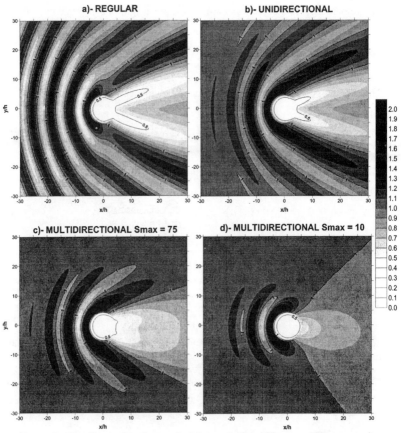

Figure 2. Nondimensional wave height evolution (Hrms$_{local}$/ Hrms$_{incident}$)

RESULTS

The solutions presented above were applied to (1) monochromatic waves, (2) spectral unidirectional waves using the standard JONSWAP spectrum, Hasselmann et al. (1973), and (3) spectral multidirectional waves again using the JONSWAP spectrum, with a directional spreading function characterized by the parameter S_{max} (as it was suggested by Mitsuyasu et al. (1975)). The latter was in turn applied with moderate ($S_{max}=75$) and large ($S_{max}=10$) spreading and to discretize the spectrum, 128 frequency and 36 angle components have been considered. In all the cases, the non-dimensional envelop is the ratio between the local and incident root mean square heights.

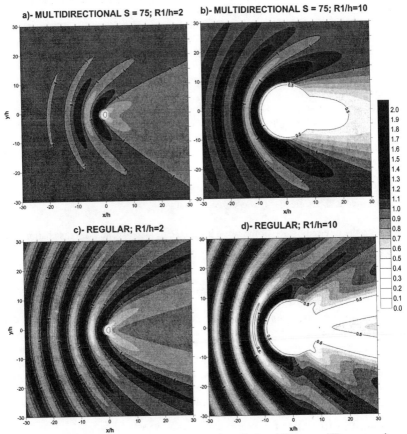

Figure 3. Nondimensional wave height evolution (Hrms$_{local}$/ Hrms$_{incident}$)

In Figures 2 and 3 show the maximum non-dimensional envelop for four different cases of incident wave spectrums (propagating from left to right), where the relative submergence of the pile $a/h = 1$, depth $h = 5m$, $R_1/h = 5$, $kh = 0.46$ and a porosity $\varepsilon = 0.4$.

We can observe that as the spectrum broadens in frequency and angle, the modulation around the structure damps faster. In all the cases, the energy inside the porous media is very small.

Figures 4 to 7 show a profile of the maximum non-dimensional wave evelop along the line $y = 0$, in the range $-30/h < x < 30/h$. Some parameters were varied to analyze their relative effects on the wave kinematics: the porosity of the pile varies from 0.3 to 0.5 (Figure 4); the relative pile radius, R_1/h, varies from 2 to 10 (Figure 5); the relative submergence, a/h, varies from 0.5 to 1.0 (Figure 6); and the relative depth, kh, varies from 0.35 to 0.68 (Figure 7). The simulation parameters are shown in the upper right corner of each figure.

CONCLUSIONS
A new mathematical model to analyze the wave kinematics in and around a submerged or emerged cylindrical permeable pile has been presented.

Using a transfer function, the monochromatic analytical solution was extended to the case of uni- and multidirectional spectral waves.

The energy dissipation depends basically on the physical characteristics of the porous media. Specifically, it was observed that:

- As the pile radius, R_1, or the wave height, H_{rms}, increase, so does dissipation and the relative maximum wave height, $H_{local}/H_{incident}$, inside the pile decreases, while outside it increases.

- As the relative submergence a/h increases, the dissipation increases.

- As the porosity, ε, or the relative depth, kh, increase, $H_{local}/H_{incident}$ inside the pile increases, while both the dissipation and $H_{local}/H_{incident}$ outside the pile decrease.

The wave modulation depends more on the characteristics of the incident wave spectrum than on the characteristics of the pile. As the spectrum broadens in frequency and angle the modulation around the structure damps faster.

ACKNOWLEDGEMENTS
This research was funded by the Mexican Consejo Nacional de Ciencia y Tecnología, CONACYT, project "Caracterización del riesgo oceanográfico en el litoral mexicano", under contract 35018-U. Highly constructive comments by Prof. Paulo Salles were received with sincere gratitude.

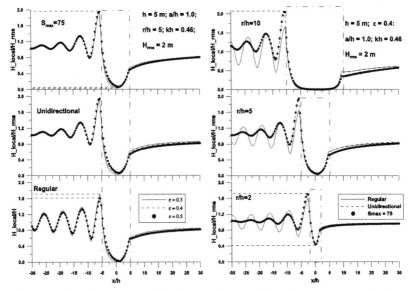

Figure 4. Maximum wave height for different porosity and incident waves

Figure 5. Maximum wave height for different relative pile radius

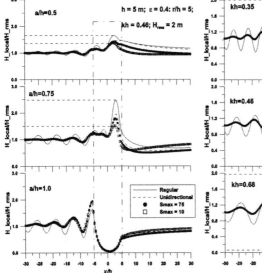

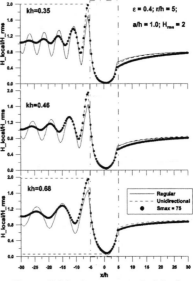

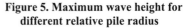

Figure 6. Maximum wave height for different relative submergences

Figure 7. Maximum wave height for different relative depths

REFERENCES

Abramowitz M. and Stegun L. E. (1964) "Handbook of Mathematical Functions with Formulas, Graphs, and Mathematical Tables" National Bureau of Standards, Applied Mathematics Series 55, U.S. Department of Commerce, Washington D.C.

Darwiche, Williams and Wang (1994) "Wave interaction with semiporous cylindrical breakwater" Journal of waterway, Port, Coastal and Ocean Engineering Vol. 120 No. 4 pp. 382-403

Govaere, G. and Silva, R. (1999) "Wave kinematics around a protected cylindrical impermeable pile" Coastal Structure Conference, Santander, Spain.

Hasselmann et al. (1973). "Measurments of wind-wave growth and swell decay during the Joint North Sea Wave Project (JONSWAP)" Deutsche Hydr. Zeit Reihe A (8°) 12 (1973).

Losada I.J., Silva R. And Losada M.A. (1996) "Interaction of non-breaking directional random waves with submerged breakwaters" Coastal Engineering Journal Vol 28 pp 249-266, Elsevier

Mitsuyasu, H. et al. (1975). "Observation of the directional spectrum of ocean waves with Mitsuyasu's directional spectrum". Tech. Note of Port and Harbour Res. Inst. 230, 45pp (In Japanese). (Reference consulted in Goda, Y. 2000. "Random seas and design of maritime structure", 2nd Edition, World Scientific Publishing Co. Pte. Ltd).

Roldan, A. (1992) "Sobre la transformación de un tren de ondas lineal, no lineal y modulado por un medio discontinuo" Ph.D. Tesis, Santander, Spain (in spanish).

Silva, R. (1995) "Transformación del oleaje debido a obras de defensa del litoral" Ph.D. Tesis, Santander, Spain (in spanish).

THE EFFECT OF WAVE ACTION ON STRUCTURES WITH LARGE CRACKS

Guido Wolters[1], Gerald Müller[2]

Abstract: Damages of coastal structures are often wave induced. A new study revealed that wave impact generated pressures can travel into water filled cracks or joints, thus damaging or destroying the structure from within. A series of experiments was conducted in QUB's wave channel to assess the characteristics of breaking wave impact induced pressure pulse propagation through partially water filled cracks. It was found that the pressure pulses generally travel at very low velocities of 45-70 m/s within completely submerged cracks, whereas velocities in partially submerged cracks are significantly higher. Pulses attenuate fast inside the fully submerged crack, but slowly in partially submerged ones. In partially submerged cracks the pressure pulse was found to travel in the air, propagating fast and with little attenuation deep into the structure, signifying that partially filled cracks are probably more dangerous than completely filled cracks. They are probably also the main cause for the *seaward* removal of blockwork in coastal engineering structures. This effect however is not limited to blockwork structures, but to every structure where cracks or joints occur, including natural rock cliffs.

INTRODUCTION

Many coastal structures, as well as natural rock cliffs, are subjected to breaking waves, which create very high and short pressure pulses. Especially older structures and soft, chalk-like rock contain *cracks* or *joints* which are exploited during storms. Typical damage mechanisms include the *seaward* removal of individual blocks in breakwaters or whole cliff sections. It is assumed that the wave impact pressures travel into the structure/cliff damaging it from within. Although it has been suspected for a long time it was only recently shown that these impact pressures can enter water filled cracks and that they propagate as compression waves, Müller (1997), Müller et al (1999).

Although in the last 20 years a lot of research effort was directed to investigate impact pressures, the mechanism of impact pressure generation is still disputed, there is still no reliable prediction formulae for such pressures and very little is known about impact pressure *propagation* and pressure pulse *characteristics*.

[1] G. Wolters, PhD Student, Queen's University Belfast, Civil Engineering Department, David Keir Building, Stranmillis Road, Belfast BT7 5AD, UK, Tel.: +44 (2890) 274001
e-mail: g.wolters@qub.ac.uk
[2] G. Müller, Lecturer, Queen's University Belfast, Civil Engineering Department, David Keir Building, Stranmillis Road, Belfast BT7 5AD, UK, Tel.: +44 (2890) 274517, Fax: +44 2890 663754
e-mail: g.muller@qub.ac.uk

The investigation of completely water filled cracks led to some new insights regarding the pulse characteristics. It was found that the pulses propagated with rather slow speeds, and that they attenuated fast. However, under natural conditions, completely filled cracks are not typical. More often, in particular e.g. in rock cliffs, crack networks with both partially and fully submerged cracks are found. Partially filled cracks are the topic of this study.

REVIEW

The authors' field of research originally centered on structures such as Alderney breakwater (Alderney being one of the Channel islands), which is a typical example of a blockwork structure with rubble filling. Damages are often caused by the seaward removal of individual blocks during storms. Fig. 1a shows a cross section of Alderney breakwater and Fig. 1b Le Havre breakwater, which is also a blockwork structure.

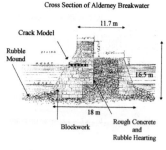

Fig. 1a: Alderney Breakwater    Fig. 1b: Le Havre Breakwater

Although Alderney is not representative for all coastal structures, it shares common features with rock cliffs and other old engineering structures: they all contain joints and are often composed of brittle materials with high compressive but low tensile strength. All are exposed to breaking wave action. The implication is that they suffer from a similar damage mechanism. As part of an EPSRC funded joint project with the Universities of Bristol, Plymouth and Belfast the authors are in the process of preparing field measurements of impact pressures and in artificial cracks on Alderney.

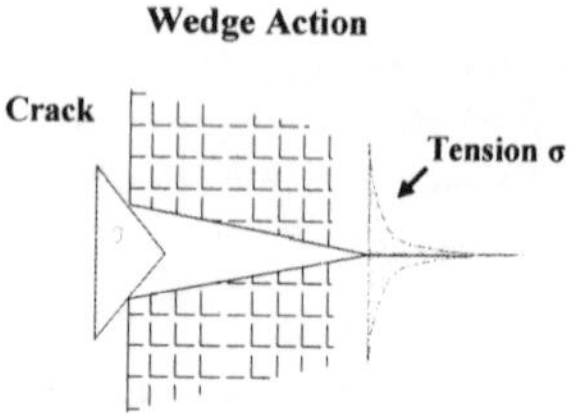

Fig. 2: Wedge Action

In rock cliffs, crack growth is caused by a process called wedge action (Fig. 2): the cracks or fissures in rock cliffs are wedge-like, with a sharp inner end. When struck by a wave the pressure inside the crack builds up rapidly, generating high tensile stresses at the crack tip, thus promoting crack growth. The material is thus loaded in tension, where its strength is weakest, at the crack tip. The sudden failure of the brittle material at this point initiates the crack growth.

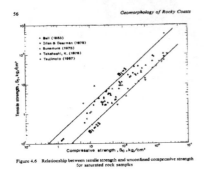

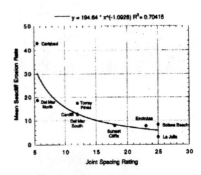

Fig. 3a) Tensile rock strength   3b) Erosion versus joint spacing

Fig. 3a, taken from Sunamura (1992), shows the tensile rock strength plotted against the compressive strength. It can be seen that the compressive strength of rock is generally 10 times higher than the tensile strength. It also shows that the tensile strength (varying between 0.01-10 MPa) is well in the range of wave impact pressures (0.7 MPa, see e.g. Rouville 1938). Fig. 3b is taken from Benumof and Griggs (1999), who measured sea cliff erosion rates at various sites in California. The graph shows cliff erosion rate plotted against joint spacing. At first glance cliff erosion rates for granite of 1mm, or for chalk of 1m per year, do not appear to pose a problem. It has however to be considered, that houses are often built very close to the cliff face, so that rock erosion can become a serious issue.

The pictures in Fig. 4 are taken at the Normandy coastline in France, near Le Havre (Etretat) in 2001: they show (a) chalk cliffs exposed to marine erosion leaving in its wake arches and spikes (cliff retreat), (b) undercutting of cliffs. The cliffs are 95 mio. years old at bottom, 70 mio years at top and about 100m high, composed of chalk layers with flint or without flint inlay and sometimes separated by darker and much harder layers of dolomite chalk. *Between* the chalk layers wide cracks can be observed (b), which extend a couple of meters into the cliff, showing that the layer boundaries are very susceptible to wave attack. Picture (c), taken at St. Pierre en Port, shows a house on top of a cliff, which faces imminent cliff collapse. In the foreground a recent rock fall can be seen.

Fig. 4a: Marine erosion

Fig. 4b: Cliff undercutting

Fig 4c: Cliff collapse

The following conclusions can be drawn from the above statements:
- Impact pressure pulses are of high magnitude and short duration.
- The integrity of blockwork structures as well as that of rock cliffs can be endangered by wave impact induced pressure pulses travelling into cracks.
- Erosion rate depends on joint width and spacing.

MODEL TESTS AT QUB

The authors model tests were conducted to address the following main objectives:
Can pressures propagate into water filled cracks? What are the characteristics of the pulse? Can blocks be removed by impact induced internal pressures?

Cracks extend generally above and below the mean water line. Those below the water line have already been researched. Here we are particularly interested in half submerged cracks.

Experimental set-up

A series of experiments were conducted in the Hydraulics Laboratory at the Queen's University of Belfast's (QUB) Civil Engineering Department in a wave tank of 17m length, 350 mm width and with a water depth of 1m. An inserted false bottom made of fibreglass brought the water depth from 1m at the deep end to 110 mm at the model sea wall with a slope of 1:10. The tank is shown in Fig. 5. At the shallow end of the sea bed, a vertical wall was installed. Waves were generated with a flap-type wave paddle in the deep water section of the tank. A single wave, with a deep water wave height of

71mm, was generated every 82 seconds. This wave resulted in a clean plunging breaker to hit the sea wall. Only the impact created by the initial breaker, which was undisturbed, was measured. Subsequent waves were affected by the reflection and turbulence left by the first breaker, and generated widely differing results. The wave tank was then allowed to settle, so that the next wave would find undisturbed conditions. This procedure allowed to generate repeatable impact pressures within acceptable limits.

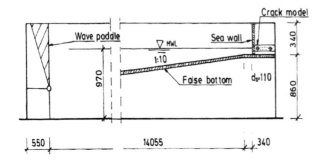

Fig. 5: Wave tank with sea wall model

Two brackets were manufactured which were inserted into the vertical seawall alternatively. The first bracket was used to record the wave impact pressures on the seawall (see Fig. 6). Four Endevco 8510B-5 pressure transducers with a maximum pressure of 5 PSI (34.5 kN/m^2) and a natural frequency of 180 kHz in air were used during the experiments. The second bracket allowed any of the crack configurations to be securely inserted into the wall. The crack widths varied between 0.5, 1 and 3mm and the height was constant at 25mm. Crack lengths of 115 and 600 mm were employed. Cracks were examined while being totally submerged as well as partially submerged. Submergence varied between 0.1 h, 0.5 h, and 0.9 h (h = crack height). Low submerged cracks (0.1 h) were not investigated for the width of 0.5mm, because the water level inside of the crack could not be controlled due to high capillary forces. The specifications of the individual cracks are given below. Fig. 7 shows the positioning of the bracket in the model wall. All the apparatus was made of stiff Perspex.

Bracket B: 145mm long and 4o mm across, with spacings from top: 12mm and then every 24mm, location: inserted 25 mm above bed level.
Crack 1 :   600mm long, with six transducer spacings from 100mm, and then every 100mm through to 600mm from the crack opening. Because of only 4 available transducers the 1st, 2nd, 4th and 6th port were used for most experiments.
Crack 2: 115mm long, with 3 transducer spacings 35mm, 65mm and 95mm from the crack opening.
Scanning frequencies were chosen as 10 kHz for the impact pressures on wall and inside of cracks.

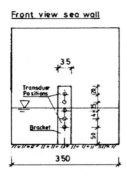

Front view sea wall

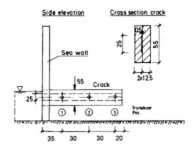

115 mm CRACK

Fig. 6: Front view of sea wall with
Bracket B and transducer

Fig.7: Crack model

EXPERIMENTAL RESULTS

Wave impact pressures

Initially, the impact pressures created by the breaking wave on the vertical wall were recorded. Fig. 8 shows a typical impact pressure record for the transducer positions 1-4 as indicated in Fig. 6; Fig. 9 shows the peak pressures for 2 series of 15 measurements.

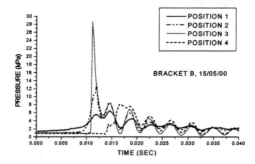

Fig. 8: Typical impact pressure records for positions 1 – 4

From Fig.'s 8 and 9 it can be seen that the impact pressures are of comparatively short duration and that – despite of the single breaker technique – a large variability of peak pressures exists. The pressure magnitude ranges from appr. 5 kPa to 40 kPa.

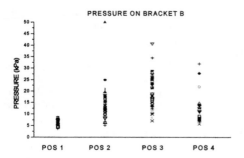

Fig. 9: Maximum peak pressures for each position and 30 waves

Pressure propagation

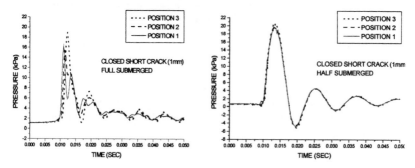

Fig. 10a: P-T-Trace of fully submerged crack   10b: P-T-Trace of half submerged crack

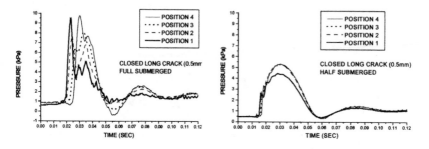

Fig. 10c: P-T-Trace of fully submerged crack   10d: P-T-Trace of half submerged crack

Fig. 10 shows typical pressure-time traces for fully and half submerged cracks.

The different colors represent the different transducer positions along the crack. Pos. 1 refers to the crack entrance and position 3 resp. 4 to the crack end.

Fully submerged crack

The time lapse between the individual pressure traces in Fig. 10a shows the pulse entering into, then propagating through the crack. Positions 1 and 2 show a partition of the initial pulse into two segments (two peaks), caused by the reflection of the pressure pulse at the crack end. Position 3 (black line) shows a higher pressure peak than positions 2 and 3, because of the superposition of the incoming and the reflected pulse. Fig. 10a shows also that the pressure pulse attenuates rapidly and that the propagation velocity is rather slow. It was found to be around 50-70 m/s. Summarizing, it can be said that pressure pulses can enter water filled cracks and travel as a compression pulse which shows *fast* attenuation and *slow* propagation speed (70 m/s). The low popagation velocity can be explained by the 2-phase character of the fluid, consisting of water and 0.2 % air, which increases significantly the compressibility and at the same time reduces the sound speed dramatically.

Partially submerged crack

The half submerged crack, Fig. 10b, shows practically no time lapse between the different transducer positions. This corresponds to a high propagation velocity, around 300 m/s, and low attenuation, giving rise to the following conclusions:

- The pressure pulse can be identified as an elastic wave *travelling in the air* at the speed of sound.
- Partially filled cracks transport wave impact pressures *fast and deep* into the inside of the structure and behind the protective blockwork.
- The high propagation velocity and the low attenuation within partially filled cracks indicates that partially filled cracks may be more dangerous than completely water filled ones.

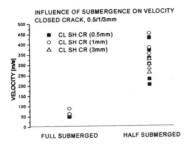

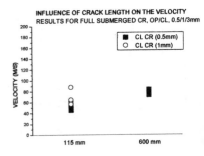

Fig. 11a: Velocity against submergence   11b: Velocity against crack length

(full submerged)

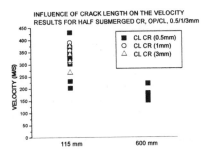

Fig. 11c: Velocity against crack length (half submerged)

Fig. 11 shows the comparison of the velocities for partially and completely submerged cracks. It can be seen that the crack length has no influence on propagation velocity. However, the data acquisition rate of 10000 Hz did not give enough points to accurately determine velocities above 250 m/s. The high variance of velocities concerning the half submerged cracks is probably due to this reason. Most of the velocities are above 200 m/s.

Originally it was thought that pressures propagate fast through water, being the denser medium, and slowly through air, but the experiments showed otherwise: low velocities were found for the full submerged crack whereas high velocities were found for the half submerged case. So it is actually the air, which behaves stiffer than the water, transporting the pressures faster and deeper into the structure.

DESIGN

These conclusions lead to the following structural requirements, which are quite contradictory: ideal structures should either be permeable or impermeable. The advantage of permeable structures is that open passages show high damping and low pressures (open end condition). On the other hand, impermeable structures also have advantages: when no cracks occur, no pressure intrusion can take place. The latest test results of the authors with cracks of varying crack width additionally indicate that smaller cracks are less dangerous, because of their slower pressure propagation velocity and their high attenuation rates.

In reality, completely im- or permeable structures are however not possible. Possible solutions include: mortar filling (or epoxy resin) of cracks could be employed to prevent pressure penetration and to minimise the crack width in *older structures*. Mortar filling is of course no permanent solution, because it will be washed out over time, but it could significantly reduce the susceptibility of the structure against impact pressures for the time being. For the design of *new structures*, permeable designing techniques should be considered. Joint reinforcement (armouring) is also possible to eliminate tension induced crack growth. As far as *rock cliffs* are concerned, mortar filling could be advantageous to prevent pressure intrusion and crack growth.

CONCLUSIONS

A series of experiments was conducted in QUB's wave channel to assess the characteristics of breaking wave impact induced pressure pulse propagation through completely and partially water filled cracks. It was found that within completely submerged cracks pressure pulses generally travel at very low velocities of 45-70 m/s, with fast attenuation of the pressure signal. The slow speed of propagation for completely submerged cracks was attributed to the fact that the water contains small amounts of air (appr. 2%) and thus constitutes a two-phase-medium with very different properties than pure water. This causes increased compressibility and dramatically reduced sound speed. In partially submerged cracks the pressure pulse was found to travel with around 300 m/s, propagating fast and with little attenuation deep into the structure, indicating that partially filled cracks may be more dangerous than completely filled cracks. It appears that certain damage types of blockwork structures and the erosion of rock cliffs can thus at least partially be attributed to wave impact induced pressures travelling into completely or partially air filled cracks.

ACKNOWLEDGEMENTS

The authors would like to gratefully acknowledge the financial support from the German Academic Exchange Board (*Deutscher Akademischer Austauschdienst*, DAAD) and from the UK's Engineering and Physical Science Research Council EPSRC under Grant GR/ M49755.

REFERENCES

Benumof, B.T., Griggs, G.B.,1999, *The Dependence of Seacliff Erosion Rates on Cliff Material Properties and Physical Processes: San Diego County, California*, Shore & Beach Vol. 67, No.4, pp. 29-41

Müller, G., 1997, Propagation of wave impact pressures into water filled cracks, *Proc Inst. Civ. Engineers, Water and Maritime,* **124**, Issue 2, 79-85.

Müller G., Allsop N.W.H., Bruce T., Cooker M., Hull P., Franco L., 1999, Impact pressure propagation into water filled joints/cracks of blockwork breakwaters and sea walls, *Proc. Coastal Structures 99*, Santander/Spain, 479-483.

Rouville, M.A., 1938, *Annales des Ponts et Chaussees*, VII, Commission des Annales a l'Ecole Nationale des Ponts et Chaussees, Paris, p.113

Shield W., 1895, *Principles and Practice of Harbour Construction*, Longmans, Green & Co., London

Sunamura, T. 1992, *Geomorphology of Rocky Coasts*, John Wiley & Sons, Chichester, 302 p.

VIOLENT WAVE OVERTOPPING: DISCHARGE THROW VELOCITIES, TRAJECTORIES AND RESULTING CROWN DECK LOADING

Tom Bruce[1], Leopoldo Franco[2], Paolo Alberti[3], Jonathan Pearson[4] and William Allsop[5]

Abstract: This paper discusses wave processes that happen *after* a wave has impacted on a coastal structure. The paper gives, for the first time, measurements of the throw of overtopping waves at vertical seawalls / breakwaters, including their velocity / trajectory, and the loadings that result when overtopping water lands back onto the deck of the structure.

1. INTRODUCTION

Pressures generated by overtopping waves on the crown deck of a vertical breakwater or seawall have a direct effect upon the serviceability of the structure, on durability of the crown pavement, and on the safety of infrastructure or equipment placed behind the crown wall. Impact of overtopping on people, vehicles (road or rail) constitute significant potential hazards, see Figure 1 which shows overtopping at a major commuter railway line at Saltcoats, on the Scottish coast south of Glasgow. Despite these dangers, and significant investigations on wave impacts on vertical / steep walls, these (post-overtopping) processes and loadings have seen almost no detailed study. One of the few references dates back to Shield (1895), who made a qualitative description of the phenomenon, and linked crown deck loadings with damage to breakwaters at Alderney and Wick.

1 Lecturer, Division of Engineering, University of Edinburgh, King's Buildings, Edinburgh, EH9 3JL, Scotland, UK; Tom.Bruce@ed.ac.uk
2 Professor, Dept Civil Engineering, University of Rome III, via V. Volterra 62, 00146 Rome, Italy
3 Research Engineer, Dept Civil Engineering, University of Rome III, via V. Volterra 62, 00146 Rome, Italy
4 Research Fellow, Division of Engineering, University of Edinburgh, Scotland, UK.
5 Professor (associate), Civil Engineering, University of Sheffield, UK / Technical Director, HR Wallingford.

Recent anecdotal evidence indicates the magnitude of the phenomenon. A steel plate covering an enclosure on the crown deck of the South Breakwater at Peterhead, Scotland was observed to have been "dished" during a storm which gave rise to violent overtopping events. The plate measured ~ 0.7 m x 0.7 m x 12 mm, and was dished by ~ 20 mm, suggesting pressures ~800 kN/m^2. (This example will be re-visited later in the paper.)

Figure 1: Overtopping of seawall at Saltcoats, Scotland (photo: HR Wallingford)

This paper reports results from physical model studies on impact pressures which, although at small scale, provides a first source of information for the estimation of wave impact pressures on the crown deck of a breakwater or seawall by overtopping waves (Sections 2 - 3). Also reported for the first time are measurements of velocity of throw discharge as it leaves the crest of the structure (Sections 4 - 5).

2. WAVE ACTION AT VERTICAL SEAWALLS / BREAKWATERS

2.1 Physical description of problem

The form of interaction of a wave with a vertical or near-vertical seawall or breakwater may be seen to fall into two broad categories: "impacting / impulsive" or "pulsating / reflecting" (Figure 2). The physics of the two cases are quite distinct and, as a result, different prediction models are required. A key outcome of the EC project PROVERBS, see Oumeraci et al (2001), was the development of a "parameter map" or "decision diagram" identifying wave / structure geometry combinations particularly susceptible to breaking wave events. Thus, it has become possible to characterise a wave / structure combination as "impacting" or "pulsating". In terms of wave overtopping phenomena, this distinction relates to green water versus violent / impulsive overtopping events.

Figure 2: Pulsating (left) and impulsive (right) wave action at a wall.

- *Green water* overtopping results from a pulsating wave running up and over the crest
- *Impulsive* or *violent* overtopping results when a wave breaks at or close to the structure, throwing the discharge in a near-vertical direction with some violence.

"Spray" overtopping, wind action on the overtopping discharge and upon the incident waves are, for reason of difficulty in scaling, not included in most laboratory studies. Studies by Ward et al (1994, 1996) suggest that very strong winds may influence run-up and overtopping, but experiments by de Waal et al (1996) indicate that the contribution by spray overtopping is small in relation to realistic design cases.

The magnitude and characteristics of the loadings on the front face of a structure under impacting conditions have been the subject of extensive research under PROVERBS resulting in prediction formulae, see Allsop & Vicinanza (1996), Allsop (2000). Similar categorisation for overtopping events is discussed by Besley et al (1998) and Besley (1999) who applied a parameter h* (Section 4).

2.2 Overtopping discharge throw and resulting forces

A wave colliding with a vertical wall is reflected both horizontally and vertically. If the wave run-up is higher than the crest, the water impinging the vertical wall is projected into the air with a vertical velocity u_{0z} and a horizontal velocity u_{0x} due to the original motion direction. Gravity and air resistance govern the flying jet's subsequent trajectory.

Considering water density ρ to be constant during the impact, and assuming dissipating phenomena and inertial forces to be negligible, the pressure generated by a jet with a velocity v impacting on a flat horizontal surface in stationary conditions would be given by $p = \rho v^2$. Thus if we knew the velocity of the landing water, we could infer a value of the generated pressure. In reality however, this phenomenon is rather more complex and is non-stationary, so cannot be described fully theoretically.

3. CROWN DECK LOADING

3.1 Experimental Setup

Model tests at the Engineering Fluids Mechanics laboratory of the University of Edinburgh used a 20m long, 0.4m wide wave flume with an operating water depth of 0.7 m. The absorbing wave maker is controlled by a PC and generated waves to a JONSWAP spectrum. Each test used approximately 1000 waves. An artificial seabed with a changeable slope modelled different water depths at the structure toe. The sidewalls and the base of the tank are glass, allowing good visualisation of the flows. The model seawall shown in Figure 3 was constructed from clear acrylic (Perspex) and mounted on the impermeable beach. It consisted of a vertical board 27 mm thick and of a horizontal board that can be set at two different heights from the toe, thus reproducing conditions with and

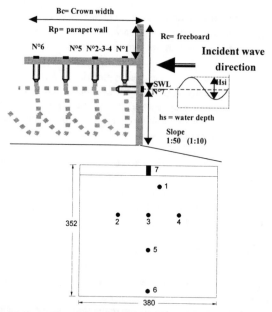

Figure 3: Model seawall and locations of pressure transducers - elevation (upper) and plan view (lower). Dimensions in mm.

without a parapet wall. Pressures were measured with seven transducers connected to a signal-conditioning unit that is itself connected to a PC. This system provides a time history of the pressure and allows sampling at rates up to 2 kHz with little distortion in order to capture the shape of the pressure impulse. The crown slab has six slots designed to hold the pressure transducers, placed with the sensor facing up to measure the pressure generated by the falling water.

3.2 Test Matrix

The experimental programme was intended to cover the widest possible range of conditions and study both low and high crested structures. From visual observation during preliminary tests, it was noticed that a low crested structure allows large green overtopping which presumably produces small impacting pressures but high total loads. Conversely, a high crested structure allows smaller overtopping rates but projects the overtopping jets up to large heights with consequent large falling impacts. It was difficult then to establish which of the two conditions was the most dangerous and also whether the presence of a parapet wall could influence the phenomenon. A total of 80 tests were performed; two front wall heights were employed and, by changing the water depth (h_s), eight different freeboards (R_c) were tested. Table 1 shows the range of structure geometries and random sea conditions tested. Measurements of wave transmission by overtopping are reported separately by Alberti et al (2001).

Table 1: Test conditions for crown deck loading experiments

R_c = 0; 50; 85; 98; 140; 150;155;198 mm	
h_s = 85; 142; 155; 190; 240 mm	60 ≤ H_{si} ≤ 113mm
0 ≤ R_c/H_{si} ≤ 2.51	1.06 ≤ d/H_{si} ≤ 3.36
0.02 ≤ s_p ≤ 0.06	T_p= 1.0, 1.3, 1.7 s
B_c= 350 mm	R_p= 0; 100 mm

As these processes are highly dynamic, rapid data sampling is essential for accurate measurements. Sampling at f_{samp} = 2 kHz was adopted to detect short duration peak pressures, noting that Schmidt (1992) measured peak pressure generated by waves on a vertical wall at f_{samp} = 2 kHz with only 2% reduction over those measured at f_{samp} = 11 kHz.

3.3 Results

Example pressure histories for two transducers are shown in Figure 4. Transducer 1 is immediately behind the crest of the structure; transducer 2 is a further 70 mm behind the crest. Two quite distinct forms are observed. At transducer 1, a very sudden rise to a high peak pressure is observed, consistent with a highly impulsive event – here, the direct impact of the falling overtopping discharge jet. The event looks somewhat different at transducer 3, where a quasi-static load is observed as the overtopping volume which landed at / near transducer 1 flows out over the surface of the deck. This form of trace would also be consistent with a *green water* overtopping event.

The pressure rise time at transducer 1 is exceedingly rapid, with the maximum reached in about 2 ms. Pressure oscillations and multiple peaks are observed after the first maximum, due to the compression of air pockets. This indicates that entrapped air has a role in the phenomenon for low crested structures, although it should be noted that this role may be exaggerated in 2-d wave channel tests such as these. The expansion of entrapped air can generate suction (negative force) which may be significant in, *eg*, dislodging elements in a blockwork wall.

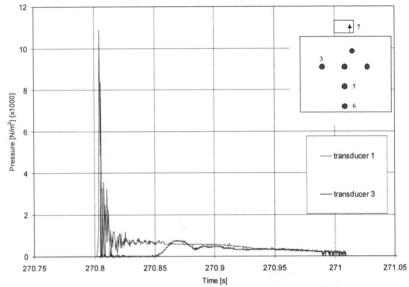

Figure 4: Example pressure histories on the crown deck.

The pressure peaks are evidently highly localised so the exact position of the transducer is of great importance to the measurements. The area of the slab where maximum pressures were measured is within a distance of 1.5 H_{si} from the seaward wall. The higher the crest elevation the closer is this distance to the front wall. For low crested structures (R_c/H_{si} < 0.5), the distance of maximum peak pressures location increases up to 2 H_{si}. Behind this area the pressure is due to the quasi-static load of the water that, after landing, is transformed in a horizontal landward flow, see Takahashi (1994). These observations are again consistent with the distinction between green water and impulsive overtopping events.

To assess their significance, the highest peak pressures measured during each test (one per test) over the crown slab are plotted in Figure 5 against the highest peak pressures measured on the front wall for that test. The two measurements do not necessarily arise from the same tests as no clear relation was found between them. The two populations have a similar order of magnitude, with pressures on the front wall reaching higher values.

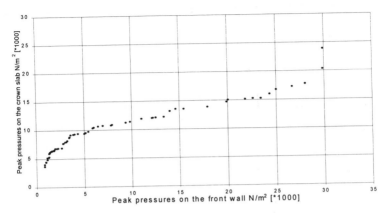

Figure 5: Maximum pressures on the crown deck plotted against (for each test) front face pressure maximum.

As expected, plunging breakers generally generate the highest pressures, at least under high-freeboard conditions. All attempts to correlate pressure maxima with crest elevations or sea state parameters (wave period and steepness) gave little insight, confirming that the process is highly stochastic, as might be expected from the complex overtopping dynamics. However, this should not to be seen as a failure in the investigation: it is well-known that pressure measurements are deeply influenced by boundary conditions and the transient and localised nature of them is frequently test specific. Even pressures generated by nominally identical conditions are only predictable in a statistical sense, see Walkden (1997).

Within the 71 tests on the 1/50 sloped beach the dimensionless pressure P* $=P_{1/250}/\rho g H_{si}$, has a maximum value of P*=17 and a mean value of P*=8. The highest value was obtained with the highest freeboard without the parapet wall, but there is no general trend with variable freeboard. Comparison of Weibull non-exceedance probabilities for pressure on the front wall (based on the number of waves) and on the crown slab (based on the number of overtopping events) in Figure 6 show broadly similar distributions.

Moving from the distribution of pressure events for a single test to seek a more general result, an equation was fitted to the Weibull distributions for the 20 largest events (at any transducer location) from each of the 71 tests. The resulting equation with correlation coefficient of 0.84 is:

$$Ln\left(-Ln\left(1-prob\left(\frac{P}{\rho g H_{si}}\right)\right)\right) = 0.21 + 0.6\,Ln\left(\frac{P}{\rho g H_{si}}\right) - 0.14\frac{R_c}{H_{si}} \qquad (1)$$

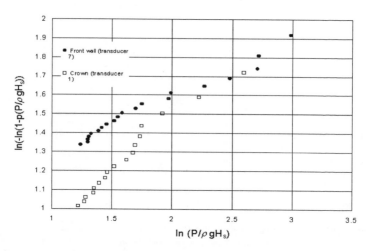

Figure 6: Weibull probability for pressures on front wall and crown deck.

Returning to the example of the dished deck plate; representative figures for the key parameters; $H_s \sim 5m$, $T_p \sim 8s$, two-hour storm peak giving ~ 1000 waves; used in equation (1) give $p_{max} \sim 850$ kN/m^2 – in broad agreement with the order of magnitude inferred from the damage observation.

Figure 7: Frames from video used in measurement of throw velocities.

4. OVERTOPPING DISCHARGE THROW VELOCITY

As noted in Section 2.2, if the velocity and trajectory of the overtopping discharge "jet" was known, it could give an estimator for the magnitude and location of the pressures. The uncertainties in this process make direct measurements, as reported above, more reliable. There is however another strong motivation for the direct measurement of velocities and trajectories – the assessment of immediate hazard to pedestrians and vehicles in the path of an overtopping event.

The vertical velocity of the thrown discharge was measured directly from video records, see Figure 7, for each of the largest 20 discharge events recorded from a 1000-

wave sequence. This was done for 14 1000-wave tests over a range of wave and structure parameters, from which Weibull distributions of throw velocities were plotted (Figure 7).

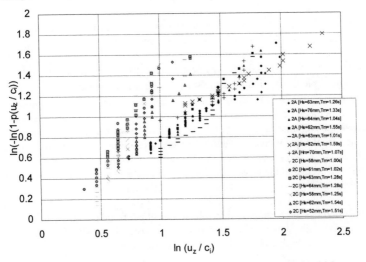

Figure 8: Weibull distribution of throw velocities associated with 20 largest discharge events; 14 tests.

It is clear that the throw velocity distributions form two distinct "populations". For the largest events (upper points on the chart), the group to the right reach a given non-exceedance probability level at velocities significantly greater than those in the left group. This observation is consistent with the qualitative distinction drawn between *green water* and *impulsive* overtopping events.

The dependence of the throw velocity upon wave parameters was investigated. For this purpose, a single measure of vertical throw velocity, u_z, was adopted to characterise each test condition. This measure was chosen as $u_{z,\ 1/25}$ – the mean of the largest 4% of throw velocities for that 1000-wave sequence, based upon the recorded number of overtopping events, $N_{OW,\ actual}$. The 1/25 level was chosen in the expectation that it would provide a more stable (arbitrary) measure than a higher non-exceedance level, eg 1% or 1/100. Comparison of $u_{z,\ 1/25}$ with the usual inshore wave parameters (zero-crossing period, significant wave height and wave steepness) gave little indicator of a possible throw velocity predictor.

Physically, what is required is a predictor which characterises the violence of wave action at the wall for a particular sea state / structure combination. Besley (1999), after Allsop *et al* (1995) presents the "h*" parameter for such a purpose:

$$h^* \equiv \left(\frac{h_s}{H_{si}}\right)\left(\frac{2\pi h_s}{gT_{mi}^2}\right) \qquad (2)$$

where h_s is the water depth at the structure. Besley suggests that *impulsive* wave action is significant for $h^* < 0.3$; *pulsating* conditions dominating for $h^* > 0.3$.

Then throw velocity $u_{z,\ 1/25}$, can be non-dimensionalised by inshore wave celerity, $c_I = (gh_s)^{0.5}$ plotted in Figure 9 against h^*. The result is striking: for $h^* > \sim 0.15$, the dimensionless throw velocity is fairly constant at $\sim 2 - 2.5$ times the inshore wave celerity, but as h^* drops below 0.15, the dimensionless throw velocity is observed to increase quite dramatically to 5 or 6 times the inshore wave celerity. This observation is in line with the qualitative distinction between *green water* and *impulsive* overtopping regimes. For $h^* > 0.15$, the wave is running up and over the structure at a velocity of the same order as that of the wave crest. For $h^* < 0.15$, the overtopping discharge is thrown violently up at speeds greatly in excess of even the wave crest velocity.

The findings here suggest that whilst transition from *impulsive* to *green water* overtopping may start at h^* around 0.3, as suggested by Besley (1999), the process becomes significantly more violent as h^* reduces to 0.15 or less.

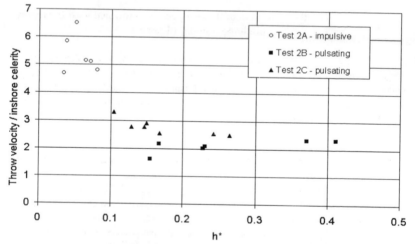

Figure 9: Dimensionless throw velocity plotted against h^*.

5. THROW TRAJECTORY

The further key element in relating overtopping discharge to actual hazards is the prediction of the loaction of the hazard. The crown deck loading measurements reported in Section 3 give some indication. Equally, knowledge of the initial throw velocity (speed *and* direction), combined with a model of the subsequent travel could give useful insights.

However, a direct approach has some merit, at least as a means of calibrating a model, which could subsequently be developed to quantify *eg* wind effects. Here, we report a first attempt at automated measurement of the trajectory of the thrown discharge, at the large wave flume at Universitat Polytechnica Catalunya, Barcelona, Spain.

The "throw board" (Figure 10, left, and visible above the left-hand end of the wall's crest in Figure 14, right) consisted of a 64 individual detector elements arranged in an 8 x 8 array at 350mm pitch. Each detector consists simply of a pair of conductors, making a switch which is closed by the presence of water. The state of all 64 switches is converted to 64 digital (on/off) signals and these signals sampled by a PC at 300 Hz. The board itself is mounted with its top and rear tilted very slightly in, in order to minimise interference to the flow of the discharge as it moves over the board.

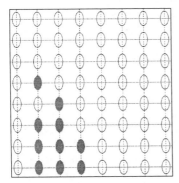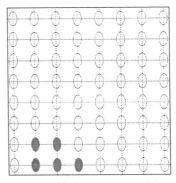

Figure 10: The trajectory detector board (left) and (right) in place above the test structure in the large flume at UPC, Barcleona.

Figure 11: "Snapshot" examples of output from overtopping trajectory device.

A real-time display of the status of the detectors was implemented to enable qualitative assessment of the operation of the devise. A macro written in "Excel" enables a visual "playback" of events from the stored data file. Visualised data during an impulsive overtopping event (left) and green water event (right) are shown in Figure 11. Trajectory data have been obatined from 30 tests at UPC, Barcelona, each test of 1000 irregular (JONSWAP) waves. Systematic analysis of these data is in progress at the time of writing.

6. CONCLUSIONS

Wave impact pressures have been measured on the crown deck of a small-scale breakwater model for 71 test conditions, with each test consisting of 1000 waves of a JONSWAP spectrum. Over all tests, the pressure at 1/250 level may be described by:

$$2 < \frac{p_{1/250}}{\rho g H_{si}} < 17 \quad \text{with a mean value of 8.}$$

For high-crested structures ($R_c / H_{si} > 0.5$) the pressure maxima were observed typically to fall within a distance of 1.5 H_{si} behind the structure crest. For lower-crested structures ($R_c / H_{si} < 0.5$), this distance was observed to increase to ~ 2 H_{si} .

An approximate Weibull distribution has been fitted to all data on crown deck pressures.

It has been observed that, for a given sea state / structure geometry condition. crown deck pressure maxima are smaller than *but of the same order of magnitude as* the pressure maxima recorded on the front face of the structure.

For wave / structure conditions giving a large number of impulsive events ($h^* <$ 0.15), it has been observed that the velocity of the thrown overtopping discharge is 5 - 7 times the inshore wave celerity. For less impulsive regimes ($h^* > 0.15$), throw velocities ~ 2 - 2.5 times the inshore wave celerity are observed.

Measurements of the trajectories followed by overtopping discharge have been carried out at large scale. Full analysis and identification to actual hazard are underway. Tests at prototype sites may be required if scale and wind effects are to be fully investigated.

ACKNOWLEDGEMENTS

Paolo Alberti's work in Edinburgh was supported by the "Blockwork Coastal Structures Network" funded by the UK EPSRC (under GR/M00893). Work on throw velocities was carried out under the UK "Violent Overtopping by Waves at Seawalls (VOWS)" project, funded by the EPSRC (GR/M42312). Access to the large wave channel at UPC Barcelona was made possible by the EC-funded "Hydralab" project under the Transnational Access to Major Research Infrastructure (TAMRI) scheme. These tests were further supported by the UK EPSRC under GR/R42306. The support of the UPC team under Prof Sanchex Arcilla and Javier Pineda is gratefully acknowledged, with special thanks for the efforts of Xavi Gironella, Quim Sospedra and Oscar Galego.

REFERENCES

Alberti P., Bruce, T. and Franco L. 2001. Wave transmission behind vertical walls due to overtopping, *Proc Coastlines, structures and breakwaters 2001*, Institution of Civil Engineers, London.

Allsop N.W.H. 2000. Wave forces on vertical and composite walls. Chapter 4 in *Handbook of Coastal Engineering*, pages 4.1-4.47, Editor J. Herbich, ISBN 0-07-134402-0, publn. McGraw-Hill, New York.

Allsop N.W.H., Besley P. and Madurini L. 1995. Overtopping performance of vertical walls and composite breakwaters, seawalls and low reflection alternatives. Paper 4.7 in MAST II *MCS Project Final Report*, publn University of Hannover.

Allsop N.W.H. and Vicinanza D. 1996. Wave impact loadings on vertical breakwaters: development of new prediction formulae. *Proc. 11th International Harbour Congress*, Antwerpen, Belgium.

Besley P. 1999. Overtopping of seawalls – design and assessment manual. R & D Technical Report W 178, ISBN 1 85705 069 X, *Environment Agency*, Bristol.

Besley P.B., Stewart T, and Allsop N.W.H. 1998. Overtopping of vertical structures: new methods to account for shallow water conditions. *Proc. Int. Conf. on Coastlines, Structures & Breakwaters '98*, pp46-57, ICE / Thomas Telford, London.

Oumeraci, H., Kortenhaus, A., Allsop, N.W.H., de Groot, M., Crouch, R., Vrijling, H. and Voortman, H. (2001) *"Probabilistic Design Tools for Vertical Breakwaters"* publn. A.A.Balkema, ISBN 90 5809 248 8.

Schmidt R., Oumeraci H. and Partenscky H-W. 1992. Impact loads induced by plunging breakers on vertical structures. *Proc. 23rd Int. Conf. on Coastal Eng, 2*, pp1545-1558 (ASCE)

Shield, W., 1885: *Principle and Practice of Harbour Construction* Longmans' *Civil Engineering Series.* London.

Takahashi S., 1994: An Investigation of the Wave Forces Acting on Breakwater Handrails. *Proc. 23rd Int. Conf. on Coastal Eng, 1*, pp1046-1060 (ASCE)

De Waal, J.P. Tonjes, P. and van der Meer, J.W. 1996. Overtopping of sea defences. *Proc 25th Int. Conf. Coastal Eng*. pp2216-2229 (ASCE)

Walkden, M.J., Crawford, A.R., Hewson, P.J. and Bullock, G.N. 1997. Scaling, *Proc. Task 1 Edinburgh technical workshop of PROVERBS project* , EC DG XII MAS3-CT95-0041

Ward, D.L., Wibner, C.G., Zhang, J. and Edge, B. 1994. Wind effects on runup and overtopping. *Proc 24th Int. Conf. Coastal Eng.*, pp1687-1699, (ASCE)

Ward, D.L., Zhang, J., Wibner, C.G. and Cinotto, C.M. 1996. Wind effects on runup and overtopping of coastal structures. *Proc 25th Int. Conf. Coastal Eng.* pp2206-2216 (ASCE)

PREDICTION OF WAVE OVERTOPPING AT STEEP SEAWALLS – VARIABILITIES AND UNCERTAINTIES

Jonathan Pearson[1], Tom Bruce[2] and William Allsop[3]

Abstract: This paper presents results from collaborative research on Violent Overtopping of Waves at Seawalls (VOWS) describing overtopping performance and processes at steep seawalls. VOWS results have been used to develop / improve prediction methods for mean overtopping discharges, wave-by-wave overtopping volumes, and overtopping throw velocities. This paper discusses variabilities and uncertainties inherent in the overtopping processes, and presents example data on uncertainties to be used in probabilistic design or hazard evaluations. The dependency of these parameter uncertainties upon the proportion of wave overtopping waves is highlighted.

1. INTRODUCTION

In rural areas, coastal defences are often formed as simple embankments. In most instances where overtopping of an embankment seawall is being assessed, the responses of most interest are time-averaged ones, such as mean overtopping discharge over 500-1000 waves, or the total overtopping volume over (the upper part of) a tide.

The situation may be quite different for urban and harbour seawalls (and some breakwaters). In these cases, consideration of personal safety and potential damage to property drives the design and will require the proportion of overtopping waves (N_{ow} / N_w) within the design storms to be very low e.g. <1%. This places a much greater requirement on the accuracy of the design tool employed, not only for the number of overtopping waves (N_{ow}), but also for the prediction of individual maximum events. A further

1 Research Fellow, Division of Engineering, University of Edinburgh, King's Buildings, Edinburgh, EH9 3JL, UK. J.Pearson@ed.ac.uk
2 Lecturer, Division of Engineering, University of Edinburgh, King's Buildings, Edinburgh, EH9 3JL, UK. Tom.Bruce@ed.ac.uk
3 Professor (associate), Civil Engineering, University of Sheffield, Technical Director, HR Wallingford, Wallingford, OX10 8BA UK. W.Allsop@shef.ac.uk

complication is that these structures are often formed by steep or vertical walls, the description of whose overtopping is more complex than for sloping embankments, for example, wave breaking may occur at the structure.

Uncertainty is a general concept which refers to the condition of being unsure about something. In coastal engineering problems, uncertainty can be defined in a number of ways, such as statistical uncertainties may be defined as the estimated amount or percentage by which an observed or calculated value may differ from the true value, or knowledge uncertainty may occur due to a lack of knowledge of all the causes and effects in the physical processes.

Quantitative estimates of uncertainties in predicted overtopping parameters, eg Q_{bar}, N_{ow} are required as a component of probabilistic and hazard assessment. This paper examines the source of these variabilities and explains the way in which the level of variability / uncertainties varies with the structure type, eg high or low allowable N_{ow}.

2. PREVIOUS WORK

2.1 Mean overtopping discharges

Most design methods for seawalls hitherto have concentrated on predicting a crest level or other aspect of wall geometry to give a (tolerable) mean discharge or overtopping volume over a storm event. Tolerable discharges have been suggested by Owen (1980), Franco et al (1994) and Besley (1999). Prediction of mean overtopping discharge are generally based on empirical formulae fitted to laboratory measurements, see for example Goda et al (1975), Owen (1982), De Waal et al (1996), Hedges & Reis (1998), Van der Meer et al (1998). These formulae mainly assume non-breaking (or "pulsating") wave conditions, but studies by Besley et al (1998) and Van der Meer et al (1998) separate non-breaking and breaking (or "impulsive") processes. Besley et al (1998) indicate configurations of vertical / composite walls for which impulsive breaking may occur, and demonstrate that simple methods may under-estimate overtopping under impact conditions.

Allsop et al (1995) demonstrated that the overtopping processes were strongly influenced by the form of the incident waves. When waves are small compared to water depth, the waves impinging on a vertical / composite wall are generally reflected back. If the waves are large relative to water depth, then they can break onto the structure, leading to significantly more abrupt overtopping characteristics. These observations led to formulation of a wave breaking parameter, $h*$, given by:

$$h* = \frac{h}{H_s} \left(\frac{2\pi h}{gT^2} \right)$$

(1)

Allsop *et al* (1995) noted that reflecting or pulsating waves predominate when $h^* > 0.3$, and that impacting waves where more likely to occur when $h^* \leq 0.3$. Under impacting wave conditions, new dimensionless discharge (Q_h) and freeboard parameters (R_h), incorporating h^* were established, and were given by:

$$Q_h = Q / (gh^3)0.5 / h^{*2} \qquad (2)$$
$$R_h = (R_c / H_s) h^* \qquad (3)$$

where Q is the mean overtopping discharge per metre run. Besley (1999) utilised the empirical studies of Allsop *et al* (1995) and de Waal *et al* (1992) for which $h^* \leq 0.3$ to derive the following relationship for vertical walls.

$$Q_h = 1.37 \times 10^{-4} R_h^{-3.24} \qquad (4)$$

2.2 Peak overtopping discharges

In terms of tolerable safety limits, a much more useful parameter is the maximum individual volume. Allsop et al (1995) followed a similar analytical procedure to Franco et al (1994), and demonstrated that peak overtopping rates under impulsive conditions could be estimated. Allsop et al (1995) showed that for a number of experimental investigations on vertical seawalls, the statistical distributions of individual wave by wave overtopping were similar. Besley (1999) adopted these observations and suggested that by given the number of overtopping events for a particular storm duration, the peak individual overtopping discharge could be estimated.

For vertical structures under impulsive conditions ($h^* \leq 0.3$), Allsop et al (1995) demonstrated empirically that the that the proportion of overtopping waves could be given by:

$$N_{ow}/N_w = 0.031 R_h^{-0.99} \qquad (5)$$

where N_{ow} is the number of overtopping waves and N_w is the number of waves. Franco *et al* (1994) and subsequent analysis by Besley (1999) demonstrated that the wave by wave individual overtopping volumes could be described by a two parameter Weibull probability distribution. Besley (1999) suggested that the expected maximum individual overtopping volume, V_{max}, in a sequence of N_{ow} overtopping waves could be described by:

$$V_{max} = a (\ln(N_{ow}))^{1/b} \qquad (6)$$

which for vertical seawalls subjected to impacting wave conditions, the Weibull scale parameter, $a = 0.92V_{bar}$, and the Weibull shape parameter, $b = 0.85$.

3. EXPERIMENTAL STUDY

Measurements have been carried out in the 2-dimensional wave flume in the Division of Engineering at the University of Edinburgh. The flume is 20m long, 0.4m wide and has an operating water depth of 0.7m. The flume is equipped with an absorbing flap-type wave maker. Experimental investigations have concentrated on a vertical seawall, in which a series of tests were undertaken to assess the repeatability, variabilities and uncertainties of the overtopping processes. For the vertical structure, a JONSWAP pseudo-random sea state spectra of varying length (128-4096 seconds, or roughly 100 – 4000 waves) was generated. For each of these conditions, approximately 10 repeat tests of different start phases have been performed. A summary of test conditions in shown in Table 1. Wave-by-wave overtopping discharge measurements have been obtained using a receiving container located behind the structure crest suspended from a load cell, the discharge measurement system was also tested to ensure consistent and repeatable results, therefore a number of preliminary experiments were performed in which known volumes of water were poured into the receiving container.

Table 1. Summary of repeatability tests *

Sequence length	Seed modification factor		Total number of tests
	Mod=3	Mod=1,2....10	
128	10	1	20
256	10	1	20
512	10	1	20
1024	10	1	20
4096	8	-	8

* Wave condition = H_s = 0.067m, T_m = 1.04s, R_c = 0.15m, h_s = 0.09m, h.= 0.071

The objective of the 2-d tests was to extend existing prediction methods into regimes in which wave breaking at or onto the structure is significant. The matrix of conditions [significant wave height at the toe of the structure (H_s), peak period (T_p), water depth at structure (h) and crest freeboard (R_c)] is shown in Table 2. Further experimental studies were undertaken for a range of wave and water level configurations, a matrix of test conditions, together with more detailed results are presented in Bruce et al (2001).

Table 2. Summary of test conditions

Structure Configuration	Test Series	Configuration	Nominal wave period T_s [s]	Significant wave height H_{si} [mm]
Vertical	2A [1:10 beach]	R_c = 150mm h = 90mm	1.0	63, 67, 69, 70, 81
			1.33	63, 70, 76, 82
			1.6	62, 77, 82
	2B [1:10 beach]	R_c = 129mm h = 247mm	1.0	71, 79, 86, 92, 100
			1.33	81,93, 94, 100, 102
			1.6	71, 74, 89, 97, 105
	2C [1:50 beach]	R_c = 85mm h = 155mm	1.0	46, 55, 58, 61
			1.33	39, 58, 63, 64
			1.6	34, 44, 52, 62
	2D [1:50 beach]	R_c = 113mm h = 127mm	1.0	55, 57, 58
			1.33	53, 57, 58
			1.6	30, 40, 59, 62

* Sequence length = 1024seconds

4. RESULTS

4.1 System accuracy

Prior to undertaking any tests the accuracy of the overtopping measurement system was tested. Two tests, each of nine simulated overtopping events were performed in which known volumes of water were 'thrown' into the measurement container. The resulting data was then passed through an algorithm to identify and quantify individual overtopping events. Figure 1 shows sample output data, the lower trace is event detector, and the upper traces are raw and processed load cell data. Software finds events from lower trace (thresholded to, eg here, 2.5V). Load cell output between events divided into two, and discharge measurement based upon second half (allowing for settling after initial discharge into container).

Figure 1: Simulated series of pre-measured discharges. Lower trace is "event detector". Upper traces are raw and processed load cell output. Table shows corresponding results of 18 simulated events, with pre-measured volumes and volumes given by discharge measurement system / software.

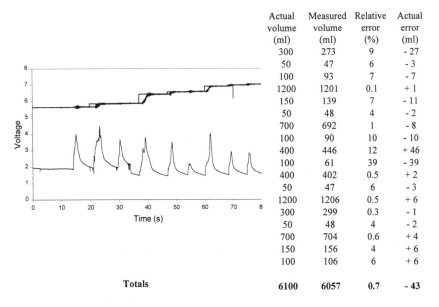

Actual volume (ml)	Measured volume (ml)	Relative error (%)	Actual error (ml)
300	273	9	- 27
50	47	6	- 3
100	93	7	- 7
1200	1201	0.1	+ 1
150	139	7	- 11
50	48	4	- 2
700	692	1	- 8
100	90	10	- 10
400	446	12	+ 46
100	61	39	- 39
400	402	0.5	+ 2
50	47	6	- 3
1200	1206	0.5	+ 6
300	299	0.3	- 1
50	48	4	- 2
700	704	0.6	+ 4
150	156	4	+ 6
100	106	6	+ 6
Totals			
6100	**6057**	**0.7**	**- 43**

From the results, a difference of 43ml (=0.7%) in total volume between the measured and actual discharge is observed. This indicates there is no significant systematic error associated with the measurement system. During the design of the test matrix, the maximum individual overtopping volume for all the tests had a predicted value of 2000ml. It was therefore concluded that any errors in the measurement system were negligible.

4.2 Mean Overtopping Discharge

Measurements of mean overtopping discharge (described here by dimensionless discharge Q_h) for the simple vertical wall show agreement with Besley *et al*'s (1998) method over the full range of test conditions studied, typically to within a factor of two, see Figure 2. The largest proportion of impulsive events was recorded under test conditions "2A", see Table 2. Adopting the procedure of Allsop *et al.* (1995), the best-fit trend line from this study was found to be

$$Q_h = 1.55 \times 10^{-4} \, R_h^{-3.03} \qquad (7)$$

with a corresponding least squares regression $R^2 = 0.92$. The results of this study when compared with the results from Besley *et al* (1998) show very similar characteristics, as shown by the two trend lines in Figure 2.

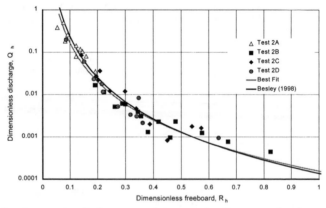

Figure 2: **Overtopping discharge on a plain vertical wall, compared to the prediction of Besley *et al* (1998).**

4.3 Number of overtopping waves/ Peak overtopping rates

A parameter which is vital in predicting peak overtopping volumes and statistics of wave-by-wave overtopping volumes is the number or proportion of overtopping waves (N_{ow}) – see Equation 5. The variation of N_{ow} with freeboard for vertical wall, as quantified from these tests, is shown in Figure 3. It is noticeable that the results of N_{ow} / N_w deviate significantly from the prediction by Equation 5, which leads to uncertainty in adopting previous prediction methods for individual maxima.

Careful analysis of the problem in defining N_{ow} precisely, reduces to a particular difficulty the exact definition of an "overtopping event". Clearly such events cannot constitute all occasions when some water passes over the wall, as that would increase with each improvement in measurement precision. Any definition must therefore generate a limit of only one overtopping event per incident wave. A method to overcome these difficulties by defining a consistent threshold level before an overtopping event is included is under development. Only then can a useful improvement be made in the prediction of N_{ow}.

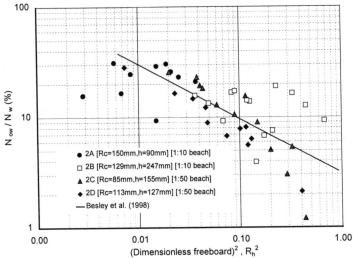

Figure 3: Percentage of waves overtopping a vertical wall, compared to Besley *et al* (1998).

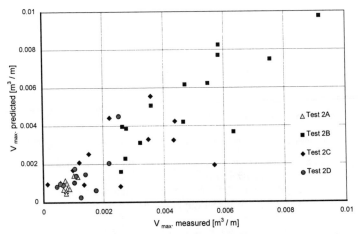

Figure 4: Comparison of measured and predicted maximum overtopping volumes

Maximum individual overtopping volumes predicted using methods by Besley (1998) are compared in Figure 4 with those measured. Despite the uncertainties in defining precisely what constitutes an overtopping event, it would appear that maxima may be slightly under-predicted. Nevertheless the results indicate a reasonable correlation between predicted and measured volumes.

4.4 Variabilities and repeatability

In order to asses the variabilities and repeatability on the overtopping characteristics, ten nominally identical repeats of each of 128s, 256s, 512s and 1024s were run. Eight nominally identical repeats of a 4096s test were also run, with equipment failure preventing the full set of ten repeats being completed. (ie, a total of 48 sequences). Figures 5 and 6 show the effect of the increase in sequence length upon four measures of overtopping: The error bars on Figures 5 & 6 represent ± one standard deviation

From Figure 5 it can be seen that, the mean discharge, Q is reasonably well settled from 512s tests and above, however it is noticeable that from the conditions tested that the mean overtopping volume varies with the selection of test sequence length, however similar characteristics were also observed in the measured wave heights, it can be therefore concluded that this variation is a function of the wave generating software rather then the overtopping characteristics. The maximum individual discharge increases slightly with the number of waves, which is as expected as the likelihood increases with sequence length.

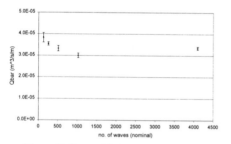

Figure 5: Q_{bar} with sequence length

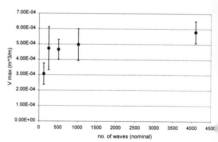

Figure 6: V_{max} with sequence length

4.5 Consistency between tests nominally identical except for JONSWAP seed

Additional experiments were performed whereby the tests reported in section 4.4, were repeated *except* that the JONSWAP seed is varied for each. Figure 7 characterises the effect upon the spread of data by looking at the way in which the standard deviation varies with sequence length for fixed and varying seeds.

From Figure 7, it can be seen that as expected the standard deviations of Q_{bar} have the lowest deviations: < 5% for same-seed, and ~ 8% for varying-seed, and the largest scatter is in V_{max} which is typically ~ 15 - 20%. For 1024s sequence (most stable data presented on this graph), there is an indication that the open symbols lie above the closed ones, ie, that scatter is increased with varying of seed. It can therefore be suggested that a typical scatter goes up from 5 - 8% for fixed seed to 10 - 15% for varying seed.

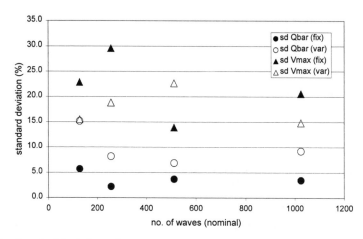

Figure 7: **Standard deviations of overtopping measured for repeated tests. For each of four sequence lengths, std. Devs. are plotted for 10 same-seed and 10 varying-seed test groups. Filled symbols = fixed-seed, open symbols = varying-seed.**

4.6 Variability and uncertainties due to definition of events of N_{ow}

Discussions on variability has so far concentrated on the variabilities due to a single wave / structure configuration, this section discusses the variabilities over the full range of tested conditions where N_{ow} varies. For each structure configuration [2A – 2D], 3 conditions have been further analysed whereby the 1000 wave test sequences have been split into ten (102s), five (204s) & two (512s) sections. The variabilities from one such test is shown in Figure 8 (2D-1).

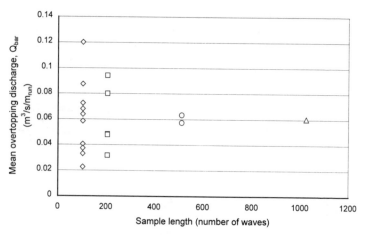

Figure 8:   Variability of mean overtopping discharge of 1000 wave test split into various equal sections (Test shown 2D-1)

For a sub-set of core tests, the variability in overtopping with respect to sequence length has been determined. The variability in mean overtopping discharge rate for different test sequence lengths is shown in Figure 9, it is of note that for cases with lower overtopping events the variability in Q_{bar} is higher.

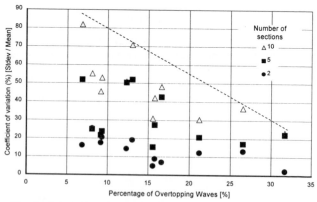

Figure 9:   Standard deviation of mean overtopping discharge of 1000 wave test sequences split into various equal sections

Although showing some scatter, it is noticeable from Figure 9, that the variability for the lower overtopping cases ($N_{ow} \sim 6\%$) are typically 3 times larger than those of the higher cases ($N_{ow} \sim 30\%$), as indicated by the upper envelope (dotted line), further work is needed to quantify the variabilities when the test sequence length is significantly extended. Only then will it be possible to give qualitative confidence limits on overtopping measurements, such as those given in Figure 2.

5. CONCLUSIONS

The prediction method of Besley *et al* (1998) for mean overtopping discharge may be used under conditions when impulsive wave action is significant or dominant, with mean discharges remaining well-predicted (typically within a factor of two). From the tested conditions, it can be seen that the results of N_{ow} deviate from the prediction of Besley (1999), but nevertheless when these results are used to compare measured and predicted maximum overtopping volumes, the results show good agreement.

For mean overtopping discharge rates, when the same wave condition is run, standard deviations of Q_{bar} have the lowest deviations: < 5% for same-seed, and ~ 8% for varying-seed. For peak overtopping rates, the scatter in V_{max} is increased with varying of seed, with a typical scatter going up from 5 - 8% for fixed seed to 10 - 15% for varying seed.

The variability in mean overtopping discharge rates for different test sequence lengths increases as the number of overtopping events reduces. The coefficient of variation of Q_{bar} increases by a factor of approximately 3 with a reduction from 30% to 10% of overtopping waves.

ACKNOWLEDGEMENTS

This work is funded by the UK EPSRC (GR/M42312), and supported by the VOWS Management Committee also including members from Manchester Metropolitan University (Derek Causon, David Ingram, Clive Mingham), Posford-Haskoning (Dick Thomas, Keming Hu), Bullen & Co (Mark Breen, Dominic Hames), HR Wallingford (Philip Besley), all of whose input and support is gratefully acknowledged. The VOWS project is also particularly pleased to acknowledge guidance and helpful supervision of their collaborative work of the EPSRC CEWE Project Manager, Michael Owen.

REFERENCES

Allsop N.W.H., Besley P. & Madurini L. (1995). "Overtopping performance of vertical walls and composite breakwaters, seawalls and low reflection alternatives" Paper 4.7 in MCS Final Report, publn University of Hannover.

Besley P. (1999) "Overtopping of seawalls – design and assessment manual " R & D Technical Report W 178, ISBN 1 85705 069 X, Environment Agency, Bristol.

Besley P.B., Stewart T, & Allsop N.W.H. (1998) "Overtopping of vertical structures: new methods to account for shallow water conditions" Proceedings of Int. Conf. on Coastlines, Structures & Breakwaters '98, Institution of Civil Engineers, pp46-57, publn. Thomas Telford, London.

Bruce, T, Allsop, N.W.H. & Pearson, J. (2001) "Violent overtopping of seawalls – extended prediction methods" Proc. "Coastlines, Seawalls and Breakwaters '01" ICE, publn Thomas Telford, London.

Franco, L., de Gerloni, M. & van der Meer, J.W. (1994), "Wave overtopping on vertical and composite breakwaters", Proc 24th Int. Conf. Coastal Eng., Kobe, publn. ASCE, New York.

Goda, Y, Kishira, Y, and Kamiyama, Y. (1975) 'Laboratory investigation on the overtopping rates of seawalls by irregular waves'. Ports and Harbour Research Institute, Vol 14, No. 4, pp 3-44, PHRI, Yokosuka.

Hedges, T.S. & Reis, M.T. (1998), "Random wave overtopping of simple sea walls: a new regression model", Proc. Instn. Civil Engrs. Water, Maritime & Energy, Volume 130, March 1998, Thomas Telford, London.

Meer van der J.W., Tonjes P. & de Waal J.P. (1998) "A code for dike height design and examination " Proceedings of Coastlines, Structures & Breakwaters '98, Int. Conf. at Institution of Civil Engineers, March 1998, pp 5-21, publn. Thomas Telford, London

Owen, M.W. (1980), "Design of seawalls allowing for wave overtopping" HR Report EX 924, June 1980, Hydraulics Research, Wallingford.
Owen, M.W. (1982), "Overtopping of Sea Defences", Proc. Intl. Conf. On Hydraulic Modelling of Civil Eng. Structures, Coventry, pp469-480, BHRA, Bedford.

Waal, de J.P., & van der Meer, J.W. (1992), "Wave run-up and overtopping on coastal structures" Proc 23rd Int. Conf. Coastal Eng., pp1758-1771, publn. ASCE, New York.

Waal, de J.P., Tonjes, P. & van der Meer, J.W. (1996), "Overtopping of sea defences" Proc 25rd Int. Conf. Coastal Eng., pp2216-2229, Orlando, publn. ASCE, New York.

WAVELENGTH OF OCEAN WAVES AND SURF BEAT AT DUCK FROM ARRAY MEASUREMENTS

A.A. Fernandes[1], H.B. Menon[2], Y.V.B. Sarma[3], Pankti D. Jog[4] and A.M. Almeida[5]

Abstract: Wavelength of ocean waves and surf beat (infra gravity waves) has for the first time been computed as a function of frequency from different combinations of non-collinear 3-gauge arrays. Data at the 15-gauge polygonal array at 8m depth at Duck, North Carolina, USA, has been used. The method used is an extension of the 3-gauge method of Esteva 1976 & 1977 for computing wave direction. Wavelength was also computed using the Singular Value Decomposition (SVD) method of Herbers et al. 1995 from (i) κ_{rms} and (ii) κ_x, κ_y. The wavelength computed from the array measurements, using both the 3-gauge as well as the SVD(κ_x, κ_y) method, is nearly identical to the wavelength predicted by the well known dispersion relation for ocean waves (DROW), viz., $(2\pi f)^2 = \kappa g \tanh \kappa d$, in the ocean wave regime (T ≤ 11.64s spectral period); and is approximately thrice the predicted value in the surf beat regime (T ≥ 32.00s). The implication of wavelength of surf beat at Duck being much larger than that predicted by DROW, is that methods for determining the directional spectrum which use DROW can erroneously indicate a surf beat direction much closer (≅ 30° closer) to the shore normal than it actually is in the field. For example the variational method of Herbers et al. 1995 and the 2-gauge array method of Borgman 1974 are unfortunately both tainted with this error due to use of DROW.

INTRODUCTION

The problem of computing the wavelength of ocean waves as a function of frequency is associated with the problem of computing the wave directional spectrum, specifically

---

1 Scientist, Physical Oceanography Division, National Institute of Oceanography, Dona Paula, Goa 403 004, India, fernan@csnio.ren.nic.in
2 Reader, Department of Marine Science, Goa University, Taleigao Plateau, Goa 403 203, India, hbmenon@unigoa.ernet.in
3 Scientist, NIO   4 Project Assistant, NIO   5 Technical Officer, NIO

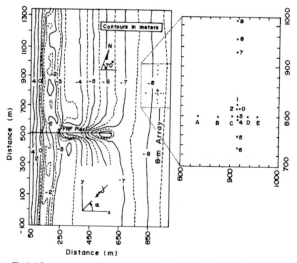

Fig 1. 15-gauge array at 8m depth at the CERC Field Research Facility at Duck, North Carolina, USA. (From Fernandes et al. 2000).

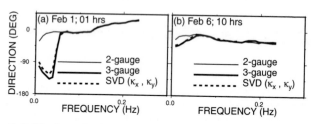

Fig 2. "Mean" wave direction computed using the 2-gauge method; 3-gauge method; and the SVD(κ_x, κ_y) method.

of computing the "mean" wave direction as a function of frequency, viz., Longuet-Higgins et al. 1963 from buoy measurements, and, Herbers and Guza 1992 from array measurements.

The present work is a fallout of our work reported in Fernandes 1999 and Fernandes et al. 2000 (referred hereafter as FERN99-2000) in consistently and accurately determining the "mean" wave direction at the 15-gauge polygonal array (Fig 1) at 8m depth at Duck, North Carolina, USA , as a function of frequency using the 2-gauge method of Borgman 1974, which suffers from an ambiguity in wave direction within a mirror symmetry; and the 3-gauge method of Esteva 1976 & 1977, which gives unambiguous wave direction.

It is generally considered as a sign of earnestness to include in a PhD thesis the same results both in form of figures as well as tables in profusion. It happened that the figures in the PhD thesis of Fernandes 1999 consistently and conspicuously showed (Fig 2) that the "mean" wave direction from 10 redundant 2-gauge arrays and from 15 redundant 3-gauge arrays was identical in the ocean wave regime (4.27s<T<14.22s spectral period);

and that the 2-gauge array estimates of wave direction were closer to the shore normal than the 3-gauge estimates by $\cong 30°$ in the surf beat (infra gravity wave) regime (T=64.00, 42.67 & 32.00s).

In our endeavour to explain the observed difference of $\cong 30°$ between the estimates of wave direction using 2-gauge arrays and 3-gauge arrays in the surf beat regime, realization suddenly dawned on us about the importance of the fact that while the 2-gauge method of Borgman makes use of the linear dispersion relation, viz., $(2\pi f)^2 = \kappa g$ tanh κd, to compute wave direction; the 3-gauge method of Esteva does not use it at all. This realization led us to derive a formula for determining the wavelength as a function of frequency from 3-gauge arrays, thus extending the scope of the method of Esteva to allow computation of the wavelength besides the direction of ocean waves.

We have computed the wavelength of ocean waves and surf beat at the 8m depth 15-gauge array at Duck using the 3-gauge method developed by us as well as by the Singular Value Decomposition method of Herbers et al. 1995. A brief description of these methods follows. An extensive review of methods used for wavelength computation from arrays, buoys etc., is given in Holland 2001.

WAVELENGTH COMPUTATION BY 3-GAUGE METHOD

Esteva 1977 derived the following formula for unambiguously determining wave direction α, from simultaneous time series measurements of surface elevation at three non-collinear gauges, i.e., a 3-gauge array or gauge triad or triangle 123:

$$\alpha = \arctan \frac{\left[\left(x_1 - x_2\right)\phi_{13} - \left(x_1 - x_3\right)\phi_{12}\right] \operatorname{sgn} p}{\left[\left(y_1 - y_3\right)\phi_{12} - \left(y_1 - y_2\right)\phi_{13}\right] \operatorname{sgn} p} \qquad (1)$$

where α is the wave direction in the range $(-\pi, \pi)$ reckoned counter clockwise positive from the positive x-axis; (x_i, y_i), i = 1,2,3 are the coordinates of the three wave gauges; ϕ_{12} and ϕ_{13} are the phase differences in the interval $(-\pi, \pi)$, between gauge pairs 12 and 13 respectively and are determined by cross spectrum analysis as a function of frequency; and $p = (x_1 - x_2)(y_1 - y_3) - (x_1 - x_3)(y_1 - y_2)$ does not vanish when the three gauges are non-collinear, and by definition sgn p=1 for p>0 and sgn p = -1 for p < 0.

By the method of Esteva just three non-collinear gauges are sufficient for determining wave direction unambiguously, i.e., the analysis is done in units of gauge triads, with the different gauge triads possible in the polygonal array giving redundant estimates of wave direction. Unfortunately Esteva was not successful in determining correctly the wave direction over the design range 25-7s of her 5-gauge polygonal array at Pt. Mugu, California, designed by Professor Leon E. Borgman.

As mentioned above FERN99-2000 determined wave direction consistently and accurately by making two modifications to the methods of Borgman and Esteva. One modification is that they performed phase *unwrapping, i.e.,* they used the *true* or *absolute* phase in the interval $(-\infty,\infty)$ obtained from the *apparent* or *relative* phase in the interval $(-\pi,\pi)$ by adding $\pm 2\pi$ to the *apparent* phase for every positive/negative jump in phase at the $\pm\pi$ pseudo-discontinuity in phase arising from circular nature of phase scale (Fig. 3); the other modification being that wave direction was registered only if the relevant gauges were coherent at 0.01 significance level following Thompson 1979.

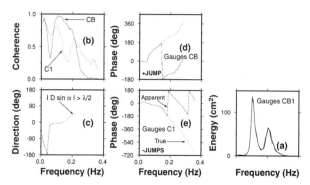

Fig 3. Analysis of wave directional spectrum using the method of Esteva for the gauge triad CB1 on 2 February 1994 at 04 hours. Note that estimates of wave direction are available even when the criterion $|D \sin \alpha| < \lambda /2$ is not satisfied (dashed line). (From Fernandes et al. 2000).

It can easily be shown that the wavelength can be determined as a function of frequency from three non-collinear gauges using the following formula:

$$\lambda = \frac{2\pi |(x_1 - x_2)(y_1 - y_3) - (x_1 - x_3)(y_1 - y_2)|}{\sqrt{[(x_1 - x_2)\phi_{13} - (x_1 - x_3)\phi_{12}]^2 + [(y_1 - y_3)\phi_{12} - (y_1 - y_2)\phi_{13}]^2}} \tag{2}$$

where λ is the wavelength, (x_i, y_i) for $i = 1,2,3$ are the coordinates of the three non-collinear gauges and ϕ_{12} and ϕ_{13} are the *true* phases determined from cross spectrum analysis between the gauges 12 and 13 respectively. An estimate of λ is registered only if the gauges 12 and 13 are coherent at 0.01 significance level – just as in case of wave direction (FERN99-2000).

The assumption made in computation of the "mean" wave direction is that at a particular spectral frequency waves approach from a *single* direction – the wave direction may be different for different frequencies. The only assumption made in wavelength computation is that at a particular spectral frequency the waves have a *single* wavelength. It may be emphasized that for computing the wavelength of ocean waves the assumption insisting that at a particular frequency waves come from a single direction *need not* be made.

WAVELENGTH COMPUTATION BY SVD METHOD

The Singular Value Decomposition (SVD) method of Herbers et al. 1995 (HERB95) consists of two parts, (1) computing from the coordinates of the gauges in the polygonal array, coefficients α_{pq} for every gauge pair pq in the array used, and, (2) using these coefficients α_{pq} and the corresponding normalized cross spectra between the different gauge pairs to obtain the wavelength as a function of frequency. When we first attempted to compute the wavelength, specifically an estimate of the root mean square of the wave number, κ_{rms}, ($\kappa = 2\pi/\lambda$), using the method of HERB95, we obtained unrealistic values (gross overestimates) of wavelength of ocean waves. Later we achieved success using a modified method described in Holland 2001, which correctly computes five different sets of α_{pq}, which can be used to compute estimates of the following five wave number parameters, (i) $<\kappa_x^2 + \kappa_y^2 > (=\kappa_{rms}^2)$, (ii) $<\kappa_x^2 - \kappa_y^2 >$, (iii) $<2\kappa_x \kappa_y >$, (iv) $<\kappa_x>$, and (v) $<\kappa_y>$; where $< >$ indicates the ensemble average. We restricted our attention to just (i), (iv) and (v), as discussed below.

In the method of Holland for estimating κ_{rms} from a Taylor series expansion (HERB95), the coefficients α_{pq} should satisfy the following equations in a least squares sense:

$$\sum_{p=1}^{N_i-1} \sum_{q=p+1}^{N_i} \alpha_{pq} \frac{(-1)^n [s(x_p - x_q)]^{2n-m} [s(y_p - y_q)]^m}{(2n-m)!m!} = \begin{cases} s^2 \text{ for } n=1, m=0; \ n=1, m=2 \\ 0 \text{ for all other } n, m \end{cases} \quad (3)$$

where N_i is the number of instruments; N_t is the number of terms in the Taylor series expansion; n=1,2,3,..., N_t; m=0,1,2,3,...,2n; s=$2\pi/D_{max}$; D_{max} being the maximum allowable distance between the gauges in the array.

In Eq 3, the number of equations is $N_t(N_t+2)$, while the number of unknowns is $N_i(N_i-1)/2$. The number of instruments, N_i, and the number of terms in the Taylor series expansion, N_t, are chosen in such a way that the number of equations is much larger than the number of unknowns, so that Eq 3 is an over determined set of equations. Then the standard method for obtaining a Least Squares solution of Eq 3, is through matrix inversion, which generally fails as the covariance matrix is often singular. Therefore the Generalized Inverse method (Menke 1984), which involves a Singular Value Decomposition (SVD) of a matrix also known as Empirical Orthogonal Function (EOF) analysis, is used as it always yields a solution (we chose $D_{max} = 80m$, $N_i = 15$, $N_t = 8$, and in the SVD only eigen values $> 10^{-4}$ times the largest were retained). In a similar situation, Professor Borgman in a paper presented at WAVES 2001 used ridge regression analysis instead of SVD.

The wavelength or rather κ_{rms} is then computed using the following formula, where the coefficients α_{pq} satisfy Eq 3; H_{pq} is the cross spectrum between the gauges pq; and H_{pp}, H_{qq} are auto spectra of gauges p and q respectively:

$$k_{rms}^2 = \sum_{p=1}^{N_i-1} \sum_{q=p+1}^{N_i} \alpha_{pq} \frac{\text{Real } [H_{pq}(f)] - 1}{[H_{pp}(f)H_{qq}(f)]^{1/2}} \quad (4)$$

The α_{pq} pertaining to κ_x are obtained by solving Eq 5, again by the Generalized Inverse method, where N_i, N_t and s are the same as in Eq 3; n=1,2,3,..., N_t; m=0,1,2,3,...,2n-1.

$$\sum_{p=1}^{N_i-1} \sum_{q=p+1}^{N_i} \alpha_{pq} \frac{(-1)^{n+1} [s(x_p - x_q)]^{2n-m-1} [s(y_p - y_q)]^m}{(2n-m-1)!m!} = \begin{cases} s \text{ for } n=1, m=0 \\ 0 \text{ for all other } n, m \end{cases} \quad (5)$$

The α_{pq} obtained from Eq 5 are then used to compute κ_x using the following formula:

$$k_x = \sum_{p=1}^{N_i-1} \sum_{q=p+1}^{N_i} \alpha_{pq} \frac{\text{Imag } [H_{pq}(f)]}{[H_{pp}(f)H_{qq}(f)]^{1/2}} \quad (6)$$

κ_y is computed in the same way as κ_x, from Eq 5 & 6, but in Eq 5, the right hand side equals s for n=1, m=1; and equals zero for all other n,m

We estimated the wavelength λ $(=2\pi/\kappa)$ directly using Eq 3 & 4, ie., SVD(κ_{rms}), and indirectly using the relation $\kappa=(k_x^2+k_y^2)^{1/2}$ from Eq 5 & 6 for k_x and k_y , i.e., SVD(κ_x, κ_y).

RESULTS AND DISCUSSION

Before discussing our results of wavelength computation, we shall give a brief description of the waves at Duck during the observation period 1-6 February 1994 determined from 3-gauge arrays (Fernandes et al. 2000). On the basis of wave conditions the observation period may be divided into the following (Fig 4):

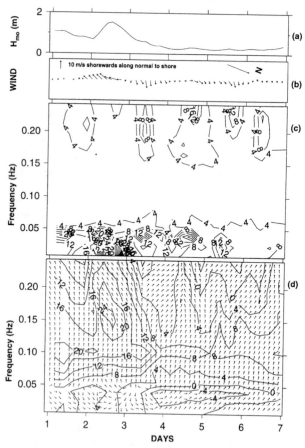

Fig 4. Wave directional spectrum for the period 1-6 February 1994, computed using the method of Esteva. (a) significant wave height, H_{mo} (b) vector mean wind (c) standard deviation of a maximum of 15 redundant estimates of wave direction (d) Mean of a maximum of 15 redundant estimates of wave direction and energy in decibels ($10*\log_{10}$ energy). Energy measured in cm^2 . Wave direction is in shore normal coordinate system, just like wind direction. (From Fernandes et al. 2000).

(i) *1 February 1994*: On 1 February surf beat (infra gravity waves, T=32.00, 42.67, 64.00s spectral period, the lowest three rows of vectors in Fig 4d) locally generated by a dominant swell (waves generated by distant storms) of ≅ 9.85s is present. The surf beat is uniformly directed up the coast away from the shore.

(ii) *2 February:* On 2 February ocean waves are locally generated (sea) by a *Northeaster* of ≅ 10 ms⁻¹, and are uniformly aligned with the wind direction.

(iii) *4-6 February:* During 4-6 February, when low wave conditions prevail (significant wave height, $H_{mo} < 0.41$m) surf beat of remote, transoceanic origin is present. The surf beat is uniformly directed up the coast towards the shore at an angle of ≅ 45° with respect to the seaward shore normal.

(iv) *1-6 February:* The full data set.

How accurate or rather consistent are the wave directions in Fig. 4d? From the standard deviation of a maximum of 15 redundant estimates of wave direction it is clear that

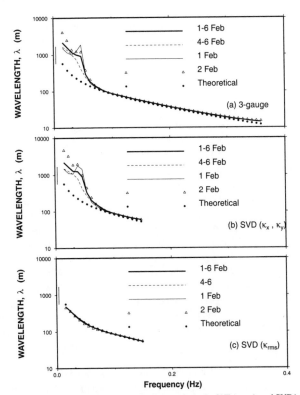

Fig 5. "Mean" wavelength computed using the 3-gauge method; the SVD(κ_x, κ_y); and SVD(κ_{rms}) method. The small vertical lines on the left indicate thrice the wavelength at 64s using the linear dispersion relationship.

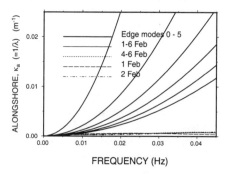

FREQUENCY (Hz)

Fig 6. "Mean" alongshore wave number k_a computed using the 3-gauge method. Edge wave modes (Huntley et al. 1981): $k_a = 2\pi f^2/[g(2n+1) \tan \beta]$, for $n = 0, 1, 2, ...,5$ (bold curves from left to right); $\beta = 0.01$ at the 8m array at Duck.

(i)   the direction of all waves with period T<20s which includes swell and sea, is very accurate throughout the observation period

(ii)   the direction of locally generated surf beat on 1 February as well as the surf beat of remote origin occurring during 4-6 February is quite accurate

indicating that for all the above cases the assumption that waves at a particular frequency come from a single direction, holds.

What is not accurate at all is the direction of surf beat on 2 February, when two rival mechanisms of surf beat generation are simultaneously at play, viz., surf beat directed up the coast away from shore generated by swell and surf beat directed down the coast away from the shore generated by locally generated waves (sea). The huge standard deviation of 15 redundant estimates of surf beat direction indicates that neither of these mechanisms is negligible on 2 February, i.e., the assumption of single direction at a particular frequency is violated. However, the directions displayed in Fig 4d indicate which of these two mechanisms is the *more* dominant.

Fig 5 shows the *mean* wavelength of ocean waves and surf beat as a function of frequency computed from array measurements during 1-6 February 1994 at Duck, North Carolina, USA using three methods, viz., the 3-gauge method developed by us; and the Singular Value Decomposition method of HERB95 using two estimates, $SVD(\kappa_{rms})$ and $SVD(\kappa_x, \kappa_y)$. The figure shows that in the ocean wave regime (T<1.64s), the wavelengths computed by these three methods are all identical to the wavelength predicted by the linear dispersion relation, $(2\pi f)^2 = \kappa g \tanh \kappa d$. However in the surf beat regime; while the wavelength computed from $SVD(\kappa_{rms})$ is also identical with the wavelength predicted by the linear dispersion relation, in lines with the original computation of $SVD(\kappa_{rms})$ of HERB95, we observe that the wavelength computed using the 3-gauge method and the $SVD(\kappa_x, \kappa_y)$ method is in contrast approximately thrice the predicted wavelength.

There is complete agreement between the 3-gauge method and the $SVD(\kappa_x, \kappa_y)$ method not just with regards to wavelength computation, but as seen from Fig 2, with regards to wave direction computation also, perhaps because both the methods essentially compute the wave vector **κ**, whereas the SVD (κ_{rms}) method computes κ^2_{rms}.

Our observation (Fig. 2) mentioned in the introduction that the wave direction computed using the 2-gauge is closer to the shore normal than that computed using the 3-gauge method by $\cong 30°$ in the *surf beat regime*, is consistent with the results of wavelength computation using the 3-gauge method and the SVD(κ_x, κ_y) method, as in the 2-gauge method, the sine of the wave direction with respect to the shore normal is directly proportional to the wavelength. Actually, D sin $\alpha = \lambda\phi/(2\pi)$, where D is the distance between the two gauges; ϕ is the phase difference between the two gauges obtained by cross spectrum analysis; and λ is the wavelength computed as a function of frequency from the dispersion relation by a process of iteration. For the record it may be mentioned that Fig. 2a exhibits the well known ambiguity in wave direction within a mirror symmetry associated with linear arrays.

Fig. 6 shows that the alongshore wave number computed from 3-gauge arrays corresponding to the above surf beat is consistently found to represent an edge wave mode of order larger than five, which is consistent with WKB derived ray trajectories reported by HERB95.

We have used in the cross spectrum analysis 80 ensembles of 256 points sampled at 2Hz giving 160 degrees of freedom, i.e. we have used the full 2.84 hour long records available at Duck thus following the present trend in using records much longer than 15 minutes in problems associated with the wave directional spectrum.

CONCLUSIONS

We found that the wavelength of surf beat at Duck is approximately thrice that predicted by the linear dispersion relation for ocean waves contrary to Herbers et al. 1995 from wavelength computation using the 3-gauge array method developed by us and the SVD(κ_x, κ_y) method. This conclusion is supported by our finding that the mean surf beat direction computed using 2-gauge arrays is $\cong 30°$ closer to the shore normal than that computed using 3-gauge arrays.

ACKNOWLEDGEMENTS

We are indebted to Dr. Charles E. Long of the CERC's Field Research Facility at Duck, North Carolina, USA for sharing with us the measured data at his 15-gauge array. We wish to thank Professor R. Mahadevan, Visiting Scientist to the National Institute of Oceanography (NIO), Dona Paula, Goa, India, for giving us some insights into the problem of wavelength computation using SVD. For our successful implementation of the SVD method we owe a debt of gratitude to Dr. K. Todd Holland of the Naval Research Laboratory, Stennis Space Center, MS, USA. We are thankful to the Director, NIO for the facilities provided. Figures 1, 3 & 4 have been reprinted from Fernandes et al. 2000 with permission from *Elsevier Science*. This is contribution No. 3713 of NIO.

REFERENCES

Barber, N. F. and Doyle, D. (1956) A method of recording the   direction of travel of ocean swell. *Deep-Sea Research* 3, 206-213.

Borgman, L. E. (1974) Statistical Reliability of computations based on wave spectral relations. *Proc. Int. Symp. on Ocean Wave Measurement and Analysis.* Vol 1, 362-378.

Esteva, D. C. (1976) Wave direction computation with three gauge arrays. *Proc. Fifteenth Coastal Engng. Conf.* Honolulu, Hawaii, Vol 1, 349-367.

Esteva, D. C. (1977) Evaluation of the computation of wave direction with three gauge arrays. U.S. Army Corps of Engineers, CERC, *Technical Paper* No 77-7.

Fernandes, A. A., 1999. Study of the directional spectrum of ocean waves using array, buoy and radar measurements. *PhD Thesis*, Goa University, Taleigao Plateau, Goa 403 203, India, June 1999.

Fernandes, A. A., Sarma, Y. V. B., and Menon, H. B. 2000. Directional spectrum of ocean waves from array measurements using phase/time/path difference methods. *Ocean Engineering*, 27(4), 345-363.

Herbers, T.H.C., and Guza, R.T. 1992. wind-wave nonlinearity observed at the sea floor. Part II: Wave numbers and third order statistics. *J. Physical Oceanography*, 22(5), 489-504.

Herbers, T. H. C., Steve Elgar and Guza, R.T. (1995) Generation and propagation of infragravity waves. *J. Geophys. Res,* 100, 24863-24872.

Holland, K.T. 2001. Application of the linear dispersion-relation with respect to depth inversion using remotely sensed data. *IEEE Journal of Geoscience and Remote Sensing* (in press).

Huntley, D.A., Guza, R.T., and Thornton, E.B., 1981. Field observations of surf beat 1: Progressive edge waves. *J. Geophys. Res.* 86, 6451-6466.

Menke, W., 1984. Geophysical data analysis: Discrete inverse theory. *Academic Press*, 260p.

Thompson, R.O.R.Y., 1979. Coherence significance levels. J. Atmosph. Sci. 36, 2020-2021.

Subject Index

Page number refers to the first page of paper

Author Index